D11723337

encyclopedia on ad hoc and ubiquitous computing

theory and design of wireless ad hoc, sensor, and mesh networks

encyclopedia on ad hoc and ubiquitous computing

theory and design of wireless ad hoc, sensor, and mesh networks

edited by

dharma p agrawal & bin xie

university of cincinnati, usa

NEW JERSEY • LONDON • SINGAPORE • BEIJING • SHANGHAI • HONG KONG • TAIPEI • CHENNAI

Published by

World Scientific Publishing Co. Pte. Ltd.

5 Toh Tuck Link, Singapore 596224

USA office: 27 Warren Street, Suite 401-402, Hackensack, NJ 07601

UK office: 57 Shelton Street, Covent Garden, London WC2H 9HE

British Library Cataloguing-in-Publication Data
A catalogue record for this book is available from the British Library.

ENCYCLOPEDIA ON AD HOC AND UBIQUITOUS COMPUTING
Theory and Design of Wireless Ad Hoc, Sensor, and Mesh Networks

Desk Editor: Tjan Kwang Wei

ISBN-13 978-981-283-348-8
ISBN-10 981-283-348-X

Typeset by Stallion Press
Email: enquiries@stallionpress.com

Printed by FuIsland Offset Printing (S) Pte Ltd, Singapore

Preface

About the Book

With an explosive growth of wireless communication devices, ad hoc and ubiquitous computing technologies have received unparallel attention from both the academia and industry. In recent years, a number of significant advances have been made in these technologies, with the target of offering more exciting and efficient services for different application scenarios, at anytime and anywhere. These advances have ranged from the development of Mobile Ad Hoc Networks (MANETs) for disaster applications, embedded wireless sensor networks for monitoring applications, and startling discovery of wireless mesh networks that promises to ensure high bandwidth Internet access to mobile users by using multi-hop, multi-channel, and multi-radio technology. These recent developments not only have kindled considerable interest in the study of different ubiquitous computing technologies, but also have fostered unusual civilian, industrial, and military applications. The intent of this book is to elaborate on the fundamental concepts, protocols, and algorithms, to provide a comprehensive understanding of this area of ubiquitous computing and to extend its usefulness to much higher heights.

Coverage of Topics

Due to diversified application scenarios, that necessitate different performance requirements in deploying these technologies, design issues for ubiquitous computing are extremely complicated and there are a number of technical challenges that need to be explored, involving every layer of the OSI (Open Systems Interconnection) protocol stack. This book has a total of 23 chapters, dealing with MANETs, wireless sensor networks, and wireless mesh networks. Design and implementation of MANET are also essential for the other two network paradigms and thus, we begin by illustrating first 8 chapters of MANETs dealing with different subject areas. Then, the design, application, and security issues in wireless sensor networks are discussed from Chapter 9 to Chapter 17. Finally, many underlying characteristics and associated issues of wireless mesh networks are covered from Chapter 18 to Chapter 23.

A. Advances in MANET

MANETs have been successfully explored for many applications such as battlefield communication and emergency response due to underlying capability of fast deployment in a scenario that resorts to quick communication, and has become a popular research area since mid 1990s. Substantial research efforts have been made in this area, and we have selected eight active topics as our prime interest, including (i) Link Quality Models, (ii) Scalable Multicast Routing Protocols, (iii) TCP, Congestion, and Admission Control Protocols for Multi-hop Transmission, (iv) Directional Antennas for MANET, (v) Peer to Peer and Content Sharing in Vehicular Ad Hoc Networks, (vi) Properties of the Vehicle to Vehicle Channel for Dedicated Short Range Communications, (vii) Radio Resource Management in Cellular Relay Networks, and (viii) Game Theoretic Tools applied to Wireless Networks. These subjects depict fundamental concepts in communication, or reflect the recent research trends in MANETs. For example, most of the designed protocols in MANETs assumed an ideal link model, with equal radio coverage in all the directions. This means that all transmitted packets can be successfully received if the receiver is located with the disk area of radius R. However, in 2003, the Berkeley mote platform has demonstrated that the radio signals to be irregular in nature, which results in variable packet losses in different directions. This has motivated many researchers to consider new link models that could accurately represent the link quality, and accordingly introduce reliable routing protocols. In the past decade, Vehicular Ad Hoc Networks (VANET), that specialize MANET to vehicle to vehicle (V2V) and vehicle to infrastructure (V2I) communications have received a great deal of attention by the research community to offer ubiquitous and pervasive network as a part of our daily life. To provide a complete understanding, we have scheduled two chapters to address the recent research activities in this emerging area.

B. Issues in wireless sensor networks

Design of wireless sensor networks was initially motivated by military applications such as battlefield surveillance. However, in the past few years, it has been the focus of many studies for numerous civilian and industrial applications, such as environment and habitat monitoring, object tracking, healthcare, home automation, and traffic control. In this book, we have included nine chapters in the area to address (i) Sensor Routing Protocols, (ii) Quality of Service (QoS) for Sensor Traffic, (iii) Mobility in wireless sensor networks, (iv) Delay-Tolerant Mobile Sensor Networks, (v) Integration of Radio frequency identification (RFID) and wireless sensor networks, (vi) Integrating Sensor Networks with the Semantic Web, (vii) Effective Multi-user Broadcast Authentication in Wireless Sensor, (viii) Wireless Sensor

Network Security, and (ix) Information Security. Unlike MANETs, routing protocols for wireless sensor networks has to be optimized such that it can not only efficiently route the packets with minimum overheads, but also consider severe power constraints. One chapter is designed to highlight architectural and operational protocols for QoS-enabled traffic in sensor networks. RFID facilitates detection and identification of objects that are not easily detectable or distinguishable by using current sensor technologies. A chapter is designed to present RFID technology and illustrates recent research works, new patents, academic products and applications that integrate RFID with sensor networks. On the other hand, semantic web represents a spectrum of effective technologies that support complex, cross-jurisdictional, heterogeneous, dynamic and large scale information systems and growing research efforts on integrating sensor networks with semantic web technologies have led to a new frontier in networking and data management research. With respect to security, three chapters are designed to illustrate security attacks, authentication, and information security in wireless sensor networks.

C. Characteristics of wireless mesh networks

Wireless mesh networks employ multi-channel and multi-radio in a static MANET with the integration of the Internet infrastructure to support cost-efficient and high bandwidth Internet accessibility for mobile users. The network coverage can span a community, an enterprise, or an entire city to support a broad range of applications such as voice, data transfer, video delivery, home networking, etc. Within a short span of time, wireless mesh network technology has stirred considerable interest in both the commercial and academic spheres, and consequently, the consideration of wireless mesh network technology is indispensable in this book. The topics addressed in the book includes: (i) Wireless Mesh Network Architecture, (ii) Multi-hop Medium Access Control (MAC) design and implementation, (iii) Multi-channel Channel Assignment, (iv) Multi-hop, Multi-path, and Load Balancing Routing, (v) Mobility Management, and (vi) Selfishness and Security. In contrast to peer to peer structure of a MANET, a wireless mesh network constitutes a wireless backhaul with static mesh routers which serves as the wireless access points for mobile users. Therefore, deployment of wireless mesh networks is substantially different from MANETs, and details are given on the chapters of wireless mesh network architecture. It may be noted that in spite of sharing some common characteristics with MANETs, wireless mesh networks have their unique design issues such as multi-channel MAC, multi-path routing, traffic patterns, and fairness, etc. We address these topics to cover significant characteristics of wireless mesh networks.

The Use of the Book

This book provides a comprehensive reference material for students, instructors, researchers, engineers, and other professionals in building their understanding of ad hoc and ubiquitous computing. The book is collective efforts of many professional experts and researchers. Each chapter on a particular topic is written in a tutorial manner and the chapters are organized so that the topics covered are easy to understand. On the other hand, each chapter is self-contained and illustrations are independent of other chapters. Therefore, the topics can be selected as per individual interest. Each chapter has about 10 questions that cover significant points of the chapter. You can find answers to selected questions at our website. In addition, slides are available at our website that can be useful for teaching and presentation by the instructors. Our publisher encouraged us to prepare this volume and we hope the efforts could prove to be useful for the readers.

Dharma Prakash Agrawal
Bin Xie
Cincinnati

Contents

PART 1
Mobile Ad hoc Networks

Chapter 1

Survey on Link Quality Models in Wireless Ad Hoc Networks

Mingming Lu* and Jie Wu†

Department of Computer Science and Engineering
Florida Atlantic University
Boca Raton, FL, 33431
**mlu2@fau.edu*
†jie@cse.fau.edu

Emerging advanced wireless ad hoc networks make it possible for network resources to be utilized anywhere and anytime. However, compared with traditional infrastructure-based networks, wireless ad hoc networks are relatively unstable and unreliable due to the underlying wireless medium and infrastructureless nature. Existing techniques compensate for the instability of wireless links by employing either packet retransmissions, network coding, opportunistic routing, thick (or multiple) paths, or route fixes (using alternative routes). This chapter seeks to compare these techniques in terms of energy cost, the increment of reliability, and various other metrics.

Keywords: Energy; link quality; reliability; wireless ad hoc networks; wireless sensor networks.

1.1. Introduction

Existing protocols proposed for wireless ad hoc/sensor networks (e.g. Refs. 21, 22, 36, 37) are mostly designed based on an ideal spherical pattern of wireless links, where perfect reception within a particular range is assumed (i.e. packet transmission is 100% reliable). However, several empirical studies[6, 15, 33, 38] on the Berkeley mote platform have shown that the coverage of a node is irregular (i.e. the radio range varies significantly in different directions), and the packet loss also varies in different directions. Therefore, the assumption of perfect reception within a particular range is unrealistic and hence may result in non-ideal performance.

The performance may improve by excluding low-quality wireless links. However, such a strategy is insufficient and recent studies[30, 33, 38, 40] have shown that the quality of communication links has a significant impact on the performance of wireless ad hoc/sensor networks, including network lifetime, network throughput, resource usage, and reliability. Thusly, link quality should be considered as a critical dimension in the design space for wireless ad hoc/sensor networks, as pointed out in Refs. 14 and 30.

The accurate measurement of link quality is very important when taking link quality in the design space of wireless ad hoc/sensor networks into account. The research results on measuring wireless link quality and patterns[6,7,10,15,26,39,41] in wireless ad hoc/sensor networks have demonstrated that the communication range of a sensor node varies temporally and spatially due to a number of factors (e.g. the varying environments where sensor networks are deployed, unreliable wireless communications, and irregular radio patterns).

Woo *et al.*[32,33] model the link quality between a pair of sensor nodes as a statistical function over time. In their model,[32,33] a sensor node is able to extract the trend of link quality changes over time by tracking the packets heard/overheard, and it can use the tracked information to estimate future link quality. Xu and Li[34] study the link quality of wireless sensor nodes by exploiting the spatial correlation in links. The intuition behind spatial correlation is that sensor nodes that are geographically close to each other may have correlated link quality. Xu and Li[34] propose a weighted regression algorithm that allows each sensor node to capture the spatial correlation in the quality of its links.

An increasing number of works have begun to model link quality and utilize it for various applications. The existence of lower quality links incurs a low packet delivery ratio, hence reducing the performance of wireless ad hoc networks.[7,38] A common fault-tolerant technique is redundancy, such as retransmissions or erasure coding. Banerjee and Misra[3] model the link cost as a function of energy consumption for a single transmission attempt across the link and the link error rate, and propose several retransmission-aware routing schemes. Banerjee *et al.*[13] extend this work by relaxing the assumption of perfect reliability (zero error rate) in the link layer. Li and Shu *et al.*[24] further extend the result[13] by integrating power control techniques into the routing problem proposed in Ref. 13.

In Ref. 25, we consider the unicast routing problem in wireless ad hoc networks with unreliable links in scenarios featuring bounded retransmission times at the link layer. We model a wireless ad hoc network as a market and introduce the concept of benefit, adopting this concept to reflect the importance of a packet. The benefit is used to balance the maximization of the packet delivery ratio and the minimization of effective energy consumption.

Kwon *et al.*[23] propose a network lifetime maximization problem under reliability constraints in wireless sensor networks. The authors first propose a retry limit allocation problem similar to the quota assignment problem.[25] They[23] propose a greedy solution to this problem. Similar to Ref. 13, Misra and Banerjee[27] model the link energy-cost as a function of energy consumption for a single transmission attempt across the link and the link error rate; however, their objective is to find the best route with the greatest residual lifetime.

Opportunistic routing[4] utilizes redundancy in omnidirectional transmissions to make up for packet loss over lossy links. Intuitively, wireless channels between a pair of nodes change over time, but if there are multiple receivers for a sender, then the probability of successful delivery will likely increase.

Besides retransmissions, network coding can also be employed to reduce the effect of lower quality links. Certain network coding methods use replication (redundancy) to send identical copies of a message simultaneously over multiple paths in an effort to mitigate the effects of a single path failure.[31] Jain *et al.*[20] apply erasure coding to delay tolerant networks (DTNs). Cui *et al.*[12] propose a jointly opportunistic source coding and opportunistic routing (OSCOR) protocol for correlated data gathering in wireless sensor networks. Cristescu and Beferull-Lozano[11] consider a sensor network that measures correlated data, where the task is to gather all data from the network nodes to a sink.

1.2. Background

Existing techniques that improve routing reliability can be classified as retransmissions, network coding, multiple paths, and alternative paths. Each of the four categories can be further classified into sub-categories as shown in Figure 1.1.

Retransmission methods can be further classified into three categories as follows: (1) Hop-by-hop retransmission: A path is chosen from source to sink, and an automatic repeat request (ARQ) is used at the link layer to request the retransmission of packets lost on every link in the path. (2) End-to-end retransmission: A path is chosen from source to sink, and packets are acknowledged by the sink, or destination node. If the acknowledgment for a packet is not received by the source,

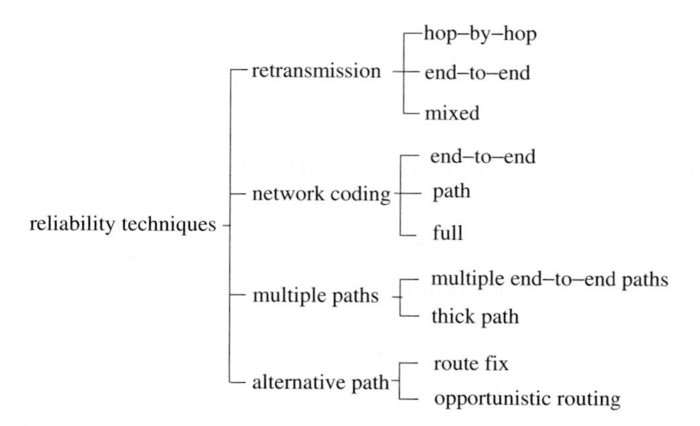

Fig. 1.1. The classification of existing techniques that can improve routing reliability.

the packet is retransmitted. (3) Mixed retransmission: Links on a path may or may not provide hop-by-hop retransmission, while end-to-end retransmission is provided at the transportation layer.

Most existing works[3,13,24,25] consider both hop-by-hop and end-to-end retransmissions. Some[9] consider only hop-by-hop retransmissions. Reference 13 further takes mixed retransmission into account.

Network coding can be classified into three sub-categories: (1) End-to-end coding: A path is chosen from source to sink, and an end-to-end forward error correction (FEC) code, such as the Reed-Solomon code, LT code, or Raptor code, is used to correct packets that become lost between source and sink. (2) Path coding: A path is chosen from source to sink, and every node on the path employs coding to correct lost packets. The most straightforward way of doing this is for each node to use one of the FEC codes for end-to-end coding, decoding, and re-encoding the packets it receives. (3) Full coding: In this case, paths are eschewed altogether. A subgraph that specifies the frequency with which every node transmits packets is chosen, and the random linear coding scheme is used.

By providing multiple paths from source to destination, multiple-path techniques can improve the bandwidth of data transmission between the source-destination pair. Two techniques are classified under this category: multiple end-to-end paths and a thick path.

The classification of link-quality-based, constraint-free routing models based on the underlying techniques is illustrated in Figure 1.1. Besides this classification, constraint-free routing models can be also classified according to their objective functions: (1) minimizing energy consumption over a single routing session;[3,13,24] (2) maximizing network life,[23,27] which is defined as the lifetime of the first node that runs out of battery power; (3) maximizing the predefined utility function,[25] which can be related to cost, link quality, and delay.

Constraint-free routing models can also be classified according to the type of application: (1) unicast[3,4,13,20,24,25,27] routing between a source-destination pair; (2) multicast[5] routing between a source and a set of receivers; (3) data gathering (reverse multicast)[12] routing between a set of sources and a sink.

1.3. Reliability-Constraint Link Quality Models

1.3.1. *Minimizing energy consumption*

Banerjee and Misra[3] argue that minimum-energy routing should not be based solely on the energy spent in a single transmission, which is usually adopted in networks using reliable links. Instead, the total energy (including energy consumed by any necessary retransmission) is the proper metric to evaluate routing optimality. Therefore, the authors model link cost as a function of energy consumption for a single transmission attempt across the link and the link error rate. Figure 1.2 gives an

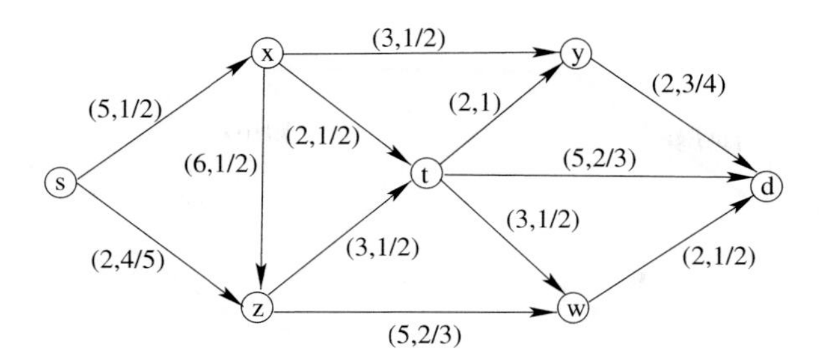

Fig. 1.2. An illustrative example where each link (i, j) is labeled with a two tuple $(e_{i,j}, q_{i,j})$ and each node is labeled with its ID.

example network graph where each link is labeled with two attributes (the energy consumption for a single transmission and the link error rate) in the form (energy, error rate).

The authors[3] consider two retransmission models: hop-by-hop and end-to-end. In the end-to-end retransmission model, data delivery is guaranteed through the retransmissions initiated by the routing source. In the hop-by-hop retransmission model, intermediate nodes provide link layer retransmissions to ensure reliable forwarding to subsequent hops.

For easy presentation, we use $e_{i,j}$ to denote the energy consumption of a single transmission across link (i, j), and $q_{i,j}$ to represent the link error rate. In the hop-by-hop retransmission model, the link cost $c_{i,j}$ is defined as

$$c_{i,j} = \frac{e_{i,j}}{1 - q_{i,j}}, \tag{1.1}$$

which is equivalent to the expected transmission count (ETX)[9] if $e_{i,j} = 1$. The ETX reflects the expected number of transmissions over link (i, j) under the condition that the number of retransmission is unlimited. Therefore, the definition of the link cost metric represents the expected energy consumption spent on a link. This link cost metric is additive for a path. Hence, the total energy cost of a path is the sum of the cost of each link on the path. The minimum cost path can be calculated through Dijkstra's algorithm. We use the example shown in Figure 1.2 to illustrate the minimum cost path in the hop-by-hop retransmission model. Figure 1.3 shows the cost of each link and the minimum cost from source s to each node of Figure 1.2. There are two minimum cost paths: $s \to x \to y \to d$ and $s \to x \to t \to w \to d$.

The authors[3] also consider total energy consumption along a path in the end-to-end retransmission model. To derive a closed form expression of the total path energy cost in the absence of hop-by-hop retransmissions, the authors make a simplified assumption that transmission errors on a link do not prohibit downstream nodes from relaying the packet. Therefore, the total energy consumption along path

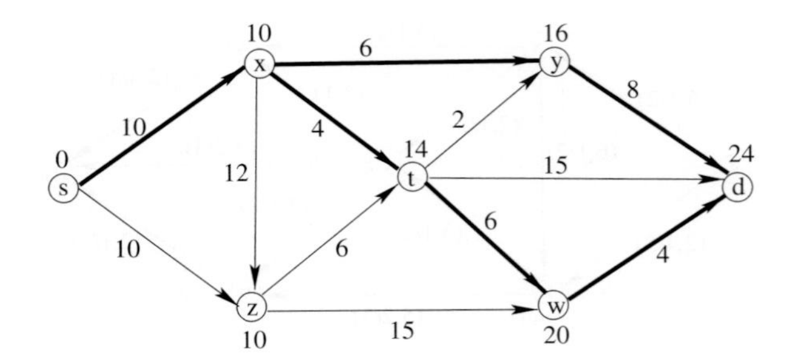

Fig. 1.3. This example illustrates the minimum cost path in the hop-by-hop retransmission model that provides link layer reliability. The topology of this example is the same as the example in Figure 1.2. The link cost $c_{i,j}$, which is labeled with each link (i, j), is calculated according to (1.1) based on transmission energy $e_{i,j}$ and link error rate $q_{i,j}$ provided in Figure 1.2. The number labeled with each node is the minimum cost from source s to the node. The bolded paths represent the minimum cost paths.

P is defined as:

$$C_P = \frac{\sum_{(i,j)\in P} e_{i,j}}{\prod_{(i,j)\in P}(1 - q_{i,j})}. \tag{1.2}$$

The above formulation cannot be expressed as a line sum of individual link costs, thereby making it inappropriate for the existing minimum-cost path computation algorithms. The authors[3] approximate Equation (1.2) by identifying all links on a path, i.e. assuming each link has the same transmission energy cost and link error rate. Therefore, Eq (1.2) can be reduced to

$$C_P = \frac{ke}{(1 - q)^k}, \tag{1.3}$$

where k is the number of links on path P, e is the transmission energy cost, and q is the link error rate. Based on Equation (1.3), the authors[3] propose a heuristic cost function for a link as follows:

$$c_{i,j}^{approx} = \frac{e_{i,j}}{(1 - q_{i,j})^L}, \tag{1.4}$$

where $L = 2, 3, \ldots$, and is chosen to be identical for all links. Hence, the total energy consumption of a path is approximated as the sum of $c_{i,j}^{approx}$ for each link on the path. In Figure 1.4, we illustrate the minimum cost path in the simplified end-to-end retransmission model. The minimum cost path when $L = 2$ for this example is $s \to x \to t \to w \to d$, which is different from that of Figure 1.3.

By analyzing the interplay between error rates, number of hops, and transmission power levels, the authors[3] concluded the following: (1) a path with multiple shorter hops is not always more beneficial than one with a smaller number of long-distance hops; (2) a routing algorithm should evaluate a candidate link (and the path) on the

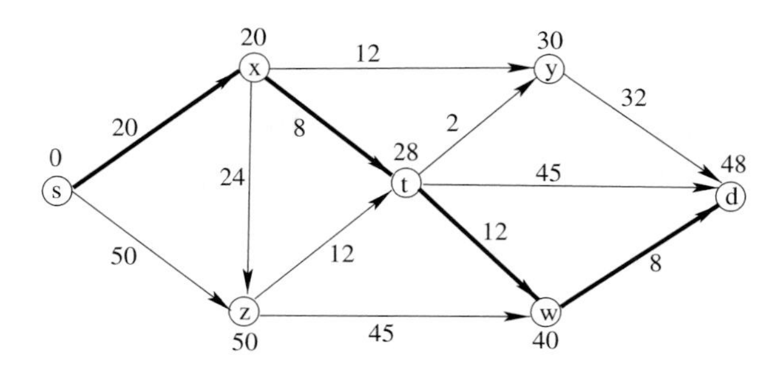

Fig. 1.4. This example illustrates the minimum cost path in the simplified end-to-end retransmission model that provides end-to-end reliability. The topology of this example is the same as the example in Figure 1.2. The link cost $c_{i,j}$, which is labeled with each link (i, j), is calculated according to (1.4) based on the transmission energy $e_{i,j}$ and link error rate $q_{i,j}$ provided in Figure 1.2. The number labeled with each node is the minimum cost from source s to the node. The bolded path represents the minimum cost path.

basis of both its power requirement and its error rate; (3) link-layer retransmission support is almost mandatory for a wireless ad hoc network because it can reduce the total energy consumption by at least one order of magnitude; (4) the advantage of using the retransmission-aware scheme is significant regardless of whether fixed or variable transmission power is used; (5) traditional Dijkstra- and Bellman-Ford-based routing algorithms can be adapted to compute the optimal energy-efficient route for the case where link layers implement perfect reliability.

Banerjee *et al.*[13] extend this work[3] by relaxing the assumption of perfect reliability in the link layer. The authors explain that all practical mechanisms achieve perfect end-to-end reliability through either end-to-end retransmission or the mixed approach (combination of hop-by-hop retransmission and end-to-end retransmission). To address the problems one at a time, they first consider the pure end-to-end retransmission model, then the mixed retransmission model.

When designing routing algorithms, the first challenge lies in defining the link cost function. In the hop-by-hop retransmission model,[3] the accumulated energy cost for a link is easy to express because it depends only on the energy spent on a single transmission across the link and the corresponding link error rate. In the pure end-to-end retransmission model, the accumulated energy consumption depends not only on the local link error rate but also on the link error rates of its downstream links. Although the authors[13] propose an expression for the link cost function in the pure end-to-end model, it is only an approximation. Banerjee *et al.*[13] improve their work by introducing a recursive expression for the accumulated energy consumption in the pure end-to-end retransmission model, which can calculate the exact energy cost of a path rather than just approximating it.

Assume that node u precedes v in the path from s to v, denoted by $P_{s,v}$. Let $P_{s,u}$ denote the pair of $P_{s,v}$ between s and u. For any path $P_{i,j}$, let $C_{P_{i,j}}$ denote the energy consumed when a packet is successfully delivered along the path from i to j. The recursive expression for the accumulated energy consumption is

$$C_{P_{s,v}} = ETX_{u,v} \cdot (C_{P_{s,u}} + e_{u,v}), \tag{1.5}$$

where $ETX_{u,v}$ is the ETX of link (u, v), and $e_{u,v}$ is the energy consumption of a single transmission attempt across link (u, v). In Dijkstra's algorithm[8] where all links are assumed to be reliable, the energy cost of a path can also be expressed in a recursive form: $C_{P_{s,v}} = C_{P_{s,u}} + e_{u,v}$. Comparing this recursive expression with Equation (1.5), it is easy to see that Dijkstra's algorithm can be used to compute the minimum accumulated energy path in pure end-to-end retransmission models by replacing the pseudo-code $C_{P_{s,v}} = \min\{C_{P_{s,v}}, C_{P_{s,u}} + e_{u,v}\}$ with the pseudo-code $C_{P_{s,v}} = \min\{C_{P_{s,v}}, ETX_{u,v} \cdot (C_{P_{s,u}} + e_{u,v})\}$.

The computation of the minimum accumulated energy path can be illustrated through the example in Figure 1.5. In the example network, each link (i, j) is labeled with a two tuple $(e_{i,j}, ETX_{i,j})$, and each node is labeled with its ID. x is the first node added except the source s, followed by its successors z and t, respectively. This process terminates after choosing d, whose predecessor is t. The minimum accumulated energy path is $s \to x \to t \to d$ with accumulated energy 87. Without considering link loss rates, a naïve shortest path is $s \to z \to w \to d$.

The authors[13] also extend the result of the pure end-to-end retransmission model to the more general mixed retransmission model where some of the links support hop-by-hop retransmissions and end-to-end retransmission is guaranteed. In Ref. 13, the authors derive a recursive expression for the accumulated energy consumption in the mixed retransmission model as follows:

$$C_{P_{s,v}} = \begin{cases} C_{P_{s,u}} + ETX_{u,v} \cdot e_{u,v}, & \text{if hop-by-hop} \\ ETX_{u,v} \cdot (C_{P_{s,u}} + e_{u,v}), & \text{otherwise} \end{cases},$$

which is based on Equations (1.1) and (1.5). In the above expression, "if hop-by-hop" means that link layer retransmission is provided to support link layer reliability.

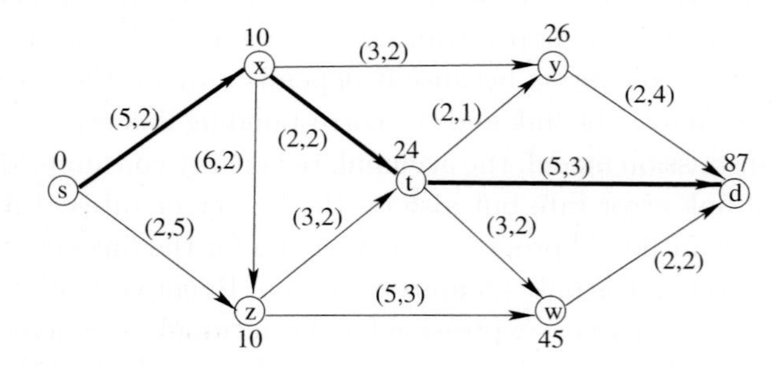

Fig. 1.5. This example illustrates the minimum cost path in the pure end-to-end retransmission model that provides end-to-end reliability. The topology of this example is the same as the example in Figure 1.2. Each link (i, j) is labeled with a two tuple $(e_{i,j}, ETX_{i,j})$. The number labeled with each node is the minimum cost from source s to the node. The bolded path represents the minimum cost path.

The minimum cost path in the mixed retransmission model can be computed by the modified Dijkstra's algorithm, which replaces the pseudo-code $C_{P_{s,v}} = \min\{C_{P_{s,v}}, C_{P_{s,u}} + e_{u,v}\}$ from the original Dijkstra's algorithm with the following pseudo-codes: $C_{P_{s,v}} = \min\{C_{P_{s,v}}, C_{P_{s,u}} + ETX_{u,v} \cdot e_{u,v}\}$ if link layer reliability is supported by link layer retransmissions; otherwise, $C_{P_{s,v}} = \min\{C_{P_{s,v}}, ETX_{u,v} \cdot (C_{P_{s,u}} + e_{u,v})\}$. In Figure 1.5, if we assume link (y, d) is reliable due to link layer retransmissions, the minimum cost path changes to $s \to x \to y \to d$ with minimum cost 34.

Based on centralized algorithms, the authors propose a non-trivial, lightweight (in terms of route exchange information), distributed algorithm. The authors also consider a multi-path routing scheme for the mixed retransmission model. The authors prove that the multi-path routing problem is NP-hard by reducing from the 3-dimensional matching problem.

Li and Shu *et al.*[24] extend previous works by integrating power control techniques into the routing problem proposed in Ref. 13. In Ref. 3, the authors model link error rates as functions of the received signal strength. In the variable-power scenario, the authors fix the received signal strength to the minimum signal strength to minimize the transmission energy across a link so that the transmitted data can be successfully decoded from the received signal. However, due to the existence of link error, the predetermination of the received signal strength, which in turn determines the transmission power, does not necessarily minimize the effective energy consumption on that link. Therefore, Li and Shu *et al.* adopt power control methods to minimize the effective energy consumption and integrate computation of the optimal transmission power into the routing algorithms[3, 13] for the reliable hop-by-hop model, the reliable end-to-end model, and the mixed model. The authors[24] also extend the unicast routing problem to single sink multiple unicast routing, multi-path (disjoint path) routing, and overlay based multi-cast routing, as well. For the first two routing problems, the authors[24] propose optimal algorithms, while for multi-path routing, the authors propose an approximation algorithm with approximation ratio 2.

Li and Shu *et al.*[24] also consider the case of bounded retransmission times in the link layer. However, they have a technical flaw in their definitions of the link cost and link error rate. Due to bounded retransmission times, the link cost is not simply the multiplication of the transmission cost across the link and the maximum allowed transmission attempts. The reason is that each transmission attempt occurs with a different probability and no retransmission is needed after a successful transmission attempt.

The bounded retransmission scheme is a generalization of the unbounded retransmission scheme because the latter can be regarded as the maximum number of transmissions, being infinity. Therefore, it is more general to study the above routing problems in case of the bounded retransmission times.

Cristescu and Beferull-Lozano[11] consider a sensor network that measures correlated data, where the task is to gather all data from the network nodes to a sink.

They consider the case where data at nodes is lossy coded with high-resolution, and the information measured by the nodes should be available at the sink within certain total and individual distortion bounds. First, they consider the problem of finding the optimal transmission structure and the rate-distortion allocations at various spatially located nodes, so as to minimize the total power consumption cost of the network. They prove that the optimal transmission structure is the shortest path tree and that the problems of rate and distortion allocation separate in the high-resolution case; namely, they first find the distortion allocation as a function of the transmission structure, and then the rate allocation is computed. Then, they study the case when the node positions can be chosen by finding the optimal node placement when two different targets of interest are considered, namely total power minimization and network lifetime extension.

1.3.2. *Maximization of network lifetime*

Kwon *et al.*[23] propose a network lifetime maximization problem under reliability and stability constraints in wireless sensor networks. The authors assume that wireless links are unreliable due to channel error alone, and modeled the link stability (per-hop probability of successful packet delivery) as a function of channel gain and transmission power. To address these problems one at a time, the authors first propose a retry limit allocation problem, similar to our quota assignment problem.[25] For a given path from a sensor to a sink, assuming the transmission power for each sensor is fixed and equivalent, the retry limit allocation problem focuses on allocating the number of transmissions to each sensor along the path. The objective of this problem is to minimize the total expected energy consumption of this path. The constraint is that the packet delivery probability along this path should be no less than a threshold probability. Kwon *et al.*[23] propose a greedy solution to this problem. This greedy solution first assigns each link one retry limit and then iteratively selects a link to increase its retry limit. The retry limit of a selected link is always increased one at a time. Each time, the selected link must be the one that can increase its link reliability the most. This process repeats until the reliability constraint is satisfied.

The authors[23] then propose a routing and power control problem. They assume that multiple paths between a particular sensor and sink pair exist. The objective of this problem is to balance the energy consumption among all possible paths from each sensor to the sink so that the network lifetime can be maximized. Besides selecting the amount of data transmitted along each path, the authors also consider the selection transmission power level and determine the retry limit on each link. Two solutions are provided to solve the problem. One is a linear programming based solution, which is optimal. The other is a cost-based heuristic solution.

The Maximum Residual Packet Capacity (MRPC) protocol proposed in Ref. 27, which considers battery charge as well as link reliability during route selection.

In Ref. 27, a node-link metric is introduced to capture the energy-lifetime of a link between nodes i (transmitter) and j, which is defined as:

$$L_{i,j} = \frac{R_i}{E_{i,j}},$$

where R_i is the residual battery charge at node i and $E_{i,j}$ is the energy required to transmit a data packet of given size over link (i, j). A suggested formulation for $E_{i,j}$ is as follows:

$$E_{i,j} = \frac{T_{i,j}}{(1 - p_{i,j})^H},$$

where $T_{i,j}$ is the energy required for one transmission attempt of the aforementioned data packet with a fixed transmission power. Also, $p_{i,j}$ is the packet error probability of the link (i, j) and $H = 1$ if unlimited hop-by-hop retransmissions are performed by the link layer. From the above formulae, it is clear that the lifetime of a link is higher when greater battery charge remains at the transmitter node, and when the reliability of the link is high, resulting in a low energy cost for correctly transmitting a packet. These formulae give an estimation for the expected number of data packets that can be transmitted over a link before the battery of the transmitter fails. If a route failure is said to occur when any single link on it fails, the lifetime of path p in number of packets is simply:

$$Life_p = \min_{(i,j) \in p} \{L_{i,j}\}.$$

MRPC considers the best route to be the one with the greatest residual lifetime. The authors suggest that the MRPC algorithm may be implemented in AODV[28] for application in MANETs. As routes are discovered, the lifetime of the path is accumulated by calculating the lifetime of each link. The next hop to a destination is always selected to be the neighbor which results in the greatest possible value for $Life_p$.

This protocol[27] results not only in load balancing, increasing the life of the network, and avoiding congestion, but also yields closer-to-optimal energy consumption per packet, as well as lower packet delay and packet loss probability due to the preference for more reliable links. It can also be implemented in an on-demand, fully distributed routing protocol, such as AODV. However, the estimation of the link reliability is not addressed.

1.4. Maximizing Reliability

1.4.1. *Opportunistic routing*

Opportunistic routing[4] utilizes redundancy in omnidirectional transmissions to make up for packet loss over lossy links. Intuitively, the wireless channels between a pair of nodes change over time, but if there are multiple receivers for a sender, then the probability of successful delivery is likely to increase.

The opportunistic routing scheme consists of two components: a routing component and a MAC component. The routing component is used to select candidate receivers for each node and determine their priorities, while the MAC component is responsible for identifying one receiver from the candidate receivers based on their priority and actual reception of packets. However, the above cross-layer routing scheme has three shortcomings. First, it does not propose an explicit optimization goal. Second, its MAC component cannot guarantee that only one receiver forwards packets. Third, the transmission range is fixed, which is unreasonable for selecting candidate receivers.

We simplified the above cross-layer routing scheme into a single-layer opportunistic routing scheme, present an explicit optimization goal, design an efficient algorithm that allows adjustable transmission ranges, prove the optimality of our algorithm, and implement the algorithm in both centralized and distributed ways. To simplify our model, we consider only one source-destination pair. Our distributed implementation provides a framework to implement routing components for all on-demand and opportunistic-based routing protocols. In our distributed implementation, only the summarized routing information, the expected network utility, is propagated from the destination to the source. Our scheme is easy to implement based on existing reactive routing protocols without introducing additional cost.

1.4.2. *Code redundancy-based reliability models*

In many transport protocols, reliability is achieved using acknowledgments and retransmissions. An alternative approach is to use replication (redundancy) and send identical copies of a message simultaneously over multiple paths to mitigate the effects of a single path failure.[31] This is in contrast to retransmission schemes which typically wait for a message to be lost before sending another copy. At the same time, erasure coding techniques have long been used to cope with partial data loss efficiently.[5]

Erasure coding[16] is a coding technique that converts a message into a set of coded packets such that any sufficiently large subset of the coded packets can be used to reconstruct the original message. In this paper, we assume that the original message has been split into k equally-sized packets. From the angle of linear algebra, each packet split from the original message can be regarded as a variable, and an erasure-coded packet is a linear combination of the k original packets. This can be expressed as a linear equation where the left-hand side of the equation is the linear combination of the k original packets and the right-hand side is the erasure-coded packet. As long as k linearly independent coded packets are given, the original message can be reconstructed by solving k linearly independent equations associated with the k coded packets.

In Ref. 20, Jain *et al.* apply erasure coding to delay tolerant networks (DTNs). Through both derivations and simulations, the authors find that there is no simple "one size fits all" answer to the question of whether erasure coding is beneficial. They outline three different regimes based on the underlying path failure probabilities and redundancy used. Using ideas from modern portfolio theory,[1] the authors propose an efficient algorithm to solve the above problem and demonstrate its efficacy as compared to simple replication and other heuristics in three different DTN scenarios.

Cui *et al.*[12] propose a jointly opportunistic source coding and opportunistic routing (OSCOR) protocol for correlated data gathering in wireless sensor networks. OSCOR improves data gathering efficiency by exploiting opportunistic data compression and cooperative diversity associated with wireless broadcast advantage. OSCOR is a cross-layer protocol. At the MAC layer, sensor nodes need to coordinate wireless transmission and packet forwarding to exploit multi-user diversity in packet reception. At the network layer, in order to achieve high diversity and compression gains, routing must be based on a metric that is dependent not only on link quality, but also compression opportunities. At the application layer, sensor nodes need a distributed source coding algorithm that has low coordination overhead and does not require that the source distributions be known. OSCOR provides solutions incorporating a slightly modified 802.11 MAC, a distributed source coding scheme based on the Lempel-Ziv code and network coding, and a node compression ratio dependent metric combined with a modified Dijkstra's algorithm for path selection. The performance of OSCOR is evaluated through simulations.

1.5. Maximizing Utility

In Ref. 25, we consider the unicast routing problem in wireless ad hoc networks with unreliable links in the scenario of the bounded retransmission times at the link layer. We first introduce a formula to calculate the expected number of transmissions given the maximum number of allowed transmissions. The expected energy consumed on a link is therefore the link transmission cost times the expected number of transmissions. We find that it is meaningless to compute the path with minimum expected energy consumption, as previous works[3,13,24] did. We observe that the minimum-expected-energy path can be a path consisting of links with high error rates. In the extreme case, the energy cost of a path can be 0 if a node on the path which does not forward packets exists. In a pure hop-by-hop model, if the maximum number of transmissions is bounded, the delivery of packets cannot be guaranteed. If we only pursue the minimization of the effective energy consumption and ignore the fundamental objective, the delivery of packets to their destination, the optimal path is meaningless. In fact, a trade-off exists between the minimization of the effective energy consumption and the maximization of packet delivery ratio.

Therefore, we model a wireless ad hoc network as a market, introduce the concept of benefit, and adopt benefit to reflect the willingness of a source node to balance the maximization of packet delivery ratio and the minimization of effective energy consumption. The link cost is no longer only dependent on its own properties: the energy cost for a single transmission across the link and its link error rate. Therefore, the introduction of the bound of the transmission times incurs interdependence between a node's expected cost and its upstream nodes' link error rates.

1.6. The Measurement of Link Quality

In networks with abundant resources, such as the Internet, link quality is usually estimated by sending passive probing packets using the Internal Gateway Routing Protocol. Link quality is measured as the ratio of arrived packets to the expected packets. However, in such networks, each packet is transmitted along a particular link, which is known a priori. Packets in wireless networks, sent via a broadcast medium, can easily get lost due to different environmental and network factors. Thus, the pattern of link quality in wireless ad hoc/sensor networks is expected to be very different from the Internet. Wireless local area network (e.g. IEEE 802.11) and sensor networks share similar qualitative communication patterns.[2] The link qualities of wireless ad hoc and local area networks are studied in Refs. 10 and 26, which demonstrate that shortest path routing may not yield a satisfactory performance due to the variance of communication links. Due to the unique characteristics of sensor networks (e.g. high node deployment density and restrictions on energy), the measurement of link quality in wireless sensor networks is different from wireless ad hoc networks.

The research results on measuring wireless link quality and patterns in wireless sensor networks[6,7,15,39,41] have demonstrated that the communication range of a sensor node varies temporally and spatially. These variations have a major impact on data acquisition, packet delivery, and reliability of the network infrastructure; hence, they affect the network performance significantly. In order to incorporate the awareness of link quality in the design and operation of wireless ad hoc/sensor networks, research projects such as Refs. 32 and 33 have modeled the pattern of link quality changes over time. Woo *et al.*[32,33] observe that the link quality between a pair of sensor nodes has a statistical relationship to time. By tracking the packets heard/overheard, a sensor node is able to extract the trend of link quality changes over time and use it to estimate future link quality.

Woo *et al.*[32,33] define link quality as: packets received in t/max (packets expected to be received in t, packets received in t), where t is a time window. Thus, for an estimator to measure the link quality, a minimum rate for message exchange between neighbor sensor nodes is required. However, the number of messages generated for link quality estimation has a direct impact on the expected lifetime of the networks. For instance, a typical sensor network may last for about one month if each sensor node broadcasts a beacon message every 30 seconds; however, the network lifetime may be increased to more than ten months if each sensor node broadcasts a beacon every 10 minutes.[18] Moreover, frequent message exchanges may cause an upsurge in network traffic when node density is increased, which may further result in a number of problems in the network, such as transmission collisions, network congestion, lower transmission throughput, high packet transmission delay, and unreliability. This is a crucial issue since high node density widely exists in many sensor networks and applications. Xu and Li[34] develop a new link quality estimator that meets the goal of estimating the quality of links without degrading network performance.

Xu and Li[34] study the link quality of wireless sensor nodes by exploiting the spatial correlation in links. The intuition behind spatial correlation is that sensor nodes geographically close to each other may have correlated link quality. The spatial correlation in link quality is observed in Ref. 38, and has been further exploited by Xu and Li,[34] where the spatial correlation in link quality of neighbor sensor nodes can be captured to estimate the link quality with substantially less transmission cost than the link quality estimators based on temporal correlation. The historical information of link quality for one node may be used for estimating not only its own link quality but also that of other neighbor sensor nodes that are geographically close. Xu and Li[34] propose a weighted regression algorithm which allows each sensor node to capture the spatial correlation in the quality of its links. By categorizing the links into classes in accordance with their quality ranges and then employing a separate regression model for each class, the link quality at a given geographical point can be estimated to a high degree of accuracy.

Predictors or estimators are widely used in various systems and applications. Raghunathan *et al.*[29] point out that sensor networks exhibit significantly correlated variations in computing and communication workloads. Thus, the energy usage can be optimized by setting the voltage accordingly if the workload of individual tasks can be predicted. Goel *et al.*[17] and Xu *et al.*[35] predict the future movements of mobile objects to reduce the network communications for reporting sensor readings to a base station. Hu and Evans[19] propose a localization algorithm for mobile sensor networks, which predicts the location of mobile sensor nodes based on their previous locations and the maximum velocity.

1.7. Conclusion and Future Work

Due to the underlying wireless medium and the infrastructure-less nature, the communication between a pair of nodes is relatively unstable. Existing techniques compensate for the instability of wireless links by employing either packet retransmissions, network coding, opportunistic routing, thick (multiple) paths, or route fixes (using alternative routes). All these methods have their own advantages and disadvantages when it comes to increasing the reliability of wireless links. Hop-by-hop retransmissions have a great impact on increasing the reliability of low-quality wireless links with relatively low expected energy cost, but hop-by-hop retransmissions cannot increase the reliability much as the maximum number of retransmissions reaches a threshold value. Meanwhile, delay caused by retransmission is a problem for hop-by-hop retransmissions. On the other hand, network coding (especially erasure-based coding) can increase the reliability almost to 1 at the expense of more energy consumption than that of the hop-by-hop retransmissions. Delay incurred encoding and decoding is also a problem.

Opportunistic routing utilizes the wireless broadcast advantage property in order to increase the bandwidth of data transmissions, which in turn can increase the reliability of data transmissions. The drawbacks of opportunistic routing are as follows: (1) its application is limited to low traffic scenarios; (2) there is a long delay caused by the requirement that each subsequent hop has to send acknowledgements in a back-off fashion. Route fixes and multiple path methods can also increase reliability, but they increase route discovery costs and introduce routing management difficulties such as data allocation problems in multiple paths.

Two important metrics for evaluating the above techniques are reliability and cost. Many works consider the problem of minimizing the energy cost by assuming that the end-to-end reliability can be guaranteed by the unlimited hop-by-hop retransmissions; this is not true because the number of hop-by-hop retransmissions is bounded in practical link layer implementation. If the end-to-end transmission is not 100%, cost alone cannot reflect the whole story because, in the presence of loss, one may incur cost and not do useful work. To balance the trade-off between reliability and cost, we could introduce the utility of packets which reflect the importance of data. If data has high importance, it is preferable to select a technique with higher reliability even with higher cost. Otherwise, it is desirable to select a technique with lower cost.

To fully utilize the advantages of the above techniques and reduce the bad impact brought by the disadvantages, it is natural to consider the combination of the techniques mentioned above. Several previous works have considered the combination of retransmission, route fixing, and erasure coding in different network models. However, as network coding continues to develop and opportunistic routing is introduced, the capacity in which reliability could be improved increases dramatically. In future work, it is desirable to consider the combination of various techniques completely. Also, metrics other than merely reliability and cost, such as utility and delay, should be considered to further identify the best solutions in various applications.

Problems

(1) What is the difference between hop-by-hop retransmission and end-to-end retransmission?
(2) What is the difference between end-to-end coding and path coding?
(3) What is the difference between multiple paths and alternative paths?
(4) What is expected transmission count (ETX)? How does it relate to the expected cost in the hop-by-hop retransmission model?

(5) In Figure 1.5, the minimum cost path is $s \to x \to t \to d$. What are the ETXs of link (s, x), (x, t), and (t, d), respectively?

(6) In Figure 1.5, if we assume that link (y, d) is reliable due to the support of link layer reliability, the minimum cost path will be $s \to x \to y \to d$. What are the ETXs of link (s, x), (x, y), and (y, d), respectively?

(7) In opportunistic routing, why should candidate receivers be prioritized? What is the drawback of this prioritization?

(8) In utility-based routing, considering a direct-connected source and sink pair, what is the expected utility of a single transmission attempt in terms of link cost, link reliability, and packet benefit?

(9) What is Kwon *et al.*'s[23] to the network lifetime maximization problem?

(10) Please briefly describe how erasure coding can increase reliability and also discuss its drawbacks.

Bibliography

1. G. J. Alexander and J. C. Francis, *Portfolio Analysis* (Prentice Hall, 1986).
2. G. Anastasi, E. Borgia, M. Conti, E. Gregori and A. Passarella, Understanding the real behavior of mote and 802.11 ad hoc networks: an experimental approach, *Pervasive and Mobile Computing* **1**(2) (2005) 237–256.
3. S. Banerjee and A. Misra, Minimum energy paths for reliable communication in multi-hop wireless networks, *Proceedings of ACM MobiHoc'02* (2002) 146–156.
4. S. Biswas and R. Morris, ExOR: opportunistic multi-hop routing for wireless networks, *Proceedings of ACM SIGCOMM'05* (2005) 133–144.
5. J. W. Byers, M. Luby and M. Mitzenmacher, A digital fountain approach to asynchronous reliable multicast, *IEEE J-SAC, Special Issue on Network Support for Multicast Communication.* **20** (2002).
6. A. Cerpa, N. Busek and D. Estrin, *SCALE: A Tool for Simple Connectivity Assessment in Lossy Environments*, Technical Report CENS-TR-03-0021, UCLA Computer Science Department (2003).
7. A. Cerpa, J. L. Wong, M. Potkonjak and D. Estrin, Temporal properties of low power wireless links: modeling and implications on multi-hop routing, *Proceedings of ACM MobiHoc'05* (2005) 414–425.
8. T. Cormen, C. Stein, R. Rivest and C. Leiserson, *Introduction to Algorithms* (McGraw-Hill Higher Education), ISBN 0070131511 (2001).
9. D. Couto, D. Aguayo, J. Bicket and R. Morris, A high-throughput path metric for multi-hop wireless routing, *Proceedings of ACM MobiCom'03* (2003) 134–146.
10. D. Couto, D. Aguayo, B. Chambers and R. Morris, Performance of multihop wireless networks: shortest path is not enough, *SIGCOMM Computer Communication Review* **33**(1) (2003b) 83–88.
11. R. Cristescu and B. Beferull-Lozano, Lossy network correlated data gathering with high-resolution coding, *IEEE/ACM Transaction Network* **14**(SI) (2006) 2817–2824.
12. T. Cui, L. Chen, T. Ho and S. Low, *Opportunistic Source Coding for Data Gathering in Wireless Sensor Networks*, Technical Report, California Institute of Technology (2007).

13. Q. Dong, S. Banerjee, M. Adler and A. Misra, Minimum energy reliable paths using unreliable wireless links, *Proceedings of ACM MobiHoc'05* (2005) 449–459.
14. Z. Fan, QoS routing using lower layer information in ad hoc networks, *Proceeding of Personal, Indoor and Mobile Radio Communications Conference* (2004) 135–139.
15. D. Ganesan, B. Krishnamachari, A. Woo, D. Culler, D. Estrin and S. Wicker, *Complex Behavior at SCALE: An Experimental Study of Low-Power Wireless Sensor Networks*, Technical Report UCLA/CSD-TR-02-0013, UCLA Computer Science (2002).
16. M. Ghaderi, D. Towsley and J. Kurose, *Reliability Benefit of Network Coding*, Computer Science Department, University of Massachusetts Amherst (2007).
17. S. Goel and T. Imielinski, Prediction-based monitoring in sensor networks: taking lessons from mpeg, *ACM Computer Communication Review* **31**(5) 2001.
18. C. Guestrin, P. Bodi, R. Thibau, M. Paski and S. Madde, Distributed regression: an efficient framework for modeling sensor network data, *Proceedings of IPSN'04* (2004).
19. L. Hu and D. Evans, Localization for mobile sensor networks, *Proceedings of MobiCom'04* (2004).
20. S. Jain, M. Demmer, R. Patra and K. Fall, Using redundancy to cope with failures in a delay tolerant network, *SIGCOMM Computer Communication Review* **35**(4) (2005) 109–120.
21. B. Karp and H. T. Kung, GPSR: greedy perimeter stateless routing for wireless networks, *Proceedings of ACM MobiCom'00* (2000) 243–254.
22. F. Kuhn, R. Wattenhofer and A. Zollinger, Worst-case optimal and average-case efficient geometric ad-hoc routing, *Proceedings of ACM MobiHoc'03* (2003) 267–278.
23. H. Kwon, T. H. Kim, S. Choi and B. G. Lee, Cross-layer lifetime maximization under reliability and stability constraints in wireless sensor networks, *Proceedings of 2005 IEEE International Conference on Communications* **5** (2005) 3285–3289.
24. X. Li, Y. Shu, H. Chen and X. Chu, Energy efficient routing with unreliable links in wireless networks, *Proceedings of the Third IEEE Mobile Adhoc and Sensor Systems (MASS'06)* (2006) 160–169.
25. M. Lu and J. Wu, Social welfare based routing in ad hoc networks, *Proceedings of the 35th International Conference on Parallel Processing (ICPP'06)* (2006) 211–218.
26. D. A. Maltz, J. Broch and D. B. Johnson, Quantitative lessons from a full-scale multihop wireless ad hoc network testbed, *Proceedings of the IEEE Wireless Communications and Network Conference* (2000).
27. A. Misra and S. Banerjee, Mrpc: maximizing network lifetime for reliable routing in wireless environments, *Proceedings of the IEEE Wireless Communications and Networking Conference* (2002).
28. C. E. Perkins and E. M. Royer, Ad hoc on-demand distance vector routing, *Proceeding of the Second IEEE Workshop Mobile Computing Systems and Applications* (1999) 90–100.
29. V. Raghunathan, C. Schurgers, S. Park and M. B. Srivastava, Energy aware wireless microsensor networks, *IEEE Signal Processing Magazine* **19**(2) (2002) 40–50.
30. K. Seada, M. Zuniga, A. Helmy and B. Krishnamachari, Energy efficient forwarding strategies for geographic routing in lossy wireless sensor networks, *Proceedings of ACM SenSys'04* (2004) 108–121.
31. A. Vahdat and D. Becker, *Epidemic Routing for Partially-Connected Ad Hoc Networks*, Technical Report CS-2000-06, Duke University (2000).
32. A. Woo and D. Culler, *Evaluation of Efficient Link Reliability Estimators for Low-Power Wireless Networks*, Technical Report UCB/CSE-03-1270, UC Berkeley (2003).
33. A. Woo, T. Tong and D. Culler, Taming the underlying challenges of reliable multihop routing in sensor networks, *Proceedings of ACM SenSys'03* (2003) 14–27.

34. Y. Xu and W.-C. Lee, Exploring spatial correlation for link quality estimation in wireless sensor networks, *Proceedings of the Fourth Annual IEEE International Conference on Pervasive Computing and Communications (PERCOM'06)* (2006) 200–211.
35. Y. Xu, J. Winter and W.-C. Lee, Dual prediction-based reporting for object tracking sensor networks, *Proceedings of International Conference on Mobile and Ubiquitous Systems: Networking and Services* (2004).
36. T. Yan, T. He and J. A. Stankovic, Differentiated surveillance for sensor networks, *Proceedings of ACM SenSys'03* (2003) 51–62.
37. Y. Yu, R. Govindan and D. Estrin, *Geographical and Energy Aware Routing: A Recursive Data Dissemination Protocol for Wireless Sensor Networks*, Technical Report UCLA/CSD-TR-01-0023, UCLA Computer Science Department (2001).
38. J. Zhao and R. Govindan, Understanding packet delivery performance in dense wireless sensor networks, *Proceedings of ACM SenSys'03* (2003) 1–13.
39. J. Zhao, R. Govindan and D. Estrin, Computing aggregates for monitoring wireless sensor networks, *Proceedings of IEEE ICC Workshop on Sensor Network Protocols and Applications* (2003).
40. G. Zhou, T. He, S. Krishnamurthy and J. A. Stankovic, Impact of radio irregularity on wireless sensor networks, *Proceedings of ACM MobiSys'04* (2004) 125–138.
41. M. Zuniga and B. Krishnamachari, Analyzing the transitional region in low power wireless links, *Proceedings of SECON'04* (2004).

Chapter 2

Scalable Multicast Routing in Mobile Ad Hoc Networks

Rolando Menchaca-Mendez* and J.J. Garcia-Luna-Aceves†

*Computer Engineering Department
University of California, Santa Cruz
Santa Cruz, CA 95064, USA
menchaca@soe.ucsc.edu

†Palo Alto Research Center, 3333 Coyote Hill Road
Palo Alto, CA 94304, USA
jj@soe.ucsc.edu

In the context of mobile and ubiquitous systems there is an increasing number of applications where data has to be transmitted, not to a single node or person, but to dynamic groups of nodes or people. This call for efficient multicast routing protocols capable of efficiently use the available bandwidth as well as the usually restricted hardware resources such memory and power. Moreover, given the current and expected sizes of this type of networks, multicast protocols have to be designed to scale up to hundreds of nodes. Here, we describe Hydra, the first multicast routing protocol for MANETs that establishes a multicast routing structure approximating the set of source-rooted shortest-path trees from multicast sources to receivers, without requiring the dissemination of control packets from each source of a multicast group. Hydra accomplishes this by (a) dynamically electing a core for the mesh of a multicast group among the sources of the group, so that at most one control packet is disseminated in the network to announce the existence of the group and (b) aggregating multicast routing state in the nodes participating in multicast meshes, so that redundant control packets are not disseminated towards the receivers of a group. We also present an improved version of PUMA which is a receiver-initiated multicast protocol that for each multicast group periodically floods a single control packet which is used to elect a core for the group and to build and maintain the multicast mesh. We present simulations results for WiFi and TDMA MAC protocols illustrating that Hydra and PUMA attain comparable or higher delivery ratios than ODMRP, but with considerably lower end-to-end delays and, in the case of Hydra, far less data overhead.

Keywords: Multicast routing; mesh-based multicast protocols; scalable multicast; state aggregation; ad hoc networks.

2.1. Introduction

The objective of a multicast routing protocol for mobile ad hoc networks (MANET) is to enable communication between a sender and a group of receivers in a network where nodes are mobile and may not be within direct wireless transmission range of each other. Any MANET node can act as traffic originator, destination, or forwarder. Hence, MANETs are well suited to applications where rapid deployment and dynamic reconfiguration are necessary. Examples of such scenarios are: military battlefield, emergency search and rescue, as well as many new emerging ubiquitous applications such as those envisioned by Mark Weiser.[13] In both civilian and military scenarios, we can find a wide range of possible applications to the multicast communication pattern, for instance, group coordination (rescue team leaders deliver instructions to a given subset of their team members), event notification (attendants of a conference receive updates regarding previously defined interests or a group of medical experts are notified of an emergency in their expertise area). Due to the nature of the underlying hardware, these multicast protocols have to be designed to efficiently use the available bandwidth as well as nodes' energy.

We can classify multicast routing protocols for MANETs by the type of routing structure they construct and maintain; namely tree-based and mesh-based protocols. A tree-based multicast routing protocol constructs and maintains either a shared multicast routing tree or multiple multicast trees (one per each sender) to deliver packets from sources to receivers. Several tree-based multicast routing protocols have been reported (e.g. Refs. 6 and 9). These approaches have proven to deliver adequate performance in wired networks. However, in the context of MANETs, establishing and maintaining a tree or a set of trees in the presence of frequent topology changes incur substantial exchange of control messages, which has a negative impact in the overall performance of the protocol.

On the other hand, a mesh-based multicast routing protocol maintains a mesh consisting of a connected sub-graph of the network containing all receivers of a particular group and the relays needed to maintain connectivity. Maintaining a connected component is far simpler than maintaining a tree and hence mesh-based protocols tend to be simpler and more robust. Three representatives of this kind of protocols are the Core Assisted Mesh Protocol (CAMP),[3] the On-Demand Multicast Routing Protocol (ODMRP),[8] and PUMA.[12] A potential concern in mesh-based schemes is that, under high channel contention, these protocols may have poor performance if too many redundant relays are involved in the forwarding of multicast traffic.

Whether multicast routing protocols for MANETs build multicast trees or meshes, all of them are based on network-wide dissemination of control packets to inform the rest of the nodes about the existence of multicast groups. In core-based or receiver-initiated schemes, only one node originates the dissemination of information about a multicast group reaching all other nodes, and receivers send explicit requests towards the core to join the group. In contrast, source-based or sender-initiated schemes have each multicast source originate the dissemination of state information that reaches all nodes in the network. Given that the sender-initiated protocols proposed to date use per source flooding, they do not scale well as the number of groups and sources increases. However, they can provide shortest paths from sources to destinations and avoid hot spots. On the other hand, core-based protocols incur far less overhead, but they do not establish shortest paths from sources to destinations, which leads to higher delays than the ideal shortest paths from sources to receivers.

The work presented here is motivated by the desirability of providing the best features from the two alternatives in the existing design space summarized above, and which we discuss in more detail in Section 2.2. Section 2.3 discuses Hydra, a multicast routing protocol that creates a multicast mesh formed by a mixture of source-specific and shared sub-trees (or sub-meshes) using as few control packets as receiver initiated schemes do. The key ideas behind Hydra are: restricting the dissemination of control packets to those regions of the network where other dynamically designated sender has previously discovered receivers, aggregation of control messages from non-core senders, and electing a sender as the core in non-destructive manner. Section 2.4 discuses PUMA,[12] a receiver-initiated mesh-based multicast routing protocol in which receivers join a multicast group using the address of a special node (core), without the need for network-wide dissemination of control or data packets from all the sources of a group. PUMA implements a distributed algorithm to elect one of the receivers of a group as the core of the group, and to inform each router in the network of at least one next-hop to the elected core of each group. Within a finite time, each router has one or multiple paths to the elected core. All nodes on shortest paths between any receiver and the core collectively form the mesh of the multicast group.

Section 2.5 describes the results of simulation experiments used to study Hydra's and PUMA's performance with that of ODMRP by considering different numbers of sources, group sizes, network density and the use of 802.11 or TDMA as the underlying MAC protocol. The results illustrate the performance benefits that should be expected from the approaches implemented in Hydra and PUMA. Both protocols provide substantial performance improvements over ODMRP even in scenarios involving relatively small networks with few multicast sources. Hydra and PUMA attain the same or better delivery ratios than ODMRP, and incurring end-to-end delays that are close to an order of magnitude smaller than in ODMRP. The simulation experiments also compare different versions of the aggregation algorithms implemented in Hydra.

2.2. Related Work

The multicast ad hoc on-demand distance vector protocol (MAODV)[10] maintains a shared tree for each multicast group consisting of receivers and relays. Sources acquire routes to the group on demand in a way similar to the ad hoc on demand distance vector protocol (AODV).[9] Each multicast group has a group leader who is the first node joining the group. The group leader is responsible for maintaining the group's sequence number, which is used to ensure freshness of routing information. The group leader periodically transmits a group hello packet to become aware of reconnections. Receivers join the shared tree by means of a special route request (RREQ) packet. Any node belonging to the multicast tree can answer to the RREQ with a route reply (RREP). A sender joins the group through the node reporting the freshest route in a RREP with the minimum hop count to the tree. Data are delivered along the tree edges maintained by MAODV. If a node that does not belong to the multicast group wishes to multicast a packet, it has to send a non-join RREQ, which is treated similar to RREQ for joining the group. As a result, the sender finds a route to a multicast group member. Once data is delivered to a group member, the remaining members receive the data along the multicast tree.

The adaptive demand-driven multicast routing protocol (ADMR)[5] maintains a source-based multicast tree for each sender of a multicast group. A new receiver performs a network-wide flood of a multicast solicitation packet when it needs to join the multicast group. Each source replies to the solicitation and the receiver sends a receiver join packet to each source that answered the solicitation. Each source-based tree is maintained by periodic keep-alive packets from the source, which allow intermediate nodes to detect link breaks in the tree by the absence of data or keep-alive packets. A new sender also sends a network-wide flood to allow existing group receivers to send receiver joins to the source. MZR[2] like ADMR, maintains source-based trees. MZR performs zonal routing; and hence the dissemination of control packets is less expensive.

In ODMRP,[8] group membership and multicast routes are established and updated by the sources. Each multicast source broadcasts *Join Query* (*JQ*) packets periodically, and these are disseminated to the entire network to establish and refresh group membership information. When a *JQ* packet reaches a multicast receiver, it creates and broadcasts a *Join Reply* (*JR*) to its neighbors stating a list of one or more forwarding nodes. Nodes receiving *JR* listing them as part of forwarding groups forward the replies with its own list of forwarding nodes. A *JR* is propagated by each forwarding group member until it reaches a multicast source via the selected paths. This process constructs (or updates) the routes from sources to receivers and builds a mesh of nodes, the forwarding group. A source can multicast data packets to multicast receivers via selected routes and forwarding groups. DCMP[1] is an extension to ODMRP that designates certain senders as cores and reduces the number of senders performing flooding. NSMP[7] is another extension to ODMRP aiming to restrict the flood of control packets to a subset of the entire network. However, DCMP and NSMP fail to eliminate entirely ODMRP's use of multiple nodes flooding control packets for each group.

CAMP[3] avoids the need for network-wide disseminations from each source to maintain multicast meshes by using one or more cores per multicast group. A receiver-initiated approach is used for receivers to join a multicast group by sending unicast join requests towards a core of the desired group. The drawbacks of CAMP are that it needs the pre-assignment of cores to groups and a unicast routing protocol to maintain routing information about the cores.

2.3. Hydra

2.3.1. *Overview*

As it is the case in ODMPR, multicast sources in Hydra periodically broadcast *Join Query* (JQ) packets to establish a partial ordering of the nodes in the network. In the case of ODMRP and Hydra, the ordering is based on the nodes' distances in hops to the sources. This ordering is further used to route *Join Reply* (JR) packets from receivers to sources, forcing intermediate nodes to join either a mesh or a tree. However, Hydra uses three mechanisms to build a routing structure as close as possible to a set of source-rooted breadth-first trees (or meshes composed of the union of breadth-first trees) spanning all the receivers while incurring as few control overhead as possible.

In contrast to ODMRP, Hydra uses an elected source as the core of the group, and this is the only source whose JQs reach the entire network. Non-core sources take advantage of the routing state established by the core to identify connected sub-graphs containing one ore more non-core sources and receivers of the group. This way, the scope of the dissemination of JQs from non-core sources is restricted to these connected regions, and other parts of the network are not flooded with unnecessary control information.

In addition, Hydra identifies regions of the network where two or more sources share common sub-graphs (meshes or trees) and performs routing-state aggregation, so that nodes located inside of those common regions only keep routing state regarding one of the aggregated sources and receive JQs and JRs only from that source. To detect the boundaries of a common sub-graph, Hydra compares the orderings established by previous sources with the ordering that is being established by the current JQ from a non-core source. If the ordering induced by the JQ is equivalent to the ordering established by a prior JQ from another source, then the current JQ is not forwarded any further and the two sources that have equivalent orderings are considered as aggregated. Two partial orderings over a graph are *equivalent* if the gradient vectors among neighbors obtained from the two orderings are the same. As the number of senders increases, the likelihood of finding equivalent regions also increases, because nothing prevents a source to share different sub-graphs with different sources, or a given sub-graph to be shared by more than two sources. This property helps the scalability of Hydra with respect to the number of sources. We also note that, while Hydra takes advantage of having a core, it is not necessary for its correct operation.

2.3.2. *Control signaling*

Hydra opportunistically groups control messages of different sources and groups into a single control packet. However, in the rest of our description, we focus on the signaling intended for a specific group.

2.3.2.1. *Join queries*

The first active multicast source for a given group considers itself to be the core for that group and states so in the *Join Query* (JQ) packets it broadcasts every *join query period.* JQs inform other nodes of the existence of the multicast group and its current core, and create a partial ordering of the network based on the distance in hops from each node to the current core. If two or more sources become active concurrently in the same partition, a distributed election is held. The details of the election algorithm are presented in Section 2.3.3.

Sources other than the core in the same multicast group are considered regular sources or *non-core senders.* Non-core senders transmit Non-Core Join Query ($JQnC$) packets to build their own trees or meshes. A JQ is composed of a packet type identifier, the address of the group, the address of the core, a TTL, the distance to the core and a sequence number. In addition, a $JQnC$ contains the address of the non-core sender, the distance to the non-core sender, and the address of the parent towards the core that is used to route $JQnC$s towards the mesh of the core.

Because non-core nodes benefit from the routing structure created by the core, the transmission of $JQnC$s is roughly synchronized with the reception of JQs. Upon receiving a JQ with a larger sequence number, non-core senders wait for a random period of time which is much smaller than the join query period. However, it is also long enough to allow the establishment of the routing structure of the core before transmitting their next $JQnC$ that refreshes the routing information for that source. $JQnC$s are also sent by non-core senders when they have data to sent but no route is known to the receivers and when the sender has not received a JQ from the core in the last two consecutive join query periods.

The objective of the combined use of JQs and $JQnC$ for a given multicast group is to order all nodes with respect to the core of the group, and to make the multicast routing structure (mesh or tree) as close as possible to the aggregation of the source trees of all the multicast sources in the group. JQs must be sent to all nodes; however, the overhead due to the dissemination of $JQnC$s is reduced using two mechanisms.

The first way of reducing the overhead incurred with $JQnC$s consist of disseminating them only to a subset of the network composed of nodes that are part of the mesh or tree established by the core, nodes that lay in the path from the non-core sender to the core, and nodes located at most k hops away from them. The set of nodes that forward $JQnC$s for a given non-core sender is called the source's *k-restricted region of interest* or simply *k-restricted region*. This way, the dissemination of $JQnC$s is carried out only among nodes that are likely to be close to receivers, and other regions of the network do not receive irrelevant control information.

The optimal value of k for a k-restricted region depends on the topology of the network as well as on the mobility of the nodes and on the length of the join query period. In our experiments, a sensitivity analysis showed that 1 is a reasonable value for k. In general, as the value of k grows, more redundancy is introduced, which helps coping with mobility. In the worst case, the k-restricted regions of interest cover the entire network, and the scheme degenerates to the case of flooding the network with control packets per sender per group as in ODMRP.

Figure 2.1 illustrates the above concepts. The figure shows two multicast groups, G_1 and G_2 with their respective cores, S_c and S_j. Each group has a non-core source; S_m for G_1 and S_i for G_2. The mesh of the core S_c of G_1 is composed of nodes labeled m_{G1} and the mesh of the core S_j of G_2 is composed of nodes labeled m_{G2}. In the figure, the 1-restricted region of interest of S_m is delimited by a dotted line. We observe that it contains the mesh constructed by the core of the group S_c, which is delimited by a solid line, as well as the nodes located one hop way from the mesh or from the path from the non-core sender to the mesh. $JQnC$s generated by S_m are forwarded only by such nodes as node x or node y, which are located inside of

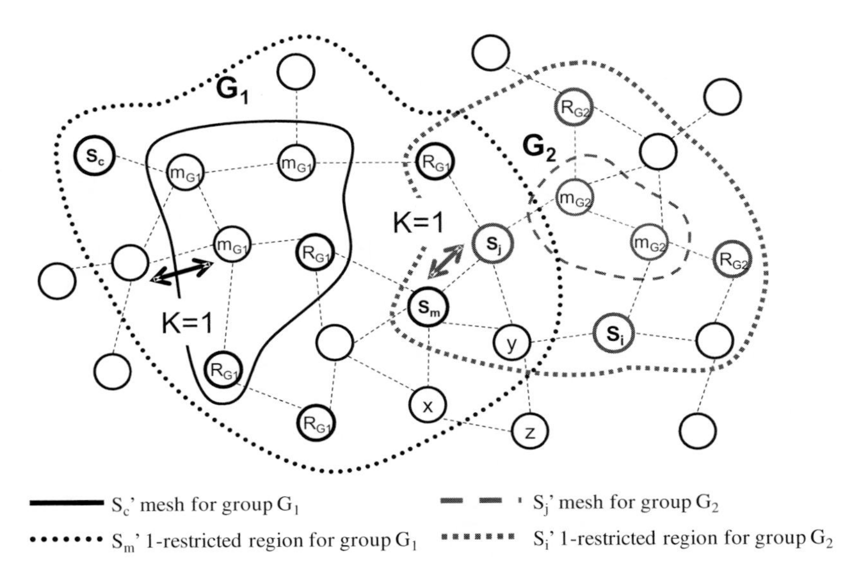

Fig. 2.1. k-restricted region. A k-restricted region of a non-core sender is a set of nodes that forward Join Queries generated by that source.

the 1-restricted region of interest, and nodes located outside of this region, such as node z, may receive the packet but does not forward it. The figure also presents a similar situation for group G_2.

The second way to reduce the number of *JQnC*s sent to the network and the state kept at nodes is to find common sub-graphs and perform multicast state aggregation on these particular regions of the network. Nodes located in a common sub-graph only receive and forward *JQ*s (or *JQnC*s) of one of the sources that share that sub-graph and keep state about the source whose join queries are forwarded. Figure 2.2 shows an example of a network in which two sources, S_c and S_2 share a common sub-graph. With this goal, we propose the *Dissemination of Multicast Aggregated-State* (*DIMAS*) algorithm. From the standpoint of message complexity, DIMAS behaves as simple flooding in the worst case. However, depending on the perceived current topology of the network, DIMAS stops disseminating control packets of non-core senders before covering the whole k-restricted region.

To do so, nodes determine if they are located in the boundary of a region that would likely be ordered by a *JQ* of a given source (say S_i) in an equivalent way as it was already ordered by a previous dissemination of *JQ*s generated by a different source (say S_j). If this is the case, then nodes stop the dissemination of control packets from S_i and mark that source as aggregated with S_j. Beyond this point, data packets generated by S_i are forwarded as if they were data packets from S_j. DIMAS (i, snd) is executed at node i when it is about to relay a *JQnC* from sender *snd* to decide whether the message is sent or the sender *snd* is aggregated at node i. The three rules used to decide when to aggregate are described below.

Rule 1: Upon reception of a *JQnC* with a larger sequence number, nodes wait a period of time equal to *FWD_DLY* to collect packets forwarded by other neighbors. Based on the distances stated in these *JQnC*, nodes compute their own distance to

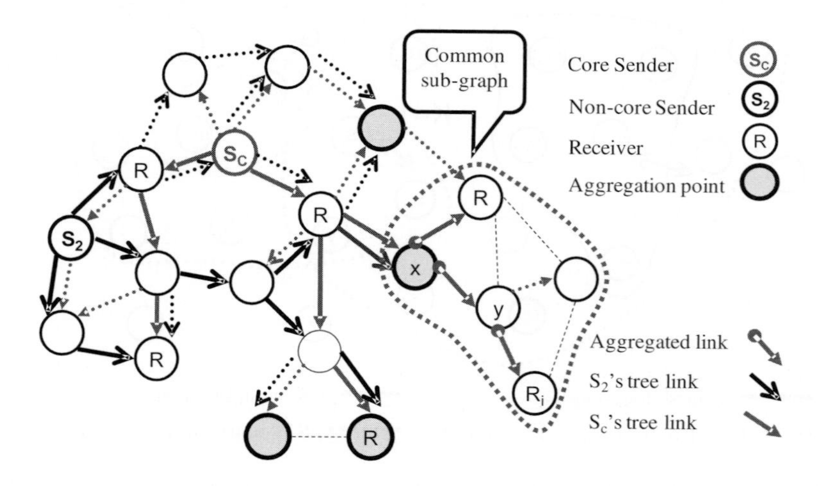

Fig. 2.2. Aggregated sub-graphs.

the source and a set of pairs $(neighbor, gradient)$ where the gradient is computed as the node's distance minus the distance reported by each neighbor. Then, nodes check if they have recently received JQs or $JQnC$s from other source, such that the set $\{(neighbor, gradient) : gradient \geq 0\}$ matches with the one computed for the current source. If that is the case, nodes do not forward the $JQnC$ and mark the senders as aggregated. If there is no match, nodes forward their own $JQnC$ (with their computed distance).

Rule 2: If a node receives $JQnC$s generated by different sources at roughly the same time (within a FWD_DLY period) and there is a match between the sets of gradient pairs, the node forwards the control packet corresponding to the source with the *largest identifier* and stops the control packets corresponding to the other sources.

Rule 3: The core source cannot be aggregated to any non-core source, because the JQs generated by the core must cover the whole network. Aggregation is allowed only either among non-core sources, or with the core aggregating non-core sources.

$DIMAS$ has two different variants. The first is called Total-DIMAS and is implemented by Hydra-TA which allows the aggregation of a non-core source with the core or any other source. The second variant is called Core-DIMAS and is implemented by Hydra-CA. This latter version only allows the aggregation of non-core senders with the core but not with other non-core senders.

2.3.2.1.1. Disseminating Aggregation Maps.

JQs and $JQnCs$ can be augmented with an *aggregation map* containing pairs of node identifiers of the form $(aggregated, aggregatedWith)$, where *aggregated* is the identifier of a source that is aggregated to other source with identifier $aggregatedWith$. The aggregation maps are stored at downstrem nodes and can be used to decide which source's routing information has to be used when forwarding a data packet of an aggregated source. Aggregation maps permit the forwarding of multicast data packets from sources whose state has been aggregated. In effect, they are routing-table extensions, as we discuss in Section 2.3.4.

2.3.2.2. *Join replies*

After receiving a JQ or a $JQnC$ with a fresher sequence number, a receiver generates a *Join Reply* (JR). A JR contains a packet type identifier, the address of the group, the address of the sender (either core or non-core), the distance to the sender, the address of the selected parent and a sequence number.

A JR is routed back to the source following the reverse direction of the gradient of the distances established by the partial ordering obtained from the diffusion of JQs. JRs travel hop-by-hop forcing intermediate notes to join the multicast routing structure, until the JRs reach the first node that is already part of the multicast routing structure or a multicast source. If a JR is generated inside of a shared sub-graph it travels along the shared region establishing routing state only about a single source. However, as soon as the JR reaches a node in which one or more $JQnC$ were stopped by the DIMAS algorithm, a new JR is generated for each $JQnC$ that was stopped in that particular node. Then, these JRs follow their own ways towards the different sources.

From the example shown in Figure 2.2, we observe that a JR generated by receiver R_i travels to node y, and establishes routing state regarding only the core S_c. However, when the JR reaches the aggregation point x, two independent JRs are forwarded towards S_c and S_1. As it is also shown in the figure, these two JRs can follow independent paths towards their respective sources.

When nodes forward JRs, they can select one or more parents to reach a source. In the first case the resulting routing structure is a tree, while in the latter case it is a mesh. In Hydra, we chose to keep state per source because we want to avoid forwarding data packets to places where they are not needed, e.g. toward other senders. This last design decision implies an increase in the state kept at the nodes. However, as our simulations show, bandwidth is a much more stringent bottleneck than memory, and spending extra memory in order to save bandwidth is a good tradeoff. Furthermore, keeping state per sender is necessary if the protocol has to support a source-specific multicast service model.[4]

2.3.3. *Non-destructive core election*

If a source has data to send to a multicast group, it first determines whether it has received a JQ from the core of that group. If the node has, it adopts the core specified in the JQ it has received, and it transmits a $JQnC$ that also advertises the same core for the group. Otherwise, it considers itself the core of the group and starts transmitting JQs periodically to its neighbors, stating itself as the core of the group and a 0 distance to itself. Nodes propagate JQs based on the best JQ they receive from their neighbors. A JQ with a higher core ID is considered better than a JQ with a lower core ID. Eventually, each connected component has only one core. If a sender becomes active for a group before other senders, then it becomes the core of the group. If several senders become active concurrently, then the one with the highest ID is elected the core of the group.

A core election is also held if the network is partitioned. The election is held in the connected component of the partition that does not have the old core. A node detects a partition if it does not receive a fresh JQ from the core for three consecutive join query intervals. Once a sender detects a partition and it has data to send, it promotes itself to the rank of core and participates in the core election. JQs from nodes with lower IDs are not discarded and the routing information regarding those senders is not destructed. Instead, JQs are just demoted to $JQnC$ and they are forwarded using the k-restricted scheme and become susceptible of being stopped by the DIMAS algorithm. This scheme contrasts with the destructive schemes used in the past core-based multicast routing protocols (e.g. PUMA) in which the partial routing structure built by the dissemination of *JQ*s of senders that contended and loose an election is eliminated when *JQ* from senders with larger IDs are received.

The way in which two partitions are merged depends on the type of join queries that traverse from one partition to the other. If a *JQ* reaches a new partition with a "better" core, then it is demoted to a *JQnC* and disseminated accordingly in that region of the network. The node or nodes that received the *JQ* from the core with a smaller ID check if they have recently forwarded a *JQ* for the current core (for instance, within the last 100 mS). If not, then they send *JQ*s that merge the partitions. While traversing the region of the network with the smaller core, *JQ*s force nodes to change to the new core but do not destroy their current routing information regarding the previously known sources. When the *JQ* is received by the senders located in the previously different partition with the smaller core, they generate new *JQnC*s with larger sequence numbers if at least one-third of the join query period has elapsed since the last transmission of a join query.

For the case of non-core join queries we have the following options. If a *JQnC* reaches a previously different partition with a better core, then the behavior is analogous to the one just described with the only difference that there is no need of demoting the message because it is already a *JQnC*. On the other hand, when a *JQnC* arrives to a previously different partition with a smaller core, nodes that first receive the message aggregate the core stated at the arriving *JQnC* with the sender that originated it, and relay the *JQnC* but now stating as a core the core with the smaller id. Nodes located at the border of the region of the network with the core with smaller ID are allowed to perform aggregation because (1) nodes in that region have not received a *JQ* from the core with larger ID (otherwise that sender would be the core), and (2) it is certain that their links are cut links between the core with the larger ID and the receivers that may be located at the region with the core with smaller ID. If the two regions remain connected long enough, then the next *JQ* generated by the core with larger ID will force all nodes in the network to have a single core.

2.3.4. *Forwarding multicast data packets*

When a source has data to send, it first checks whether it has received at least one *JR* with the same sequence number as the last transmitted *JQ* or *JQnC*. If it is the case, the source considers the node from which it received the *JR* a child and transmits the data packet. If the source does not have any child, then it checks if has elapsed *ALLOW_NEXT_JQ* time since the last time it sent either a *JQ* or a *JQnC*. If so, it piggybacks the data packet in a *JQ* (or *JQnC*) with a newer sequence number and transmits it. Otherwise, the packet is silently dropped.

A multicast data packet received from a sender s_i is discarded by a node if a hit is found in the packet cache at the node based on the packet's sender and

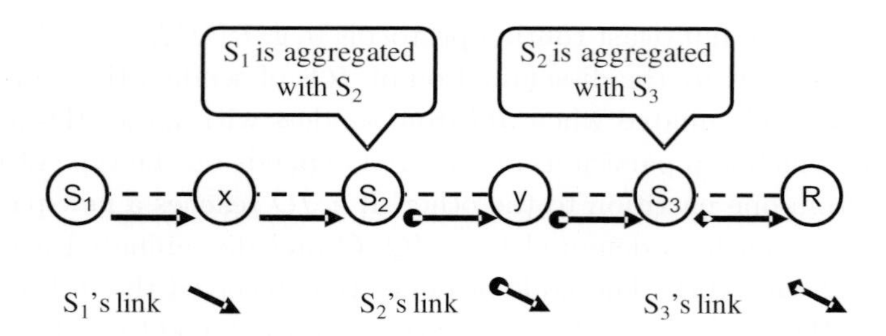

Fig. 2.3. Aggregated path. The $S_1 \rightsquigarrow R$ path is an aggregated path composed of the $S_1 \rightsquigarrow S_2$, $S_2 \rightsquigarrow S_3$ and $S_3 \rightsquigarrow R$ sub-paths.

its sequence number. Otherwise, the receiving node inserts the (*sender's address, sequence number*) pair in its packet cache and determines the *address of the effective source* (es) of the packet, which is the one used to decide whether the packet has to be forwarded or not. The address of the effective source for a given packet is the original source s_i, or s_j if s_i was aggregated with s_j at the current node. Once the node determines the value of es, it forwards the data packet if it has at least one child for es. The effective source is obtained from the aggregation map maintained by each node.

Figure 2.3 shows a simple example of a path composed of three concatenated paths, each of which corresponds to a path established for an aggregated source. In the figure, data packets generated by S_1 are routed by x using S_1 as the address of the effective source (es). When a S_1's packet reaches S_2, it determines from its aggregation map that the effective source for S_1 is itself (S_2) and forwards the packet accordingly using its routing table. Nodes along the subpath from S_2 to S_3 similarly determine that S_2 is the effective source for S_1. At node S_3, the effective source becomes S_3 itself, and the same is true for the relays in the subpath from S_3 to R.

If aggregation maps are not communicated among neighbors as part of *JQnCs* and are maintained only locally, the original multicast data packet can be encapsulated in another multicast data packet with the aggregating source as the sender and the same group as the destination. At each hop, a relay decapsulates and encapsulates the packet before forwarding it.

2.4. PUMA

As well as Hydra, PUMA supports the IP multicast service model of allowing any source to send multicast packets addressed to a given multicast group, without having to know the constituency of the group. And sources need not join a multicast group in order to send data packets to the group. Like CAMP and MAODV, PUMA uses a receiver-initiated approach in which receivers join a multicast group using the address of a special node (core in CAMP or group leader in MAODV), without the need for network-wide flooding of control or data packets from all the sources of a group. PUMA implements a distributed algorithm to elect one of the receivers of a group as the core of the group, and to inform each router in the network of at least one next-hop to the elected core of each group.

Every receiver can be connected to the elected core along all shortest paths between the receiver and the core. Nodes on shortest paths between any receiver and the core are candidate mesh members. The actual number of shortest paths used to establish the mesh depends on the number of next hops to the core that are forced to join the mesh. A sender sends a data packet to the group along any of the shortest paths between the sender and the core. When the data packet reaches a mesh member, it is flooded within the mesh. Nodes maintain a packet ID cache to drop duplicate data packets.

PUMA uses a single control message for all its functions, the multicast announcement. Each multicast announcement specifies a sequence number, the address of the group (group ID), the address of the core (core ID), the distance to the core, a mesh member flag that is set when the sending node belongs to the mesh, and a parent that states the preferred neighbor to reach the core. With the information contained in such announcements, nodes elect cores, determine the routes for sources outside a multicast group to unicast multicast data packets towards the group, notify others about joining or leaving the mesh of a group, and maintain the mesh of the group.

2.4.1. *Connectivity lists and propagation of multicast announcements: mesh establishment and maintenance*

A node that believes itself to be the core of a group transmits multicast announcements periodically for that group. As the multicast announcement travels through the network, it establishes a data structure known as connectivity list at every node in the network. A node stores the data from all the multicast announcements it receives from its neighbors in the connectivity list. Fresher multicast announcements from a neighbor (i.e. one with a higher sequence number) overwrite entries with lower sequence numbers for the same group. Each entry in the connectivity list, in addition to storing the multicast announcement, also stores the time when it was received, and the neighbor from which it was received. The node then generates its own multicast announcement based on the best entry in the connectivity list: for the same core ID, only multicast announcements with the highest sequence number are considered valid. For the same core ID and sequence number, multicast announcements with smaller distances to the core are considered better. When all those fields are the same, the multicast announcement that arrived earlier is considered better. The distance to the core of the current node is one plus the distance to core reported in the best multicast announcement. The parent field is filled with the address of the neighbor from which the best multicast announcement was received. The mesh membership flag indicates if the current node is a mesh member. In general a node is a mesh member if it has recently heard about at least one neighbor with a larger distance to the core and that is either a receiver or a mesh member.

After receiving a multicast announcement with a fresh sequence number, nodes wait for a short period (e.g. 100 ms) to collect multicast announcements from multiple neighbors before generating their own multicast announcement. When multiple groups exist, nodes group all the fresh multicast announcements they receive into a single control packet, and broadcast them periodically every multicast announcement interval. However, multicast announcements representing groups being heard for the first time, resulting in a new core, or resulting in changes in mesh membership status are forwarded with a much smaller delay. This is to avoid large delays in critical operations, like core elections and mesh establishment which could lead to a delay in establishing the correct mesh, and could lead to packet drops as well as unnecessary transmissions of data packets. However, announcements of this type are not sent immediately, this, to avoid oscillations in the mesh establishment process and to allow opportunistic grouping of two or more announcements in a single control packet.

A node also generate new multicast announcement when it detects a change in its mesh member status. This could occur when a node detects a mesh child for the first time, or when a node that previously had a mesh child detects that it has no mesh children. This way, only nodes lying in shortest paths from receivers to the core are forced to become mesh members.

2.4.2. *Core election*

When a receiver needs to join a multicast group, and similar to Hydra, it first determines whether it has received a multicast announcement for that group. If the node has, it adopts the core specified in the announcement it has received, and it starts transmitting multicast announcements that specify the same core for the group. Otherwise it considers itself the core of the group and starts transmitting multicast announcements periodically to its neighbors stating itself as the core of the group and a 0 distance to itself. Nodes propagate multicast announcements based on the best multicast announcements they receive from their neighbors. A multicast announcement with higher core ID is considered better than a multicast announcement with a lower core ID. Eventually, each connected component has only one core. If one receiver joins the group before other receivers, then it becomes the core of the group. If several receivers join the group concurrently, then the one with the highest ID becomes the core of the group.

A core election is also held if the network is partitioned. The election is held in the partition which does not have the old core. A node detects a partition if it does not receive a fresh core announcement for 3 × *multicast announcement interval*. Once a receiver detects a partition, it behaves in exactly the same way it would upon joining the group, and participates in the core election.

2.4.3. *Forwarding multicast data packets*

The parent field of the connectivity list entry for a particular neighbor corresponds to the node from which the neighbor received its best multicast announcement. This field allows nodes that are non-members to forward multicast packets towards the mesh of a group. A node forwards a multicast data packet it receives from its neighbor if the parent for the neighbor is the node itself. Hence, with no need of packet encapsulation, multicast data packets move hop-by-hop, until they reach mesh members. The packets are then flooded within the mesh, and group members use a packet ID cache to detect and discard packet duplicates.

2.5. Performance Results

In this section we present simulation results in which we compare PUMA and three different variants of Hydra against ODMRP. The three variants of Hydra consist of using the Total-Aggregation algorithm (Hydra-TA), the Core-Aggregation Algorithm (Hydra-CA) or no aggregation (Hydra-NA). We chose ODMRP for our comparison because it is a representative of the state of the art in multicast protocols for MANETs, and because it constructs its forwarding mesh as the union of the source-specific trees of the senders of a particular group. Given that Hydra also builds a structure that is close to a set of source-specific trees or meshes, comparing Hydra against ODMRP allows us to highlight the benefits obtained by the signaling used in Hydra. On the other hand, the results obtained by PUMA allow us to identify the tradeoffs of further reducing the control overhead but at the expense of increased data overhead caused by having meshes that are not composed of shortest paths from senders to receivers.

We use packet delivery ratio, average end-to-end delay and average number of packets relayed per packet received at receivers as our performance metrics. The latter metric can be seen as the cost in terms of bandwidth that the protocol pays to achieve a given delivery ratio.

The multicast protocols are tested with IEEE 802.11 and TDMA as the underlying MAC protocols. The former is the most commonly used MAC in the literature of multicast protocols for MANETs and the latter allows us to isolate the multicast signaling and construction of routing structures from the effects of collisions at the MAC layer. We used the discrete event simulator Qualnet[11] version 3.9. The software distribution of Qualnet itself has the ODMRP code, which was used for the ODMRP simulations. Each simulation was run for ten different seed values. To have meaningful comparisons, all the protocols use the same join query intervals (or multicast announcement interval in the case of PUMA) of 3 seconds for 802.11 and 60 seconds for TDMA. For ODMRP, the forwarding group timeout was set to three times the value of the join query interval, as advised by its designers. Figure 2.4 lists the details of the simulation environment. For Hydra, we set the value of k of

Simulation Environment					
Total Nodes	50	Node Placement	Random	Data Source	MCBR
Transmission Power	15 dbm	Mobility Model	Random Waypoint	Pkts. sent per src.	1000
Channel Capacity	2000000 bps	Simulation Area	1400x1400m		
MAC Protocol 802.11					
Simulation Time	150s	Pause Time	10s	Min-Max Vel.	1-20m/s
MAC Protocol TDMA (1 slot per node in round-robin)					
Simulation Time	30min	Pause Time	50s	Min-Max Vel.	1-2m/s

Fig. 2.4. Simulation environment.

the *k-restricted* region to 1. Hence, only nodes that are at most one hop away from the routing structure of the core disseminate $JQnC$s. For PUMA, each node selects two parents (if availabe) to reach the core of the group.

2.5.1. *Results using 802.11 MAC*

We first focus our attention on an experiment in which the number of concurrent active senders changes. Each sender transmits 20 packets of 256 bytes per second and the group is composed of 20 nodes. We observe from Figure 2.5 that for up to 6 concurrent sources, all the versions of Hydra (TA, CA and NA) consistently outperform ODMRP and PUMA, but, starting from 9 sources, PUMA performs

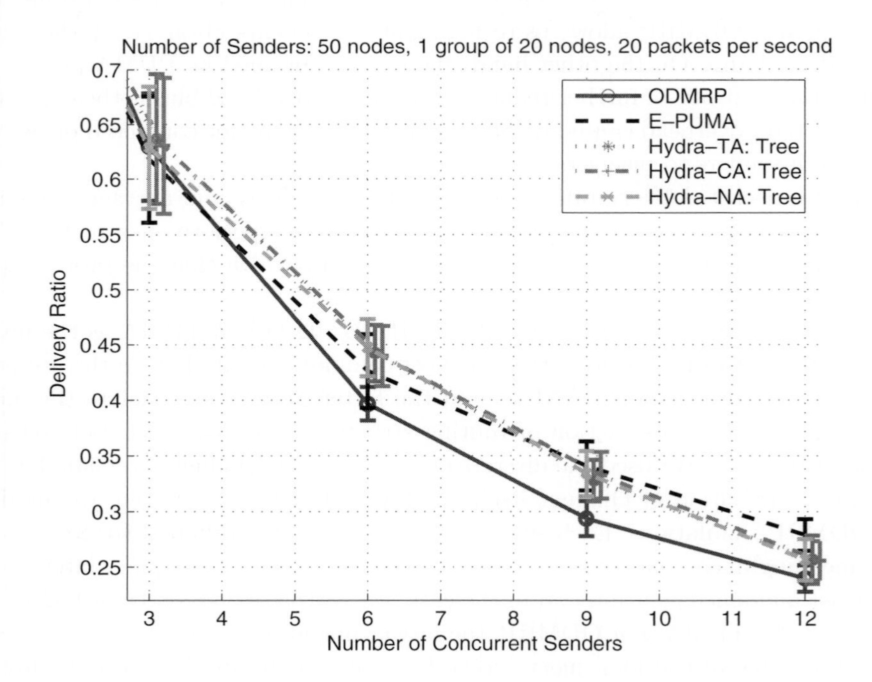

Fig. 2.5. Delivery ratio when varying the number of concurrent active sources with 802.11 MAC.

similar or better than all the other protocols. These results show that the reduced amount of control overhead induced by PUMA allows it to scale better than the other protocols that induce control overhead per source per group. A detailed analysis of the simulations reveals that ODMRP and Hydra reach first the point where the combined overhead induced by control and data traffic does not allow the correct establishment of the routing structures. On the other hand, Hydra performs better than PUMA for up to 6 senders because the routing structures established by Hydra which are close to a set of source specific breath-search trees are more efficient than the shared-trees rooted at an arbitrary receiver built by PUMA.

Among the Hydra variants we can observe that the versions that use aggregation are capable of attaining delivery ratios equivalent to the ones attained by the versions that do not use aggregation, but incurring less control overhead. For these experiments, the tree and mesh versions of Hydra-TA transmitted an average of 77.89% and 78.73% of the *Join Queries* ($JQ + JQnC$) transmitted by the tree and mesh versions of Hydra-NA, respectively.

Figure 2.6 shows the average number of data packets relayed per packet received by receivers. We can observe that in general, PUMA employs more transmissions to cover the receivers of the group. This is due to the fact that the PUMA's mesh-establishment process does not take into account the location of the sources. Among

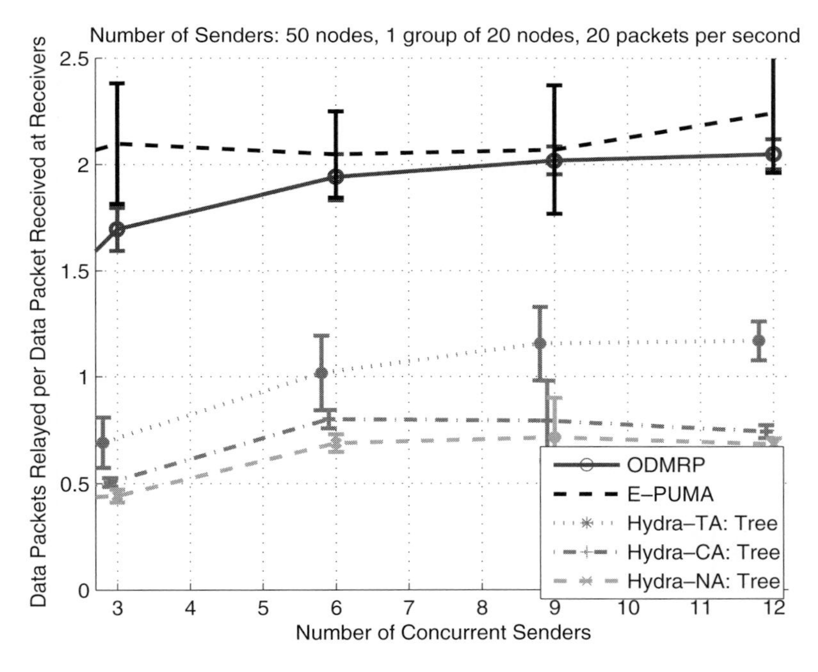

Fig. 2.6. Data packets relayed per data packet received at receivers when varying the number of concurrent active sources with 802.11 MAC.

the protocols that use source-rooted trees, we can see that ODMRP incurs considerably more redundancy than the Hydra variants. As we would expect, the variant of Hydra that uses less redundancy is the tree version with no aggregation, while the one that uses more redundancy is the mesh version with total aggregation. In general, the versions that use total aggregation generate more redundancy than the core aggregation versions, which in turn generate more redundancy than the versions that do not use aggregation. This is due to the fact that the no-aggregation versions try to establish source-specific shortest-path trees or meshes, while the structures created by the aggregated versions are not necessarily composed of shortest paths. If we analyze these two metrics, we can notice how the versions of Hydra are able to attain higher delivery ratios at a much smaller cost than ODMRP. This is a strong indication that the routing structures built by Hydra are more efficient than the ODMRP's mesh that is composed of the union of the source-rooted trees of the sources of the group. We can also observe that even when PUMA has more data overhead its reduced control overhead pays off as the number of senders increases.

Figure 2.7 shows the average end-to-end delay attained by the protocols. We observe a situation similar to the delivery ratio, namely, that the versions of Hydra perform similar or better than the other protocols for up to 6 sources and that PUMA attains the lowest end-to-end delay for more than 6 sources. There are two important factors behind this behavior. The first one is the queueing delay, which

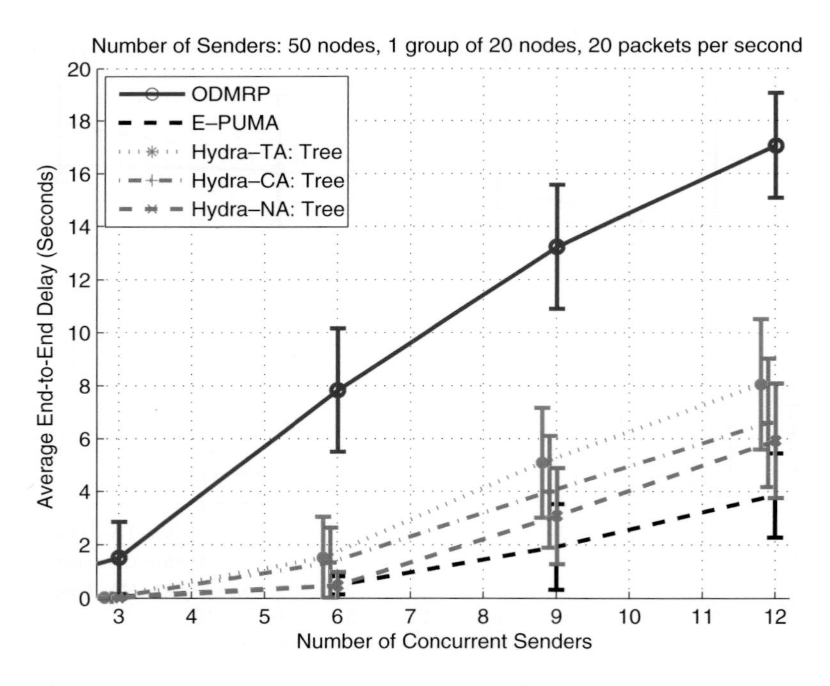

Fig. 2.7. Averge end-to-end delay when varying the number of concurrent active sources with 802.11 MAC.

is strongly related to the amount of control overhead and redundancy used by the protocol. Here is also important to point out that in the context of IP, the queue used for control traffic has higher priority than the data queue, hence the control overhead tends to have a considerable effect in the end-to-end delay. Therefore, protocols that incur less control overhead and redundancy when forwarding data packets tend to attain lower end-to-end delays than the ones with increased level of control overhead or redundancy. The other factor is the length of the paths followed by the data packets.

Figure 2.8 summarizes a number of simulation results for three concurrent sources in different scenarios: node mobility, node density and group size. The results clearly show that Hydra and PUMA render similar delivery ratios and with close to an order of magnitude improvement in delay. Hydra also has the advantage of incurring far less data overhead (about half).

2.5.2. *Results using TDMA MAC*

The results shown in Figures 2.9–2.11 present a picture of the behavior of the protocols when the transmission of control information (by assigning control packets to a higher priority queue) is isolated from the effects of the data traffic. For these experiments, each sender transmits 1 packet of 256 bytes per second and the group is composed of 20 nodes. TDMA allocates 1 slot of 10 mS per node in round-robin fashion by address. For one sender, all the protocols have equivalent performance in terms of delivery ratio and end-to-end delay, as should be expected. As in the case of 802.11 MAC, PUMA incurs more transmission overhead because its mesh is not rooted at the sender. However, as the number of sources increases, the performance of ODMRP drops sharply, which is due to the per source per group flooding strategy it uses. For six or more sources, the Hydra family and PUMA provide an improvement of around 30% in delivery ratio and up to 40% in end-to-end delay. The decreased amount of overhead shown by ODMRP for 12 sources is due to the fact that much more data packets are being dropped early. For 12 sources, the reduced control overhead of PUMA allows it to scale better than the Hydra family showing an improvement of around 5% in delivery ratio and a more considerable improvement of close to 30% in end-to-end delay. The reason of the increased improvement in delay is that PUMA generates much less high-priority control traffic that is sent before the low-priority data traffic.

2.6. Conclusions

We described Hydra and an improved version of PUMA which are two mesh-based multicast routing protocols for MANETs. Hydra is the first multicast routing protocol that establishes routing structures that approximate those built with sender-initiated approaches, but incurring the communication overhead of a

Mobility (pause time)	0S		5S		10S		15S		20S	
	Avg	SD	Avg	SD	Avg	SD	Avg	SD	Avg	SD
	Delivery Ratio									
ODMRP	0.632	0.037	0.639	0.055	0.630	0.049	0.629	0.045	0.630	0.050
PUMA	0.635	0.066	0.619	0.063	0.618	0.057	0.62	0.055	0.62	0.057
Hydra-TA	0.656	0.057	0.646	0.048	0.644	0.057	**0.646**	0.056	0.635	0.054
Hydra-CA	0.649	0.053	**0.651**	0.060	0.637	0.049	0.641	0.057	**0.643**	0.059
Hydra-NA	**0.657**	0.046	0.649	0.057	**0.653**	0.057	0.645	0.060	**0.643**	0.059
	Number of Data Packets Relayed per Data Packet Received at Receivers									
ODMRP	1.714	0.125	1.699	0.090	1.696	0.101	1.727	0.113	1.695	0.120
PUMA	1.918	0.229	1.976	0.251	2.099	0.282	1.999	0.255	2.064	0.276
Hydra-TA	0.986	0.140	1.064	0.171	1.020	0.146	1.001	0.184	1.034	0.162
Hydra-CA	1.005	0.182	0.948	0.113	0.958	0.158	0.940	0.176	0.923	0.167
Hydra-NA	**0.864**	0.155	**0.866**	0.122	**0.822**	0.084	**0.889**	0.146	**0.848**	0.085
	End-to-End Delay (S)									
ODMRP	1.691	1.374	1.570	1.308	1.513	1.349	1.594	1.490	1.291	1.232
PUMA	**0.038**	0.021	**0.046**	0.029	0.056	0.053	**0.063**	0.084	0.070	0.077
Hydra-TA	0.166	0.360	0.180	0.288	0.101	0.164	0.172	0.427	0.134	0.247
Hydra-CA	0.179	0.311	0.049	0.048	0.107	0.174	0.078	0.146	0.068	0.115
Hydra-NA	0.092	0.143	0.076	0.107	**0.026**	0.004	0.082	0.114	**0.036**	0.022
Terrain Dimensions	1200x1200m		1300x1300m		1400x1400m		1500x1500m		1600x1600m	
	Delivery Ratio									
ODMRP	0.659	0.029	0.646	0.033	0.630	0.049	**0.619**	0.049	**0.597**	0.070
PUMA	0.669	0.053	0.648	0.063	0.618	0.057	0.594	0.059	0.564	0.065
Hydra-TA	0.685	0.048	0.670	0.052	0.644	0.057	0.618	0.057	0.586	0.062
Hydra-CA	0.703	0.052	0.672	0.049	0.637	0.049	0.618	0.062	0.587	0.059
Hydra-NA	**0.709**	0.053	**0.680**	0.052	**0.653**	0.057	0.617	0.064	0.589	0.062
	Number of Data Packets Relayed per Data Packet Received at Receivers									
ODMRP	1.660	0.119	1.699	0.128	1.696	0.101	1.663	0.137	1.690	0.093
PUMA	1.771	0.342	1.763	0.359	2.099	0.282	2.149	0.165	2.172	0.146
Hydra-TA	1.050	0.258	1.017	0.194	1.020	0.146	0.994	0.132	1.092	0.082
Hydra-CA	0.888	0.225	0.967	0.246	0.958	0.158	0.982	0.096	1.010	0.112
Hydra-NA	**0.802**	0.231	**0.846**	0.187	**0.822**	0.084	**0.851**	0.087	**0.917**	0.094
	End-to-End Delay (S)									
ODMRP	2.875	1.531	2.098	1.575	1.513	1.349	0.854	0.972	0.464	0.602
PUMA	**0.123**	0.163	**0.039**	0.022	**0.056**	0.053	0.064	0.084	0.036	0.014
Hydra-TA	0.667	0.765	0.267	0.382	0.101	0.164	0.058	0.071	0.038	0.016
Hydra-CA	0.384	0.598	0.313	0.542	0.107	0.174	0.033	0.009	0.031	0.007
Hydra-NA	0.345	0.645	0.175	0.304	0.026	0.004	**0.027**	0.005	**0.027**	0.004
Group size (receivers)	10		15		20		25		30	
	Delivery Ratio									
ODMRP	**0.656**	0.062	0.636	0.052	0.630	0.049	0.625	0.034	0.620	0.045
PUMA	0.589	0.065	0.61	0.071	0.601	0.062	0.618	0.065	0.612	0.069
Hydra-TA	0.650	0.063	0.651	0.055	0.644	0.057	0.639	0.058	0.634	0.050
Hydra-CA	**0.656**	0.055	**0.654**	0.048	0.637	0.049	**0.649**	0.048	0.645	0.054
Hydra-NA	0.647	0.059	**0.654**	0.055	**0.653**	0.057	**0.649**	0.058	**0.653**	0.061
	Number of Data Packets Relayed per Data Packet Received at Receivers									
ODMRP	2.857	0.248	2.145	0.158	1.696	0.101	1.368	0.095	1.200	0.067
PUMA	2.459	0.854	1.979	0.571	1.756	0.264	1.412	0.205	1.213	0.160
Hydra-TA	1.464	0.215	1.158	0.206	1.020	0.146	0.855	0.182	0.745	0.094
Hydra-CA	1.304	0.207	1.080	0.124	0.958	0.158	0.829	0.115	0.741	0.094
Hydra-NA	**1.250**	0.140	**0.972**	0.122	**0.822**	0.084	**0.734**	0.116	**0.647**	0.126
	End-to-End Delay (S)									
ODMRP	0.723	0.75	1.298	1.274	1.513	1.349	1.53	1.381	1.79	1.411
PUMA	0.031	0.038	0.036	0.042	0.027	0.007	**0.03**	0.018	**0.03**	0.021
Hydra-TA	0.033	0.027	0.086	0.17	0.101	0.164	0.271	0.488	0.286	0.402
Hydra-CA	0.022	0.003	0.034	0.021	0.107	0.174	0.185	0.252	0.219	0.361
Hydra-NA	**0.02**	0.002	**0.02**	0.004	**0.03**	0.004	0.097	0.152	0.14	0.286

Fig. 2.8. Mobility, node density, and group size with 802.11 MAC.

receiver-initiated approach. This is accomplished by limiting the dissemination of control information from non-core sources to small regions of the network, and aggregating routing information by establishing common sub-trees (or sub-meshes) that are shared by two or more senders. Hydra can work in either mesh or tree mode

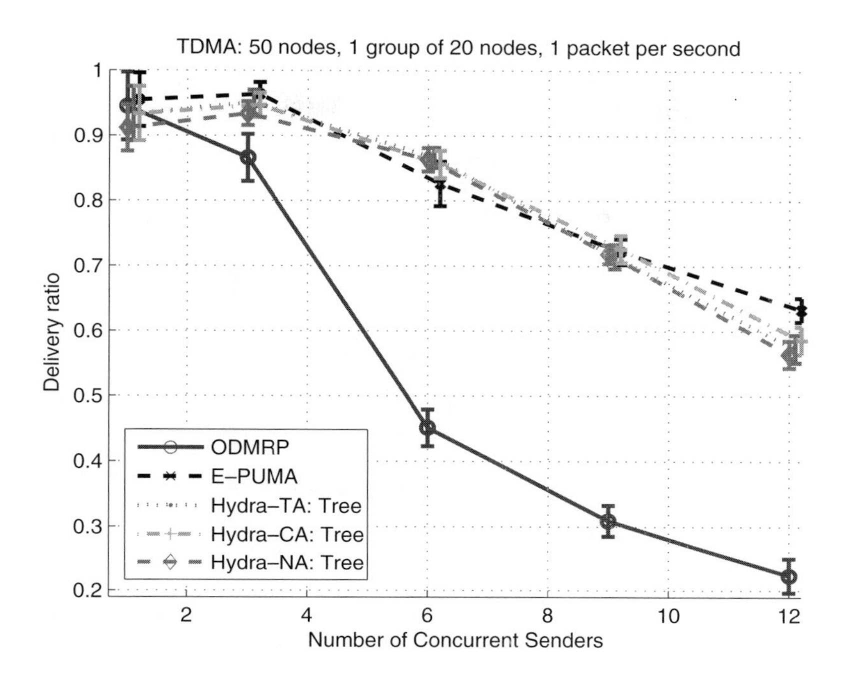

Fig. 2.9. Delivery ratio when varying the number of concurrent active sources with TDMA MAC.

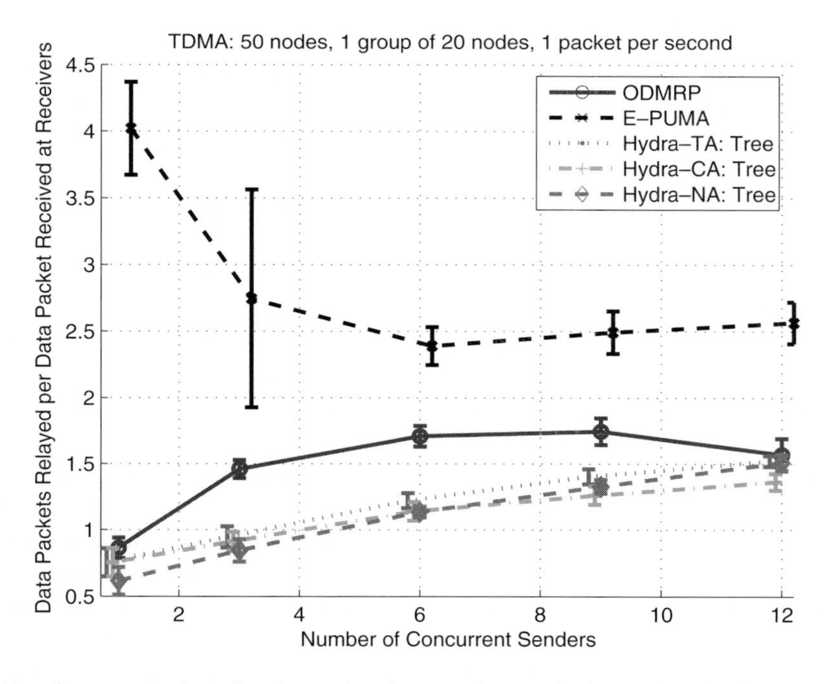

Fig. 2.10. Average number of packets relayed per packet received at receivers when varying the number of concurrent active sources with TDMA MAC.

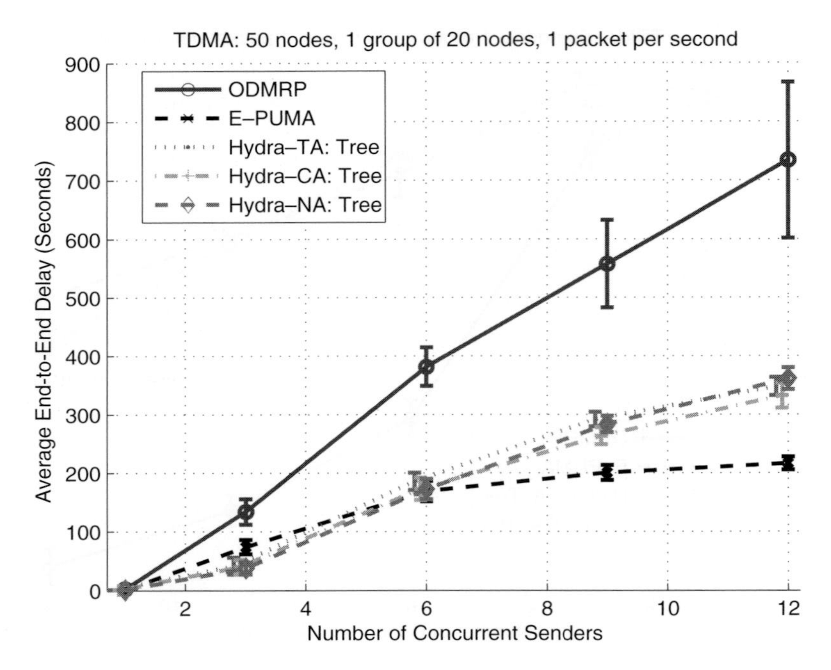

Fig. 2.11. Average end-to-end delay when varying the number of concurrent active sources with TDMA MAC.

by restricting the number of parents that are forced to join to the routing structure. Hydra takes advantage of the work carried out by a core but a core is not necessary for correctness. Cores are elected using a non-destructive core election protocol that does not destroy recently established trees (or meshes). To save bandwidth and take advantage of the broadcast nature of the wireless medium, Hydra opportunistically combines control messages from different groups and sources into a single control packet. PUMA uses a receiver-initiated approach in which receivers join a multicast group using the address of a special node (core), without the need for network-wide dissemination of control or data packets from all the sources of a group. PUMA implements a distributed algorithm to elect one of the receivers of a group as the core of the group, and to inform each router in the network of at least one next-hop to the elected core of each group. Within a finite time, each router has one or multiple paths to the elected core. All nodes on shortest paths between any receiver and the core collectively form the mesh of the multicast group. A sender node can send packets to the multicast group by encapsulating them in unicast packets to the core along any of the paths to the core. The version of PUMA presented here opportunistically combines multicast announcement of different multicast group in a control packet which enables it to react faster than the original PUMA protocol to changes in the topology of the network. We also presented the results of a series of simulation experiments using 802.11 and TDMA MAC protocols illustrating that

Hydra and PUMA attain higher delivery ratios and considerably lower end-to-end delays than ODMRP, and in the case of Hydra, inducing far less data retransmission overhead, even in relatively small networks with few multicast sources. Even though PUMA and Hydra scale better than ODMRP, we are far from having a really scalable multicast protocol for MANETs, hence further research is needed.

Acknowledgments

This work was partially sponsored by the US Army Research Office (ARO) under grants W911NF-04-1-0224 and W911NF-05-1-0246, by the National Science Foundation (NSF) under grant NS-0435522, by the Defense Advanced Research Projects Agency (DARPA) through Air Force Research Laboratory (AFRL) Contract FA8750-07-C-0169, by the Baskin Chair of Computer Engineering, by the CONACyT-UC MEXUS program, and by the Mexican National Polytechnic Institute. The views and conclusions contained in this document are those of the authors and should not be interpreted as representing the official policies, either expressed or implied, of the Defense Advanced Research Projects.

Problems

(1) Explain the mechanism used by ODMRP, PUMA and Hydra to discard duplicate packets. Why is this necessary?
(2) Mention some of the advantages of selecting a core or leader dynamically instead of having it predetermined.
(3) Mention some of the advantages and disadvantages of mesh-based protocols against their tree-based counterparts?
(4) Why are core-based multicast routing protocols more prone to create contention hot-spots than their source-specific counterparts?
(5) Why does the control overhead induced by source-specific multicast routing protocols such as ODMRP grow faster than the overhead induced by core-based protocols such as PUMA?
(6) Describe how the average end-to-end delay attained by multicast routing protocols that run on top of contention-based MAC protocols is related with the control and data overhead they incur.
(7) In the case of ODMRP, PUMA and Hydra, what is the function of the sequence numbers included in the control packets?
(8) Describe some of the differences between the core election protocol implemented in PUMA and the one implemented in Hydra.
(9) Describe the mechanisms implemented in Hydra to reduce the number of control packets used to establish and maintain the multicast mesh.

Bibliography

1. S. K. Das, B. S. B. S. Manoj and C. S. R. Murthy, A dynamic core based multicast routing protocol for ad hoc wireless networks, *MobiHoc '02: Proc. 3rd ACM Int. Symp. Mob. Ad Hoc Net. Comput. (ACM)*, ISBN 1-58113-501-7, pp. 24–35, doi: http://doi.acm.org/10.1145/513800.513804 (2002).
2. V. Devarapalli and D. Sidhu, Mzr: a multicast protocol for mobile ad hoc networks, *Proc. IEEE Int. Conf. Commun. ICC 2001.* **3** (2001) 886–891, vol. 3, doi: 10.1109/ICC.2001.937365.
3. J. J. Garcia-Luna-Aceves and E. L. Madruga, The core-assisted mesh protocol, *IEEE J. Selected Areas Commun.* **17**(8) (1999) 1380–1394.
4. H. Holbrook and B. Cain, Source-specific multicast for ip. internet-draft (2000).
5. J. G. Jetcheva and D. B. Johnson, Adaptive demand-driven multicast routing in multi-hop wireless ad hoc networks, *MobiHoc '01: Proc. 2nd ACM Int. Symp. Mob. Ad Hoc Net. Comput. (ACM)*, ISBN 1-58113-428-2 (2001) 33–44.
6. L. Ji and M. S. Corson, Differential destination multicast-a manet multicast routing protocol for small groups, *Proc. INFOCOM 2001. Twentieth Ann. Joint Conf. IEEE Comput. Commun. Soc.* **2** (2001) 1192–1201.
7. S. Lee and C. Kim, Neighbor supporting ad hoc multicast routing protocol, *MobiHoc '00: Proc. 1st ACM Int. Symp. Mob. Ad Hoc Net. Comput. (ACM)*, ISBN 0-7803-6534-8 (2000) 37–44.
8. S.-J. Lee, M. Gerla and C.-C. Chiang, On-demand multicast routing protocol, *Proc. IEEE Wireless Commun. Net. Conf. (WCNC, 1999)* **3** (1999) 1298–1302.
9. C. E. Perkins and E. M. Royer, Ad-hoc on-demand distance vector routing, *Proc. Second IEEE Workshop Mob. Comput. Syst. Appl. (WMCSA '99)*, doi: 10.1109/MCSA.1999.749281 (1999), 90–100.
10. E. M. Royer and C. E. Perkins, Multicast operation of the ad-hoc on-demand distance vector routing protocol, *MobiCom '99: Proc. 5th Ann. ACM/IEEE Int. Conf. Mob. Comput. Net. (ACM)*, ISBN 1-58113-142-9, pp. 207–218, doi: http://doi.acm.org/10.1145/313451.313538 (1999).
11. SNT Qualnet 3.9, scalable network technologies: http://www.scalablenetworks.com (2005).
12. R. Vaishampayan and J. J. Garcia-Luna-Aceves, Efficient and robust multicast routing in mobile ad hoc networks, *Proc. IEEE Conf. Mob. Ad Hoc Sensor Syst.*, (2004) 304–313.
13. M. Weiser, The computer of the 21st century, *Sci. Am.* **265**(3) (1991) 66–75.

Chapter 3

TCP, Congestion and Admission Control Protocols in Ad Hoc Networks

Amitabh Mishra*, Baruch Awerbuch† and Robert Cole‡

Department of Computer Science
The Johns Hopkins University
**amitabh@cs.jhu.edu*
†baruch@cs.jhu.edu
‡Robert.cole@jhuapl.edu

Wireless ad hoc networks evolved as a wireless extension of the Internet which is a packet switched data network characterized by TCP/IP suite of protocols for packet routing and transport. It was generally believed that TCP/IP will seamlessly extend the reach of Internet to mobile devices. IP which is at the heart of the proliferation and popularity of wired portion of the Internet was found to be not suitable for wireless environments without significant modifications, as a result considerable research efforts were expended in developing several new routing protocols since the beginning of development of PRNET[30] and that trend still continuing with several new routing proposals presented even today. Similarly, the TCP protocol which provides the reliable end-to-end delivery of data even on unreliable wired networks has been found to have several major shortcomings while transmitting data streams on the wireless part of the network due to significant differences at the physical layer in wired and wireless networks. In this book chapter we provide the basics of TCP protocol, discuss several proposals which make it suitable for wireless environments and then present the recent efforts in providing admission control over ad hoc networks.

This chapter is divided into four major sections, namely (i) TCP and congestion control for the Internet. (ii) congestion control for ad hoc wireless networks and TCP based solutions. (iii) admission control for ad hoc networks. (iv) conclusions and further readings.

This work is partly based on the recent survey papers[22, 35] that have appeared in the literature on the topics of congestion control in ad hoc networks with significant enhancements to make the chapter more readable.

Keywords: TCP; congestion control; admission control; ad hoc wireless networks; protocols and algorithms.

3.1. TCP Overview

TCP is a connection oriented reliable data transport protocol for the Internet. The full description of TCP is available in RFC 793, RFC 1122, RFC 1323, RFC 2018, and RFC 2581 which are available from IETF.[24] In this section we briefly go over the operation of TCP and how it takes care of congestion control using different mechanisms.

TCP provides full duplex service for a point to point connection. The connection establishment process between two entities on different hosts that intend to communicate with each other requires a three way handshake. Three-way handshake results in establishment of send and receive buffers, agreements on initial sequence and acknowledgement numbers between the two communicating processes.

TCP uses cumulative acknowledgements for retransmission of missing TCP segments. TCP employs a roundtrip time (RTT) defined as the average elapsed time at the sender between transmitting a data segment and receiving an acknowledgement for it from the receiver, estimation process to provide it with the initial waiting time interval (timeout period) before it can retransmit the segment if required due to loss of the segment, data corruption etc. TCP also implements several modifications that allow it to double the time-out period if the data acknowledgement is not received within the specified time-out period. Whenever the time out event occurs TCP transmits the not yet acknowledged segments with the smallest sequence number. But each time TCP retransmits, it sets the next time interval to be twice of the previous value. The time-out intervals grow exponentially after each retransmission. When TCP sender is in this situation it believes that the acknowledgements from the receiver are getting delayed due to a minor congestion in the network on the path these data segments and their acknowledgements are traversing. So the TCP modification that leads to doubling the timeout period can be considered as a limited form of congestion control.

One of the issue with the retransmissions that are triggered due to time-out is that the time-out period after a few retransmission attempts becomes relatively long which contributes to long end-to-end delays. TCP also implements a solutions to the problem of long delays by using a fast retransmit procedure which gets triggered as follows: when a sender receives a duplicate ACK which means that it has already received an earlier ACK for this particular segment. When a sender receives a third duplicate ACK, it retransmits the segment even before the expiry of the time-out period. In this scenario TCP believes that the receipt of the triple ACK is due to missing a particular segment due to reasons other than congestion.

In addition to providing congestion control which is our focus in this chapter, it also provides flow control. The TCP uses end-to-end congestion control and believes that a TCP sender be allowed to transmit data according to perceived network congestion. If congestion is less, more data can be transmitted and conversely if congestion is heavy the sender's transmission rate has to be lower. TCP uses a variable called *congestion window* (*CW*) to implement congestion control. The congestion window in essence determines the rate at which a TCP sender can transmit segments into the network. It turns out a TCP sender roughly transmits data at a CW/RTT rate into the network.

The congestion window grows or shrinks as a result of network congestion which is determined due to occurrence of a loss event at the TCP sender. A loss event occurs when the sender timeouts or receives three duplicate ACKs for a particular segment. The congestion control algorithm of TCP has three components, namely (i) AIMD (Additive Increase and Multiplicative Decrease) — Congestion Avoidance (CA). (ii) Slow Start (SS). (iii) Reaction to time out events. In AIMD congestion window grows additively i.e. linearly when the path is congestion free and shrinks to a multiplicative decrease — half of previous value, when there is a loss event. In the SS phase congestion window doubles every RTT starting from a value of 1 and growing till a loss event occurs when it reduces to half of its existing value and again follows AIMD procedure.

The TCP reacts differently to loss events that are detected by timeout than those detected by receipts of triple duplicate ACKs. After receiving a triple duplicate ACK, the TCP reduces its congestion window to half and then increases it linearly. But when TCP sender encounters a timeout, it sets its congestion window to 1 and then grows the window exponentially. The exponential growth continues until the congestion window becomes half of the value it had before the timeout event. After reaching this point, the congestion window grows linearly as it will be for the triple duplicate ACK case.

TCP uses a variable called *threshold* to determine the window size at which slow start will end and congestion avoidance will begin. The threshold is initially set to a large value and it immediately is set to half the value of the current congestion window upon detecting a loss event. So starting from the slow start phase which the TCP sender entered due to timeout event, it increases the CW exponentially till it reaches threshold and then it starts congestion avoidance phase in which the window grows linearly as stated above. The details of this algorithm are described very well in Ref. 48 and readers should refer it.

3.2. Congestion in Mobile Ad Hoc Networks (MANETs)

As we know TCP works very well for wired Internet but in the wireless environment in which mobile ad hoc networks operate it suffers through several impairments which arise primarily due to its congestion control algorithm. In this section we briefly discuss these impairments.

The two key impairments arise: (i) due to the mobility of nodes of the ad hoc network and (ii) the shared nature of wireless multi-hop channel among several nodes. Frequent route changes due to node mobility as well as unreliable nature of wireless medium result in packet delays as well as packet losses that vary very widely. It is imperative that these losses should not be considered as losses due to congestion.

In a wireless multi-hop ad hoc network only one transmission can take place in a transmission range of a node in a neighborhood to avoid interference. Therefore links that are geographically close are not independent of each other. Because of this reason, the congestion in ad hoc network manifests itself on several nodes in a region as compared to wired Internet in which a congested router or a congested incoming or outgoing link may be an isolated, very local issue.

Packet losses in ad hoc networks are very frequent due to the use of wireless medium which is inherently lossy and these losses have nothing to do with congestion. As it happens on the Internet, the losses are mainly due to congestion. The fluctuations in the RTT on the Internet are not as rapid or severe as encountered in the wireless world because of multi-path propagation and corresponding delays. Both of these factors affect the working of the congestion control algorithm of the TCP. Some of the observations that researchers have made in the context of TCP running on the ad hoc networks:

1. Route failures trigger inappropriate TCP congestion control reactions
2. The TCP transmission time grows too fast in MANET
3. Rerouting leads to paths with dissimilar characteristics
4. Wireless links have random losses
5. Unfairness among TCP flows in MANET
6. Long feedback paths
7. Data and acknowledgements can interfere with each other
8. TCP may over-saturate the network
9. Presence of intra-flow contention
10. TCP generates lots of acknowledgements
11. TCP is not suitable for MANET environments.

These observations are further discussed in different sections of this chapter. There have been several proposed approaches in the literature that try to address the adverse effects of the observations listed above on the TCP performance and improve the transport layer performance in MANET. We discuss some of these approaches in brief in Section 3 of this chapter. The new transport protocols that have been proposed for ad hoc networks which are not based on TCP are discussed in Section 4.

3.3. TCP for Ad hoc Networks — Enhancements & Solutions

In this section we review several proposals that have been proposed to make TCP friendly to ad hoc networks. These enhancements are merged in subsections according to the observations listed in Section 2.

3.3.1. *Route failures and TCP*

In wireless ad hoc networks, routes among source and destination pairs break more often than in a wired network primarily due to the time varying nature of the wireless links as well as mobility of nodes. The route failures result into excessive delays as the network needs to initiate a new route discovery process for each of the source and destination pairs before beginning to transmit and retransmit any missing packets. During route discovery process no application data could be delivered and no corresponding acknowledgements are received as a result the TCP congestion window gets significantly reduced even though there is no sign of congestion in the network. In this section, we review some of the approaches that have been proposed in the literature to deal with this situation.

An approach called TCP-F (TCP Feedback)[7] provides congestion control for ad hoc networks which is based on the premise that TCP congestion control should be disabled for non-congestion related symptoms such as delayed ACKs or timeouts which are entirely caused by route failures.

TCP-F utilizes two control messages called Route Failure Notification (RFN) and Route Re-establishment Notification) which inform the sender of route failure and route establishment respectively. When a sender receives an RFN from an intermediate node on the path to destination or from the destination itself, it enters in a state called "snooze state" in which it freezes the TCP related variables such as timers and congestion window sizes. Any intermediate node can generate RFN or RRN if it detects the imminent route failures or hears the new route to the destination and sends these messages to the source node. A source node, upon receiving a RRN, restarts the TCP session using the previously frozen state variables.

Explicit Link Failure Notification (ELFN) technique was proposed in Ref. 23 for MANET which relies on the lower layers such as routing and MAC to inform the TCP regarding the occurrence of route failures. Upon receiving a notification from lower layer, a TCP sender freezes its state and enters a "standby mode". ELFN uses the probe packet which it sends periodically towards the destination as part of route discovery process. Upon receiving an acknowledgement from the receiver the sender becomes active again. ELFN uses ICMP (Internet Control Message Protocol) message to inform nodes about the route failures and has not proposed any new control message. ELFN has been shown to improve the TCP performance significantly but it is more prone to MAC layer collisions which lead to false broken link detection[38] and wasteful new route discoveries as a result.

TCP-BuS (TCP Buffering Capabilities and Sequence Number) in Ref. 26 is another attempt to improve TCP performance during route failures. This approach extends the ideas of ELFN by proposing to require intermediate nodes to buffer packets in case of route failures. This prevents retransmissions from the same node once the route gets established in future. In order for this enhancement to work, initial timeout values are increased.

Approach of ELFN with minor variations has been further examined in Refs. 15 and 50 and TCP-RC[60] and reported.

ATCP[34] considered several TCP related issues arising from wireless environments and proposed a solution that handles route failures due to non-congestion and longer network disconnection times. To deal with non-congestion related losses from others, it uses ELFN like probe packet. Congestions are handled by ECN message (Explicit Congestion Notification), so when a loss occurs without an ECN message, it is attributed to non-congestion related. ATCP is compatible with TCP and implemented in an additional layer below TCP.

TCP-DOOR[52] which means TCP with Detection of Out-of-Order and response is an end-to-end solution to packet loss due to route failures. In TCP-DOOR when data packet or acknowledgements are delivered out of order, it is interpreted that the routing has changed. This approach utilizes two innovative mechanisms to deal-without of order events. Upon detecting that out of order packets are being delivered to receivers, a sender temporarily disables TCP's congestion control mechanisms by freezing its state variables which are kept constant. One of the conclusions of this work is that out-of-order packet delivery detection either at sender or the receiver is sufficient and duplicating this functionality at both places does not yield significant benefits. TCP-DOOR performs the best when it disables TCP congestion control and reverts to a pre-congestion state known as "instant recovery".

The idea of preemptive routing[21] is that to foresee the route failures by looking at the received signal strengths at each node on a given path and start the route discovery process so that disconnection time can be reduced. The atra framework[20] also deals with route failure in MANET by proposing and incorporating three new mechanisms called Symmetric Route Pinning, Route Failure Prediction, and Proactive Route Errors. These mechanisms have been tailored for the DSR routing protocol. The Route Pinning forces TCP acknowledgements to use the same route as corresponding data so that data protocol does not have to deal with two route failures at the same time. It needs to restore only one. The Route Failure Prediction is analogous to preemptive routing scheme discussed earlier. When Route Prediction scheme does not work, it uses the third mechanism Proactive Route Failure which notifies all nodes that are using the broken link. This approach is shown to result in a better performance than ELFN.

Table 3.1 provides the summary of the approaches that were discussed in this section. The approaches are compared along several dimensions such as (1) Compatibility with TCP i.e. whether the approach will work with standard TCP. (2) interworking with wired Internet network via a gateway or any suitable infrastructure i.e. whether the approach will work or not or its not clear. (3) reliability and Congestion Control i.e. whether approach uses the TCP features of reliability and congestion control or has a separate implementation. (4) Data Rate — constant bit rate (smooth) or bursty. (5) Network layer — whether the approach is dependent on a particular routing protocol or independent. (6) MAC layer whether approach

Table 3.1 Summary of approaches related to route failures.

Proposed Approach	TCP Friendly	Internet Compatible	Reliability & Congestion Control	Data Rate	Network Layer	MAC Layer	Reference
TCP-F	Yes	Yes	TCP	Bursty	Dependent	Independent	7
ELFN	Yes	Yes	TCP	Bursty	Dependent	Independent	23
Fixed RTO	Yes	Yes	TCP	Bursty	Independent	Independent	15
ENIC	Yes	Yes	TCP	Bursty	Dependent	Independent	50
TCP-RC	Yes	Yes	TCP	Bursty	Dependent	Independent	60
ATCP	Yes	No	TCP	Bursty	Inependent	Independent	34
Cross-Layer	Yes	No	TCP	Bursty	Dependent	Independent	56
TCP-DOOR	Yes	Yes	TCP	Bursty	Inependent	Independent	52
Preemptive routing	Yes	Unknown	TCP	Bursty	Dependent	Independent	21
Atra	Yes	Unknown	TCP	Bursty	Dependent	Independent	47
Signal strength based link management	Yes	Unknown	TCP	Bursty	Dependent	Dependent	27
TCP-Bus	Yes	Yes	TCP	Bursty	Dependent	Independent	26
Multipath-TCP	Yes	No	TCP	Bursty	Dependent	Independent	31

is dependent or independent of any medium access control layer. (7) Reference that has all the detailed and pertinent information about the approach.

Multi-path TCP[31] explored use of discovering multiple paths between source and destination and using them in parallel by distributing packets to them. The paper concludes that it is not very beneficial to have multiple parallel routes to address route failure because it leads to frequent out-of-order packet delivery at the destination/source triggering unnecessary congestion control reaction. Having discussed the proposed TCP solutions for the frequent route failure problem in MANET we now review emerging literature that targets schemes dealing with wireless losses in the next section.

3.3.2. *TCP and wireless losses*

As we have mentioned earlier the wireless link is much more prone to random packet loss as compared to a wired link. Such losses affect the TCP performance if TCP treats them wrongly as losses due to congestion in the network. We discuss some of the prominent approaches to deal with these losses in this section.

One of the approaches is due to Vlahovic[51] which advocates introducing three states in TCP senders and is called TCP with Restricted Congestion Window Enlargement (TCP/RCWE). The approach uses ELFN mechanism which handles link breakages and losses associated with them. The new idea in the scheme is a heuristic based observation of RTO. If the RTO increases, the congestion window size is not increased. If RTO decreases or remains constant, the congestion window is increased as per TCP congestion control rule such as AIMD/SS. RCWE is reported to have much smaller congestion window which results in a higher goodput as compared to standard TCP without ELFN. What is the contribution of ELFN in achieving higher goodput in RCWE is not reported in the paper.

ADTCP[18] is an approach that uses multiple metrics for bit error loss detection instead of relying on single one. The first new metric is called inter-packet-delay difference (jitter) at the receiver and the second is called short-term throughput in the immediate past. Authors make the observation that during the congestion the jitter is higher and the throughput is lower. Both of these metrics are combined to derive a single more robust metric. In ADTPC out-of-order packet arrivals and the packet loss ratio are used to detect route changes and channel errors. The receiver is tasked to determine the current network state and provide this feedback to sender so that both entities take appropriate actions at their end by distinguishing losses due to congestion or channel errors. This scheme is compatible with standard wired TCP. ADTFRC[20] is another approach that builds on already proposed approach[17] TFRC for wired TCP which has the smooth rate adaptation properties for multimedia applications.

De Oliveira *et al.* propose an edge based approach which distinguishes congestion losses from the media losses by using measured round trip times. The route failures are detected when a time out occurs and no packets are received for extended period of time. When this happens the approach starts to use ELFN probes so that a route is reestablished.

We now discuss the third major issue that arises due shared media and unfairness that is caused by TCP.

3.3.3. *Shared medium and TCP*

In an ad hoc network the medium is shared by all nodes in a particular geographical location. So while performing congestion control, it is important to consider the location wide congestion instead of a single node-based congestion. This observation makes congestion control more complex. In this section, we briefly review some of the well known approaches.

Reference 19 shows that for a given topology and traffic pattern there exists an optimal TCP window size but TCP is unable to find it, and usually resorts to using a much larger congestion window. A larger congestion window leads to more transmissions thereby increasing the number of packets that get dropped due to the link layer contention. This paper presents two mechanisms to improve TCP by reacting to link layer overload, the first is called Link RED (LRED) and an *adaptive pacing* strategy. LRED follows the principles of RED mechanisms proposed for wired networks, which allows the routers to drop packets with a probability that grows with the size of queue length. In LRED, the probability of packet drop is related to number of transmission attempts at the MAC layer which is increased with the number of retries. The adaptive pacing is primarily used to regulate the retransmission traffic in LRED, which is enabled after reaching a certain threshold. When enabled the adaptive pacing adds an additional packet transmission time to the backoff timer thus allowing medium to be free for a longer period in which packet is likely to get transmitted from the next hop neighbor also.

TCP unfairness problem in a locally shared medium is addressed in Ref. 53 which, proposes NRED (neighborhood RED). In this approach every node in the neighborhood estimates the total number of packets waiting to be transmitted in the neighborhood. If this number exceeds a threshold the packets are dropped with a certain probability. NRED is designed to have three steps (i) The neighborhood queue size estimation — this is done by looking at the channel utilization. (ii) If the utilization is above a certain threshold, the neighborhood is presumed to be congested and every node in the neighborhood is advised of a drop probability, (iii) Using suggested drop probability figure of the neighborhood, each node computes its own drop probability and drops the packets accordingly as a result the entire neighborhood reduces the congestion collectively.

COPAS[12] (Contention Based Path Selection) addresses the path selection issue from the node capture point of view which leads to unfairness. In this approach all routes between a source and destination nodes are discovered and then two node disjoint paths are chosen — one for packet transmission and the other for acknowledgements. COPAS calculates the backoff times for each of the nodes on the discovered paths continuously and chooses the two with the lower numbers. Authors claim that their method addresses the path capture issues where one path has access to media for significantly longer times.

Split TCP[29] is a congestion control approach which is quite different from other proposals. This scheme advocates separation of congestion control and end-to-end reliability functionalities of the TCP. Using the notion of TCP proxies in the intermediate nodes, the path is split into several independent segments. Each proxy stores the packets and transmits them to the next proxy or to the destination. Acknowledgements are used for local data transmission within a segment. Additionally, end-to-end cumulative ACKs are used for reliable data transmission due to proxy failures.

Split TCP solves the mobility related link breakage problem to some extent by separating the path into different number of segments and letting most of the links functioning while recovery of an individual link is being worked. Path capture effects are minimum because of data transmission occurs from segment to segment.

There are two schemes[5, 58] that deal with alleviating self contention on contention between packets belonging to the same flow in both directions, and reducing the impact of inter-flow and intra-flow contention on the throughput and fairness. The first approach advocates adding an extra packet transmission time as part of RTS/CTS exchange and the second relies on assigning higher priority for the medium access to a node that has just received a packet.

3.3.4. *ACK traffic in TCP*

It has been widely reported[2,14] that packets using the same route in opposite directions, for example, TCP segments and the corresponding acknowledgements create intra-flow contention affecting the throughput of each other as a result studies have started to look into the minimization of negative impact of ACK traffic on the upstream traffic. Use of delayed ACKs (DACKs) and selective ACKS (SACKs) instead of cumulative ACKs has been propose earlier[15, 50] to reduce the ACK traffic.

Dynamic Delayed ACK proposal extends the use of DACKS by allowing receiver to send acknowledgement after k packets where $d \geq 2$. The scheme has shown significant performance improvements for $d = 2$. The scheme has been experimented with d increasing up to 4 and remaining constant thereafter.

As an extension to Dynamic Delayed ACK, De Olivera[14] proposes a new scheme called *dynamic adaptive acknowledgement* by applying concepts proposed in RFC 2581 which describes TCP congestion control. This scheme argues against having a fixed number of packets handled by a single ACK as well as keeping the ACK timeout to be fixed.

RFC 2581 advocates sending an acknowledgement immediately if the receiver receives an out-of-order packet or a packet that was missing. The dynamic adaptive acknowledgement scheme suggests that the receiver should wait additional time till it receives the second packet before timeout. The receipt of the second packet will determine what action the receiver has to take. For instance, if the receiver receives an in order second packet which may be delayed, it should not timeout. If the received second packet is out of order then an immediate acknowledgement should be sent. The authors also propose growing the parameter d dynamically in increments of one up to a value of 4. Upon receiving an out-of-order packet it is set back to 2.

Sugano[46] presents the *preferred ACK retransmission* scheme by using ELFN, DACK on a new radio network called Flexible Radio Network (FRN) which is a proprietary network from Fuji and is different from IEEE 802.11 networks. FRN is a time-slot based network. In this scheme the repeated collisions of ACKs with the data streams is avoided by giving a higher priority to ACK packets over the data packets only after repeated ACK collisions. The higher priority is given by reducing the ACK retransmission interval at the MAC layer. The preferred ACK retransmission scheme proposed for FRN is further enhanced in Yuki[57] which proposes to combine data and ACK packets into one packet at intermediate nodes if data and ACK packets are present. This minimizes the potential contention between data and ACK streams. The combined packet has two destinations as part of the header which the receiving nodes can decode.

3.3.5. *Throttling packet rate in TCP*

The congestion window of TCP grows following AIMD as discussed in previous sections and very soon it saturates the network. Limiting the size of the congestion window (smaller) particularly for MANETS have shown significant benefits also discussed earlier. In this section, we discuss the approaches that investigate the variations on the congestion window sizes for MANETs.

A proposal to establish a congestion window or a limit on congestion window (CWL) dynamically is presented in Chen[8] which is based on the bandwidth-delay product of the on-going connection and it cannot exceed the round-trip hop-count (RTHC) in a wireless multi-hop network. Using the DSR routing, this scheme requires sources to determine the path-length using which CWL can be computed. The simulation results presented in their paper are indicative of significant performance gains which were achieved by using the maximum retransmission timeout of TCP to 2 seconds instead of 240 recommended in RFC 1122.

Slow Congestion Avoidance (SCA) is a scheme that is critical of the previous scheme because of the use of lower maximum retransmission time. The basic idea of this scheme is to grow the congestion window by one after a given number of round trip times. This scheme saw an enhancement in Nahm[39] which is known as fractional window increment (FeW). The basic premise of FeW is that the TCP operates at a higher loss rate for networks that have low bandwidth-delay product which are mainly attributed to link-layer losses. These losses can potential affect the routing. This scheme tries to lower the operational range of the TCP in which window does not grow by an integer per RTT and is more or less identical to SCA.

The idea of a low congestion window has been tried in a network in which ad hoc network interfaces with a wired backbone in Ref. 53 with an observation that in such scenarios the performance is significantly degraded because of small number of packets present in the wired part of the network. The solution to this problem is a scheduling algorithm also proposed by the authors called as non-work-conserving scheduling which is mainly applied to the wireless interface. The scheduling scheme works as follows: a sender after transmitting a data packet sets a timer. Until this time expires, node is not allowed to send any other data packets. The value of timer keeps on increasing if there are more packets are waiting to be transmitted as a result congested nodes are throttled more. This algorithm provides better fairness at the cost of reduced throughput. The non-work-conserving algorithm needs to be used by all intermediate nodes which interface with the wired part of the network.

The approach we just discussed implements rate control via scheduling algorithm to achieve fairness. Other proposals argue to incorporate another rate control mechanism right at the TCP output, so TCP output to a lower layer is adequately regulated. The Rate Based Congestion Control (RBCC)[59] is one such proposal which adds a leaky bucket to the TCP output queue. RBCC is implemented via creating a field in the header of the packet which is examined and modified by every intermediate node to regulate the maximum rate and provide feedback to others. In addition, every node on a given route is required to compute the percentage of time the channel is busy locally and add this information in the packet. When the source node ultimately gets to see this information for all the nodes on the path, it chooses the appropriate rate.

Extending the ideas of RBCC further Kliazovich[28] in a cross-layer congestion control scheme a field is added to the header of the link layer that stores the measurements from all the intermediate nodes on the current path. When the destination is generating an ACK to be sent to the sender, the feedback field is copied in the ACK so the sender knows what is the desirable window size receiver is advertising and in a way can change the sender's congestion window. A hybrid approach that uses rate based as well congestion based approaches is described in ElRakabawy[16] and is known as TCP-AP (TCP with Adaptive Pacing). The central theme of this approach is a definition of a new parameter called 4-hop propagation delay which is estimated elapsed time between transmission of a packet by the source and its reception by a node which is 4 hops away from it. The 4-hop propagation delay is estimated using the RTT of packets. The reason behind choosing of the 4-hop propagation delay comes from the observation that a transmission currently in progress is assumed to interfere within a range of four hops.

3.3.6. *Non-TCP transmission protocols*

The previous sections described approaches to provide congestion control in ad hoc networking environments by adapting and modifying TCP in ways so that it performs better. In parallel there have been schemes proposed that advocated entirely new transport protocols for ad hoc networks that are significantly different from TCP. We review a few of the major ones in this section. The one major drawback of these approaches is that these may not work with TCP in environments such as at the interface of wired and wireless networks.

EXACT protocol[8,9] is a transport protocol that is designed for ad hoc networks. In this scheme every node maintains the state of all flows passing through it to neighboring nodes. Using these state variables nodes are able to allocate bandwidth fairly across all flows. Explicit rate information is inserted in all packets that are passing through nodes, which allows determination of the node which has the minimum flow and this value is used in subsequent transmissions. This mechanism is used twice via having two fields in the header. One field contains the current rate of the sender and the other one is the requested rate by the application running on the sender. This feature allows changing the rate dynamically by intermediate nodes by not allowing more than necessary bandwidth consumed by a flow. As part of this scheme a parameter called *safety window* is proposed, which restricts the sender from overloading the network in case of link failures. Safety window limits the number of unacknowledged packets in the flow as their number cannot exceed this window.

EXACT has a TCP like version for reliable data transmission and as well as UDP like version which is not so reliable. Retransmissions are implemented with the help of selective acknowledgements (SACKs). Concerns have been raised in the literature regarding scalability of this scheme because of maintenance of state of each flow in a node.

Ad hoc Transport Protocol (ATP)[47] is a variation on the theme proposed by EXACT to overcome the issue of low throughput in TCP and high packet loss in UDP by arguing that UDP packets should be acknowledged as well if needed as an option. ATP delegates the option of retransmissions to the applications which are informed if and when an acknowledgement is received or not. In ATP congestion control is not part of the transport protocol and is left to applications to decide and implement congestion control of their choice. Lack of a network wide congestion control can be an issue with ATP.

Wireless Explicit Congestion Control Protocol (WXCP)[49] is a protocol that is variation on XCP[25] which, was proposed as a transport protocol for wired networks with high bandwidth-delay product. By design XCP is not compatible with TCP. WXCP uses additional set of metrics such as locally available bandwidth, local queue lengths, and the average number of link layer retransmissions which are used by the intermediate nodes to prevent the necessity of network probing to discover higher available bandwidth. These three metrics are combined in a way to produce a feedback known as aggregate feedback. Link layer retransmission metric has a special significance in WXCP as it allows detection of presence of self-interference within a flow in a particular neighborhood.

Fairness and congestion are treated separately in WXCP. For fairness, it uses time fairness criteria, which means each flow is given same amount of medium access time against the throughput fairness which may not be achievable in wireless environment. A new feature of WXCP is called *discovery state* using which it switches between window based approach which is a default to a rate based approach which is typically slow in the event that small congestion window is not allowing additional packet transmissions and ACKs are missing. Another new feature of WXCP is that WXCP switches from window based scheme to rate based scheme by a mechanism which it calls pacing. Pacing is used when number of packets to be transmitted exceeds the window threshold and then the packets are evenly distributed per RTT and transmitted as per this pace.

TPA (Transport Protocol for Ad Hoc Network) is another transport protocol for ad hoc networks that incorporates TCP like congestion control mechanisms but suggests different features for retransmission minimizations. In TPA packets are transmitted in blocks using a window based scheme. A block is defined as a set of N number of packets with N being variable. No other packets from the next block can be transmitted until all packets of the current block are acknowledged which may entail retransmissions of missing packets. TPA discovers route failures if it encounters consecutive time-outs or it can employ ELFN mechanisms to detect route failures. In response to route failures, TPA reduces the congestion window to one and freezes its state. RTT is used as a means to fix the retransmission timeouts and newer values of RTT are given higher weights when route changes are encountered.

To implement congestion control, TPA uses a large congestion window of 2 or 3 segments and a small congestion window of 1 segment which are used in normal and under congestion operations respectively. TPA shows evidence that a simple end-to-end protocol without relying on intermediate node intelligence can still be superior in throughput as compared to TCP in small and simple networks. It remains to be seen if such benefits can still be achieved for large, complex and dynamic networks.

Table 3.2 provides the summary of the features of other alternative transport protocols which are independent of TCP which are discussed above.

3.4. Admission Control

Wireless ad hoc networks are envisioned to support multi-media applications such as streaming audio and video and even voice which require resource guarantees in terms of delay and jitter. Both of these constraints translate to provisioning of sufficient bandwidth for the application to work correctly. The role of admission control becomes very important when the network is to provide performance guarantees. To

Table 3.2 Summary of alternative transport protocols.

Proposed Approach	TCP Friendly	Internet Compatible	Reliability & Congestion Control	Data Rate	Network Layer	MAC Layer	Reference
EXACT	No	Yes	Separate	Smooth	Independent	Dependent	9
ATP (Sundaresan)	No	No	Separate	Smooth	Independent	Independent	47
ATP (Liu)	No	Unknown	Unknown	Unknown	Independent	Independent	34
WXCP	No	No	No	Bursty	Independent	Dependent	49
TPA	No	Unknown	No	Bursty	Independent	Independent	3

make an admission control decision, the network must first accurately estimate the resources a flow will consume if admitted and the network is able to provide these. In addition, the network must also ensure that this admitted flow will not affect the performance of already admitted flows. The estimation of available resources such as bandwidth is relatively easy in wired networks as individual nodes are able to accurately monitor it. It is relative hard in wireless environments as it is an open and shared medium. A wireless node's transmission consumes bandwidth shared with other nodes in the vicinity since these nodes cannot simultaneously access the shared medium. In a wireless medium, the transmission consumes bandwidth at all nodes within the carrier sensing range of a transmitting node. Presence of multiple nodes within the carrier sensing distance on a multi-hop path leads to the contention of the medium among them thereby preventing simultaneous transmissions which is known to cause intra-flow contention.[45] An intra-flow contention is defined as contention among the packets belonging to the same flow that are forwarded at different hops along a multi-hop path.

The MANETs require very different perspective on network admission control management. In the literature, Time Division Multiple Access (TDMA) based approaches[11,32,61]and several others have been proposed which require synchronization among all nodes. Providing synchronization in ad hoc network is difficult even in static case because of the absence of a base station and much harder when nodes move. A single channel MAC layer scheduling have been proposed in Ref. 36 which deal with fair resource allocation at individual nodes for QoS but have no features of admission control which is an end-to end concept like QoS. Some solutions that have mission control such as Ahn,[1] Maltz,[37] and Barry[4] but these approaches do not consider intra-flow contention in a neighborhood as a result provide optimistic results. In this section we only cover admission control approaches that incorporate neighborhood contention in their proposals. Here we discuss estimation of intra-flow contention, and present the prominent admission control procedures proposed using this contention for ad hoc networks. Also, our focus is only on the prominent schemes that are CSMA/CA based and do not involve any infrastructure such as IEEE 802.11 in ad hoc mode.

3.4.1. *Intra-flow contention*

To calculate the intra-flow contention of nodes along the path, it is required that we know the contention count for each node. The contention count (CC) at a node is the number of nodes on the multi-hop path that are located within carrier sensing range of the given node. The effective bandwidth consumed by a flow at each node is the CC times the single hop flow bandwidth required by the application. Figure 3.1 shows the ranges of IEEE 802.11 wireless communication.[45] The maximum separation between a sender and receiver for successful packet reception is called Reception Range (RR). Nodes within RR of a particular sender can directly

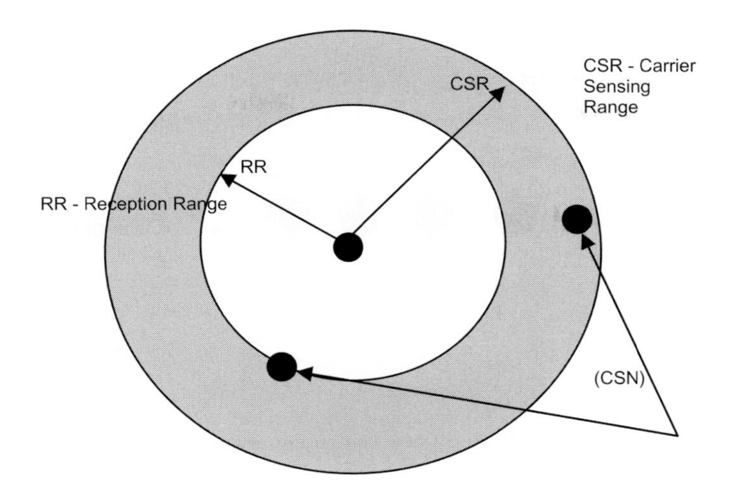

Fig. 3.1. Ranges of IEEE 802.11 wireless communications.

communicate with the sender and are considered its neighbor. The maximum distance that a node can detect an ongoing packet transmission is called carrier sensing range (CSR). This range is usually larger than the reception range. The nodes that are within the CSR of a node are known as its carrier sensing neighbors (CSN). The CSN nodes can detect a transmission but may not be able to decode the packet if they are outside of the reception range.

The CC at a node is defined as an intersection of a set of nodes that lie on the multi-hop path (NoP) with the carrier sensing neighbors CSNs. So the CC for a node i can be written as

$$CC_i = |CSN_i \cap N_oP| + 1. \tag{3.1}$$

The first term is the number of competing CSN, and one is added to account for the impact of node i itself. Lets use Figure 3.2 to calculate CC. Here the path is A-B-C-D, and CSN for node A is B and C. So intersection gives us 2 and when add 1, we get CC as 3. Same is the case for all other nodes on the path. In Figure 3.2, the solid circle represent radio transmission range and the dotted circles represent the carrier sensing range.

3.4.2. *CACP — contention aware admission control*

CACP[55] has been specifically designed to provide admission control for ad hoc networks and it computes the CC using Equation (3.1). When making a decision to admit a flow, it computes the locally available bandwidth by using Equation (3.2) given below:

$$B_{available} = (1 - U) * B_{\max}. \tag{3.2}$$

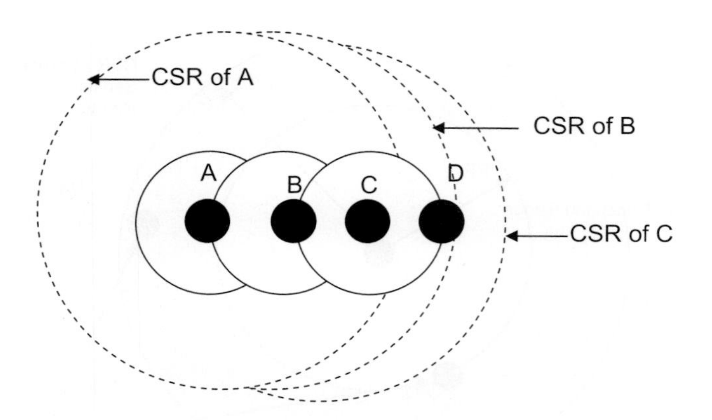

Fig. 3.2. Intra-flow contention illustration.

Table 3.3 Available bandwidth.

Time/Node	W	X&Y	Z
T1	100%	50%	50%
T2	74%	25%	50%

Here $B_{\max}$ is a measure of the maximum available bandwidth, and U is the channel utilization. U can be computed by using Equation (3.3) which gives us the total transmission time for a data packet including the MAC layer overheads.

$$T_{data} = T_{difs} + T_{rts} + T_{cts} + \frac{L+H}{B_{\max}} + T_{ack} + 3T_{sifs}. \tag{3.3}$$

Here T, Times for *difs, sifs, rts, cts*, and *ack* are related to IEEE 802.11 MAC protocol. L is the length of the data in the packet and H is the size of the header.

Other proposals for estimating the channel utilization are based on measurements which may be more accurate and reflect the true traffic conditions are described in Chakeres.[6]

There are two main operations in CACP — admission control decision and a multi-hop routing protocol. The admission control decision involves computation of available bandwidth on hop by hop basis using the CC. Before admitting a new flow over one hop, it is necessary to check if the requested bandwidth is available. Since the available bandwidth calculation does not include all nodes that may be impacted by a new flow, a query message must be sent to all nodes within carrier sensing range. If all CSN detect enough available bandwidth then the flow is admitted.

CACP considers two methods to query the available bandwidth at the CSN of a node prior to flow admission. The first method is multi-hop approach that uses flooding of query messages with a pre-specified hop count assuming that all nodes that are in CSN are reachable. The second method involves sending bandwidth availability query using a high power packet transmission so that all nodes within

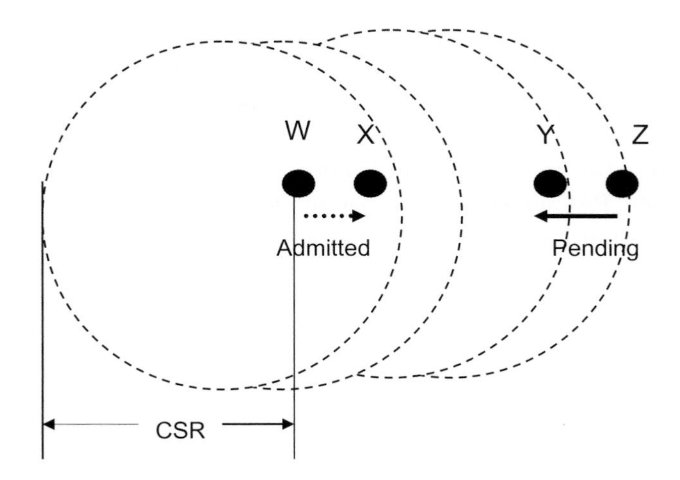

Fig. 3.3. Single hop admission control decision — Example.

carrier sensing range of a sender are contacted. If any node that does not have enough bandwidth to support the new flow, it sends a rejection message also using high power transmission. This approach takes care of the nodes which are not reachable.

CACP operation can be better described by an example which is given in Ref. 6 and briefly described here. Figure 3.3 shows a network with an admitted traffic flow between nodes Z and Y that consumes 50% of the bandwidth. Table 1 shows the current operating conditions of the network at time t_1. At this moment nodes X, Y, and Z are able to detect the current flow, but W cannot because it is outside of carrier sensing range. Now W intends to initiate a flow that will consume 25% of the bandwidth. Node W checks and it finds that it can support this flow and now sends a query to the nodes in the CSR i.e. to X, and Y, whether they can support this flow. Both nodes check and having found that they indeed can, and do not send a rejection message to node W. W admits the flow. At time t_2, W intends to initiate yet another flow that will consume 50% of the bandwidth. W checks with itself and with the neighbors. This time W receives a rejection from both X and Y.

CACP is shown to work well. However, it has a few issues such as those arising from the control packet loss which may lead to erroneous admissions and no mobility support. Mobility is delegated to the routing protocol DSR which leads to a excessive control traffic. Spatial reuse is also limited in CACP as a result the aggregate throughput has been found to be lower. Other details of this protocol are in Yang[55].

3.4.3. *Perceptive admission control (PAC)*

Some of the issues that CACP has been found to exhibit e.g. reservation of excessive bandwidth than necessary in a large geographical area, and use of negative acknowledgements which leads to denial of admission to legitimate flows are overcome in perceptive admission control scheme.[6] PAC reduces the protocol overhead and the risk of making erroneous admission by using a new method of determining available bandwidth which does not require sending any control message to its neighbors. PAC also uses channel utilization metric like CAPC but does it in a different manner. PAC extends the range of channel utilization measurements by increasing the distance required for two simultaneous transmissions to occur without a collision. Admission of a flow that requires bandwidth B_{req}, the inequality in Equation (3.4) should be satisfied.

$$B_{available} - B_{rsv} > B_{req}, \tag{3.4}$$

where B_{rsv} is a small fraction of the bandwidth that is reserved to better handle bandwidth fluctuations and prevent the network from becoming congested.

It has been found that mobility can cause congestion even with a initial proper admission control. Through continuous monitoring of bandwidth by nodes in PAC, this situation is avoided. If node is involved in an ongoing traffic and encounters a situation of bandwidth lower than threshold, then the source node should stop the flow and wait according to any fair backoff scheme and try readmitting the flow. Bandwidth monitoring coupled with temporary blocking of flow, allows PAC to keep the network uncongested.

Both CACP and PAC work in the multi-hop environments but their performance requires improvements when dealing with mobile nodes to support admission control. MACMAN[33] Multi-Path Admission Control for Mobile Ad Hoc Networks is a new protocol that implements the admission control for mobile networks. The key idea of this approach is the discovery of multiple paths between the source and the destination and subsequently the switching of the flow from one path to the next if one becomes unavailable.

3.4.4. *MACMAN*

The *route discovery* phase of MACMAN is analogous to that used by PAC for calculating the local available bandwidth but adapted for multiple routes. In this scheme no state information is kept for alternate routes, which is an advantage as source keeps the complete routes in its cache. When destination receives the route requests it replies to first as well as subsequent requests. When source receives multiple route replies from the destination it chooses the best route for the data transmission and keeps other routes in cache for future use depending on its needs.

MACMAN continuously monitors alternative routes by sending a message called Route Capacity Query (RCQ) which has bandwidth requirement as a parameter to assess changing capacity on all the paths as well as preventing the stale routing problem encountered in ad hoc networks. RCQ also has information about the current node that is being used. When a node on an alternate route receives the

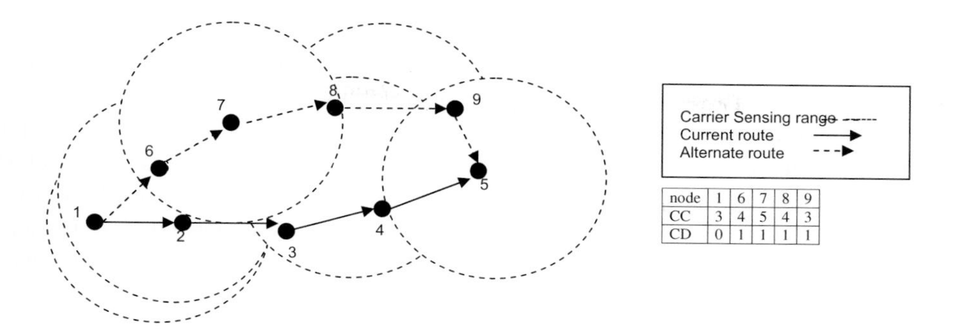

node	1	6	7	8	9
CC	3	4	5	4	3
CD	0	1	1	1	1

Fig. 3.4. Calculation of contention difference.

RCQ it determines if it can support the requested bandwidth and a new CC called CD (Contention Difference) is calculated using Equation (3.5) which takes care of the differences in contentions between the two paths if a decision to switch the path is taken

$$CD = |CSN \cap R_{new} \backslash \{D\}| - |CSN \cap R_{curr} \backslash \{D\}|, \tag{3.5}$$

where R_{new} is the alternate route on which RCQ was sent, R_{curr} is the current route, and D is the destination. Using CD, a flow can be admitted to the alternate route, if equality of Equation (3.6) holds. Figure 3.4 shows the CD calculation for an example network[33]. In the diagram the route $1 \to 2 \to 3 \to 4 \to 5$ is currently used. The alternate path consists of node $1 \to 6 \to 7 \to 8 \to 9$ and the RCQ is sent along this path. Since nodes 1, 2, and 3 are already involved in the transmission along the current path the CD is equal to 1:

$$B_{avail} - Brsv > CD \cdot B_{req}. \tag{3.6}$$

If a node on the alternate route finds that this inequality does not hold, it can send a Route Capacity Failed Message (RCF) back to the source of the flow, otherwise it sends RCQ to next hop which is towards the destination.

MACMAN uses the route maintenance to keep fresh routes in the cache and discards the routes which are stale or do not meet the QoS. Interested readers should consult Lndgren[33] for additional details.

3.5. Conclusions & Future Readings

In this chapter, we presented the various issues that arise when TCP protocol is used in wireless ad hoc networking environments. The congestion control features of TCP need significant modifications which we have discussed in this chapter albeit briefly. In the literature there are several other proposals for data transport which are not TCP like we also highlighted these. With the increasing popularity of wireless communication, provisioning ad hoc networks to support real-time traffic such as audio and video has also gaining prominence in the literature. In this context role of admission control algorithms specially tailored for ad hoc networks becomes very important as the traditional admission control approaches for Internet require significant modifications. In this chapter, we have presented some of the major admission control schemes proposed for the ad hoc networks also. Wireless TCP and its variants as well as wireless admission control in infrastructure less environments continues to be active area of research and it is suggested that readers should look into IEEE/ACM conference proceedings for interesting ideas and newer results.

Problems

1. Consider an i-hop IEEE 802.11 network in which nodes are equidistant and are in a straight line. Assume capacity of links between the nodes to be identical to be C bits/second. Express the throughput bound for the network when (a) $1 \leq i \leq 3$, (b) $i > 3$.
2. Intuitively, TCP throughput should monotonically decrease as the speed of the nodes in the ad hoc networks increases. But for some scenarios, the TCP throughput has been found to increase at higher speeds. Describe such scenarios.
3. Explain the role of intra-flow contention in determining the correct admission control algorithm for ad hoc networks. Also argue the role this contention plays in providing QoS guarantees for ad hoc networks.
4. Describe the desirable properties in the media access control protocol which can improve TCP performance over ad hoc networks.
5. Which characteristics of routing protocol will have significant performance impact on TCP over ad hoc networks? Rank order these and substantiate your arguments.

Bibliography

1. G. S. Ahn, A. Campbell and A. Veres, SWAN: service differentiation in stateless wireless ad hoc networks, *Proceedings of IEEE Infocom'02* (2002).
2. E. Altman and T. Jiminez, Novel delayed ACK techniques for improving TCP performance in multi-hop wireless networks, *Proceedings of PWC'03* (September 2003) 237–250.
3. G. Anastasi, E. Ancillotti, M. Conti and A. Pasarella, A transport protocol for ad hoc networks, *Proceedings of IEEE International Symposium on Computers and Communications* (June 2005) 51–56.
4. M. G. Barry, A. T. Campbell and A. Veres, Distributed control algorithms for service differentiations in wireless packet networks, *Proceedings of IEEE Infocom'01* (2001).
5. D. Berger, Z. Ye, P. Sinha, S. Krisnamurthy and S. K. Tripathi, TCP-friendly medium access control for ad hoc wireless networks: alleviating self-contention, *Proceedings of MASS*, Fort Lauderdale (2004) 214–223.

6. I. D. Chakeres and E. M. Belding-Royer, PAC: perceptive admission control for mobile wireless networks, *Proceedings of QShine*, Dallas (2004).
7. K. Chandran, S. Raghunathan, S. Venkatesan and S. Prakash, A feedback based scheme for improving TCP Performance in ad hoc wireless networks, *Proceedings of International Conference on Distributed Computing System* (1998) 472–479.
8. K. Chen, Y. Xue and K. Nahrstedt, On setting TCP congestion window limit in mobile ad hoc networks, *Proceedings of IEEE ICC'03*, Anchorage Alaska (2003).
9. K. Chen and K. Nahrstedt, EXACT: an explicit rate based flow control frame work in MANET, Technical Report UIUCDCS-R-2002-2286'UILU-ENG-2202-1730 (2002).
10. K. Chen, K. Nahrstedt and N. H. Vaidya, The utility of explicit rate based flow control in mobile ad hoc networks, *Proceedings of IEEE WCNC'04* (2004) 1921–1926.
11. T. W. Chen, J. T. Tsai and M. Gerla, QoS routing performance in multi-hop multimedia wireless networks, *Proceedings of the IEEE International Conference on Universal Personal Communication (ICUPC)* (1997).
12. M. De Cordeiro, S. R. Das and D. P. Agrawal, COPAS: dynamic contention balancing to enhance performance of the TCP over multi-hop wireless networks, *Proceedings of ICCCN* (2002) 382–387.
13. R. De Oliveira, T. Braun and M. Heissenbutel, An edge based approach for improving TCP in wireless mobile ad hoc networks, *Proceedings of DADs* (2003).
14. R. De Oliveira and T. Braun, An edge based approach for improving TCP in wireless mobile ad hoc networks, *Proceedings of IEEE Infocom* (2005) 1863–1874.
15. T. D. Dyer and R. V. Bopanna, A comparison of TCP performance over three routing protocols for mobile ad hoc networks, *Proceedings of MobiHoc* (2001) 56–66.
16. S. M. ElRakabawy, A. Klemm and C. Lindermann, TCP with adaptive pacing for multi-hop wireless networks, *Proceedings of ACM Mobihoc'05* (2005) 288–299.
17. S. Floyd, M. Handley, J. Padhey and J. Windmar, Equation based congestion control for unicast applications, *ACM SIGCOMM, CCR* (2000) 43–56.
18. Z. Fu, B. Greenstein, X. Meng and S. Lu, Design and implementation of a TCP friendly transport protocol for ad hoc wireless networks, *Proceedings of ICNP'02* (2002) 216–255.
19. Z. Fu, X. Meng, S. Lu and M. Gerla, The impact of multi-hop wireless channel on TCP throughput and loss, *Proceedings of INFOCOM* (2003) 1744–1753.
20. Z. Fu, X. Meng and S. Lu, A transport protocol for supporting multimedia streaming in mobile ad hoc networks, *IEEE Journal on Selected Areas in Communication* **21**(3) (2003) 1615–1626.
21. T. Goff, N. B. Abu-Ghazaleh, D. S. Phatak, R. Kahvecioglu, Preemptive routing in ad hoc networks, *Journal of Parallel and Distributed Computing* **63**(2) (2003) 123–140.
22. A. Al Hanbali, E. Altman and P. Nain, A survey of TCP over ad hoc networks, *IEEE Communications Surveys and Tutorials* **7**(3) (2003) 22–36.
23. G. Holland and N. H. Vaidya, Analysis of TCP performance over mobile ad hoc networks, *Proceedings of ACM Mobicom* (1999) 219–230.
24. Internet Engineering Task Force homepage, http://www.ietf.org.
25. D. Katabi, M. Handley and C. Rohrs, Congestion control for high bandwidth delay product networks, *Proceedings of SIGCOMM'02* (2002) 89–102.
26. D. Kim, C. K. Toh and Y. Choi, TCP-bus: improving TCP performance in wireless ad hoc networks, *Proceedings of IEEE ICC* (2000) 1707–1713.
27. F. Klemm *et al.*, Improving TCP performance in ad hoc networks using signal strength based link management, *Ad hoc Networks* **3**(2) (2005) 175–191.

28. D. Kliazovich and F. Granelli, Cross-layer congestion control in ad hoc wireless networks, *Ad Hoc Networks* **4**(6) (2006) 687–708.
29. S. Kopparty, S. Krishnamurthy and S. K. Tripathi, Split TCP for mobile ad hoc networks, *Proceedings of Globecom'02* (November 2002) 138–142.
30. B. Leiner *et al.*, The past and future history of the Internet, *Communications of the ACM* **40**(2) (1997) 102–108.
31. H. Lim, K. Xu and M. Gerla, TCP performance over multipath routing in mobile ad hoc networks, *Proceedings of IEEE ICC 2003* **2** (2003) 1064–1068.
32. C. R. Lin and J. S. Liu, QoS routing in ad hoc wireless networks *IEEE Journal of Selected Areas in Communication* **17**(8) (1999) 1426–1438.
33. A. Lindgren and E. Belding-Royer, Multi-path admission control for mobile ad hoc networks, *Proceedings of IEEE Mobiquitous* (2005).
34. J. Liu and S. Singh, ATCP: TCP for mobile ad hoc networks, *IEEE Journal on Selected Areas in Communication* **9**(7) (2001) 1300–1315.
35. C. Lochert, B. Scheuermann and M. Mauve, A survey on congestion control for mobile ad hoc networks, *Wireless Communications and Mobile Computing* **7** (2007) 655–676.
36. H. Luo, S. Lu, V. Bhargavan, J. Cheng and Z. Zhong, A packet scheduling approach to QoS support in multi-hop wireless networks, *ACM Journal of MONET* (2002).
37. D. Maltz, Resource management in multi-hop ad hoc networks, Technical Report, CMU, CS-00-150, School of Computer Science, Carnegie Mellon University (July 2000).
38. J. P. Monks, P. Sinha and V. Bhargavan, Limitations of TCP ELFN for ad hoc networks, *Proceedings of MoMuc'00* (2000).
39. K. Nahm, A. Helmy and C.-C. J. Kuo, TCP over multihop 802.11 networks: issues and performance enhancements, *Proceedings of the ACM Mobihoc'05* (2005) 277–287.
40. J. Postel, *Transmission Control Protocol* (1981).
41. R. Braden, *Requirements for Internet Hosts — communication Layers*, http://www.rfc-editor.org/rfc/rfc1122.txt (1989).
42. V. Jacobson, S. Braden and D. Borman, *TCP Extension for High Performance*, http://www.rfc.-editor.org/rfc/rfc1323.txt (May 1992).
43. M. Mathis, J. Mahdavi, S. Floyd and A. Romanow, *TCP Selective Acknowledgement Options*, http://www.rfc.-editor.org/rfc/rfc2018.txt (October 1996).
44. M. Allman, V. Paxson and W. Stevens, *TCP Congestion Control*, http://www.rfc.-editor.org/rfc/rfc2581.txt (April 1999).
45. K. Sanzgiri, I. D. Chakeres and E. M. Belding-Royer, Determining intra-flow contention along multi-hop paths in wireless networks, *Proceedings of IEEE BroadNets*, San Jose, California (October 2004).
46. M. Sugano and M. Murata, Performance improvement of TCP on a wireless ad hoc network, *Proceedings of IEEE VTC* (2003) 2276–2280.
47. K. Sundaresan, V. Ananthraman, H. Y. Hsieh and R. Sivakumar, ATP — a reliable transport protocol for ad hoc networks, *Proceedings of MobiHoc'03* (June 2003) 64–75.
48. W. R. Stevens, *TCP/IP Illustrated*, Vol. 1: The Protocols (Addison, Wesley, 1994).
49. Y. Su and T. Gross, WXCP — explicit congestion control for wireless multi-hop networks, *Proceedings of 12th International Workshop on Quality of Service* (*IWQoS'05*), (June 2005).
50. D. Sun and H. Man, ENIC — an improved reliable transport scheme for mobile ad hoc networks, *Proceedings of IEEE Globecom* (2001) 2852–2856.
51. M. Gunes and D. Vlahovic, The performance of the TCP/RCWE enhancement for ad hoc networks, *Proceedings of International Symposium on Computers and Communications* (*ISCC*) (2002) 43–48.

52. F. Wang and Y. Zhang, Improving TCP performance over mobile ad hoc networks with out of order detection and response, *Proceedings of MobiHoc* (2002) 217–255.
53. K. Xu, M. Gerla, L. Qi and Y. Shu, TCP unfairness in ad hoc wireless networks and a neighborhood RED solution, *Wireless Networks* (2005) 389–399.
54. L. Yang, W. K. G. Seah and Q. Yin, Improving fairness among TCP flows crossing wireless ad hoc and wired networks, *ACM MobiHoc'03*, New York (2003) 57–63.
55. Y. Yang and R. Kravets, Contention aware admission control for ad hoc networks, *IEEE Transactions on Mobile Computing* **4**(4) (2005) 363–377.
56. X. Yu, Improving TCP performance over mobile ad hoc networks by exploiting cross-layer information awareness, *ACM Mobicom'04* (2004) 231–244.
57. T. Yuki, T. Yamamoto and M. Sugano *et al.*, Improvement of TCP throughput by combination of data and ACK throughput by combination of data and ACK packets in ad hoc networks, *IEICE Transactions on Communications* (2004) 2493–2499.
58. H. Zhai and J. Wang *et al.*, Alleviating intra-flow and inter-flow contentions for reliable service in mobile ad hoc networks, *IEEE Milcom'04* (October 2004) 1640–1646.
59. H. Zhai, X. Chen and J. Wang *et al.*, Rate based transport control for mobile ad hoc networks, *Proceedings of IEEE WCNC'05* (March 2005) 2264–2269.
60. J. Zhou, B. Shi and L. Zou, Improve TCP performance in ad hoc network by TCP-RC, *Proceedings of IEEE PIMRC* (September 2003) 216–220.
61. C. Zhu and M. S. Corson, QoS routing in ad hoc wireless networks, Technical Report CSHCN TR 2001-18, Institute of System Research, University of Maryland (2001).

Chapter 4

Wireless Ad Hoc Networks with Directional Antennas

Basel Alawieh* and Chadi Assi†

Faculty of Engineering and Computer Science
Concordia University, Montreal, Quebec, Canada
** b_alawi@encs.concordia.ca*
† assi@encs.concordia.ca

Hussein Mouftah

School of Information Technology and Engineering
University of Ottawa, Ottawa, Ontario, Canada
mouftah@site.uottawa.ca

Currently, the IEEE 802.11 standard and its distributed coordination function (DCF) access method have gained global acceptance and popularity both in wireless LANs and wireless multihop ad hoc environment. It has been shown that the DCF access method does not make efficient use of the shared channel due to its inherent conservative approach in assessing the level of interference. To date, various mechanisms have been proposed to improve the capacity of IEEE 802.11-based multi-hop wireless networks. These mechanisms can be broadly classified as temporal and spatial approaches depending on their focus of optimization on the channel bandwidth. The temporal approaches attempt to better utilize the channel along the time dimension by optimizing or improving the exponential bakeoff algorithm. On the other hand, the spatial approaches try to find more chances of spatial reuse without significantly increasing the chance of collisions. These mechanisms include the tuning of the carrier sensing threshold, the data rate adaptation, the transmission power control, and the use of directional antennas. This chapter is aimed specifically at providing a comprehensive survey on schemes that deploy the use of directional antenna and the schemes that consider coupling of directional antenna with power control.

Keywords: IEEE 802.11; directional antenna; power control.

4.1. Introduction

The rapid evolution of the mobile Internet technology has provided incentives for building efficient multihop ad hoc networks. A wireless ad hoc network precludes the use of a wired infrastructure and allows hosts to communicate either directly or indirectly over radio channels. These networks are applicable to environments in which a prior deployment of network infrastructure is not possible. Wireless ad hoc networks have enabled the existence of various applications ranging from the monitoring of herds of animals to supporting communication in military battlefields and civilian disaster recovery scenarios as well as emergency warning system for vehicles. Due to the scarce wireless channel resources, an effective medium access control (MAC) protocol which regulates the nodes access to the shared channel is required.

Currently, the distributed coordination function (DCF) of the IEEE 802.11[5] is the dominant MAC protocol for both wireless LANs and wireless multihop ad hoc environment due to its simple implementation and distributed nature. However, it has been shown that the DCF access method does not make efficient use of the shared channel due to its inherent conservative approach in assessing the level of interference. To date, various mechanisms have been proposed to improve the capacity of IEEE 802.11-based multi-hop wireless networks. These mechanisms can be broadly classified as temporal and spatial approaches depending on their focus of optimization on the channel bandwidth. The temporal approaches attempt to better utilize the channel along the time dimension by optimizing or improving the exponential bakeoff algorithm. On the other hand, the spatial approaches try to find more chances of spatial reuse without significantly increasing the chance of collisions. These mechanisms include the tuning of the carrier sensing threshold, the data rate adaptation, the transmission power control, and the use of directional antennas. This chapter discusses the use of directional antennas.

Directional antennas offer numerous benefits, such as extended communication range, improved capacity and suppressed interference. Due to the great potentials that directional antennas have shown in cellular networks, it is expected that using directional antennas in a multi-hop WLAN environment could also lead to a better performance in terms of spatial reuse and energy consumption. However in order to take full advantage of these potential benefits, efficient MAC protocols that are directional antenna-friendly need to be designed.

The rest of the chapter is as follows. In Section 2, we give a brief background on limitations of the implementation of IEEE 802.11[1] with Omnidirectional, we proceed further to introduce the concept of power control. The merit behind using directional antenna is presented in Section 4.3. Operational problems facing medium access protocol design when using directional antennas is presented in Section 4.4. The literature that specifies directional antenna implementations is surveyed in Section 4.5. Section 4.6 presents the gains when coupling directional antenna with power control and then existing work. Conclusion and future research directions in this area are listed in Section 4.7.

4.2. Background

4.2.1. *Omnidirectional antennas*

The omnidirectional antenna radiates or receives equally well in all directions; i.e. a 360 degrees mode. It is also called the "non-directional" antenna because it does not favor any particular direction. Figure 4.1 shows the coverage pattern for an omnidirectional antenna. This type of pattern is commonly associated with verticals, ground planes and other antenna types in which the radiator element is vertical with respect to the Earth's surface. The radiated signal has the same strength in all directions. Nodes equipped with omnidirectional antennas are constrained with limited transmission distance range. Only nodes lying in the vicinity of this distance range are able to receive and decode the packets correctly. In ad hoc networks the range of these antennas will be a few 100's of meters. Omnidirectional antennas are practical for broadcasting and multicasting applications.

4.2.2. *Directional antennas*

A directional antenna receives or sends signals most effectively in a particular direction. The advantages of directional antennas over omnidirectional antennas can be summarized by the increase of the signal quality (SNR) through beam forming gain, reduced interference through null steering, spectral reuse through spatial multiple reuse and robustness to multipath fading through spatial diversity. Directional antennas exist in various forms. Two of them are mostly used in literature: the switched beam antenna and steering antenna. The switched beams antenna system as shown in Figure 4.2 is divided into a fixed number of identical sectors. Hence for an N-sectored, there are n antenna elements covering with a beam angle of $\frac{2*\pi}{N}$

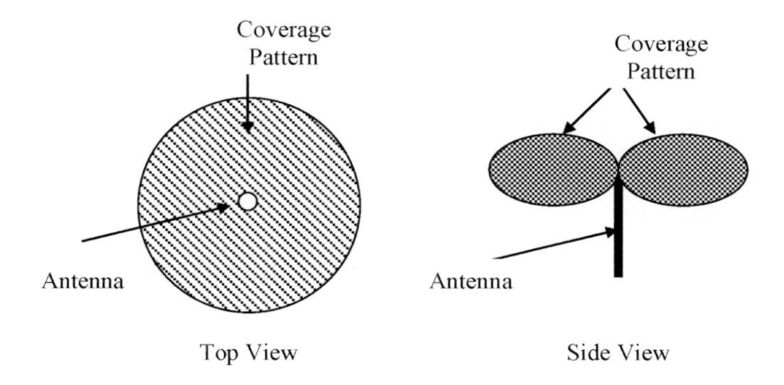

Fig. 4.1. Omnidirectional antenna coverage pattern.

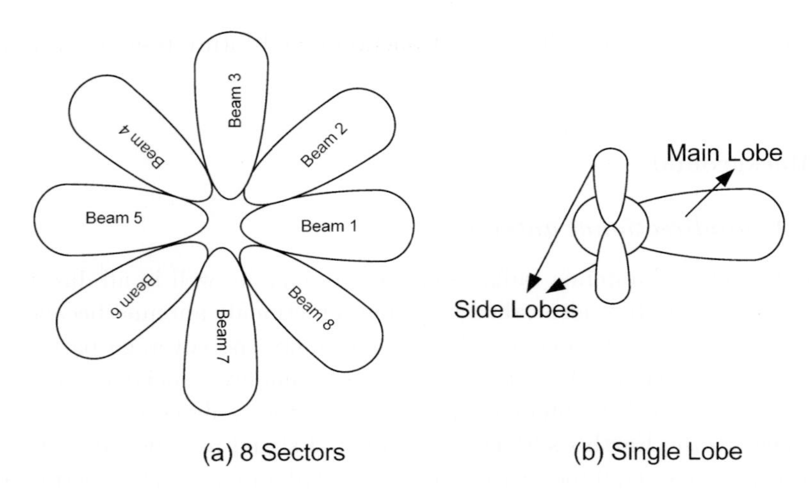

Fig. 4.2. Switched beam antenna.

radians. The transmission pattern of each sector forms the main lobe. For every main lobe, there are two side lobes. The lobes represent both the transmission and reception gains.

The steering type antenna can direct the beam to the desired receiver direction. Thus, it is considered more intelligent than the Switched beam antennas. This antenna consists of a number of antenna elements. The antenna system logic combines the antenna elements in such a way that the beam is directed towards any given angle. Through Null steering,[2] these kinds of intelligent antennas are able to minimize interference from unnecessary nodes. This is performed by integrating the antenna elements in such a mode that mainlobe, side lobes and tail lobe are not aimed at the interferer, the antenna minimize the interference. For better illustration of this technique, consider the example in Figure 4.3. The example shows a steerable antenna listening to the sender by mixing its antenna elements such that the main lobe points to the sender and the side lobes and the tail lobes avoid the interferer. Steerable antennas are more complex and expensive than switched beam antennas. Four possible connectivity scenarios exist when using omni and directional antennas; these are depicted in Figure 4.4. Clearly, the maximum connectivity reach is when the transmit and receive modes of the antennas are both directional.

4.2.3. *IEEE 802.11*

The IEEE 802.11 DCF relies on carrier sensing multiple access with collision avoidance (CSMA/CA). DCF employs two different channel access modes for data packet transmission; the default 2-way (basic access) and the optional four-way handshaking (request-to-send (RTS) clear-to-send CTS)) access scheme. Here, the optional (RTS/CTS) scheme assumes the transmission of RTS and CTS control packets

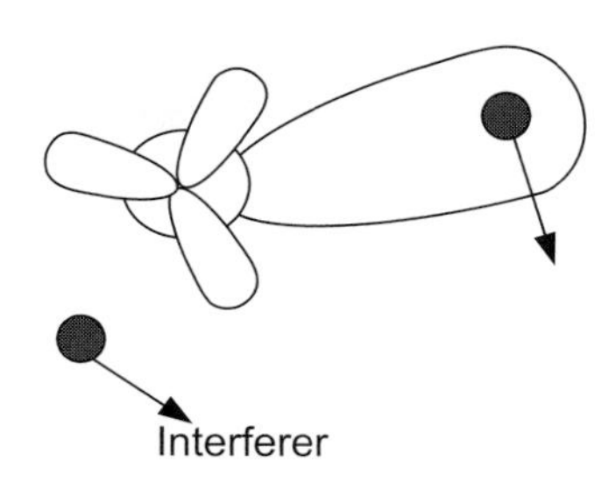

Fig. 4.3. Steering antenna.

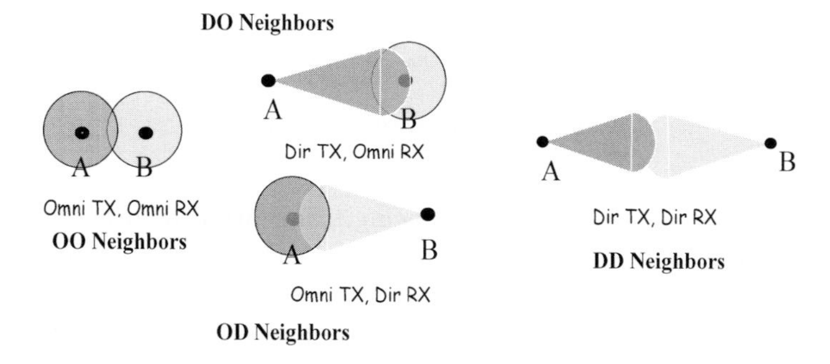

Fig. 4.4. Transmission and reception mode of antenna.

prior to the data packet transmission. In the (RTS/CTS) access scheme, a node with packets to transmit first senses the medium through physical carrier sensing (PCS). If the medium is idle for at least a certain period DIFS (Distributed Interframe Space), it will immediately request the channel by sending a short control frame RTS to the receiver node. If the receiver correctly receives the RTS, it will reply with a short control frame CTS after waiting a SIFS (Short Interframe Space) period. Once the sender receives the CTS, it will start to transfer DATA. After the successful reception of DATA, the receiver sends an ACK to the sender. SIFS duration is the shortest of the interframe spaces and is used after the RTS, CTS, and DATA frames to give the highest priority to CTS, DATA and ACK, respectively. Nodes implementing IEEE 802.11 maintains a NAV (Network Allocation Vector) which shows the remaining time of the on-going transmission sessions. Using the duration information in the RTS, CTS, and DATA packets, nodes adjust their NAVs whenever they receive a packet. This is shown in Figure 4.5. For the basic scheme, nodes upon sensing the DATA or ACK packet will refrain from transmitting any packet for EIFS. Moreover, if the channel is sensed busy, the station has to wait until the channel is sensed idle for a DIFS time. At this point, the station generates a random backoff time interval before transmission, in order to minimize

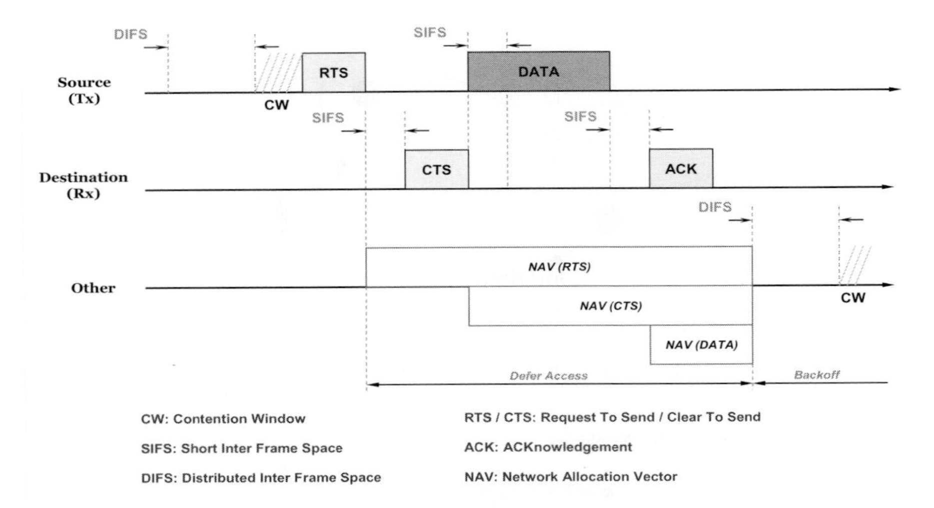

Fig. 4.5. IEEE 802.11. DCF.

the probability of collisions with packets being transmitted by other stations. DCF adopts an exponential back off scheme. At each packet transmission, the back off time is uniformly chosen in the range [0, CW]. The value CW is called contention window, and depends on the number of failed transmissions for the packet. At the first transmission attempt, CW set to the minimum value, CW(min). After each retransmission, CW will be exponentially increased until the maximum value. Upon successful transmission, CW will be reset to CW(min).

4.2.4. *Transmission power control merits*

In IEEE 802.11, all packets (RTS/CTS/DATA/ACK) are transmitted at the maximum power. It has been shown that this kind of handshake communication affects the spatial reuse and thus decreases capacity throughput and additionally yields unnecessary energy consumption. We illustrate the benefits of transmission power control through the following example. Consider the scenario shown in Figure 4.6, where Node A uses its maximum transmission power (TP) to send packets to Node B. Now Nodes C and F will not be able to initiate a communication with Nodes D and E respectively since they lie in the transmission vicinity of Nodes A and B respectively. Here, Nodes C and F are termed as exposed terminals. However, it is easy to show that the three transmissions A $\rightarrow$ B, D $\rightarrow$ C, and E $\rightarrow$ F can overlap in time if nodes are able to select their transmission power appropriately as shown in Figure 4.6. Accordingly, spatial reuse is enhanced, which yields an increase in the network throughput and ultimately reduces the overall energy consumption. Thus, adjusting the transmitted power is extremely important in ad hoc networks due to the major three reasons:

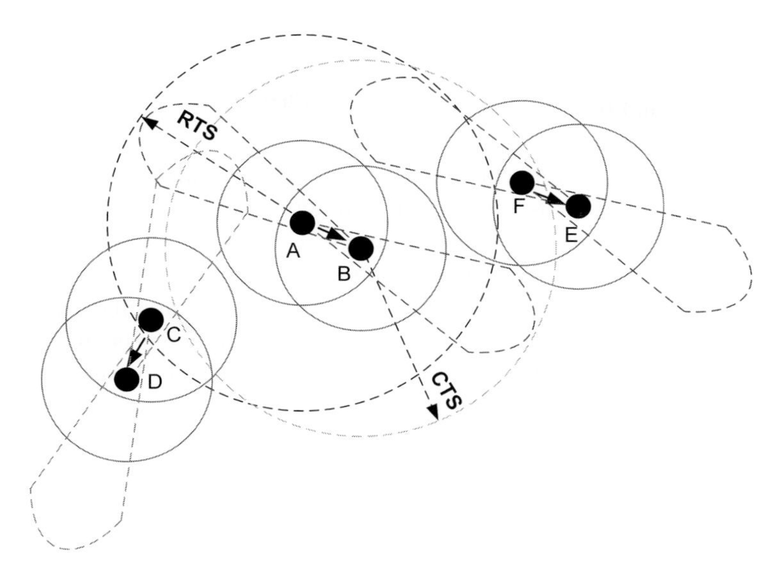

Fig. 4.6. Power control and directional antenna merits.

- The transmitted power of the mobile node determines the network topology. The network topology in turn has considerable impact on the throughput performance of the network.
- Mobile nodes in multihop ad hoc networks are usually energy constrained, and they have to be as energy efficient as possible. Hence, the power control should adjust the transmitted power to be the least power required to send the data packet to meet the required Signal to Interference plus Noise Ratio (SINR) threshold.
- Transmitting at high power can degrade other ongoing transmissions and in the meantime can prevent future transmission as illustrated in Figure 4.6. Thus, reducing the transmission power will reduce interference on other nodes communication and may enhance the overall network throughput as well as energy consumption.

4.3. Directional Antenna Evolution

One aspect for performing power conservation is the use of directional antenna. Directional antenna offers clear advantages for improving the network capacity by increasing the potential for spatial reuse. Allowing antennas to direct their transmissions in the direction of the intended receiver clearly reduces the level of contention with other nodes, thereby allowing for more simultaneous transmissions. Moreover, directional antennas can increase the signaling range without spending extra power (as opposed to omnidirectional) and accordingly, some receivers outside the omnidirectional range may be reached in one hop transmission. This longer range results in a smaller number of hops on end-to-end paths, yielding an increase in connection throughput. To elaborate more on the merits of directional antenna, we consider again the scenarios shown in Figure 4.6; Nodes A and B, Nodes C and D, and Nodes E and F will not be able to communicate simultaneously with omnidirectional operation whereas with directional antenna, this can be done; thus spatial reuse is enhanced. Further, Nodes E cannot reach Node A with omnidirectional antenna but using directional antenna, Node E reaches Node A. Since a directional antenna has a higher gain, a transmitter using directional antenna requires a lower amount of power to transmit to the same distance as would be needed with an omnidirectional antenna. Hence, transmitting nodes can conserve power by adequately reducing the transmit power when using directional transmissions. If Node A uses an omnidirectional antenna, the power sent by A towards Node B will be radiated uniformly in all directions at the same time instance resulting in a circular transmission/reception pattern. Although the transmitted packet is intended for a specific receiver, all nodes that lie within the vicinity of this circular transmission receive the signal, this results in unnecessary power wastage. If Node A transmits using a directional antenna, the power will be radiated in specific direction towards Node B and thus achieves much little energy consumption. For a given transmission distance, the power required by a transmitter using directional antenna is proportional to its beamwidth. This implies that Node A using a directional antenna with 1/6 beamwidth will approximately requires 1/6 the amount of power an omnidirectional antenna needs to transmit to Node B.

4.4. Directional Antenna Implementation Challenges

One of the first modifications of the distributed coordination function that aims at directional antennas was proposed in the Directional Medium Access Control (D-MAC) protocol.[3] Here, RTS/CTS/DATA/ACK frames are all transmitted directionally. However DMAC suffers from operational problems such as the deafness problem, the hidden terminal problem, the exposed terminal problem, and the Head Of Line (HOL) blocking problem. We identify the origin of each problem and evaluate its impacts on the network performance.

4.4.1. *Deafness problem*

The deafness problem[4] results when an intended receiver fails to reply to an RTS message initiated by an intended sender since the intended receiver beam is set in a direction away from the intended sender. Figure 4.7(a) illustrates this problem; an ongoing directional communication is occurring between Nodes B and C. Node A attempts to initiate a communication with Node B by sending Directional RTS towards Node B after sensing the channel in that direction to B idle. As a result,

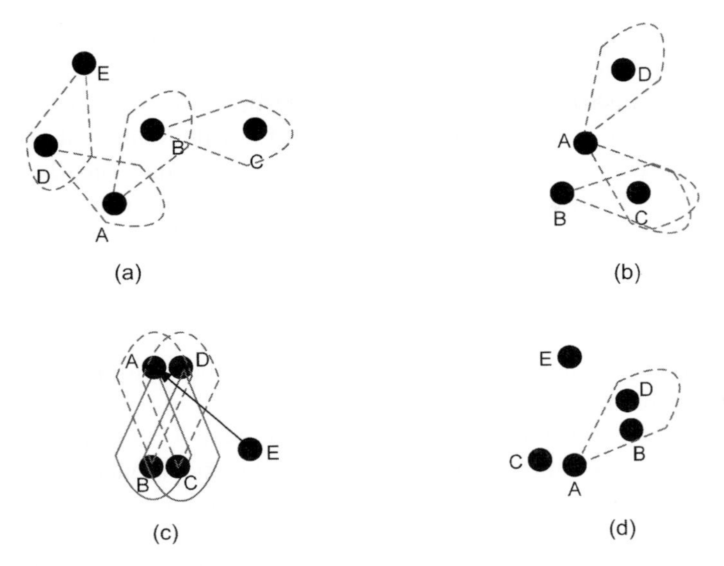

Fig. 4.7. Directional antenna operational challenges. (a) Node B is communicating with C, A tries to communicate with B, at the same time, D tries communicating with A, the same for E; result is deadlock; (b) Node A unware of communications taking place between Nodes B and C, sends its RTS message; collision will occur at C; (c) Node E tries to communicate with Node A but Node A is unable to respond since it is in the vicinity of directional transmission of Node C; (d) Node C can win the channel towards Node E but it has a packet for Node D, so more packet delivery delays.

Node B will be unable to respond to Node A's RTS message, since its beam is directed towards Node C. This is termed as deafness. Moreover, Node A is ignorant about the cause Node B is not replying for. Node A assumes packet collision and exhaust its RTS retry limit and in the case where the communication between Nodes B and C takes longer times than the time needed for RTS retry limit, node A will drop the packet. Now suppose there are more than such scenario in the network, the network will be jammed and the overall performance in terms of energy savings and throughput will definitely decrease. Now, consider the same scenario as before where Node D is trying to communicate with Node A while Node A is attempting to communicate with Node B. Another Node, say Node E, initiates a communication towards Node D while Node D is trying to communicate with Node A. Deadlock is reached and severe consequences may result. To conclude, deafness is a severe phenomenon which may hinder to a great extent the merits of using directional antennas if left unsolved.

4.4.2. *Hidden terminal*

The hidden terminal problem is due to the combination of inefficient timing criteria of the IEEE 802.11 and the directional transmission of RTS/CTS/DATA/ACK messages. To elaborate more on this, let us consider the simple scenario depicted in Figure 4.7(b). While Node A was communicating with Node D, it was not able to hear the RTS/CTS messages between Node B and C. If Node A has a packet to send to Node C, Node A sends an RTS message towards Node C at the same time Node B starts sending DATA message towards Node C; consequently collision will occur at Node C. Collision has high effect on increasing energy consumption and declining capacity throughput.

4.4.3. *Exposed terminals*

Exposed Terminals in multi-hop wireless network reduce spatial reuse and consequently the throughput performance. Figure 4.7(c) illustrates this problem; Node A cannot initiate any transmission to Node B while Node C is communicating with Node D and vice versa since all Nodes lies in the directional vicinity of each other. Another problem that is built on the exposed terminal problem is the jointly exposed terminal problem and the receiver blocking problem. We illustrate this further in Figure 4.7(c); Node A cannot reply with a CTS message to an initiated RTS message from Node B, since Node A lies in the vicinity of the ongoing directional communication of Node C. Node B will keep sending an RTS message according to the number of retries allowed and configured in RTS retry limit, thinking that the previous messages collided. This term of problem has the same effect on the network as deafness.

4.4.4. *Head of line blocking*

Networks implementing First-In-First-Out (FIFO) queuing service rule suffers from Head-of-Line blocking Problem. Networks with such queuing schemes suffers to a great extent when using directional antenna. Nodes contend to win a channel before transmitting their packets. Suppose Nodes A in Figure 4.7(d) has a packet to transmit to Node B, Node C has a packet to transmit to Node D. Node A wins the channel and transmits the data message to Node B. On the other hand, Node C can win the channel in the direction of Node E and transmit any intended packet to Node E, but it is unable to do so since it has a packet for Node D. This is termed as HOL blocking problem. This problems adds more delay for packet delivery and thus affect the overall network throughput.

4.5. Directional MAC Literature

Another protocol[5] assumes that it is not necessary to know each node's location information. Each node is equipped with beam switched directional antenna. The sender and receiver estimate the location of each other by noting the antenna that received the maximum power when sending the omnidirectional RTS/CTS packets.

Then receiver and the sender points their antenna beams in the direction towards each other for data transmission to occur. Mobility is considered in the implementation of the protocol.

Directional Virtual Carrier Sensing (DVCS) scheme with steerable antenna system was introduced in Ref. 6. Estimated Angle of Arrival (AoAs) of nodes neighbors is cached when it hears any signal, regardless of whether the signal is targeted at the node. A node initiating a transmission will check its AoAs, if it exists; the node uses a directional RTS to that node otherwise omnidirectional RTS is sent. Directional Network Allocation Vector (DNAV) a directional version of IEEE 802.11 NAV (Network Allocation Vector) operates in reserving the channel for others only in a range of directions.

The Multihop Medium Access Control (MMAC) protocol[7] integrated with DVCS is based on the basic DMAC. For Directional transmission, a steerable single beam antenna is used. MMAC was proposed to make use of the extended communication range advantage of directional antenna while maintaining the spatial reuse of the basic DMAC. Two communications aspects are required for correct implementation of MMAC the Direction-Omni (DO) Neighbors and Direction-Direction (DD) Neighbors. A node attempts to establish a DD link by sending a directional RTS and waiting for the CTS in that direction for predefined timeout interval. If timeout occurs, it forwards multiple hops RTS along the DO-neighbor route to the destination. As noted, the MAC depends on the routing layer in getting the DO-neighbours. The destination node will reply by directional CTS, directly after which data transfer occurs with the sender and the receiver beams pointing towards each other. Thus a DD link is established.

A Circular-DMAC protocol[8] was proposed in an attempt to fully eliminate deafness and to completely utilize the benefits of directional antennas. No omnidirectional mode was considered in the four-way handshake (RTS/CTS/DATA/ACK). RTS packets are directionally sent via sweeping sequentially the antenna beams to cover the whole space using switched beam antenna. This is termed as circular directional RTS. CTS and Data frames are sent directionally afterwards. Nodes locations tables containing the nodes neighbours identities, associated with the beam index on which neighbor nodes can be reached, and the corresponding beam index used by the neighbour is proposed with the DNAV mechanism to combat the hidden terminal and deafness problems. Although, deafness and hidden terminal problems were decreased in a significant manner, circular directional RTS degrades performance due to the excessive overhead it imposes on the network. Moreover, the location table update for every packet reception is not practical for highly dynamic networks.

The Dual Busy Tone Multiple Access with Directional Antennas (DBTMA/DA) protocol[9] is based on the Dual Busy Tone Multiple Access (DBTMA) protocol with omnidirectional antennas. This idea was motivated by the aspect of performance deterioration due to RTS/CTS collision. In this scheme, busy tones, RTS/CTS and Data packets are sent directionally. Two busy tones: transmission busy tone and reception busy tone are assigned two separate single frequencies in the control channel. This has shown to be an efficient scheme to decrease the possibility of hidden and exposed terminal problems. Thus spatial reuse and channel capacity are increased which was verified by significant advantageous effects on throughput and end-to-end delay via simulations. Upon hearing a transmission/reception busy tone the node will reschedule its transmission, which alleviates both the hidden and exposed terminal problems. Moreover, a comparison between sending omnidirectional busy tones and directional busy tone is studied. It is shown that the directional transmission of transmit busy tones favors over the omnidirectional transmission, since it avoids the exposed terminal problem, whereas a tradeoff exists when comparing a directional receive busy tone with an omni one. Obviously, omnidirectional receive busy tones degrade the spatial reuse in favor of reducing hidden terminal problems where as directional receive busy tones enhance spatial reuse, but introduce new hidden terminal problems. With such scheme, the DNAV does not provide advantageous influence to MAC and can be avoided if wanted.

The Tone-based directional MAC (ToneDMAC) protocol[10] has been proposed to alleviate Deafness; in other words to get nodes to understand that other nodes are deaf and are unable to hear their transmission. In this scheme, tones are sent omnidirectionaly after the directional DATA/ACK exchange has been completed. By this, a neighboring node can conclude deafness if it overhears a tone from its targeted destination. Upon hearing the tone, the node concludes that this node is deaf and decreases its contention window to a minimum value and has thus a fair chance to win the next channel contention. Moreover, the authors divided the channel into two sub-channels: a data channel and a narrow control channel. RTS/CTS, Data and ACK packets are transmitted on the data channel and tones are transmitted on the control channel. Every node is assigned a unique identifier by which its neighbour can recognize it through. Identifying a node is performed via a hash function that contains both the tone frequency and duration. Multiple Tones usage in the network was an obstacle and was rectified via frequency tone reuse through the hash function implementations. Furthermore, the location information regarding the neighboring nodes was used to reduce the probability of tone mismatching.

Smart-Aloha protocol, proposed in Ref. 11, is based on the slotted aloha and follows a Tone/Packet/ACK handshake. A source node beamforms towards the receivers and sends a tone. Upon receiving the tone, the receiver in omnidirectional mode runs a Detection of Angle algorithm to locate the best direction for the received tone. DATA/ACK exchange occurs afterwards. The drawback in this scheme is that the sent tone does not identify the destinations address. Thus, a receiver may receive a data packet that was not intended for it. In such a case, the packet will be discarded and the sender enters backoff. Smart-802.11b,[12] a mix of IEEE 802.11b standard and the Smart-Aloha protocol implementation, avoids the collisions resulted by its predecessor Smart Aloha by using sender-tone and receiver-tone as well. In such a scheme, the transmitter waits for a receiver-tone before sending a packet. After the packet is sent, an ACK is required. If the sender fails to receive the ACK, it will enter backoff like IEEE 802.11b. Destination fields are not encapsulated inside the tone which is not the case for RTS/CTS. So, Smart-802.11b suffers from the same drawback as Smart-Aloha but with less packets collisions.

DOA-MAC is another directional MAC proposed in Ref. 13. Each slot in DOAMAC is divided into 3 mini-slots resembling the TONEsent/Tonereceived/ DATA/ACK handshake. The tones handshake occurring in the first mini-slot follows the same steps as in Smart-Aloha protocol. Data and ACK transmissions follow in the next two mini-slots. If the transmitter does not receive the ACK, it will retransmit the packets later in another slot. Furthermore, since it follows the same principle as in the smart 802.11, receivers of the packets may be not the targeted ones, in such a case, no ACK will be sent in the third mini-slot. Randomness dominates this approach. The probability of unintended packet receptions and the probability of deadlocks have been addressed in evaluation of this protocol. As an enhancement of DOA-MAC the authors implemented a single-entry cache scheme. It allows a receiver to beamform toward the second strongest signal if the receiver was not the intended receiver of a packet transmitted by a node providing the strongest signal.

Space Division Multiple Access (SDMA) is proposed in Ref. 14 to improve the throughput of nodes by synchronizing the packet receptions from other nodes. This scheme employs time division duplex (TDD) between transmission and reception. A node informs all its neighbours its readiness for packet reception by initiating an omnidirectional Ready-to-Receive (RTR) packet periodically. Senders with a packet to transmit will reply with a directional RTS packet, after which directional CTS/DATA/ACK sequences follows. Synchronization among node is achieved via training sequences. This enables the receiver to train switch its beam to receive all signals. RTR by itself is a large control packet and can degrade the performance due to extra overhead it puts on the network. Thus the protocol performance is unacceptable under low loads.

Another scenario was considered in Ref. 15. In this scheme, omnidirectional RTS, Directional CTS and DATA packets are considered. The authors deduce that omnidirectional RTS reduces the hidden terminal problem, whereas a directional transmission decreases the exposed terminal problem. The authors propose two corresponding schemes (1) sending an omnidirectional RTS and (2) sending directional RTS. They verify via simulations that Directional transmission of RTS outperforms the omnidirectional scheme. The authors further consider the effect of the carrier sensing threshold on the network throughput. The hidden terminal problem domination over the exposed terminal problem when carrier detection is not activated was their conclusion.

The authors in Refs. 16, 17 and 18 propose various MAC protocols for MANET with directional antennas. In Ref. 16, they provide enhancements to the popular DMAC to reduce deafness. One scheme is to intelligently handle the RTS/CTS handshake; nodes hearing an RTS would block their beams towards the source nodes. Nodes hearing a CTS frame should not block their antennas in that direction since it may miss important control or broadcast packets in that direction. The implementation was straightforward; a flag in the DNAV table was added. This flag shows if the DNAV presently set was originally due to an RTS or not. Another enhancement was built on an assumption that the location of a node is known, number of antenna beams is identical for all nodes, the same radiation patterns profile exists for directional antennas and addresses an efficient way to estimate destination status. With such a technique, consider, for example, Node A will check if it is reliable to initiate an RTS to neighbour Node B while an ongoing data transmission is taking place in the region. This is done by checking in advance the associated beam of the Node B if it is blocked or not due to an RTS from the ongoing transmission. The author further investigates short retry limit implementations and argues that sender can deduce deafness from its intended receiver by continuously listening to the packets in its neighborhood. For instance, if Node B receives an RTS from S, and if it has a packet to sent to S, it resets its SRL as well as its contention window. Various Directional MACs scenarios have been considered for performance evaluations. Simulation results showed that the achieved performance gain is dependent of the type of directional MAC protocol and on the network topology as well as the traffic pattern between nodes. Directional Antenna Medium Access protocol (DAMA)[17] designed to effectively pass the limitations incurred by DMAC and CRM. Directional four-way handshake is considered to take the benefit of increased gain obtained by directional antennas. DAMA employs a circular transmission of RTS and CTS to prevent the problems if deafness and hidden terminal problems. In order not to overwhelm the network with these control packets, DAMA performs an optimized transmission of RTS/CTS. A node discovers all its neighbors and consequently sends its RTS through the antenna beams associated with these neighbors. The worst case scenario is the dense network topology where the node has to use all of its antennas beams to reach all its neighbors, thus performance issue regarding overhead pops out. The mentioned technique showed a performance gain in case of sparse network. DAMA employs a three way handshake in a way Directional RTS is sent then followed by directional CTS then at the same DRTS/DCTS are sent directly via sweeping through all the selected beams. Finally ACK is sent. DAMA implements an adaptive mechanism where it learns and caches information about those sectors with neighbors. Various topologies have been considered for performance evaluation. Simulation showed that DAMA performs better than IEEE 802.11 DMAC and CRM in all scenarios except in the linear topology. The linear topology case is particularly degrading to all directional MAC protocols, but DAMA is still observed to perform best in terms of all directional MAC protocols considered, while IEEE 802.11 performs best overall. MAC for directional Antenna (MDA)[18] another fruitful work by the same authors consider enhancing DAMA by employing an efficient sweeping procedure for sending circular RTS and CTS called the Diametrically Opposite Directional (DOD).

SYN-DMAC a directional MAC protocol for ad hoc networks with Synchronization is investigated in Ref. 19. Nodes are assumed to be equipped with GPS receivers for synchronization purposes. A switched-beam antenna is adopted for this protocol. SYN-DMAC proposes a timing-structure different than that of the IEEE 802.11 to alleviate the hidden and deafness problems. The timing structure in each cycle is made up of three phases. The first phase is the random channel access multiple nodes contend to win out a channel. Neighbor discovery is carried along with channel contention in the same phase. One or more node-pair may win out a channel on a condition these node-pairs should not collide with each other. Simultaneous Data transmissions occur in second phase. Based on the variation of the channel, each node-pair modify its power or rate accordingly. Finally parallel contention-free ACKs are transmitted in phase three. The protocol also addresses the HOL problem. Simulations showed significantly improvement gain over the IEEE 802.11 MAC protocol.

To fully exploit the benefits of directional antennas, DBSMA,[20] a MAC protocol for Multi-hop Ad Hoc Networks with Directional Antennas, considers all transmissions, receptions, and idle listening to be directional. The authors describe a set of requirements that should be met by directional MAC protocol and then proposed DBSMA as a voted MAC. DBSMA uses novel concepts of idle directional listening, beam sweeping, Invitation Signal and directional back-off windows. Listening is performed via sweeping to cover the whole space. Every beam sector for each direction in a switch beam antenna implements its own back off window. Implementing the same back off window would result in unfairness to channel contention. Increasing a backoff window of a node due to collision in one direction affects the node's ability to content for a channel in the remaining directions due to the fact the node implements the same backoff window for all directions. If such a collision occurs in an uncongested area, this will deteriorate the node's capability to send in congested area. Thus, the network condition becomes more unstable. Independent backoff window for every direction rectifies this problem. An Invitation signal is introduced before an RTS control packet. This signal is sufficiently long to make all neighbors hear it. Upon hearing it, all nodes stop listening in all directions and lock their antenna beam in the direction of IS signal and waits for RTS message. After receiving an RTS, a destination node starts transmitting a busy tone till the end of the DATA and ACK transmissions. The busy signal is a narrow bandwidth, out of band signal. Simulations showed that implementing independent backoff window for every direction has enhanced the throughput by factor greater than 20%. The author observed the that the minimum back-off window using directional antennas should be smaller compared to the single back-off window due to the reality that the number of nodes existing in within the directional transmission range is smaller than the number of all possible neighbors. It is also shown that IS duration highly affects the throughput. Finally, DBSMA achieved higher performance gain over the popular DMAC in the case the control packet is much smaller than the data packets.

Smart Antenna based Wide-Range Access MAC protocol (SWAMP) is presented in Ref. 21. All nodes made us of GPS to locate the direction of the target nodes. The protocol is effectively based on location information and uses two type of modes in communication: the Omnidirectional area Communication access mode (OC-mode) and (Extend area communication access mode) EC-mode. EC-mode means that all control and Data packets are sent directionally where as OC-mode, only Data and ACK are sent directionally. The protocol makes use of Next Hop Direction Information table (NHDI) to switch between these two accesses modes. Start of frame control packet (SOF) is added between CTS and DATA in order to forward the NHDI to neighbors of both the source and destination. The source node inserts the NHDI of the destination node into the SOF. Also, destination nodes should insert the NHDI of the source node into the CTS that is an extended traditional control frame. If destination node is registered in the NHDI table, EC-mode is selected otherwise the OC-mode is selected. Comparison is carried with IEEE 802.11 and enhancements in throughput are recorded. The author concluded the paper with a qualitative evaluation with DMAC and MMAC.

4.6. Power Control and Directional Antennas

4.6.1. *Preliminaries*

With the increase of the number of concurrent communications with the use of directional antenna, the overall interference impact may increase. Additionally, a higher gain of directional transmissions leads to more interferences at nodes receiving in omnidirectional mode. Interference, in turn, will increase the probability of packet delivery error since it decreases the signal to interference ratio (SIR), and this may have a strong negative effect on the overall network throughput and energy gains. This motivates the use of power control with directional antenna.[22]

The integration of directional antenna with transmission power control scheme can give more benefits than anticipated.[22] Figure 4.8 illustrates this very easily by comparing the four combinations assuming an angle of 10 degrees and applying the path loss law: (a) no power or directional, (b) power control without using directional antenna, (c) directional antenna, (d) power control with directional antenna. Neglecting the side lobes of directional antenna and taking into account the ratio of the areas resulting from applying the four combinations, we can argue the merits achieved in terms of spatial reuse from integrating power control with directional antenna. With power control and without the use of directional antenna, the area is reduced by a factor of 4 if the distance between the sender and the receiver is half the distance of the maximum range. With directional antenna and without power control, the same area is reduced by a factor of 6 if the beamwidth is 10 degrees.

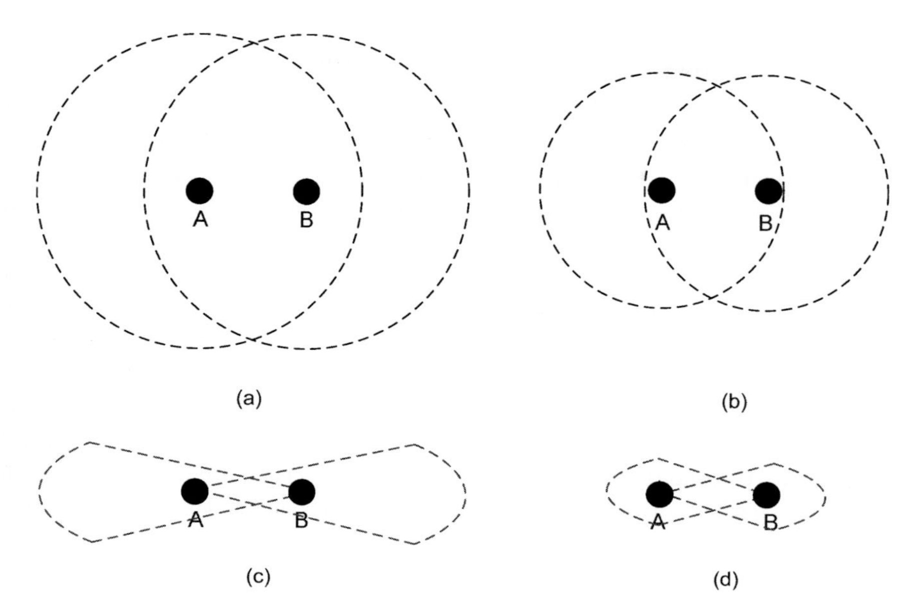

Fig. 4.8. Coupling directional antenna and power control merits. (a) No power control and no directional antenna (Area = A); (b) power control and no directional antenna (Area = A/4); (c) Directional antenna (Area = A/6) if using beamwidth of 10 degrees; (d) directional antenna with power control (Area = A/144) if using beamwidth of 10 degrees.

Finally using directional antenna with power control, the area is reduced by a factor of 144% with the same distance and beamwidth. It can be seen that coupling directional antenna is far better than using each technique separately. According to the path-loss, energy is directly proportional to the distance between the sender and receiver raised to power of path-loss factor. The path loss factor can take values between 2 and 6 depending on the wireless medium.

With the use of directional antenna, energy is directly proportional to the beam width angle and the distance between the sender and the receiver. Applying power control with directional antenna can enhance the energy savings by 144%,[22] if using path loss free space model. Nevertheless, one cannot expect to match the theoretical potentials of using directional antenna with power control due to the fact that directional antenna has its own drawbacks when implemented, but the question of how difference the incorporation of power control with directional antenna makes merit attention with the existing operational burdens. Potential MAC proposals that make use of directional antenna and power control are discussed in the next section.

4.6.2. *Survey of directional MAC with power control*

The use of directional antennas for single-hop packet radio network was first proposed in Ref. 23 where a slotted ALOHA packet radio network was considered.

Authors derived an equation model to calculate the performance improvement that can be obtained in a slotted ALOHA channel by the use of directional antennas and multiple receivers. The idea was then reformulated to multi-hop networks but using directional antenna with power control. The throughput performance was investigated through a derived equation model. The derived model showed that the throughput increases dramatically if power controlled directional antennas are used for transmission. Moreover, the authors argue that using narrow beams antenna, the risk of destructive packet collision is reduced and nodes will be able to communicate with higher transmission probability. Thereafter, several studies have been carried in to benefit from the controlled transmission power gains of directional antennas.

Performance evaluation of directional antenna with power control was studied in Ref. 24. The RTS message is sent at a predetermined power — the maximum power. The receiver will find the difference between the received power of the RTS message and its threshold power. The threshold power is the minimum power needed to decode the packet correctly. The value of the difference is sent within the CTS message. The source node will use power value that is equal to maximum power minus the difference value. A simulation experiment consisted of 40 static nodes equipped with directional antenna were randomly distributed in an specific area. Since the packets considered in simulation are large packets, the delay metric is a better indicator of performance. Adding power control with directional antenna dramatically reduced the delay by up to 28% whereas with only directional antenna the factor of delay is around 2% to 3%. Throughput enhancement of 118% was recorded.

The authors in Ref. 25 proposed the use of adaptive antenna arrays. RTS/CTS messages are sent omnidirectional with maximum power P_{max} whereas DATA/ACK frames are sent directionally with controlled power. A *SHORT_{NAV}* term is used to alleviate the exposed terminal problem. Two power control schemes were introduced: (1) global power control (GPC), (2) local power control (LPC). DATA/ACK power values in GPC are determined based on a factor k such that power of the DATA and ACK packet is equal to $k.P_{max}$; whereas the power of DATA/ACK packets in LPC is set for each transmission so the Signal to Noise ratio (SNR) is a pre-determined value. This can be done by using the values of the received RTS/CTS power levels to compute how much power reduction is required. Performance evaluation of GPC and LPC showed the following; normalized system capacity for GPC was 475% over IEEE 802.11, LPC was 525% over IEEE 802.11, whereas with only the use of directional antenna it is 260% over IEEE 802.11.

Based on the omnidirectional BASIC power control protocol, a similar scheme but with the use of directional antenna was investigated in Ref. 26 They name it directional antenna based MAC protocol with power control (DMACP). Here, RTS/CTS/DATA/ACK frames are sent directionally; the RTS and CTS messages are sent with maximum power but the data packets are transmitted with power control. Through the RTS-CTS handshake, the power value for transmitting the data packet is assigned. Moreover, a destination node upon receiving an RTS packet, it calculates the difference between the values of signal to interference plus noise ratio (SINR) of the RTS packet and the minimum SIR threshold. This difference value is encapsulated in the CTS message sent to the source. Based on this value, the source reduces the power value needed for ensuing the data packet by an amount that is equal to this difference minus a margin 6 dB, not exceeding the maximum power level of the transmitter. The performance evaluation of DMACP showed that integrating power control with use of directional antenna does not have a significant impact on the throughput but on energy consumption.

A distributed power control (DPC) protocol has been introduced for ad hoc nodes with smart antennas in Ref. 27 In this protocol, the receivers measures the local interference information and send it to the transmitters; upon receiving this information, the transmitter use it together with corresponding minimum SINR (signal to interference plus noise) to estimate the power reduction factors for each activated link. DATA and ACK transmissions are in (beamformed) array-mode since smart antennas are used at both ends of the link. In DPC protocol, the interference information is collected during both omnidirectional RTS/CTS transmission and the beamformed DATA/ACK transmission. RTS/CTS packets are always transmitted with full power in omnidirectional mode, and the power level of DATA/ACK transmission is determined by a power reduction factor which is determined by the maximum interference. Protocol performance evaluation showed significant performance improvement has been achieved when compared the conventional IEEE 802.11 protocol.

A directional medium access protocol with power control (DMAP) was presented in Ref. 28. RTS message sent omnidirectionally while CTS/DATA/ACK messages are sent directionally. The main target of DMAP was to alleviate some of the problems associated with directional antenna use. Moreover, DMAP minimizes the energy consumption by integrating transmission power control with the use of directional antennas. Separate data and control (RTS/CTS/ACK) channels were used to rectify the hidden terminal problem due to unheard RTS/CTS messages. In DMAP, a transmitter sends an omnidirectional RTS. The receiver, before replying with directional CTS (D-CTS), will sense the data channel towards the transmitter and measures the interference. A power control factor is encapsulated within the D-CTS packet for the transmitter to read so as to assign a power value for data packets. The CTS message is sent with a power that is multiplied by directional gain factor as if the RTS message is sent directionally. The author argues that deafness would be eliminated due to the power scaling of D-CTS. Performance evaluation of DMAP when compared with IEEE 802.11b showed throughput enhancement by a factor of 200% and energy consumption reduction by a factor of 82%.

A load-based concurrent access protocol (LCAP) was proposed in Ref. 29. LCAP aims at increasing spatial reuse by allowing interference-limited, simultaneous transmissions to take place within the same vicinity by using transmission power control.

RTS messages are sent omnidirectionally with maximum power, CTS/DATA/ACK messages are sent directionally. Similar to its predecessor (DMAP), LCAP uses separate data and control channel to alleviate the hidden terminal problem due to unheard RTS/CTS messages. LCAP uses the same procedure in DMAP for scaling and finding the power of CTS to solve deafness. Moreover, the receiver uses a *load control* technique to determine the power value of the data packets and encapsulate it within the CTS message. This data power value is determined so as to ensure a balance between energy consumption and spatial reuse. Furthermore, upon finding the data power value, the receiver calculates the difference between this value and the minimum power value needed to decode the packet correctly. The difference is also encapsulated in the CTS packet and is used by the nodes hearing the CTS messages to find in case they have to initiate any communication, the amount of interference they can put on the receiver. Thus the difference value is interference margin that nodes decide on the maximum value of their future interfering transmissions. LCAP showed interesting performance metrics when compared with IEEE 802.11b for different network topologies.

Three channels power control scheme with the use of directional antennas is presented in Ref. 30. The protocol proposed uses one channel for data packets, a second channel for control packets and a third channel for busy tone. Busy tones are sent directionally, RTS message is sent omnidirectionally with maximum power, CTS/DATA/ACK sent directionally with CTS/ACK messages with maximum power. The busy tone is used to solve the deafness problem. An Interference model is calculated to estimate the interference around the receiver. Based on this interference calculated, a proposed power control scheme is designed. A node receiving an omnidirectional RTS will calculate the maximum interference using the mention model then decide on the power value the source node should send its DATA message by. Data power value is advised as follows. The receiver computes the difference between the maximum interference calculated by the model and subtracts from it the total measured noise power; then adds to this difference the minimum power needed to transmit the packet. The minimum power to transmit the packet is simply the minimum power needed to decode the packet correctly upon reception times the channel loss gain. Two busy tones are defined: transmit busy tone and receive busy tones. Both of these busy tones are sent at maximum power and with the RTS/CTS messages respectively. The protocol showed enhancement in the channel utilization and energy consumption when compared with IEEE 802.11 performances.

A new transmission power control scheme for enhancing the throughput and energy efficiency of the directional IEEE 802.11 based MAC protocol has been proposed in Ref. 31. It is noted that the use of single channel for the 4-way handshake communication is a necessary condition for this scheme to be adopted. The scheme considers the hidden terminal problem, deafness and side lobe interference effects caused by directional antennas. The scheme estimates the interference around the sending node and based on the predicted interference, it determines the relation that exists between the transmitted power of the RTS/CTS/DATA/ACK frames in order to achieve a target signal to interference (SIR) at both the transmitter and receiver. In other words, given the transmission power of RTS frame, the receiver can derive the power needed for transmitting the CTS frame; further, given the transmission power of CTS frame, the sender can determine the power for sending the DATA packet and finally we deduce the power needed for transmitting the ACK message based on the power of the DATA packet. Simulation results that correlated power control scheme is efficient in terms of throughput and energy consumption. The proposed scheme outperforms the IEEE 802.11 by 60% factor. At the same time, 55% reduction in energy consumed over the IEEE 802.11 is achieved by the scheme.

Table 4.1 illustrates the comparison between different directional MAC protocols presented in this chapter.

4.7. Summary and Future Work

In this chapter, we have presented an overview on the benefits and limitations of the integration of directional antenna in the multi-hop wireless ad hoc environment. The majority of the chapter has been devoted to the survey of state-of-the-art MAC protocols deploying the directional antenna and the union of directional antenna and power control. The idea of directional antenna with power control will continually receive considerable attention and will open good trends and innovations in the coming future for implementing an efficient multi-hop wireless ad hoc environment.

Reduction in power is a key-factor in improving the capacity as well as energy consumption. This is what we have figured out from the survey we carried. Beamforming when combined with power control is highly synergistic; i.e. when integrated together is far better in performance than carrying each alone. Although there have been some advances in the integration of directional antenna with power control, a lot more opportunities remain uninvestigated. We summarize them as follows:

- Convertness — What is known as low probability of detection. As explained before, the narrower the beam, the less energy needed in transmitting the packet. With this minimal leakage of energy characteristics, beamforming with narrow beams does not provide the chance for eavesdroppers to intercept packets. This is an interesting topic that needs to be addressed and implemented.
- Most of the work carried out using directional antenna with power control solves most of the operational problems associated with directional antenna using a second channel or simply a busy tone channel. An interesting research is to rectify all these problems using single channel.

Table 4.1 Comparisons of directional MAC protocols.

MAC Reference	Antenna Type	Side Lobes	Power Control	Hidden Terminal	Exposed Terminal	Deafness	Idle Listening	Channels
3	Switch	No	No	No	No	No	Omni	1
5	Switch	No	No	No	No	No	Omni	1
6	Adaptive Array	Yes	No	No	No	No	Omni	1
7	Steer Able	Yes	No	No	No	No	Omni	1
8	Switch	No	No	No	Solved	Solved	Omni	1
9	Switch	No	No	No	Solved	Does not exist	Omni	2
10	Switch	Yes	No	No	No	Solved	Omni	2
11	Adaptive Array	Yes	No	No	No	No	Omni	1
12	Adaptive Array	Yes	No	No	No	No	Omni	1
13	Adaptive Array	Yes	No	No	No	No	Omni	1
14	Switch	No	No	Yes	Yes	Yes	Omni	1
15	Switch	No	No	Yes	Yes	Yes	Omni	1
16,17,18	Switch	No	No	Yes	Yes	Yes	Omni	1
19	Switch	No	No	Yes	Yes	Yes	Omni	1
20	Switch	Yes	No	Yes	Yes	Yes	Directional	3
21	Switch	Yes	No	Yes	Yes	Yes	Omni	2
24	Switch	No	Yes	No	No	No	Omni	1
25	Adaptive Array	No	Yes	No	No	No	Omni	1
26	Switch	No	Yes	No	No	No	Omni	1
27	Adaptive Array	No	Yes	No	No	No	Omni	1
28	Switch	Yes	Yes	Yes	Yes	No	Omni	2
29	Switch	Yes	Yes	Yes	Yes	No	Omni	2
30	Switch	Yes	Yes	Yes	Yes	Yes	Omni	3
31	Switch	Yes	Yes	Yes	No	No	Omni	1

- Exploitation of beamforming with power control for other functions, such as routing policies, multicasting. A potential direction is to use directional antennas with power control to reduce the query of flooding overhead.
- Cross layer design of MAC protocols especially with the physical layer will open several new frontiers in research. An example of which is integrating modulation techniques with directional antenna with power control.
- No theoretical work on limits of wireless ad hoc capacity throughput and energy consumption with nodes equipped with directional antenna exists.
- To date, various mechanisms have been proposed to improve the capacity of IEEE 802.11-based multi-hop wireless networks. These mechanisms can be broadly classified as temporal and spatial approaches depending on their focus of optimization

on the channel bandwidth. The temporal approaches attempt to better utilize the channel along the time dimension by optimizing or improving the backoff algorithm[32,33] of the DCF protocol. On the other hand, the spatial approaches try to find more chances of spatial reuse without significantly increasing the chance of collisions. These mechanisms include the tuning of the carrier sensing threshold,[34–36] the data rate adaptation,[37] the transmission power control, and the use of directional antennas. Integrating Directional Antennas with spatial (not power control) and temporal has not been further investigated. For instance, the design of MAC protocol should consider the effect of idle listening to be directional and integrate with it a directional power control back off mechanism.

Problems

(1) What are the two techniques used to save energy?
(2) How a node using IEEE 802.11 DCF RTS/CTS transmits a packet?
(3) What are the two major problems associated with the implementation of TPC in an IEEE 802.11 based ad hoc networks?
(4) Consider the network of four Nodes A, B, C, D shown in Figure 4.9. Node A is communicating with Nodes B and C is communicating with Node D. If the distances between terminals are given as $d(A, B) = 95$ meters, $d(A, D) = 98$ meters, $d(C, B) = 65$ meters, and $d(C, D) = 57$ meters. Given that the target SINR for both connections is 10 dB. Assume that the target SINR is defined as the minimum required SINR when receiving a packet so as the node will be able to correctly receive and decode a packet. The additive average noise power is -90 dBw (10^{-10} mW). The transmitted power is 24 dB (250 mW) for both transmitting Nodes A and C. Moreover, each node is equipped with a transceiver of processing gain $= 10$; the height and the antenna Gains are assumed equal 1. The channel model is a two-ray model.

   (a) Assume no power control is applied; check if the transmission from Nodes A to B and from Nodes C to D will be successful.
   (b) Repeat Part (a) but now with applying power control. Assume that the transmitted power of Node A is 20 mW and that of Node C is 100 mW.

   Here, the SINR is calculated according to the following equation:

$$SINR = G * \left(\frac{P_r}{P_n + I} \right)$$

   where P_r is the good received signal, P_n is the average noise power, I is the interference from other communicating nodes. G is the processing gain and is assumed to be equal to 10. We adopt the two-ray channel model.

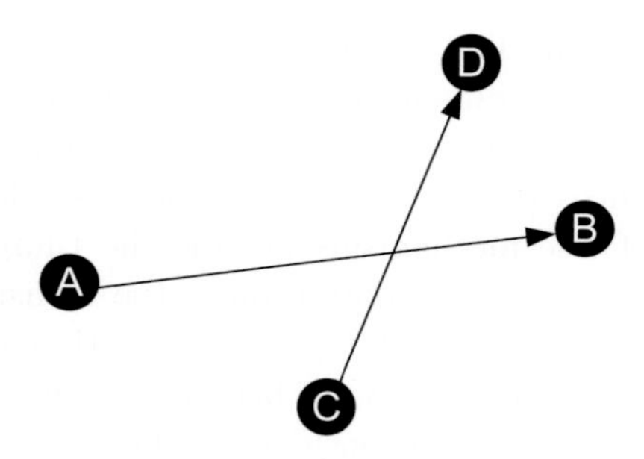

Fig. 4.9. System configuration of Question 4.

Thus the received power at a receiver from a transmitter is modeled as:

$$P_r = \frac{P_t * G_t * G_r * h_t * h_r}{d^4},$$

where P_t is the transmitted power, G_t, G_r, h_t, h_r are the antenna gains and heights of the transmitter and receiver respectively.

(5) What are the most two used types of Directional Antenna used in literature?

(6) How many Possible connectivity scenarios exist when using omni (O) and directional antennas (D) for receiving and transmitting?

(7) What are the operational problem associated with the use of directional antennas?

(8) The example in Section 4.6.1 has presented the benefits of coupling directional antenna with power control. Repeat the gain of using directional antenna for angle of 90 degrees.

(9) A generic model of directional antenna for determining the interference is shown in Figure 4.10, where R denotes the maximal permission range of Node A, R″ is the maximal range of the side lobe of Node A, R′ is the constraint range of the side lobe, and θ is the beam width of the main lobe. Here, nodes lying in the area of constrained range and in the area formed by the intersection of the node's main beam (the white region in Figure 4.10) with that of side lobe of radius R″ are refrained from transmission in any direction since their transmission may highly affect the ongoing communication. Two types of interference result from the application of directional antennas, namely the potential interference and the indirect interference. All nodes outside the main lobe and outside the side lobe range (the dotted shaded region in Figure 4.10) are considered potential interferences, and may turn their directional antenna in any direction. All nodes inside the main beam of A within range R and greater than R″ or inside the side lobe of A with a range between R′ and R″ are considered indirect interferences (the shaded region in Figure 4.10) since they will refrain from transmission in the direction of Node A and they will not cause any direct interference to Node A. These nodes are free to be engaged in any communication towards other directions. Let Pt be the transmission

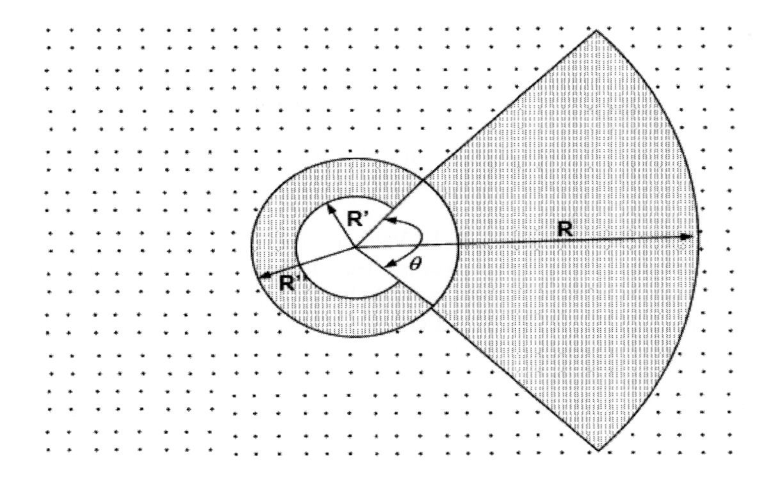

Fig. 4.10. Generic directional antenna interference region.

power, Gt be the transmitter gain of the main lobe, Gr be the receiving gain. Gs be the gain of the main side lobe represented by R$''$, and h is the antenna height. The receiver is able to receive and decode the packet correctly if the received power Pr of a frame from a transmitter node in its transmission zone is higher than or equal to κ (the reception sensitivity). Furthermore, using the two-ray propagation model and with the exponential attenuation factor equal to 4, Determine the relations between R, R$'$ and R$''$.

(10) What are other techniques to enhance spatial reuse in an IEEE 802.11 based multi-hop ad hoc networks?

Terminology

MANET — Mobile Ad Hoc Networks
TPC — Transmission Power Control
IEEE 802.11 — The adopted MAC standard for Wireless LAN
SINR — Signal to interference plus noise ratio
Deafness problem — Intended receiver is unable to reply to a message because it is beam is directed in other direction
Spatial reuse — Allowing simultaneous transmissions to take place in the same vicinity
HOL — Head of Line Blocking
DVCS — Directional Virtual Carrier Sensing
DNAV — Directional Network Allocation Vector
DD neighbours — Both nodes use directional transmission and receptions

Bibliography

1. IEEE 802.11 wireless LAN media access control (MAC) and physical layer (PHY) specifications (1999).
2. G. Li, L. L. Yang, W. S. Conner and B. Sadeghi, *Opportunities and Challenges for Mesh Networks Using Directional Antennas*, www.cs.ucdavis.edu/~prasant/WIMESH/p13.pdf.
3. Y. B. Ko, V. Shankarkumar and N. H. Vaidya, Medium access control protocols using directional antennas in ad hoc networks, *Annual Joint Conference of the IEEE Computer and Communications Societies (INFOCOM)* (2000) 13–21.
4. R. Choudhury and N. H. Vaidya, Performance of ad hoc routing using directional antennas, *Elseiver Journal of Ad Hoc Networks* **3**(2) (2005) 157–173.
5. A. Nasipuri, S. Ye and R. E. Hiromoto, A mac protocol for mobile ad hoc networks using directional antennas, *IEEE Wireless Communications and Networking Conference, WCNC* (2000).
6. R. Bagrodia, M. Takai, J. Martin and A. Ren, Directional virtual carrier sensing for directional antennas in mobile ad hoc networks, A. Press (ed.), *3rd ACM International Symposium on Mobile Ad Hoc Networking and computing (MobiHOC2002)*, Lausanne, Switzerland (December 2002) 183–193.
7. R. R. Choudhury, X. Yang, N. H. Vaidya and R. Ramanathan, Using directional antennas for medium access control in ad hoc networks, *MOBICOM*, Atlanta, Georgia, USA (September 2002) 183–193,
8. T. Korakis, G. Jakllari and L. Tassiulas, A mac protocol for full exploitation of directional antennas in ad-hoc wireless networks, *ACM International Symposium on Mobile Ad Hoc Networking and Computing*, Annapolis, Maryland, USA (September 2003) 98–107.
9. C. S. Z. Huang, C.-C. Shen and C. Jaikaeo, A busy-tone based directional mac protocol for ad hoc networks, *Military Communication Conference, MILCOM* (2002).
10. R. R. Choudhury and N. H. Vaidya, Deafness: a mac problem in ad hoc networks when using directional antennas. *ICNP 2004. Proceedings of the 12th IEEE International Conference on Network Protocols*, Annapolis, Maryland, USA (September 2004) 283–292.
11. H. Singh and S. Singh, Smart-802.11bmac protocol for use with smart antennas, *IEEE International Conference on Communications (ICC2004)*, Paris, France (June 2004) 3684–3688.
12. H. Singh and S. Singh, A mac protocol based on adaptive beamforming for ad hoc networks, *IEEE International Symposium on Personal, Indoor and Mobile Radio Communications, PIMRC* (2003).
13. H. Singh and S. Singh, Tone based mac protocol for use with adaptive array antennas, *IEEE Wireless Communications and Networking Conference (WCNC2004)*, Paris, France (2004).
14. D. Lal, R. Toshniwal, R. Radhakrishnan, D. P. Agrawal and J. Caffery. A novel mac layer protocol for space division multiple access in wireless ad hoc networks, *ICNP* (2002).
15. M. Sanchez, T. Giles and J. Zander, Csma/ca with beam forming antennas in multihop packet radio, *Swedish Workshop on Wireless Ad hoc Networks* (2001).
16. H. Gossain, C. Cordeiro and D. P. Agrawal, Minimizing the effect of deafness and hidden terminal problem in wireless ad hoc networks using directional antennas, *Wireless Communications and Mobile Computing* **6**(7) (2006) 917–931.

17. H. Gossain, C. M. Cordeiro, D. Cavalcanti and D. P. Agrawal, The deafness problems and solutions in wireless ad hoc networks using directional antennas, *IEEE Workshop on Wireless Ad Hoc and Sensor Networks in Conjunction with IEEE Globecom* (2004).
18. C. M. Cordeiro, H. Gossain and D. P. Agrawal, *A Directional Antenna Medium Access Control Protocol for Wireless Ad Hoc Networks.*
19. J. Wang, Y. Fang and D. Wu, SYN-DMAC: a directional Mac protocol for ad hoc networks with synchronization, *IEEE Military Communications Conference* (2005).
20. S. S. Kulkarni and C. Rosenberg, DBSMA: a Mac protocol for multi-hop ad-hoc networks with directional antennas, *International Symposium on Personal Indoor and Mobile Radio Communications, PIMRC* (2005).
21. K. Nagashima, M. Takata and T. Watanabe, Evaluations of a directional mac protocol for ad hoc networks, *International Conference on Distributed Computing Systems Workshops (ICDCSW)*, Tokyo, Japan (2004).
22. R. Ramanathan, Antenna beamforming and power control for ad hoc networks, *Mobile Ad Hoc Networking*, S. Basagni *et al.* (eds.), Chapter 5, IEEE Press/Wiley (2004).
23. J. Zander, Slotted Aloha multihop packet radio networks with directional antennas, *IEE Electronics Letters* **26**(25) (1990) 2098–2100.
24. R. Ramanathan, On the performance of ad hoc networks with beamforming antennas, *IEEE International Symposium on Mobile Ad Hoc Networking and Computing (MobiHoc)*, Long Beach, CA, USA (2001) 95–105.
25. N. S. Fahmy, T. D. Todd and V. Kezys, Ad hoc networks with smart antennas using IEEE 802.11-based protocols, *IEEE International Conference on Communication (ICC)*, New York, USA (2002) 3144–3148.
26. A. Nasipuri, K. Li and U. Sappidi, Power consumption and throughput in mobile ad hoc networks using directional antennas, *IEEE International Conference on Computer Communications and Networks*, Miami, FL, USA (2002) 620–626.
27. N. S. Fahmy, T. D. Todd and V. Kezys, Distributed power control for ad hoc networks with smart antennas, *IEEE Vehicular Technology Conference (VTC)*, Vancouver, BC, Canada (2002) 2141–2144.
28. A. Arora, M. Krunz and A. Muqattash, Directional medium access protocol DMAP with power control for wireless ad hoc networks, *IEEE Global Telecommunications Conference (GLOBECOM)*, Dallas, USA (2004) 2797–2801.
29. A. Arora and M. Krunz, Interference-limited MAC protocol for MANETs with directional antennas, *IEEE International Symposium on a World of Wireless, Mobile and Multimedia Networks (WoWMoM)*, Taormina, Italy (2005) 2–10.
30. Z. Xiaodong, L. Jiandong and Z. Dongfang, A novel power control multiple access protocol for ad hoc networks with directional antennas, *IEEE International Conference on Advanced Information Networking and Applications (AINA)*, Vienna, Austria (2006) 822–826.
31. B. ALawieh, C. Assi and W. Ajib, A power control scheme for directional MAC protocols in MANET, *IEEE Wireless Communications and Networking Conference (WCNC)*, Hong Kong, China (2007).
32. K. Medepalli and F. Tobagi, On optimization of CSMA/CA based wireless LANs: Part I — impact of exponential backoff, *Proceedings of IEEE ICC*, Istanbul, Turkey (2006) 2089–2094.
33. K. Medepalli, F. Tobagi, D. Famolari and T. Kodama, On optimization of CSMA/CA based wireless LANs: Part I — mitigating efficiency loss, *Proceedings of IEEE ICC*, Istanbul, Turkey (2006) 4799–4804.
34. J. Deng, B. Liang and P. K. Varshney, Tuning the carrier sensing range of IEEE 802.11 MAC, *Proceedings of the IEEE International Global Telecommunications Conference (GLOBECOM)*, Texas, USA (2004) 2987–2991.

35. K. Jamieson, B. Hull, A. Miu and H. Balakrishnan, Understanding the real-world performance of carrier sense, *Proceedings of ACM Workshop on Experimental Approaches to Wireless Network Design and Analysis (E-WIND-05)*, Philadelphia, PA, USA (2005) 1321–1329.
36. J. Zhu, B. Metzler, X. Guo and Y. Liu, Adaptive CSMA for scalable network capacity in high-density WLAN: a hardware prototyping approach, *Proceedings of IEEE INFOCOM*, Barcelona, Spain (2006) 1–10.
37. T. Kim, J. Hou and H. Lim, Improving spatial reuse through tuning transmit power, carrier sense threshold, and data rate in multihop wireless networks, *Proceedings of ACM Mobicom*, CA, USA (2006) 366–377.

Chapter 5

Peer-to-Peer and Content Sharing in Vehicular Ad Hoc Networks

Mahmoud Abuelela* and Stephan Olariu†

Department of Computer Science, Old Dominion University
Norfolk, VA 23529–0162, USA
**eabu@cs.odu.edu*
†olariu@cs.odu.edu

Recently, it was realized that Vehicular Ad hoc Networks can support multimedia applications, including peer-to-peer content provisioning. Not surprisingly, a number of peer-to-peer systems for VANET have been recently proposed in the literature. Many of these systems are relying on preinstalled roadside infrastructures and/or using some of the existing techniques like network coding and gossip protocol. In this chapter, we are cover some of the most noticeable peer to peer systems developed for VANETs showing motivation, overview and shortcomings of each. We also provide a discussion and comparison between these systems showing a possible future work.

Keywords: Vehicular ad hoc networks; peer-to-peer systems; multimedia-streaming; content-sharing.

5.1. Introduction

In the past decade, Vehicular Ad Hoc Networks (VANETs), that specialize Mobile Ad Hoc Networks (MANET), to vehicle to vehicle (V2V) and vehicle to infrastructure (V2I) communications have received a great deal of attention in the research community and, with good reason, vehicular communications promise to integrate driving into an ubiquitous and pervasive network that is already redefining the way we live and work.

The potential societal impact of VANET was confirmed by the proliferation of consortia and initiatives involving car manufacturers, government agencies and academia including, among others, the Car-2-Car Communication Consortium, the Vehicle Safety Consortium, the Networks-on-Wheels Project, and the Vehicle Infrastructure Integration Program, and the Advanced Safety Vehicle Program.

While the original impetus for VANET was traffic safety, more recent concerns involve privacy and security. It was recently noticed that allocation of 75 MHz spectrum in the 5.9 GHz band for Dedicate Short Range Communications (DSRC) in North America opens VANET to multimedia applications including peer-to-peer (P2P) content provisioning and the fast-growing mobile infotainment industry.

In spite of their close resemblance to MANET, with which they share the same underlying philosophy, VANET networks have a number of specific characteristics that set them apart from MANET. First, while most MANET networks are deployed in support of special-purpose operations including disaster relief, search-and-rescue, law-enforcement and multimedia classrooms, all of which are intrinsically short-lived and involve a small number of nodes, VANET networks may involve thousands of fast-moving vehicles over tens of miles of roadways and streets. Second, it has been recently noticed that while MANET networks may experience transient periods of loss of connectivity, in VANET, especially under sparse traffic conditions, extended periods of disconnection are the norm rather than the exception.

Most P2P file sharing systems (e.g. Gnutella, BitTorrent) are developed targeting wired IP networks and thus hardly work as intended in MANETs without modification. Recently, several P2P schemes targeting MANET, MANET-optimized version of existing P2P schemes as well as clean-slate designs, have been proposed.

Recently, it was realized that VANETs can support multimedia applications, including peer-to-peer content provisioning. Not surprisingly, a number of peer-to-peer systems for VANET have been recently proposed in the literature. However, most of these systems rely on either the existing cellular system or else on some form of roadside infrastructure installed every few miles.

Although people can download files from road-side access points (APs) that provide Internet connections, the conventional client-server model will not work neither scale well for the following reasons. First, due to the high mobility, the actual contact time to an AP is short. For example, assuming that the WiFi range is 300 m, when driving at the speed of 45 mph, we can have 30 seconds of contact period. With the overhead of association, Dynamic Host Configuration Protocol (DHCP), and Internet connections, the actual contact period is shorter than 30 seconds. Second, in real environments signal strength is mainly a function of distance; i.e. as the distance from the AP increases, the signal strength decreases. This increases the packet error rate; consequently, the effective throughput that one can achieve is much less than expected. Third, it is neither practical to install APs every 300 m, nor feasible to stop in the middle of roads to download a file. Thus, we conclude that in reality, the contact period is short, and its good put is low. To effectively handle this situation, we advocate the use of peer-to-peer file swarming in which users out of AP range can still download parts of files from others.[5]

Cooperative networking is a challenge in VANET because of rapid node movement and intermittent disconnection of the nodes. If cooperative downloading is employed, then every vehicle can share their data with each other and every one can be benefited by that. For example, if the traffic condition is to be shared among a group of vehicles heading towards a particular direction, cooperative downloading can significantly speedup the process of dissemination of data.

In this chapter, a survey about different peer-to-peer systems developed for VANETs is introduced showing strengths and weakness of each.

5.2. Related Background

Before introducing some of the most noticeable P2P systems developed for VANETs in the literature, we think that it is very important to provide an overview of some concepts that are used by many of these systems.

5.2.1. *Network coding*

In 2000, Ahlswede *et al.*[3] demonstrated that allowing intermediate nodes to process the information can increase the achievable rate in a multicasting scenario with respect to simple routing. The proposed approach termed *network coding* basically requires intermediate nodes to perform combinations of the incoming packets. The main idea behind network coding is to allow and encourage mixing of data at intermediate network nodes. Routing itself can be viewed as a special case of coding wherein the outputs of a node are permutations of the inputs.

A very simple scenario for network coding is shown in Figure 5.1. Assume that A and B are not in the communication range of each other while C is in the communication range of both. Let us assume further that A wants to send a single bit b_1 to B and B wants to send a single bit b_2 to A. Using any routing protocol, such communications will be performed in $four$ transmissions through C. However, Figure 5.1 shows how network coding can reduce these to only three transmissions. First, A may send b_1 to C. Then B may send b_2 to C. Finally, C transmits $b_1 \oplus b_2$ to both.

In general, network coding can be described as follows. Given n bits b_1, b_2, b_n as input, a node may generate a coded frame as a linear combination of these n bits, $c = \sum_{i=1}^{n} e_i * b_i$ where e_i's are random numbers. After collecting enough coded frames, at least n of them are linealy independent, along with the encoding vectors e's, a node can easily reconstruct the n bits using simple matrix inversion.

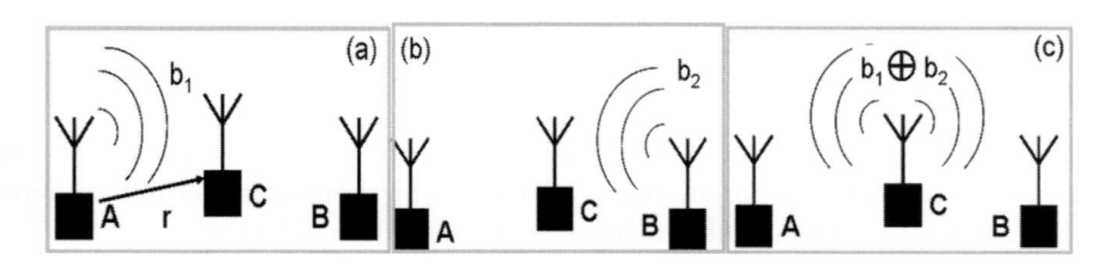

Fig. 5.1. Illustrating the idea of network coding.

The main advantage behind network coding was to design networks being able to achieve the maxflow bound on the information transmission rate in a multicast scenario. On the other hand, the main disadvantage is that the loss of one packet could affect many other packets and renders some information useless at the receiver.

5.2.2. *GOSSIP*

A gossip protocol is a style of computer-to-computer communication protocol inspired by the form of gossip seen in social networks. Modern distributed systems often use gossip protocols to solve problems that might be difficult to solve in other ways, either because the underlying network has an inconvenient structure, is extremely large. A gossip protocol is simply a protocol designed to mimic the way that information spreads when people gossip about some fact.

Suppose that we want to find the object that most closely matches some search pattern, within a network of unknown size, but where the nodes can communicate with each other and are running a small agent program that implements a gossip protocol. To start the search, a user would ask the local agent to begin to gossip about the search string. Periodically, at some rate (let's say ten times per second), each agent picks some other agent at random, and gossips with it. Search strings known to a Node A will now also be known to another Node B, and vice versa. In the next "round" of gossip A and B will pick additional random peers, may be C and D. This round-by-round doubling phenomenon makes the protocol very robust, even if some messages get lost, or some of the selected peers are the same or already know about the search string. It should be easy to see that within logarithmic time in the size of the network (the number of agents), any new search string will have reached all agents.

The mapping for VANETs is very obvious where initially cars know nothing about each other. Then every car broadcasts information about itself as well as its information about others. After some determined time, every car should know about every car else in the local region.

5.3. VANETCODE

Ahmed *et al.*[8] proposed a network coding based co-operative content distribution scheme called VANETCODE that is based on preinstalled gateways every few miles.

5.3.1. *Motivation*

The main motivation of VANETCODE is to eliminate the need of peer selection, content selection and neighbor discovery, which take up significant time and resource in other cooperative downloading mechanism proposed for VANETs.

So, VANETCODE was developed to overcome these problems by using network coding in conjunction with preinstalled gateways along the highway.

5.3.2. *Overview*

In VANETCODE, stationary gateways are assumed to be installed along the freeways at regular intervals of about 2–10 miles where the gateways are expected to be co-located with traffic lights, gas stations and rest areas. The communication between the vehicles and the static gateways can use one of the multitudes of access technologies currently available such as Dedicated Short-Range Communication (DSRC), IEEE 802.11 or the newly emerging 802.16 WiMaX standard.

In a typical scenario, a car is expected to be within the communication range of a gateway for a short duration of the order of a minute.[7] Further, the WLAN connectivity with the gateway will be intermittent; usually the vehicles have short periods of connectivity with the gateway with alternate longer periods of non-connectivity rendering traditional MANET peer-to-peer protocols do not work effectively. VANETCODE assumes that files are to present in their entirety at the static gateways which could be proactively distributed by the content providers amongst the gateways, similar to a content distribution network (CDN) or could be downloaded on demand.

The main idea of VANETCODE can be described as follows. The gateways, which act as the servers, split the original file into k blocks. Whenever a request is received, from one or more cars, for a file, the gateway generates k linearly independent frames from the k blocks and sends them to the requesting cars, if possible, or to their neighbors. After being out of the gateway range, cars may cooperate with each other to exchange frames. If a car collects n independent frames, it can reconstruct the entire file.

5.3.3. *Shortcomings*

VANETCODE is afflicted with a number of problems. Firstly, the gateways are the only sources of data and we should expect all data of interest to be exist and replicated at all gateways. Secondly, there is no sharing of the contents stored at different vehicles as gateways are the only source of information that violates one of the basics of any P2P systems. Thirdly, the system does not scale well; adding or modifying data to all gateways is an issue and require much effort. Finally, although network coding is efficient in maximizing the throughput, it is not practical for multimedia streaming where a driver cannot wait to collect the entire file before he can watch or listen to it. Moreover, if, for any reason, a car could not collect the required n independent frames, the collected frames have no meaning as the car could not reconstruct any part of the file from them.

5.4. PAVAN

PAVAN[9] was introduced by Ghandeharizadeh *et al.* as a policy framework for predicting the availability of a movies (files) in local regions using the cellular network to advertise about files and car-to-car communication to download files from nearby cars.

5.4.1. *Motivation*

Most traditional peer-to-peer systems are based on the initiation of a search request from users. However, in VANETs this may not work well because of the limited resources and high dynamics of the system. By limited resources, we mean that even if thousands of cars do exist, practically routing from far cars is not possible and hence we may assume that they are not practically exist. Hence, one of the basic motivations behind PAVAN is that the vehicle's entertainment system must somehow predicatively determine which titles are available either immediately or within the future δ time, so that the user can select a title to view. The available title list must seek to satisfy the user by striking a delicate balance between showing far fewer titles than can actually be accessed and showing too many titles that cannot be accessed.

5.4.2. *Overview*

A vehicular ad hoc network may potentially cover a large geographical area, such as a metropolitan city. At such large distances, discovering titles available for viewing becomes a very challenging problem indeed. Thus, it is easy to see that on-demand flooding/simple query-based approaches to resource discovery within the ad hoc network will not scale well. So, the authors in PAVAN proposed a hierarchical architecture that also leverages the existing large scale heterogeneous wired-wireless cellular network infrastructure that aids in the collection of localized aggregate information that can be used to distribute the decision making. Every car is assumed to be equipped with Car-to-Car Peer-to-Peer (C2P2) devices where each device has a wireless interface, gigabytes storage and a fast processor. It may also use its local storage to cache different titles.

As shown in Figure 5.2, PAVAN is based on a two-tiered architecture that consists of separate data and control networks. The data network (edges labeled 3) consists of the C2P2 vehicular ad hoc network, facilitating the video streaming and inter-node data exchanges. The control network (edges labeled 2) is a low data rate cellular network infrastructure, with base stations dividing a large geographical area into localized cells. This architectural framework also localizes the communication in the data network mostly to within a cell. Since the display of a typical title would take a C2P2 through several cells, the availability problem essentially consists of determining if sufficient replicas of the title will be present in C2P2s in each cell in this intended path. This in turn requires sufficient information about nearby (regional) cells to be made available to the availability policy module, which resides in each individual C2P2 device. More details about both types of networks can be described as follow.

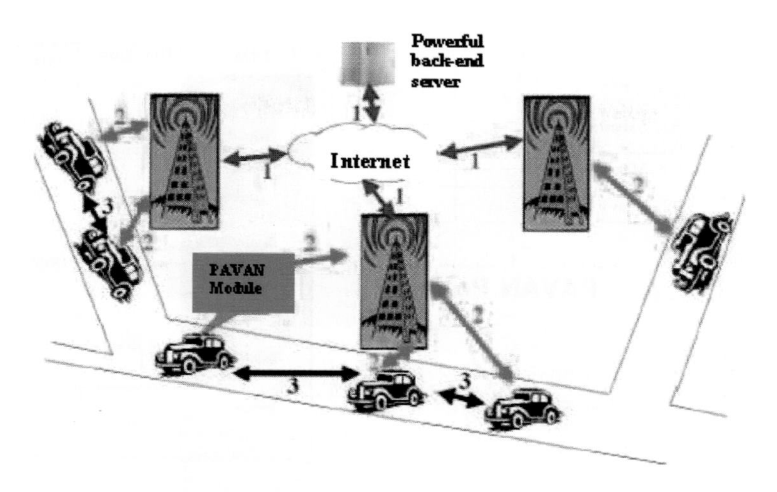

Fig. 5.2. PAVAN architecture.

- Data network: Consists of the vehicular ad hoc network of car-to-car devices. The system storage of content is distributed among the various C2P2 devices within this network. The links in this network are all high bandwidth (tens to hundreds of Mbps each) for streaming of different titles. At each instant, the communication is among nodes that are moving within the same region. We assume that every C2P2 in the same cell is network connected. A typical path between two devices in the same cell may be multi-hop. This is because the range of a cellular base station is almost certainly much larger than the range of high bandwidth network devices (e.g. 802.11a[7]) employed by C2P2 devices. The number of hops is expected to be short, on the order of 3 to 4 hops.
- Control network: Provides three key functionalities: first, monitoring and collection of pertinent content and mobility information from individual car's C2P2 devices to the base station; second, regional consolidation and storage of this information into maps, mobility models and content information by nearby base stations and remote servers within the cellular network infrastructure. The third function is to periodical update of pertinent regional map, mobility, and content information of C2P2 devices within each cell. A base station may perform the last step by broadcasting information. Control messages are typically small and require a low data rate in the order of tens of Kilo bits per second (Kbps).

An example output for PAVAN is shown in Figure 5.3. As shown in the figure, the program tells users some important imformation about movies like titles, duration and when will these movies be available. The prediction and presentation of the availability latency empowers users to make informed decisions.

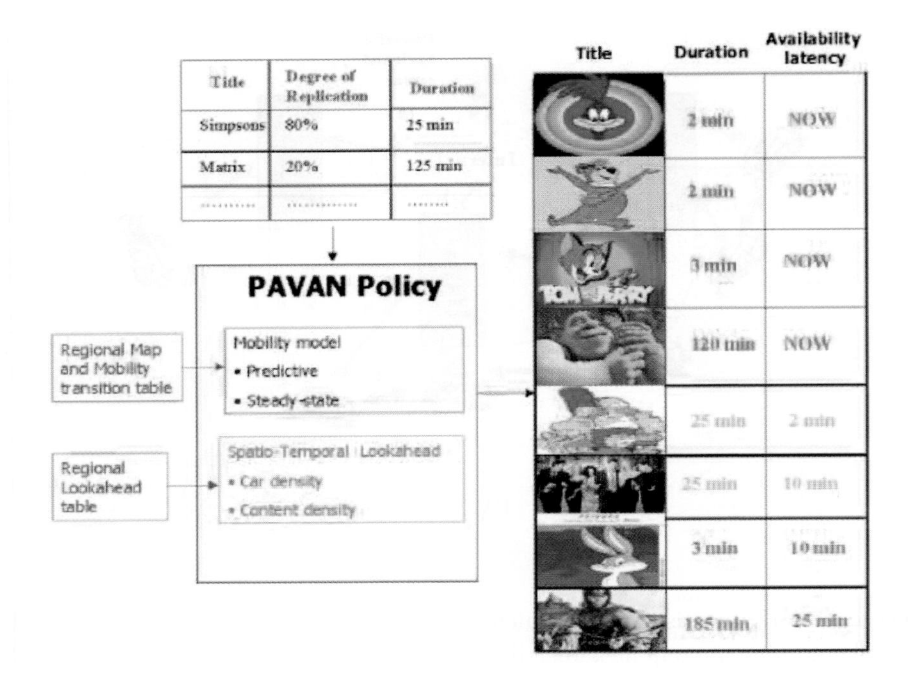

Fig. 5.3. PAVAN output.

5.4.3. *Shortcomings*

If number of vehicles transmitting their files description increases, the cellular network that supports only a few tens of kilo bits per seconds will not be able to carry such load. Thus, we are running in a scalability problem in PAVAN. Also billing is a very important issue here as we do not expect cellular network providers to saturate their network for free. Hence, the real deployment of PAVAN will have many problems to solve.

5.5. CarTorrent

The goal of CarTorrent[5] is to test the feasibility of in-vehicle content sharing. Toward this goal, Lee *et al.* have implemented CarTorrent and measure its performance in a real VANET testbed which was the first of its kind.

5.5.1. *Motivation*

People can download files from road-side access points (APs) that provide Internet connections, which is known as Wardriving. The conventional client-server model will not work neither scale well for the following reasons. First, due to the high mobility, the actual contact time to an AP is short. For example, assuming that the WiFi range is 300 m, when driving at the speed of 45 mph, we can have 30 seconds of contact period. With the overhead of association, DHCP, and Internet connections, the actual contact period is shorter than 30 seconds. Second, in real environments signal strength is mainly a function of distance; i.e. as the distance from the AP increases, the signal strength decreases. This increases the packet error rate; consequently, the effective throughput that one can achieve is much less than expected. Third, it is neither practical to install APs every 300 m, nor feasible to stop in the middle of roads to download a file.

Because of all of the above reasons, Lee *et al.* has implemented CarTorrent. CarTorrent was implemented and deployed on a real VANET testbed to affirm the feasibility of the peer-to-peer file sharing application tailored to VANET. The deployment of such a content sharing application on a real VANETs testbed is the first of its kind.

5.5.2. *Overview*

The basic CarTorrent protocol can be described as follow. For a given file, CarTorrent clients disseminate their piece availability information via gossiping (i.e. by *k-hop* limited scope broadcasting). Each gossip message is forwarded until it reaches to nodes located *k-hop* away from the originator. Thus, peers can gather statistics such as local topology and piece availability. Statistics are then used to select a piece/peer that is preferably close in proximity. In other words, given that two peers A and B own a rarest piece that C desires, C would choose A because A has a shorter hop count to C than B does.

For example, if client A wishes to share a file F. F is split into pieces by FileSplitterThread. SendGossipThread periodically sends gossips that contains the originator of a file, a sequence number, a filename, a piece availability bit vector, a Time-To-Live (TTL), and hop count.

Upon sharing F, SendGossipThread sends gossips regarding the existence of file F. Client B sees the file F being shared as its RcvGossipThread receives gossips about file F. The gossips are then fed to CarTorrent File Manager which runs the engine of the piece selection algorithm. Based on the algorithm, SendPacketThread requests a particular piece from Client A through *AODV*. Client A's ListenThread serves the incoming requests. Upon receiving a request for a particular piece, the connection is immediately processed by ReceivePacketThread. It retrieves the particular piece from its stable storage and sends the piece through *AODV*. The difference between SendPacketThread and ReceivePacketThread is SendPacketThread only sends out requests and ReceivePacketThread receives requests and pieces for the requests. Figure 5.4 shows the designed layout for CarTorrent.

5.5.3. *Shortcomings*

CarTorrent has some disadvantages that may be stated as follow. First, each car has to gossip its files information even if no driver is interested in it, which makes

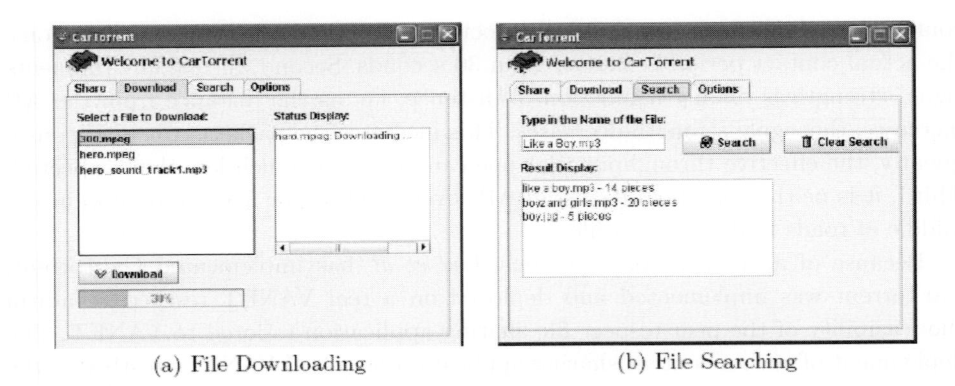

(a) File Downloading (b) File Searching

Fig. 5.4. CarTorrent layout.

unnecessary messages and consumptions of resources. Of course, that will not scale in dense traffic where many cars have some files to advertise about. Also, installing roadside infrastructure every few miles may be deployed in a small scale for some developed countries. However, it cannot be deployed in a large scale in such countries or in any developing country.

5.6. CodeTorrent

Lee *et al.*[6] introduced CodeTorrent as zero infrastructure system that applies network coding as a basis for peer-to-peer communications in VANETs.

5.6.1. *Motivation*

The main motivation behind CodeTorrent may be summarized as follows. The wireless channel is error prone. So, if a protocol is designed without considering errors, the performance of the protocol in real deployment will be seriously degraded. For example, TCP connections usually die out in multi-hop networks with lossy channel but most P2P protocols simply assume that TCP offers reasonable bandwidth. User density is a very important factor to be considered. In an urban scenario, VANETs can scale up to tens of thousands of nodes, and theoretically, all of the nodes can be users running P2P protocols. Even with any cross-layer optimization, no conventional MANET routing protocols is expected to support such big networks.

The second motivation behind CodeTorrent is that most MANET protocols are designed based on the assumption of node cooperativeness. Actually, multi-hop routes can only be established when there are nodes willing to serve as relays for the sake of data sender. In some MANET scenarios, such as a military tactical network, nodes can easily be forced to cooperate to achieve a common goal. But in VANETs, it is very likely that nodes are operated by different entities for their own good and thus it may not possible to force every node to cooperate each other.

Finally, IP addressing is non-trivial in VANETs. It is not clear how each node will be assigned an IP address in VANETs. Random MAC address is a candidate idea for privacy protection, which violates the static and unique MAC address assumption that every MANET routing protocol is built atop.

5.6.2. *Overview*

A car which intends to share a file, a seed car, creates and broadcasts to its neighbors in a single hop the description of that file. This description contains information about the file like identification number, name, size, number of pieces, etc. At the seed car, a file F is divided into n pieces $p_1, p_2, \ldots, p_n$. where nodes apply the idea of network coding by exchanging coded frames instead of file pieces. A coded frame c can be defined as a linear combination of file piece $p_k s$ $c = \sum_{k=1}^{n} e_k * p_k$ where e_k is a certain element in a certain finite field F over which every arithmetic operation is.

Whenever the seed car is requested to exchange a coded frame, the car transmits a newly generated a coded frame c and when generating c, each e_k is drawn randomly from F, hence the name of random linear coding. In the header of a coded frame, the encoding vector $e = [e_1 \cdots e_n]$ is stored for the purpose of later decoding. A node learns of a file from receiving the files description transmitted from neighbors. If the node finds the file interesting, it broadcasts a request containing the identification of that file. Upon receiving such a request, every node possessing any file piece or coded frame of the requested file responds with a newly generated coded frame. A node keeps requesting neighbors to send coded frames until it collects n coded frames carrying encoding vectors that are linearly independent of each other.

Any seed node of a file or any node which possesses any coded frame of the file and willing to share, it periodically broadcasts (at a very low rate) to its 1-hop neighbors the description of the file.

If a node has multiple files to share, multiple descriptions are packed into the least number of packets that can carry all of them and then transmitted.

A car may apply caching by listening to packets even if the car is not the designated receiver, so that it can use them if possible or even to deliver it to another car if a request is overheard.

After collecting n independent frames, a car can safely reconstruct the file blocks using the encoding vectors. Of course the use of network coding here is very efficient if many cars are interested in the file which is true for most peer-to-peer systems. However, if only one node is interested in the file, network coding cannot help and is just waste of time to encode and decode packets.

5.6.3. *Shortcomings*

In spite of its appeal, CodeTorrent has some disadvantages that may be stated as follow. First, each seed car should broadcast its file description even if no driver is interested in it, which makes unnecessary messages and consumptions of resources.

Second, if many cars possess more data to share, the wireless medium will saturate because every car has to broadcast information about every single file it possesses that requires a perfect MAC protocol to manage this heavy load. It is very important here to mention that this heavy load is because of broadcasting files descriptions not even files downloading. Thirdly, although network coding is efficient in maximizing the throughput, it is not practical for multimedia streaming where a driver cannot wait to collect the entire file before he can watch or listen to it. Finally, if for any reason, a car could not collect any block, the remaining collected frames have no meaning as the car could not reconstruct any part of the file from them.

5.7. ZIPPER

Abuelela *et al.* proposed ZIPPER as a zero infrastructure peer-to-peer system for VANETs.[2] In addition to being zero infrastructure, one of the targets of ZIPPER is to allow multimedia streaming between cars at which a driver may enjoy watching (listening) to the first part of a file while downloading the remaining parts.

5.7.1. *Motivation*

Most of the P2P systems proposed for VANETs in the literatures are not zero-infrastructure. Instead they rely on various instances of vehicle to infrastructure communications. The cost of installing and maintaining this infrastructure, which by all accounts are likely to be exorbitant, casts a long shadow of doubt on the feasibility and scalability of such systems especially in developing countries. Also many of the proposed protocols rely on the idea of network coding that is not suitable for media streaming in which a driver may enjoy the first block while still downloading the remaining. Another problem of network coding occurs if the file blocks are rare and most cars in neighborhood have just part of the file, in that situation a car would not be able to reconstruct the file. Finally, for a large files, movies, for example, P2P may be categorized as a delay tolerant network at which the response time is not the most important. Instead, after downloading the first block, a driver may enjoy it while still looking for the remaining blocks.

5.7.2. *Overview*

In ZIPPER, files are stored as a collection of *blocks* and a vehicle may not possess the entire set of blocks of a certain file, as it may be in the process of downloading now.

ZIPPER requires no preinstalled infrastructure along the road so it can be implemented with no extra cost. The following are assumed about vehicles.

- Vehicles are assumed to have virtually infinite power supply and storage.
- Each vehicle has a powerful on-board computer.

- As mandated by DSRC, every 300 m/s each vehicle sends a beacon message with a range of about 200–300 m. These beacons contain information that allows vehicles to handshake and synchronize.
- Each vehicle is equipped with a GPS system and a digital map that helps estimating its accurate position.

ZIPPER operations are performed by four main threads: the Initiator, the ReceivePacket, the ReceiveRequest and the SendPacket threads while implementing TAPR[1] as the underlying packet relaying protocol. Multi-threaded implementation was chosen because it enables a vehicle to perform many operations at the same time. So, a vehicle may initiate a request while searching for and/or relaying some blocks for some other vehicles which means that at any moment of time, zero or more instances of any thread may be in execution.

For better understanding of ZIPPER, let us consider the application scenario illustrated in Figure 5.5. Suppose that the driver at vehicle V_1 initiates a request for a certain movie that consists of four blocks A, B, C and D. Let us also suppose that vehicle V_2 possesses block A and vehicle V_3 possesses block C where V_2 and V_3 are within communication range of V_1.

Figure 5.5(b) shows a sequence diagram illustrating basic control, dotted lines, and data messages, solid lines, exchanged between V_1,V_2 and V_3 that can be

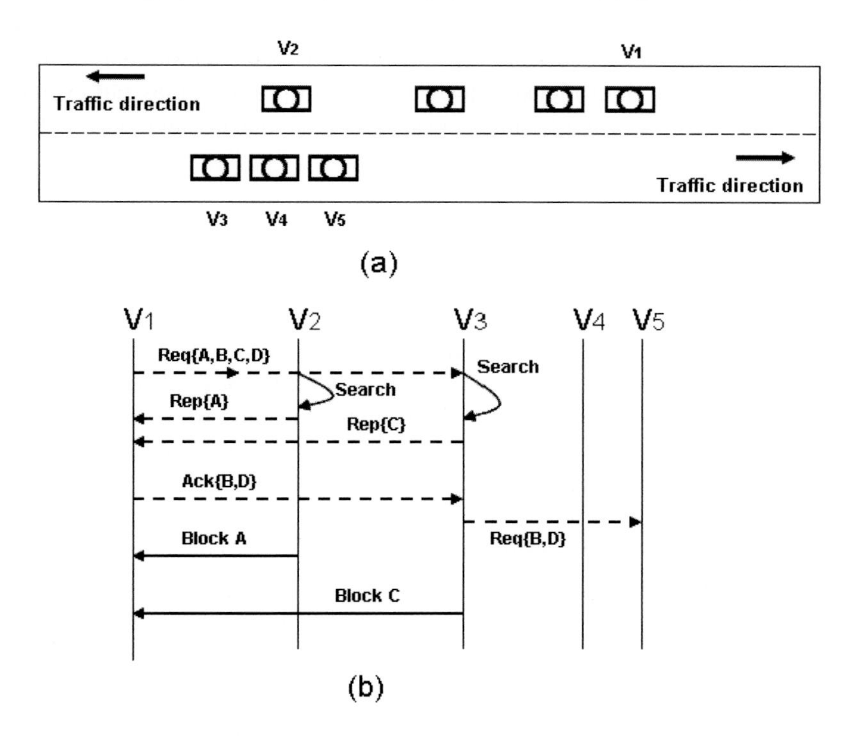

Fig. 5.5. ZIPPER example and sequence diagram.

described as follows. First, V_1 broadcasts its request for A, B, C and D to both V_2 and V_3 since both are 1-hop away from V_1. Both V_2 and V_3 start searching their databases for the required blocks. So, V_2 replies with a control message that it possesses block A while V_3 replies that it possesses block C. Then V_1 sends the third control message $Ack\{\text{B, D}\}$ that contains the new remaining blocks to V_3 which is the farthest vehicle from V_1. Now V_1 can download blocks A and C from vehicles V_2 and V_3 respectively.

Upon receiving $Ack\{\text{B, D}\}$, V_3 can impersonate V_1 in propagating the request and receiving control messages from V_4 and V_5 as follows. V_3 will send $Req\{\text{B, D}\}$ to both V_4 and V_5, assuming that they are the only physical neighbors of V_3 then, V_4 will send $Rep\{\text{D}\}$ to V_3 while V_5 will send an empty reply as it does not possess any of the required blocks. Finally V_3 sends $Ack\{\text{B}\}$ to V_5 which is the farthest vehicle. Of couse, block D will be routed from V_4 to V_1 using TAPR or in general if any vehicle that is not a physical neighbor of V_1 possesses any block, TAPR will be used to relay that block to V_1.

To illustrate the importance of these control messages, let us imagine that they do not exit. So, the application scenario in Figure 5.5 could be explained as follows. V_1 broadcasts its request for blocks A, B, C and D to both V_2 and V_3. After searching their databases, V_2 may send block A with low adjusted power to V_1 and V_3 sends block C to V_1. In order to propagate the request to the next vehicles on the road, V_3 will update the remaining blocks to be {A, B, D} as it did not hear V_2 and V_2 would propagate the new request for blocks {B, C, D} to next vehicles on the road. Obviously, many redundant blocks will be routed to V_1 and we may not even be able to decide when to stop propagating requests when all blocks have been collected.

5.7.3. *Coupon collector's problem*

Although this problem was introduced in ZIPPER, it may be applied for any P2P system for VANETs. The classic Coupon Collector's Problem[4] can be stated as follows. Suppose that there are n distinct types of coupons. At each trial, we draw a coupon whose type is uniformly distributed among the n possible types. Indeed, it is well known[4] that the expected number of steps needed to collect the entire set of coupons is nH_n, where H_n is the nth harmonic number. Since $H_n = \ln n + \gamma + O(\frac{1}{n})$, it follows that nH_n is close to $n \ln n + \gamma n + O(1)$, where $\gamma = 0.57721\ldots$ is the well known Euler constant.

There is an obvious connection between the coupon collector's problem and our interest in evaluating the number of cars that need to be probed to ensure that we collect all blocks of interest. However, the problem at hand involves an additional difficulty, since not every car needs to have a block of the desired file. Think, for example, that we are interested in collecting the various blocks of the popular "sesame street". Since this is a show of immense interest to children, but seldom to adults, most cars are expected not to store any of its blocks.

Abuelela *et al.*[2] proved that with probability exceeding $1 - e^{-c}$ at the end of $m \ln m + cm$ probes, all the m blocks have been collected. For example, if we are probing $\frac{m}{p} \ln m + 5\frac{m}{p}$ vehicles. The probability of having obtained all the blocks exceeds 99.36%.

5.7.4. *Shortcomings*

Although ZIPPER looks very promising, it will not scale well if many cars are running it. This is because it applies routing to deliver blocks from source to the requesting car. Also, in sparse traffic, this routing will not be even possible without the help of some infrastructure.

5.8. Discussion

In this chapter, we presented an overview of different peer-to-peer systems proposed in the literature for VANETS. These protocols may be categorized according to different parameters as follows.

Some of them are relying on preinstalled roadside infrastructure like CarTorrent and VANETCODE. Some are zero infrastructures like ZIPPER and CodeTorrent while PAVAN relies on the existing cellular network. Zero infrastructures protocols are scalable and cheaper to deploy in real environments especially in developing countries. However, in sparse traffic, these protocols would not work as frequent disconnection would be the norm rather than the exception rendering the need for some infrastructure to help connecting different cars.

Another fundamental difference is whether to apply network coding or not. CodeTorrent and VanetCode do apply network coding while the others do not. As we described throughout the chapter, network coding is a very efficient approach when many cars are interested in downloading a certain file or when each of them has some frames and looking for downloading the remaining. However, network coding is not suitable for media streaming as a driver may need only to download and watch the first block while waiting for the remaining blocks. So, fast response time for all blocks may not be necessary. Also, if a car could not download n independent frames, whatever it had in hand has no value as it cannot reconstruct the file from them.

A third fundamental difference is how cars search for and learn about files. Some protocols like CarTorrent and CodeTorrent apply the gossip protocol while some others like ZIPPER assumes that the driver will start his request knowing nothing about the current local available files. Actually, gossip may be very important in an ad hoc environment like VANETs where each node initially has no knowledge about others. However it has many shortcomings that may be stated as follows. First, the system in VANETs is highly dynamic and a car may leave the local region that it just learnt about before taking advantage of this knowledge, as the car may take and exit, speed up or slow down leaving current region and joining another one. So, if the cannot learn fast enough about its neighbors, the gossip has no meaning and is just waste of resources. Second, a car has to broadcast information about itself and others even if no driver is interested to download these files. So, file selection, which file to advertise, and number of files is a very critical parameters that needs to be chosen very carefully.

Problems

(1) Several P2P systems have been developed for MANETs. Explain why do we need a P2P systems developed specially for VANETs?
(2) Network coding idea is illustrated in 5.2.1, explain when network coding be an efficient idea to save bandwidth and when it is not useful at all and may even harm the performance.
(3) Some protocols mentioned in this chapter rely on preinstalled infrastructure and some others do not. Explain the advantage and disadvantage of each?
(4) Consider a new approach in which cars rely only on the cellular network to exchange information and files. For example, a car may send its request to all cars covered by the same cellular base station and if any of them has the required data, it sends it back to the requester over the cellular network. What can you say about the advantage and disadvantage of this approach? Is there any situation at which this approach work? Show also an example at which this approach does not work.
(5) Is there any difference between P2P in city (streets traffic lights and stop signs) and highway scenarios? If your answer is yes, show these differences? If your answer is no, explain why?
(6) GOSIIP is a protocol that allows nodes in ad hoc network to learn about each other. Under what condition is it appropriate for VANETs?
(7) In this chapter, we have studied two approaches for P2P in VANETs. Some protocols at which cars advertise about their contents and whoever is interested may select from what is available. Some other protocols require that the car interested in a file should send a request to its neighbors to search for that file. Discuss the advantage and disadvantage of each by giving an example on each approach?

Bibliography

1. M. Abuelela and S. Olariu, Traffic-adaptive packet relaying in vanet, *VANET '07: Proceedings of the 4th ACM International Workshop on Vehicular Ad Hoc Networks* (ACM, New York, NY, USA), ISBN 978-1-59593-739-1, doi:http://doi.acm.org/10.1145/1287748.1287764 (2007a) 77–78.

2. M. Abuelela and S. Olariu, Zipper: a zero-infrastructure peer-to-peer system for-vanet, *WMuNeP '07: Proceedings of the 3rd ACM Workshop on Wireless Multimedia Networking and Performance Modeling* (ACM, New York, NY, USA), ISBN 978-1-59593-804-6, doi:http://doi.acm.org/10.1145/1298216.1298218 (2007b) 2–8.
3. R. Ahlswede, N. Cai, S.-Y. Li and R. Yeung, Network information flow, *Information Theory, IEEE Transactions* **46**(4) doi:10.1109/18.850663 (2000) 1204–1216.
4. W. Feller, *An Introduction to Probability Theory and Its Applications*, Vol. I, 3rd edn. (John Wiley and Sons, 2000).
5. K. Lee, S.-H. Lee, R. Cheung, U. Lee and M. Gerla, First experience with cartorrent in a real vehicular ad hoc network testbed, *2007 Mobile Networking for Vehicular Environments*, doi:10.1109/MOVE.2007.4300814 (2007) 109–114.
6. U. Lee, J.-S. Park, J. Yeh, G. Pau and M. Gerla, Code torrent: content distribution using network coding in vanet, *MobiShare '06: Proceedings of the 1st International Workshop on Decentralized Resource Sharing in Mobile Computing and Networking* (ACM), ISBN 1-59593-558-4, doi:http://doi.acm.org/10.1145/1161252.1161254 (2006) 1–5.
7. J. Ott and D. Kutscher, Drive-thru Internet: IEEE 802.11b for "automobile" users, *Proceedings of IEEE INFOCOM* (2004).
8. A. Shabbir and S. Salil, Vanetcode: network coding to enhance cooperative downloading in vehicular ad-hoc networks, *IWCMC '06: Proceedings of the International Conference on Wireless Communications and Mobile Computing* (ACM, New York, NY, USA), ISBN 1-59593-306-9, doi:http://doi.acm.org/10.1145/1143549.1143654 (2006) 527–532.
9. G. Shahram, K. Shyam and N. Bhaskar, Pavan: a policy framework for content availabilty in vehicular ad-hoc networks, *VANET '04: Proceedings of the 1st ACM International Workshop on Vehicular Ad Hoc Networks* (ACM, New York, NY, USA), ISBN 1-58113-922-5, doi:http://doi.acm.org/10.1145/1023875.1023885 (2004) 57–65.

Chapter 6

Properties of the Vehicle-to-Vehicle Channel for Dedicated Short Range Communications

Lin Cheng*, Benjamin E. Henty† and Daniel D. Stancil‡

Carnegie Mellon University, Pittsburgh, PA, 15213
* *lincheng@andrew.cmu.edu*
† *henty@andrew.cmu.edu*
‡ *stancil@andrew.cmu.edu*

Fan Bai

General Motors Research, Warren, MI, 48090, USA
fan.bai@gm.com

Ad hoc networks involving vehicles on roads and highways are of interest because of the potential for supporting a wide range of services and applications, including safety, business, and infotainment. Technology and standards for Dedicated Short Range Communications (DSRC) are presently under development for frequencies near 5.9 GHz in the United States, and near 5.8 GHz in Europe and Japan. To best design equipment and protocols for these applications, the nature of the vehicle-to-vehicle channel in these bands must be understood. This article presents a survey of recent research into both narrowband and wideband channel characteristics, as well as the performance of data transmissions using orthogonal frequency division multiplexing (OFDM). Specific emphasis is placed on investigating the detailed behavior of the channel through on-road measurements in suburban, rural, and highway driving in and near Pittsburgh, PA. The influence of driver behavior on the channel properties is discussed, and this influence is illustrated with the use of the effective speed-separation diagram. Potential implications for vehicle-to-vehicle network implementation and performance are also discussed.

Keywords: Vehicular networks; intelligent transportation systems; propagation models.

6.1. Introduction

The wide usage of mobile computing devices and the scaling of device technologies have made possible significant increases in the embedding of computing devices in modern vehicles. Embedded mobile computing devices have been an indispensable part in manufactured vehicles. The new trend, however, is the networking of these devices and their ubiquity not only in traditional embedded applications, but to allow vehicles to remain in communication with each other and roadside units.

A trend deserving particular attention is that in which large numbers of simple, inexpensive on-board units are embedded in moving vehicles on the road, forming a Vehicular Ad Hoc Network (VANET) using dedicated short range communications (DSRC) at 5.9 GHz assigned by the Federal Communications Commission (FCC). VANET brings significant potential for supporting a wide range of services and applications, including safety, business, and infotainment. Safety applications include collision avoidance and cooperative driving, to name a few. In addition to safety applications, VANETs will provide convenience and commercial applications to reduce time on the road and to improve the driving experience.

The highly safety-sensitive nature of VANETs and the reliability associated with these diverse applications make the design issues of VANET complicated and there are a number of technical challenges that need to be explored. In particular, this article intends to answer the following questions:

(1) What is the detailed behavior of the V2V channel in various on-road scenarios?
(2) What is the influence of driver behavior on the channel properties?
(3) What are the potential implications for V2V network implementation and performance?

The remainder of this chapter continues with a description of the background and related work in Section 6.2. The measurement methodology and apparatus are given in Section 6.3. The environments studied are described in Section 6.4. In Section 6.5, we present measured channel properties and their implications. Influence from driver behavior is discussed in Section 6.6. Section 6.7 discusses the challenges for future research. We conclude the chapter in Section 6.8.

6.2. Background and Related Work

The starting point for vehicular communications is the intelligent transport system (ITS). Interest in ITS arises from the issues induced by traffic congestion and the emerging new technologies for real-time control, simulation and communications, aiming to construct "the person, the road and the vehicle" as one system.[1] The applications include improving traffic flow and safety efficiently (reduce the probability of accidents), alleviation of congestion, and increasing commuter awareness of traffic and weather conditions in real-time. As a sample project,[2] described the TrafficView device that can be embedded in the next generation of vehicles to provide the drivers with a realtime view of the road traffic far beyond what they can physically see.

Besides using standard available wireless systems for communications, a number of emerging implementation strategies have been suggested. For example, Ref. 1 proposed to use a wireless Bluetooth enabled system to minimize the costs of equipment. Using access point (AP) areas, the Internet could be accessed from inside the vehicle, and information such as news and weather information could be downloaded.

Another relatively long term example application is to enable the flow of multimedia between different traveling vehicles, with the possibility to turn tiring long journeys into pleasant driving times that can be shared with friends and family. This information flowing between vehicles would likely be multimedia: data, images, video, and voice. A good summary of possible applications can be found in Ref. 3.

One of the major research focuses at present is the implementation of vehicle-to-vehicle communications using vehicular ad hoc networks. These Ad Hoc Networks extend the scope of vehicle-to-vehicle communications from a fixed number of a priori specified vehicles to an unlimited numbers of vehicles along the road. Vehicular ad hoc networks also remove the dependency on conventional cellular networks for communication between vehicles. This is important since in densely populated areas, the current cellular architecture can become a bottleneck for basic cellular calls, so transferring the transmissions of vehicle-to-vehicle communications onto other reliable networks would save vehicle-to-vehicle communication users both time and money. Public safety (PS) applications may also be enabled by employing vehicle-to-vehicle communications.[4]

Recently, a standard[5] for vehicle-to-vehicle communication in the 5.9 GHz Unlicensed National Information Infrastructure (UNII) band has been proposed, aiming to extend the IEEE 802.11a application environment. Perhaps the most important goal for these communication services is to reduce the number of accidents and eliminate all fatal consequences. Therefore it is imperative that reliable transfer of information be enabled by the vehicle-to-vehicle applications. As described before, to accurately predict the vehicle-to-vehicle system performance, it is essential to test the wireless systems under realistic channels before their actual deployment. Hence providing a thorough understanding of the vehicle-to-vehicle channel will greatly benefit researchers in evaluating the performance of their systems under realistic conditions.

The challenges for Vehicular Ad Hoc Networks, through a layered design perspective, was addressed in Refs. 6 and 7 where the authors provide a detailed network layered approach for assessing the feasibility and performance of VANET. The studies in Ref. 8 provide an overview of DSRC-based vehicular safety communications and propose a coherent set of protocols to address these requirements. Example initiatives concerning the implementation of VANET include the California PATH project in the US,[9] and the Fleetnet project[10] and its successor, the network-on-wheels project[11] in Europe. These projects have investigated a number of issues relating to the radio devices, the medium access (MAC) layer, and routing protocols.

Several authors have also considered physical layer issues in the mobile-to-mobile context. In one of the earliest theoretical works,[12] a mobile-to-mobile model was proposed in which each mobile node is surrounded isotropically by dense scatterers, and there is no line-of-sight. In Refs. 13 to 16, the authors extend the work presented in Ref. 12, detailing methods by which items such as autocorrelation and simulation methodology can be constructed and compared with theory.

Other works that are more empirical in nature have been conducted, though many of these used frequencies outside the 5.9 GHz DSRC band targeted for nationwide deployment in the United States. Although the principles of channel impairment are similar, scattering and obstruction by various objects in the environment can vary significantly with frequency. The authors in Ref. 17 reported vehicle-to-vehicle RF propagation measurements in the 900 MHz band between parked cars in a roadway environment. Reference 18 reported vehicle-to-vehicle measurements at 915 MHz in a mobile environment, and joint Doppler-delay power profile measurements at 2.4 GHz were described in Ref. 19. A number of studies reported measurements in the 5.2–5.3 GHz band. Reference 20 conducted flat-fading narrow-band measurements of inter-vehicle transmission at 5.2 GHz, and Refs. 21 and 22 reported measurements in the indoor low-mobility environment at 5.3 GHz. More recently, Ref. 23 presented measurements at 5.9 GHz for a chosen highway site. A method to develop a channel emulator model for a doubly selective V2V wireless channel at 5.9 GHz was presented in Ref. 23. References 21 and 22 reported Doppler spectra from IEEE 802.11 type devices in the indoor low-mobility environment at 5.3 GHz.

6.3. Measurement Methodology and Apparatus

The challenges of V2V systems on dynamic on-road scenarios bring V2V channel evaluation to the forefront of the design process. At the same time, since V2V channels are coupled with on-road scenarios, these evaluations must take into account issues such as vehicle traffic conditions, driving behavior and wireless communication characteristics, etc.

While theoretical analysis and simulation provide perhaps the most convenient evaluation method, many on-road scenarios are too complicated to adequately satisfy the assumptions made in theoretical models. We believe experimental evaluation on-road from multiple dimensions is needed in understanding the behavior of the V2V channel. These experiments not only provide a way to cover the dynamic and rapidly changing V2V channel, they also help to derive a meaningful understanding of V2V system behavior from realistic settings, enabling the study of detailed behaviors that are difficult to capture with pure analytic models.

An experimental RF platform that meets the above criteria was developed with the purpose of performing location-aware investigations of the V2V channel. This platform was also employed to investigate the impact of driver behavior and correlations among various mobility and environmental variables on system performance,

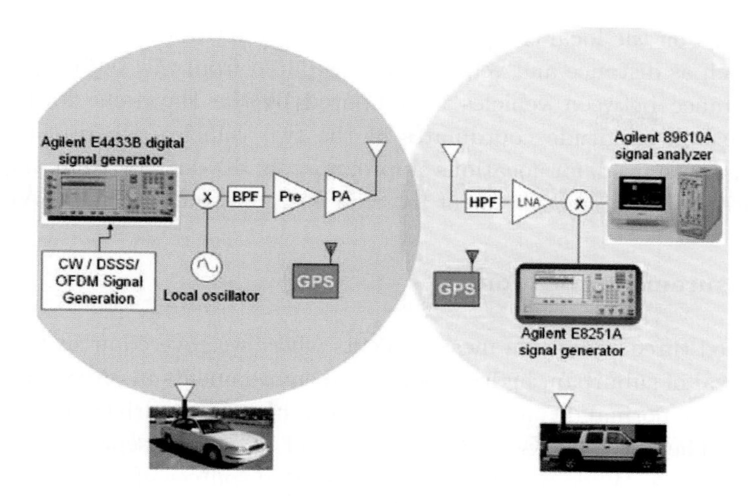

Fig. 6.1. System setup diagram.

as well as the feasibility of using the draft IEEE 802.11p standard to achieve dedicated short range communications among vehicles.

As shown in Figure 6.1, the platform we constructed is designed to provide us the flexibility of performing measurements with many signaling schemes (e.g. continuous wave, spread spectrum sequences, orthogonal frequency-division multiplexing (OFDM) signals employing DSRC standards, etc.). The platform consists of a number of laboratory grade instruments installed in two vehicles. At the transmitter side, an Agilent E4438B Digital Signal Generator (DSG) with dual base-band arbitrary waveform generators is used to repeatedly send the desired waveform in the V2V channel.

At the receiver side, a broadband, low noise amplifier is used to maximize the sensitivity of the receiver. A mixer is used to down-convert the received RF signal into the 0 to 40 MHz range that can be analyzed by the Agilent 89600 Vector Signal Analyzer (VSA). The VSA samples the signal with a high-speed, wide-dynamic-range analog-to-digital converter (ADC). Automation software is used to record the data capture of the signal. Data frames captured from the VSA serve as the basis for our analysis.

In addition to the RF components, we describe how our system measures and calculates separation distance and relative speed. To ensure the precision of our measurements, both vehicles are equipped with a CSI wireless Differential GPS (DGPS) receiver and a Linux laptop computer that logs GPS data. The accuracy of DGPS is on the order of 1 m. Thus, we believe that the location information provided by the DGPS receiver is accurate enough for our analysis. The DGPS receivers also help us overcome the difficulties in properly synchronizing measurements performed while the two vehicles are in motion during field tests. All of the transmitted data packets are also stored on the local computer along with the recorded GPS data. Location statistics such as distance and velocity are computed from raw GPS recordings.

The distance between vehicles is computed by the Haversine formula, using the latitude and longitude coordinates of the two vehicles, where the radius of the earth is optimized for locations approximately 40 degrees from the equator (i.e. the latitude of the measurement location near Pittsburgh, PA, USA).

6.4. Measurement Locations

We performed three sessions of measurements where we covered areas that we considered typical of suburban, highway and rural environments in or near Pittsburgh, PA. To preserve normal driving behavior, both vehicles were driven at each driver's prerogative. The two vehicles under study travel at various separations and with different velocities. While on road, there are several sources of reflection and scattering — buildings, trees, hills, other vehicles, etc. — in a typical link between the two mobile vehicles. We refer to these objects generically as scatterers. While moving down the road, the cars pass through multiple kinds of local scatterers. Some of these scatterers such as buildings and trees are stationary, while others such as vehicles and pedestrians are in motion.

In suburban areas, measurements were made in neighborhoods near Carnegie Mellon University, with routes consisting primarily of 2-lane suburban streets. The width of these 2-lane suburban streets in Pittsburgh is on the order of 8 to 10 m, with trees, houses, apartment and office buildings set back 10–12 m from the curb. In these two-lane streets, the experimental vehicles did not pass each other during the measurement.

On the highway, measurements were made while the two sounding vehicles traversed I-79 between Pittsburgh, PA and Grove city, PA. This segment of interstate highway consists of two to three lane segments with moderate traffic. During the experiments, the two vehicles were moving steadily at high speed, along with other vehicles with similar speeds. The two vehicles occasionally changed lanes and overtook each other. There were in general no nearby buildings, but occasional overpasses were observed.

Rural field data was acquired as the vehicles traversed the rolling country side north of Pittsburgh, PA. It features sections of high speed interspersed with towns and intersections. The traffic was very light on these 2-lane roads, and the vehicles did not pass one another. While the close vicinity consisted of different kinds of low-height vegetation, remote trees and hills were commonly observed.

6.5. V2V Channel Properties

Using the aforementioned platform, a direct-sequence spread spectrum (DSSS) signal was transmitted into the V2V channel. The de-spread received signals were then used to form an estimate of the channel impulse response. The channel response can also be obtained from swept frequency measurements using a vector or scalar network analyzer. However, this method normally requires a reference line connection between the transmitting and receiving systems. Since the V2V channel is between two high-speed vehicles with separation up to a few hundred meters, we adopt the time-delay domain signal excitation approach.

The V2V channel is a multi-path channel. The transmitted signal arrives at the receiver in various paths of different length. The multiple versions of the signal may interfere with each other, causing inter-symbol interference (ISI). The common representation of the multi-path channel is the channel impulse response which is the signal at the receiver if a single pulse is transmitted. In the field implementations, the multi-path delay profile is typically used to describe the channel impulse response over a small area. Statistical characterizations that are typically performed on the channel measurements include statistical metrics based on the channel power delay profile (PDP). For example, the RMS delay spread σ_τ can be derived from the first and second moment of a measured multi-path delay profile.[24]

While the abovementioned metrics can be applied to describe the V2V channel properties from the multi-path delay perspective, there is no guarantee that the V2V channel's communication performance will be the same for the same values of delay spread. Many other impacting factors must be taken into account. For example, the transmitting and receiving vehicles pass various kinds of scatterers, making the V2V mobile radio channel time variant. The propagation-path also changes owing to the relative motion between transmitter and receiver vehicles, as well as the motion of the scatterers in the vicinity of the vehicles. This makes it necessary to investigate channel coherency over time and Doppler spreading in the frequency domain as well. We describe V2V channel properties along these dimensions in the following sections.

6.5.1. *Time delay domain properties*

Representative statistical characterizations in the time domain are typically estimated from moments of the multi-path delay profile. The two statistics we describe here are RMS delay spread and maximum excess delay. While the RMS delay spread gives an indication of how long (in the RMS sense) the multiple propagating copies of the transmitted signal remain in the channel given an energy level, the maximum excess delay captures the extreme time period beyond which multi-path replicas are unlikely to appear.

The RMS delay spread σ_τ can be derived from the first and second moment of a measured multi-path delay profile[24,25]

$$\sigma_\tau = \sqrt{\overline{\tau^2} - (\bar{\tau})^2}, \tag{6.1}$$

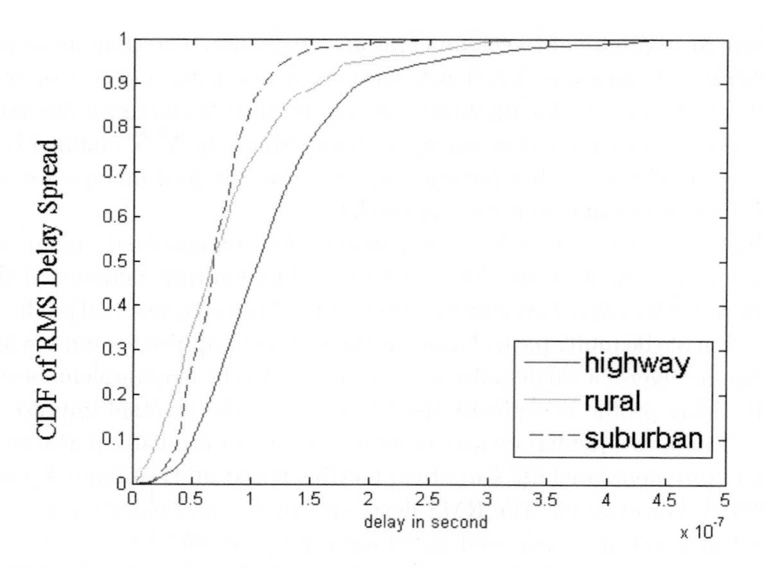

Fig. 6.2. CDF of RMS delay spread using a 15 dB threshold.[26]

where $\bar{\tau}$ and $\bar{\tau}^2$ are computed using (for sampled experimental waveforms)

$$\bar{\tau} = \frac{\sum_k \tau(k) p_r(k)}{\sum_k p_r(k)}, \tag{6.2}$$

and

$$\bar{\tau}^2 = \frac{\sum_k \tau^2(k) p_r(k)}{\sum_k p_r(k)} \tag{6.3}$$

where $p_r(k)$ represents the discrete measured multi-path delay profile p_r at index k, and $\tau(k)$ is the discrete time delay τ at index k.

Evaluating the RMS delay spread for each of the obtained power delay profiles allows investigators to explore the probability distribution of the RMS delay spread in several different driving environments of the V2V channel. Figure 6.2 compares the cumulative distribution of the RMS delay spread for suburban, highway and rural environments. If we draw a horizontal line at a y value, the corresponding x_i ($i = 1, 2, 3$) value for each of the three environments implies that the probability $Pr(\sigma_\tau \leq x_i) = y$. The minimum values of σ_τ for which $y = 0.9$ were determined from the CDF by computing the level below which σ_τ stays for 90% of the time. This means the RMS delay spread stays below these x values for 90% of the time. For $y = 0.9$, it is observed that the suburban environment has the smallest RMS delay spread, and the highway environment has the largest. Further, we note that the median values of RMS delay spread ($y = 0.5$) for the rural and suburban environments are very similar, though the rural distribution is broader. The highway environment also exhibits the largest median value (about 110 ns).

The larger values of 90% RMS delay spread for the rural and highway environments can be explained by the presence of larger open areas along with wider separation between vehicles. It also implies there are objects both very close (other vehicles) and far away (remote hills, trees, etc.). In contrast, the limited distance to buildings in the suburban environment results in a narrower range of echo times. In view of these differences, the similarity in the median values of the suburban and rural distributions is interesting.

In contrast to the RMS delay spread, the maximum excess delay captures the time period during which the power delay profile falls to T dB below its maximum, where T is the threshold applied. Figure 6.3 shows the cumulative distribution of the empirically measured maximum excess delay over all environments under investigation. The maximum excess delay is useful as an estimate of the upper bound for the temporal dispersion.

Although Figure 6.3 shows a behavior similar to the RMS delay spread (but with larger values), the rural case does appear to saturate more slowly in the high probability region. Unlike the suburban case with no observed maximum excess delay value larger than 1 μs, values of several μs are observed in the rural and the highway cases.

While the RMS delay spread describes the weighted second order central moment of the delay profile, the maximum excess delay is more of a bound. Since the objects observed in the rural case are typically distributed more remotely compared to the suburban and highway environments, the larger extension of the maximum excess delay in the rural case is likely due to reflections from remote hills, etc. as discussed in connection with the RMS delay spread. Again, the suburban distribution is seen

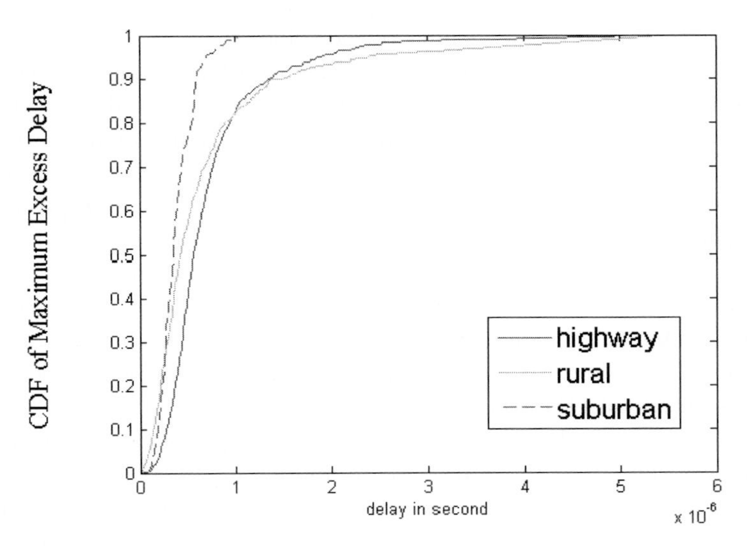

Fig. 6.3. CDF of maximum excess delay using a 15 dB threshold.[26]

to saturate faster, owing to the tighter distribution of scattering objects in the vicinity compared to the rural and highway environments.

It is also interesting to observe again that though the distributions are very different, the median values of the rural and suburban distributions are very similar. The median value of the highway distribution is again the largest, with a value of about 600 ns.

From a design perspective, one of the major implications from the above studies is for the transmission scheme. Since many applications of V2V networks are likely to have different data rate and reliability requirements, it is important to understand how the multi-path propagation affects the transmission scheme. One of the key impacts of multi-path propagation on digital communications is the introduction of inter-symbol interference (ISI). If the symbol periods of the deployed VANET system are small such that copies of a transmitted symbol arrive during the subsequent symbol period, interference is then introduced and must be solved using mitigation techniques (for example, using an equalizer) to ensure a reliable system realization.

6.5.2. *Channel coherency over time*

In the V2V scenario, the transmitter and receiver vehicles travel through different environments with changing velocities. Consequently, the vehicles pass various kinds of scatterers in their vicinity. Representative scatterers include stationary objects (trees, houses, overhead bridges, etc.) as well as other vehicles in movement. All of the above make the V2V mobile radio channel time variant; the propagation-path changes owing to the motion of the transmitter and receiver vehicles, as well as the motion of the scatterers in the vicinity of the vehicles.

To be concrete, if a narrowband signal is transmitted through the channel, the receiver vehicle would experience severe changes in the amplitude and phase of the received signal owing to such motion. Moreover, even a stationary parked vehicle would experience some changes in the propagation path owing to the movement of cars (scatterers) around it. One metric to characterize this time variation is the coherence time, which describes the time over which the channel can be considered unchanged. It can be readily converted into the frequency domain as well. In the frequency domain, the received spectrum would be spread by signals arriving from all directions. One of the commonly used metrics in the frequency domain is the Doppler spread (discussed in the next section). The interference between frequency components determines the coherence time.

In our study here, although the time correlation is the Fourier transform of the Doppler power spectrum function,[27] we compute the coherence time directly by evaluating the autocorrelation of the time-domain signal measured by the VSA. The autocorrelation function $\rho(\tau)$ of the received envelope describes the correlation between the channel response to a sinusoid sent at a time t_i, and the response to a similar signal sent at a time t_j, with $\tau = t_j - t_i$. While the coherence time can be calculated for different levels of correlation of $\rho(\tau)$, here we use the 90% coherence time. (There is some judgment involved in selecting criteria; for example, another commonly used value is the 50% coherence time. The values we have used we believe to be reasonable but somewhat on the conservative side.) The 90% coherence times are estimated from the time offsets over which the autocorrelation function drops to 90% of its peak value.

The measured cumulative distribution of the 90% coherence time in each environment is shown in Figure 6.4. If we draw a horizontal line across at $y = y_0$, we can estimate the coherence time at $100(1 - y_0)\%$ of the time, for each environment. To be concrete, we illustrate using the suburban curve: if the horizontal line is specified at $y_0 = 0.1$, we can find the corresponding Tc, y_0 value on the horizontal axis for the suburban environment, which reads to be about 1 ms based on the cumulative function. This means the coherence time stays above 1 ms for $(1 - y_0) = 90\%$ of the time, i.e. the suburban V2V channel is likely to stay invariant over a time duration of 1 ms for 90% of the time. By the same token, we can obtain the values for the rural and highway environments from their cumulative functions. Table 6.1 summaries the minimum 90% coherence time Tc for all three environments.

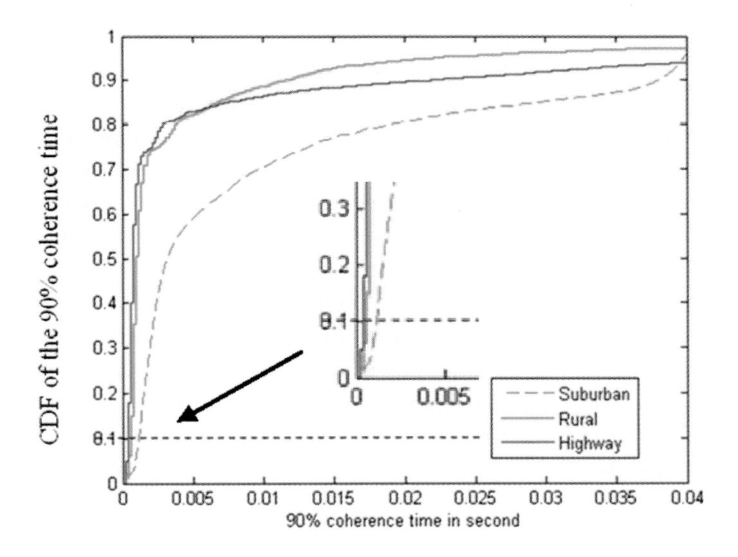

Fig. 6.4. CDF of the 90% coherence time (overall) and a close-up of the region near the origin.

Table 6.1 Comparison of minimum 90% coherence time.

Environment	Suburban	Highway	Rural
Minimum 90% coherence time	1 ms	0.3 ms	0.4 ms

6.5.3. *Doppler analysis*

In V2V scenarios, signals from multiple paths are received by a moving vehicle, causing the observed frequency of each signal to be Doppler shifted depending on the component of the vehicle velocity along the direction of arrival of the path. Doppler shifts also result from the motion of the transmitter and other nearby vehicles. All of the received signals are summed together at the receiver resulting in a broadened spectrum compared to the transmitted signal. This phenomenon is referred to as Doppler spread.

Doppler spread is an important parameter for characterizing the time-variant behavior of the V2V channel. With the dynamic behavior in each environment, we start by slicing the observed Doppler spread with regard to mobility. After that, we present the explanation details and the implications.

Among the vehicle mobility variables, the effective speed (defined by $v_{eff} = \sqrt{v_T^2 + v_R^2}$, where v_T and v_R are ground speeds for the transmitter and receiver, respectively) is shown experimentally to be correlated with the Doppler spread in our earlier studies,[28] as well as in theoretical mobile-to-mobile model derivations.[15]

While the Doppler and effective speed dependence has been studied, joint analysis of Doppler combining the mobility and environmental impacts on V2V is not widely understood. Moreover, environmental assumptions are made in several theoretical mobile-to-mobile studies[12] that need to be validated. On-road empirical investigations in different environments are complementarily to these studies, serving to validate and extend these environmental assumptions.

In the following paragraphs, we compare the Doppler spread as a function of effective speed across environments. Spectral estimation is used to characterize the Doppler spread from frequency domain recordings of the VSA. Specifically, we use the square root of the second order central moment of a spectrum to quantify the empirical Doppler spread fE.

Figure 6.5 depicts the dependence of Doppler spread on the effective speed for three different environments. The error bars in the plot represent the 95% confidence intervals. Linear regression results shown in solid lines are also plotted for each environment respectively. The parameters from the linear regression are summarized in Table 6.2. The dashed line represents the theoretical Doppler spread curve given by theoretical models with isotropic scatterers and no line-of-sight[12,15] computed using

$$f_D = \left(\frac{1}{\lambda}\right)\sqrt{\frac{v_R^2 + v_T^2}{2}} = \left(\frac{1}{\lambda\sqrt{2}}\right) v_{eff}. \tag{6.4}$$

While the data confirms the linear dependence for different environments, several observations are made from the empirical results. First, the empirical slopes are consistently about 40% of the theoretical value in all three environments. Second,

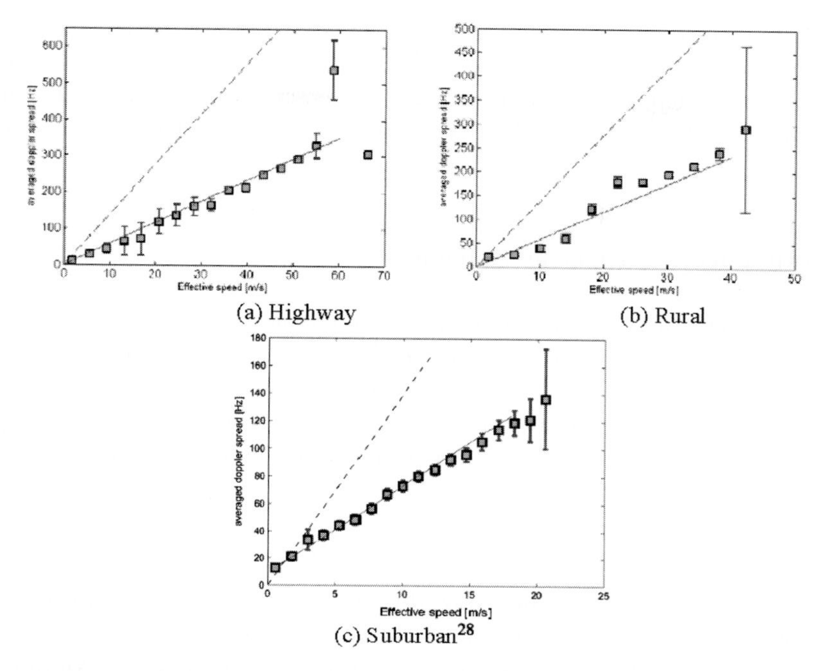

Fig. 6.5. Doppler spread versus effective speed. (a) Highway; (b) rural; (c) suburban. The solid lines represent the linear regression.

Table 6.2 Comparison descriptions.

Environment	Rural	Highway	Suburban[28]
Offset (Hz)	0.5	0.2	11.2
Slope	$0.420/(\lambda\sqrt{2})$	$0.414/(\lambda\sqrt{2})$	$0.428/(\lambda\sqrt{2})$
Plot	Figure 6.5(a)	Figure 6.5(b)	Figure 6.5(c)

all of the environments show an offset at zero effective speed. This offset is the largest in the suburban environment.

We believe the primary reason for the smaller observed slopes is the fact that the theoretical two-ring models do not include a line-of-sight (LoS) path, whereas our on-road data almost always does have such a path. The presence of a strong component with narrower width will reduce the computed second central moment of the spectrum. Another possible contributing factor is the possibility of angles-of-arrival and departure at each vehicle that are not uniformly nor densely distributed. If these paths occur over a smaller range of angles, then the range of Doppler frequencies would also be reduced.

The remnant Doppler spreads at zero effective speed in each environment are due to other moving objects in the environment. Perhaps the reason for the lower offsets in the highway and rural environments is that when the vehicles are stationary, they are likely to be further away from moving objects than in the suburban environment. On the highway, this may be because of the location being on an exit ramp, while in the rural environment it may simply reflect the sparse traffic conditions. In contrast, in stop-and-go suburban traffic, there are often vehicles in nearby lanes that are still in motion.

6.6. Influence from Driver Behavior

One of the interesting studies our RF platform enables is the influence from driver behavior. As indicated in the previous section, the GPS position and velocity of each vehicle is logged for each VSA measurement sweep. The separation between the vehicles is readily obtained from these GPS logs. The velocities of the vehicles determine the Doppler spread, and according to Table 6.2, the Doppler spread is given by

$$f_D \approx \frac{0.42}{\lambda\sqrt{2}} v_{\mathit{eff}} + f_{D\phi} \tag{6.5}$$

where $f_{D\phi}$ is the residual Doppler spread from motion in the environment. Sample scatter plots of v_{eff} versus separation in suburban environments are shown in Figure 6.6. We refer to these plots as Speed-Separation (S-S) diagrams. An interesting observation is that the data from the two different suburban experiments exhibit similar patterns, which we interpret in terms of driver behavior. Beyond a minimum separation of about 5 m, the speed increases with separation up to the speed limit. The correlation of speed with distance reflects the natural tendency of drivers to allow greater separation as the speed increases. Both data sets also show a decrease in speed beyond about 100 m. The driving factors giving rise to this driving behavior are presently under study, but one of the reasons may be due to a desire of drivers to stay together.

This underlying correlation between separation and speed must be taken into account when analyzing the data for dependence on speed and separation. This led us to develop a phenomenological model to describe the relationship between the separation d and the maximum effective speed $v_{\mathit{eff}\,\max(d)} = \max(\sqrt{(v_R^2(d) + v_T^2(d))})$

$$v_{\mathit{eff}\,\max}(d) = \frac{v_{limit}}{1 + e^{-\alpha(d-d_o)}}. \tag{6.6}$$

Here v_{limit} is the maximum value of v_{eff} observed during a data run, and α and d_0 are adjustable parameters. In our analysis, we tune α and d_0 to control the steepness and position of the function to obtain the best match with the top edge of the actual measured data (here $\alpha = 0.18\,\mathrm{m}^{-1}$, $d_0 = 16\,\mathrm{m}$ are found to fit well). As shown in Figure 6.6, this phenomenological model bounds the points in the S-S diagram with reasonable accuracy for both data sets.

Using the observation that Doppler spread is proportional to the effective speed $V_{\mathit{eff}}(d) = \sqrt{v_R^2(d) + v_T^2(d)}$, as given in Equation (6.5), an approximation to the

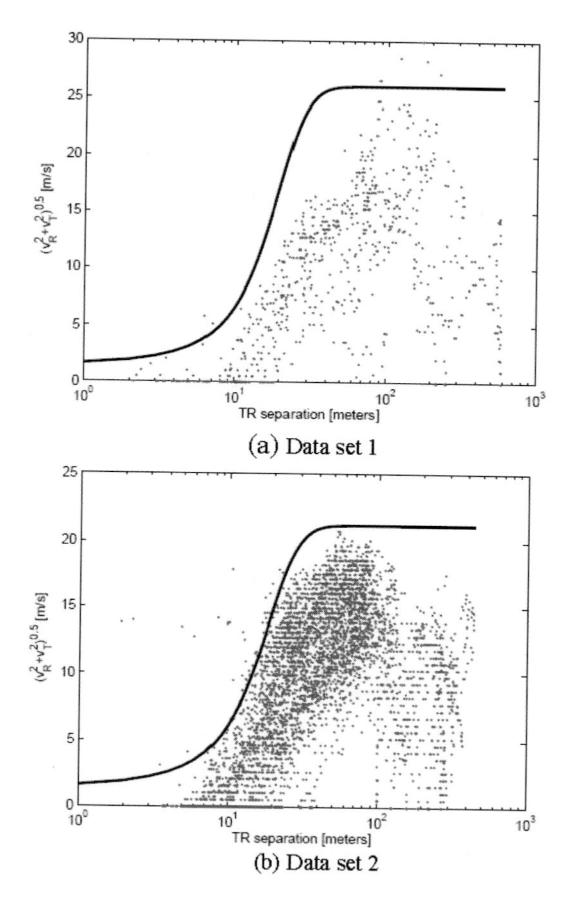

Fig. 6.6. Doppler spread versus effective speed. (a) Highway; (b) rural; (c) suburban. The solid lines represent the linear regression.

Doppler spread for a given distance bin can be constructed using the S-S diagram. Consider the points that lie in a particular range of distances in either Figure 6.6(a) or (b). Since the Doppler spread is proportional to the effective speed, each point in the bin contributes to a particular value of spread. By averaging all of the resulting Doppler spreads (or equivalently averaging the effective speeds) in a particular distance bin, an estimate of the Doppler spread for that distance can be obtained. The Doppler spread versus distance curves calculated in this way for the two data sets are shown as the curves marked with "*" in Figure 6.7. The agreement with the values of Doppler spread calculated from the experimental spectra is good, demonstrating the validity and usefulness of the S-S diagram.

The S-S diagram allows others to compute the expected Doppler spectrum from just measurements of vehicle separation and speed. It should be noted that while an S-S diagram can be based on measurements, this is not required. In particular, it

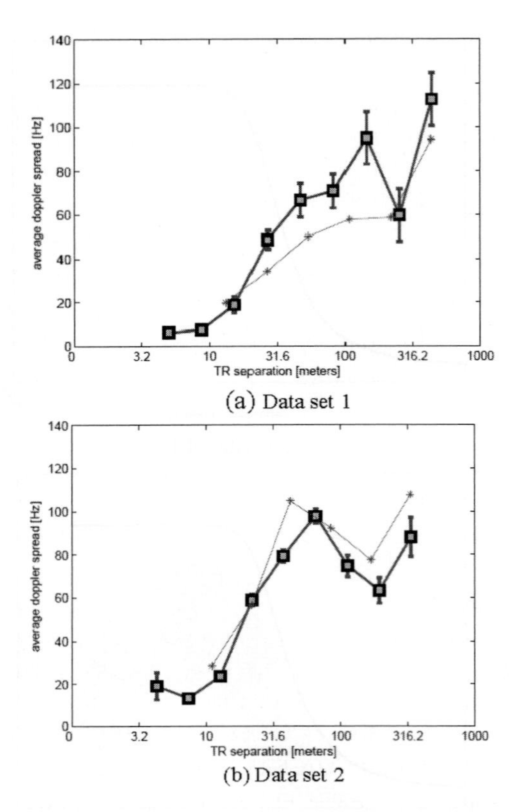

(a) Data set 1

(b) Data set 2

Fig. 6.7. Doppler spread versus distance for the suburban environment. (a) Data set 1, (b) data set 2. Curves marked with "*" result from the S-S model.[28]

would be natural to simulate driver behavior in order to generate an S-S diagram and use that diagram to compute what the expected Doppler spread would be in such an environment. This would also allow the calculation of small-scale fading statistics and coherence time.

This technique would be an extremely convenient way to generate accurate small-scale fading using only a model of driving behavior, with or without actual measurements.

6.7. Challenges for Future Research

While important advances have been made in V2V research, there still exist many open research challenges owing to the unique characteristics of V2V networks. For example, the V2V communication channel can vary from a simple point-to-point microwave link for vehicles in open rural areas, to severe Rayleigh fading between cars within busy cities. To make things worse, the V2V channel could change considerably every few seconds owing to, for example, the intermittent blockage of line of sight owing to trucks, etc. The ideal physical channel is expected to provide high data rates even though the communication time can be limited to a few seconds, and under diverse channel conditions. Therefore it is desirable to have adaptive and efficient channel estimation algorithms. Diversity techniques to overcome fading effects are also of great interest.

Antenna design for VANET DSRC applications is constrained by practical considerations for deployments, including height, placement, and packaging. Further, while a cell tower could be in any direction with respect to the vehicle, other participants in V2V communications are most likely to be in front of or behind the vehicle. Owing to these different requirements, it is difficult to directly apply the extensively-studied antenna design for cellular networks to V2V. Therefore it is important to consider the effect of antenna mounting placement and diversity on V2V communications.

There could exist a plethora of potential cyber-attacks to VANET including bogus traffic information, attacks on a vehicle platoon, etc. Securing vehicular networks against cyber-attacks is a major issue due to the criticality of many of their applications in Intelligent Transportation Systems. It is expected that the security solution should also meet the diverse needs of the applications. At the same time, it is desirable to take into consideration the processing capabilities of the onboard units. In addition, the network should require minimum human interaction to avoid diverting the driver's attention from the road.

Efficient broadcasting algorithms are essential for delivery of safety and routing messages. There still exists a need to further investigate routing protocols in VANET to ascertain their stability and effectiveness when few cars are on the road as well as in congested areas.

Other challenges for future research include specifications for interfaces between driver and on-board units, the need for new hand-off schemes intended to support seamless handoff at high speeds, spectrum demand and interference management, and the need for traffic-adaptive protocols. As operating conditions change (traffic, environmental, etc.), the system has to adapt and reconfigure on-the-fly to achieve better functionality.

Together with what has already been addressed in this chapter, these considerations are expected to work together in an efficient fashion to reach the overall objective of safer, more comfortable and information rich trips while driving on the road.

6.8. Conclusions

In this chapter, we have discussed some fundamental research issues concerning both physical propagation and communication aspects in V2V networks. We have shown that as applications become more complex, the properties of the V2V channel play increasingly important roles in the design process. In particular, we describe our recent research on investigating the detailed behavior of the V2V channel through on-road measurements in suburban, rural, and highway driving environments. We illustrate the influence of driver behavior on the channel properties with the use of the effective speed-separation diagram. Finally, we also summarize potential implications for V2V system designers to ensure the transmission channel operates reliably in realistic and dynamic environments and road traffic.

Problems

(1) What is VANET, and what new capabilities will it provide over today's state-of-the-art vehicles?
(2) In what ways is the vehicle-to-vehicle channel different from the traditional fixed-to-mobile channel?
(3) Why is the V2V channel time-variant? Use an example to illustrate.
(4) How and why does the V2V channel behave differently in different environments?
(5) Will the Doppler spread potentially be impacted by other adjacent vehicles with high speed? If so, how?
(6) What is the influence of driver behavior on the Doppler spread?
(7) An equalizer can be used to correct for various channel impairments. According to Table 6.1, over what time interval could a particular equalizer setting be used in a suburban environment with the expectation that it would be effective 90% of the time?
(8) Two vehicles are traveling on the highway such that $v_R = 40\,\mathrm{m/s}$ and $v_T = 50\,\mathrm{m/s}$. What is the expected Doppler spread of the channel if the frequency is 5.9 GHz?
(9) Consider two vehicles communicating in a suburban environment. From Figure 6.7(a), estimate the expected Doppler spread if the separation is (a) 10 m, (b) 100 m.
(10) What is the median RMS delay spread for the following environments: (a) suburban, (b) rural, (c) highway?

Bibliography

1. A. Sugiura and C. Dermawan, In Traffic Jam IVC-RVC System for ITS using bluetooth, *IEEE Transactions on Intelligent Transportation Systems* **6** (2005) 302–313.
2. S. Dashtinezhad, T. Nadeem, B. Dorohonceanu, C. Borcea, P. Kang and L. Iftode, Trafficview: a driver assistant device for traffic monitoring based on car-to-car communication, *Vehicular Technology Conference* (2004).
3. W. Chen and S. Cai, Ad hoc peer-to-peer network architecture for vehicle safety communications, *IEEE Communication Magazine* (2005) 100–107.
4. T. L. Doumi, Spectrum considerations for public safety in the United States, *IEEE Communication Magazine* **44** (2006) 30–37.

5. Standard specification for telecommunications and information exchange between roadside and vehicle systems — 5GHz band dedicated short range communications (DSRC) medium access control (MAC) and physical layer (PHY) specifications, ASTM e2213-03 (2003).
6. J. Yin, T. ElBatt, G. Yeung, B. Ryu, S. Habermas, H. Krishnan and T. Talty, Performance evaluation of safety applications over DSRC vehicular ad hoc networks, *VANET '04: Proceedings of the 1st ACM International workshop on Vehicular Ad Hoc Networks* (2004) 1–9.
7. J. J. Blum, A. Eskandarian and L. J. Hoffman, Challenges of inter-vehicle ad hoc networks, *IEEE Transactions on Intelligent Transportation Systems* **5** (2004) 347–351.
8. D. Jiang, V. Taliwal, A. Meier, W. Holfelder and R. Herrtwich, Design of 5.9 GHz DSRC-based vehicular safety communication, *IEEE Wireless Communications* **13** (2006) 36–43.
9. The California PATH Project, http://www.path.berkeley.edu.
10. The Fleetnet Project, http://www.et2.tu-harburg.de/fleetnet/.
11. Network-on-wheels, http://www.network-on-wheels.de/.
12. A. S. Akki and F. Haber, A statistical model of mobile-to-mobile land communication channel, *IEEE Transactions on Vehicle Technology* (1986).
13. R. Wang and D. Cox, Double mobility mitigates fading in ad hoc wireless networks, *IEEE Antennas and Propagation Society International Symposium* **2** (2002) 16–21.
14. R. Wang and D. Cox, Channel modeling for ad hoc mobile wireless networks, *IEEE Vehicular Technology Conference* **1** (2002) 21–25.
15. C. Patel, G. Stuber and T. G. Pratt, Simulation of Rayleigh-faded mobile-to-mobile communication channels, *IEEE Transactions on Communications* **53** (2005) 1876–1884.
16. L.-C. Wang and Y.-H. Cheng, A statistical mobile-to-mobile Rician fading channel model, *IEEE Vehicular Technology Conference* (2005).
17. J. S. Davis and J. P. M. G. Linnartz, Vehicle-to-vehicle Rf propagation measurements, *28th Asilomar Conference on Signals, Systems and Computers* **1** (1994) 470–474.
18. R. J. Punnoose, P. V. Nikitin, J. Broch and D. D. Stancil, Optimizing wireless network protocols using real-time predictive propagation modeling, *1999 Radio and Wireless Conference* (*RAWCON 99*) (1999) 39–44.
19. G. Acosta, K. Tokuda and M. A. Ingram, Measured joint Doppler-delay power profiles for vehicle-to-vehicle communications at 2.4 GHz, *GLOBECOM '04. IEEE* **6** (2004) 3813–3817.
20. J. Maurer, T. Fugen and W. Wiesbeck, Narrow-band measurements and analysis of the inter-vehicle transmission channel at 5.2 GHz, *IEEE Vehicular Technology Conference* **3** (2002) 1274–1278.
21. X. Zhao, J. Kivinen, P. Vainikainen and K. Skog, Characterization of Doppler spectra for mobile communications at 5.3 GHz, *IEEE Transactions on Vehicular Technology* **52** (2003) 14–23.
22. X. Zhao, J. Kivinen, P. Vainikainen and K. Skog, Propagation characteristics for wideband outdoor mobile communications at 5.3 GHz, *IEEE Journal on Selected Areas in Communications* **20**(3) (2002) 507–514.
23. G. Acosta and M. A Ingram, Doubly selective vehicle-to-vehicle channel measurements and modeling at 5.9 GHz, *Wireless Personal Multimedia Communications Conference* (2006).
24. T. S. Rappaport, *Wireless Communications: Principles and Practice* (Prentice Hall, 1999).

25. W. G. Newhall, *Simulation Models, and Processing Techniques for a Sliding Correlator Measurement System*, M.S. thesis, Virginia Polytechnic Institute and State University (1997).
26. L. Cheng, B. Henty, D. Stancil and F. Bai, Multi-path propagation measurements for vehicular networks at 5.9 GHz, *IEEE Wireless Communications and Networking Conference* (2008).
27. B. Sklar, Rayleigh fading channels in mobile digital communication systems, *IEEE Communications Magazine* (1997) 102–109.
28. L. Cheng, B. Henty, D. Stancil, F. Bai and P. Mudalige, Mobile vehicle-to-vehicle narrow-band channel measurement and characterization of the 5.9 GHz dedicated short range communication (DSRC) frequency band, *IEEE Journal on Selected Areas in Communications* **25**(8) (2007) 1501–1516.

Chapter 7

Radio Resource Management in Cellular Relay Networks

Ki-Dong Lee

LG Electronics Mobile Research, San Diego, CA 92131, USA

Victor C. M. Leung

University of British Columbia, Vancouver, BC V6T-1Z4, Canada

As the networking architecture evolves toward low-cost high-utility services, the development cost-effective techniques for reliable and efficient operations of the network with inherently increased complexity is require. This chapter provides the reader (specifically for engineering students) with a holistic view on resource management in the cellular relay network, ranging from system modeling, optimization problem formulation and solution. It deals with very popular solution methods for resource allocation problems and also deals with two case examples commonly arising in cellular relay networks: fair scheduling and energy efficient scheduling. Based on the optimization techniques introduced in the former part of this chapter, we also deal with several problems associated with their applications in the latter part.

Keywords: Relay network; resource management; fair scheduling; energy efficient scheduling; cooperative diversity technique; optimization technique.

7.1. Cellular Relay Networks

7.1.1. *Introduction*

As wireless applications are getting more popular and demanding, one of the hot issues is how to provide higher utility for the subscribers at the lowest possible cost, which leads to the maximization of the long term revenue of the service provider. In a wireless network, the volume of traffic successfully delivered to the destination over a certain period of time, i.e. the throughput, is a good measure of determining the utility. This utility is obtained at a certain cost, most commonly characterized by the usage of radio resources. In time-division multiple access, the usage of each timeslot is considered to be a certain amount of cost; in code-division multiple access, the power consumption is considered to be a cost. One of the most important problems for the service provider in system engineering is to provide the best possible utility using the least cost.

In multi-class service networks, the utility is not a simple notion but is a function of a variety of affecting parameters, such as call blockage (for real-time calls), jitter (video streaming), delay, carried traffic, battery lifetime, etc. It is interesting to note that as the available radio resources is increased, the effects of these parameters tend to diminish and the utility tends to be improved.

One of the methods that are being widely considered to achieving utility improvements is to use *relays*. Using a relay for a connection provides power savings and coverage extension. Also, using relays in a multiple-node networking environment may provide even more performance enhancements obtained by cooperation diversity and network coding. However, use of relays may bring about some, such as additional delays due to multiple hops. Nevertheless, these drawbacks can be considered as constraints to system design and pose interesting topics for future investigations.

In this chapter, we present advanced techniques for designing wireless networks with relays. We also present a brief introduction on optimization techniques that can be applied to solve problems emerging from networking with relays.

7.1.2. *Recent standardization activities*

The use of relays in wireless networking has a long history in the research literature[1] However, the wireless industry has only initiated discussions on related issues recently. The IEEE 802.16 Wireless Broadband Access Working Group consists of several Task Groups (TGs). The Relay Task Group, TGj, is completing a specification for multi-hop relay support. However, the IEEE 802.16j specification[1] has a limited scope; specifically, it does not include operations of the mobile stations (MSs) and is therefore not to be applied on the handset side. The specification defines two types of relay stations (RSs): non-transparent RS and transparent RS. A non-transparent RS transmits the same downlink (DL) synchronization and overhead information as a base station (BS), whereas a transparent RS does not transmit such information but instead relay the information sent by the BS. A multi-hop relay BS (MR-BS) is a generalized equipment set providing connectivity, management, and control of RSs. The transparent mode frame structure consists of two subframes: DL subframe and UL subframe. Each subframe consists of "access zone" and "relay zone". In the DL subframe, the access zone is used for MR-BS to transmit and for RS to receive whereas the relay zone is used for RS to relay downward to other RSs or MSs. In the UL subframe, the access zone can be used for MS to MR-BS/RS transmissions and the relay zone is used for RS to MR-BS/RS transmissions.

Recently, IEEE 802.16 TGm[3] has begun to work on an advanced air interface. The system requirement and evaluation methodology documents have been completed but the system description document is still under development. More flexible relay operations are expected to be specified in IEEE 802.16m.

7.2. Radio Resource Allocation

7.2.1. *Optimization criteria — economic aspects*

7.2.1.1. *Revenue*

In general, revenue refers to business income, and profit is the total revenue minus total expenses. In wireless networking, revenue is often defined as "carried traffic". For example, we can define a revenue function as the amount of traffic carried through a certain network. Also, we can define the time average of that amount as a revenue function in a time-average sense. In the layered architecture of communication networking, the revenue functions may have different forms and characteristics depending on the layer in which it is defined. For example, for a packet sent through the same path in the network, the revenues quantified in the physical layer and in media access control (MAC) layer are not necessarily equivalent to each other.

No matter how different the physical meanings are, revenue is a means to take care of the service provider's concerns, not the users' concerns, even if the two are related. Therefore, it is not guaranteed that the service provider can achieve the long term optimum in revenue even if a resource allocation is selected that maximizes the revenue in the short term.

7.2.1.2. *Utility*

One of key difference of functional characteristics between revenue and utility is that there is usually an upper bound in utility; i.e. there is a saturation point after which the increase in utility is asymptotically zero when the traffic is increased. The blocking probability modeled by an M/M/c/c queue is a typical example for the diminishing return in utility. As the offered traffic is increased, the blocking probability approaches 1 and no the amount of traffic actually accommodated becomes saturated.

7.2.1.3. *Fairness*

In a wireless network, not all the users are in the same situation: their channel conditions and the stochastic behavior of the channels are very different from each other due to various factors, such as different locations, source-destination distances, speeds of movement, and so on. To achieve the best possible throughput, we can choose the users that have the best channel quality (see "opportunistic scheduler"[32]). However, if fairness is not realized in resource allocation, some users may monopolize the resources whereas other users are starving for resources.

Max-min fairness: A scheduling policy is said to be max–min fair if the minimum of throughputs of multiple users is maximized. For example, suppose that out of all options we have two feasible policies (policies A and B) and that we have four users. For policy A, the achievable throughput vector for four users is (1, 1, 1, 1); for policy B, it is (2, 0.5, 1, 1). Policy B is better than policy A in terms of system throughput ($4.5 > 4$) but policy A is fair and, furthermore, max-min fair since the minimum throughput is maximized.

Proportional fairness: The key point of the "proportional fair" scheduler is to take advantage of the fact that the data rate (or throughput) of each user is changing over time. So we can choose the best performing user at any time. The difference between the opportunistic scheduler and the proportional fair scheduler is that the user with the maximum throughput is chosen in the opportunistic scheduler whereas the user with the maximum of throughput relative to its own average throughput is chosen in the proportional fair scheduler. In Figure 7.1, user 1 is always chosen if we use the opportunistic scheduler. This is not fair at all. However, in the case of the proportional fair scheduler user 1 is chosen in "timeslot a" because its throughput relative its own average is greater than that of user 2 in the timeslot and user 2 can be chosen in the other timeslot, says "timeslot b", because its throughput relative to its own average is greater than that of user 1 in the timeslot. By doing so, the proportional fair scheduler can achieve better system performance than a round-robin scheduler because there is no chance for users that are not best performing relative to their own average throughput to send traffic, whereas this may happen in the round-robin scheduler and degrade the system performance.

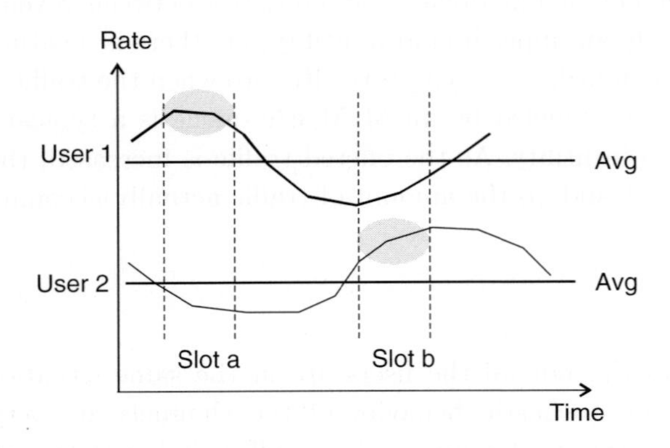

Fig. 7.1. An example on proportional fairness.

7.2.2. *Real-time solution techniques*

7.2.2.1. *Problem decomposition*

The main purpose of problem decomposition is to reduce the computational complexity for solving a certain problem of interest. In this subsection, we briefly introduce how a linear or non-linear programming problem is decomposed. More mathematical details, such as Dantzig-Wolfe Decomposition, can be found in books on optimization theory such as Ref. 5.

7.2.2.1.1. Decomposition of a linear program

Consider a following problem:

$$(\text{P}) \quad \begin{aligned} \max \quad & a^T x \\ s.t. \quad & Ax \leq b. \\ & x \geq 0 \end{aligned}$$

Suppose that the computational complexity of this problem is $O(N^M)$ where N is the cardinality of vector x and M is a certain number determined by the algorithm we use. If the coefficient matrix A has the following characteristic:

$$A = \begin{pmatrix} A_1 & 0 \\ 0 & A_2 \end{pmatrix} \quad \text{or} \quad A = \begin{pmatrix} 0 & A_2 \\ A_1 & 0 \end{pmatrix},$$

then problem (P) can be completely decomposed into two subproblems:

$$(\text{P1}) \quad \begin{aligned} \max \quad & a_1^T x_1 \\ s.t. \quad & A_1 x_1 \leq b_1, \\ & x_1 \geq 0 \end{aligned}$$

and

$$(\text{P2}) \quad \begin{aligned} \max \quad & a_2^T x_2 \\ s.t. \quad & A_2 x_2 \leq b_2 \\ & x_2 \geq 0 \end{aligned}$$

where

$$a = \begin{pmatrix} a_1 \\ a_2 \end{pmatrix}, \quad b = \begin{pmatrix} b_1 \\ b_2 \end{pmatrix}, \quad x = \begin{pmatrix} x_1 \\ x_2 \end{pmatrix}.$$

The computational complexity of the decomposed problem is $O(\max\{N_1, N_2\}^M)$ where N_i is the cardinality of vector x_i, which results in a reduction in computational burden.

7.2.2.1.2. Program decomposition by Lagrange relaxation

Even if the coefficient matrix is not in the form given above, i.e. there are constraints that prevent us from direct application of the above method, we are still able to decompose a problem using Lagrange relaxation.

$$
(\mathrm{P}) \quad
\begin{aligned}
\max \quad & \begin{pmatrix} a_1 & a_2 \end{pmatrix}^T \begin{pmatrix} x_1 \\ x_2 \end{pmatrix} \\
s.t. \quad & \begin{pmatrix} A_1 & 0 \\ 0 & A_2 \end{pmatrix} \begin{pmatrix} x_1 \\ x_2 \end{pmatrix} \le \begin{pmatrix} b_1 \\ b_2 \end{pmatrix} \\
& \begin{pmatrix} c_1 & c_2 \end{pmatrix}^T \begin{pmatrix} x_1 \\ x_2 \end{pmatrix} \le b_3 \qquad (const\ 1). \\
& \begin{pmatrix} x_1 \\ x_2 \end{pmatrix} \ge 0
\end{aligned}
$$

If we do not have the constraint $(c_1\ c_2)^T \binom{x_1}{x_2} \le b_3$, which is called a *coupling constraint*, then we can completely decompose the problem using the procedure shown above.

In this situation, we can relax the *coupling constraint*, for example (const1), using Lagrange relaxation:

$$
\begin{aligned}
\max \quad & \{a_1 x_1 + a_2 x_2\} + \lambda \{c_1 x_1 + c_2 x_2 - b_3\} \\
s.t. \quad & \begin{pmatrix} A_1 & 0 \\ 0 & A_2 \end{pmatrix} \begin{pmatrix} x_1 \\ x_2 \end{pmatrix} \le \begin{pmatrix} b_1 \\ b_2 \end{pmatrix} \\
& \begin{pmatrix} x_1 \\ x_2 \end{pmatrix} \ge 0, \quad \lambda \ge 0.
\end{aligned}
$$

Thus, we have two subproblems:

$$
\begin{aligned}
\max \quad & (a_1 + \lambda c_1) x_1 \\
s.t. \quad & A_1 x_1 \le b_1 \qquad , \\
& x_1 \ge 0, \quad \lambda \ge 0
\end{aligned}
$$

and

$$
\begin{aligned}
\max \quad & (a_2 + \lambda c_2) x_2 \\
s.t. \quad & A_2 x_2 \le b_2 \qquad . \\
& x_2 \ge 0, \quad \lambda \ge 0
\end{aligned}
$$

Then we can take on each subproblem that is much easier to solve. It is very common to solve the subproblems on λ to find $x_1^*(\lambda)$ and $x_2^*(\lambda)$. With $x_1^*(\lambda)$ and $x_2^*(\lambda)$, we can calculate the Lagrangean dual function to find the dual optimal solution λ^*. This will give the primal solution in regard to the dual optimal solution. In the linear program, there is no duality gap; therefore, the dual optimal solution gives the primal optimal solution.

7.2.2.1.3. Decomposition of a non-linear program

The above technique can be used in non-linear program cases. The following gives an illustrative example:

$$\begin{aligned} \min \quad & f = x_1^2 + x_2^2 \\ s.t. \quad & -x_1 - 2x_2 + 4 \leq 0. \\ & x_1,\ x_2 \geq 0 \end{aligned}$$

Verification of the Optimum: From the property of the objective function and the constraint, it is obvious that the constraint is "binding" (i.e. equality holds) at the optimum. Substituting the objective function with $x_1 = 4 - 2x_2$, we have

$$f = (4 - 2x_2)^2 + x_2^2,$$

with $x_2 \leq 2$. This becomes

$$\begin{aligned} f &= 5x_2^2 - 16x_2 + 16 \\ &= 5\left(x_2 - \frac{8}{5}\right)^2 + 16 - \frac{8^2}{5} \\ &\geq \frac{16}{5} \left(= \text{holds if } x_2 = \frac{8}{5}\right). \end{aligned}$$

Therefore, the optimum is $(x_1^*, x_2^*) = \left(\frac{4}{5}, \frac{8}{5}\right)$.

Solution by Decomposition: The Lagrangean function is

$$\begin{aligned} L(x_1, x_2, \lambda) &= f + \lambda \cdot (-x_1 - 2x_2 + 4) \\ &= x_1^2 + x_2^2 + \lambda \cdot (-x_1 - 2x_2 + 4) \end{aligned}$$

for $\lambda \geq 0$. Thus, the dual problem is

$$\begin{aligned} q(\lambda) = \min \quad & x_1^2 + x_2^2 + \lambda \cdot (-x_1 - 2x_2 + 4) \\ s.t \quad & x_1, x_2 \geq 0. \end{aligned}$$

Then we have two subproblems:

$$\text{(sub1)} \quad \begin{matrix} \min & x_1^2 - \lambda x_1 \\ s.t & x_1 \geq 0 \end{matrix},$$

where the optimal solution is $x_1^* = \min\left(4, \frac{\lambda}{2}\right)$;

$$\text{(sub2)} \quad \begin{matrix} \min & x_2^2 - 2\lambda x_2 \\ s.t & x_2 \geq 0 \end{matrix},$$

where the optimal solution is $x_2^* = \min(2, \lambda)$.

For $\lambda \leq 2$,

$$
\begin{aligned}
q(\lambda) &= -\frac{5}{4}\lambda^2 + 4\lambda \\
&= -\frac{5}{4}\left(\lambda - \frac{8}{5}\right)^2 + \frac{16}{5}. \\
&\leq \frac{16}{5} \quad \left(\text{equality holds if } \lambda = \frac{8}{5}\right)
\end{aligned}
$$

This yields $(x_1^*, x_2^*) = \left(\frac{4}{5}, \frac{8}{5}\right)$.

This example demonstrates how the decomposition method works. In practical cases where the number of variables is large, it is very attractive to use such decomposition techniques to yield a smaller computational burden with a smaller memory space requirement (i.e. it is attractive in terms of both *computational complexity* and *memory complexity*).

7.2.3. *Case study on fair scheduling in a cellular relay network*

Advances in information and communications technologies over the past decade have caused a remarkable increase of traffic demand especially for Internet access, which requirements for high data rate, quality of service (QoS) support and reliability have motivated the development of advanced wireless networks. Recently, wireless mesh networks (WMNs) have emerged as a promising technology for next-generation wireless networks[6,7] with flexible and reconfigurable architectures. To realize the potential of this new technology, research is rapidly progressing to develop high-performance techniques for advanced radio transmissions, medium access control and routing. Because of the challenges presented by multi-hop communications over WMNs, techniques to improve basic network performance, such as end-to-end transmission delay, end-to-end data rate and fairness, remain important subjects of investigation. Recently, adaptive resource management for multiuser orthogonal frequency-division multiple-access (OFDMA) systems has attracted enormous research interest.[8–14] OFDMA is a very promising solution to provide a high-performance physical layer for emerging wireless networks, as it is based on orthogonal frequency-division multiplexing (OFDM) and inherits its desirable properties of immunity to inter-symbol interference and frequency selective fading.

In Ref. 8, the authors studied how to minimize the total transmit power while satisfying a minimum rate constraint for each user. The problem was formulated as an integer programming problem and a continuous-relaxation based suboptimal solution method was studied. In Ref. 9, computationally inexpensive methods for power allocation and subcarrier assignment were developed, which are shown to achieve comparable performance, but do not require intensive computations. However, these approaches did not provide a *fair* opportunity for users so that some users may become dominant in resource occupancy even when the others starve.

In Ref. 10, the authors proposed a fair scheduling scheme to minimize the total transmit power by allocating subcarriers to the users and then to determine the number of bits transmitted on each subcarrier. Also, they developed suboptimal solution algorithms by using the linear programming technique and the Hungarian method. In Ref. 11, the authors formulated a combinatorial problem to jointly optimize the subcarrier and power allocation, subject to the constraint that resources are to be allocated to user according to predetermined fractions. By using the constraint, the resources can be fairly allocated. A novel scheme to fairly allocate subcarrier, rate and power for multiuser OFDMA systems was proposed in Ref. 13, which introduces a new fairness criterion for generalized proportional fairness based on Nash bargaining solutions (NBS) and coalitions. This study is very different from the previous OFDMA scheduling studies in the sense that the resource allocation is performed with a game-theoretic decision rule. They proposed a very fast near-optimal algorithm using the Hungarian method and showed by simulations that their fair-scheduling scheme provides a similar overall rate to that of a rate-maximizing scheme. However, the studies reviewed above have focused on wireless *single-hop* networks.

We propose a new fair-scheduling scheme, called distributed hierarchical scheduling, for a WMN that is connected to the Internet via a mesh router (MR), and which consists of a group of mesh clients (MCs) that can potentially relay each other's packets to/from the MR. Hierarchical scheduling has been widely used in computer systems[15] and computer networks.[16]

The goal of the proposed hierarchical scheduling scheme is to fairly allocate the subcarriers and power to multiple users; i.e. the MCs. To achieve this goal in a both reliable and efficient manner, we employ *hierarchical decoupling* for the WMN scheduling between the MR level and the MC level. At the MR level (Level-0 scheduling performed by the MR), the objective is to fairly allocate subcarriers to MCs so that the NBS fairness criterion is maximized. At the MC level (Level-1 scheduling performed by each MC in a distributed manner) the objective is to fairly allocate the total transmit power to the available subcarriers assigned to each outgoing link so that the NBS fairness criterion is maximized. Distributed hierarchical scheduling is very useful for WMNs because in practice it is difficult if not impossible to obtain perfect information about every node and every link in the WMN at a central location (e.g. MR). The proposed hierarchical decoupling method allows fair scheduling to be contemplated based on the use of a suitable subset of the network information at the MR for Level-0 scheduling, and local information available at each MC for Level-1 scheduling. Since a limited subset of network information can be provided to the MR more frequently, the scheduling input based on this is potentially more reliable than full network information that is hard to update, thus resulting in more reliable scheduling. Using distributed hierarchical scheduling also reduces the signaling overhead and distributes the processing loads of scheduling among the network nodes so that both the signaling and scheduling efficiencies can be improved.

In the hierarchically decoupled scheduling problem, we formulate the first problem (Level-0 scheduling) as a nonlinear integer programming problem and develop an efficient near-optimal solution algorithm. Also, we formulate the second problem (Level-1 scheduling) as a nonlinear mixed integer programming problem. To solve the latter problem more efficiently, we transform this problem into a time-division scheduling problem and develop a closed-form solution for subcarrier and power allocation. Extensive simulation results demonstrate that the proposed scheme provides fair opportunities to the MCs and an overall end-to-end rate comparable to existing max-rate schemes[13] when the number of users increases.

7.2.3.1. *The proposed hierarchical scheduling*

There are extensive studies on routing algorithms in general wired and wireless networks.[17–19] Developing routing algorithms is not the focus of this section. Instead, this section focuses on developing fair resource-allocation algorithms in an OFDMA WMN given that routing has been determined. The resources to schedule are defined as a set of subcarriers, and the total transmit power available at each node. We consider the following model for distributed hierarchical scheduling, where the MR only performs a rough scheduling with limited information (because not all the information is always available) and the MCs perform more refined scheduling with full information that is available locally.[4]

7.2.3.1.1. MR scheduling

- Scheduling Input: The traffic demand of each MC in terms of the data rate, R_i; the average channel gain of all outgoing links at MC i, $\overline{G_i}$ (abbreviated information).
- The Role of MR: To determine the number of subcarriers to be assigned to each node. However, the MR does not allocate power levels on the respective subcarriers because the MR does not know G_{ij}^k. Instead, power allocation is performed by the MC.[a]

7.2.3.1.2. MC scheduling

- Scheduling Input: The traffic demand for each outgoing link (i, j) in terms of the data rate, R_{ij}; the channel gain of subcarrier k on its outgoing link (i, j) G_{ij}^k (full information available).
- The Role of MC: To allocate its available subcarriers that have been assigned to it by the MR and to allocate its available transmit power among these subcarriers.

[a]The spatial reuse of radio resources is not considered in this chapter.

7.2.3.2. *Model description*

We consider an OFDMA WMN that consists of N nodes: one mesh router (MR) and $N-1$ mesh clients (MCs). Figure 7.2 illustrates the network model considered in this section. The set of nodes is denoted by $\mathbf{N} = \{0, 1, \ldots, N-1\}$, where node 0 represents the MR that delivers traffic between the MCs and the Internet. Some MCs, denoted by set $\mathbf{N}_0$, are located in the effective radio coverage of the MR whereas the others, denoted by set $\mathbf{N}_1 = \mathbf{N} - \mathbf{N}_0$, are located outside the MR's effective radio coverage and need to access the Internet via multi-hops (e.g. MCs 4, 5, and 6 in Figure 7.2).

MC i, where $i \in \mathbf{N}_0$, may alternately access the Internet via multi-hops if there exists a path to the MR (node 0) that cost less than a direct transmission to the MR. Also, MC i may have multiple paths to the MR (e.g. paths $4 \to 2 \to 0$ and $4 \to 3 \to 0$ in Figure 7.2). A multi-path routing policy, where a portion of traffic of MC i is delivered to the destination via a neighboring MC and another portion is delivered via a different neighboring MC, is useful compared to single-path routing because it can provide more flexibility in network resource allocation, potentially increasing the network capacity. For example, suppose that neither MC 2 nor MC 3 in Figure 7.2 has enough capacity to deliver the traffic from MC 4 but the sum of their capacity is sufficient. Another possible scenario is that both MC 2 and MC 3 have enough capacity to deliver the traffic from MC 4 but the use of multi-path routing saves the overall delivery cost compared to single-path routing. Such scenarios are very common in OFDMA WMNs because of the strict concavity of the Shannon capacity of each subcarrier. The routing information is denoted by a matrix y_{ij}, where y_{ij} is the fraction of traffic that MC i sends to MC j (in the case of $j = 0$, MC 0 denotes MR) so that $\sum_{\forall j} y_{ij} = 1$.

There are a total of C subcarriers in the system. Each subcarrier has a band width of W. The channel gain of subcarrier k on link (i, j), which connects MC i to MC j, is denoted by G_{ij}^k and the transmit power of MC i on subcarrier k is denoted by p_i^k. Each MC i has a transmit power limit of $\overline{p_i}$ and a minimal rate requirement

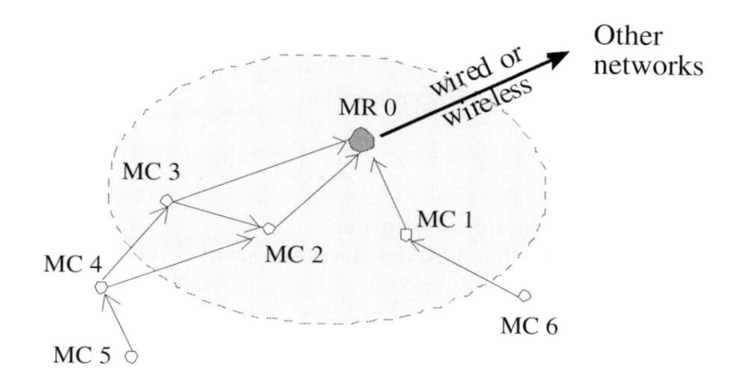

Fig. 7.2. Wireless relay network model.

of R_i. M-ary quadrature amplitude modulation (MQAM) is adopted in our system. We consider a slow-fading channel such that the channel is stationary (channel gain is constant) within each OFDM frame. This is consistent with contemporary applications of OFDMA WMNs.

7.2.3.3. *Subcarrier allocation at level-0 scheduling*

7.2.3.3.1. MR scheduling

In Level-0 scheduling, the MR determines how many subcarriers to allocate to the respective MCs (nodes $i = 1, \ldots, N-1$). The scheduling inputs are traffic demand vector $\mathbf{R}_i^T$ and the channel gain vector $\overline{\mathbf{G}_i}^T$. The MR does not need any routing information for Level-0 scheduling. The scheduling method consists of two phases (refer to Figure 7.3): (1) the MC optimizes the vector $\mathbf{x}^* = (x_i^*)^T$, which element x_i^* is the decision variable denoting the number of subcarriers to allocate to MC i, using only information available at the MR; (2) the MR randomly generates the set of subcarriers for each MC i, say $\mathbf{C}_i$, $i \in \mathbf{N} - \{0\}$, using the vector $\mathbf{x}^*$ $(x_i^* = |\mathbf{C}_i|)$. The main reason that the MR randomly determines $\mathbf{C}_i$ based on the vector x_i^* is that exact and complete information needed for deterministic channel assignments to the MCs is not usually available at the MR. The scheduling output, $\mathbf{C}_i$, $i \in \mathbf{N} - \{0\}$, is sent to the respective nodes via multi-hops if necessary.

We formulate the Level-0 scheduling as a nonlinear integer programming problem and suggest a simple and efficient solution algorithm in the following two subsections.

7.2.3.3.2. MR problem formulation

Each MR solves problem (P1) as defined below to determine $\mathbf{x} = (x_1, \ldots, x_{N-1})^T$, given R_i and $\overline{G}_i$, and sends the information regarding the number of subcarrier x_i that is allocated to MC i to the respective MC.

Step 1: Find an optimal real-number solution
If $\lambda^* < 0$, then (P1) is not feasible (Stop);

$$\overline{x}_i^* = g_i^{-1}(\lambda^*);$$

Step 2: Find an initial integer solution

$$\tilde{x}_i = \lfloor \overline{x}_i^* \rfloor; \quad s = \sum_{i=1}^{N-1} (\overline{x}_i^* - \tilde{x}_i);$$

Step 3: Improve the initial solution
Choose $\tilde{x}_k$ that has the steepest ascent direction on f_1;

$$\tilde{x}_k + = 1; \quad s - = 1;$$

Repeat the above until $s < 1$;

$$x^* = \tilde{x};$$

Fig. 7.3. Iterative steepest-ascent direction search for (P1) performed by MR.

$$\text{(P1)}\quad \begin{array}{ll} \max & f_0(\mathbf{x}) = \prod_{i=1}^{N-1} \{r_i(x_i) - R_i\} \\ \text{subject to} & -C + \sum_{i=1}^{N-1} x_i \leq 0 \\ & -r_i(x_i) + R_i \leq 0 \\ & x_i : \text{non-negative integer} \end{array},$$

where $r_i(x_i)$, the rate realized at MC i from allocation x_i^*, is given by

$$r_i(x_i) = x_i W \log_2 \left(1 + a\overline{G}_i \frac{\overline{p}_i}{x_i}\right).$$

Since we consider MQAM in this section, we have $a \approx -\frac{1.5}{\sigma^2 \cdot \log(5 \cdot BER)}$ where σ^2 is the thermal noise power at the receiver over the bandwidth of each subcarrier, and BER is the desired bit-error rate. We assume that σ^2 is constant at all receivers over all subcarrier channels.

Proposition 7.1 (Proportional Fairness). *Suppose that $R_i = 0$. Given that the optimal solution to problem* (P1) *exists, the optimal solution provides* proportionally fair *resource allocation.*

Proof. See Exercise. □

7.2.3.3.3. Proposed solution procedure

The problem (P1) is combinatorial in nature. By relaxing the integer control variables into real variables, we can obtain the unique global-optimal real-number solution. Then, a near-optimal integer solution can be searched around this real-number solution.

Taking advantage of the strictly increasing property of a logarithm function, we can transform the problem (P1) into the same problem with the objective function $\log f_0(\mathbf{r})$. First, we define a Lagrangean function

$$L(\mathbf{x}, \lambda) = \log f_0(\mathbf{r}) - \lambda \cdot \left(\sum_{i=1}^{N-1} x_i - C\right),$$

where $\mathbf{x} = (x_1, \ldots, x_{N-1})^T$, λ is the associated Lagrange multiplier, $\mathbf{R}$ and $\mathbf{R}_0$ denote the set of real numbers and the set of non-negative real numbers, respectively. Let

$$g_i(x_i) = \frac{\partial \log f_0(\mathbf{x})}{\partial x_i},$$

and let $\delta_i \equiv a\overline{G_i}\overline{p}_i$, then we have

$$g_i(x_i) = \frac{\log_2(1 + \delta_i/x_i) - \delta_i/(\delta_i + x_i)/\log 2}{x_i W \log_2(1 + \delta_i/x_i) - R_i}.$$

Proposition 7.2. *Suppose that the maximum transmit power of MC i is positive, i.e. $\overline{p}_i > 0$. Then, $\log f_0(\mathbf{x})$ is strictly increasing for all x_i's such that $x_i W \log_2(1 + \delta_i/x_i) > R_i$ for $i = 1, \ldots, N-1$.*

Proof. See Exercise. □

From the Karush-Kuhn-Tucker (KKT) theorem,[5,19] we have

$$\frac{\partial L(\mathbf{x}, \lambda)}{\partial x_i} = 0, \text{ i.e. } g_i(x_i) = \lambda, \forall i.$$

By differentiating $g_i(x_i)$ with respect to x_i, we can simply verify that $dg_i/dx_i < 0$ for all $x_i > x_i^{\min}$, i.e. $g_i(x_i)$ is a strictly decreasing function.[b]

Thus, the inverse function of $g_i(x_i)$, $g_i^{-1}(x_i)$, exists for $x_i > 0$. Because $g_i^{-1}(\lambda) = x_i$, $\forall i$, we have

$$\sum_{i=1}^{N-1} g_i^{-1}(\lambda) = C.$$

Let

$$h(\lambda) \equiv \sum_{i=1}^{N-1} g_i^{-1}(\lambda).$$

Proposition 7.3. *$h(\lambda)$ is invertible.*

Proof. From the fact that $dg_i/dx_i < 0$, we have

$$\frac{dh}{d\lambda} = \sum_{i=1}^{N-1} \frac{dg_i^{-1}(\lambda)}{d\lambda} = \sum_{i=1}^{N-1} \frac{1}{\left(\frac{dg_i}{dx_i}\right)} < 0,$$

namely, $h(\lambda)$ is a strictly decreasing function. □

[b]Recall that $x_i^{\min}$ is the unique solution of $x_i W \log_2(1 + \delta_i/x_i) = R_i$, which is positive.

From Proposition 7.3, we obtain

$$\lambda^* = h^{-1}(C)$$

and finally

$$x_i^* = g_i^{-1}(\lambda^*).$$

This is the optimal real-number solution of (P1). The near-optimal algorithm to find an integer solution based on this optimal real-number solution is presented as Phase 2 in Figure 7.3.

Even though the integer solution obtained by the proposed algorithm (Figure 7.3) is based on searching the neighbors of the optimal real-number solution for an integer solution, the optimality of the integer solution is not always guaranteed. To show the performance of the proposed algorithm, we evaluate the gap between the optimal real-number solution, which is obtained by the proposed algorithm in Phase 1 of Figure 7.3, and our integer solution. The experimental results on the gap are shown in Figure 7.5, where it is observed that among 2000 randomly configured test problems, more than 96% of them have a gap smaller than 0.002 (0.2%).

7.2.3.4. *Subcarrier and power allocation at level-*1 *scheduling*

7.2.3.4.1. MC scheduling

In Level-1 scheduling, each MC i performs scheduling to determine which subcarriers (out of the subcarriers allocated to node i) to allocate to the respective links (i, j), $j \in \mathbf{A}_i$, and how much power to allocate to the respective subcarriers. The scheduling inputs for MC i are: the set of allocated subcarriers $\mathbf{C}_i$, a vector traffic demands, including forwarding traffic, from MC i to MCs $j \in \mathbf{A}_i$, R_{ij}^t ($R_{ij} \equiv R_i y_{ij}$), and the channel gains of the respective links (i, j), $j \in \mathbf{A}_i$, on subcarriers $k \in \mathbf{C}_i$, G_{ij}^k. The objective of Level-0 scheduling is to fairly allocate subcarriers to the respective MCs whereas that of Level-1 scheduling is to fairly allocate the subcarriers, which are allocated to each MC, to the respective links and to allocate the powers for the respective subcarriers. We formulate an optimization problem for the Level-1 scheduling and suggest an alternative scheduling problem with an exact solution method in the following subsections.

7.2.3.4.2. MC problem formulation

Each MC $i \in \mathbf{N} - \{0\}$ should solve problem (P2) below to determine (a) the number of subcarriers, $\sum_{k \in \mathbf{C}_i} x_{ij}^k$ to allocate to each link (i, j), where $x_{ij}^k = 1$ if (i, j) is assigned subcarrier k and $x_{ij}^k = 0$ otherwise, and (b) the power p_i^k for each subcarrier k assigned to link (i, j). Problem (P2) is solved at MC i using the current values of $\mathbf{C}_i$ ($x_i^* = |\mathbf{C}_i|$) obtained from the MR, and R_{ij}'s and G_{ij}^k's obtained locally.

This problem can be considered as a generalized case of proportionally fair resource allocation.

$$
\text{(P2)} \quad
\begin{aligned}
\max \quad & \prod_{j \in \mathbf{A}_i} \left\{ -R_{ij} + \sum_{k \in \mathbf{C}_i} x_{ij}^k W \log_2(1 + aG_{ij}^k p_i^k) \right\} \\
\text{subject to} \quad & -\overline{p}_i + \sum_{k \in \mathbf{C}_i} p_i^k \le 0 \\
& -\sum_{k \in \mathbf{C}_i} x_{ij}^k W \log_2(1 + aG_{ij}^k p_i^k) + R_{ij} \le 0 \\
& -x_i^* + \sum_{k \in \mathbf{C}_i} \sum_{j \in \mathbf{A}_i} x_{ij}^k \le 0 \\
& -1 + \sum_{j \in \mathbf{A}_i} x_{ij}^k \le 0 \\
& p_i^k : \text{ non-negative real-number} \\
& x_{ij}^k : \text{ binary integer}
\end{aligned}
.
$$

7.2.3.4.3. The proposed solution using problem transformation

As shown above, (P2) is a non-linear mixed integer programming problem. To improve the tractability of the solution technique, such as criteria for checking feasibility and methods for finding the optimal solution in real-time computations, we suggest an alternate problem formulation (P3) and solution method that can give better performance in terms of both the computational efficiency and the optimum objective value. (P3) given below is a non-linear real-number programming problem. The alternate problem formulation is possible not due to forced relaxation but due to the benefit of time allocation scheduling where the control variable of allocation time is a real-number. The characteristics of (P2) and (P3) are compared in Table 7.1.

In the alternate problem formulation (P3), the problem (P2) is reduced to a time allocation problem in which the allocated time fraction τ_j is optimized, where τ_j is defined as the fraction of time when link (i, j), $j \in \mathbf{A}_i$, occupies the full capacity that MC i can achieve using all the subcarriers assigned to it.

Table 7.1 Comparison of (P2) and (P3).

Item	(P2)		(P3)
Problem type	Nonlinear program		Nonlinear program
Integer variable	Included		None
Real variable	Included		Included
Feasibility check	Difficult		Simple
Achievable rate		$\le$	
Complexity	High		Low

$$
\text{(P3)} \quad
\begin{aligned}
\max \quad & \prod_{j \in \mathbf{A}_i} \left\{ \tau_j \sum_{k \in \mathbf{C}_i} W \log_2(1 + aG_{ij}^k p_i^k) - R_{ij} \right\} \\
\text{subject to} \quad & -\overline{p}_i + \sum_{k \in \mathbf{C}_i} p_i^k \le 0 \\
& -\tau_j \sum_{k \in \mathbf{C}_i} W \log_2(1 + aG_{ij}^k p_i^k) + R_{ij} \le 0 \\
& -1 + \sum_{j \in \mathbf{A}_i} \tau_j \le 0 \\
& sp_i^k, \tau_j \text{: non-negative real-number}
\end{aligned}
\quad .
$$

The value 1 on the right-hand side of the second constraint above is the normalized scheduling interval of this time allocation scheduling. For any desired scheduling interval T, the optimal amount of time at MC i scheduled for link (i, j) is $\tau_j^* \cdot T$.

Proposition 7.4. *Suppose that* (P3) *is feasible. For any given feasible* τ_j*, the power allocation vector* $\mathbf{p}_i^* = (p_i^{k*})^T$ *is optimal if and only if* p_i^{k*} *is the (not necessarily unique) maximizer of* $\sum_{k \in \mathbf{C}_i} W \log_2(1 + aG_{ij}^k p_i^k)$ *subject to the constraint* $-\overline{p}_i + \sum_{k \in \mathbf{C}_i} p_i^k \le 0$.

Proof. Note that all the subcarriers that have been allocated to MC i are used during each time interval τ_j. The amount of resource assigned to each respective link (i, j)'s is dependent on τ_j's only. □

The following shows more details.

($\leftarrow$ "sufficient condition" for optimality)

For any given feasible τ_j, if $\mathbf{p}_i^*$ is the maximizer of $\sum_{k \in \mathbf{C}_i} W \log_2(1 + aG_{ij}^k p_i^k)$ subject to the constraint $\sum_{k \in \mathbf{C}_i} p_i^k \le \overline{p}_i$, then it is also the maximizer of $\tau_j \sum_{k \in \mathbf{C}_i} W \log_2(1 + aG_{ij}^k p_i^k) - R_{ij}$.

($\rightarrow$ "necessary condition")

For any given feasible τ_j, the maximizer of $\tau_j \sum_{k \in \mathbf{C}_i} W \log_2(1 + aG_{ij}^k p_i^k) - R_{ij}$ is the maximizer of $\sum_{k \in \mathbf{C}_i} W \log_2(1 + aG_{ij}^k p_i^k)$ subject to the constraint $-\overline{p}_i + \sum_{k \in \mathbf{C}_i} p_i^k \le 0$.

Proposition 7.5 (Optimum Power Allocation). *Suppose that* (P3) *is feasible. For any given MC* i*, the elements of the optimal power allocation vector* $\mathbf{p}_i^* = (p_i^{k*})^T$ *for* (P3) *are given by*

$$
p_i^{k*} = \left(\frac{1}{|\mathbf{C}_i|} \left\{ \overline{p}_i + \sum_{l \in \mathbf{C}_i} \frac{1}{aG_{ij}^l} \right\} - \frac{1}{aG_{ij}^k} \right)^+,
$$

where $x^+ = \max(0, x)$.

Proof. See Exercise. □

Note that the existence of the optimum power allocation given by Proposition 7.4 does not imply the feasibility of (P3). We establish the feasibility of (P3) as follows.

Corollary 7.1 (Feasibility Criterion for (P3)). *For any given MC i, (P3) is feasible if and only if*

$$\sum_{j \in \mathbf{A}_i} \frac{R_{ij}}{\sum_{k \in \mathbf{C}_i} W \log_2(1 + aG_{ij}^k p_i^{k*})} \leq 1,$$

where p_i^* is the optimal power allocation vector.

Proof. With the optimal power allocation vector p_i^*, the rest of constraints above result in this criterion. □

Proposition 7.6 (Optimum Time Allocation). *Suppose that (P3) is feasible. For any given MC i, the elements of the optimal time allocation vector $\tau^* = (\tau_j^*)^T$ for (P3) are given by*

$$\tau_j^* = \frac{1}{|\mathbf{A}_i|} \cdot \left\{ 1 - \sum_{l \in \mathbf{A}_i} \frac{R_{il}}{\sum_{k \in \mathbf{C}_i} W \log_2(1 + aG_{ij}^k p_i^{k*})} \right\} + \frac{R_{ij}}{\sum_{k \in \mathbf{C}_i} W \log_2(1 + aG_{ij}^k p_i^{k*})}.$$

Proof. See Exercise. □

Based on the above properties of (P3), we summarize the exact solution algorithm for the problem in Figure 7.4.

Step 1: Find an optimum power vector

$$p_i^* = (p_i^{1*}, \ldots, p_i^{|C_i|*})^T;$$

where $p_i^{k*} = \left\{ \overline{p}_i + \sum_{l \in C_i} (aG_{ij}^l)^{-1} \right\} / C_i - (aG_{ij}^k)^{-1}$;

Step 2: Check the feasibility

$$\text{If } \sum_{j \in \mathbf{A}_i} \frac{R_{ij}}{\sum_{k \in \mathbf{C}_i} W \log_2(1 + aG_{ij}^k p_i^{k*})} > 1,$$

then (P3) is not feasible (Stop);

Step 3: Find an optimum time-fraction vector

$$\frac{1}{\lambda^*} = \frac{1}{|\mathbf{A}_i|} \left\{ 1 - \sum_{l \in \mathbf{A}_i} \frac{R_{il}}{\sum_{k \in \mathbf{C}_i} W \log_2(1 + aG_{ij}^k p_i^{k*})} \right\};$$

$$\tau_j^* = \frac{1}{\lambda^*} + \frac{R_{ij}}{\sum_{k \in \mathbf{C}_i} W \log_2(1 + aG_{ij}^k p_i^{k*})};$$

Fig. 7.4. An exact algorithm for (P3) performed by MC i.

7.2.3.5. *Performance evaluations*

To evaluate the performance of the proposed schemes, we consider network scenarios with two MCs and multiple MCs in simulations. Two different optimization criteria (end-to-end maximal rate and generalized proportional fairness) are compared. Here, the end-to-end maximal rate scheme is to maximize the overall end-to-end rate with the same constraint of (P1),[13] whereas the generalized proportional fairness denotes the fair scheme proposed in this section. We simulated an OFDMA WMN with 128 subcarriers over a 3.2-MHz band, where all parameters except those related to single-hopping are the same as those used in Ref. 13. Each transmitter is synchronized with respect to the receiver's clock reference to make the tones orthogonal. The maximal power at MC i is $\overline{p}_i = 50\,\text{mW}$, the thermal noise power is $\sigma^2 = 10^{-11}$ W, and two desired bit-error rates (*BERs*): 10^{-2} and 10^{-5} are considered. The propagation path loss exponent is three.

In the two MC scenario, the three nodes are located on a straight line with the MR. The distance between MC 1 and the MR is fixed at $D_1 = 10\,\text{m}$, whereas D_2, the distance between MC 1 and MC 2, varies from 110 to 300 m. Each MC requires a minimum rate of $\overline{R}_i = 100\,\text{Kbps}\ \forall\ i$. For $i, j \in \mathbf{N} = \{0, 1, 2\}$, the routing information is fixed as: $y_{10} = y_{21} = 1$; $y_{ij} = 0$ otherwise. The routing information in the reverse direction is $(y_{ij})^T$.

In Figure 7.6, the end-to-end rates of both MCs for the proposed scheme and maximal rate scheme are shown versus D_2. The end-to-end rate of each MC includes only traffic generated by that MC, but not forwarded traffic from adjacent MCs. For the maximal rate scheme, the MCs closer to the MR has a higher rate and the rate increases as D_2 increases. This is because allocating as much resources as possible to the MC that has a better channel gain increases the overall rate. For the proposed scheme, the difference between the rates of two MCs is smaller than the difference observed in the maximal rate scheme. However, the difference increases

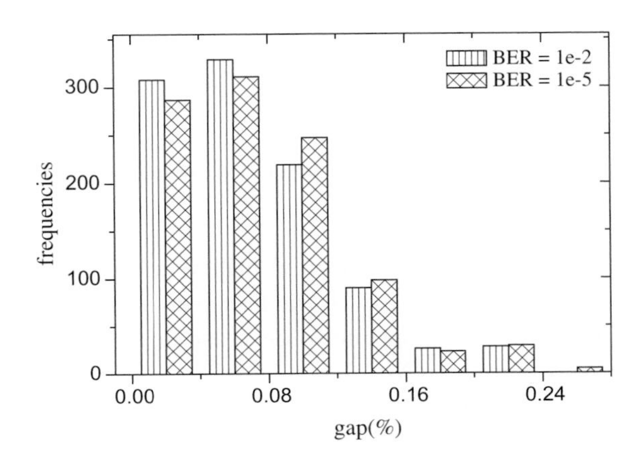

Fig. 7.5. Gap performance of the proposed algorithm for (P1). Gap is less than 0.2% (0.002) for 96.6% of randomly configured 2000 test problems for BER $= 10^{-2}$, 10^{-5}.

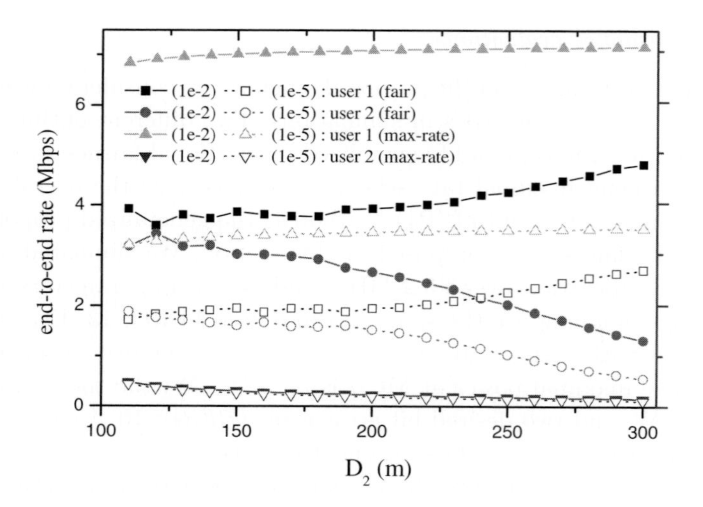

Fig. 7.6. Each user's end-to-end rate (Mbps).

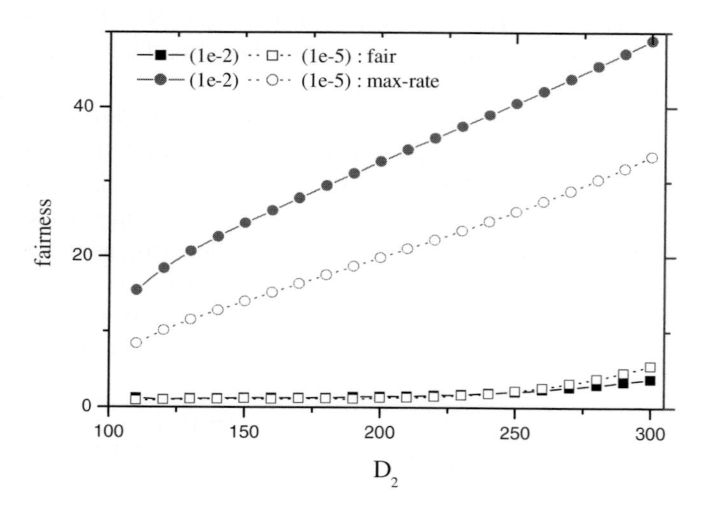

Fig. 7.7. Fairness comparison. In this two-MC simulation setup, the end-to-end rate of MC 1 is $r_1 - r_2$ whereas that of MC 2 is r_2.

as D_2 increases. This is because the marginal utility in the case of allocating more resources is decreasing in accordance with the fast decrease of channel gain with D_2, i.e. the fast decrease of channel gain due to decrease in D_2 cannot be compensated with allocating more resources to MC 2.

In addition, the fairness ratio between the two nodes' end-to-end rates is shown in Figure 7.7, where the end-to-end rates of MC 1 and MC 2 are $r_1 - r_2$ and r_2, respectively, and the fairness ratio is given by $(r_1 - r_2)/r_2$. For the maximal rate scheme, it is observed that the fairness ratio changes greatly at the rate of about +3.1 per ten meters as D_2 increases and that the ratio is much greater than unity. This shows that the maximal rate scheme is very unfair. Even though allocating resources as much as possible to the MC that has a better channel gain increases the overall rate, this results in the MC with a better channel gain being dominant in resource usage, causing the other MC(s) to starve for resource. For the proposed scheme, the ratio $(r_1 - r_2)/r_2$ changes very slowly at the rate of changes about +0.027 per ten meters, which is less than 1% of the rate of change in the maximal rate scheme. This shows the much better fairness of the proposed resource allocation scheme.

Figure 7.8 shows the overall (total) end-to-end rate of the two MCs for the maximal rate and proposed schemes versus D_2. It is observed that the maximal rate scheme achieves an almost constant overall rate. This means that the scheme does not care for MC 2 even though the MC starves for resource under bad channel conditions, which shows that the maximal rate scheme is very unfair. In contrast, the proposed scheme trades off the overall rate for fair support of MC 2 under bad channel conditions.

In Figure 7.9, the rates of both MCs for the proposed and maximal rate schemes are shown versus D_2. The rate of each MC includes both traffic generated by the MC and the forwarded traffic coming from neighboring MCs. For the maximal rate scheme, the MC closer to the MR has a higher rate, and the difference increases, with an upper bound, as the distance D_2 increases. For the proposed scheme, however, the rate of MC 1 decreases as the distance of MC 2 from the MR D_2 increases. That is, the rates of both MC 1 and MC 2 decrease as the average channel gain decreases.

In Figure 7.10, we show the overall (total) rate of the two MCs for the two schemes versus D_2. Note that the rate shown in this figure includes all the rates in

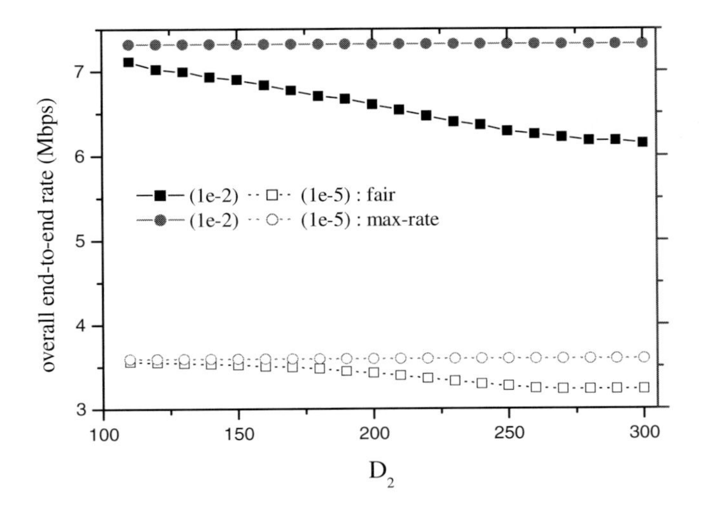

Fig. 7.8. Overall end-to-end rate (Mbps) versus D2.

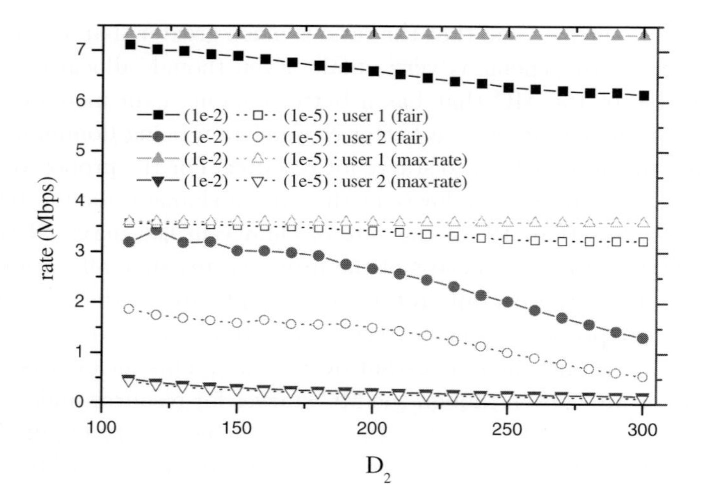

Fig. 7.9. Each user's rate (Mbps) versus D2.

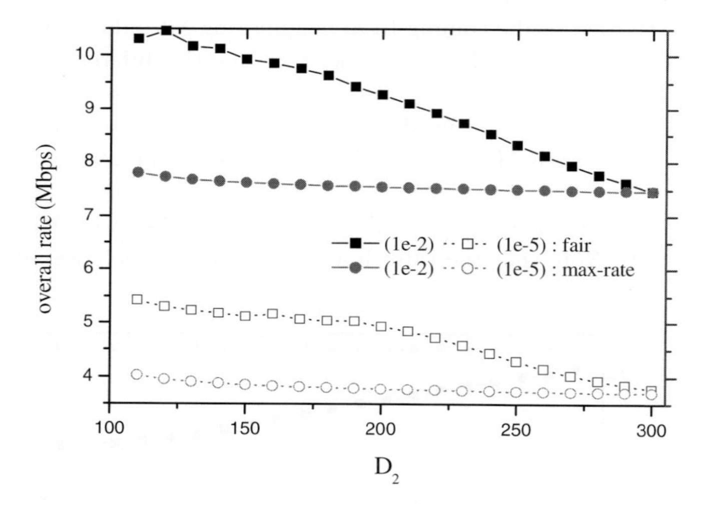

Fig. 7.10. Overall rate (Mbps) $r_1 + r_2$.

all the links. For example, suppose that 20 bps is allocated to link $(1, 0)$ and 10 bps is to link $(2, 1)$. Then the overall rate mentioned here is $20 + 10 = 30$ bps, which is not necessarily equal to the overall end-to-end rate. In the figure, we observe that the overall rate of the proposed scheme is greater than that of the maximal rate scheme. Note that the maximal rate scheme does not maximize the overall rate but it maximizes the overall end-to-end rate instead. This shows that a greater overall rate does not always give a greater overall end-to-end rate in WMNs. Thus, the overall rate is not a performance metric that has significant implications in WMNs.

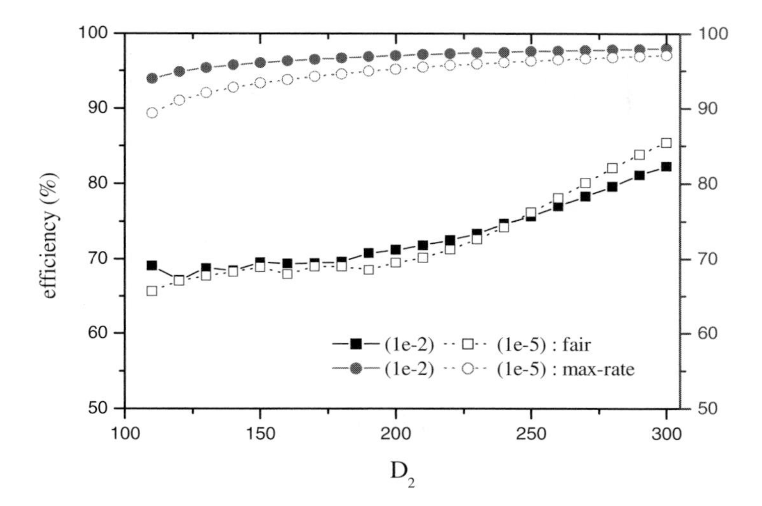

Fig. 7.11. Efficiency of resource utilization (%). The fraction of overall end-to-end rates out of overall rates.

Here, we just consider this metric to show the efficiency of the schemes. In this section, we define the efficiency as the relative ratio of the overall end-to-end rate to the overall rate that a network consumes. Figure 7.11 shows the efficiency of two schemes versus D_2, where we observe that the efficiency increases as D_2 increases. This is because the rate of MC 2 gets smaller as it gets farther from MC 1, making the rate of decrease in the overall rate greater than the rate of decrease in the overall end-to-end rate.

Next we set up the simulations with a total of 19 MCs to evaluate the proposed scheme at $BER = 10^{-2}$. We consider a circular cell with a 150 m radius. The MR is located at the center of the cell, 6 MCs (called *near MCs*) are randomly located with the cell and the other 12 MCs (called *far MCs*) are randomly located within the co-center circle of radius 300 m but outside the cell so that the 12 far MCs cannot access the MR without relaying. Also, the far MCs may not access the MR if there are no relaying MCs near themselves. To evaluate the schemes in a multi-hop environment, the routing matrices for the respective layouts of MCs are chosen to form minimum spanning trees with the MR at the root. The other settings are the same as those of the two-MC scenarios.

Figure 7.12 shows the overall end-to-end rate versus the number of MCs connected to the MR (regardless of the number of hops). It is observed that the two schemes have better performance when the number of connected-MCs increases. This is due to multiuser diversity. The performance improvement decreases gradually. The proposed scheme has comparable performance to that of the maximal rate scheme. Also, the performance gap between the proposed scheme and the maximal rate scheme decreases when the number of connected-users increases. This is

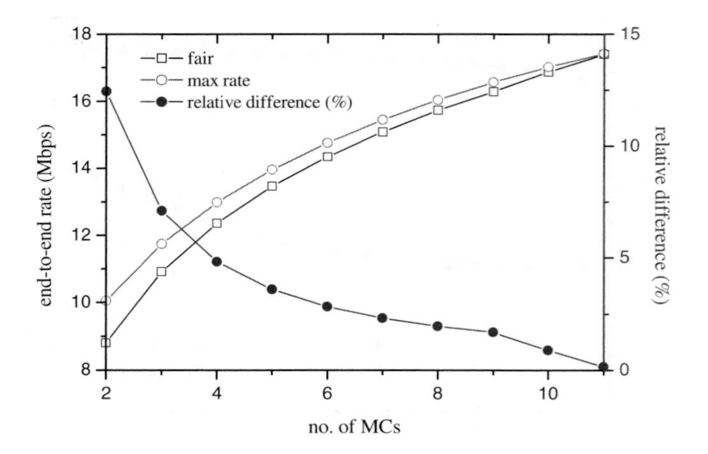

Fig. 7.12. Overall end-to-end rate (Mbps) versus number of MC's.

because both schemes search for feasible solutions which do not violate the constraint that the allocated rate must be greater than or equal to the minimum rate requirement, $r_i \leq R_i$. As the number of connected-MCs increases, the feasible region of rate allocation decreases and, finally, it converges to a single point, $\mathbf{r}^* = \mathbf{R}^*$. If there are further MCs trying to join the WMN, then the two problems (maximal rate problem and fairness problem) become infeasible. This demonstrates that the maximal rate scheme and the proposed scheme go to the same rate allocation vector $\mathbf{r}^* = \mathbf{R}^*$ as far as the problems are feasible, making the gap converge to zero.

7.2.3.6. *Summary*

In this section, we have proposed a new fair-scheduling scheme for wireless mesh networks using orthogonal frequency-division multiple access (OFDMA), with the objective of fairly allocating subcarriers and power so that the NBS fairness criterion is maximized. Instead of solving a single global control problem, the subcarrier and power allocation problem has been decoupled into two subproblems that can be solved hierarchically in a distributed manner. This decoupling is useful and necessary because not all the information necessary for centralized scheduling is available at the mesh router in a wireless mesh network. By decoupling the global control problem, we have proposed a distributed fair-scheduling scheme, where the mesh router solves only the subcarrier allocation problem online based on the limited available information, and each mesh client solves the reduced subcarrier and power allocation problem based on the limited available information online. We have formulated the two problems into non-linear integer programming problems. Also, we have developed simple and efficient solution algorithms for the mesh router's problem and developed a closed form solution by transforming the mesh client's problem into a time-division scheduling problem. Extensive simulation results have demonstrated that the proposed scheme provides fair opportunities to the mesh clients and an overall end-to-end rate comparable to maximal rate scheduling when the number of mesh clients increases.

7.3. Energy Efficient Techniques

7.3.1. *Case study on energy efficiency of cooperative relaying*

The complexity in wireless networks is increasing as the subscriber nodes or source nodes (SNs) become increasingly smaller in size and numerous in number. However, the large number of nodes will enable such networks to benefit from space diversity, or multi-antenna diversity. The advantages of multiple-input multiple-output (MIMO) systems have been intensively studied recently. Even though space diversity is known advantageous, it may not be feasible in practical cellular and/or ad hoc networks to have multiple antennas installed in a very small node because of size and power limitations.[22–24] To overcome this problem and to reap some of the benefits introduced by MIMO systems, the concept of cooperative communications has been introduced and several cooperative diversity techniques have been studied.[22–29]

The benefits of cooperative communications result from *cooperative diversity*,[25,26] which can be achieved by *relaying*. Although both cooperative communication networks and multi-hop networks employ data relaying by network nodes, the former are distinguished from the latter through the use of cooperative diversity realized from multiple virtual links established in the process of relaying. Cooperative diversity was introduced by Sendonaris *et al.*[22,23] In the papers, the authors implemented a two-user cooperative CDMA cellular system, where both users (or SNs) are active at the same time and use orthogonal codes to avoid multiple access interference. Also, for ad hoc networks, Laneman *et al.* proposed various cooperative diversity protocols and analyzed their performance in terms of outage probability.[25,26]

The objective is to evaluate the performance of a cooperative cellular network with two heterogeneous classes of nodes, namely SNs and relay agents (RAs), focusing on the average achievable rate and power consumption.[1] When a SN is connected to the base station (BS) through one or more RAs, the BS may receive the signals from both the cooperating RAs and the SN, resulting in a performance gain via cooperative diversity. In most previous studies, the characteristics of nodes are assumed identical, or homogeneous, and it is considered that nodes act as both sources and relays. However, in practice, different kinds of nodes with diverse mobility characteristics, battery lifetime, and so on, may exist in the network. For example, a portable handset has a relatively short battery lifetime whereas a radio transceiver onboard a vehicle has a much longer battery lifetime.

It is commonly considered that a SN needs to be small in size (e.g. it is preferred to have as small a handset as possible). However, it is not necessary that RAs should be small in size. As mentioned above, this is because we may have different kinds of

RAs such as dedicated devices that have little concern about power consumption. Such RAs can be installed in vehicles, as mobile relays, or on top of buildings, as fixed relays. Assuming that RAs are available in a cellular network, the SNs can potentially save a lot of power because they can take advantage of cooperative diversity offered by RAs to reduce their own transmit power, and they need not act as relays. This heterogeneous cooperative cellular architecture is not only useful to improve the average achievable rate and reduce power consumption at SNs, but the reduction of SN power emission may also offer benefits in allaying health concerns with regard to electromagnetic radiations, and reducing probability of interception of signals by unintended receivers (e.g. in military applications).

Figure 7.13 illustrates the proposed cooperative cellular network architecture with the two types of nodes mentioned above. Based on this network architecture, we propose a dual agent relaying mechanism (DARM) in which up to two RAs, when available, are employed in parallel for cooperative relaying of data from the source node. While additional RAs could be employed as first relays, with the minimum number of first relays that enable cooperative diversity, DARM gives a performance lower bound for cooperative diversity relaying with heterogeneous nodes.

In DARM, we consider no more than two parallel RAs as the first relays from the SN. However, the first relays may each employ multiple RAs in the remaining multiple paths to the destination. We present analytical results obtained under the worst case condition that only the parallel first RAs exist between the SN and the destination BS, second RAs are not available between the first RAs and the BS. Results show that DARM can increase the rate by 10–80% in most cases of practical population densities, compared to the baseline scenario where no more than one RA is used for relaying a SN's data to the BS. DARM also achieves a reduction in power consumption up to 15–30% compared to the baseline scenario. In most cases, the probability that an arbitrary SN finds at least two RAs for relaying, and this probability is nearly constant relative to the population density or the ratio between the transmission range of a SN and the cell area.

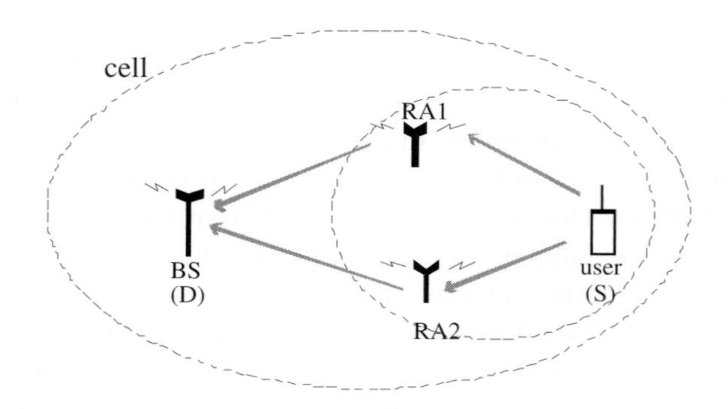

Fig. 7.13. Cooperative relaying architecture with dual agents.

7.3.1.1. *The proposed relaying mechanism*

We consider a cellular network employing orthogonal frequency division multiple access (OFDMA) over the air interface. OFDMA is one of the most promising transmission technologies that are gaining popularity in wireless networks.[13] In each cell, we consider uplink transmissions from the SNs to the BS via RAs acting as relays if available. Also, we consider that each transmitter (BS or RA) uses orthogonal code to reduce the interference level and we assume that inter-cell interference is negligible. In cooperative networks, it is usually assumed that SNs and RAs have full channel state information (CSI).[30,31] We consider a series of time frames where each frame consists of two timeslots: the SN can transmit in the first timeslot whereas the relay(s) can forward in the second timeslot.[25,29] When a SN has two relays for DARM, the two relays are time-synchronized with respect to their receiver.

Figure 7.14 illustrates DARM for the aforementioned relaying architecture with two heterogeneous classes of nodes. Consider a user (SN) as shown in the figure. In the figure, the transmission range of this SN is shown. If there exists at least one RA in the transmission range, the SN may request the RA for relaying. In this case, the BS needs to receive signals from both the SN and the RA to achieve cooperative diversity. To do this, the SN needs to maintain its transmission power level comparable to the level in the case that he/she is connected to the BS without relaying.

On the other hand, if there are multiple RAs in the transmission range, the SN selects two best RAs, where the criterion can be based on signal-to-noise ratio (SNR) or signal-to-interference ratio (SIR), and requests them for parallel relaying.[27] In the case of parallel relaying with two RAs, the SN can reap the benefits of cooperative diversity big enough even though the two agents do not have any other relaying agents in the residual path toward the destination. It is possible that the SN has more than two RAs in the first stage of relaying (i.e. the number of first RAs is greater than two) and that the SN has more RAs in the next stage of relaying (i.e. the number of second RAs is greater than zero). As studied in Ref. 32, these types of relaying are known to be better in terms of capacity but there is a tradeoff between capacity gain and computational burden: the more relays are involved in relaying, the more signaling and control burdens the network nodes may have. Investigation on the tradeoff is an interesting further work but is not the major focus of this section. The major objective of this study is to analyze how much gains we can achieve through the use of the cooperation diversity with heterogeneous types of nodes in the relaying architecture.

The procedure of DARM is detailed in Figure 7.14. If a SN needs a new action, such as handover or new call setup requests, then the SN checks both the pilot signal(s) coming from the associated BS(s) and the pilot signal(s) coming from neighboring RA(s), including the channel gain information between the respective RA(s) and the associated BS.

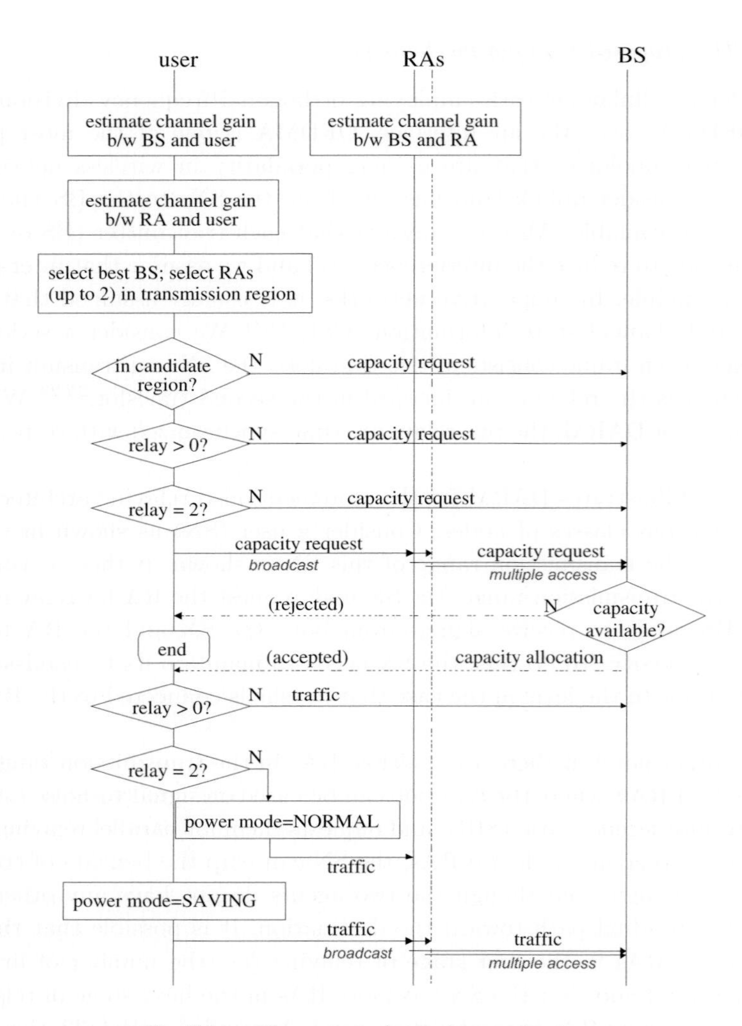

Fig. 7.14. The proposed relaying mechanism.

BSs periodically broadcast the pilot signal so that SNs and RAs can estimate their channel gain between them and the associated BS.[c] RAs periodically broadcast the pilot signal so that SNs can estimate the channel gain between them and the neighboring RAs. Also, RAs can send out the channel gain information between them and the associated BS. Each SN takes advantage of the information to make decisions required in the relaying architecture.

If the distance between a SN and its best BS is small enough, then the SN may be connected to the BS directly without any relaying. The cell region excluding

[c]Developing the estimation techniques can be referred to the literature but is not the focus of this section.

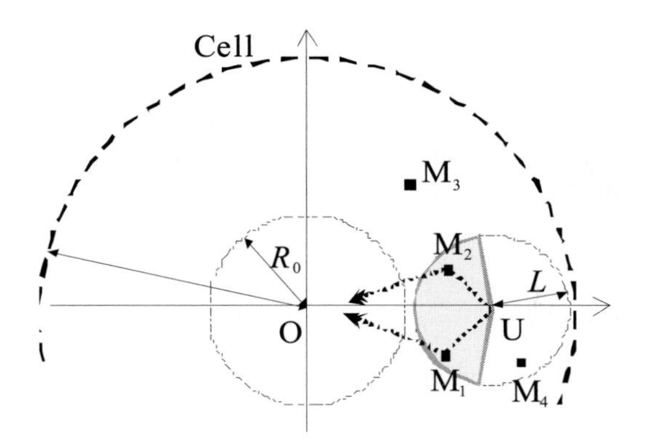

Fig. 7.15. Geometry for analytical evaluation.

such a region is called candidate region in this analysis. For example, the region $\{(r,\theta)|R_0 \leq r \leq R,\ 0 \leq \theta < 2\pi\}$ denotes the candidate region (see Figure 7.15).

The SN selects the best BS and best two RAs (selection is completed by handshaking). If RAs are unavailable (i.e. *relay* $= 0$[d]), the SN sends capacity request (CR)[e] to the BS directly. Otherwise, the SN checks if two RAs are available. If one RA is available (i.e. *relay* $= 1$), the SN sends CR to the BS. Otherwise (i.e. *relay* $= 2$), the SN sends CR to both RAs (broadcast), and the RAs deliver the CR to the destination BS (multiple access).

Once the BS received the CR,[f] the BS checks if the required capacity is available according to the admission control policy it uses. If there is no capacity available for the SN that has originated the CR, then the BS rejects the request. Otherwise, the SN is allocated a certain amount of capacity and starts transmitting information in accordance with the relay mode. Before transmitting information through the use of cooperative diversity, the SN can optimize the transmission power level accordingly. That is to say, if a SN can transmit information using a cooperative diversity technique, it is reasonable that the SN should optimize/adjust the transmit power level because there is no need to consume the same level of transmit power.

The transmission power of a SN needs to be continuously adjusted according to channel conditions. However, the transmission power of a SN in DARM is roughly divided into two modes: NORMAL mode and SAVING mode. In the event of parallel dual agent relaying (i.e. *relay* $= 2$), the SN needs not transmit information at the normal power level (NORMAL mode). This is because cooperative diversity can be achieved with two parallel links if two agents are relaying in parallel (i.e. *relay* $= 2$). In this case, the SN may reduce its power level as far as its signal may reach both

[d]The parameter "relay" is defined as the number of selected RAs. The possible values of "relay" are 0, 1, and 2.
[e]Every CR message includes a data field to indicate the priority level. For example, the BS knows if a CR is intended for a handover or a new call setup.
[f]A duplicated copy of the same CR is discarded.

RAs (SAVING mode). However, in the case of no parallel dual agent relaying (i.e. *relay* < 2), it is necessary for the SN to transmit information at the normal power level so that its signal can reach the BS directly. In the case of single relaying (i.e. *relay* $= 1$), the BS can experience cooperative diversity if the BS receives the SN's signal directly. In the case of no relaying (*relay* $= 0$), it is imperative for the BS to receive the SN's signal directly.

7.3.1.2. *Analytical model*

In this section, we build an analytic model to evaluate the performance of DARM. We assume that the cell is a circle of radius R. Even though the shape of an actual cell is irregular, a circle-shaped cell is commonly used in analytical methods for performance evaluation.

We consider Rayleigh fading and, for analysis purposes, we assume that the channel gain with respect to the distance between transmitter (e.g. SN) and receiver (e.g. BS) is known. We consider a BS and a SN located at $O(0,0)$ and at point $U(a,0)$, respectively, in the polar coordinate system as shown in Figure 7.15. In the figure, we can find two co-center circles with radii R and R_0, respectively, and a circle with radius L. The circle with radius R denotes the cell. If a SN is located within the region defined by $\{(r,\theta)\,|\,R_0 \le r \le R,\ 0 \le \theta < 2\pi\}$, then the SN needs not be served by relaying agents because the power consumption in the region is small enough. The SN tries to find RAs within its transmission range to make relayed connections to the base station. For example, in the figure, RAs M_1 and M_2 located within the transmission range can make two parallel virtual links for the SN.

7.3.1.3. *Evaluation of the average achievable rate*

In this section, we analyze (the upper bound of) the average achievable rate per subcarrier for an arbitrary SN. Even though RAs may experience multiple relaying toward the destination BS as mentioned (see Figure 7.13), we analyze the average transmit rate in the worst case that the first RA(s) from the source (e.g. RA 1 and RA 2 in the same figure) do not have next RAs to the destination. We compare the achievable rate when DARM is not used with that by using DARM. To compare these two rates, we consider the same transmit power level of the SN.

7.3.1.3.1. The case without the proposed mechanism

Consider a SN located a distant from the base station located at the origin $O(0,0)$. Suppose that DARM is not used (single relaying or no relaying), then the average rate per subcarrier of bandwidth W at power level p is given by

$$C_0(a) \le I_0(a) = W \cdot \log_2\{1 + S_0(pG(a))\}, \quad 0 \le a \le R$$

where $S_0(\cdot)$ denotes the scaled SNR[g] in terms of transmit power and channel gain and is given by

$$S_0(x) = \kappa x,$$

and $G(a)$ is the channel gain for a given distance a. Here, $\kappa \approx -1.5/(\sigma^2 \cdot \log(5 \cdot BER))$ for MQAM[13] (σ^2 is the thermal noise power for each subcarrier, and BER is the desired bit-error rate). Thus, the average rate per subcarrier of an arbitrary SN without using DARM is given by

$$C_{no} \leq I_{no} = \int_0^R I_0(a) \cdot \frac{2a}{R^2} da.$$

7.3.1.3.2. The case with the proposed mechanism

First, let us consider the average achievable rate for a SN located within the candidate region $\{(r,\theta) \,|\, R_0 \leq r \leq R, 0 \leq \theta < 2\pi\}$. Without loss of generality, we can assume that the SN's location is $U(a, 0)$. For this SN, the average achievable rate per subcarrier of bandwidth W at power p when an RA, which is relaying for this SN, located at (r,θ)[h] with respect to the location of the SN is given by

$$C_1(r,\theta; a) \leq I_1(r,\theta; a) = \frac{1}{2} W \log_2\{1 + S_1(pG(a), p_1 G(d(r,\theta)))\},$$

where p_i denotes the transmit power level of the ith relaying agent, $i = 1, 2$, and $d(r,\theta)$ denotes the distance between the BS and RA and is given by

$$d(r,\theta) = \sqrt{a^2 + r^2 + 2ar\cos\theta},$$

and $S_1(\cdot,\cdot)$ denotes the (scaled) SNR in terms of transmit powers and channel gains when an RA is involved in cooperation and is characterized by the associated cooperative diversity protocol.[25]

For the transmission range, the average achievable rate per subcarrier at power p is given by

$$C_1(a) \leq I_1(a) = \int_0^L \int_{\theta_L}^{\theta_U} I_1(r,\theta; a) \frac{r}{A_e(a)} d\theta dr,$$

where $A_e(a)$[1] is the area of the transmission range of the SN located at (a, θ), $0 \leq \theta < 2$ and

$$(\theta_L, \theta_U) = \left(\pi - \cos^{-1}\frac{L}{a},\ \pi + \cos^{-1}\frac{L}{a}\right).$$

[g]This means SNR multiplied by the square of the associated fading coefficient.
[h]The RA's location is equal to $(a + r\cos\theta, r\sin\theta)$ in the Cartesian coordinate.

Next, let us consider the same case as above but there are two relaying agents with a SN. With the same coordinate of the reference SN above, we denote the locations of two relaying agents as (r_1, θ_1) and (r_2, θ_2), respectively, with respect to the coordinate of the SN. Then the average achievable rate can be expressed as

$$\begin{aligned}
&C_2(r_1, \theta_1, r_2, \theta_2; a) \\
&\quad \leq I_2(r_1, \theta_1, r_2, \theta_2; a) \\
&\quad = \frac{1}{2} W \log_2\{1 + S_2(pG(a), p_1G(d(r_1, \theta_1)), p_2G(d(r_2, \theta_2)))\},
\end{aligned}$$

where $S_2(\cdot, \cdot, \cdot)$ denotes the (scaled) SNR in terms of transmit powers and channel gains when two RAs are involved in cooperation and is characterized by the associated cooperative diversity protocol.[25] By unconditioning, we can obtain the average achievable rate per subcarrier at transmit power p of the SN as

$$\begin{aligned}
&C_2(a) \leq I_2(a) \\
&\quad = \int_0^L \int_0^L \int_{\theta_L}^{\theta_U} \int_{\theta_L}^{\theta_U} I_2(r_1, \theta_1, r_2, \theta_2; a) \frac{r_1 r_2}{A_e(a)^2} d\theta_2 d\theta_1 dr_2 dr_1.
\end{aligned}$$

Finally, let us consider the average achievable rate for a SN located within the region $\{(r, \theta) \,|\, 0 \leq r < R_0,\ 0 \leq \theta < 2\pi\}$. As mentioned above, since these SNs do not make use of relaying agents, the average rate per subcarrier at power p is given by

$$C_0(a) \leq I_0(a) = W \cdot \log_2\{1 + S_0(pG(a))\}, \quad 0 \leq a \leq R_0.$$

When DARM is used with the transmit power of each relaying agent approximately equal to that of the SN (i.e. $p_i \approx p$), the average achievable rate of a SN whose location is (a, θ), $0 \leq \theta < 2\pi$, is given by

$$\begin{aligned}
\overline{C}(a) &\equiv \phi_0 C_0(a) + \phi_1 C_1(a) + (1 - \phi_1 - \phi_2) C_2(a) \\
&\leq \overline{I}(a) \equiv \phi_0 I_0(a) + \phi_1 I_1(a) + (1 - \phi_1 - \phi_2) I_2(a),
\end{aligned}$$

where ϕ_k denotes the limiting probability that an arbitrary SN finds k RAs in its transmission range.

By unconditioning with respect to a over the candidate region, we can obtain the average achievable rate per subcarrier of a SN residing in the candidate region as

$$I = \int_{R_0}^{R} \overline{I}(a) \frac{2a}{R^2 - R_0^2} da,$$

and that of a SN residing in the inner co-center circle is given by

$$I_0 = \int_0^{R_0} I_0(a) \frac{2a}{R_0^2} da.$$

Finally, the average achievable rate per subcarrier of an arbitrary SN is given by

$$\overline{I} = I \cdot \left(1 - \frac{R_0^2}{R^2}\right) + I_0 \cdot \frac{R_0^2}{R^2}.$$

The (relative) improvement in the average achievable rate per subcarrier is obtained by

$$f_c \equiv \frac{\overline{I} - I_{no}}{I_{no}},$$

where f_c is greater than zero if DARM increases the average achievable rate. This quantity means the ratio of instantaneous rates in the time domain. By taking integration of the respective portion of rates with respect to the associated fading coefficient squares, we can obtain the relative improvement averaged in the time domain. The following proposition will show the quantity of instantaneous relative improvement is greater than or equal to zero, which implies the time-average improvement is also greater than or equal to zero.

Proposition 7.7 (Improvement in the Average Achievable Rate Upper-Bound). *For any three or four nodes within a single set of cooperation that achieves $I_0 \leq I_1 \leq I_2$, the maximum achievable rate by using DARM is not less than that without using DARM, i.e. $f_c \leq 0$.*

Proof. See Exercise. □

Remark 7.1. The improvement of an upper bound of a reference metric (i.e. the average achievable rate) does not necessarily imply the improvement of the reference metric. In our case, we can extend the proof in terms of the improvement of the reference metric by considering Rayleigh fading. However, it yields the same result. Instead of tedious procedural proof, we show numerical examples for wide range of affecting parameter values in the following subsection.

7.3.1.3.3. Numerical results

We consider the OFDMA cellular system with 128 subcarriers over the 3.2-MHz band. Each transmitter is synchronized with respect to the receiver's clock reference to make the tones orthogonal. The transmission power of a SN per subcarrier is $p = 50\,\text{mW}$, the thermal noise power is $\sigma^2 = 10^{-10}\,\text{W}$, and the desired $BER = 10^{-3}$. The path loss exponent α is four.

In Figure 7.16, the average achievable rates per subcarrier versus the fraction of the transmission range area and the cell area, denoted by f_e. Here, it is observed that the average achievable rate of DARM increases when the area fraction of the transmission range and the cell, f_e, increases. In the figure, it is observed that the average achievable rate is improved by more than 20% when the area fraction f_e is greater than $10^{-6}\%$ ($= 10^{-8}$) and that the improvement is approximately 100%

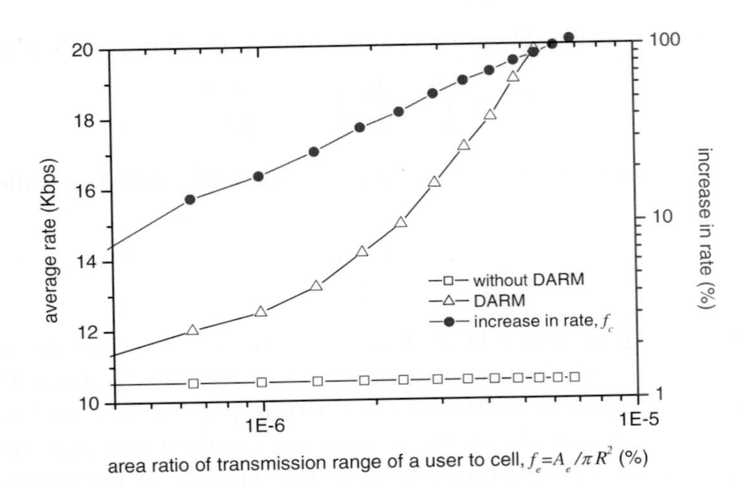

Fig. 7.16. The average achievable rate of a SN per subcarrier versus the area ratio of transmission rate of a user to cell.

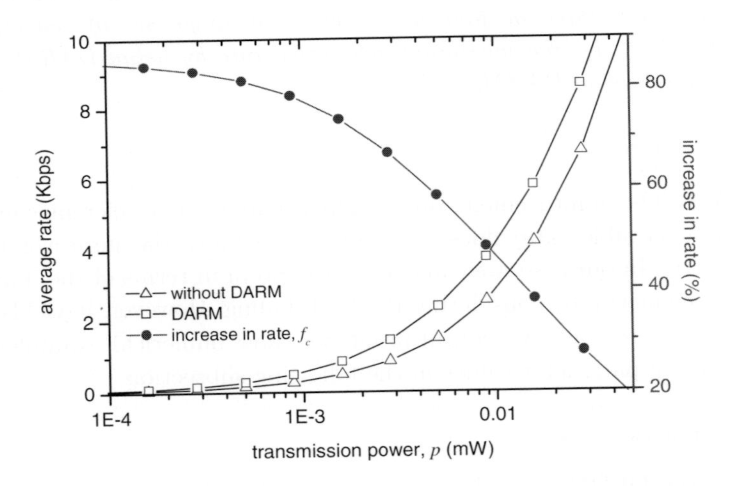

Fig. 7.17. The average achievable rate of a SN per subcarrier versus p.

when $f_e = 6 \cdot 10^{-6}\%$. It is also observed that the relative increase f_c is more than 20% and 60% when $L/R \geq 10$ % and $L/R \geq 20\%$, respectively. This demonstrates that the average achievable rate is remarkably improved by DARM even though the radius of the transmission range L is small relative to the cell radius R.

In Figure 7.17, the average achievable rates per subcarrier versus transmission power are shown. It is observed that the average achievable rates of both mechanisms are increasing with a decreasing speed of increase, i.e. concave increasing,

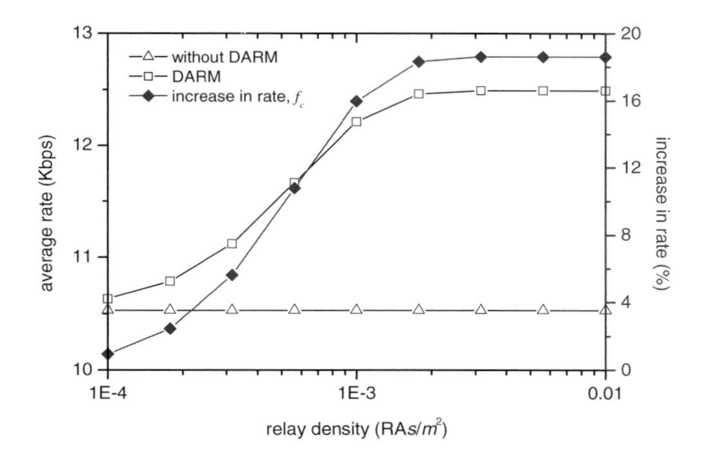

Fig. 7.18. The average achievable rate of a SN per subcarrier versus the RA density.

with respect to power increase. For small transmission power levels, even though the amount of rate increase is small, the relative increase f_c is very large. As the transmission power increases, the relative increase f_c decreases, slowly approaching 0%. However, this does not imply that the net increase converges to zero: this means that the net increase diverges with the speed of its divergence (increase) smaller than that of rate increase due to increasing transmit power.

In Figure 7.18, the effect of the density of RAs in the cell site on the average achievable rates per subcarrier is shown. We test the effect with RA densities ranging from 0.00001 to 0.01 (RAs/m^2). The density of 0.01 is chosen for representing the situation of population density in urban area (for example, the population density in New York City is approximately 0.01/m^2). Under the assumption that everyone has an auto that can act as a relaying agent, this value represents densely populated situations. On the other hand, the value 0.00001 (RAs/m^2) is used for representing the rural population situation. In the figure, it is observed that the average rate of DARM steeply increases as the density of RAs increases and satiates around 0.002 (RAs/m^2). Also, for population densities 0.00032/m^2 and 0.00100/m^2 (suburban areas), the relative increase f_c is 5.6% and 18.6%, respectively. This demonstrates that DARM is much more useful for most cases of population density. Even though the subscription ratio is considered, we believe that the performance has practical meaning for financial and/or shopping districts, where the density is much higher than any other area.

Figure 7.19 shows the trends of $\phi_0 + \phi_1$ versus the radius of the transmission range, the distance between the BS and the SN, and the density of relaying agents, respectively. Here, the quantity $\phi_0 + \phi_1$ denotes the probability that an arbitrary SN finds at most one RA in its transmission range. In this case, the SN cannot be served in parallel by two relaying agents residing in its transmission range. In Figure 7.19, it is observed that the probability $\phi_0 + \phi_1$ decreases rapidly. Note that

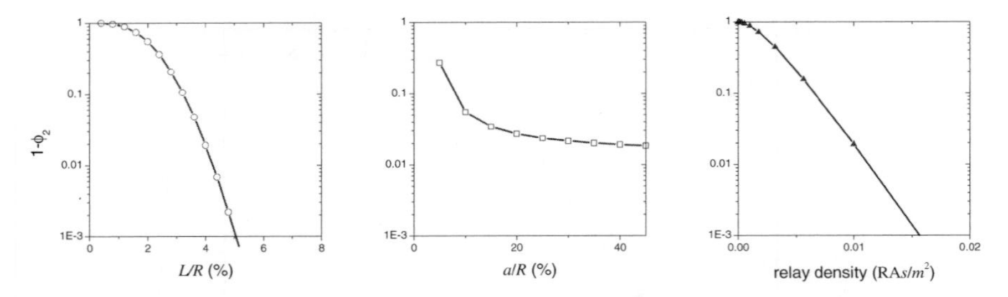

Fig. 7.19. The probability that a SN does not have two RAs versus L/R, a/R, and the density of RA.

population densities of most cities lie in the interval (0.00, 0.02). This demonstrates that the probability that an arbitrary SN is served in parallel by two relaying agents is very high in practical situations.

7.3.1.4. *Evaluation of power consumption*

The average power consumed by an arbitrary SN to experience a given transmission rate x per subcarrier between the user (source) and BS (destination) is analyzed. Here, the consumed power does not include the power that RAs (relays) consume. In the literature,[25,26] it is already known that cooperation diversity may improve the achievable rate as well as the power consumption. According to this fact, it is obvious that the average power consumption over the two types of nodes in our model is not greater than that in the model without using DARM. Since we consider a system with two different types of node where relaying nodes do not have any concerns about power consumption, we focus on the average power consumption of the source node side.

For clarity of the notion of power consumption, we recall the transmission power modes of a SN used in DARM. If a SN is connected to the BS via parallel dual agent relaying (i.e. $relay = 2$), then the SN reduces the transmission power level as much as possible because its signal only has to reach two RAs residing in its transmission range. However, if the SN is connected to the BS via single agent relaying or directly (no relaying), he/she should consume a normal level of transmission power so that its signal can reach the BS directly.

7.3.1.4.1. The case without the proposed mechanism

Let x denote the target transmission rate of x per subcarrier. Then the average power consumed to achieve the target transmission rate per subcarrier is given by

$$P_{no} = \min\{p : I_{no} \geq x\}.$$

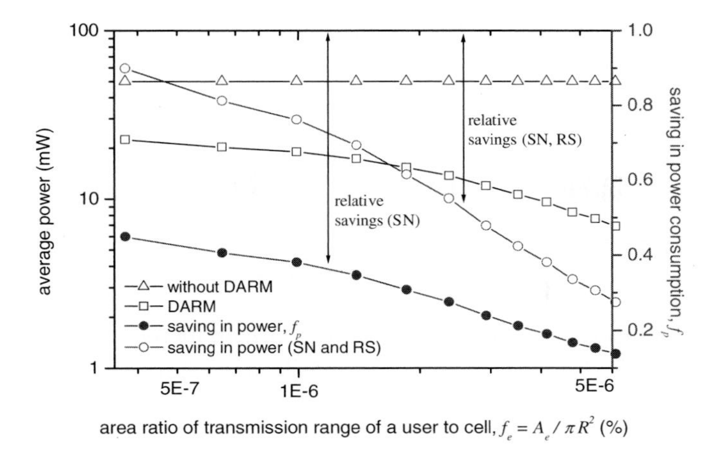

Fig. 7.20. The average power consumption of a SN per subcarrier versus the area ratio.

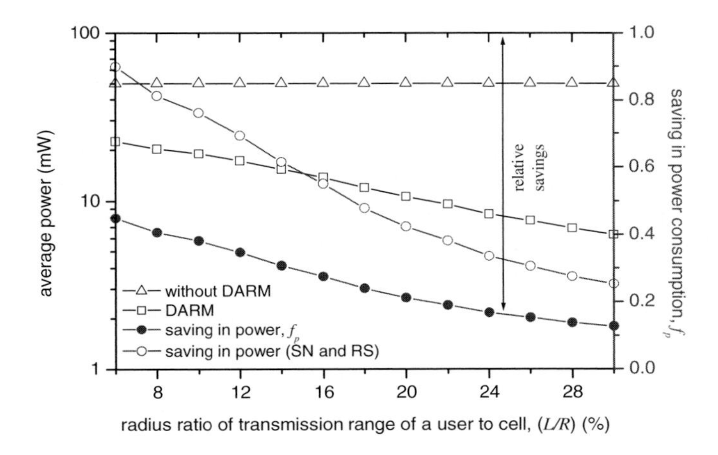

Fig. 7.21. The average power consumption of a SN per subcarrier versus L/R.

7.3.1.4.2. The case with the proposed mechanism

Similarly, when DARM is used, the average power consumed to achieve the target transmission rate per subcarrier is given by

$$\begin{aligned}\overline{P} &= \phi_0 \cdot \min\{p : I_{no} \geq x\} \\ &\quad + \phi_1 \cdot \left\{ \frac{1}{2} \cdot \min\left\{ p : \int_0^R I_1(a)da \geq x \right\} + \frac{1}{2} \cdot 0 \right\} \\ &\quad + (1 - \phi_0 - \phi_1) \cdot \frac{1}{2} \cdot \min\left\{ p : \int_0^R I_2(a)da \geq x \right\}.\end{aligned}$$

The (relative) improvement in the average power consumption per subcarrier is obtained by

$$f_p = \frac{\overline{P}}{P_{no}},$$

where f_p is less than unity if DARM saves the power consumption.

Proposition 7.7 (Power Saving). *The power consumed by a SN per subcarrier by DARM is not less than the power when DARM is not employed, i.e.* $f_p \leq 1$.

Proof [Abbreviated]. Note that $\overline{C}$ and C_{no} are an increasing function of transmit power at the SN. From the previous proposition, we have $\overline{C} \geq C_{no}$ for any given transmission power level p at the SN. Therefore, for the inverse functions, we have $\overline{P} \leq P_{no}$. □

7.3.1.4.3. Numerical results

The same parameters setup shown in former examples is also used in power consumption examples. We consider a various spectrum of target rate. In Figure 7.16, the transmission power consumptions of a SN per subcarrier versus the fraction of the transmission range area and the cell area f_e. The results are plotted again with respect to L/R in the next figure. As defined above, f_p is the relative ratio of the average transmission power level of DARM to that of a baseline scheme without using DARM. Thus, the difference between unity and f_p value denotes the amount of relative savings: for example, for $f_e = 5 \cdot 10^{-6}$, we have $f_p = 0.28$, which means 72% of savings in transmit power.

In the logarithmic scale of f_e, a rapid increase in improvement of power consumption is observed for small L/R. For $L/R > 16\%$, the power saving amount is more than 50%. As shown in the figure, DARM provides an excellent power efficiency (roughly 75% savings when $L/R = 0.3$, which is 4-fold efficient: $75\% = 1 - 1/4$) comparing to the previous mechanism in most practical cases.

Figure 7.23 shows the power consumptions versus the desired BER. A steady improvement in power consumption for the test values of BER is observed. A remarkable reduction in power is observed. The saving in power consumption f_p is approximately 17–30%. Also, it is observed that DARM is more effective to a higher reliability transmission in terms of energy efficiency. Figure 7.22 shows the power consumptions versus required rates. DARM is more effective to a low rate transmission than to high rate transmission in terms of the relative saving in transmit power. In the logarithmic scale of the average transmit power, however, it is observed that the saving in transmit power increases as the required rate (target rate) increases.

Figure 7.24 shows the power consumptions versus the density of RAs in cell. It is observed that the saving in power consumption f_p is rapidly increasing until density approaches 0.001 (see the rapid decrease of f_p in the logarithmic scale of

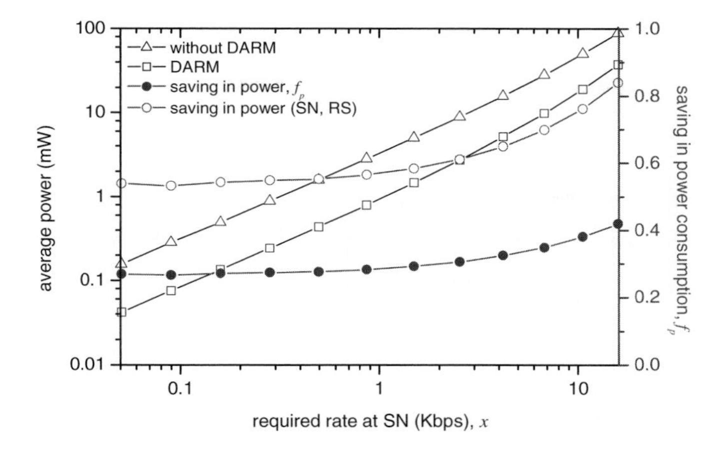

Fig. 7.22. The average power consumption of a SN per subcarrier versus required rate.

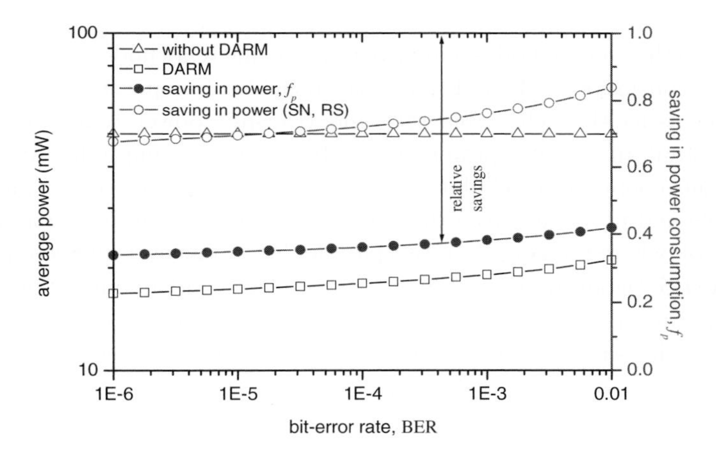

Fig. 7.23. The average power consumption of a SN per subcarrier versus BER.

density). According to the results, the relative saving in power consumption in a site of population density $0.00037/\text{m}^2$, such as in Montreal, is 15% whereas that of density $0.0015/\text{m}^2$, such as in San Diego, is 24%. These results demonstrate that DARM can provide a remarkable power saving effect in suburban and urban areas.

7.3.1.5. *Summary*

In this section, we have examined the average achievable rate and the average power consumption in cooperative cellular networks with two classes of nodes. Unlike previously proposed architecture employing homogeneous nodes that function as both sources and relays, we propose a functional differentiation between source and

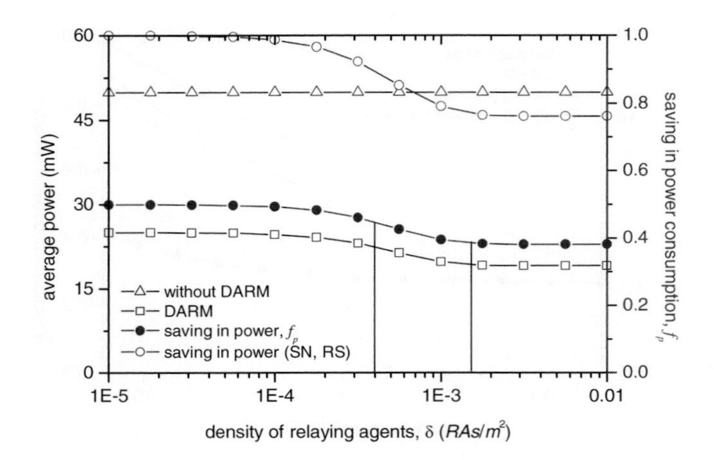

Fig. 7.24. The average power consumption of a SN per subcarrier versus density of RA.

relay nodes so that source nodes may save power while relay nodes have little concern about power consumption. In addition to the benefits of cooperative communication, such as increasing the achievable rate and transmit power reductions, SN's can achieve substantially greater power savings by not performing relay functions. The heterogeneous cooperative cellular network architecture proposed in this section is both feasible and practical. According to analysis results obtained under the worst case condition where no more than two parallel RAs exist over an SN's transmission paths to the BS, the proposed architecture achieves a good amount of increase in the average rate and a good amount of reduction in the average power consumption, compared to the situation where at most one RA is available between each SN and the BS.

Problems

(1) Show that a proportional fair scheduler performs better than a random round-robin scheduler in terms of system throughput.
Hint: See 7.2.1.3.

(2) Consider the following problem.

$$\begin{aligned} \min \quad & f = x_1^2 + x_2^2 \\ s.t. \quad & -x_1 - x_2 + 1 \leq 0 \\ & x_1,\, x_2 \geq 0 \end{aligned}$$

(3) Describe the pros and cons of distributed scheduling in multi-hop relay networks.
Hint: See 7.2.3.1.

(4) Prove Proposition 7.1.
(5) Prove Proposition 7.2.
(6) Prove Proposition 7.5.
(7) Prove Proposition 7.6.
(8) Prove Proposition 7.7.

Bibliography

1. K.-D. Lee and V. C. M. Leung, Evaluations of achievable rate and power consumption in cooperative cellular networks with two classes of nodes, accepted for publication in *IEEE Trans. Veh. Technol.* (May 2007).
2. IEEE 802.16j/D2, Draft Amendment to IEEE Standard for Local and Metropolitan Area Networks — Part 16: Air Interface for Fixed and Mobile Broadband Wireless Access (Multihop Relay Specification) (2007).
3. IEEE 802.16 Task Group m (TGm) website: http://www.ieee802.org/16/tgm/.
4. X. Liu, E. K. P. Chong and N. B. Shroff, A framework for opportunistic scheduling in wireless networks, *Computer Net.* **41** (2003) 451–474.
5. D. Bertsekas, *Nonlinear Programming* (Athena Scientific, Belmont, MA, 1995).
6. I. F. Akyildiz and X. Wang, A survey on wireless mesh networks, *IEEE Commun. Mag.* **43**(9) (2005) 23–30.
7. S. Ghosh, K. Basu, and S. K. Das, What a mesh! An architecture for next-generation radio access networks, *IEEE Network* **19**(5) (2005) 35–42.
8. C. Y. Wong, R. S. Cheng, K. B. Lataief and R. D. Murch, Multiuser OFDM with adaptive subcarrier, bit, and power allocation, *IEEE J. Sel. Areas Commun.* **17**(10) (1999) 1747–1758.
9. D. Kivanc, G. Li and H. Liu, Computationally efficient bandwidth allocation and power control for OFDMA, *IEEE Trans. Wireless Commun.* **2**(6) (2003) 1150–1158.
10. M. Ergen, S. Coleri and P. Varaiya, QoS aware adaptive resource allocation techniques for fair scheduling in OFDMA based broadband wireless access systems, *IEEE Trans. Broadcast.* **49**(4) (2003) 362–370.
11. C. Mohanram and S. Bhashyam, A sub-optimal joint subcarrier and power allocation algorithm for multiuser OFDM, *IEEE Commun. Lett.* **9**(8) (2005) 685–687.
12. Y. J. Zhang and K. B. Letaief, Multiuser adaptive subcarrier-and-bit allocation with adaptive cell selection for OFDM systems, *IEEE Trans. Wireless Commun.* **3**(5) (2004) 1566–1575.
13. Z. Han, Z. J. Ji and K. J. Ray Liu, Fair multiuser channel allocation for OFDMA networks using Nash bargaining solutions and coalitions, *IEEE Trans. Commun.* **53**(8) (2005) 1366–1376.
14. Y. Yao and G. B. Giannakis, Rate-maximizing power allocation in OFDM based on partial channel knowledge, *IEEE Trans. Wireless Commun.* **4**(3) (2005) 1073–1083.
15. Y. Li and W. Wolf, Hierarchical scheduling and allocation of multirate systems on heterogeneous multiprocessors, *Proc. Eur. Des. Test Conf.* (1997) 134–139.
16. B. Li and Y. Qin, Traffic scheduling in a photonic packet switching system with QoS guarantee, *J. Lightwave Technol.* **16**(12) (1998) 2281–2295.
17. M. Zorzi and R. R. Rao, Geographic random forwarding (GeRaF) for ad hoc and sensor networks: multi-hop performance, *IEEE Trans. Mobile Comput.* **2**(4) (2003) 337–348.

18. J.-H. Song, V. W. S. Wong and V. C. M. Leung, Efficient on-demand routing for mobile ad-hoc wireless access networks, *IEEE J. Sel. Areas Commun.* **22** (2004) 1374–1383.
19. K. G. Murty, *Linear and Combinatorial Programming* (John Wiley and Sons, New York, 1976).
20. K.-D. Lee and V. C. M. Leung, Fair allocation of subcarrier and power in an OFDMA wireless mesh network, *IEEE J. Sel. Areas Commun.* **24**(11) (2006) 2051–2060.
21. F. P. Kelly, A. K. Maulloo and D. K. H. Tan, Rate control in communication networks: shadow prices, proportional fairness and stability, *J. Oper. Res. Soc.* **49** (1998) 237–252.
22. A. Sendonaris, E. Erkip and B. Aazhang, User cooperation diversity, Part I: system description, *IEEE Trans. Commun.* **51**(11) (2003) 1927–1938.
23. A. Sendonaris, E. Erkip and B. Aazhang, User cooperation diversity, Part II: implementation aspects and performance analysis, *IEEE Trans. Commun.* **51**(11) (2003) 1939–1948.
24. A. Nosratinia, T. E. Hunter and A. Hedayat, Cooperative communication in wireless networks, *IEEE Commun. Mag.* **42**(10) (2004) 74–80.
25. J. N. Laneman, D. Tse and G. W. Wornell, Cooperative diversity in wireless networks: efficient protocols and outage behaviour, *IEEE Trans. Inform. Theory* **50**(12) (2004) 3062–3080.
26. J. N. Laneman and G. W. Wornell, Distributed space-time coded protocols for exploiting cooperative diversity in wireless networks, *IEEE Trans. Inform. Theory* **49** (2004) 2415–2425.
27. B. Schein and R. Gallager, The Gaussian parallel relay network, *Proc. IEEE ISIT* (2000) 22.
28. W. Su, A. K. Sadek and K. J. R. Liu, SER performance analysis and optimum power allocation for decode-and-forward cooperation protocol in wireless networks, *Proc. IEEE WCNC* (2005).
29. T. C.-Y. Ng and W. Yu, Joint optimization of relay strategies and resource allocations in cooperative celluler networks, *IEEE J. Sel. Areas Commun.* **25**(2) (2007) 328–339.
30. A. Goldsmith, Rate limits and cross-layer design in cooperative communications, *WICAT Workshop on Cooperative Communications*, Polytechnic University, Brooklyn, New York (2005).
31. A. Host-Madsen and J. Zhang, Capacity bounds and power allocation for wireless relay channels, *IEEE Trans. Inform. Theory* **51**(6) (2005) 2020–2040.
32. P. Gupta and P. R. Kumar, Towards an information theory of large networks: an achievable rate region, *IEEE Trans. Inform. Theory* **49**(8) (2003) 1877–1894.

Chapter 8

Game Theoretic Tools Applied to Wireless Networks

Hua Liu[*,‡], Bhaskar Krishnamachari[*,†,§], Shyam Kapadia[*,||]

[*]*Department of Computer Science*
[†]*Department of Electrical Engineering*
Viterbi School of Engineering
University of Southern California
Los Angeles, CA 90089, USA
[‡]*hual@usc.edu*
[§]*bkrishna@usc.edu*
[||]*kapadia@usc.edu*

Wireless networks always consist of equipments which belong to different organizations/persons. Hence, nodes in wireless networks are self-interested. Game theory, a set of tools originally studied in microeconomics to predict the behavior of selfish individuals in free market, now becomes a valuable tool to handle various problems for routing and media access in wireless networks with selfish and rational users. This chapter focuses on the applications of game theoretic tools in wireless networks. In each subsection of this chapter, we will introduce important concepts in game theory using definitions/theorems and followed by their applications on important papers published in wireless networks.

The chapter starts with strategic form games, which are the most widely used game formulations in wireless research. Pure Nash equilibrium and mixed equilibrium are discussed as two major branches of strategic form games. Supermodular games and potential games, two classes of the games with desired property to improve system performance, are presented as special cases of strategic form games. Subsequently, repeated games (a.k.a. iterated games) are discussed with its application in reputation system design. This is followed by a brief discussion of Bayesian games to model scenarios with incomplete information. The chapter concludes with a discussion of truthful auctions and their application in pricing based wireless routing problems.

Keywords: Game theory; Nash equilibrium; wireless networks; strategic form games; Bayesian games; repeated games; auction.

8.1. Introduction

Game theory is a set of tools that originated from microeconomics and is employed to predict the behavior of selfish individuals in a free market. The tool set enables capturing the behavior of individuals in strategic situations where an individual's success is dependent on the choices made by the other individuals. At the core of the development of game theoretic tools lies the concept of "equilibrium" — a state where rational individuals are unlikely to change their behavior. While many definitions of equilibria have been proposed in the game theoretic literature, the most popular and widely used one is that given by Nash in 1950 called the *Nash* equilibrium.[14,22,28]

In recent years, game theory has been gaining the attention of the wireless research community since the tool set is powerful in analyzing rational selfish player behavior. This stems from the fact that, self-interested users are considered as basic components for some wireless networks, especially ad hoc networks. In this chapter, we will introduce important concepts in game theory followed by its applications in wireless networks some of which have appeared in recent research literature. A wide variety of problem spaces have been described in which game theory has been successfully applied like data routing, power control, wireless security systems, topology control, medium access control and reliable routing.

Game theory like other fields of applied mathematics has matured over the past 50 years. The 1944 classical work called *Theory of Games and Economic Behavior* by Neumann and Morgenstern served in initially popularizing the field of game theory. In the decades to follow, many scholars helped developing this theory further. Some notable contributions besides those made by Nash are described below. In 1957, Luce and Raiffa published the first popular game theory textbook[21] where they introduced the concept of repeated games. Vickrey provided the first formalization of auctions in 1961.[35] Subsequently, in 1967, Harsanyi developed the concept of a "Bayesian Nash Equilibrium" for Bayesian games.[15]

We have organized this survey of game theory in wireless networks based on different types of game formulations. First, we discuss the strategic form game, which is the most widely used formulation in wireless contexts. Then, we define pure Nash equilibrium and mixed equilibrium. Subsequently, we introduce a special class of games called supermodular games. Then, we describe repeated games and related applications in reputation system designs. This is followed by a discussion of Bayesian games that involve the use of probabilistic methods in dominant strategy calculations when players have incomplete information. The last part of the chapter provides a brief description on truthful auctions.

8.2. Strategic Form Game

Strategic form (or normal form) game is a basic component in game theory. A strategic form game[14] is a triplet $\langle I, (S_i)_{i\in I}, (u_i)_{i\in I}\rangle$ with:

- A finite set of players $i \in I$; $I = 1, \ldots, I$.
- A set of available actions $(S_i)_{i\in I}$ where S_i is a non-empty set of actions for player i. A tuple $s = \{s_1, \ldots, s_I\}$ is called an action profile.
- A payoff function (also called as a utility function) u_i for each profile $s = \{s_1, \ldots, s_I\}$.

In most of the notations, authors always refer to all players other than some given player i as "player i's opponents" and denote them as "$-i$". Note that a strategic form game is a simultaneous decision game. In this section, we consider strategic games with complete information, which implies that each player has knowledge about all the other players' utility functions.

There are two kinds of strategies for a player: *pure strategy* and *mixed strategy.* Given a set of strategies, if a player chooses to take one action with probability 1 then that player is playing a pure strategy. A mixed strategy is a probability distribution over pure strategies. Each player's randomization is statistically independent of those of his opponents, and the payoffs to a profile of mixed strategies are the expected values of the corresponding pure strategy payoffs.

8.2.1. *The Nash equilibrium concept*

Nash equilibrium is named after its inventor John Nash (1950), though a similar concept was introduced by Cournot[9] in 1897. Generally speaking, Nash equilibrium is a steady state where no player in the game would unilaterally change his strategy if he is selfish and rational. If a game converges, it must converge to a Nash equilibrium. There might be several Nash equilibriums existing in a game. Since players in the game are self-interested, the Nash equilibrium is not necessary the same as the social welfare. In fact, in most cases, Nash equilibrium is not same as the global optimal solution. The global optimal solution is a solution where players completely cooperate with each other to maximize the system throughput regardless individual gain. Hence, the existence and uniqueness of Nash equilibrium for a game is an important topic for researchers.

8.2.1.1. *Background*

A Nash equilibrium is a set of strategies, one for each player, such that no player has an incentive to **unilaterally** change his action. In other words, players are in Nash equilibrium, if a change in strategies by any one of them would result in lower gains for that player than if he had remained with his current strategy. Mathematically, a pure strategy Nash equilibrium is a pure strategy profile s^* such that for all player i,

$$u_i(s_i^*, s_{-i}^*) \geq u_i(s_i, s_{-i}^*),$$

for all $s_i \in S_i$. A mixed strategy Nash equilibrium profile σ^* is defined when for each player i and for each $\sigma_i \in \Sigma_i$

$$u_i(\sigma_i^*, \sigma_{-i}^*) \geq u_i(\sigma_i, \sigma_{-i}^*),$$

where Σ_i is the probability distribution over player i's pure strategies. That is, each element $\sigma_i \in \Sigma_i$ indicates how frequently every pure strategy is chosen. In this case, the utility function value is the expected value of the corresponding pure strategy values.

Nash equilibrium is an important concept in game theory since if every player is rational in a game, the Nash outcome is a player's best response when given other players' strategy. In many network game theory contexts, only pure strategies are reasonable or desirable because of practical and easy implementation. However, according to Nash, a mixed Nash equilibrium always exists in finite games, but a pure strategy Nash equilibrium might not exist. Therefore, many studies focus on the existence and complexity of deciding the existence of pure Nash equilibria.

In a game formulation, if a Nash equilibrium exists, we need a metric to measure the performance of the system at Nash equilibrium in comparison to the system performance yielded at the global optima. Two widely used metrics in the literature are Price of Anarchy (POA)[18] and Price of Stability (POS).[4] For maximization problems, POA and POS are defined as follows:

$$POA = \frac{Value\ of\ Worst\ Equilibrium}{Value\ of\ Optimal\ Solution}, \quad POS = \frac{Value\ of\ Best\ Equilibrium}{Value\ of\ Optimal\ Solution}.$$

For minimization problems, the definitions are turned over. In both cases, $POA \geq 1$ and $POS \geq 1$.

In the following subsection, we will discuss some papers which focus on the existence of a pure strategy Nash equilibrium and investigate the system performance on Nash equilibrium.

8.2.1.2. *Topology control*

Eidenbenz *et al.*[11] consider a problem of topology control in wireless ad hoc networks. In their formulation, the players are all the nodes (refered to as points in the network graph) in the network. The strategy for each player is to decide its own transmission radio range. The authors consider 3 kinds of games: connectivity game, strong connectivity game and reachability game. In the connectivity game, a set of pairs of source and destination nodes (in the form of (s_i, t_i)) is given. In each pair, the source node s_i needs to choose a radius to get connected to destination t_i (possibly through multiple intermediate nodes) while keeping the radius as small as possible.

Figure 8.1 gives an example of the connectivity game. Given a source-destination pair $\{s, t\}$, Figure 8.1(a) shows a pure Nash equilibrium result while Figure 8.1(b) depicts a result that is not a Nash equilibrium. The strong connectivity game is a

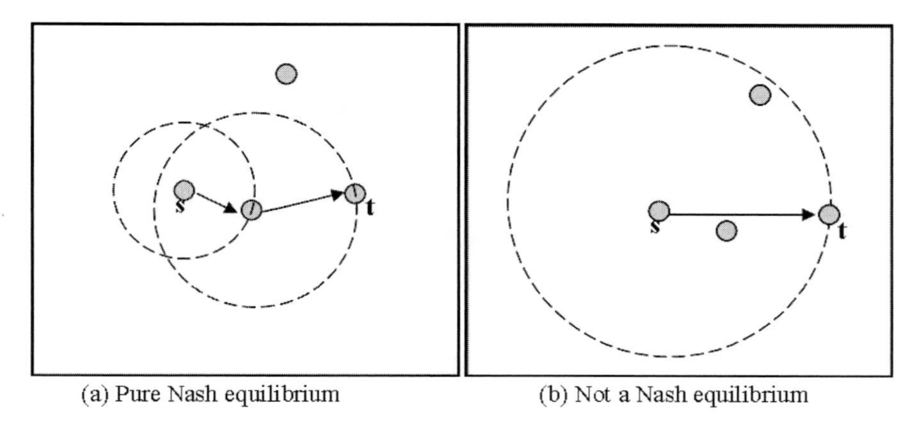

Fig. 8.1. An example of connectivity game.

special case of the connectivity game, where each node needs to be connected with every other node. In both connectivity games, the utility functions are designed to penalize a node relatively heavily if it cannot satisfy the connectivity constraints. Otherwise, a node's utility increases with decreasing radius. In the reachability game, each node tries to reach as many other nodes as possible while minimizing its radius. The utility function of the reachability game is defined as the total number of connected nodes minus the power cost to set up the radio range. Mathematically, given radius vector[a] $\bar{r}$, utility of user i is $U(i) = f_{\bar{r}}(i) - r_i^{\alpha}$, where $f_{\bar{r}}(i)$ is the number vertices reachable from i and α is the distance power gradient, usually being 2.

Major conclusions of the work are three-fold. First, the authors claim that in the strong connectivity game, a polynomial time algorithm can find Nash equilibria with costs at most a factor 2 of the minimum possible cost. Second, for the connectivity game, when a given network graph satisfies the triangle inequality, the problem of finding the existence of a Nash equilibrium is NP-hard. Finally, the authors also state that for the reachability game, $1 + O(1)$ approximate Nash equilibria exist for graphs characterized by a random distribution of nodes.

A related topology control game is studied by Kesselman *et al.*[17] Considering every node in the network as a player, they consider two games: a broadcast game and a convergecast game. In the broadcast game, each node has to establish a directed path from a pre-defined root node to itself. In the convergecast game, each node needs to deliver their packets to the root node. They formulate the game with both cost sharing and non-sharing models. In cost sharing model, the cost of each node can be paid by *several* other nodes. In the cost model without sharing, each node pays its own transmission cost. A strategy of a player u is a payment function Pay_v^u, which indicates how much player u would pay for player v. The transmit

[a]Each element in the vector denotes the transmitting radius of each node in the network. r_i is node i's radius.

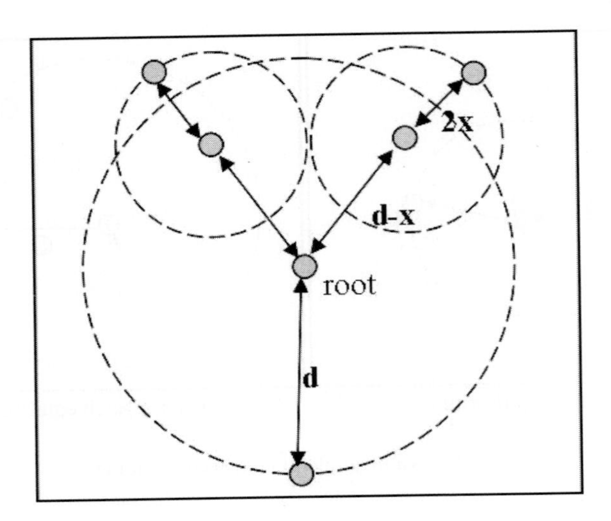

Fig. 8.2. An example of a unique pure Nash equilibrium in a broadcast game.

power for a node is proportional to its total income obtained from all other nodes in the network. The social cost of the outcome of the game is the total transmission power of all the nodes in the network.

Figure 8.2 gives an example of a broadcast game with a unique Nash equilibrium. In this case, there are rays emanating from the root node and the angle between two adjacent rays is 120 degrees. On the upper left and upper right rays, there are two nodes at distance $d-x$ and $d+x$ ($x < d$) respectively from root node. The ray at the bottom has a node with distance d from the root node. The unique Nash equilibrium occurs when the bottom node pays the root node for a transmission range d and the furthest upper nodes pay the relay nodes for transmission range $2x$.

The authors also study the pessimistic and optimistic price of anarchy (PoA). By denoting the cost of the Nash equilibrium as C_p and the cost of the optimal solution as C_p^*, the authors define the pessimist PoA as $\max_p\{\max\{C_p/C_p^*\}\}$ and the optimistic PoA as $\max_p\{\min\{C_p/C_p^*\}\}$. The conclusions highlight three findings of the study. First, in a broadcast game, a pure Nash equilibrium may *not* exist and the optimistic PoA is bounded away from 1 while the pessimistic PoA is $\theta(n)$. Secondly, for a convergecast game without sharing, there always exists a pure Nash equilibrium. The optimistic PoA for this game is 1 and the pessimist PoA is $\theta(n^{\alpha-1})$ where $2 \Leftarrow \alpha \Leftarrow 8$ is the distance-power gradient.[b] Finally, for a convergecast game with sharing, the optimistic PoA is 1 and the pessimist PoA is $\theta(n)$. They also provide a heuristic greedy algorithm based on a minimum spanning tree to calculate the Nash equilibrium for a hop-bounded broadcast game.

[b]In their modeling, the neighbors of a node u are determined by its transmission power P_u, which means node u can reach all nodes within its transmission range $R_u = P_u^{1/\alpha}$.

8.2.1.3. *Fair bandwidth allocation*

Fang and Bensaou formulate and solve a game theoretic problem pertaining to fair bandwidth/rate allocation for flows in a wireless multi-hop network setting.[12] They start with a general formulation in which the goal is to maximize the weighted sum of some given convex functions of the rates for each flow, subject to bandwidth consumption constraints that are modeled using a clique-resource approach.[c] The general problem formulation is given as follows:

$$\begin{aligned} P: \quad & \max_{x_i} \sum_i w_i f_i(x_i) \\ \text{s.t} \quad & \sum_{i \in S(j)} x_i \leq c_j, \quad j = 1, 2, \ldots, M \\ & x_i \geq 0, \qquad\quad i = 1, 2, \ldots, N. \end{aligned}$$

From this general problem, they derive an unconstrained Lagrange relaxation based optimization formulation in which the goal is to maximize the sum of a derived utility function V_i for each flow that is a function of all the flow rates. This is used to model the problem as a non-cooperative strategic form game where the players are the individual flows, the strategy space consists of the continuous flow rate for each flow, and the utility function is V_i which is proved to be strictly concave. The authors show that this game has a unique pure Nash equilibrium. Furthermore, they employ a result pertaining to games with concave utility functions[30] to show that the equilibrium can be reached by changing the strategy for each flow at a rate proportional to the gradient of its payoff function with respect to its strategy. This fact yields a simple distributed algorithm for the problem, whose effectiveness is verified through numerical solutions.

8.2.1.4. *Power control*

Wireless networks in contrast to wired networks are characterized by higher signal interference. Together with fading, multi-path and other impairments, signals are distorted as they travel from hop to hop. Signal to interference noise ratio (SIR) is a common denominator employed to characterize the quality of signals. Generally, improving the signal transmit power level leads to a higher SIR. However, wireless devices always have limited resources in terms of battery power. Hence, it is important to optimize the usage of the limited power resource while striving to meet the user's required signal quality. In this section, we will discuss some

[c]This approach uses a graph to capture the contention among flows. In this graph, every vertex represents an active flow/link and each edge denotes interference. Flows in the same maximal clique cannot transmit simultaneously. A flow can succeed in transmission if and only if all flows that share at least one clique with this flow do not transmit.

research related to this power control problem within the context of a game theoretic framework.

A significant piece of work which is widely cited was conducted by Alpcan *et al.*[2] In this study, the authors model the power control of Code Division Multiple Access (CDMA) uplinks as a non-cooperative strategic game. The players are the users sharing a single cell. The strategy space for each user is a set of available power levels. The players are trying to minimize their own cost while maximizing their utility. The cost function is defined as the difference of a pricing function and a utility function. For player i, the pricing function P_i is proportional to the transmit power level p_i, and the utility function is the logarithmic function of the user's SIR γ_i:

$$J(p_i, p_{-i}) = \lambda_i p_i - u_i \ln(1 + \gamma_i),$$

where λ_i is a positive number and u_i is a user-specific weight parameter.

The major contribution of this study is that the authors not only prove that there exists a unique Nash equilibrium for this game, but they also capture and characterize the Nash equilibrium. To obtain the Nash equilibrium point, they employ the following steps. First, for each user i, they define a user specific parameter $a_i = (u_i h_i/\lambda_i) - (\sigma^2/L)$ where $0 < h_i < 1$ is the channel gain from user i to the base station, σ^2 is the interference and $L > 1$ is the throughput. Then, they index array a in a descending manner, i.e.

$$a_i \leq a_j \Rightarrow i > j.$$

Let $M^* \leq M$ (where M is the total number of users) denote the largest integer $\tilde{M}$ which satisfies:

$$a_{\tilde{M}} > \frac{1}{L + \tilde{M} - 1} \sum_{i=1}^{\tilde{M}} a_i.$$

The conclusion is that all users $M^* + 1, \ldots, M$ have **zero** power levels and the users $1, 2, \ldots, M^*$ have power levels which can be calculated by

$$p^* = \frac{1}{h_i} \left\{ \frac{L}{L-1} \left[a_i - \frac{1}{L + M^* - 1} \sum_{j \in M^*} a_j \right] \right\}.$$

If no such $\tilde{M}$ exists, then the Nash equilibrium solution just assigns **zero** power levels to all M players. Furthermore, the authors provide two power updating algorithms, namely parallel update algorithm (PUA) and random update algorithm (RUA) to converge to the Nash equilibrium. In PUA, nodes update their power level in each iteration (discrete time intervals) according to the reaction function $p_i = \max\left\{0, \frac{1}{h_i}\left[a_i - \frac{1}{L}\tilde{y}_{-i}\right]\right\}$ where $\tilde{y}_{-i}$ is the summation of power levels of other users as received at the base station. In RUA, users update their power levels with a pre-defined probability. Stability, robustness, and convergence of each algorithm have been discussed in the paper.

8.2.1.5. *Reliable routing*

Wireless networks usually consist off equipments that belong to different owners. Consequently, in such contexts, selfish behavior is inherent, especially in routing problems. There is a lack of incentive for intermediate nodes to participate in the routing process. Pricing can be introduced into the system to encourage node participation.

Liu and Krishnamachari[19] consider a destination-driven price-based reliable routing problem in wireless networks with selfish users. In their formulation, if a user in a network, denoted as an undirected graph $Graph(V, E)$, wants to get a piece of information, he will grant a price G to the information provider (a.k.a. source node) to initiate a routing request. The source node then will announce a fixed price p for all the other nodes in the network to encourage nodes to participate in the routing. In this network, each node v_i in V is characterized by a probability of reliably forwarding a packet R_i, and each link (v_i, v_{i+1}) in E is characterized by a cost $C_{v_i,v_{i+1}}$ of transmission. The promised prices are paid only after the destination receives the requested information. The objective of this game is to form a stable and reliable routing path between a given source and destination pair.

In the game formulation, the source node and all the intermediate nodes constitute the player space. Strategy for each node involves the selection of another node as it is a next hop.[d]

The utility function for the source node is defined as follows:

$$u_{src} = \begin{cases} 0 & \textit{if no path exists} \\ (G - hp) \prod\limits_{v_i \in Path} R_i - C_{src,\, v_i} & \textit{otherwise} \end{cases}$$

where h is the number of hops in the route.

The utility function for intermediate node v_i is defined as:

$$u_{v_i} = \begin{cases} 0 & \textit{if no path exists or if } v_i \notin Path \\ p \prod\limits_{v_i \in Path \textit{ from } v_i \textit{ onward}} R_i - C_{v_i,v_{i+1}} & \textit{otherwise} \end{cases}$$

The authors present a polynomial time algorithm which gives a Nash equilibrium path and evaluate the performance of the game with respect to different parameters such as price and source behavior (selfish source or cooperative source with respect to destination) for various network settings.

[d]If a node does not participate in the routing, its next hop is set as empty.

8.2.1.6. *Security: intrusion detection*

Security is a critical concern in wireless networks. Over the years, researchers have applied various kinds of reinforcement learning and data mining techniques to detect intrusions and thereby defending wireless equipment from attackers. Application of game theoretic tools toward security related problems is still at a very early stage.

Agah *et al.* propose a static game theoretic approach for security management in mobile wireless sensor networks.[1] The game is based on a cluster-based sensor network comprising trusted base station with infinite energy. This game has two players: an attacker and a cluster-protector. The attacker is trying to launch a denial of service (DoS) attack on a certain cluster head. The protector (say Intrusion Detection System (IDS)) is trying to protect the cluster head before the attack is launched. An important assumption of this game is that the attacker is rational and not malicious. Note that a rational player's goal is to maximize his own payoff as opposed to a malicious player whose goal may be to destroy the system without any payoff considerations. With respect to one fixed cluster, say the kth cluster, the attacker employs one of the following 3 strategies: (AS_1) attack cluster k; (AS_2): do not attack at all; (AS_3): attack a cluster other than cluster k. The IDS has two strategies: (SS_1): defend cluster k or SS_2: defend a different cluster. The payoff table for the combination of strategies is shown in Table 8.1.

Here, the notation k' denotes a certain cluster other than cluster k and k'' denotes a cluster which is neither cluster k nor cluster k'. Other symbols are defined as follows: $U_{ij}(t)$ is the utility of ongoing sessions in the sensor network; AL_k is the average loss incurred if cluster k is lost (i.e. the loss if the attacker attacks cluster k while the defender defends a cluster other than k); C_k is defined as the average cost of defending cluster k; CW is the cost of waiting and deciding to attack in the future; CI denotes the cost of intrusion for the attacker; and $PI(t)$ is the average profit for each attack.

Using this payoff table, the authors carefully define the formulae to calculate CI, C_k, CW, and $PI(t)$. Briefly, the cost of intrusion for attacker, CI, is defined proportional to the complicity of the attack. The cost of intrusion for IDS, C_k, is defined as a linear combination of the previous attack history of cluster k and the cluster size of k; the history has a weight that is significantly higher than the cluster size. The cost of waiting for the attacker, CW, is defined as proportional to the number of previous unsuccessful attack attempts on this cluster plus a small

Table 8.1 Payoffs of players in intrusion detection game.

	AS_1	AS_2	AS_3
SS_1	$((PI(t) - CI),$ $(U_{ij}(t) - C_k))$	$CW, U_{ij}(t) - C_k)$	$((PI(t) - CI),$ $(U_{ij}(t) - C_k - \sum AL_{k''}))$
SS_2	$((PI(t) - CI),$ $(U_{ij}(t) - C_{k'} - \sum AL_{k''}))$	$CW, U_{ij}(t) - C_{k'})$	$((PI(t) - CI),$ $(U_{ij}(t) - C_{k'} - \sum AL_{k''}))$

constant. The profit for the attacker to successfully attack the cluster equals the IDS loss of the cluster, which contains the values of all the processes running in this cluster.

The major contribution of this piece of work is that, with the well designed utility table, the authors prove that this game has a unique pure strategy Nash equilibrium at strategy pair (AS_1, SS_1). This implies that the equilibrium occurs when both the attacker and the protector target the same cluster. In other words, if the attacker is selfish (not malicious) and plays by the game rules, theoretically, the defender will always defend the attacked cluster successfully.

8.2.2. *Pareto improvement and pareto optimal*

Nash equilibrium only considers the strategy profile when no rational player will unilaterally change his strategy. Sometimes a game is characterized by the existence of multiple Nash equilibria and among them, one Nash equilibrium will result in all players having better or equal payoffs (and at least one player having a better payoff) than in another Nash equilibrium. In such cases, the former Nash equilibrium is preferred over the latter one(s). The former one is known as a Pareto improvement to the latter one. In the next section, we provide a brief discussion on how the concept of Pareto improvement can be utilized to achieve a desirable Nash equilibrium.

8.2.2.1. *Background*

If a strategy profile can make at least one player better off, without making any other player worse off, we call this a Pareto improvement or Pareto optimization in comparison to the previous strategy profile. That is, for player strategy profiles s' and s, if for every player i in the game, we have

$$u_i(s'_i, s'_{-i}) \geq u_i(s_i, s_{-i}),$$

and there exists a player j such that

$$u_j(s'_j, s'_{-j}) > u_j(s_j, s_{-j}),$$

then (s'_i, s'_{-i}) is a Pareto improvement over (s_i, s_{-i}). A strategy profile is Pareto efficient or Pareto optimal if no further Pareto improvements can be made.

The simplest example to explain the relationship between the Nash equilibrium point and the Pareto optimal point is the famous prisoner's dilemma problem. The classical prisoner's dilemma (PD) is as follows[14]:

Two suspects, A and B, are arrested by the police. The police have insufficient evidence for a conviction, and, having separated both prisoners, visit each of them to offer the same deal: if one testifies for the prosecution against the other and the other remains silent, the betrayer goes free and the silent accomplice receives the full 10-year sentence. If both stay silent, the police can sentence both prisoners to only one year in jail for a minor charge. If each betrays the other, each will receive

Player 1 / Player 2	Stay Silent	Betray
Stay Silent	(-1,-1)	(0,-10)
Betray	(-10,0)	(-2,-2)

Fig. 8.3. Payoff matrix for prisoner's dilemma.

a two-year sentence. Each prisoner must independently make the choice of whether to betray the other or to remain silent. However, neither prisoner knows for sure what choice the other prisoner will make.

Let negative integers denote the years of sentence a prisoner gets. The payoff matrix for the prisoner's dilemma problem is described in Figure 8.3.

In this case, {Betray, Betray} is a Nash equilibrium, while {Silent, Silent} is a Pareto improvement to the Nash equilibrium; in fact, it is the Pareto optimal strategy profile for both prisoners. In general, Pareto optimal performs no worse than Nash equilibrium as an operation point. Hence, many game theoretic studies have proposed techniques to achieve Pareto improvement and Pareto optimality if the Nash equilibrium is Pareto inefficient.

8.2.2.2. *Power control*

Power control problems are a set of problems seeking to intelligently select transmit power in a communication system to achieve good performance. This performance is achieved by optimizing metrics of interest such as link data rate, network capacity, geographic coverage and range, and life of the network and network devices etc. Power control problems are considered in many contexts, including cellular networks, sensor networks, wireless LANs, and DSL modems. A power control game is a power control problem form in a game theoretic context. Concerned metrics in such game map to appropriate utility functions. In a power control game, a solution is Pareto-optimal if there exists no other power allocation for which one or more users can improve their utilities without reduction in the utilities of the other users. In this section, we describe a few studies that have investigated power control games in wireless networks.

Shah *et al.*[33] discuss design issues in developing a game theoretic framework for a wireless power control problem in Code Division Multiple Access (CDMA) systems. The authors formulate the power control problem as a non-cooperative strategic game where the players are all the users in the network who are transmitting data to a common base station. The strategy space for each player is a set of values that the user's power is restricted to. The strategy space for each player is assumed to be continuous.

The authors also provide a discussion on the desired properties a utility function of a realistic wireless network should possess. There are five such properties. Let $u_i(p_i, \gamma_i)$ denote the utility function of player i where p_i is its transmit power and γ_i denotes the user's signal to interference noise ratio (SIR) for a fixed transmit power p_i. First, for a fixed transmission power, the utility should monotonically increase with SIR, i.e.

$$\frac{\partial u_i(p_i, \gamma_i)}{\partial \gamma_i} > 0 \quad \forall \gamma_i, \ p_i > 0.$$

Second, the utility should obey the law of diminishing marginal utility[e] when SIR is large enough. That is,

$$\lim_{\gamma_i \to \infty} \frac{\partial u_i(p_i, \gamma_i)}{\partial \gamma_i} = 0 \quad p_i > 0.$$

Third, for a fixed SIR, the utility should be a monotonically decreasing function of the user's transmitter power. Mathematically,

$$\frac{\partial u_i(p_i, \gamma_i)}{\partial p_i} < 0 \quad \forall \gamma_i, \ p_i > 0.$$

Fourth, when transmit power approaches zero, the utility should go to zero.

$$\lim_{p_i \to 0} u_i = 0.$$

Finally, when transmit power goes to infinity, the utility should go to zero,

$$\lim_{p_i \to \infty} u_i = 0.$$

As an example, the authors employ the following utility function which satisfies all the five properties mentioned above:

$$u_i(p_i, \gamma_i) = \frac{ER}{p_i}(1 - e^{-0.5*\gamma})^L,$$

where E is the remaining power of the node, R is the data transmit rate, and L is the length of the data packet. This quasi-concave utility function[f] yields a unique Nash equilibrium point as per Debreu's theorem.[10] However, this Nash equilibrium point is Pareto inefficient since the authors prove that if all the users are willing to reduce their transmit power by a small amount; they all will improve their utility.

In order to achieve Pareto improvement, the notion of pricing is introduced into the game. The basic idea of pricing is that if more users are competing in the network, each of them will have to utilize higher transmit powers. Hence, the price

[e] Law of diminishing marginal utility is the customer's view of law of diminishing returns in economics.

[f] Definition: A function $f(x)$ is quasi-concave if it satisfies: $f(\lambda x_1 + (1-\lambda)x_2 \geq \min[f(x_1), f(x_2)]$, $\forall x_1, x_2$ in the domain where $0 \leq \lambda \leq 1$.

of using the network should be higher, which in turn should discourage users from using the network. Therefore, the price function should be monotonically increasing with transmit power. The authors model the price function as $F_i = tRp_i$ where t is a positive constant. Incorporating the notion of pricing yields a new utility (payoff) function which is described as the difference of user's satisfaction and the price i.e. $u_i - F_i$. With pricing, the new game has a unique Nash equilibrium and is a Pareto improvement in comparison to the original one. However, the authors have not been able to conclusively state whether the Pareto improved Nash equilibrium is Pareto optimal or not. The main results of this study are used by many of the following studies.

A closely related study on power control is conducted by Saraydar *et al.*[31] A non-cooperative power control game is proposed where as before the players are mutually interfering users. The strategy space is a set of power levels that a player can choose for data transmission. The utility function is defined as

$$u_i(p_i, p_{-i}) = \frac{LR}{Mp_i} f(\gamma_i),$$

with unit bits/joule. Here, R is the data rate, $\frac{L}{M}$ is the total number of packets to be transmitted, p_i is the transmit power level, and $f(\gamma_i)$ is called the efficiency function which indicates the probability of successful packet reception. As before,[33] the authors prove the existence of a unique Nash equilibrium point with this utility function; however, the equilibrium is shown to be Pareto inefficient. With the goal of providing Pareto improvement, a pricing scheme is introduced in the problem framework. The usage-based price is defined as a linear function of the transmit power. The authors claim that this enhanced price-involved game is a supermodular game.[34] Subsequently, they prove the existence of a Nash equilibrium for this game using supermodularity theory (see Section 8.2.3). However, they also point out that the Nash equilibrium is not unique. Furthermore, they conclude that when the pricing function is defined as

$$c_i = -\left(\frac{1}{p_i}\right) \sum_{j=1, j\neq i}^{n} p_j u_j,$$

the Nash equilibrium obtained is Pareto optimal. But achieving this equilibrium needs a centralized mechanism since the utility value for each node u_i needs to be known globally.

Meshkati *et al.* have performed several extensive studies in the power control game theoretic context with different scenarios/constraints.[24–26] The main distinction of these works from prior studies is that the network comprises multiple base stations which may be chosen by the users for data transmission.[26] The game G is defined as $G = [K, \{A_k\}, \{u_k\}]$ where $K = \{1, \ldots, K\}$ are the users, $A_k = [0, P_{\max}]$ ($P_{\max}$ is the maximum allowed power for transmission) is the strategy space for the kth user, and the utility function u_k is defined as the ratio of throughput to transmit power level with units of bits per joule. Here, throughput is the number of reliable (without error or correctable) transmitted bits.

With these definitions, the authors show that there exists a unique Nash equilibrium. This can be achieved when all users pick the minimum mean-square error (MMSE) detector as their uplink receiver, and choose their transmit powers such that their output SINRs are all equal to γ^* where γ^* is the solution to $\frac{\partial f(\gamma)}{\partial \gamma} = 0$ while $f(\gamma)$ is an efficiency function[g] which denotes the packet success rate (PSR). Furthermore, the authors state that the obtained Nash equilibrium is not Pareto optimal. They also illustrate that generally the difference between the two is not significant. A similar formulation and results by same authors have been represented in Ref. 25 with more details and concrete examples. The authors also provide a distributed greedy algorithm to achieve the Nash equilibrium. Note that with Refs. 25 and 26, the goal of each user is to maximize its throughput while employing minimum transmission power.

With a similar game setting, taking delay requirements into consideration, Meshkati *et al.* prove the existence and uniqueness of Nash equilibrium for a delay constraint problem.[24] In this study, the authors assume that if there is any error during data transmission, the nodes will retransmit the packet. The transmission delay for a packet is then directly proportional to the number of transmissions required for a packet to be received without any errors (denoted as X). Assuming efficiency function $f(\gamma)$, the probability that exactly m transmissions are required for successfully transmitting a packet is given by

$$\Pr\{X = m\} = f(\gamma)(1 - f(\gamma))^{m-1}.$$

Hence, the delay requirement of a particular user is given as a pair (D, β) where

$$\Pr\{X \le D\} \ge \beta,$$

i.e. the delay requirement is modeled as a CDF which states "the number of transmissions be at most D with a probability larger than or equal to β". Since $f(\gamma)$ is an increasing function of γ, this constraint provides a lower bound on the value of γ. Let $\tilde{\gamma}$ denote the γ value which satisfies the equality of the delay requirement. A delay constrained utility function for player k is defined as:

$$\tilde{u}_k = \begin{cases} u_k & \text{if } \gamma_k \ge \tilde{\gamma}_k \\ 0 & \text{if } \gamma_k < \tilde{\gamma}_k \end{cases}.$$

With this modified utility function, the unique Nash equilibrium is derived as $\tilde{p}_k^* = \min\{p_k^*, P_{\max}\}$ for each user k. Here, p_k^* is the transmit power that yields an SIR equal to γ_k^* at the output of the receiver with $\overset{\sim}{\gamma_k^*} = \max\{\tilde{\gamma}_k, \gamma^*\}$.[h] As part of the

[g] Generally, γ represents SIR and $f(\gamma)$ is a monotonically non-decreasing function of γ.
[h] γ^* is the optimal γ value derived from the original problem without considering delay requirement.

validation process, the authors show the efficiency of the Nash equilibrium point via numerical results.

8.2.3. *Supermodularity*

Supermodular games were developed by Topkis in 1979.[34] The main distinguishing idea in a supermodular game is that when a player takes a higher action according to a defined order, the other players are better off if they also take the higher action, i.e. the game has increasing best responses. Moreover, supermodular games are particularly well behaved. For example, they have pure strategy Nash equilibria and are characterized by the existence of polynomial time algorithms to identify these equilibria.

8.2.3.1. *Background*

Supermodularity is defined through the concept of increasing differences.[14]

Definition (***Increasing differences***). Function $u_i(s_i, s_{-i})$ has *increasing differences* in (s_i, s_{-i}) if, for all $(s_i, \tilde{s}_i) \in S_i^2)$ and $(s_{-i}, \tilde{s}_{-i}) \in S_{-i}^2$ such that $s_i \geq \tilde{s}_i$ and $s_{-i} \geq \tilde{s}_{-i}$,

$$u_i(s_i, s_{-i}) - u_i(\tilde{s}_i, s_{-i}) \geq u_i(s_i, \tilde{s}_{-i}) - u_i(\tilde{s}_i, \tilde{s}_{-i}).$$

Definition (***Supermodular***}. $u_i(s_i, s_{-i})$ is supermodular in s_i if for each s_{-i}

$$u_i(s_i, s_{-i}) - u_i(\tilde{s}_i, s_{-i}) \leq u_i(s_i \wedge \tilde{s}_{-i}, s_{-i}) - u_i(s_i \vee \tilde{s}_{-i}, s_{-i}),$$

for all $(s_i, \tilde{s}_i) \in S_i^2$. Here, the operator "meet" $(x \wedge y)$ and operator "join" $(x \vee y)$ are defined in Euclidean space R^k as:

$$x \wedge y \equiv (\min(x_1, y_1), \ldots, \min(x_k, y_k))$$
$$x \vee y \equiv (\max(x_1, y_1), \ldots, \max(x_k, y_k)).$$

Definition (***Supermodular game***). A supermodular game $G = \langle I, S, U \rangle$ is such that, for each i, S_i is a sub-lattice of R^{m_i}, u_i has increasing differences in $(s_i,\ s_{-i})$ and u_i is supermodular in s_i.

If the utility function u_i is twice continuously differentiable, u_i is supermodular if and only if, for any two components s_l and s_k of s,

$$\frac{\partial^2 u_i}{\partial s_l \partial s_k} \geq 0.$$

8.2.3.2. *Power control*

In this section, we describe an application of supermodular games to a power control problem in wireless data networks. Recall that in a supermodular game, players can iteratively modify their strategy in an ordered direction to achieve a desired

Nash equilibrium. Saraydar *et al.* discuss modifying their uplink power control game to a supermodular game by changing the player strategy space.[31] Note that the original uplink power control game formulation involving pricing has been described earlier in Section 8.2.2.2. This original game is not a supermodular game since the utility function does not satisfy the non-decreasing difference condition (see Equation (8.2.19)) in that strategy space. Instead of allowing the lower bound of the power level to be any non-negative number, the modified strategy space for player i is defined as $P_i = [\underline{p}_i, \overline{p}_i]$ where $\underline{p}_i$ is the minimum transmit power which satisfies $\gamma_i \geq 2 \ln M$ where M is the frame length and $\overline{p}_i$ is the maximum transmit power limited by system constraints. Transmission power for player i: $p_i \geq \underline{p}_i$ ensures $\frac{\partial u_i(p)^2}{\partial p_i \partial p_j} \geq 0$ for all (i, j) pairs where $i \neq j$. This game with the enhanced strategy space, by the definition in Section 8.2.3, is a supermodular game.

With this game formulation, the authors provide an iterative algorithm to obtain the smallest (i.e. most power-efficient) equilibrium in the set of Nash equilibria. The iterative algorithm initializes the power for each node at time $t = 0$ as $p(0) = \underline{p}$. In each iteration, $t = k$, each node calculates the best transmit power at that time instant which maximizes its utility, given all other node strategies at time $t = k - 1$ over the modified strategy space. The authors prove that this algorithm converges to the most power efficient equilibrium by using the non-decreasing difference property of supermodular games.

8.2.4. *Mixed Nash equilibrium*

So far we have considered applications of pure strategy Nash equilibrium. However, pure Nash equilibrium does not always exist. In this section, we will consider mixed Nash equilibria, which exist for every finite game. A mixed strategy is a strategy in which a player performs his available pure strategies with certain probabilities. In an n-player game, if the game has finite number of pure strategies, then there exists at least one equilibrium in mixed strategies. This existence of a mixed equilibrium is proved by a famous theorem by Nash.[14] Furthermore, if no pure strategy Nash equilibria exist, then there *must* exist a unique mixed equilibrium. A famous example is the game of rock-paper-scissors with payoff table shown in Figure 8.4.

	Rock	**Paper**	**Scissors**
Rock	(0,0)	(-1,1)	(1,-1)
Paper	(1,-1)	(0,0)	(-1,1)
Scissors	(-1,1)	(1,-1)	(0,0)

Fig. 8.4. Payoff table for rock-paper-scissors game.

Mixed strategies are best understood in the context of repeated games (see Section 8.3), where each player's goal is to keep the other players guessing. However, the mixed strategy concept can be applied to any kind of game.

8.2.4.1. *Background*

In the pure strategy context, a player can only choose to perform one strategy among many given strategies. In the context of mixed strategies, a player i chooses to play each pure strategy in his strategy space ($S_i = \{s_i^1, s_i^2, \ldots\}$) with a certain probability given by

$$p_{s_i^j} \quad (0 \leq p_{s_i^j} \leq 1).$$

The payoff for each player in the mixed strategy game is the expected payoff calculated given the probabilistic distribution of all the other player strategies. A mixed strategy Nash equilibrium is defined on the expected payoff of all the players.

8.2.4.2. *Power saving*

In this section, we discuss an example of applying mixed Nash equilibrium to improve the system power saving in a query system with server broadcasting the replies.

In their study,[36] Yeung and Kwok consider the following scenario for a wireless data access problem: several clients are interested in a set of data items kept at a server. One client sends a query request to the server for a desired data item, and the server responds by broadcasting the requested item so that all clients receive it. In this scenario, the authors show that it is not necessary for each client to send query requests to the server for each data item. Instead, each client determines data item request probability without any explicit communication with each other. The probability distribution on the strategy space forms a mixed Nash equilibrium.

In this non-cooperative game, players are the clients numbered $\{1, 2, \ldots, n\}$. Each player determines its request probability across all items as its mixed strategy. The strategy space of player i is given by $S_i = \{s_i \,|\, 0 \leq s_i \leq 1\}$. The strategy space for the game is the Cartesian product of all the player's strategy space, $S = \times_{i \in N} S_i$. The utility of each player i is expressed as the number of queries that can be completed given a fixed energy source. Mathematically, client i's utility is given by

$$U_i = \frac{E_{total}}{E_{UL}^i + E_{DL}^i},$$

where E_{total} is defined as the amount of energy available for the client; E_{UL}^i and E_{DL}^i are the expected amounts of energy consumption for data request and data download respectively for client$i$; E_{DL} is proportional to the size of the requested data item.

With this formulation, the authors provide a solution for a two player game. Let the energy cost for sending a request signal be E_s and the energy dissipated while waiting be $E_w = \alpha E_s$. Note that if in a pre-fixed duration t nobody sends a request; all players will try again with the strategy specified probability. In order to avoid waiting infinitely, after two broadcast cycles,[i] if a player still has not obtained the data, it will be forced to send out a request with probability 1. Hence, the expected energy cost for request transmission is

$$E_{UL}^{i} = s_i E_s + (E_s + \alpha E_s) \prod_{j=1}^{n} (1 - s_j) + (1 + 2\alpha) E_s \prod_{j=1}^{n} (1 - s_j)^2.$$

Substituting this into the utility function specified by Equation (8.2.24) and plotting the best strategy curve, the authors claim that the symmetric equilibrium strategy for each player in this game is given by:

$$s^* = \frac{2(1 + 2\alpha)(1 - s^*)^{2(n-1)} - (1 - \alpha)(1 - s^*)^{n-1} - 1}{2(1 + 2\alpha)(1 - s^*)^{2(n-1)} - 2(1 - s^*)^{n-1}}.$$

8.3. Repeated Games

All the discussions so far have been related to one-shot games, also known as single stage games or single shot games. As the name implies, these fall under the category of non-repeated games. During a game, if a player's actions are observed periodically, it becomes possible for other players to condition their play based on the past play actions of their opponents. This can lead to equilibrium outcomes that do not arise in one-shot games. A repeated game consists of a series of one-shot games. In game theory, a repeated game (or an iterated game) is an extensive form game which consists of a number of repetitions of some stage game.

8.3.1. *Background*

Repeated games[14] may be repeated finitely or infinitely many times. The most widely studied repeated games are those repeated possibly infinite number of times. Finite repeated games have some inherit defects. Since the players are inherently selfish, if they know when the game is about to end, players can take advantage of this information and may tend to cheat in the final stages of the game. There are several alternate specifications of payoff functions for infinitely repeated games. The most widely used involves applying a discount factor ($\delta < 1$) to each subsequent game stage. This discount factor has two primary interpretations. First, it captures the fact that at each stage there is some finite probability that the game will end.

[i]Broadcast cycle is the estimated/known-beforehand time for server to broadcast reply to whole network.

Second, it also indicates that each individual cares slightly less about each successive stage.

Repeated games need to keep track of the historic behavior of each player in the game to judge if a player violates a cooperation agreement. Normally, a reputation system is designed to detect misbehavior of selfish players.

There are some well-known policies employed in repeated games to punish detected misbehaving players. Grim Trigger[5] is one of the famous policies.[j] Initially, a player using Grim Trigger will cooperate, but as soon as his opponent misbehaves (thus satisfying the detect trigger condition), the player using Grim Trigger will betray the agreement for the remainder of the iterated game. Since a single defect by the opponent triggers continuous defection for all subsequent stages, Grim Trigger is the most strictly unforgiving policy in an iterated game. Compared to Grim Trigger, Tit-For-Tat is a more effective forgiving strategy. Intuitively, Tit-For-Tat obeys the simple rule of repeating the action (to cooperate or not) of the opponent in the previous game. That is, if previously the opponent was cooperative, the player is cooperative; if not, the player is non-cooperative.

8.3.2. *Routing security and reputation systems*

In wireless ad hoc network routing problems, *attackers* refer to nodes that gain an edge over other nodes by means of malicious behavior. Two important related problems defined in this space are: misleading and selfish behavior. Misleading refers to cases when a node after prior agreement to forward packets does not actually do so. Selfish refers to node behavior where nodes do not forward packets from any other nodes but occasionally send forwarding request for its own packets. To thwart attacks from misleading and selfish nodes, systems can either use prevention, or detection with response. According to Schneier,[32] a prevention-only strategy only works if the prevention mechanisms are **perfect**. Otherwise, odes will find ways to get around them. In real systems, it is extremely difficult to design a perfect prevention system. On the other hand, a relatively easier and more effective option is to design detect with response systems, using reputation systems.

Marti *et al.* propose a simple while effective reputation system to handle the misleading problem. The system comprises of two components: *watchdog* and *path-rater*[23] that are executed on each node in the network. Watchdog provides a mechanism to track the behavior of the nodes in a network. In wireless networks, as opposed to their wired counterparts, nodes can "overhear" transmissions from neighboring nodes. Watchdog takes advantage of this characteristic and records the forwarding attempts of a given node's next-hop node. That is, a node's watchdog can detect misbehavior if it does not overhear its neighbors forwarding packets. Global behavior record exchange helps set up the reputation system of each node. A node's

[j] Grim Trigger is originally credited to James Friedman because he used the concept in a 1971 paper titled "A non-cooperative equilibrium for super games".

reputation is based on the merge of all the other nodes' observations in the network. The path-rater on each node combines knowledge of misbehaving nodes with link reliability data to pick the route most likely to be reliable. After the global reputation message exchange, the reputation record of each node should be kept consistent. Hence the path-rater on each node will induce the same path in the network. Simulation results show that reputation systems can improve throughput by up to 17% with the presence of 40% malicious nodes in the network. This improvement comes at the cost of increasing overhead due to message transmissions by 9–17%.

Bounchegger and Bouder propose a more comprehensive reputation and path selection system for wireless routing, called CONFIDENT.[8] CONFIDENT is an extension to a reactive source-routing protocol based on DSR (Dynamic Source Routing.[16] The authors show that CONFIDENT can handle both problems of misleading and selfishness by having each node periodically monitor behavior of neighboring nodes. In this system, each node runs CONFIDENT locally in a distributed manner. The CONFIDENT protocol consists of 4 major parts: (a) a monitor, which keeps track of the behavior of nodes in the neighborhood of a given node; (b) a trust manager, which handles the misbehavior reports and sends warnings of malicious nodes if necessary; (c) a reputation system, which manages a table consisting of entries for nodes and their corresponding ratings. The table follows along with packet path as a reference for other nodes; (d) a path manager (similar to the path-rater in Ref. 23), which ranks paths according to how the reputation system reacts to misbehaved nodes' routing requests.[k] Simulation results indicate that such a reputation based system performs well even when the fraction of malicious nodes in a network is as high as 60%.

All the reputation systems mentioned so far require global message exchange, which not only increases the system overhead but also generates additional problems. Although these designs have been proved to be effective, exchanging "second-hand information" (merging reputation information from other nodes) is not guaranteed to be secure. Nodes may maliciously accuse other nodes of misbehavior. It is difficult to decide whether an accusation is true, especially so when there is a collusion of several nodes. Bansal and Baker[6] propose a reputation system called OCEAN (Observation-based Cooperation Enforcement in Ad Hoc Networks). OCEAN is a sub-layer that resides between the network and data link layers of the network protocol stack and it handles the problem of misleading. A critical characteristic of OCEAN is that it totally abandons employing second-hand information when setting up node reputations. Instead, nodes make routing decisions based purely on *direct* observations of its neighboring nodes' exchange history with themselves. Surprisingly, the authors find that this system works well in terms of throughput, considering its simplicity compared to the other global message exchange based

[k]Ignoring onward packet forwarding requests from malicious nodes is the most widely used punishment.

schemes. The OCEAN idea works partially based on the fact that in a low mobile network, a node normally only needs to interact with a relatively small set of neighboring nodes while selecting a forwarding path. However, performance of OCEAN has not been evaluated for networks with high mobility where a nodes neighbor may change frequently over shorter durations.

8.3.3. *Incentives in routing*

Rather than building reputation systems and employing punishment policies to coerce nodes to be cooperative, some researchers claim that enough incentives to node utilities can automatically prevent selfish node behavior effectively.

Dasilva and Srivastava analytically model node behavior in voluntary resource sharing networks and quantify cost-benefit tradeoffs that lead to nodes voluntarily participating in resource sharing (for example, forwarding packets of other nodes). In each one-shot game, the players are the N user nodes in the network. The strategy s for each player j is either sharing ($s_j = 1$) or not sharing ($s_j = 0$) the resource. The utility function is a linear combination of the benefits from all other nodes' action and the cost for sharing its own resource with others. This is given by,

$$u_j(s) = \alpha_i \left(\sum_{i \in N, i \neq j} s_i \right) + \beta_i(s_j),$$

where α_i and β_i are the weights for benefit and cost parts respectively. Note that $\beta_i(s_j)$ can be negative (if sharing is considered as a cost) or positive (if the user can derive some benefit or satisfaction from sharing with others). Similar to the prisoner's dilemma problem discussed in Section 8.2.2.1, this one-shot game might reach non-optimal Nash equilibriums. However, better equilibria (cooperation enhanced equilibria) are achievable when the game is repeated. Specifically, the authors assume the repeated game plays K times where K is a geometrically distributed discrete random variable with parameter $p(0 < p < 1)$. The authors consider a grim-trigger strategy adopted by all nodes: each node shares as long as *all* other nodes share; the node does not share if *any* of the other nodes have deviated in the previous round. If everybody shares, at any time a player's expected payoff from that point onward is

$$\frac{\alpha_i(N-1) + \beta_i(1)}{p}.$$

If a node decides to unilaterally deviate at the beginning of one round, its expected payoff from that point forward is just $\alpha_i\ (N-1)$. Therefore, it is a Nash equilibrium for all nodes to participate and share as long as

$$\alpha_i(N-1) > \frac{-\beta_i(1)}{1-p}.$$

As mentioned before, Grim Trigger is considered a "tough" policy since it never forgives and it requires all nodes to participate in resource sharing which is not a very practical scheme in real scenarios. As an alternate strategy, the authors propose that in the kth round of the game, a rational node will participate in sharing if the following condition is satisfied:

$$\alpha_i \left(\sum_{i \in N, i \neq j} s_i^{k-1} \right) > \frac{-\beta_i j}{1-p}.$$

Intuitively, this formula implies that if the expected share gain for a node in future rounds from all the share volunteers is greater than its cost in the current round, it should decide to participate in resource sharing. The expected share gain is measured from the historic sharing status. One major achievement of this strategy is that desirable equilibria can be achieved even if nodes think there is a low likelihood of the game being repeated for additional rounds in the future. The authors also analyze the risk of having rogue nodes by modeling the participant problem as a Bayesian game which is introduced in the next section.

Fabio *et al.* present policies to achieve cooperation among selfish users without a central authority using a repeated game formulation.[27] This study models packet relaying as a game between two neighboring nodes, A and B. In the basic game, if B forwards one packets for A, A hears the transmission and gains utility α. B consumes its resource and gains utility $-\beta$. This game G can be defined as $G = \langle N, \{p_i\}, \{u_i\} \rangle$ where $N = \{1, 2\}$ is the set of players, p_i in $[0, 1]$ is the probability of dropping packets for player i, and $u_i = \beta p_i - \alpha p_{-i}$ is the payoff of player i.

In this game, the Nash equilibrium is mutual defection, i.e. $p_i = 1$ for $i = \{1, 2\}$ is the unique Nash equilibrium for this two player game G. However, the authors show that if this game is repeatedly played, cooperation can be achieved under certain conditions. The authors consider two cases: cooperation without package collisions and cooperation with package collisions.

In the cooperation without collision case, the repeated packet relaying game Γ is defined as a multistage game $\Gamma = \langle N, \{s_i\}, \{U_i\} \rangle$ where $N = \{1, 2\}$ is the set of players, $p_i^{(k)}$ is the dropping probability of player i at stage k, and $u_i^{(k)} = \beta p_i^{(k)} - \alpha p_{-i}^{(k)}$ is the payoff of player i at stage k. The strategy for player i, s_i, is given by $(P^{(0)}, \ldots, P^{(k-1)}) \to p_i^{(k)}$. U_i is the discounted payoff of player i, defined as, $U_i = \sum_{k \geq 0} \delta^k u_i^{(k)}$. The discount parameter δ in $[0, 1]$ represents the subjective evaluation of the future by the player. The simplest strategy to achieve cooperation is Tit-For-Tat (TFT).

In the cooperation with collision case, A wants to send a packet to C through B and, at the same time D wants to a send packet to E. A potential collision between these transmissions prevents A from correctly overhearing B's transmission to C (See Figure 8.5).

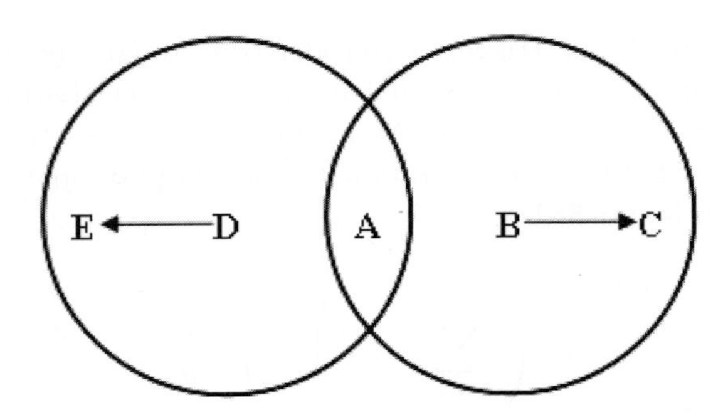

Fig. 8.5. Illustration of relaying model with collisions.

In other words, A is unaware of whether B transmitted its packet to C. The authors employ λ to represent the probability with which each node attempts a transmission in each time instant[1] to capture the distortion introduced by packet collision. Hence, $\hat{p}_i^{(k)} = \lambda + (1-\lambda)p_i^{(k)}$ represents the perceived defection of player i at stage k. Similarly, the perceived payoff of player i at stage k is $\hat{u}_i^{(k)} = \beta\hat{p}_i^{(k)} - \alpha\hat{p}_i^{(k)}$ and perceived discounted payoff of player i is $\hat{U}_i = \sum_{k\geq 0}\delta^k \hat{u}_i^{(k)}$. The authors show in this game, TFT is no longer sufficient to sustain mutual cooperation. The solution is to add a tolerance threshold γ_i to the TFT to accommodate a limited number of defections.

8.4. Bayesian Games

A player is assumed to have global knowledge of all the other players' payoff attributes in all the game formulations described so far. This is a simplifying assumption. In the real world, many times players do not have this global information; they may only have partial global information. When some players do not know the payoffs of the other players, the game is said to have *incomplete information*. A Bayesian game is a game with incomplete information.

8.4.1. *Background*

In a Bayesian game, it is necessary to specify the strategy spaces, type spaces, payoff functions, and beliefs for every player. A strategy for a player is a complete plan of actions that covers every contingency that might arise for every type for a player. A strategy must not only specify the actions of the player given the type that he is, but must also specify the actions that he would take if he were of another type. Strategy spaces are defined as below. A type space for a player is just the set of all possible types of that player. The beliefs of a player describe the uncertainty of

[1]The authors employ a discrete time model for this relay game.

that player about the types of the other players. Each belief is the probability of the other players having particular types, given the type of the player with that belief (i.e. the belief is p (types of other players | type of this player)). A payoff function is a 2-place function of strategy profiles and types. If a player has payoff function $U(x, y)$ and he has type t, the payoff he receives is $U(x^*, t)$, where x^* is the strategy profile played in the game.

A Bayesian Nash equilibrium is defined as a combination of a (i) strategy profile and (ii) belief set for each player about the types of the other players that maximizes the expected payoff for each player given their beliefs about the other player's types and given the strategies played by the other players.

8.4.2. *Security mechanism*

Intrusion detection is an effective security tool widely employed in both wired and more recently wireless networks. Within this context, Liu *et al.* propose a game theoretic framework to analyze interactions between pairs of attacking/defending nodes using a Bayesian formulation.[20] The attacker/defender game is more suitably modeled as an incomplete information game, when the defender is uncertain about the attacker. Two types of games are defined: a static game and a dynamic game. In a static game, the defender does not take game evolution into account while deciding its strategy. The defender always uses his prior belief of his opponent. In contrast, in a dynamic game, the defender adjusts his monitoring strategy based on the new observation of his opponent.

In a static game, there are two players, i is the potential attacker, j is the defender. θ_i represents the player type, $\theta_i = 0$ means regular and $\theta_i = 1$ means malicious. Note $\theta_j = 0$ is always true.

The malicious type of i has two strategies: to attack or to not attack. The defender, player j, has two strategies, to monitor or to not monitor. These strategies have different payoff values. i and j choose their strategies simultaneously at the start of the game. The goal of both players is to maximize their payoff.

μ_0 is defined as a common prior, i.e. player i knows defender j's belief of μ_0. The conclusion for a static game is that if μ_0 is high enough, a mixed-strategy Bayesian Nash equilibrium exists for which defender j plays monitor and i plays attack. A pure-strategy Bayesian Nash equilibrium exists for which the defender j plays his pure strategy to not monitor, and the player i plays his pure strategy to attack if malicious, and to not attack if regular.

The dynamic game is a multi-stage game, which is an extension of the Bayesian static game. The static Bayesian game is repeatedly played. The authors make several assumptions: (1) the game has an infinite horizon; (2) the payoffs of the players in each stage of the game are the same as in the previous stage; (3) there is no discount factor for payoff to the players; (4) the identities of the players remain consistent throughout the game. The authors show that this multi-stage game has a perfect Bayesian equilibrium.

8.4.3. *Medium access control*

Benammar and Baras propose a new medium access control protocol that provides incentives for each of the entities in a wireless network to optimize the overall utility using a Bayesian game formulation.[7] An n player Bayesian game is described as: $\Gamma = \{S_1, \ldots, S_n, T_1, \ldots, T_n, p_1, \ldots, p_n, U_1, \ldots, U_2\}$ where S_i denotes a set of strategies for player i, T_i is the set of the types of the player i, $p_i = p(t_{-i} \mid t_i)$ is the player's belief about other the player types $t-_i$ given his own type t_i, and U_i is the player utility and is a function of the player types and their strategies.

In this game, the authors assume that all users are of the same type. Hence, a strategy profile $\delta = (\delta_1, \ldots, \delta_n)$ is a Bayesian equilibrium of Γ if

$$U_i[\delta(t), t] \geq U_i[\delta_{-i}(t), s_i, t], \quad \forall\ s_i \in S_i.$$

The authors model a three station game and then generalize the game to consider n stations. The strategy for each station is to transmit or wait, $S_i = \{T,\ W\}$. In this formulation, u_s, u_f, u_i denote the payoffs for a successful transmission, failed transmission, and no transmission, respectively. The transmission probabilities of station 1, 2, 3 are x, y, z respectively. $U_{\{i \mid X\}}$ denotes the expected utility of station i that follows strategy X.

If station 1 is a mix between transmitting and waiting, he must have the same expected payoff i.e.

$$U_{1|T} = U_{1|W}.$$

With $x = y = z$, the above equation can be derived as

$$(u_s - u_f)x^2 + 2(u_f - ux)x + u_s - u_i = 0.$$

This has a unique solution given by,

$$x^* = 1 - \sqrt{\frac{u_i - u_f}{u_s - u_f}}.$$

The solution is a mixed Nash equilibrium point. Generalizing the game to consider n stations, we have,

$$x_n^* = 1 - \left(\frac{u_i - u_f}{u_s - u_f}\right)^{\frac{1}{n-1}}.$$

Note that in Equations (8.4.4) and (8.4.5), $u_i - u_f$ is the transmission cost and $u_s - u_f$ is the payoff for a successful transmission.

8.4.4. *Routing*

In this section, we briefly describe some studies exploring power constrained routing using Bayesian game formulations. Nurmi *et al.* present a repeated Bayesian game formulation for energy constrained routing in wireless ad hoc networks with selfish participants.[29] In this game, each repetition is a stage. A new stage starts at time step t_k^i when some node i generates $g(t_k)$ packets to the network $\langle N, E\rangle$, where N and E denote the node and the edge sets in a connected graph. $s(t_k)$ denotes the actual number of packets sent by the node i. θ_i denotes the remaining energy of i and the set of all possible values of θ_i is Θ. While θ_i is private to i, $\mu_{t_k}^j$ denotes i's belief about the remaining energy of at node j at time step t_k and this value is independent over the nodes.

A route discovery phase is performed at the beginning of the game. J is the set of nodes on the routing path. Therefore, the set of players are $i \cup J$. The belief system is $B = \mu_{t_k}^j$ and the action space of i is $A(t_k) = s(t_k) \mid s(t_k) \leq g(t_k)$. $f_j(t_k)$ denotes the number of packets j forwards to i. The action space of a forwarding node j is $A_j(t_k) = f_j\ (t_k) \mid 0 \leq f_j(t_k) \leq g(t_k)$. u_i and u_j are used to denote the utility functions of the source i and the forwarder j. $\gamma(\theta_j)$ denotes the probability of forwarding with energy level θ_j. The authors assume that with more energy levels, the node has a higher probability of forwarding a packet. $\hat{\gamma}_{j,i}$ denotes node j's estimate of the cooperation probability of node i. The probability, $\delta_{j,i}$, with which j forwards to i is a combination of the energy dependent probabilities and the probabilities given by the cooperation mechanism, i.e.

$$\delta_{j,i} = \alpha_j \hat{\gamma}_{j,\,i} + \beta_i \gamma(\Theta_j),$$

where α and β are adjustable importance parameters.

$\phi_j(t_k)$ denotes the belief of the source node i's interest of forwarding to node j. We have $\phi_j(t_k) = \{\phi_j(\hat{t}_k), \hat{\theta}_j\}$. Node i will select a path P from the path set PP. The probability of selecting a routing path $P \in PP$ is $\mu(P) = \prod_{j \in P}[\alpha_i \gamma_{i,\,j} + \hat{\beta}_i \gamma(\theta_j)]$. The utility function of node i depends on which path is employed and how many packets are sent. Thus, it is defined as:

$$u(s(t_k), P) = \mu(P)[h(s(t_k)) - h_\gamma(s(t_k))],$$

where $h(.)$ indicates the number of packets to send to i.

The authors have a two-fold conclusion. First, every perfect equilibrium point is a sequential equilibrium.[m] Second, the proposed model has at least one sequential equilibrium point and the learning sequence converges to a sequential equilibrium. If there is only one sequential equilibrium, the sequence converges to that point. Otherwise, the learning sequence converges to some sequential equilibrium point.

[m] A sequential equilibrium specifies both the strategy for each of the players and a corresponding belief vector. The belief vector gives a probability distribution on the nodes in each player's information set.

8.5. Auctions

An auction depicts a specific scenario of games with imperfect information. Generally, we consider a *symmetric, private-value, sealed-bid, standard* auction game on a single object. The information gathered for each player is symmetric. Private-value means that each bidder knows his own valuation at the time of bidding. He also has some idea of the expectations of other bidders. A sealed-bid is one that is submitted simultaneously by the bidders. A standard auction implies that the participant that submits the highest bid gets the object.

There are two major models of an auction: the first-price auction and the second-price auction. In a first-price auction, the winner pays the price that he submitted in his bid (i.e. the highest price among all bids). In a second-price auction, the winner pays the second highest price across all the submitted bids.

8.5.1. *Background*

Specifically, in an n bidders first-price auction, each bidder submits a sealed bid b_i. With the given bids $b_1, \ldots, b_n$, the payoffs of the players are:

$$u_i = \begin{cases} x_i - b_i & \textit{for } b_i > \max\limits_{j \neq i} b_j \\ 0 & \textit{for } b_i < \max\limits_{j \neq i} b_j \end{cases}.$$

In a first-price auction, the players tend to bid a little less than their valuation in order to increase their payoff.

In an n bidders second-price auction, the payoff for each player is:

$$u_i = \begin{cases} x_i - \max\limits_{j \neq i} b_i & \textit{for } b_i > \max\limits_{j \neq i} b_j \\ 0 & \textit{for } b_i < \max\limits_{j \neq i} b_j \end{cases}.$$

In a second-price sealed bid auction, a weakly dominant strategy is for players to bid as per their valuation price. Therefore, we can say that a second-price sealed bid auction enforces truthfulness. The second-price sealed bid auction is also known as a VCG (Vickrey-Clarke-Groves) auction.

8.5.2. *Routing*

In wireless networks, recall that equipments generally belong to individual users. If the users are selfish, they will only relay packets for a certain price which can at least compensate for their cost of energy consumption. This raises new questions with regards to the selection of the most cost-efficient route for data transfer while obtaining the true price information from individual users (i.e. how to prevent cheating).

Anderegg and Eidenbenz propose a routing protocol for wireless ad hoc networks with selfish users based on the VCG protocol. This is named as ad hoc VCG.[3] The basic idea of the ad hoc VCG mechanism is to employ the second highest sealed price bid (VCG mechanism) while paying the intermediate nodes an extra price in addition to their actual costs to ensure truthfulness. In this non-cooperative game, the source nodes and intermediate nodes are the players. There are two phases in this ad hoc VCG protocol: a route discovery phase followed by a data transmission phase. In the first phase, the nodes send out emission signals to determine the energy expended for transmission between every pair of nodes. This information is employed to create a weighted graph, where each edge is the labeled with a cost value that depends on the energy consumed for transferring unit data using that link. The destination node collects information about all the edge weights and calculates the shortest path from the source to itself. In the second phase, packets are forwarded along the shortest path route and payments are made to the intermediate nodes along the selected route.

There are two alternatives for payments in the data transmission phase: (a) source model and (b) central-bank model. In a source model system, the source will pay both, the actual cost and the extra price of an intermediate node. In a central-bank model system, the source only pays the actual cost while a central-bank (a special node that does not participate in routing) pays the extra price for all the intermediate nodes. Recall that the extra price is given to ensure node truthfulness. It makes cheating unattractive by providing payments as high as a node could possibly attain by cheating. Specifically, in an ad hoc VCG, this extra price for an intermediate node v_i is defined as the difference between shortest path calculated with and without node v_i. Mathematically, the VCG payment M_i for intermediate node v_i is defined as:

$$M_i = \left|SP^{-i}\right| - |SP| + c_i P_{i,i+1}^{\min},$$

where $|SP^{-i}|$ is the cost of the shortest path without node v_i, $|SP|$ is the cost of the shortest path with node v_i, c_i is the cost for unit power consumption, and $P_{i,i+1}^{\min}$ is the minimum transmission energy from node v_i to node v_{i+1}, calculated using the emission signal information and the information in the acknowledge packets from neighbors that contain received signal strength.

The authors also prove that with this mechanism, the resulting total overpayment is always bounded by a factor of $2^{\alpha+1}\frac{c_{\max}}{c_{\min}}$ where α is the signal loss exponent and $c_{\max}$ ($c_{\min}$) is the maximum (minimum) cost-of-energy declared by the nodes on the most cost-efficient path (i.e. shortest path calculated by destination).

Zhong *et al.* propose a protocol, called Corsac, that gives nodes incentives to route and forward packets in wireless ad hoc networks with selfish users.[37] Corsac is also based on the VCG mechanism. Again, as before, the players in this routing game are all the nodes in the network. The strategy for each player is a finite, discrete set of transmit power levels l. The utility function for player i is defined as

$U_i = -l\alpha + p_i$ where p_i is the credit paid for participating in the forwarding process and α is the energy consumption parameter.

The authors argue that there exists no dominant forwarding protocol for such an ad hoc routing game. In order to give further incentives to the players to participate in the forwarding process, the authors split the game into two stages as suggested by Feigenbaum and Shenker.[13] The game now has two parts: the routing decision part and the forwarding part. In the routing stage, the nodes take a routing decision, which specifies what each node is "supposed to do" in the forwarding stage. In the forwarding stage, a node's routing decision and the forwarding sub-action (the action whether nodes should forward or not) jointly decides each node's utility. Now the strategy for each node is separated into two sub-actions: $S_i = (S_i^{(r)}, S_i^{(f)})$ where $S_i^{(r)}$ and $S_i^{(f)}$ are the player's strategy for routing and forwarding, respectively. The utility function is defined as: $U_i' = U_i(R, a^{(f)})$. The goal of the players is to maximize the joint utility U_i'. Specifically, the optimal solution can be achieved if routing decision is sub-stage optimal and the incentive for nodes to follow the routing decision in forwarding stage leads to sub-stage optimal too.

In the routing stage, the routing decision is based on the VCG mechanism. Instead of doing a link cost estimation of the previous work,[3] the authors combined the VCG and the cryptographic technique in order to prevent cheating. In this routing protocol design, before sending out data packets, the source sends out test packets to give information on the destination to choose the cheapest energy cost path. Each intermediate node forwards the test packet while recording its own MAC address, in an encrypted manner, into the routing information and the test packet. Upon collecting all the routing information and cost information, the destination checks the routing information using a symmetric key mechanism and picks the minimum energy cost path as the routing path. The authors show that this routing protocol is a routing-dominant-protocol.[n]

8.6. Challenges for Future Research

Game theory is a powerful tool to model interactions among self-interested users and predicting their behavior. However, a lot of challenges still exist in the process of applying game theoretic tools to network contexts. Here, we list some of the main challenges:

- Modeling a realistic problem to a game with rational selfish users and a reasonable strategy space such that the problem is tractable.
- Definition of suitable utility functions that are reasonable and possess "good" properties from the game theoretic point of view (for example, can yield a unique Nash equilibrium or make the game supermodular).

[n]Definition (from Ref. 13): a routing protocol is a routing-dominant protocol to the routing stage if following the protocol is a dominant sub-action of each potential forwarding node in the routing stage.

- Finding distributed techniques to attain the converged equilibrium point and reducing the computational complexity of these approaches.
- Nash equilibrium is defined as a state when no player is willing to unilaterally change his strategy. However, it does not consider cases where two or more players may change their strategies at the same time. The problem of two or more players cooperating to gain higher benefits as compared to the other players is known as *collusion*. Although there has been some pioneering work related to the collusion problem,[38] further investigation is needed to better understand the collusion problem, how to effectively avoiding such problems and what kind of policies need to be in place to detect collusion and respond appropriately.

8.7. Conclusions

In this chapter, we briefly introduced some core concepts employed in game theory and then provided a variety of recent results highlighting various applications of game theoretic tools in wireless networks. The growing interest of the research community in applying game theoretic tools toward newer problems in wireless networks promises exciting results in the upcoming years.

To the reader interested in learning more about game theory, we suggest some further reading. Fudenberg and Tirole's[14] book on game theory gives a comprehensive graduate-level treatment of game theory and its many formulations. MacKenzie and DaSilva's[22] book titled "Game Theory for Wireless Engineers" is a more friendly introduction to the basics of game theory explicitly aimed at wireless engineers.

Problems

1. **Pure Nash Equilibrium and Dominant Strategy Calculation:** Given a 3×3 game as depicted in Figure P1, find all the **dominated** strategies. Also, find all the pure Nash equilibria.

		Player 2		
		Left	Middle	Right
Player 1	N	(1, 1)	(3, 4)	(2, 1)
	C	(2, 4)	(2, 5)	(8, 1)
	S	(3, 3)	(0, 4)	(0, 9)

Fig. P1. Payoff table for Problem 1.

2. **Pure Strategy Equilibrium and Dominant Strategies:** Assumptions and payoff table are given by the description of security attacker/IDS paper.[1] Prove that AS1/SS1 is the unique pure Nash equilibrium for this 2-player game.
3. **Mixed Strategy Equilibrium Calculation:** In the power saving problem[36] formulation with multiple requesting clients, calculate the mixed strategy equilibrium for a 2-player game and draw the best response curves for both players in a single graph.
4. **Nash Equilibrium Calculation:** Given the payoff table in Figure P4, calculate all the pure and mixed Nash equilibria of this game.

I \ II	*l*	*c*	*r*
T	0 2	2 2	3 1
B	5 0	4 4	2 3

Fig. P4. Payoff table for Problem 4.

5. **POA and POS calculation:** Given the table in Figure P5, calculate the price of anarchy and price of stability in terms of maximize payoffs.

		Player 2	
		X	**Y**
Player 1	**A**	**1, 1**	**1, 2**
	B	**3, 1**	**1, 1**

Fig. P5. Payoff table for Problem 5.

6. **Routing and Selfishness (Braess's paradox):** Consider the following network routing graph labeled with link delay l as shown in Figure P6.

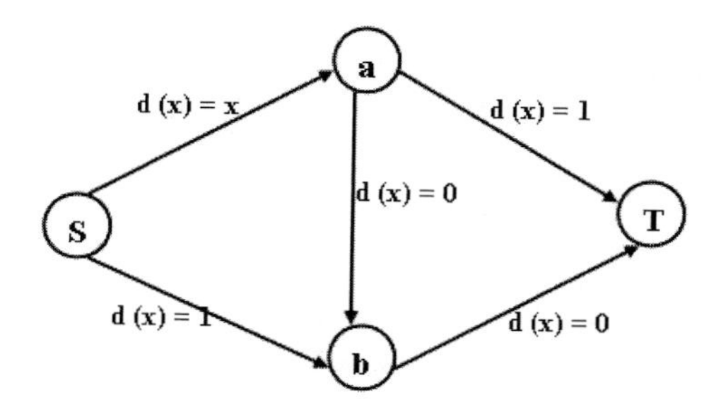

Fig. P6. Illustration of Braess's paradox.

Let's normalize the whole set of users of this network as 1. In the figure, $d(x) = x$ indicates that the delay of the link is proportional to the user size x choosing that link. Suppose all nodes picking path from S to T selfishly choose to minimize the path delay (if there exist multiple paths that are indifferent for the user, he will pick one randomly). Find out the Nash equilibrium path for the users. Calculate path delay for each user at the Nash equilibrium. Is this Nash same as the optimal solution?

Now we assume that for some reason, link $a \to b$ is broken and this information is known to all the users in the network. With deletion of such a "super link", is the delay of each user worse or better than before?

7. **Supermodularity:** In the power control game,[31] prove the non-decreasing differences property of the game with the modified strategy space.
8. **Supermodularity:** Consider an $M/M/1$ queue with service rate μ. There are two groups of users (modeled as two players) with arrival rate λ_1 and λ_2 respectively. For the two players, the strategy is to selfishly choose arrival rate λ_i so as to maximize their own utility function $U(i) = \lambda_i\ (\mu - \lambda_1 - \lambda_2)^\alpha$ where $\alpha < 1$ is a given constant. We also assume the queue will not grow to infinity, i.e. $\lambda_1 + \lambda_2 < \mu$. Investigate if this game is a supermodular game.
9. **Truthful Auction:** In the ad-hoc VCG mechanism design,[3] source node (or the centralized bank) will overpay each of the intermediate routing nodes so that they do not have incentive to cheat. How much extra price is paid? Prove that the total over-payment with this mechanism is bounded.
10. **Truthful Auction (General):** We already know that a second price sealed auction ensures truthfulness. Consider kth price sealed auction where the winner(s) will pay the price of kth highest bids. Does this mechanism ensure truthfulness in a single item auction? Does this mechanism ensure truthfulness in a multiple items (i.e. exists multiple winners) auction?

Bibliography

1. A. Agah, S. K. Das, K. Basu and M. Asadi, Intrusion detection in sensor networks: a non-cooperative game approach, *NCA '04: Proceedings of the Third IEEE*

International Symposium on Network Computing and Applications, IEEE Computer Society, Washington, DC, USA (2004) 343–346.
2. T. Alpcan, T. Basar, R. Srikant and E. Altman, CDMA uplink power control as a noncooperative game, *Wireless Networks* **8**(6) (2002) 659–670.
3. L. Anderegg and S. Eidenbenz, Ad hoc-VCG: a truthful and cost-efficient routing protocol for mobile ad hoc networks with selfish agents, *MobiCom '03: Proceedings of the 9th Annual International Conference on Mobile Computing and Networking* (ACM Press, New York, NY, USA, 2003) 245–259.
4. E. Anshelevich, A. Dasgupta, J. Kleinberg, E. Tardos, T. Wexler and T. Roughgarden, The price of stability for network design with fair cost allocation, *FOCS '04: Proceedings of the 45th Conference on Foundations of Computer Science* (2004) 295–304.
5. R. M. Axelrod, *The Evolution of Cooperation* (Basic Books, New York, 1984).
6. S. Bansal and M. Baker, Observation-based cooperation enforcement in ad hoc networks, *The Computing Research Repository*, cs.NI/0307012, Informal publication (2003).
7. N. BenAmmar and J. S. Baras, Incentive compatible medium access control in wireless networks, *GameNets '06: Proceedings of the 2006 Workshop on Game Theory for Communications and Networks* (ACM Press, New York, NY, USA, 2006) 5.
8. S. Buchegger and J. L. Boudec, Performance analysis of the CONFIDANT protocol, *MobiHoc '02: Proceedings of the 3rd ACM International Symposium on Mobile Ad Hoc Networking & Computing* (2002) 226–236.
9. A. Cournot, *A Research Into the Mathematical Principles of the Theory of Wealth* (Libraire des sciences politiques et sociales, 1897).
10. G. Debreu, Valuation equilibrium and pareto optimum, *Proceedings of the National Academy of Sciences* **40** (1954) 588–592.
11. S. Eidenbenz, V. S. A. Kumar and S. Zust, Equilibria in topology control games for ad hoc networks, *Mobile Networks and Applications* **11**(2) (2006) 143–159.
12. Z. Fang and B. Bensaou, Fair bandwidth sharing algorithms based on game theory frameworks for wireless ad-hoc networks, *Proceedings of the IEEE Infocom* (2004).
13. J. Feigenbaum and S. Shenker, Incentives and internet computation, *PODC '03: Proceedings of ACM Symposium on Principles of Distributed Computing* (2003).
14. D. Fudenberg and J. Tirole, *Game Theory* (MIT Press, 1991).
15. J. C. Harsanyi, Games with incomplete information played by Bayesian players, *Management Science* **14**(3) (1967) 159–182.
16. D. Johnson and D. Maltz, Dynamic source routing in ad hoc wireless networks, *Mobile Computing* (1996) 153–181.
17. A. Kesselman, D. Kowalski and M. Segal, Energy efficient communication in ad hoc networks from user's and designer's perspective, *SIGMOBILE Mobile Computing and Communications Review* **9**(1) (2005) 15–26.
18. E. Koutsoupias and C. Papadimitriou, Worst-case equilibria, *Lect. Notes Comput. Sci.* **1563** (1999) 404–413.
19. H. Liu and B. Krishnamachari, A price-based reliable routing game in wireless networks, *GameNets '06: Proceeding from the 2006 Workshop on Game Theory for Communications and Networks* (ACM Press, New York, NY, USA, 2006).
20. Y. Liu, C. Comaniciu and H. Man, A Bayesian game approach for intrusion detection in wireless ad hoc networks, *GameNets '06: Proceeding from the 2006 Workshop on Game Theory for Communications and Networks* (ACM Press, New York, NY, USA, 2006).
21. R. D. Luce and H. Raiffa, *Game and Decisions — Introduction and Critical Survey* (John Wiley & Sons, Inc., New York, 1957).

22. A. MacKenzie and L. DaSilva, *Game Theory for Wireless Engineers, Synthesis Lectures on Communications* (Morgan & Claypool Publishers, 2006).
23. S. Marti, T. J. Giuli, K. Lai and M. Baker, Mitigating routing misbehavior in mobile ad hoc networks, *Proceedings of Mobile Computing and Networking* (2000) 255–265.
24. F. Meshkati, M. Chiang, H. V. Poor and S. C. Schwartz, A game-theoretic approach to energy-efficient power control in multicarrier CDMA systems, *IEEE Journal on Selected Areas in Communications* **24**(6) (2006) 1115–1129.
25. F. Meshkati, H. V. Poor and S. C. Schwartz, A non-cooperative power control game in delay-constrained multiple-access networks, *IEEE International Symposium on Information Theory* (2005).
26. F. Meshkati, H. V. Poor, S. C. Schwartz and N. B. Mandayam, An energy-efficient approach to power control and receiver design in wireless data networks, *IEEE Transactions on Communications* **53**(11) (2005) 1885–1894.
27. F. Milan, J. J. Jaramillo and R. Srikant, Achieving cooperation in multihop wireless networks of selfish nodes, *GameNets '06: Proceeding from the 2006 Workshop on Game Theory for Communications and Networks* (ACM Press, New York, NY, USA, 2006) 3.
28. J. F. Nash, Equilibrium points in N-person games, *Proceedings of the National Academy of Sciences of the United States of America* **36**(1) (1950) 48–49.
29. P. Nurmi, Modeling energy constrained routing in selfish ad hoc networks, *GameNets '06: Proceeding from the 2006 Workshop on Game Theory for Communications and Networks* (ACM Press, New York, NY, USA, 2006) 3.
30. J. B. Rosen, Existence and uniqueness of equilibrium points for concave n-player games, *Econometrica* **33** (1965) 550–534.
31. C. U. Saraydar, N. B. Mandayam and D. J. Goodman, Efficient power control via pricing in wireless data networks. *IEEE Transactions on Communications* **50**(2) (2002) 291–303.
32. B. Schneier, *Secrets and Lies: Digital Security in a Networked World* (Wiley, New York, 2000).
33. V. Shah, N. B. Mandayam and D. J. Goodman, Power control for wireless data based on utility and pricing, *The 9th IEEE International Symposium on Personal, Indoor and Mobile Radio Communications* **3** (1998) 1427–1432.
34. D. M. Topkis, Equilibrium points in nonzero-sum n-person submodular games, *SIAM Journal on Control and Optimization* **17**(6) (1979) 773–787.
35. W. Vickery, Counterspeculation, auctions and competitive sealed tenders, *The Journal of Finance* **16**(1) (1961) 8–37.
36. M. Yeung and Y. Kwok, Game theoretic power aware wireless data access, *WOWMOM '05: Proceedings of the 6th IEEE International Symposium on a World of Wireless Mobile and Multimedia Networks*, IEEE Computer Society, Washington, DC, USA (2005) 324–329.
37. S. Zhong, L. E. Li, Y. G. Liu and Y. R. Yang, On designing incentive-compatible routing and forwarding protocols in wireless ad-hoc networks: an integrated approach using game theoretical and cryptographic techniques, *MobiCom '05: Proceedings of the 11th Annual International Conference on Mobile Computing and Networking* (ACM Press, New York, NY, USA, 2005) 117–131.
38. S. Zhong, Y. Yang and J. Chen, Sprite: a simple, cheat-proof, credit-based system for mobile ad hoc networks, *Proceedings of IEEE Infocom* (2003).

PART 2

Wireless Sensor Networks

Chapter 9

Wireless Sensor Networks — Routing Protocols

Abbas Jamalipour* and Mohammad A. Azim
University of Sydney, NSW 2006, Australia
* *a.jamalipour@ieee.org*

One of the most important issues in wireless sensor networks (WSNs) is the design of optimized routing protocols that cannot only route the packets in an efficient way and with minimum overheads but also consider severe power restriction of those networks. Correct design of routing protocols can prolong the lifetime of these networks and therefore much research has been carried out to develop new and customized routing protocols for the WSNs. In this chapter, we will first look at different topologies of the WSNs, say flat and hierarchical networks and compare the two topologies from the routing efficiency point of view. We will then review the most important routing protocols designed for the WSNs for flat and hierarchical topologies. We will further look at some other routing protocols in those networks that aim at providing quality of service in sensor networks and can possibly control the power consumption and therefore the lifetime of the networks. Finally we will review data-centric and geographic routing approaches to complete our survey on routing protocols in WSNs.

Keywords: Wireless sensor networks; routing protocols; flat routing; hierarchical routing; quality-of-service based routing; data-centric routing; geographic routing; energy efficiency; quality of service; network lifetime.

9.1. Introduction

Wireless sensor networks (WSNs) have become the focus of many studies in the past few years. The protocol stack used by WSN nodes may consist of application layer, transport layer, network layer, data link layer and physical layer along with the power, mobility and task management planes.[6] With such capabilities, WSN has the potential for deployment in the real-world for a large number of diversified applications. So far a number of such applications have been developed and are highlighted below.

- Environmental applications[73,74,107]
- Wildlife monitoring[51,69]
- Military, security and rescue applications[76,77,92,106]
- Health applications[14,89]
- Agriculture and farming[15,22,23]
- Consumption monitoring and retail management[53,84]
- Industry and home applications.[9,85,94]

In any of these applications, sensors generate the desired data required by the user, while the sink (a high end node) acts as a gateway to the infrastructure for collecting the information from the nodes and forwarding the queries from the user. For end-to-end data traversal (i.e. from source to sink), the intermediate nodes forward the data and sometimes perform additional processing such as data aggregation (data is aggregated based on certain deterministic characteristics). The participation of these different types of nodes in the routing process incorporates the following functionalities with some specific patterns.

- **Event detection:** Source detects the events that match the query and reports back to the sink with the sensed information.
- **Function approximation:** To map the area approximately, sensor nodes are used to approximate a function of location that estimates the change of the physical value from one point to another.
- **Periodic measurement:** Sources report the sensed value to the destination sink periodically according to the set interval or as per application requirements.
- **Tracking:** As an object of interest, information is obtained regarding the mobility of a sensor node. In fact, sensor nodes generate meaningful information by interacting with each other.

9.1.1. *Challenges in the WSNs*

Due to the limitation of resource-limited sensor nodes, development and deployment of WSN for a variety of applications pose a number of challenges.[2,3,6] These are briefly summarized below.

- **Lifetime:** It is likely that the provider/operator of the WSN intends to operate the network at least for the time necessary to fulfill the given task. Particularly the lifetime will be application specific and of course will depend on the energy efficiency of various mechanisms employed (such as routing protocol, sleep strategy, aggregation mechanism etc.).
- **Quality of service (QoS):** The sensor network needs to provide some QoS to its applications which define the level of service granularity. The QoS functions are therefore translated into a set of cost metrics such as delay, jitter, bandwidth, packet loss, and so on.

- **Scalability:** In cases where a large number of sensor nodes need to be deployed in the sensor bed, scalability becomes a key design parameter. Here the underlying assumption is that the presence of a large number of nodes is not detrimental to the proper functioning of the protocols and algorithms that have been incorporated to support efficient network functionality.
- **Fault tolerance:** Due to the inherent unreliable characteristics of the WSN (on account of wireless interface), the network needs to incorporate fault tolerant capability so as to prevent network partitioning. A simple and accepted way of providing such a mechanism is to deploy sensors redundantly.
- **Cooperation:** It is necessary for the sensor nodes to interact with each other in a cooperative manner to detect an event, to route the sensed data effectively, to participate in the aggregation process, and so on.
- **Sustainability:** A topology change due to the mobility or due to the death of a sensor node (resulting in network partitioning) must not act as a deterrent for the regular service in the network. A WSN must be robust and adaptive enough to tackle such a situation.
- **Programmability:** Sensors must be able to adjust or modify the processing options dynamically and without any manual intervention based on the perceived circumstances.

From the above discussion, it is evident that a huge research effort is necessary to address these issues. Not only from the perspective of the academia but a concerted effort by the industry alike is warranted to develop a truly energy-efficient network, necessitating the clear identification of the research areas. According to literature, WSN research can be loosely subdivided into several categories i.e. routing,[6] localization,[82,104] MAC,[35,102] TCP and security.[21,36,100,103] Energy-efficient routing constitutes one of the most important design parameters in WSNs.

9.1.2. *Routing in the WSNs*

Routing protocols dictate the path that a packet follows from source to destination. The ultimate goal of such a protocol is to minimize the cost of the route. Fundamentally definition of cost varies from one network to another. For example, in the reliable infrastructure-based network such as Internet, hop distance defines the fundamental cost of the route. In wireless networks with unreliable interface (e.g. ad hoc networks, sensor networks, etc.) a markedly different approach is warranted. Heuristically this depends on the corresponding routing perspective of individual networks and their related applications. In sensor networks, although mobility imposes restrictions on the routing constraints (similar to ad hoc networks), the primary focus of a routing protocol design relies in defining an energy efficient network. This stems from the fact that due to the structure of the sensor networks and the limitations of the battery power of a sensor node, it is almost impossible to replace a network once it is deployed, necessitating attempts by the network designers in extending the network lifetime. After the energy issue the second most important thing is the application itself for which the routing protocol is deployed. Several solutions to address these energy efficient routing issues will be reviewed in this chapter.

The following section provides a detailed discussion of the existing routing protocols, and highlights their pros and cons. The review of the routing protocols is presented primarily based on the application specific usage of the sensor networks and their logical structure. The review clearly identifies the need for developing energy efficient routing protocols to extend the lifetime of the network.

9.2. Background

For the past few years, WSN has witnessed phenomenal research efforts into developing routing protocols to address the growing number of diversified applications. This section presents the-state-of-the-art routing protocols that have been proposed so far in WSN. A good survey of the routing protocols can also be found in Refs. 3, 8, 82.

The routing protocols in WSN can be classified in several ways.[3,99] Here, the routing protocols are classified in terms of their logical structures flat, hierarchical (shown in Figure 9.1), and QoS based routing. Besides these, a brief overview of data centric routing, geographic routing, in-network aggregation and network flow based approaches are also presented.

9.2.1. *Flat routing*

In flat network (in the absence of any hierarchy), each and every node assumes the same responsibility and performs the routing task. The most significant flat routing protocols in literature are discussed below.

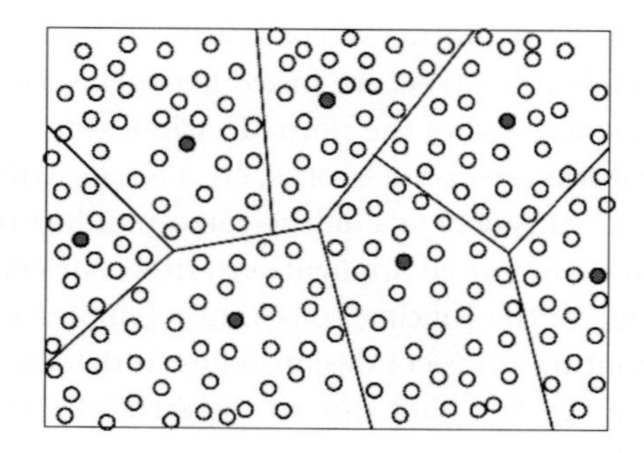

Fig. 9.1. WSN architectures.

9.2.1.1. *Reactive routing with controlled flooding*

Reference 101 proposes a reactive routing protocol that provides a reliable transmission mechanism with low energy consumption. A controlled broadcasting technique has been employed to flood the queries from sink to the source. The broadcast is controlled such that only the energy rich nodes with high distance (from the node it receives packet) is used to rebroadcast. When the query finds the source the reply message follows the reverse path to get back to the sink. However flooding may often result in limited wireless resource wastage from data implosion thereby impending the key design requirement of energy efficient protocol.

9.2.1.2. *Minimum transmit energy (MTE)*

Minimum transmit energy (MTE)[75] proposes a multi-hop routing protocol for flat sensor networks, where it attempts to minimize the transmit energy cost from source to destination through multiple hops. The transmit energy cost is conceived by taking the cost function based on the distance squares over multiple hops. Figure 9.2 illustrates the operating principle of MTE. Assuming node C and D are the candidate next hop nodes for node A en route to the sink B, the cost function for C and D are derived as $costC = distance_{AC}^2 + distance_{CB}^2$ and $costD = distance_{AD}^2 + distance_{DB}^2$, respectively, and the optimal selection is defined by the minimum cost between the two.

Since the protocol only considers the distance, its focus is limited to minimizing the transmit cost and accordingly fails to consider load distribution across the sensor field in a uniform and efficient manner. By incorporating metrics such as energy and link usage it is possible for the protocol to effectively extend the network lifetime.

9.2.1.3. *Energy and mobility aware geographical multi-path routing protocol (EM-GMR)*

Another flat routing protocol termed as geographical multi-path routing (GMR) is introduced in Ref. 50. This location based algorithm uses only the location information to select the next hop node. Reference 63 introduces an improvement of the

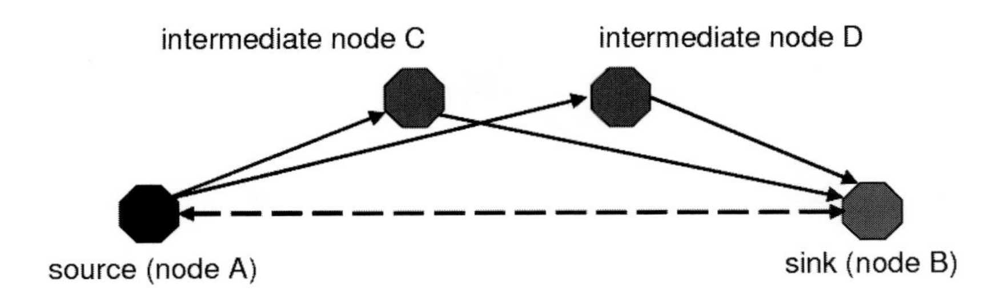

Fig. 9.2. MTE forwarding scheme.

GMR approach termed as energy and mobility aware geographical multi-path routing protocol (EM-GMR). EM-GMR utilizes two additional metrics i.e. energy and mobility alongside the location information. These metrics are optimized through fuzzy inference systems. Although mobility is taken into account as a metric to resolve the scenario under consideration i.e. a highly mobile sensor network, for cases where the sensors are less mobile and the network is moderately loaded, other metrics can also be considered, in particular the link usage may become more important an issue to be considered.

9.2.1.4. *Sensor protocol for information via negotiation (SPIN)*

Sensor protocol for information via negotiation (SPIN),[59,60] disseminates all the data to every nodes under the assumption that each and every node can serve as a base station. Under this assumption user can query any node to get the information readily. SPIN is based on the simple idea that the nodes residing in the close proximity can have the same data. Therefore forwarding the redundant data to that particular node is eliminated through negotiation and resource adaptation. SPIN uses high-level naming scheme to completely describe their collected data called metadata. Metadata negotiation is carried out before performing any particular data transmission. Through this negotiation SPIN family of protocols overcome the limitations of classic flooding approach.

A number of protocols have been defined in the SPIN family of protocols; among them the most important two are SPIN-1 and SPIN-2. The negotiation is incorporated to ensure only to transfer the useful information. SPIN-1 is a three stage protocol that negotiates before communications. Three types of messages are exchanged i.e. ADV, REQ and DATA. The ADV message is broadcasted to advertise the message the node possesses. A REQ message is only initiated when a neighboring node needs to have the message advertised. Finally the DATA is transferred in accordance to the REQ received from the initiator node. Here, a messaging of the SPIN protocol is depicted in Figure 9.3. Here, the source advertises the metadata to all

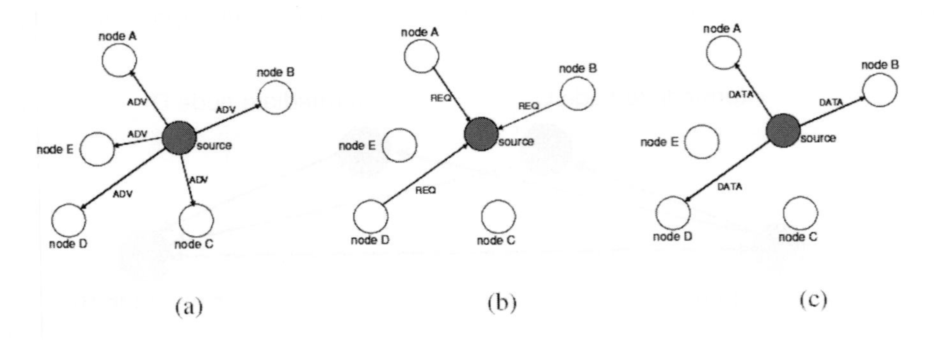

Fig. 9.3. Messaging in SPIN protocol.

the neighbors (shown in Figure 9.3(a)). Among the neighbors only node A, B and D request for the content (in Figure 9.3(b)) and thereby served by the source (in Figure 9.3(c)). In an extension to SPIN-1, SPIN-2 incorporates a threshold based resource awareness mechanism. When the energy of the node is plenty, it works as SPIN-1. However, when the energy goes down to a low threshold value node's participation to forwarding is also reduced to conserve energy. This participation is only when it believes it can complete all the stages of the protocol without going below the low energy threshold.

As a result, SPIN protocol conserves energy through metadata negotiation and thereby limits the redundant data. In addition, as the topology changes are localized, nodes only need to know its single hop neighbors.

Unfortunately, SPIN family of protocols suffers from few disadvantages. Most importantly this protocol does not guarantee the delivery of the data. If the intermediate nodes are not interested about a particular data, then the data will not be delivered to the destination at all. Moreover, a replication like SPIN might be considered as wasteful usage of internal resources where the network has too many copies of the same data.

9.2.1.5. *Directed diffusion (DD)*

Reference 49 introduces directed diffusion, a hybrid data centric routing approach for WSN. Here, sinks generate interests that propagate through the network looking for nodes with matching event records. The events are forwarded to the originators of the interests through multiple paths. The interests contain the interval attribute field that indicates the frequency with which data related to these interests should be sent. The longevity of the communication allows the protocol to discover the good paths. The network reinforces one or a small number of these paths during the communications.

Figure 9.4 shows the three steps of the DD, where the sink broadcast the interest (shown in Figure 9.4(a)), upon receiving it the source broadcast the events to the sink (in Figure 9.4(b)) and finally the best path is reinforced by the sink (in Figure 9.4(c)). Nodes in the networks are application aware, therefore the diffusion achieves energy efficiency by selecting good paths empirically. It also caches and processes data in the networks. This caching increases the efficiency, robustness and scalability of the coordination among the nodes. By this directed diffusion approach, spontaneous propagating can be achieved in some sections of the network. It can be used to present queries where the requesting nodes are not expecting data that satisfy query for a specific duration. In addition, the algorithm is resilient to node failure as it can reach the destination through a number of routes.

However since the protocol floods the network with interests, for the one time queries, directed diffusion is considered too costly because of the limited worth in establishing the gradients for queries that use the path only once.

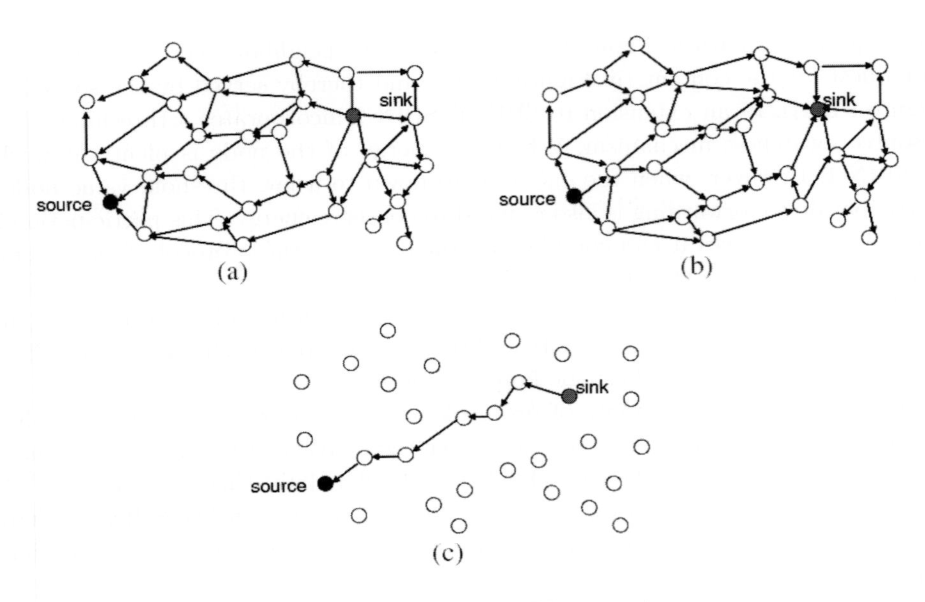

Fig. 9.4. Messaging in DD protocol.

9.2.1.6. *Fuzzy directed diffusion*

In Refs. 12 and 13, a fuzzy diffusion is introduced as an improvement of the directed-diffusion (DD) protocol. As DD uses flooding to propagate interest and flooding is undesirable in the context of sensor networks due to its high cost, this protocol is incorporated (to limit the cost). The protocol proposes a fuzzy optimization base probabilistic interest forwarding technique instead of flooding and thereby makes the system more efficient. According to this approach, interest is propagated to the nodes with a higher probability of having more available energy and queue. Another approach of limiting the DD flooding is proposed in Ref. 93 where the node decides about whether to participate in the query retrieval process by fuzzy logic, based on the data and the remaining battery power.

9.2.1.7. *Trajectory based forwarding (TBF/TBF+)*

Trajectory based forwarding (TBF)[81] offers a technique to disseminate packets from the sink to the set of nodes along a predefined curve. The curve represents the trajectory of the packet where intermediate node forwards the packet based on their distance from the defined trajectory.

As TBF does not define how the curves will be selected or generated, in TBF+[67] the trajectory is defined by the energy map of the network. Here, 50% of the nodes with high remaining energies are selected to start with the trajectory selection algorithm while the curves are defined by utilizing multiple regression analysis.[78]

9.2.1.8. *Rumor routing*

In an alternative to directed diffusion (DD), rumor routing approach (alternatively known as gossip based routing) is introduced in Ref. 20. This stems from the fact that when there is no geographical criterion to diffuse the tasks, directed diffusion uses flooding throughout the region to inject the queries. Rumor routing avoids the expensive flooding, especially in cases where the number of events is small and the number of queries is large. Here, the underlying idea is to pass the queries to those particular nodes that have observed a particular event. In this protocol both the source and the sink spread information. When these two meets in a node, the path between the source and the sink has been established.

This protocol also reduces the communication cost by employing agents in the networks. However, it suffers from the limitation that it requires a number of ongoing events. If there are not many events of interest, then the cost of managing the agents becomes uneconomical, resulting in unnecessary exhaustion of energy. In addition, the routing may cause a large amount of delay while delivering the interest to a desired destination.

9.2.1.9. *COUGAR*

An additional query layer has been proposed in COUGAR[110] for the abstraction of complex query and in-network data aggregation to satisfy the ultimate goal of energy conservation. Here, the sink is responsible to generate the query plan. This plan incorporates the necessary information regarding the data flow, in-network computation, selection of the leader of the query as well as the delivery of the query to the relevant nodes.

Although the protocol provides a solution to the query problem independent of the network layer, it runs the risk of becoming a burden due to its overhead from the additional layer. Energy consumption arising from such overhead can be reduced at the cost of additional delay (not very uncommon, such as PEGASIS).[64]

Intelligent selection of the leader node is also necessary to ensure that the leader node is not static, rather dynamic and therefore is not susceptible to early death from excessive burden.

9.2.1.10. *ACtive QUery forwarding In sensoR nEtworks (ACQUIRE)*

ACtive QUery forwarding In sensoR nEtworks (ACQUIRE)[86] is proposed for the networks where queries are complex and consists of a number of sub-queries.

According to the protocol, the queries (complex) are initiated by the sink, and nodes based on their observed/stored value, tries to answer the query (possibly partially). If the query is not fully answered the node then gathers information from its h_0 hop neighbors. After resolving the query, the final and aggregated response is sent to the query generator sink. The neighborhood here is largely affected by the selection of h_0. To be effective in querying, the value of h_0 is varied to make the query region dynamic. In the worst case scenario, when this value becomes as big as the network size, it is interpreted as flooding the whole networks with the query.

9.2.1.11. *Periodic, event driven and query based protocol (PEC)*

Periodic, event driven and query based protocol (PEC)[18] uses public/subscribe paradigm to disseminate data across the networks. The protocol also provides a fault tolerant technique based on the acknowledge schemes. The public/subscribe provides full decoupling of the participant that makes the system scalable and flexible.[37] This decoupling is in the perspective of time, space and flow. Decoupling in time means the publisher and the subscriber do not need to be active at the same time. Decoupling in space and flow relate to the fact that they do not need to know each other and do not need to be synchronized to interact, respectively. In PEC the public/subscribe is used to perform routing for the periodic, query-based and event driven data. The protocol starts with sink initiating a flooding process to build a hop tree in the network. After building the tree, the sink initiates subscription messages. Upon receiving the message, the intended recipient (through the location match) stores the message, otherwise it rebroadcasts the packet. In case of an event match, the query node verifies the sender ID of the subscription and forwards the data. As such, through successive hops the query reaches the sink. To make the system robust, the protocol employs the ACK message. When the ACK is not reached, the node initiates a local search to reestablish a path.

9.2.1.12. *Energy-aware routing*

Energy-aware routing (EAR)[91] is a reactive approach to sensor network routing protocol. Similar to DD approach, EAR maintains a set of paths instead of a single path. These paths are used for forwarding data with a certain probability. Depending on the energy consumption to each specific path the probability is set. Utilizing the set of paths and using all the paths the protocol increases the network lifetime. The operation of the protocol is initiated through the localized flooding and discovers all the existing paths from the source to destination. From the discovered paths the protocol then eliminates the paths having very high costs. Among the remaining paths, data is forwarded according to the probability which is inversely proportional to the cost of that particular path. Destination node utilizes localized flooding to keep the derived paths alive. This protocol achieves a better performance with respect to the directed diffusion through a more complicated route setup procedure.

9.2.1.13. *Minimum cost forwarding algorithm (MCFA)*

In minimum cost forwarding algorithm (MCFA),[111] a node maintains its least cost estimated from the node itself and the sink. To maintain the estimation, every node in the initial time sets the cost to infinity. First, sink broadcasts the cost establishment message with the cost field set to zero. When the node receives the message, it first calculates the sum of the message cost field and the link on which it is received. If the calculated sum is less than the old estimate, the estimate is updated with the new value, otherwise the node disregards the broadcast and the estimate remains unchanged. As this update procedure suffers from having multiple updates in the distant nodes (from the sink), a backoff algorithm is incorporated where a node does not send an update message before the backoff timer expires. Here the backoff timer is set to $a \times l_c$ where a is a constant and l_c is the link cost at the time when the message is received. After this initial establishment of the cost, data is forwarded to the sink through successive broadcasting. Neighbor nodes receive the data and check whether it is on the forwarding path of the source and the sink. If yes, it rebroadcasts the data. Accordingly the data from the source reaches to the sink over multiple hops.

9.2.1.14. *Random walk*

Random walk routing technique[90] is used for load balancing in a sensor network with very limited mobility while randomly turning the sensor nodes on and off. The protocol uses multi-path routing and achieves load balancing in a statistical sense. Although the topology can be irregular, the nodes are so arranged that each node falls in the intersection of a grid on a plane. This intersection, i.e. the lattice coordinate is obtained by computing distances between the nodes using distributed asynchronous Bellman-Ford algorithm. Nodes in the neighborhood are selected as next hop according to the probability in relation to the closeness of the node to the destination. Unfortunately the uniform grid structure limits its usage in real-life scenarios.

9.2.1.15. *Constrained anisotropic diffusion routing (CADR)*

Reference 31 provides two techniques namely information-driven sensor querying (IDSQ) and constrained anisotropic diffusion routing (CADR). Only nodes closest to the event are activated and generate data, while the routes are dynamically adjusted. Unlike directed diffusion not only is the communication cost considered, information gain is also taken into account. The goal is therefore to maximize the information gain as well as minimize the latency and bandwidth. In doing so, instead of gradient of the cost/height as in Refs. 25 and 91, CADR uses information gain per unit cost for routing while IDSQ selects the optimal order of sensor for maximal information gain. Unfortunately, the protocol (i.e. CADR) does not specify the path to route the query and the data to and from the source and the sink, necessitating an additional routing protocol design where IDSQ can be regarded as complementing the routing technique.

9.2.2. *Hierarchical routing*

Whereas in flat routing architecture every node operates in the same logical layer, in hierarchical routing nodes are categorized according to their criterion such as battery power, mobility condition and occupy different logical layers within the WSN architecture. Grouped into clusters, nodes perform various operations in relation to their corresponding layers. A brief overview of the most significant hierarchical routing protocols is presented here.

9.2.2.1. *Low energy adaptive clustering hierarchy (LEACH)*

In a solution to the cluster based network architecture, low energy adaptive clustering hierarchy (LEACH)[45] nodes collect data via the cluster heads (CHs). In LEACH, a predefined percentage of nodes work as CHs where the selections of the CHs are randomly defined by the nodes themselves. After defining themselves as CHs, nodes broadcast the decisions, and the rest of the non-CH nodes join the CHs based on the received signal strength to form the clusters. Following this cluster formation, nodes send the sensed data to the CHs which then aggregate the collected data and send it to the sink. Here all the communications are carried out through direct communications as shown in Figure 9.5. However, because of the random cluster formation process, the shape of the clusters can become irregular with different number of participating nodes or different number of clusters.

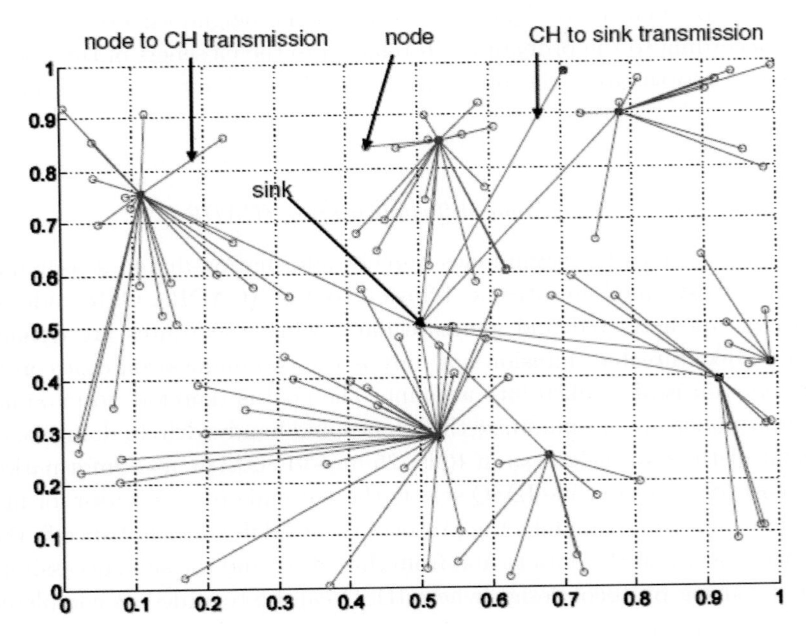

Fig. 9.5. Transmissions in LEACH protocol.

9.2.2.2. *Centralized LEACH (LEACH-C)*

To address the irregular formation of clusters in LEACH (a decentralized approach), LEACH-C[46] (a centralized cluster formation scheme) is proposed, where the CH selection is carried out by the sink. Here, the information about the location and energy of the sensor nodes is delivered to the sink prior to the CH selection. Using this information the sink then calculates the average energy of the nodes. Nodes having energy level below the average energy are not considered as a prospective CH. Accordingly the cluster is formed in such a way that the total amount of energy needed to transmit data from the non-CHs to the CHs is minimized. Apart from this cluster formation scheme, the aggregation and forwarding in LEACH-C is same as LEACH.

9.2.2.3. *Maximum energy cluster head (MECH)*

In order to improve the random rotation of the CHs and to limit the irregular cluster formation in LEACH, a routing protocol titled maximum energy cluster head (MECH)[25] is proposed that provides improvement over LEACH in different aspects. MECH forms the clusters in such a way that a maximum number of nodes inside the cluster as well as a maximum size of the cluster is maintained. Here, the node with the highest battery power is selected as a CH in rotation. Moreover, a hierarchical tree based routing approach is utilized to reduce the energy expenditure in relation to the data transfer. As a result, despite the high set up cost of the network, the system conserves energy in the long run.

9.2.2.4. *Base-station controlled dynamic clustering protocol (BCDCP)*

Reference 80 proposes a centralized cluster based routing approach titled base-station controlled dynamic clustering protocol (BCDCP). This protocol calculates the average energy of the nodes, and only the nodes having energy above the average are selected as CHs. Based on the remaining power of the nodes the sink finds the set of nodes S that has battery power more than the average. The sink then finds two nodes those have maximum separation distance. Remaining nodes then join with these two with respect to the distance. The protocol then refines this formation by balancing the nodes and making groups having approximately the same number of nodes known as sub-clusters. The same procedure continues and results in further division of the sub-clusters into smaller clusters. After the completion of the clusters, one randomly chosen CH forwards the collected data to the sink. This particular CH collects the data from other CHs through the minimum-spanning tree constructed by all the CHs.

9.2.2.5. *TEEN*

Threshold sensitive energy efficient sensor network protocol (TEEN)[70] is a time critical routing application that saves energy by limiting the generating packets.

To limit the number of packets the protocol introduces two thresholds of sensed value namely a hard threshold (T_h) and a soft threshold (T_s). Data is only sent to the sink if the current value is greater than T_h or if the difference between the current and the previous value is greater or equal to T_s.

9.2.2.6. *APTEEN*

As mentioned earlier, LEACH and TEEN are different types of protocols where the former collects the periodic data proactively while the latter reports time critical data reactively. LEACH sends data to the sink even if there is no significant importance of the sensed values thereby resulting in wastage of resources. Conversely, if the thresholds are not reached, TEEN may not deliver data to the sink for a long period of time. To resolve these issues, APTEEN,[71,72] a hybrid protocol, has been protocol that can respond to the queries on the historic data, forward the periodic data and also respond to any sudden change on the sensed value.

9.2.2.7. *Fuzzy centralized routing*

In Refs. 42 and 114, a cluster based hierarchical routing protocol is proposed where the gateways are responsible for construction and maintenance of a centralized routing table for the next hop for each sensor node in a particular cluster. The costs of communication between any two nodes are calculated by the fuzzy logic. After finding the costs, the routes are determined using the Dijkstra's shortest path algorithm. However, such a centralized mechanism incorporates a single point of failure and may result in partitioning of the network.

9.2.2.8. *Two-tier data dissemination (TTDD)*

A grid structure model of data dissemination is provided by the protocol two-tier data dissemination (TTDD).[66,112] The protocol offers a technique for collection of data through the multiple mobile sinks. In TTDD, data sources construct the grid structures proactively. This structure enables the continually mobile sink to gather information through flooding queries within the local cells only, thereby avoiding a network wide flooding of the query and thus saving the energy.

9.2.2.9. *Energy conserving passive clustering (ECPC)*

The data centric routing protocol directed-diffusion is disadvantaged due to its interest propagation with flooding. Reference 43 proposes passive clustering directed diffusion (PCDD) mechanism that improves this particular problem by utilizing the concept of passive clustering.[41] Passive clustering is a technique that uses on-demand clustering to avoid potential set-up time and reduce massive overhead. But this passive clustering suffers from the shortcoming of generating a large number of gateways. In addition, such clustering approach does not incorporate the node energy during the cluster set up mechanism. Energy conserving passive clustering (ECPC)[83] approach is an enhancement to the PCDD approach that considers distance as well as energy to form the cluster thereby improving the efficiency of the routing scheme.

9.2.2.10. *GA-clustering*

Another improvement of LEACH is proposed in Ref. 48, which introduces cluster formation algorithm where the genetic algorithm[55] is used as a tool to optimize the among various metrics. This algorithm uses distance to the sink, distance to the cluster, standard deviation of the distance to the cluster, transfer energy, and number of transmissions as fitness functions to be minimized.

9.2.2.11. *HEAR-SN*

Routing approaches that utilize single-hop communication suffer from connectivity problem in large networks. The most popular routing protocols such as LEACH, LEACH-C, TEEN, APTEEN, and so on, use single long hops for the communication and as a result are limited to smaller networks. However this problem is addressed by a heterogeneous network proposed in Ref. 47 where the clusters are fixed and the CHs are high end nodes with higher transmission and battery power. Unfortunately such straightforward approach with high end nodes incurs significant monitory cost a very undesirable situation for large scale network deployment. Despite these cost issues, it is a shared belief among the research community that with the advent of technology, sensors will become very cheap in the near future as low end devices.

9.2.2.12. *Power efficient gathering in sensor information systems (PEGASIS)*

Power efficient gathering in sensor information systems (PEGASIS)[64] addresses the energy saving problem by constructing a near optimal chain. The chain is constructed so that the nodes can only communicate with their nearest neighbors. PEGASIS reduces the communication cost efficiently; as a result it increases the life of the network. But this encouraging result comes with the cost of increasing delay in data delivery (approximately 4.3 times more than LEACH).[98] Instead of this very high delay sometimes a moderate delay is required for the applications.

9.2.2.13. *E-PEGASIS*

To distribute the energy dissipation evenly, PEGASIS rotates the leader nodes and ensures that all the nodes in turn participate in the last hop data delivery (to the sink). But this leader selection results in redundant data transmission where the selection is independent of the location information. Reference 52 proposes enhanced-PEGASIS (EPEGASIS) that eliminates the aforementioned problem and enhances the energy efficiency for sensor networks. Here, a concentric cluster formation scheme is proposed with the sink located in the center of the circles. The levels of the clusters are so defined that the distant cluster posses the highest level. For every cluster a head node (as CH) is selected through rotation. This head node is responsible for the collection of the data from the respective cluster through a chain. After collecting the data from all the sensors inside the cluster the head node passes the data down to the lower level of the clusters, and it reaches the sink hop-by-hop.

9.2.2.14. *Tree based energy efficient protocol for sensor information (TREEPSI)*

Tree based energy efficient protocol for sensor information (TREEPSI)[87] is a close variant of PEGASIS where an equivalent node functions as the leader node in collecting the data from all the nodes through a minimum tree routing path. Data is aggregated in every hop along the paths toward the sink. After reaching the leader node over multiple hops, it sends the collected and aggregated data to the sink. This process continues until the leader dies and the node with the next ID takes on the role of the leader, thereby maintaining the data flow.

9.2.2.15. *Event driven clustering (EDC)*

In event driven clustering algorithm (EDC),[115] authors propose selection of CHs depending on the highest remaining energy where CHs aggregate and route data to the sink through the mobile gateways. Similar to LEACH these mobile gateways gather data from the nodes aggregate the data and send it to the sink. The transmissions (both from node to gateways and gateways to sink) are carried out through large single hops.

9.2.2.16. *Two-phase clustering (TPC)*

For energy saving and delay adaptive data gathering in WSN, two-phase clustering (TPC) scheme[30] is proposed. In phase-1, direct links are set up from nodes to the CHs. For time critical data aggregation, the protocol uses the direct links established by this phase. On the other hand, in phase-2 nodes search for other nodes those are closer than the CHs and set up small relay links. For non-time critical data gathering, nodes use the relay links to save energy.

9.2.2.17. *Hierarchical battery aware routing (H-BAR)*

In Ref. 79, authors propose a hierarchical battery aware routing (H-BAR) protocol. H-BAR works as LEACH except the clusters are not reformed after the initial formation. Only the role of the CH changes in every round. Here, node having the highest remaining energy becomes a CH. In terms of lifetime consideration the protocol outperforms LEACH.

9.2.2.18. *Hierarchical-PEC (HPEC)*

Though PEC is fast, reliable and fault tolerant, it is designed for the flat networks. A hierarchical variant of the protocol known as hierarchical periodic, event-driven and query-based (HPEC) protocol is proposed in Ref. 18. In addition to PEC advantages, HPEC is energy-aware and can support aggregation. Protocol starts with flooding and construction of the hop gradient throughout the networks. The aggregator, also known as CHs, send the aggregated data to the sink using this gradient. The CHs are selected based on the high-energy reserve. Each node generates a random number [0, 1]. When the number is less than the predefined value the node broadcasts a packet requesting the neighboring node's energy level. Based on the highest energy, the node is selected as an aggregator. The aggregator then sends a broadcast message putting a time-to-live (TTL) field. This TTL is used to maintain the maximum size of the cluster. Nodes join the aggregator that is closest among all the aggregators from which it receives the broadcast. HPEC follows the data routing and path repair mechanism incorporated in the PEC algorithm.

9.2.2.19. *Virtual grid architecture (VGA) routing*

In VGA routing,[7] the sensor bed is divided in square shaped clusters where a CH takes the responsibility of that particular cluster. Data aggregation is carried out inside the network in two levels, (i) local and (ii) global. The CH performing local aggregations are known as local aggregators (LAs). Among the LAs, a subset of them works as global aggregators known as master aggregators (MAs). The virtual grid architecture is presented in the Figure 9.6.[6] An optimal and three near optimal solutions to the routing problem with data aggregation is presented in Ref. 6. The exact solution to the problem of finding the number of MAs among the LAs while maximizing network lifetime is achieved through the integer linear program formulation (ILP) approach. The approximate solutions are achieved by algorithms such as genetic algorithm based heuristics, k-mean heuristic and greedy-based heuristics. An alternative solution to the same problem is proposed in Ref. 7. Known as the clustering based aggregation heuristic (CBAH), the solution functions similar to the classical bin packing problem.

9.2.2.20. *Hierarchical power aware routing*

Reference 62 proposes hierarchical power aware routing. Here, the network is divided into clusters with geographic proximity and termed as zones where the zones work as an entity. Zones take the routing decision to forward data hierarchically through other zones and maximizes the battery life of the system. The algorithm relates to the dilemma of selection between a higher remaining powered node with more expensive path and a lower powered node with lesser routing cost. The approximate algorithm, called $\max - \min zP_{\min}$, is proposed as a solution to resolve this problem. The proposal finds a trade-off between minimizing the total power consumption and

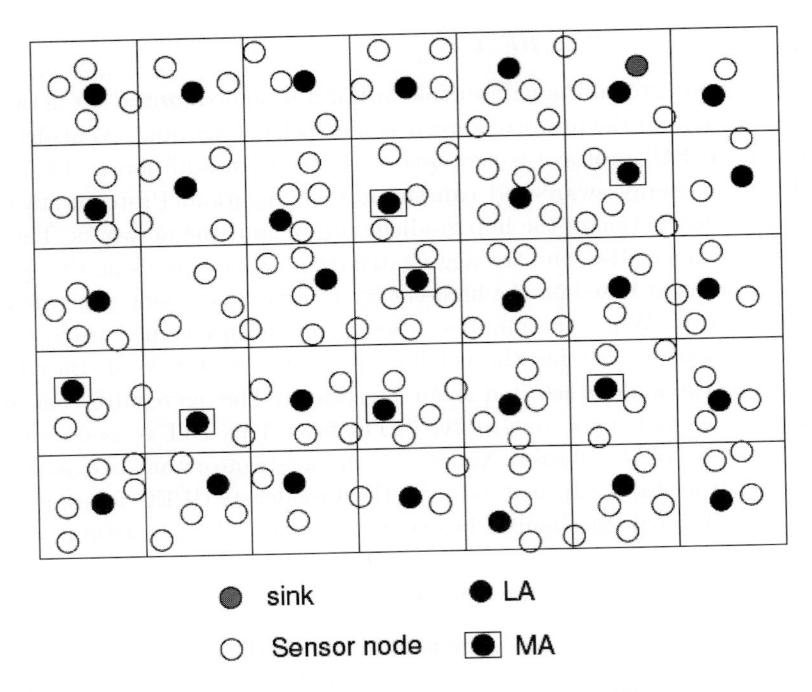

Fig. 9.6. Virtual grid architecture in a WSN.

maximizing the minimal residual power of the network. The protocol uses Dijkstra algorithm to find the path with least power consumption $P_{\min}$. Then it finds a path that maximizes the minimum residual power in the network. These two criteria are then optimized through the relaxation technique. The minimum power consumption is then relaxed to $zP_{\min}$ for the message where $z > 1$. The algorithm therefore maximizes the minimal residual power while consuming the most $zP_{\min}$.

In another routing solution of sending a message across the entire region, a global path from zone to zone is found. Zones take the routing decision based on the power level estimates of the zones independently. A modified Bellman-Ford algorithm is used in the decision process in order to select the optimal path.

9.2.2.21. *Sensor aggregates routing*

Focused on target tracking application, Ref. 38 proposes a set of algorithms that construct and maintain sensor aggregates. The nodes in the network that satisfy the grouping predicate for a collaborative processing task create the aggregates. The predicates depend on the task and its resource requirements. The clusters are formed such that there exists only one peak (sensed signal) in a cluster. Nodes having highest sensed signal is elected as CHs through the neighbor information exchange. Three algorithms are proposed.

- Distributed aggregate management: The node determines whether it belongs to an aggregate, based on the result of applying the predicate and information from the other nodes.
- Energy-based activity monitoring: The sensed signal strength at each node is estimated by deriving the signal impact area.
- Expectation-maximization like activity monitoring: Estimates target positions and signal energy.

The system only functions well when the targets do not interfere with each other.

9.2.2.22. *Self-organizing protocol*

In the scenario where some of the sensors are mobile and the rest of them are stationary,[97] proposes a self-organizing protocol for hierarchical networks with some of the stationary nodes acting as routers and forming the backbone of the communications. A fault tolerant broadcasting is achieved by incorporating random walks on spanning trees of graphs (known as local Markov loops (LML)). In the first phase neighbor nodes are discovered, while in the second phase (also known as organization phase) the hierarchical groups are made and broadcast spanning trees are formed. In the maintenance phase, energy levels and routing trees are updated. In case of a failure or death, regrouping and reorganization takes place (self-reconfiguration).

9.2.2.23. *Ant colony optimization*

As the complexity of the routing algorithm often increases with the size of the network, some of the routing algorithms incorporate the technique that mimic the real biological world. Ant colony optimization[16] is a technique that finds an approximate solution to the combinational optimization. According to this algorithm, artificial ants move inside the problem graph. These ants leave artificial pheromones on the graph. Based on these pheromones, future ants find a better solution. Based on this concept AntNet[24] routing protocol was proposed. Performing in an adaptive and distributed fashion, this protocol shows robustness and also reaches a stable condition rapidly. To optimize different valuable criteria such as energy, energy efficiency, latency and success rate, Refs. 1 and 116 utilize ant colony based reinforcement learning for the WSN and demonstrate the corresponding performance enhancements.

9.2.3. *QoS aware routing*

A number of QoS based routing protocols have been introduced.[28] Among them Ref. 108 investigates the mean link quality and its variances with respect to the various node distances. Authors uses window mean exponential weighted moving average (WMEWMA) for link estimation. They also provide a frequency based table management and cost based routing in the paper.

9.2.3.1. *Reliable information forwarding using multi-path (ReInForM)*

In reliable information forwarding using multi-path (ReInForM)[34] end-to-end reliability in multi-hop routing is addressed by delivering data through redundant paths. Desired reliability is set according to the information content in the data being sent. Based on this desired reliability and the local channel error rates, the data is sent from the source to the destination through a thin band of paths having low deviation from the optimal path. Load is also distributed by incorporating randomization in the next hop selection.

9.2.3.2. *Reliability constrained optimized routing (RC-OR)*

Reference 61 provides an optimal power allocation and routing algorithm to maximize the lifetime of the sensor networks under constrained end-to-end success probability of data delivery. Here the optimized power allocation algorithm minimizes the total power required by a path (source to destination) while maintaining the designated successful delivery ratio. The optimal routing algorithm, on the other hand, applies integer programming attempts to maximize the lifetime by balancing the energy consumption of all the paths. Reference 61 also provides a suboptimal minimum cost path routing with a heuristic based approach with less complexity.

9.2.3.3. *Stateless protocol for real-time communication in sensor networks (SPEED)*

Stateless protocol for real-time communication in sensor networks (SPEED)[44] attempts to enforce a minimum data delivery time for packets to its destination. An explicit soft real-time delay is set and maintained by a uniformly maintained speed. In case the speed is not achieved, the maximum allowable delay is evidently not attainable and consequently the packet is automatically dropped.

9.2.3.4. *Multi-path multi-SPEED (MMSPEED)*

Multi-path multi-SPEED (MMSPEED)[40] provides an extension of SPEED and offers QoS provisioning in the domain of reliability and timeliness i.e. it supports different classes of packets with the requirements of different velocities and reliabilities. In the reliability domain, the protocol employs probabilistic multi-path forwarding depending on the end-to-end success probability. On the other hand, in the timeliness domain, the protocol provides multiple speed options where the traffic dynamically chooses the proper speed and meets their end-to-end deadlines. Figure 9.7 presents the messaging of the MMSPEED protocol. Here, Figures 9.7(a) and (b) present the reliability domain and Figures 9.7(c) and (d) present the timeliness domain. Messaging in Figure 9.7(b) has a better reliability than messaging in Figure 9.7(a) as it uses a redundant path for the data delivery. On the other hand,

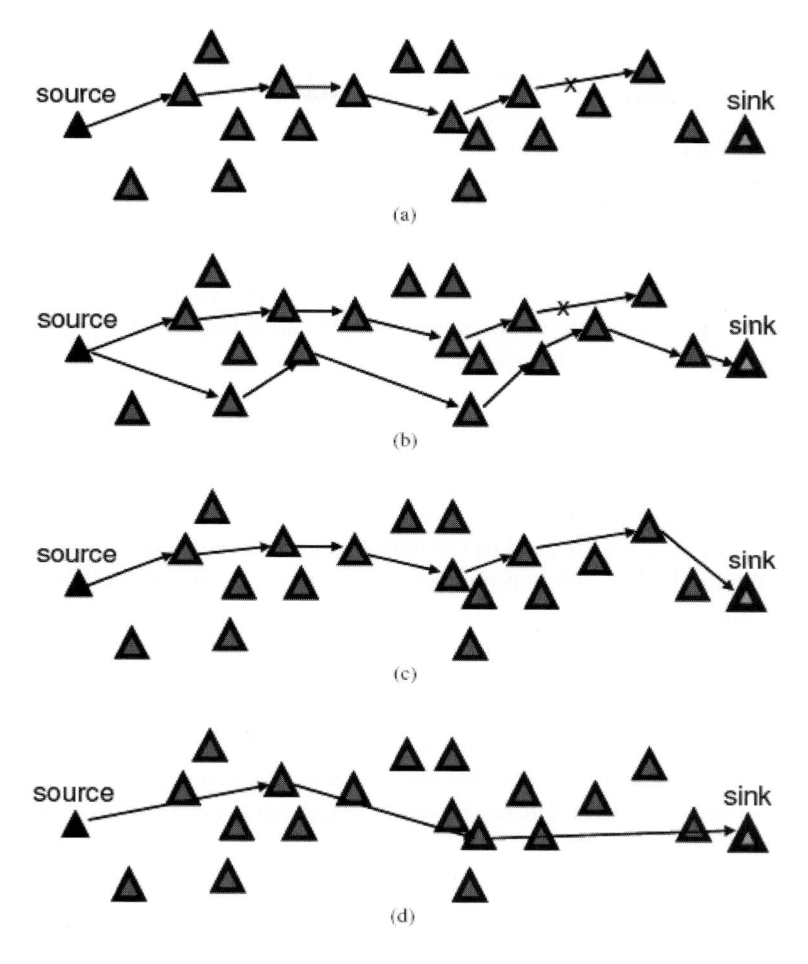

Fig. 9.7. Messaging in MMSPEED protocol.

Figure 9.7(d) uses a higher speed path than the path in Figure 9.7(c) to support more time critical data.

9.2.3.5. *Cluster-based energy-aware routing (CER)*

Reference 5 proposes a cluster-based energy aware routing protocol for WSNs that can support real-time and non-real-time applications coexisting in a given time. The protocol tend to minimize the cost of transmissions by considering distances, remaining energy, expected time of the battery level and link error rate. To find a minimum cost path, Dijkstra's algorithm[33] is enforced. The protocol then evaluates if the path is capable of delivering data with the end-to-end delay constrains. In case of failure, k_c different paths with minimum cost are found through modified Dijkstra's algorithm, followed by selection of the suitable path with the delay requirements. For non-real-time applications, throughput is maximized by adjusting the service rate for both the traffic classes.

9.2.3.6. *Just-in-time scheduling (JiTS)*

Reference 65 proposes a scheduling algorithm called just-in-time scheduling (JiTS) that allocates the available slack intelligently among the hops. The protocol implements the policy where the packets are non-uniformly delayed at the intermediate nodes as the load, resulting in higher contention as the packet gets closer to the sink. The deadline miss ratio or the packet drop ratio is improved by exploiting the non-uniform contention of the networks.

9.2.3.7. *QoS and energy aware routing protocol (QoSERP)*

Reference 68 proposes a QoS and energy aware routing protocol for real-time traffic in WSN. The problem to the packet loss is addressed by packet duplication at the source node and data is delivered to the sink by two different paths. The underlying idea of balancing the two different metrics of energy and delay is such that if a data is sent over a small hop then the energy is saved but this is at the expense of extended delay from multi-hop propagation. So in a careful consideration with respect to tolerable delay the next hop distance is adjusted and data is sent to the destination. For example, data having three different timeliness levels as low, medium and high has three paths depicted in Figures 9.8(a), (b) and (c) respectively. Data paths represented in Figure 9.8(a) energy is conserved with small hops whereas in Figure 9.8(c) no such option left as the required data delivery time is extremely low.

9.2.3.8. *Multiple-objective fuzzy decision-making based information-aware (MOFD-MIA) routing*

Reference 105 proposes a multiple-objective fuzzy decision-making based information-aware (MOFDMIA) routing protocol for the WSN environment where multiple data types coexist. The interested network is composed of a number of clusters where the gateways act as CHs. This protocol defines how the data will flow from the nodes to these gateways. Multiple trees are generated rooted from the single hop neighbors of the gateways. The trees grow outwards through branching while avoiding nodes with undesirable characteristics such as low energy, low bandwidth, and high delay. Three different types of data are supported and their requirements are characterized. The protocol selects the best paths for the different types of the data according to the fuzzy decision making algorithm.

9.2.3.9. *QoS-based adaptive clustering algorithm (QAC)*

Reference 29 proposes a QoS-based adaptive clustering algorithm (QAC) where the formation of the cluster relates to both energy and reliability. Unlike a regular

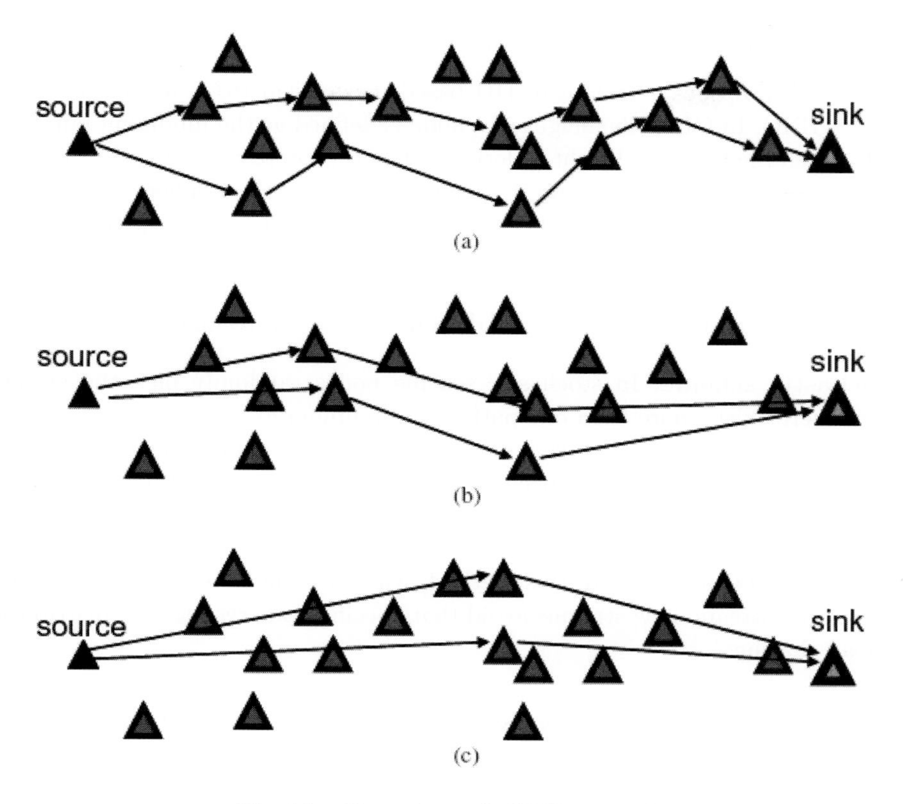

Fig. 9.8. Messaging in QoSERP protocol.

clustering technique, this cluster formation algorithm offer dual CH model, and thereby distributes the load of the CH as well increases reliability. When the number of nodes inside a cluster is beyond a predefined threshold value k, the two CHs are needed to support the cluster. The master-CH designates a second CH named slave-CH to handle the additional workload. The slave-CHs aggregate the data from its designated nodes and send the aggregated data to the master-CH. The master-CHs aggregate the data from the nodes within its cluster (including the slave-CH) and sends it to the sink.

9.2.4. *Data centric routing*

In data centric routing, the sink sends queries to the regions and waits for the data from the location of the specified region. Although this chapter discussed many of the protocols in flat networks they also fall in this specific category. For example, SPIN, DD, rumor routing are such kind of approaches. Other examples include the gradient-based routing (GBR) is briefly discussed below.

9.2.4.1. *Gradient-based routing (GBR)*

Reference 88 proposes a variant of DD based mechanism titled as gradient-based routing (GBR). In GBR, the height of a node is defined as the minimum number of hops to reach the BS. This height is derived when the interest is diffused throughout the network. From the node's and its neighbor's height, the gradient of the height is calculated by the difference between these two. Upon establishing the gradient, a packet is forwarded through the link possessing the greatest gradient. Nodes in the multiple routing paths combine i.e. aggregate the data with certain functions. GBR proposes three different types of data dissemination techniques/schemes:

- **Stochastic scheme:** In stochastic scheme nodes randomly pick the gradient when there exists more than one path with minimum gradient.
- **Energy-based scheme:** In this method, when the remaining energy of a node goes down from a predefined specific value, a threshold node artificially increases its height. Such increment in turn discourages nodes from sending data to that particular node.
- **Stream-based scheme:** In this method, when a node is located in the path of an existing stream, new streams avoid that specific node resulting in distribution of load sharing.

9.2.5. *Geographic routing approaches*

Geographic routing approaches are based on the location of the nodes in the networks. Greedy forwarding is one of such routing approaches that provide a simple and efficient routing mechanism. The remainder of this subsection mainly focuses on the different geographical routing approaches namely geographic adaptive fidelity (GAF) and geographic and energy aware routing (GEAR) along with a group of protocols that handle the routing hole in the geographical forwarding arena.

9.2.5.1. *Geographic adaptive fidelity (GAF)*

Geographic adaptive fidelity (GAF)[109] utilizes location information and turns OFF the redundant nodes to conserve energy. Though primarily developed and proposed for the ad hoc networks, it is also and probably more suitable for energy constrained sensor networks. GAF forms a virtual grid for the covered area where the nodes find out their presence to the associated grid. Nodes located in the same grid are considered equivalent with respect to the routing cost. Therefore, some of the nodes can go to sleep while the rest can do the forwarding job on behalf of the sleeping nodes inside the grid. Nodes change their states (sleep to active) to balance the energy consumption among the nodes. Besides these two states, an additional state named discovery is required to discover the neighbors in the grid. This protocol is able to handle both mobile and static sensor beds where the mobility is supported by the estimation of possible time of a node leaving the specific grid and informing the neighborhood about the departure. To keep the routing fidelity, the sleeping nodes adjust their sleeping time accordingly. GAF can also be considered as a hierarchical protocol as a leader node inside a particular grid is responsible for the data forwarding to the other nodes. One of the biggest disadvantages of this protocol is that it does not provide any aggregation (which is not very common in the hierarchical networks), though it conserves energy without increasing the latency and the packet loss.

9.2.5.2. *Geographic and energy aware routing (GEAR)*

Geographic and energy aware routing (GEAR)[113] utilize the location information while disseminating queries to the appropriate regions. An energy aware geographic neighbor selection heuristics is applied here. Therefore the queries are routed to the target region effectively. Nodes maintain a repository to keep an estimated cost and a learning cost with respect to every neighbor in order to reach the destination. Here the estimated cost is defined as the combination of the cost of distance to destination and residual energy. Conversely the learning cost is defined as a refined cost that reflects the cost of a detour for a packet due to the presence of a hole somewhere in the path. To forward a packet, a node therefore first checks if there exist a neighbor which is geographically closer to the current location. If it is so, the node passes the packet to the node having the minimum estimated cost. Otherwise it uses the second type of cost i.e. the learning cost to forward the data. Once the packet reaches the designated region, data is diffused around the region. If the network is densely deployed then the recursive geographic flooding is used for the diffusion. However for the less dense or sparse network, restricted flooding is used. These two different flooding approaches have been developed because they are cheaper for the corresponding node distributions with respect to the energy usage. In recursive geographic flooding the whole region is divided into four sub-regions and four copies of packets are created. This process continues until there exists a single node in the sub-region.

Compared to GPSR, this protocol performs better both in terms of energy efficiency and delivery ratio.

9.2.5.3. *Void avoidance in the geographic routing arena*

Data packets always take a positive progression towards the destination in the geographic routing algorithms. If a negative progression is ever allowed, it is highly likely that the packets may become trapped into routing loops. Unfortunately, when a negative progression is not permitted and no node in the forwarding path exists, it is called a void in the geographic forwarding. It is therefore necessary to tackle this issue through the algorithms titled void handling algorithms. Figure 9.9 shows such a void where a packet is stuck in the dead node. Therefore a void handling technique is invoked to forward the data along the path shown by the arrows.

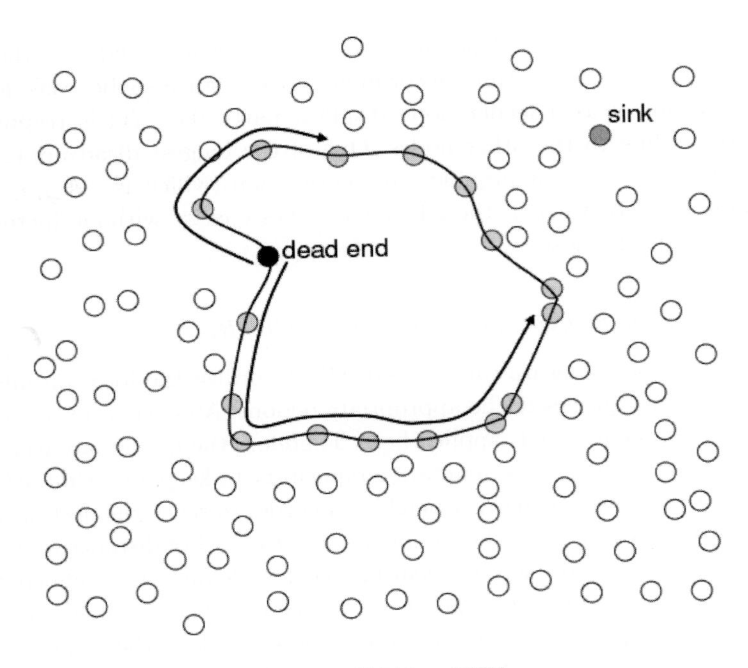

Fig. 9.9. Void in a WSN.

A number of void handling approaches are in the literature due to the importance of tackling the issue.[28] Planner-graph based approach is incorporated in a few algorithms that handle this problem based on the right-hand-rule, of maze exit (a very ancient idea). According to this rule an entrapped person can exit the maze walking forward by placing their right hand to the wall and exit at the end.[28] Using this technique, the routing algorithms avoid loops in the path by extracting the planner sub-graph from the network graph. It uses different graph algorithms such as graph traversing algorithm[56–58,117] and distributed planarization algorithms.[54] Having computational cost involved in these process planner graph techniques guarantee packet delivery[17] (of course there must need to exist a topologically valid path from source to destination).

In geometric void handling,[39] voids are tackled by identifying the voids proactively or reactively, depending on the applications. After finding the void the information is stored along the boundaries of the hole. This boundary (connecting the void nodes and the non-void nodes) forms a cycle where the packets are routed along the boundary to circumvent the hole.

One of the crude ways of tackling the void issue is to flood the whole network. But due to its enormous resource consumption this technique is not well accepted. Rather a limited flooding based approach is sometimes adopted. Reference 95 proposes a one-hop flooding approach where the void node initiates the flooding limited to a single hop. After reception, the neighboring nodes use the original greedy forwarding to forward the packet. In case the packet goes to the original void node that initiated the broadcast, the void node sends a packet to notify the particular neighbor node. The neighbor node then selects the next best node from its neighboring table. If no such node exists then this node becomes a new void node and these processes continue until the packet comes out of the trap. Reference 96 proposes an on demand routing that starts upon a packet reaching in a void. The void node then starts a graph search algorithm known as the breadth first search. This process continues until the packet comes out of the dead end. The proposed protocol in Ref. 28 introduces an expanded ring search for the problem where the node widens the search area by sending out the discovery packet to two hops, three hops and so on, until it overcomes the void.

Cost based void handling techniques are proposed in Refs. 118 and 119. Generally cost based routing is directed by the rule that a packet is forwarded from the lower cost node to the higher cost node. Here, cost is defined by the Euclidean distance from the destination. Node identifies itself as a void node increasing its cost to a higher value. Therefore the packet, directed by the same rule (high cost to low cost), eventually reaches the destination through an efficient path by getting around the void.

Reference 44 proposes a solution where a viod node actively propagates the information to the upstream nodes about the existence of the void (alternately known as backpressure beaconing). The upstream node then attempts to reroute the packet through an alternative node. In case such a node does not exist, a further backpressure beaconing is activated until a new path is found.

9.3. Challenges for Future Research

Based on the earlier discussion of the WSN routing protocols, it can be concluded that in order to extend the network lifetime through conservation of battery energy and distribution of data traffic across the sensor field in an equitable manner, new energy-efficient routing protocol designs are warranted. This will be even more important as the network size grows beyond the capabilities of existing proposals. In addition, node mobility and the problem with offering energy-efficient routing becomes manifold. Although traditional proposals can address some of the issues to some extent, a lot more is desired.

With the advent of new applications, WSN has evolved into a more diversified network and often varies in size (small and large) and shape. As such, a single architecture may not suffice and protocol may need to support both flat and hierarchical networks, unlike the traditional support of a specific architecture. For example, while MTE and GMR are specific to the flat networks, their incorporation of distance based routing fails to provide an energy conservation mechanism in cases when the network is heavily loaded.

Table 9.1 Open issues for WSN routing protocols.

	Multihop	Multiple routing metrics	Optimal energy efficiency	Application specific (Real-time: RT/ Non-real-time: NRT)
Flat				
GMR	√	X	X	NRT
MTE	√	X	X	NRT
EM-GMR	√	√	√	NRT
Hierarchical				
LEACH	X	X	X	NRT
C-LEAXH	X	X	√	NRT
TEEN	X	X	X	RT
QoS aware				
ReInForM	√	X	X	NRT
SPEED	√	X	√	RT
MMSPEED	√	√	√	RT

Similarly well referenced routing protocols such as LEACH family of protocols (LEACH, LEACH-C, TEEN, APTEEN) suffer from scalability problem where they limit their corresponding network size due to the adaptation of single long hops. On the other hand, deploying a high cost CH (as proposed in Ref. 47) is also undesirable for the high monitory cost involved. Moreover longer hops are not suitable for conservation of battery power. Another approach is to incorporate a centralized next hop decision making algorithm (proposed in Refs. 5, 42 and 144). Unfortunately, such an approach facilitates a single point failure and can lead to network partitioning.

In order to resolve these issues, optimal routing algorithms that can offer energy efficiency to extend the network lifetime similar to those introduced in Refs. 10 and 11 are required. Table 9.1 highlights the routing algorithms discussed in this chapter and their corresponding open issues that govern the development/design of an energy efficient routing protocol.

9.4. Conclusions

In this chapter, a comprehensive review of the available routing protocols for a variety of topologies in WSNs has been provided. Table 9.2 summarizes the large number of routing protocols that have already been proposed for the WSNs and discussed in this chapter. We have divided the routing protocols based on the topology of the network; that is, flat versus hierarchical, and then provided summary of most important routing techniques available in the literature. The routing protocols discussed in this chapter are designed for different scenarios and different applications. Each routing protocol also tries to focus on one or more characteristics such as data delivery delay, energy efficiency, routing overhead, and so on. It is very difficult to

Table 9.2 Routing techniques in WSNs.

Flat routing	Hierarchical routing	QoS aware routing	Data centric routing	Geographic routing
• Reactive routing with controlled flooding	• LEACH	• ReInForM	• GBR	• GAF
• MTE	• LEACH-C	• RC-OR	• SPIN	• GEAR
• EM-GMR	• MECH	• SPEED	• Directed diffusion	
• SPIN	• BCDCP	• MMSPEED		
• Directed diffusion	• TEEN	• CER	• Rumor routing	
• Fuzzy directed diffusion	• APTEEN	• JiTS		
• TBF/TBF+	• Fuzzy centralized routing	• QoSERP		
• Rumor routing	• TTDD	• MOFD-MIA		
• COUGAR	• ECPC	• QAC		
• ACQUIRE	• GA-Clustering			
• PEC	• HEAR-SN			
• Energy-aware routing	• PEGASIS			
• MCFA	• E-PEGASIS			
• Random walk	• TREEPSI			
• CADR	• EDC			
	• TPC			
	• H-BAR			
	• HPEC			
	• VGA routing			
	• Power aware routing			
	• Sensor aggregates routing			
	• Self-organizing protocol			
	• Ant colony optimization			

specify a routing protocol that can be useful for all sorts of the WSNs as this is very much dependent on the specification of a particular network and applications associated with it. The information provided in this chapter however should be useful in giving the interested readers an insight to what sort of routing technique can be adopted in their own network.

Problems

(1) Specify five potential applications for the wireless sensor networks and explain each briefly.
(2) Explain the general functions of the nodes in a wireless sensor network.
(3) Explain the importance of lifetime and sustainability consideration in a wireless sensor network.
(4) Explain the difference between the two major routing architecture, flat and hierarchical routing techniques in wireless sensor networks and give one example from each one.
(5) Briefly explain the direct diffusion routing scheme in a flat wireless sensor network.
(6) Load balancing is considered as an effective method in increasing the lifetime of a wireless sensor network. Give an example of a routing protocol where the load balancing is employed and explain how it works.
(7) What are the differences between the LEACH and LEACH-C?
(8) What are the differences between the TEEN and APTEEN routing techniques?
(9) Explain the main characteristics of the PEGASIS routing technique.
(10) What are the main characteristics of geographic routing approaches in wireless sensor networks?
(11) List and explain some of the main challenges in routing for future wireless sensor networks.
(12) List the main factors that differentiate the routing techniques in wireless sensor networks and those of ad hoc networks.

Bibliography

1. R. G. Aghaei, M. A. Rahman, W. Gueaieb and A. E. Saddik, Ant colony-based reinforcement learning algorithm for routing in wireless sensor networks, *IEEE International Conference on Instrumentation and Measurement Technology (IMTC)*, Warsaw, Poland (2007) 1–6.
2. D. P. Agrawal, R. Biswas, N. Jain, A. Mukherjee, S. Sekhar and A. Gupta, Sensor systems: state of the art and future challenges, *Handbook on Theoretical and Algorithmic Aspects of Ad hoc and Sensor Networks* (Auerbach Publications, 2006).
3. K. Akkaya and M. Younis, An energy-aware QoS routing protocol for wireless sensor networks, *23rd International Conference on Distributed Computing Systems*, Rhode Island, USA (2003) 710–715.

4. K. Akkaya and M. Younis, A survey on routing protocol for wireless sensor networks, *Elsevier Ad Hoc Networks* **3** (2005) 325–349.
5. K. Akkaya and M. Younis, An energy-aware QoS routing protocol for wireless sensor networks, *23rd International Conference on Distributed Computing Systems*, Rhode Island, USA (2003) 710–715.
6. I. F. Akyildiz, W. Su, Y. Sankarasubramaniam and E. Cayirci, A survey on sensor networks, *IEEE Communications Magazine* **40**(8) (2002) 102–114.
7. J. N. Al-Karaki and A. E. Kamal, On the correlated data gathering problem in wireless sensor networks, *IEEE International Symposium on Computers and Communications (ISCC)*, Alexandria, Egypt **1** (2004) 226–231.
8. J. N. Al-Karaki and A. E. Kamal, Routing techniques in wireless sensor networks: a survey, *IEEE Wireless Communications* **11**(6) (2005) 6–28.
9. S. Antifakos, F. Michahelles and B. Schiele, Proactive instructions for furniture assembly, *International Conference on Ubiquitous Computing (Ubicomp)*, Gothenburg, Sweden (2002) 351–360.
10. M. A. Azim and A. Jamalipour, Optimized forwarding for wireless sensor networks by fuzzy inference system, *IEEE International Conference on Wireless Broadband and Ultra Wideband Communications*, Sydney, Australia (2006).
11. M. A. Azim and A. Jamalipour, Performance evaluation of optimized forwarding strategy for flat sensor networks, *IEEE Global Communications Conference (GLOBECOM)*, Washington, DC, USA (2007).
12. M. Balakrishnan and E. E. Johnson, Fuzzy diffusion for distributed sensor networks, *IEEE Military Communications Conference (MILCOM)*, Atlantic City, NJ **5** (2005) 1–6.
13. M. Balakrishnan and E. E. Johnson, Fuzzy diffusion analysis: decision significance and applicable scenarios, *IEEE Military Communications Conference (MILCOM)*, Washington, DC (2006) 1–7.
14. H. Baldus, K. Klabunde and G. Muesch, Reliable set-up of medical body-sensor networks, *European Conference on Wireless Sensor Networks (EWSN)*, Gothenburg, Sweden **2920** (2004) 353–363.
15. R. Beckwith, D. Teibel and P. Bowen, Pervasive computing and proactive agriculture, *ACM Conference of Pervasive Computing*, Vienna, Austria (2004).
16. E. Bonabeau, M. Dorigo and G. Theraulaz, *Swarm Intelligence: From Natural to Artificial Systems* (Oxford University Press, 1999).
17. P. Bose, P. Morin, I. Stojmenovic and J. Urrutia, Routing with guaranteed delivery in ad hoc wireless networks, *ACM Wireless Networks* **7**(6) (2001) 609–616.
18. A. Boukerche, R. W. N. Pazzi and R. B. Araujo, A fast and reliable protocol for wireless sensor networks in critical conditions monitoring applications, *IEEE MSWIM*, Venice, Italy (2004).
19. A. Boukerche, R. W. N. Pazzi and R. B. Araujo, HPEQ — a hierarchical periodic, event-driven and query-based wireless sensor network protocol, *IEEE Local Computer Networks*, Sydney, Australia (2005) 560–567.
20. D. Braginsky and D. Estrin, Rumor routing algorithm for sensor networks, *ACM Workshop on Sensor Networks and Applications*, Atlanta, GA (2002) 22–31.
21. T. Braun, T. Voigt and A. Dunkels, TCP support for sensor networks, *IEEE Wireless on Demand Network Systems and Services (WONS)*, Austria (2007) 162–169.
22. J. Burrell, T. Brooke and R. Beckwith, Vineyard computing: sensor networks in agricultural production, *IEEE Pervasive Computing* **3**(1) (2007) 38–45.
23. Z. Butler, P. Corke, R. Peterson and D. Rus, Networked cows: virtual fences for controlling cows, *ACM Workshop on Applications of Mobile Embedded Systems (WAMES)*, Boston, USA (2004).

24. G. D. Caro and M. Dorigo, AntNet: distributed stigmergetic control for communications networks, *Artificial Intelligence Research* **9** (1998) 317–365.
25. R. Chang and C. Kuo, An energy efficient routing mechanism for wireless sensor networks, *IEEE International Conference on Advanced Information Networking and Applications (AINA)*, Vienna, Austria **2** (2006).
26. D. Chen and P. K. Varshney, QoS support in wireless sensor networks: a survey, *International Conference on Wireless Networks (ICWN)*, Las Vegas, Nevada, USA (2006).
27. D. Chen and P. K. Varshney, On demand geographic forwarding for data delivery in wireless sensor networks, *Elsevier Computer Communication Journal, Special Issue on Network Coverage and Routing Schemes for Wireless Sensor Networks* (2006).
28. D. Chen and P. K. Varshney, A survey of void handling techniques for geographic routing in wireless networks, *IEEE Communications Surveys and Tutorials* **9**(1), (2007) 50–67.
29. W. Chen, W. Li, H. Shou and B. Yuan, A QoS-based adaptive clustering algorithm for wireless sensor networks, *IEEE International Conference on Mechatronics and Automation*, Bangkok, Thailand (2006b) 1947–1952.
30. W. Choi, P. Shah and S. K. Das, A framework for energy-saving data gathering using two-phase clustering in wireless sensor networks, *International Conference on Mobile and Ubiquitous Systems: Networking and Services*, Boston, MA (2004) 203–212.
31. M. Chu, H. Haussecker and F. Zhao, Scalable information driven sensor querying and routing for ad hoc heterogeneous sensor networks, *The International Journal of High Performance Computing Applications* (2002) 293–313.
32. C. D. M. Cordeiro and D. P. Agrawal, *Ad Hoc and Sensor Networks: Theory and Applications* (World Scientific Publishing, 2006).
33. T. H. Cormen, C. E. Leiserson, R. L. Rivest and C. Stein, *Introduction to Algorithms* (MIT Press and McGraw-Hill, 2001).
34. B. Deb, S. Bhatnagar and B. Nath, ReInForM: reliable information forwarding using multiple paths in sensor networks, *IEEE International Conference on Local Computer Networks*, Bonn, Germany (2003) 406–415.
35. I. Demirkol, C. Ersoy and F. Alagoz, MAC protocols for wireless sensor networks: a survey, *IEEE Communications Magazine* **44**(4) (2006) 115–121.
36. M. Eltoweissy, M. Moharrum and R. Mukkamala, Dynamic key management in sensor networks, *IEEE Communications Magazine* **44**(4) (2006) 122–130.
37. P. T. Eugster, R. Guerraoui and J. Sventek, *Distributed Asynchronous Collections: Abstractions for Publish/Subscribe Interaction* (Lecture Notes in Computer Science, Vol. 1850 (Springer-Verlag, 2000) 252–276.
38. Q. Fang, F. Zhao and L. Guibas, Lightweight sensing and communication protocols for target enumeration and aggregation, *ACM/IEEE Mobile and Ad Hoc Networking and Computing (MobiHoc)*, Maryland, USA (2003) 165–176.
39. Q. Fang, J. Gao and L. J. Guibas, Locating and bypassing routing holes in sensor networks, *IEEE Computer and Communications Conference (INFOCOM)*, Hong Kong **4** (2004) 2458–2468.
40. E. Felemban, C. Lee and E. Ekici, MMSPEED: multipath multi-SPEED protocol for QoS guarantee of reliability and timeliness in wireless sensor networks, *IEEE Transactions on Mobile Computing* **5**(6) (2006) 738–757.
41. M. Gerla, T. J. Kwon and G. Pei, On-demand routing in large ad hoc wireless networks with passive clustering, *IEEE Wireless Communications and Networking Conference (WCNC)*, Chicago, IL **1** (2000) 100–105.

42. T. Haider and M. Yusuf, FPGA based fuzzy link cost processor for energy — aware routing in wireless sensor networks — design and implementation, *IEEE International Multitopic Conference (INMIC)*, Karachi, Pakistan (2005) 1–6.
43. V. Handziski, A. Kopke, H. Karl, C. Frank and W. Drytkiewicz, Improving the energy efficiency of directed diffusion using passive clustering, *1st European Workshop on Wireless Sensor Networks (EWSN)*, Berlin, Germany (2004) 172–187.
44. T. Hee, J. A. Stankovic, C. Lu and T. Abdelzaher, SPEED: a stateless protocol for real-time communication in sensor networks, *IEEE Distributed Computing Systems*, Rhode Island, USA (2003) 46–55.
45. W. Heinzelman, A. Chandrakasan and H. Balakrishanan, Energy-efficient routing protocols for wireless micorsensor networks, *33rd Hawaii International Conference on System Sciences (HICSS)*, Maui (2000) 3005–3014.
46. W. Heinzelman, A. Chandrakasan and H. Balakrishanan, An application-specific protocol architecture for wireless microsensor networks, *IEEE Transactions on Wireless Communications* **1**(4) (2002) 660–670.
47. M. Hempel, H. Sharif and P. Raviraj, HEAR-SN: a new hierarchical energy-aware routing protocol for sensor networks, *38th Annual Hawaii International Conference on System Sciences*, Hawaii.
48. S. Hussain, A. W. Matin and O. Islam, Genetic algorithm for energy efficient clusters in wireless sensor networks, *IEEE International Conference on Information Technology (ITNG)*, Las Vegas, Nevada (2007) 147–154.
49. C. Intanagonwiwat, R. Govindan, D. Estrin, J. Heidemann and F. Silva, Directed diffusion for wireless sensor networking, *IEEE/ACM Transactions on Networking* **11**(1) (2003) 2–16.
50. R. Jain, A. Puri and R. Sengupta, Geographical routing using partial information for wireless ad hoc networks, *IEEE Personal Communications* **8**(1) (2001) 48–57.
51. P. Juang, H. Oki, Y. Wang, M. Martonosi, L. S. Peh and D. Rubenstein, Energy-efficient computing for wildlife tracking: design tradeoffs and early experiences with Zebranet, *ACM Architectural Support for Programming Languages and Operating Systems (ASPLOS)*, San Jose, CA **36** (2002) 96–107.
52. S. Jung, Y. Han and T. Chung, The concentric clustering scheme for efficient energy consumption in the PEGASIS, *IEEE International Conference on Advanced Communication Technology (ICACT)*, Gangwon-Do, Korea **1** (2007) 260–265.
53. C. Kappler and G. Riegel, A real-world, simple wireless sensor network for monitoring electrical energy consumption, *European Conference on Wireless Sensor Networks (EWSN)*, Berlin, Germany **2920** (2004) 339–352.
54. B. Karp and H. T. Kung, GPSR: greedy perimeter stateless routing for wireless networks, *IEEE Mobile Computing and Networking*, Boston, MA (2000) 243–254.
55. A. Konak, D. W. Coit and A. E. Smith, Multi-objective optimization using genetic algorithms: a tutorial, *Elsevier Reliability Engineering and System Safety* **91**(9) (2006) 992–1007.
56. E. Kranakis, H. Singh and J. Urrutia, Compass routing on geometric networks, *11th Canadian Conference Computational Geometry*, Vancouver, Canada (1999) 51–54.
57. F. Kuhn, R. Wattenhofer and A. Zollinger, Worst-case optimal and average-case efficient geometric ad-hoc routing, *ACM Mobile Ad Hoc Networking and Computing (MobiHoc)*, Annapolis, MD (2003a) 267–278.
58. F. Kuhn, R. Wattenhofer, Y. Zhang and A. Zollinger, Geometric ad hoc routing: of theory and practice, *ACM Principles of Distributed Computing (PODC)*, Boston, MA (2003b) 62–73.

59. J. Kulik, W. Rabiner and H. Balakrishnan, Adaptive protocols for information dissemination in wireless sensor networks, *ACM/IEEE International Conference on Mobile Computing and Networking*, Seattle, WA **2** (1999) 174–185.
60. J. Kulik, W. R. Heinzelman and H. Balakrishnan, Negotiation-based protocols for disseminating information in wireless sensor networks, *Wireless Networks* **8**(2–3) (2002) 169–185.
61. H. Kwon, T. H. Kim, S. Choi and B. G. Lee, Lifetime maximization under reliability constraint via cross-layer strategy in wireless sensor networks, *IEEE Wireless Communications and Networking Conference (WCNC)*, New Orleans, LA **3** (2005) 1891–1896.
62. Y. Li, J. Chen, R. Lin and Z. Wang, A reliable routing protocol design for wireless sensor networks, *IEEE International Conference on Mobile Ad Hoc and Sensor Systems*, Washington, DC (2005).
63. Q. Liang and Q. Ren, Energy and mobility aware geographic multipath routing for wireless sensor networks, *Wireless Communications and Networking Conference (WCNC)*, New Orleans, LA **3** (2005) 1867–1871.
64. S. Lindsay and C. Raghavendra, PEGASSIS: power efficient gathering in sensor information systems, *IEEE Aerospace Conference*, Big Sky, Montana **3** (2002) 1125–1130.
65. K. Liu, N. Abu-Ghazaleh and K.-D. Kang, JiTS: just-in-time scheduling for real-time sensor data dissemination, *4th IEEE International Conference on Pervasive Computing and Communications* (2006).
66. H. Luo, F. Ye, J. Cheng, S. Lu and L. Zhang, TTDD: two-tier data dissemination in large-scale wireless sensor networks, *Springer Wirless Networks* **11**(1–2) (2005) 161–175.
67. M. V. Machado, O. Goussevskaia, R. A. F. Mini, C. G. Rezende, A. A. F. Loureiro, G. R. Mateus and J. M. Nogueira, Data dissemination using the energy map, *IEEE International Conference on Wireless On-Demand Network Systems and Services (WONS)*, St. Moritz, Switzerland (2005) 139–148.
68. A. Mahapatra, K. Anand and D. P. Agrawal, QoS and energy aware routing for real-time traffic in wireless sensor networks, *Elsevier Journal for Computer Communications* **29**(4) (2006) 437–445.
69. A. Mainwaring, J. Polastre, R. Szewczyk, D. Culler and J. Anderson, Wireless sensor networks for habitat monitoring, *ACM Wireless Sensor Networks and Applications (WSNA)*, Atlanta, USA (2002) 88–97.
70. A. Manjeshwar and D. P. Agarwal, TEEN: a routing protocol for enhanced efficiency in wireless sensor networks, *IEEE Parallel and Distributed Processing Symposium*, San Francisco (2001) 2009–2015.
71. A. Manjeshwar and D. P. Agrawal, APTEEN: a hybrid protocol for efficient routing and comprehensive information retrieval in wireless sensor networks, *IEEE International Parallel and Distributed Processing Symposium*, Florida, USA (2002) 195–202.
72. A. Manjeshwar, Q. Zeng and D. P. Agrawal, An analytical model for information retrieval in wireless sensor networks using enhanced APTEEN protocol, *IEEE Transactions on Parallel and Distributed Systems* **13**(12) (2002) 1290–1302.
73. I. W. Marshall, C. Roadknight, I. Wokoma and L. Sacks, Self-organizing sensor networks, *UbiNet*, London, UK (2003).
74. K. Martinez, R. Ong and J. Hart, GLACSWEB: a sensor network for hostile environments, *IEEE Sensor and Ad Hoc Communications and Networks (SECON)*, Berlin, Germany (2004) 81–87.
75. T. Meng and R. Volkan, Distributed network protocols for wireless communication, *IEEE ISCAS*, Monterey, USA **4** (1998) 600–603.

76. W. M. Meriall, F. Newberg, K. Sohrabi, W. Kaiser and G. Pottie, Collaborative networking requirements for unattended ground sensor systems, *IEEE Aerospace Conference*, Big Sky, Montana **5** (2003) 2153–2165.
77. F. Michahelles, P. Matter, A. Schmidt and B. Schiele, Applying wearable sensors to avalanche rescue, *Elsevier Computers and Graphics* **27**(6) (2003) 839–847.
78. D. C. Montgomery, E. A. Peck and G. G. Vining, *Introduction to Linear Regression Analysis* (John Wiley and Sons, 2001).
79. R. Musunuri and J. A. Cobb, Hierarchical-battery aware routing in wireless sensor networks, *IEEE Vehicular Technology Conference (VTC)*, Dallas, TX (2005) 203–212.
80. S. D. Muruganathan, D. C. F. Ma, R. I. Bhasin and A. O. Fapojuwo, A centralized energy-efficient routing protocol for wireless sensor networks, *IEEE Communications Magazine* **43**(3) (2005) 8–13.
81. D. Niculescu and B. Nath, Trajectory-based forwarding and its applications, *IEEE International Conference on Mobile Computing and Networking*, San Diego, CA, USA (2003) 260–272.
82. N. Patwari, J. N. Ash, S. Kyperountas, A. O. Hero, R. L. Moses and N. S. Correal, Locating the nodes: cooperative localization in wireless sensor networks, *IEEE Signal Processing Magazine* **22**(4) (2005) 54–69.
83. O. Rashid, M. Mamun, M. M. Alam and C. S. Hong, Energy conserving passive clustering for efficient routing in wireless sensor network, *IEEE International Conference on Advanced Communication Technology*, Phoenix Park, Korea **2** (2005) 982–986.
84. R. Riem-Vis, Cold chain management using an ultra low power wireless sensor network, *ACM Workshop on Applications of Mobile Embedded Systems (WAMES)*, Boston, USA (2004) 18–20.
85. A. Rytter, *Vibration Based Inspection of Civil Engineering Structures*, PhD Dissertation, Department of Building Technology and Structural Engineering, University of Aalborg, Denmark.
86. N. Sadagopan, B. Krishnamachari and A. Helmy, The ACQUIRE mechanism for efficient querying in sensor networks, *IEEE Sensor Network Protocols and Applications*, Anchorage, Alaska (2003) 149–155.
87. S. S. Satapathy and N. Sarma, TREEPSI: tree based energy efficient protocol for sensor information, *IEEE International Conference on Wireless and Optical Communications Networks*, Bangalore, India (2006).
88. C. Schurgers and M. B. Srivastava, Energy efficient routing in wireless sensor networks, *IEEE Military Communications Conference (MILCOM)*, McLean, VA **1** (2001) 357–361.
89. L. Schwiebert, S. K. S. Gupta and J. Weinmann, Research challenges in wireless networks of biomedical sensors, *ACM/IEEE Mobile Computing and Networking (MobiCom)*, Rome, Italy (2001) 151–165.
90. S. Servetto and G. Barrenechea, Constrained random walks on random graphs: routing algorithms for large scale wireless sensor networks, *ACM Wireless Sensor Networks and Applications*, New York, USA (2002) 12–21.
91. R. Shah and J. Rabaey, Energy aware routing for low energy ad hoc sensor networks, *IEEE Wireless Communications and Networking Conference (WCNC)*, Orlando, FL **1** (2002) 350–355.
92. G. Simon, A. Ledezczi and M. Maroti, Sensor network-based countersniper system, *ACM Workshop on Applications of Mobile Embedded Systems (WAMES)*, Baltimore, USA (2004) 1–12.

93. T. Srinivasan, R. Chandrasekar and V. Vijaykumar, A fuzzy, energy efficient scheme for data centric multipath routing in wireless sensor networks, *IEEE International Conference on Wireless and Optical Communications Networks*, Bangalore, India (2006).
94. M. B. Srivastava, R. R. Muntz and M. Potkonjak, Smart kindergarten: sensorbased wireless networks for smart developmental problem-solving enviroments, *ACM/IEEE Mobile Computing and Networking (MobiCom)*, Rome, Italy (2001) 132–138.
95. I. Stojmenovic and X. Lin, Loop-free hybrid single-path/flooding routing algorithms with guaranteed delivery for wireless networks, *IEEE Transactions on Parallel and Distributed Systems* (2001).
96. I. Stojmenvoic, M. Russell and B. Vukojevic, Depth first search and location based localized routing and QoS routing in wireless networks, *Computers and Informatics* **21**(2) (2002) 149–165.
97. L. Subramanian and R. H. Katz, An architecture for building self-configurable systems, *IEEE Mobile and Ad Hoc Networking and Computing*, Boston, MA (2000) 63–73.
98. N. Tabassum, Q. E. K. Mamun and Y. Urano, COSEN: a chain oriented sensor network for efficient data collection, *IEEE Information Technology*, Las Vegas, Nevada (2006) 262–267.
99. S. Tilak, N. B. Abu-Ghazaleh and W. Heinzelman, A taxonomy of wireless microsensor network models, *ACM Mobile Computing and Communications Review (MC2R)*, **6**(2) (2002) 28–36.
100. P. Traynor, R. Kumar, H. Choi, G. Cao, S. Zhu and T. L. Porta, Efficient hybrid security mechanisms for heterogeneous sensor networks, *IEEE Transactions on Mobile Computing* **6**(6) (2007) 663–677.
101. R. Vidhyapriya and P. T. Vanathi, Energy aware routing for wireless sensor networks, *IEEE International Conference on Signal Processing, Communications and Networking (ICSCN)*, Chennai, India (2007) 545–550.
102. M. C. Vuran and I. F. Akyildiz, Spatial correlation-based collaborative medium access control in wireless sensor networks, *IEEE/ACM Transactions on Networking* **14**(2) 316–329.
103. C. Wang, K. Sohraby, Y. Hu, B. Li and W. Tang, Issues of transport control protocols for wireless sensor networks, *IEEE Communications, Circuits and Systems*, Hong Kong **1** (2005) 422–426.
104. C. Wang and L. Xiao, Sensor localization under limited measurement capabilities, *IEEE Network* **21**(3) (2007) 16–23.
105. L. Wang, S. Chen and X. Li, A multiple-objective fuzzy decision making based information-aware routing protocol for wireless sensor networks, *IEEE Wireless Communications, Networking and Mobile Computing (WiCOM)*, Rome, Italy (2006b) 1–4.
106. Web, *ESF Exploratory Workshop on Wireless Sensor Networks*, www.vs.inf.ethz.ch/events/esf-wsn04 (2004).
107. Web, ARGO — Global Ocean Sensor Network, www.argo.ucsd.edu (2007b).
108. A. Woo, T. Tong and D. Culler, Taming the underlying challenges of reliable mullti-hop routing in sensor networks, *Embedded Networked Sensor Systems* (ACM SenSys, 2003), Los Angeles, CA (2003) 14–27.
109. Y. Xu, J. Heidemann and D. Estrin, Geography-informed energy conservation for ad hoc routing, *ACM/IEEE International Conference on Mobile Computing and Networking*, Rome, Italy (2001) 70–84.
110. Y. Yao and J. Gehrke, The Cougar approach to in-network query processing in sensor networks, *ACM SIGMOD Record* **31** (2002) 3–8.

111. F. Ye, A. Chen, S. Lu and L. Zhang, A scalable solution to minimum cost forwarding in large sensor networks, *IEEE International Parallel and Distributed Processing Symposium*, Arizona, USA (2001) 304–309.
112. F. Ye, H. Luo, J. Cheng, S. Lu and L. Zhang, A two-tier data dissemination model for large-scale wireless sensor networks, *ACM/IEEE International Conference on Mobile Computing and Networking*, Atlanta GA (2002) 148–159.
113. Y. Yu, D. Estrin and R. Govindan, Geographical and energy-aware routing: a recursive data dissemination protocol for wireless sensor networks, UCLA Computer Science Department Technical Report, UCLA-CSD TR-01-0023.
114. M. Yusuf and T. Haider, Energy-aware fuzzy routing for wireless sensor networks, *IEEE Emerging Technologies*, Richardson, TX (2005) 63–69.
115. Z. Zeng-Wei, W. Zou-Hui and L. Huai-Zhong, An event-driven clustering routing algorithm for wireless sensor networks, *IEEE Intelligent Robots and Systems*, Japan **2** (2004) 1802–1806.
116. Y. Zhang, L. D. Kuhn and M. P. J. Fromherz, Improvements on ant routing for sensor networks, *International Workshop on Ant Colony Optimization and Swarm Intelligence*, Warsaw, Poland (2004).
117. Q. Zhao and L. Tong, Quality-of-service specific information retrieval for densely deployed sensor networks, *IEEE Military Communications Conference (MILCOM)*, Boston, MA **1** (2003) 591–596.
118. L. Zou, M. Lu and Z. Xiong, PAGER-M: a novel location based routing protocol for mobile sensor networks, *IEEE Broadband Wireless Services and Applications (BroadWISE)*, San Jose, CA (2004).
119. L. Zou, M. Lu and Z. Xiong, Distributed algorithm for the dead end problem of location based routing in sensor networks, *IEEE Transactions on Vehicular Technology* **54**(4) (2005) 1509–1522.

Chapter 10

Handling QoS Traffic in Wireless Sensor Networks

Mohamed Younis
University of Maryland at Baltimore County, USA
younis@cs.umbc.edu

Kemal Akkaya
Southern Illinois University, USA
kemal@cs.siu.edu

Moustafa Youssef
Nile University, Egypt
mayoussef@nileu.edu.eg

Many new routing and MAC layer protocols have been proposed for wireless sensor networks tackling the issues raised by the resource constrained unattended sensor nodes in large-scale deployments. The majority of these protocols considered energy efficiency as the main objective and assumed data traffic with unconstrained delivery requirements. However, the growing interests in applications that demand certain end-to-end performance guarantees and the introduction of imaging and video sensors have posed additional challenges. Transmission of data in such cases requires both energy and QoS aware network management in order to ensure efficient usage of the sensor resources and effective access to the gathered measurements. In this chapter, we highlight the architectural and operational challenges of handling of QoS traffic in sensor networks. We report on progress make to-date and discuss sample of our work in more details. Finally, we outline open research problems.

Keywords: Ad hoc networks; sensor networks; managing quality of service traffic; energy efficient design; energy-aware routing; energy-aware MAC protocols; quality of service issues.

10.1. Introduction

Recent technological breakthroughs in ultra-high integration and low-power electronics have enabled the development of tiny battery-operated sensors.[1–4] In addition to the sensing circuitry, a sensor typically includes a signal-processor and a radio. The sensing circuitry measures ambient conditions, related to the environment surrounding the sensor and transforms them into an electric signal. Processing such a signal reveals some properties about objects located and/or events happening in the vicinity of the sensor. The sensor sends such collected data via the radio transmitter, to a command center (sink) either directly or through a data concentration center (a gateway). The gateway can perform fusion of the sensed data in order to filter out erroneous data and anomalies and to draw conclusions from the reported data over a period of time.

The continuous decrease in the size and cost of sensors has motivated intensive research addressing the potential of collaboration among sensors in data gathering and processing via an ad hoc wireless network. Networking unattended sensor nodes is expected to have significant impact on the efficiency of many military and civil applications, such as disaster management, combat field surveillance, and security. In disaster management situations such as earthquakes, sensor networks can be used to selectively map the affected regions directing emergency response units to survivors. In military situations, sensor networks can be used in surveillance missions and can be used to detect moving targets, chemical gases, or presence of micro-agents. Sensors in such environments are energy constrained and their batteries cannot be recharged. Therefore, designing energy-aware algorithms becomes an important factor for extending the lifetime of sensor networks.[5,6]

The signal processing and communication activities are the main consumers of sensor's energy. Since sensors are battery-operated, keeping the sensor active all the time will limit the battery's lifetime. Therefore, optimal organization and management of the sensor network are crucial in order to perform the desired function with an acceptable level of quality and to maintain sufficient sensor energy for the required mission. Mission-oriented organization of the sensor network enables the appropriate selection of only a subset of the sensors to be turned on and thus avoids wasting the energy of sensors that do not have to be involved. Energy-aware network management will ensure a desired level of performance for the data transfer while extending the life of the network.

Energy constraints combined with a typical deployment of large number of sensor nodes have necessitated energy-awareness at most layers of networking protocol stack including network and link layers. Current research on routing in wireless sensor networks mostly has focused on protocols that are energy-aware to maximize the lifetime of the network, are scalable to accommodate a large number of sensor nodes, and are tolerant to sensor damage and battery exhaustion.[7–10] In addition, medium access is a major consumer of sensor energy, especially when the radio receiver is turned on all time. Energy consumed for radio transmission is directly proportional to the distance squared and can significantly magnify in a noisy environment. Energy-aware routing can optimize the transmission energy, while collision avoidance and minimization of energy consumed by the receiver can be achieved via energy-efficient medium access control (MAC) mechanisms.[11–13]

Since such energy consideration has dominated most of the research in sensor networks, the concepts of latency, throughput and delay jitter were not primary concerns in most of the published work on sensor networks. However, the increasing interest in real-time applications along with the introduction of imaging and video sensors has posed additional challenges. For instance, the transmission of imaging and video data requires careful handling in order to ensure that end-to-end delay is within acceptable range and the variation in such delay is acceptable. Such performance metrics are usually referred to as quality of service (QoS) of the communication network. Therefore, collecting sensed imaging and video data requires both energy and QoS aware network protocols in order to ensure efficient usage of the sensors and effective access to the gathered measurements.

QoS protocols in sensor networks have several applications including real-time target tracking in battlefield environments, emergent event triggering in monitoring applications, etc. Consider the following scenario: in a battlefield environment in order to identify a target, we should employ imaging sensors. After detecting and locating a target using contemporary types of sensors, e.g. acoustic, imaging sensors can be turned on to capture a picture of such a target periodically for sending to the gateway. Since, it is a battlefield environment; this requires a real-time data exchange between sensors and controller in order to take the proper actions. Delivering such time-constrained data requires certain bandwidth with minimum possible delay and thus a service differentiation mechanism will be needed in order to guarantee timeliness.

Energy-aware QoS routing in sensor networks will ensure guaranteed bandwidth (or delay) through the duration of a connection as well as providing the use of the most energy efficient path. To the best of our knowledge, little attention has been paid by the research community to addressing QoS requirements in sensor networks. In this chapter, we analyze the challenges of supporting QoS in traffic at the network and link layers, survey current state of the research, and point out open issues.

In the balance of this chapter, we will briefly summarize the system architecture design issues for sensor networks and their implications on data routing and medium access control in Section 2 along with an example of the very recent work and discusses it elaborately. We analyze the complexity of handling QoS requirements in sensor networks, report on progress made in the research community, and outline open research issues and directions for future research in Section 3. Finally, we conclude the chapter in Section 4.

10.2. Background

In this section, we give a background related to QoS in wireless sensor networks. We start by the different design issues for sensor networks that affect QoS traffic. Following that, we discuss how QoS is currently handled in sensor wireless networks. We end with an example illustrating the various concepts covered in the section.

10.2.1. *System architecture and design issues*

Depending on the application, different architectures and design goals/constraints have been considered for sensor networks. Since the performance of routing and MAC protocols are closely related to the architectural model, in this section we strive to capture architectural design issues and highlight their implications on the sensor network performance. Later we will analyze the complexity of supporting QoS traffic in light of these design variations. A summary of design issues is given in Table 10.1.

10.2.1.1. *Network dynamics*

There are three main components in a sensor network. These are the sensor nodes, sink, and monitored events. Aside from the very few setups that utilize mobile sensors,[14] most of the network architectures assume that sensor nodes are stationary. On the other hand, supporting the mobility of sinks or cluster-heads (gateways) is sometimes deemed necessary.[15] Routing messages from or to moving nodes is more challenging since route stability becomes an important optimization factor, in addition to energy, bandwidth, etc. The sensed event can be either mobile or stationary depending on the application.[16] For instance, in a target detection/tracking application, the event (phenomenon) is mobile whereas forest monitoring for early fire prevention is an example of stationary events. Monitoring mobile events is more challenging than stationary events.

10.2.1.2. *Node deployment*

Another consideration is the topological deployment of nodes. This is application dependent and affects the performance of the routing protocol. The deployment

Table 10.1 A summary of the architectural design issues.

Design issue	Primary factors
Network Dynamics	Mobility of node, target, and sink
Node Deployment	Deterministic or Ad Hoc
Node Communications	Single-hop or multi-hop
Data Delivery Models	Continuous, event-driven, query-driven, or hybrid
Node Capabilities	Multi- or single function, Homogeneous or heterogeneous capabilities
Data Aggregation/Fusion	In-network (partially or fully) or out-of-network

is either deterministic or self-organizing. In deterministic situations, the sensors are manually placed and data is routed through pre-determined paths. In addition, collision among the transmissions of the different nodes can be minimized through the pre-scheduling of medium access. However in self-organizing systems, the sensor nodes are scattered randomly creating an infrastructure in an ad hoc manner.[5,10,17,18] In that infrastructure, the position of the sink or the cluster-head is also crucial in terms of energy efficiency and performance. When the distribution of nodes is not uniform, optimal clustering becomes a pressing issue to enable energy efficient network operation.

10.2.1.3. *Energy considerations*

During the creation of an infrastructure, the process of setting up the routes is greatly influenced by energy considerations. Since the transmission power of a wireless radio is proportional to the square of the distance or even higher order in the presence of obstacles, multi-hop routing will consume less energy than direct communication. However, multi-hop routing introduces significant overhead for topology management and medium access control. Direct routing would perform well enough if all the nodes were very close to the sink.[18] Most of the time sensors are scattered randomly over an area of interest and multi-hop routing becomes unavoidable. Arbitrating medium access in this case becomes cumbersome.

10.2.1.4. *Data delivery models*

Depending on the application of the sensor network, the data delivery model to the sink can be continuous, event-driven, query-driven, or hybrid.[16] In the continuous delivery model, each sensor sends data periodically. In event-driven and query-driven models, the transmission of data is triggered when an event occurs or a query is generated by the sink. Some networks apply a hybrid model using a combination of continuous, event-driven and query-driven data delivery. The routing and MAC protocols are highly influenced by the data delivery model, especially with regard to the minimization of energy consumption and route stability. For instance, it has been concluded by Heinzelman[18] that for a habitat monitoring application, where data is continuously transmitted to the sink, a hierarchical routing protocol is the most efficient alternative. This is due to the fact that such an application generates significant redundant data that can be aggregated on route to the sink, thus reducing traffic and saving energy. In addition, in a continuous data delivery model time-based medium access can achieve significant energy saving since it will enable turning off sensors' radio receivers.[11] Carrier Sense Multiple Access (CSMA) medium access arbitration is a good fit for event-based data delivery models since the data is generated sporadically.

10.2.1.5. *Node capabilities*

In a sensor network, different functionalities can be associated with the sensor nodes. In early work on sensor networks,[4,7,9] all sensor nodes are assumed to be homogenous, having equal capacity in terms of computation, communication and power. However, depending on the application, a node can be dedicated to a particular special function such as relaying, sensing and aggregation since engaging the three functionalities at the same time on a node might quickly drain the energy of that node. Some of the hierarchical protocols proposed in the literature designate a cluster-head different from the normal sensors. While some networks have picked cluster-heads from the deployed sensors,[18–20] in other applications a cluster-head is more powerful than the sensor nodes in terms of energy, bandwidth and memory.[10,14] In such cases, the burden of transmission to the sink and aggregation is handled by the cluster-head.

10.2.1.6. *Data aggregation/fusion*

Since sensor nodes might generate significant redundant data, in some applications similar packets from multiple nodes can be aggregated so that the number of transmissions would be reduced. Data aggregation is the combination of data from different sources by using functions such as *suppression* (eliminating duplicates), *min, max* and *average*.[21] Some of these functions can be performed either partially or fully in each sensor node, by allowing sensor nodes to conduct in-network data reduction.[7,19,22] Recognizing that computation would be less energy consuming than communication, substantial energy savings can be obtained through data aggregation.[9] This technique has been used to achieve energy efficiency and traffic optimization in a number of routing protocols. In some network architectures, all aggregation functions are assigned to more powerful and specialized nodes.[14] Data aggregation is also feasible through signal processing techniques. In that case, it is referred as *data fusion* where a node is capable of producing a more accurate signal by reducing the noise and using some techniques such as *beamforming* to combine the signals.[9] Data aggregation makes medium access control complex since redundant packets will be eliminated and such elimination will require instantaneous medium access arbitration. In such case, only CSMA and Code Division Multiple Access (CDMA)-based MAC protocols are typically applicable leading to an increase in energy consumption. In addition, data aggregation becomes more challenging when dealing with multimedia data, such as video and images, as described in Section 10.3.

10.2.2. *Supporting QoS in sensor networks*

The network and link layers of the communication protocol stack have been the focus of researchers for improving energy utilization. Especially in wireless sensor networks, new energy-conscious routing algorithms and medium access arbitration have been designed. However, little research has been done on supporting QoS constrained traffic. Although there has been some research on QoS routing for mobile ad hoc networks, it has not been studied in the context of wireless sensor networks.

Before getting to the detailed analysis of the QoS issues in wireless sensor networks, it is important to differentiate between QoS objectives and constraints. Having design or operational goals in terms of QoS attributes, e.g. minimizing end-to-end delay (response time), is very common in all types of networks. Supporting traffic that is subject to QoS requirements is generally more difficult. Meeting QoS requirements in a resource-constrained environment, such as sensor networks, is exceptionally challenging.

In this section, we analyze the technical issues for handling QoS constrained traffic in wireless sensor networks and report on the state of the research. First we outline the research effort related to energy-aware QoS in general mobile ad hoc networks and comment on the appropriateness of the developed techniques to wireless sensor networks. Section 3.2 lists the main challenges of supporting QoS traffic in wireless sensor networks. Finally we survey published and on-going research on routing and MAC layer protocols for QoS sensor data.

10.2.2.1. *QoS in general wireless networks*

While contemporary best-effort routing approaches address unconstrained traffic, QoS routing in a connection-oriented communication is usually performed through resource reservation in order to meet the QoS requirements for each individual connection. Although many mechanisms have been proposed for routing QoS constrained multimedia data in wire-based networks,[23–27] they cannot be directly applied to wireless networks due to the inherent characteristics of wireless environments affecting link quality and to the limited resources, such as bandwidth. Therefore several new protocols have been proposed for QoS routing in wireless ad hoc networks taking the dynamic nature of the network into account.[28–32]

Some of the proposed protocols considered to work with imprecise state information[28,29] since it is difficult to capture precise state information due to dynamic nature of mobile ad hoc networks. CEDAR[30] is another QoS aware protocol, which uses the idea of core nodes (dominating set) of the network while determining the paths.[30] Using routes found through the network core, a QoS path can be found. However, if a node in the core breaks down, it will cost too much in terms of resource usage to reconstruct the core. Lin[31] and Zhu *et al.*[32] have proposed QoS routing protocols specifically designed for Time Division Multiple Access (TDMA)-based ad hoc networks. Both protocols can build a QoS route from a source to destination with reserved bandwidth. The bandwidth calculation is done hop-by-hop in a distributed fashion. For wireless sensor networks, the use of reservation based protocols for supporting QoS constrained traffic will be impractical unless the network follows a continuous data delivery model.

On the other hand, applications that need regular delivery of QoS constrained data are not expected to employ sensor networks due to the lack of sufficient resources, in particular energy and bandwidth, to handle such demanding QoS traffic. In addition, the nature of sensor networks poses unique challenges compared to general wireless networks and thus requires special attention. The next subsection discusses such challenges.

10.2.2.2. *Survey of QoS routing*

While having QoS objectives has not been uncommon for data routing in sensor networks,[5,10] very little attention has been paid to QoS constrained traffic. Recently few research projects have started to emerge addressing the support of QoS requirements in wireless sensor networks. In this subsection, we survey the state of the research summarizing published work and highlighting the subset of QoS issues being addressed. In summary, the work published so far falls into two categories. The first category focuses on the energy and delay trade-off without much consideration to the other issues. The second category strives to spread traffic in order to effectively boost the bandwidths and lower the delay. The following is a more elaborate summary of these approaches.

SAR: Sequential Assignment Routing (SAR) is the first protocol for sensor networks that includes a notion of QoS in its routing decisions.[5] It is a table-driven multi-path approach striving to achieve energy efficiency and fault tolerance. The SAR protocol supports only QoS objectives. It creates trees rooted at one-hop neighbors of the sink by taking QoS metrics, available energy resources on each path and priority level of each packet into consideration. By using the created trees, multiple paths from sink to sensors are formed, of which only one is actually used keeping the rest as backup. Failure recovery is done by enforcing routing table consistency between upstream and downstream nodes on each path. Any local failure causes an automatic path restoration procedure. Simulation results show that SAR offers less power consumption than the minimum-energy metric algorithm, which focuses only on the energy consumption of each packet without considering its priority. SAR maintains multiple paths from nodes to sink. Although, this allows fault-tolerance and easy recovery, the protocol suffers from the overhead of maintaining the tables and states at each sensor node especially when the number of nodes is huge. SAR also does not use redundant routes to split the load and effectively boost the bandwidth.

Energy-Aware QoS Routing Protocol: A fairly new QoS aware protocol for sensor networks is proposed by Akkaya and Younis.[33] They consider a network architecture in which sensors are partitioned into clusters. A gateway node that is less resource constrained than sensors manages the route setup within a cluster. The proposed protocol finds a least cost and energy efficient path that meets an end-to-end delay bound. The link cost used is a function that captures the nodes' energy reserve, transmission energy, error rate and other communication parameters.[10]

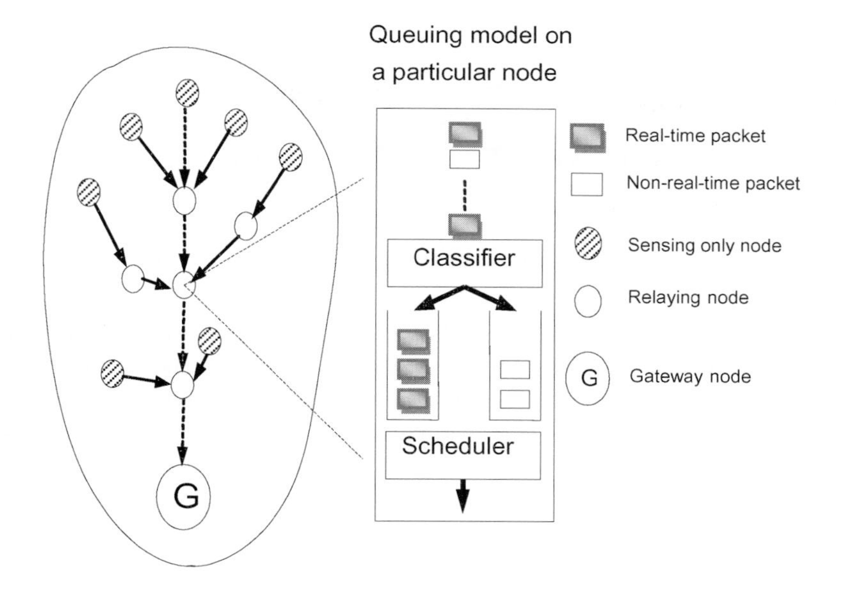

Fig. 10.1. A class-based queuing model is employed at each sensor node.[33]

In order to support both best effort and real-time traffic at the same time, a class-based queuing model is employed. The queuing model allows service sharing for real-time and non-real-time traffic. The bandwidth ratio r, is defined as an initial value set by the gateway and represents the amount of bandwidth to be dedicated both to the real-time and non-real-time traffic on a particular outgoing link in case of a congestion. As a consequence, the throughput for normal data does not diminish by properly adjusting such "r" value. The queuing model is depicted in Figure 10.1. The protocol finds a list of least cost paths by using an extended version of Dijkstra's algorithm and picks a path from that list which meets the end-to-end delay requirement.

Simulation results show that the proposed protocol consistently performs well with respect to QoS and energy metrics. However, the same r-value is set initially for all nodes, which does not provide flexible adjusting of bandwidth sharing for different links. The protocol is further extended[34] by assigning a different r-value for each node in order to achieve a better utilization of the links.

SPEED: A QoS routing protocol for sensor networks that provides soft real-time end-to-end guarantees is proposed by He *et al.*[35] The protocol requires each node to maintain information about its neighbors and uses geographic forwarding to find the paths. In addition, SPEED strive to ensure a certain speed for each packet in the network so that each application can estimate the end-to-end delay for the packets by considering the distance to the sink and the speed of the packet before making the admission decision. Moreover, SPEED can provide congestion avoidance when the network is overloaded.

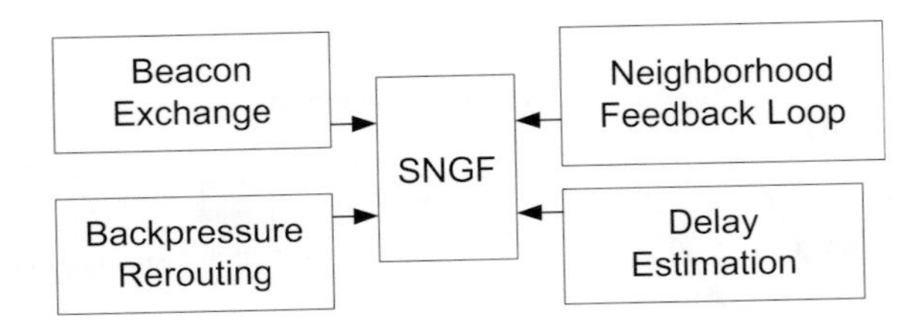

Fig. 10.2. Description of the routing module of SPEED.[35]

The routing module in SPEED is called Stateless Non-Deterministic Geographic Forwarding (SNGF) and works with four other modules at the network layer, as shown in Figure 10.2. The beacon exchange mechanism collects information about the nodes and their location. Delay estimation at each node is basically made by calculating the elapsed time when an ACK is received from a neighbor as a response to a transmitted data packet. By looking at the delay values, SNGF selects the node, which meets the speed requirement. If such a node cannot be found, the *relay ratio* of the node is checked. The Neighborhood Feedback Loop module is responsible for providing the relay ratio, which is calculated by looking at the miss ratios of the neighbors of a node (the nodes which could not provide the desired speed) and is fed to the SNGF module. If the relay ratio is less than a randomly generated number between 0 and 1, the packet is dropped. And finally, the backpressure-rerouting module is used to prevent voids, when a node fails to find a next hop node, and to eliminate congestion by sending messages back to the source nodes so that they will pursue new routes.

When compared to Dynamic Source Routing (DSR)[36] and Ad hoc on-demand vector routing (AODV),[37] SPEED performs better in terms of end-to-end delay and miss ratio. Moreover, the total transmission energy is less due to the simplicity of the routing algorithm, i.e. control packet overhead is less, and to the even traffic distribution. Such load balancing is achieved through the SNGF mechanism of dispersing packets into a large relay area. SPEED does not consider any energy metric in its routing protocol.

10.2.2.3. *MAC level support*

Several MAC protocols were proposed for wireless networks based on contention and carrier sense.[38,39] However, such protocols aim at maximizing the throughput and do not provide any real-time guarantee. In order to provide QoS guarantees such as bounded delay, many protocols have employed special real-time packet scheduling mechanisms.[40–43] However, these protocols introduce a significant amount of control packet overhead, which can be a burden for the limited energy resources of a sensor node and therefore are not applicable to sensor networks.

While many energy-aware MAC protocols have been proposed for sensor networks,[13,44,45] very little research has been conducted to combine real-time scheduling techniques and energy-awareness. Recently, Caccamo *et al.* have proposed a prioritized access protocol for sensor networks and employed the Earliest Deadline First (EDF) scheduling algorithm in order to ensure timeliness for real-time traffic.[46] The idea is to take advantage of the periodic nature of the sensor data traffic to create a schedule rather than using control packets for channel reservation. A sensor network architecture composed of several hexagonal cells is considered. In order to avoid channel interferences, 7 different frequency channels are used. Within each cell, all the nodes are assumed to be fully connected so that there will be no hidden channel problem. Enabling multicast within each cell provides elimination of redundant data since only one message is transmitted out of the cell after intra-cell message exchanges. Simulation results show that the protocol performs better in terms of throughput and average delay in heavy load conditions when compared to CSMA/CA with RTS/CTS option disabled.

RAP[47] is another project that considers a real-time scheduling policy for sensor networks. RAP is a communication architecture for sensor networks that proposes velocity-monotonic scheduling in order to minimize deadline miss ratios for packets. Each packet is put to a different FIFO queue based on their requested velocity, i.e. the deadline and closeness to the sink. This ensures a prioritization in the MAC layer. An extension of IEEE 802.11[48] is used along with such prioritization.

Power-aware reservation based MAC (PARMAC)[49] is an energy-aware protocol primarily designed for mobile ad hoc networks. However, this MAC protocol can be applied to sensor networks as well. The network is divided into grids and in each grid each node is assumed to reach all the other nodes within the grid area. Time is divided into fixed frames where each frame is composed of Reservation Period (RP) and Contention Free Period (CFP). In each RP, nodes within a grid cell exchange 3 messages to reserve the slots for data transmission and reception. Data is then sent in the CFP. If the reservation can be made before the deadline of real-time packets, a delay bounds can be provided. The protocol saves energy by minimizing the idle time of the nodes and allowing the nodes to sleep during a CFP. Moreover, control packets overhead and packet retransmissions are minimal, achieving significant energy savings.

10.2.3. *Example: energy-aware WFQ-based routing*

In order to describe the routing problem, we consider the following scenario: in a battlefield environment it is crucial to locate, detect and identify a target. In order to identify a target, we should employ imaging sensors. After locating and detecting the target without the need of imaging sensors, we can turn on those sensors to get, for instance, an image of the target periodically and send to the gateway. Since, it is a battlefield environment; this requires a real-time data exchange between sensors and controller in order to take the proper actions. In that case, we should deal with real-time imaging data, which requires certain bandwidth with minimum possible delay. Therefore, a service differentiation mechanism is needed in order to guarantee the reliable delivery of such data.

The goal in such an application will be to find an optimal path to the gateway in terms of energy consumption and error rate while meeting the end-to-end delay requirement. End-to-end delay requirement is associated only with the real-time data. The described QoS routing problem is very similar to typical path constrained path optimization (PCPO) problems, which are proved to be NP-complete.[50] We are trying to find a least-cost path, which meets the end-to-end delay path constraint. Our approach is based on associating a cost function for each link and using a K least cost path algorithm to find a set of candidate routes. Such routes are checked against the end-to-end constraints and the one that meets the requirements is picked. In the balance of this section, we will briefly describe the underlying network operation model, give some background on WFQ and explain the details of proposed algorithm.

10.2.3.1. *Network operation*

In the system architecture, gateway nodes assume responsibility for sensor organization based on missions that are assigned to the network. Mission-oriented organization of the sensor network enables the appropriate selection of only a subset of the sensors to be turned on and thus avoids wasting the energy of sensors that do not have to be involved. Thus the gateway will control the configuration of the data processing circuitry of each sensor. Assigning the responsibility of network management to the gateway can increase the efficiency of the usage of the sensor resources. The gateway node can apply energy-aware metrics to the network management guided by the sensor participation in current missions and its available energy. Since the gateway sends configuration instructions to sensors, the gateway has the responsibility of managing transmission time and establishing routes for the outgoing messages.

The sensor nodes can be in one of four main states: sensing only, relaying only, sensing-relaying, and inactive. In the *sensing state*, the node sensing circuitry is on and it sends data to the gateway in a constant rate. In the *relaying state*, the node does not probe the environment but its communications circuitry is on to relay the data from other active nodes. When a node is both sensing the target and relaying messages from other nodes, it is considered in the *sensing-relaying state*. Otherwise, the node is considered *inactive* and can turn off its sensing and communication circuitry. The decision for determining the node's state is done at the gateway based on the current sensor organization, node battery levels, and desired network performance measures. It should be noted that our routing approach is transparent to the method of selecting the nodes that should sense the environment.

The gateway will use model-based energy consumption for the data processor, radio transmitter and receiver in order to track the life of the sensor battery. This model is used in the routing algorithm as explained later. The gateway updates the sensor energy model with each packet received by changing the remaining battery capacity for the nodes along the path from the source sensor node to the gateway. The typical operation of the network consists of two alternating cycles: data cycle and routing cycle. During the data cycle, the nodes, which are sensing the environment sends their data to the gateway. In the routing cycle, the state of each node in the network is determined by the gateway and the nodes are then informed about their newly assigned states and how to route the data. Rerouting is triggered by an application-related event requiring different set of sensors to probe the environment including the activation of imaging sensors or the depletion of the battery of an active node.

10.2.3.2. *Background on WFQ*

In our system model, each node employs a packet scheduling discipline that approximates Generalized Processor Sharing (GPS). GPS achieves exact weighted max-min fairness by dedicating a separate FIFO queue for each session (flow) and serving an infinitely small amount of data from each queue in a weighted round robin fashion. Before explaining how GPS works in details, we introduce the following notation:

σ_i : Maximum burst size for leaky bucket on flow i
ρ_i : Average data rate of the flow i
$D(i)$: End-to-end delay for flow i
C : Link bandwidth
$P_{\max}(i)$: Maximum packet size for flow i
$P_{\max}$: Maximum packet size allowed in the network
g_i^m : Service rate on node m for flow i
$g(i)$: Minimum of all service rates for flow i
M : The number of nodes on path of flow i.

A GPS server m serves n sessions on a link by giving each session a share of the link based on n positive real numbers, $\Phi_1^m, \Phi_2^m, \ldots, \Phi_n^m$. These numbers denote the relative amount of service to each flow on the server m. Note that this sharing is only for backlogged connections since non-backlogged connections already receive what they ask for. The GPS server ensures that backlogged connections share the remaining bandwidth in proportion to the assigned weights. As analyzed in Ref. 51, each backlogged connection "i" receives a service rate of:

$$g_i^m = \frac{\Phi_i^m}{\sum_{j=1}^{n} \Phi_j^m} C. \tag{10.1}$$

However, GPS cannot be implemented in practice due to its ideal fluid model. Therefore, packet approximation algorithms of GPS were proposed. Weighted Fair Queuing (WFQ)[52] and Packetized Generalized Processor Sharing (PGPS)[53] are two identical disciplines developed independently and do not require GPS's infinitely small service assumption. They serve the incoming packets according to their service times under GPS. Therefore, for each flow the packet with the earlier service time is served first. Throughout this paper, we use the term WFQ for the packet-based version of GPS.

WFQ has two important features: first, it can provide fair allocation of bandwidth among all backlogged sessions as long as the total service rate of all sessions is less than the link bandwidth. Second, when combined with traffic regulation, it has been shown to provide an upper bound for the end-to-end delay.[53] Such regulation is done through the use of leaky bucket mechanism at the sources in order to ensure a constant data rate and to restrict the burst size for the traffic reaching the next relay nodes. As shown in Ref. 51, if a session "i" is leaky bucket constrained, the amount of session i traffic entering the network during interval $(\tau,\, t]$ will be:

$$A_i(\tau, t) \leq \sigma_i + \rho_i(t - \tau), \quad \forall t \geq \tau \geq 0.$$

Assuming flow i is constrained by a leaky bucket with parameters $(\sigma_i,\, \rho_i)$, the maximum end-to-end delay (transmission + queuing delay) for a packet of flow i under WFQ, given in Ref. 53 is:

$$D(i) \leq \frac{\sigma_i}{g(i)} + \sum_{m=1}^{M-1} \frac{P_{\max}(i)}{g_i^m} + \sum_{m=1}^{M} \frac{P_{\max}}{C}. \tag{10.2}$$

10.2.3.3. *Proposed queuing model*

The queuing model is specifically designed for the case of coexistence of real-time and non-real-time traffic in each sensor node. The model we employ is inspired from class-based queuing.[54] We use different queues for the two different types of traffic. Basically, we have real-time traffic and non-real-time (normal) traffic whose packets are labeled accordingly. On each node there is a classifier, which checks the type of the incoming packet and sends it to the appropriate queue. There is also a scheduler, which determines the order of packets to be transmitted from the queues according to the bandwidth ratio "r" of each type of traffic on that link. The model is depicted in Figure 10.1. The bandwidth ratio r, is actually a value set by the gateway and is used in allocating the amount of bandwidth to be dedicated to the real-time and non-real-time traffic on a particular outgoing link. As indicated in Figure 10.1, this r-value is also used to calculate the service rate for each type of traffic on that particular node, with $r_m\mu$ and $(1 - r_m)\mu$ being respectively the service rate for real-time and non-real-time data on sensor node m.

Since in WFQ each flow has its own queue, we consider each imaging sensor node as a source of different real-time flow, however only one real-time queue is used in the node to serve the data coming from these multiple flows. This mechanism is an approximation to flow-based WFQ approach and is used due to two reasons: first, having a different queue for each real-time flow will be inefficient in terms of the storage capacity of a sensor node. Second, the real-time flows are generated dynamically depending on the number of active imaging sensors. Since the number

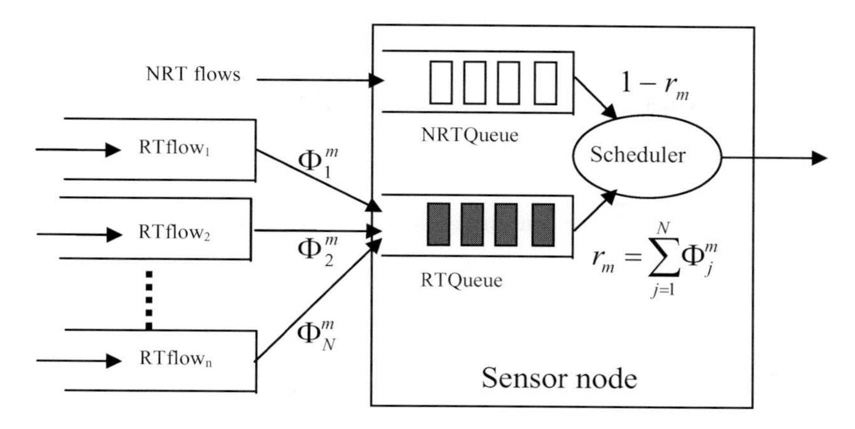

Fig. 10.3. Queuing model on a sensor node.

of such flows can change during the sensing activity, having one queue will reduce the maintenance overhead. In our model, we dedicate another separate queue to serve non-real-time data coming from different sources. The model is depicted in Figure 10.3. In this model, the service ratio r_m for the real-time queue on a node will be summation of link shares of all real-time flows passing through that node.

10.2.3.4. *Calculation of link costs*

We consider the factors for the cost function on each particular link separately except the end-to-end delay requirement, which should be for the whole path, i.e. all the links on that path. We define the following cost function for a link between nodes i and j:

$$\cos t_{ij} = \sum_{k=0}^{2} CF\mathrm{k} = c_0 \times (dist_{ij})^l + c_1 \times f(energy_j) + c_2 \times f(e_{ij}),$$

where
$dist_{ij}$ is the distance between the nodes i and j,
$f(energy_j)$ is the function for finding current residual energy of node j,
$f(e_{ij})$ is the function for finding the error rate on the link between i and j.
Cost factors are defined as follows:

CF_0 (*Communication Cost*) $= c_0 \times (dist_{ij})^l$, where c_0 is a weighting constant and the parameter l depends on the environment, and typically equals to 2. This factor reflects the cost of the wireless transmission power, which is directly proportional to the distance raised to some power l. The closer a node to the destination, the less its cost factor CF_0 and the more attractive it is for routing.

CF_1 (*Energy Stock*)$= c_1 \times f(energy_j)$. This factor reflects the remaining battery lifetime (energy usage rate) favoring nodes with more energy. The more energy the node maintains, the better the stability of the formed routes.

$CF_2(Error\ rate) = c_2 \times f(e_{ij})$ where f is a function of distance between nodes i and j and buffer size on node j (i.e. $dist_{ij}/\ buffer\_size_j$). Links with high error rate will increase the cost function, thus will be avoided.

10.2.3.5. *Estimation of r-values*

In order to find a QoS path for sending real-time data to the gateway, first we obtain a set of energy-efficient paths. Then, we further check them to identify the one that can meet the end-to-end delay requirement by trying to find an r-value for each node on that path. Therefore, for each flow the necessary service rate at each node should be estimated.

Let $T_{required}$ be the required end-to-end delay for the application. Thus, we should find r-values so that $D(i) \leq T_{required}$. Using (10.2),

$$\frac{\sigma_i}{g(i)} + \sum_{m=1}^{M-1} \frac{P_{\max}(i)}{g_i^m} + \sum_{m=1}^{M} \frac{P_{\max}}{C} \leq T_{required}.$$

In order to find g_i^m from the above equation, we assume that the service rate is same for all the nodes on the path of a particular flow, i.e. $g(i) = g_i^m$. Once g_i^m is calculated, then the link share can be calculated directly from (1),

$$\Phi_i^m = \frac{g_i^m * \sum_{j=1}^{N} \Phi_j^m}{C}.$$

As mentioned in Section 2.3, the service ratio for real-time data on a sensor node will be the summation of the link shares of all flows passing through that node, i.e. $r_m = \sum_{j=1}^{N} \Phi_j^m$, where N is the number of real-time data flows passing through node m and Φ_j^m is the link share for a real-time flow j on node m. This r-value is calculated for each node at the gateway and then sent to the corresponding node by the gateway.

By considering the above calculations, we propose the algorithm shown in Figure 10.4, to find a least-cost path that meets the constraints for real-time data. The algorithm calculates the cost for each link, line 0 of Figure 10.4, based on the cost function defined in Section 10.2.3.4. Then, for each flow from imaging sensors, the least cost path to the gateway is found by running a K-shortest path algorithm in line 3. Here, the least cost path is taken among K alternatives by setting K to 1 in line 1. Between lines 4–5, appropriate r-values are calculated for each real-time data flow. In lines 6 and 7, each node on the path of a flow calculates its final r-value. If that value is not between 0 and 1, alternative paths with bigger costs are tried by increasing the K value (line 2 and 3). As soon as a proper r-value is found, the loop exits (line 11). If there is no such r-value, the connection request of that node to the gateway is rejected (line 12–13).

In order to find the K shortest paths, i.e. K least-cost paths, we modified an extended version of Dijkstra's algorithm given in Ref. 55. Since the algorithm finds

```
0 Calculate cost_ij, ∀i, j ∈ V
1 k ← 1
2 Repeat
3     Find kth least cost path for each node
4     for each flow from imaging sensor i do
      begin
5         Compute Φ_i for all the nodes from D(i) ≤ T_required
6         for each node m (on the path of flow i) do
7                 r^m ← r^m + Φ_i
8                 if ( r^m is not in range [0,1)) then
9                         break;
      end
10    k ← k + 1
11 until k > K or a proper r^m for each node is found
12 if no proper r^m is found then
13    Reject the connection
```

Fig. 10.4. Pseudo-code for setting r-values.

a set of paths with similar nodes and links, we modified the algorithm in such a way that each time a new path is searched for a particular node; only node-disjoint paths are considered during the process. This ensures simplicity and helps in finding a proper r-value more easily since that node-disjoint path will not inherit the congestion of paths whose costs are less and for which the end-to-end delay requirements could not be met.

10.3. Challenges for Future Research

While sensor networks inherit most of the QoS issues from the general wireless networks described in Section 10.2.2.1, their characteristics pose unique challenges. This section discusses different challenges that have to be addressed by future research in order to materialize QoS traffic handling in wireless sensor networks.

10.3.1. *Bandwidth limitation*

A typical issue for general wireless networks is securing the bandwidth needed for achieving the required QoS. Bandwidth limitation is going to be a more pressing issue for wireless sensor networks. Traffic in sensor networks can be burst with a mixture of real-time and non-real-time traffic. Dedicating available bandwidth solely to QoS traffic will not be acceptable. A trade-off in image/video quality may be necessary to accommodate non-real-time traffic. In addition, a simultaneous use of multiple independent routes will be sometimes needed to split the traffic and allow for meeting the QoS requirements. Setting up independent routes for the same flow can be very complex and challenging in sensor networks due to the energy constraints, the limitation of computational resources and the potential increase in collisions among sensor transmissions.

10.3.2. *Removal of redundancy*

As discussed in Section 10.1, sensor networks are characterized with high redundancy in the generated data. For unconstrained traffic, elimination of redundant data messages is somewhat easy since simple aggregation functions would suffice. However, conducting data aggregation for QoS traffic is much more complex. Comparison of images and video streams is not computationally trivial and can consume significant energy resources. A combination of system and sensor level rules would be necessary to make aggregation of QoS data computationally feasible. For example, data aggregation of imaging data can be selectively performed for traffic generated by sensors pointing to same direction since the images may be very similar. Another factor of consideration is the amount of QoS traffic at a particular moment. For low traffic it may be more efficient to cease data aggregation since the overhead would become dominant. Despite the complexity of data aggregation of imaging and video data, it can be very rewarding from a network performance point-of-view, given the size of the data and the frequency of the transmission.

10.3.3. *Energy and delay trade-off*

Since the transmission power of radio is proportional to the distance squared or even higher order in noisy environments or in the non-flat terrain, the use of multi-hop routing is almost standard in wireless sensor networks. Although the increase in the number of hops dramatically reduces the energy consumed for data collection, the cumulative packet delay magnifies. Since packet queuing delay dominates its propagation delay, the increase in the number of hops can, not only slow down packet delivery but also complicate the analysis and the handling of delay-constrained traffic. Therefore, it is expected that QoS routing of sensor data would have to sacrifice energy efficiency to meet delivery requirements. In addition, redundant routing of data may be unavoidable to cope with the typical high error rate in wireless communication, further complicating the trade-off between energy consumption and delay of packet delivery.

10.3.4. *Buffer size limitation*

Sensor nodes are usually constrained in processing and storage capabilities. Multi-hop routing relies on intermediate relaying nodes for storing incoming packets for forwarding to the next hop. While a small buffer size can conceivably suffice, buffering of multiple packets has some advantages in wireless sensor networks. First, the transition of the radio circuitry between transmission and reception modes consumes considerable energy[6] and thus it is advantageous to receive many packets prior to forwarding them. In addition, data aggregation and fusion involves multiple packets. Multi-hop routing of QoS data would typically require long sessions and buffering of even larger data, especially when the delay jitter is of interest. The buffer size limitation will increase the delay variation that packets incur while traveling on different routes and even on the same route. Such an issue will complicate medium access scheduling and make it difficult to meet QoS requirements.

10.3.5. *Support of multiple traffic types*

Inclusion of heterogeneous set of sensors raises multiple technical issues related to data routing. For instance, some applications might require a diverse mixture of sensors for monitoring temperature, pressure and humidity of the surrounding environment, detecting motion via acoustic signatures and capturing the image or video tracking of moving objects. These special sensors are either deployed independently or the functionality can be included on the normal sensors to be used on demand. Readings generated from these sensors can be at different rates, subject to diverse quality of service constraints and following multiple data delivery models, as explained earlier. Therefore, such a heterogeneous environment makes data routing more challenging.

10.3.6. *Handling mobility*

Most of the current protocols assume that the sensor nodes and the sink are stationary. However, there might be situations, such as battlefield environments, where the sink and possibly the sensors need to be mobile. In such cases, the frequent update of the position of the sink and the sensor nodes and the propagation of that information through the network may excessively drain the energy of nodes. Clever QoS routing and MAC protocols are needed in order to handle the overhead of mobility and topology changes in such energy-constrained environment.

10.3.7. *Supporting different QoS metrics*

End-to-end delay bounds have been the main QoS requirement considered so far, leaving out the consideration of delay jitter. Meeting delay jitter constraints is a much tougher problem that yet to be investigated.

10.3.8. *Integration with IP-based networks*

Other possible future research for routing protocols includes the integration of sensor networks with IP-based networks (e.g. Internet). Most of the applications in security and environmental monitoring require the data collected from the sensor nodes to be transmitted to a server so that further analysis can be done. On the other hand, the requests from the user could be made to the sink through Internet. Since the routing requirements of each environment are different, further research is necessary for handling QoS requirements under these kinds of situations.

10.4. Conclusions

Several new routing protocols have been proposed for wireless sensor networks in recent years. Almost all of these routing protocols considered energy efficiency as the ultimate objective since energy is a very scarce resource for sensor nodes. However, the introduction of imaging and video sensors has posed additional challenges. Transmission of imaging data and video streams requires both energy and QoS aware routing in order to ensure efficient usage of the sensors and effective access to the gathered measurements. In this chapter, we have analyzed the technical issues for supporting QoS constrained traffic in wireless sensor networks. In addition we have reported on the state of the research in energy-aware QoS network and link layer protocols for sensor networks.

Overcoming bandwidth limitation, removal of redundancy, effective energy and delay trade-off, handling buffer size limitation, supporting multiple traffic types, handling mobility, supporting different Qos metrics, and integration with IP-based networks have been identified as the main technical challenges for supporting QoS requirements in wireless sensor networks. Few research projects have started to tackle only a subset of these issues, leaving lots of room for future research.

Problems

(1) What are the different sensor network applications that require QoS support?
(2) How do network dynamics affect QoS in sensor networks?
(3) What is the disadvantage of using CEDAR for quality of service in sensor networks?
(4) Mention examples of routing protocols that support QoS in sensor networks.
(5) How does the MAC protocol of Caccamo *et al.* address the hidden node problem?
(6) How does RAP provide real-time scheduling?
(7) Why cannot we use the Generalized Processor Sharing for providing QoS in sensor networks?
(8) What are the quality of service challenges in sensor networks?
(9) Why is data fusion a challenge for quality of service in sensor networks?
(10) Why is buffer size limitation a challenge for QoS in sensor networks?
(11) What is the main issue that needs to be addressed for providing QoS when integrating sensor networks with the Internet?
(12) How can a routing protocol for QoS support in sensor networks take into effect both the energy and QoS constraints?

Bibliography

1. I. F. Akyildiz, W. Su, Y. Sankarasubramaniam and E. Cayirci, Wireless sensor networks: a survey, *Computer Networks* **38** (2002) 393–422.
2. D. Estrin, R. Govindan, J. Heidemann and S. Kumar, Next century challenges: scalable coordination in sensor networks, *Proceedings of the 5th IEEE/ACM Annual Conference on Mobile Computing and Networks (MobiCOM'99)*, Seattle, WA (1999).
3. G. J. Pottie and W. J. Kaiser, Wireless integrated network sensors, *Communications of the ACM* **43**(5) (2000) 51–58.
4. J. Rabaey, M. Ammer, J. Silva Jr., D. Patel and S. Roundy, PicoRadio supports ad hoc ultra low power wireless networking, *IEEE Computer* **33** (2000) 42–48.
5. K. Sohrabi, J. Gao, V. Ailawadhi and G. J. Pottie, Protocols for self-organization of a wireless sensor network, *IEEE Personal Communications* **7**(5) (2000) 16–27.
6. R. Min, M. Bhardwaj, S. Cho, E. Shih, A. Sinha, A. Wang and A. Chandrakasan, Low power wireless sensor networks, *Proceedings of International Conference on VLSI Design*, Bangalore, India (2001).
7. C. Intanagonwiwat, R. Govindan and D. Estrin, Directed diffusion: a scalable and robust communication paradigm for sensor networks, *Proceedings of the 6th IEEE/ACM Annual Conference on Mobile Computing and Networks (MobiCOM'00)*, Boston, MA (2000).
8. R. Shah and J. Rabaey, Energy aware routing for low energy ad hoc sensor networks, *Proceedings of IEEE Wireless Communications and Networking Conference (WCNC)*, Orlando, FL (2002).
9. W. Heinzelman, A. Chandrakasan and H. Balakrishnan, Energy-efficient communication protocols for wireless microsensor networks, *Proceedings of the Hawaii International Conference on System Sciences (HICSS)* (2000).
10. M. Younis, M. Youssef and K. Arisha, Energy-aware routing in cluster-based sensor networks, *Proceedings of the 10th IEEE/ACM Symposium on Modeling, Analysis and Simulation of Computer and Telecommunication Systems (MASCOTS'02)*, Fort Worth, TX (2002).
11. G. Jolly and M. Younis, Energy efficient arbitration of medium access in sensor networks, *Proceedings of the IASTED Conference on Wireless and Optical Communications (WOC 2003)*, Banff, Canada (2003).
12. S. Singh and C. S. Raghavendra, PAMAS: power aware multi-access protocol with signaling for ad hoc networks, *ACM Computer Communications Review* (1998).
13. S. Xu and T. Saadawi, Does the IEEE 802.11 MAC protocol work well in multihop wireless ad hoc networks? *IEEE Communications Magazine* (2001).
14. L. Subramanian and R. H. Katz, An architecture for building self configurable systems, *Proceedings of IEEE/ACM Workshop on Mobile Ad Hoc Networking and Computing*, Boston, MA (2000).
15. F. Ye, H. Luo, J. Cheng, S. Lu and L. Zhang, A two-tier data dissemination model for large-scale wireless sensor networks, *Proceedings of the 8th IEEE/ACM Annual Conference on Mobile Computing and Networks (MobiCOM'02)*, Atlanta, GA (2002).
16. S. Tilak, N. B. Abu-Ghazaleh and W. Heinzelman, A taxonomy of wireless microsensor network models, *ACM Mobile Computing and Communications Review (MC2R)* (2002).
17. A. Manjeshwar and D. P. Agrawal, TEEN: a protocol for enhanced efficiency in wireless sensor networks, *Proceedings of the Workshop on Parallel and Distributed Computing Issues in Wireless Networks and Mobile Computing*, San Francisco, CA (2001).

18. W. Heinzelman, Application specific protocol architectures for wireless networks, Ph.D. thesis (MIT, 2000).
19. S. Lindsey and C. S. Raghavendra, PEGASIS: power efficient gathering in sensor information systems, *Proceedings of the IEEE Aerospace Conference*, Big Sky, Montana (2002).
20. S. Lindsey, C. S. Raghavendra and K. Sivalingam, Data gathering in sensor networks using the Energy*Delay Metric, *Proceedings of the Workshop on Parallel and Distributed Computing Issues in Wireless Networks and Mobile Computing*, San Francisco, CA (2002).
21. B. Krishnamachari, D. Estrin and S. Wicker, Modeling data centric routing in wireless sensor networks, *Proceedings of IEEE INFOCOM'02*, New York (2002).
22. Y. Yao and J. Gehrke, The cougar approach to in-network query processing in sensor networks, *SIGMOD Record* (2002).
23. W. C. Lee, M. G. Hluchyj and P. A. Humblet, Routing subject to quality of service constraints integrated communication networks, *IEEE Network* (1995).
24. Z. Wang and J. Crowcraft, QoS-based routing for supporting resource reservation, *IEEE Journal on Selected Area of Communications* (1996).
25. Q. Ma and P. Steenkiste, Quality-of-service routing with performance guarantees, *Proceedings of the 4th IFIP Workshop on Quality of Service* (1997).
26. L. Zhang, S. Deering, D. Estrin, S. Shenker and D. Zappala, RSVP: a new resource reservation protocol, *IEEE Network* (1993).
27. E. Crowley, R. Nair, B. Rajagopalan and H. Sandick, A framework for QoS based routing in the Internet, *Internet Draft*, draft-ietf-qosr-framework-06.txt (1998).
28. R. Querin and A. Orda, QoS-based routing in networks with inaccurate information: theory and algorithms, *Proceedings of IEEE INFOCOM'97*, Japan (1997) 75–83.
29. S. Chen and K. Nahrstedt, Distributed quality-of-service routing in ad hoc networks, *IEEE Journal on Selected Areas in Communications* **17**(8) (1999).
30. R. Sivakumar, P. Sinha and V. Bharghavan, Core extraction distributed ad hoc routing (CEDAR) specification, *IETF Internet Draft*, draft-ietf-manet-cedar-spec-00.txt (1988).
31. C. R. Lin, On demand QoS routing in multihop mobile networks, *IEICE Transactions on Communications* (2000).
32. C. Zhu and M. S. Corson, QoS routing for mobile ad hoc networks, *Proceedings of IEEE INFOCOM'02*, New York (2002).
33. K. Akkaya and M. Younis, An energy-aware QoS routing protocol for wireless sensor networks, *Proceedings of the IEEE Workshop on Mobile and Wireless Networks (MWN 2003)*, Rhode Island, USA (2003).
34. K. Akkaya and M. Younis, Energy and QoS aware routing in wireless sensor networks, *Journal of Cluster Computing, Special Issue on Ad Hoc Networks* (to appear).
35. T. He, J. A. Stankovic, C. Lu and T. Abdelzaher, SPEED: a stateless protocol for real-time communication in sensor networks, *Proceedings of the International Conference on Distributed Computing Systems (ICDCS)*, Rhode Island, USA (2003).
36. D. B Johnson and D. A. Maltz, Dynamic source routing in ad hoc wireless networks, *Mobile Computing*, eds. T. Imielinski and H. Korth, Chap. 5 (Kluwer Academic Publishers, 1996) 153–181, ISBN: 0792396979.
37. C. E. Perkins, E. M. Belding-Royer and S. Das, Ad hoc on-demand distance vector (AODV) routing, *IETF RFC 3561* (2003).
38. V. Bharghavan, A. Demers, S. Shneker and L. Zhang, Macaw: a media access protocol for wireless LAN, *Proceedings SIGCOMM'94 Conference, ACM* (1994) 212–225.

39. L. Kleinrock and F. Tobagi, Packet switching in radio channels, Part 1 — carrier sense multiple access modes and their throughput characteristics, *IEEE Transactions on Communications* **23**(12) (1975) 1400–1416.
40. J. J. Garcia-Luna-Aceves and C. Fullmer, Floor acquisition multiple access (FAMA) in single-channel wireless networks, *ACM Mobile Networks and Applications Journal, Special Issue on Ad Hoc Networks* **4** (1999) 157–174.
41. D. J. Goodman, R. A. Valenzuela, K. T. Gayliard and B. Ramanurthi, Packet reservation multiple access for local wireless communications, *IEEE Transactions on Communications* **37** (1989) 885–890.
42. M. Adamou, S. Khanna, I. Lee, I. Shin and S. Zhou, Fair real-time traffic scheduling over a wireless LAN, *Proceedings of the 22nd IEEE Real-Time Systems Symposium*, London, UK (2001).
43. J. L. Sobrinho and A. S. Krishnakumar, Quality of service for ad hoc carrier sense multiple access networks, *IEEE Journal on Selected Areas in Communications* **17**(8) (1999) 1353–1368.
44. A. Woo and D. Culler, A transmission control scheme for media access in sensor networks, *Proceedings of 7th ACM/IEEE Annual Conference on Mobile Computing and Networks (MobiCOM'01)*, Rome, Italy (2001).
45. W. Ye, J. Heidemann and D. Estrin, An energy-efficient MAC protocol for wireless sensor networks, *Proceedings of the IEEE INFOCOM'02*, New York (2002).
46. M. Caccamo, L. Y. Zhang, L. Sha and G. Buttazzo, An implicit prioritized access protocol for wireless sensor network, *Proceedings of the IEEE Real-Time Systems Symposium* (2002).
47. C. Lu, B. Blum, T. Abdelzaher, J. Stankovic and T. He, RAP: a real-time communication architecture for large-scale wireless sensor networks, *Proceedings of the IEEE Real-Time and Embedded Technology and Applications Symposium (RTAS 2002)*, San Jose, CA (2002).
48. I. Aad and C. Castelluccia, Differentiation mechanisms for IEEE 802.11, *Proceedings of IEEE INFOCOM'01*, Anchorage, Alaska (2001).
49. M. Adamou, I. Lee and I. Shin, An energy efficient real-time medium access control protocol for wireless ad hoc networks, *WIP Session of IEEE Real-Time Systems Symposium*, London, UK (2001).
50. G. Feng *et al.*, Performance evaluation of delay-constrained least-cost routing algorithms based on linear and nonlinear Lagrange relaxation, *Proceedings of the IEEE ICC'2002*, New York (2002).
51. A. K. Parekh and G. Gallager, A generalized processor sharing approach to flow control in integrated services networks: the single-node case, *IEEE Transactions on Networking* **1**(3) (1993) 344–357.
52. A. Demers *et al.*, Analysis and simulation of a fair queuing algorithm, *Journal of Internetworking Research and Experience* (1990) 3–26.
53. A. K. Parekh and G. Gallager, A generalized processor sharing approach to flow control in integrated services networks services networks: the multiple node case, *Proceedings of IEEE* (1993b).
54. S. Floyd and V. Jacobson, Link sharing and resource management models for packet networks, *IEEE/ACM Transactions on Networking* **3**(4) (1995) 365–386.
55. Q. V. M. Ernesto *et al.*, *The K Shortest Paths Problem*, Research Report, CISUC (June 1998).
56. R. Sivakumar, P. Sinha and V. Bharghavan, CEDAR: a core-extraction distributed ad hoc routing algorithm, *Selected Areas in Communications, IEEE Journal* **17**(8) (1999) 1454–1465.

Chapter 11

Mobility in Wireless Sensor Networks

Arun Asok* and Krishna M. Sivalingam*,†

*Indian Institute of Technology Madras
Chennai — 600036, India
arunasok@gmail.com

†University of Maryland, Baltimore County
Baltimore, MD 21250, USA
kristhri@gmail.com

Prathima Agrawal

Auburn University, Auburn, AL 36849, USA
pagrawal@eng.auburn.edu

Wireless sensor networks (WSNs) have been proposed as a solution to several monitoring, tracking, environmental and other applications. In such networks, a large number of small and simple sensor devices communicate over short range wireless interfaces to deliver observations over multiple hops to central locations called sinks. WSNs have been considered for several critical application scenarios including battlefield surveillance, habitat monitoring, healthcare, manufacturing, traffic monitoring, homeland defense and security applications. Initial studies and deployments of WSNs considered static sensor nodes and static data sinks. In the past few years, there has been substantial interest in introducing limited levels of mobility in the network architecture. This chapter will present a comprehensive survey of such mobility approaches studied in the literature. Mobility has been proposed with several objectives, including improved data dissemination capacity for high-bandwidth sensor traffic, balanced energy consumption among sensor nodes, improved locationing and improved security services. This chapter will describe schemes with mobile data sinks (or base station), mobile data collection entities, mobile relay nodes, and mobile security/locationing support nodes.

Keywords: Wireless sensor networks; base station; information gateway; mobile gateway; mobile base station; mobile sink; mobile data collector.

---

†Contact author.

11.1. Introduction

Wireless Sensor Networking (WSN) is a fast growing and exciting research area that has attracted considerable research attention in the recent past.[43,44] This has been fueled by the recent tremendous technological advances in the development of low-cost sensor devices equipped with wireless network interfaces. The creation of large-scale sensor networks interconnecting several hundred to a few thousand sensor nodes opens up several technical challenges and immense application possibilities.

Energy efficiency is an important factor in the design of WSNs. Since the sensor nodes are energy constrained, WSNs use a multi-hop communication model for communication within the network. The sensor nodes around the sink node in the network forward the data received from all other nodes to the sink, and thus deplete their energy very quickly. It has been found that the use of mobile sink for target tracking applications can maximize the network lifetime. Using mobile sinks is one of the many approaches used to exploit mobility in WSNs for the cause of increase in lifetime. A mobile sink can change its position during operation time and hence collect data from the sensor nodes in a more efficient way there by reducing the power consumption at the sensor nodes.

The network nodes are mostly small battery powered sensor devices, each equipped with one or more sensors, an embedded processor and low power radio for wireless communication. The base station (BS) or information gateway (GW) collect the information from the nodes and process it further for the particular application. An example network consisting of 10 sensor nodes and a GW node is shown in Figure 11.1. The sensor nodes can be deployed randomly on the ground and

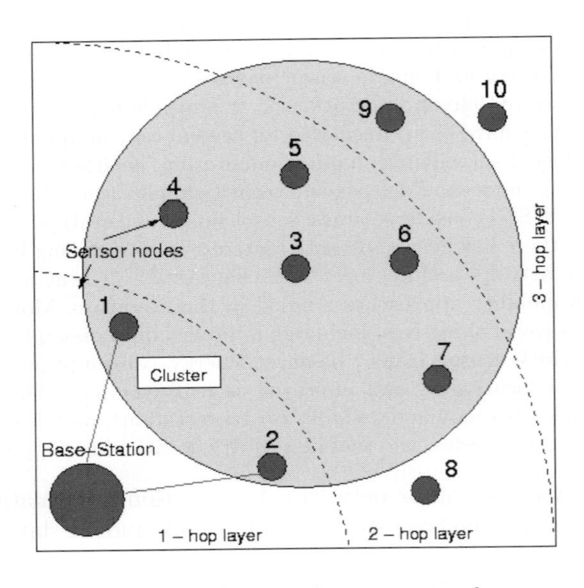

Fig. 11.1. An example sensor network.

operated unattended or can be operated in a more controlled setting. Different types of sensors such as seismic, magnetic, thermal, visual, infrared, acoustic and radar, are used to monitor a wide variety of ambient conditions viz. temperature, pressure, humidity, vehicular movement, noise level. Sensor network applications involve continuous sensing, event detection, event identification, location sensing. The sensors working collaboratively offer numerous valuable capabilities: larger system coverage with advanced properties such as self-configuration, redundancy, tolerance and intelligent sensing with higher accuracy.

Sensor networks can be used in applications spanning several domains including environmental monitoring, strategic, military, medical, industrial, and home networks. WSNs have moved from the research domain into the real world with the commercial availability of sensors with networking capabilities. Companies such as Crossbow, Sentilla, and Intel have emerged as suppliers/designers of the necessary hardware and software building blocks.

The current generation sensors' limited power and communication capabilities lead to fundamental problems such as bandwidth and operational lifetime of the sensor network. The battery lifetime of such nodes is of the order of a few months and perhaps a year or so. Narrowband RF-based schemes typically offer very low data rate of the order of few Kbps and result in poor performance as data and network sizes increase. Some of the key challenges are: scalability of network protocols to large number of nodes, design of simple and efficient protocols for different network operations, design of power-conserving protocols, design of data handling techniques including data querying, data mining, data fusion and data dissemination, localization techniques, time synchronization and development of exciting new applications that exploit the potential of WSNs.

The first generation of sensor nodes were predominantly static nodes and the applications were based on purely data-gathering models. In the past few years, there has been increasing interest in including a small set of mobile nodes in WSNs. This has been made possible by the development of robots with multiple sensing capabilities and by attaching sensor nodes to simple off-the-shelf mobile components.[33–36] This chapter will present a comprehensive survey of such mobility approaches studied in the literature. Mobility has been proposed with several objectives, including improved data dissemination capacity for high-bandwidth sensor traffic, balanced energy consumption among sensor nodes, improved locationing and improved security services. This chapter will provide a comprehensive survey of some of these mobility based schemes developed for WSNs.

The typical sensor network architecture consists of static sensor nodes (SN) and gateway or sink nodes (GW). The sensor nodes send the sensed data to the GW node through multi-hop paths directly to the GW. It is possible to group a subset of sensor nodes into a cluster; in this case, data is sent by the SN to its clusterhead in one or multiple hops; the clusterhead then forwards the data to the GW using single-hop transmission or using multi-hop transmission via other clusterheads. The GW node usually collects, analyzes and stores the data and can also answer queries on the sensor data. It is also usually connected to the Internet, and enables remote querying and possibly control of the sensor networks. In some architectures as in directed diffusion, there is also a mobile querying node that receives data from sensor nodes; however, these nodes do not normally transmit data back to the GW.

In the context of these architectures, mobility in sensor networks can be categorized as follows. In the first category, the GW is mobile and moves around the network, changing its positions periodically while spending time at each stopping point in the network.[26] The main objective of this approach is to reduce the bottleneck region around the GW. When the sensors forward data to the GW, it has to be ultimately routed to nodes closer to the GW. Hence, these closer nodes end up consuming a lot of battery power and hence die sooner than other remote network nodes. By moving the GW around the network, it is possible to balance the energy consumed by the different nodes due to data forwarding. In the second category, the network makes uses of specialized data collector nodes that are mobile and move around the network collecting data from the sensor nodes or from specialized clusterhead nodes (that have collected data from nodes in their respective cluster using regular multi-hop wireless transmissions). The mobile nodes then deliver the collected data to the GW at a later point in time. The speed and buffer capacity at the collector node and the additional delay due to waiting for the collector node are important issues to be considered for this approach.

Mobility has also been considered in other contexts such as relay nodes to improve connectivity and reduce energy, location determination using mobile reference nodes, and security in order to verify location of nodes. In some articles, mobility of the sensor nodes themselves has been considered, in the context of relocation of sensor nodes to meet certain network requirements. This requires that most of the sensor nodes are mobile. The notion of "mobile agents", a software program that is sent from the GW node to the sensor nodes, has also been considered.[22–24] These agents help in data processing, data aggregation and other functionality, adapting the sensor functionality as needed. In this chapter, we primarily describe mechanisms that use mobile GW nodes, followed by mechanisms that use mobile collector nodes.

11.2. Mobile Gateway Node Based Mechanisms

In this section, we present some of the main mechanisms where the mobile gateway approach is used. The basic idea is that the gateway or sink node (GW) moves around the network, stopping at various locations in the network. An example of this mechanism is presented in Figure 11.2. The expectation of using gateway mobility is that it can increase network lifetime by balancing energy consumption among the network nodes. This reduces the problem of nodes dying in concentrated pockets of the network which in turns leads to a disconnected network. However,

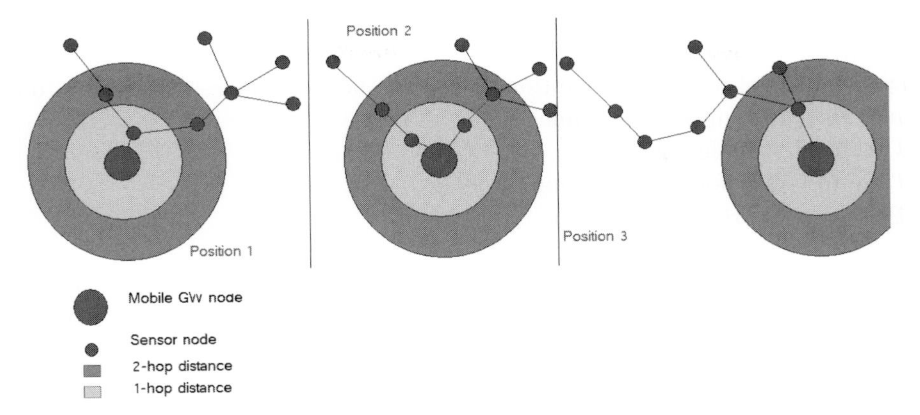

Fig. 11.2. The Mobile GW node, is shown here moving to three different locations. The routing tree is updated after each move.

it introduces additional challenges related to updating the network with the sink's current location and appropriately updating the routing tables of the sensors.

Some of the important challenges to be addressed in designing such mechanisms are:

- What is the path selected by the GW node? What will be the stopping points along this path?
- How long will the GW spend at each of the stopping points?
- How will the sensor nodes be informed about the GW node's current location and period of stay at that location?
- How will the sensor nodes update their routing tables to correctly forward data to the GW?
- Is the network architecture flat or is there a hierarchical structure? For example, the commonly cluster based communication is a two-tier architecture. Here, sensor nodes send data to cluster-head nodes (that are periodically changed by some election process) which in turn send data to the GW node. In this context, is it adequate for the cluster-head nodes to know the GW node's current location?
- Is the GW node's path, stopping points and duration of stay dependent upon network events? For example, if there is a hotspot area in the network, where substantially large amount of data is being generated compared to other areas, does the GW node move there?
- Will the speed and buffer capacity of the GW node affect the system performance?
- Can the system use more than one mobile GW node? If so, how do the sensor nodes determine to which mobile GW node their data has to be transmitted?

The various mechanisms proposed in literature address one or more of the above questions. In terms of determining the performance benefits using a mobile GW node, the commonly used metrics are:

- Distribution of power consumption among the network nodes (the goal is to balance the consumption over time).
- Time to first node failure.
- Time to failure for some fraction of nodes.
- Time to network disconnection.

In the following sections, we describe a subset of the mechanisms, highlight the main solution approach and describe the important observations about each mechanism.

11.2.1. *Energy consumption and network lifetime*

In Ref. 25, the authors consider a network with multiple mobile GW nodes (or sinks) in a sensor network, as shown in Figure 11.3. The objective was to minimize the energy consumed at each node and the total energy consumed over a period of time. Using an Integer Linear Programming formulation (ILP) that considered the traffic flow characteristics of the sensor nodes, a set of feasible stopping points is determined. The mobile GW nodes move to a subset of these stopping points, stay there for a period of time (called a "round") and then move to new locations. The ILP formulations were shown to yield a feasible set of locations that the GW nodes could move to, while achieving the objectives of energy minimization.

However, the authors in Ref. 3 claim that minimizing energy consumption does not necessarily optimize network lifetime. The work presented in Ref. 3 attempts to increase network lifetime, i.e. the instance at which the first node failure occurs, using the mobile sink approach. A single sink is considered and the joint problem of sink movement and sink sojourn time at stopping points is formulated using linear programming. The results show that the network lifetime can be increased

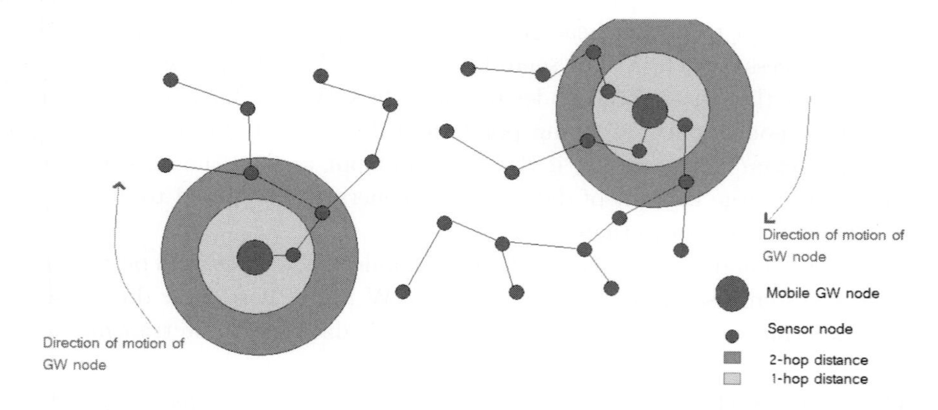

Fig. 11.3. A network with two mobile GW nodes.

by almost 500% for a 256-node network as compared to the network lifetime in a WSN with static sink, provided that the stopping points are the four corners and the central location (for a rectangular deployment area).

In Ref. 4, a joint mobile sink movement strategy and routing algorithm is studied. For network lifetime maximization, the routing protocol and the moving strategy should be jointly designed, and hence, both of them become constraints for the maximization of network life time. In this work, the load of a sensor node is defined as the power spent in sending and receiving data. The optimization formulation attempts to balance the load of each node by solving a max-min problem. In solving the joint problem, the proposed approach first assumes a shortest path routing strategy based on which the optimum mobility strategy is determined. Then, for the optimum mobility strategy, a routing strategy that performs better than shortest path routing is determined. The paper shows that for a circular network deployment area, the optimum mobility strategy is along the periphery of the network. For routing, the circle is divided into an inner circle and the annulus around this circle. For nodes within the inner circle, shortest path routing is used; for nodes in the annulus, the concept of "round routing", where a path in parallel with the circle's periphery until it is in line with the sink location (at the periphery), followed by shortest-path routing to the sink. The simulation based results show an increase in lifetime of up to 400% for an 800-node network, using the mobile sink approach when compared to the static sink system.

In Ref. 29, the movement of the GW node is based on measuring network performance and towards regions generating higher amount of traffic. When a selected metric (e.g. packet delay) is increasing beyond some threshold, the GW node attempts to find a suitable location to move to, that has the potential to improve performance. The protocol also provides for relaying of packets when the GW node is moving to its new location. Overall, this is a complex problem that at present is best solved using heuristic techniques.

In Ref. 30, an algorithm for routing in sensor networks using mobile GW nodes and an implementation on TOSSIM is described. The mobile GW approach is compared to a system with a static GW that is positioned at the center or at the border. After considering interactions with medium access layer, the results show that using a mobile sink is more appropriate for larger networks and that the medium access control protocol should be free of over-hearing for the mobile sink approach to have maximum effect. The paper also presents an adaptive algorithm that changes the sojourn time of the mobile node at each stopping point based on the power consumption profile of the network nodes.

11.2.2. *Static and mobile GW node architectures*

In order to reduce the extra cost incurred for updating the mobile sink's current location, a Dual-Sink method that uses two sinks, one mobile and the other static, is

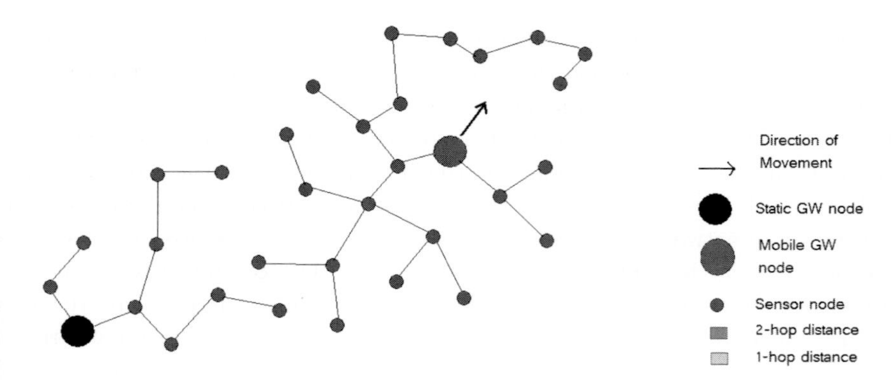

Fig. 11.4. A network with a static and a mobile sink.

proposed in Ref. 11. Figure 11.4 presents an example network to illustrate the dual-sink concept. The static sink broadcasts its location to the network nodes, at the beginning of network establishment. When the mobile sink reaches a new location, it broadcasts its location information to only a subset of nodes (small number of hops away) and not the entire network. Thus a sensor node sends data to the nearest sink (either static or mobile) by maintaining a routing table that indicates the number of hops to each sink type. It is shown through simulations that the method has advantage over using either a single static or a single mobile link. A hop limit of 3 to 4 hops is shown to be suitable for limiting the propagation of the mobile node's location update message.

The mobile sink mechanisms have been extended to consider multiple mobile sinks in order to support large scale WSNs. In Ref. 12, three different protocols to exploit multiple mobile sink movement are presented, with the objective of improving network lifetime. The first protocol employs random walk model for all mobile sinks. The mobile sinks move around randomly, independent of each other. When a sink reaches a new location, it sends a local beacon. Sensor nodes that can hear the beacon send the data to the sink with the strongest beacon, using single-hop transmission (assuming symmetric wireless links). The second protocol uses the dual-sink architecture, where the network has both static and mobile sinks. A sensor node sends packets to the closest sink using a multi-hop path. In both these protocols, the mobile sink movements and locations are not coordinated. The third approach uses the same dual-sink approach, but requires the sink nodes to cooperate so that the same region is not served by more than one sink. For a 200×200 m network with 300 randomly deployed nodes, around 3 randomly moving mobiles is shown to provide almost 100% network coverage. The success rate for the other two mechanisms is not as high as the random walk mechanism; with careful tuning of the number of mobile networks, it is possible to achieve high success rates and low packet delays.

Clustering the sensor nodes can enhance the effect of using mobile sinks. Multi-hop forwarding, which helps form a cluster in the vicinity of the expected position of mobile sink, can accomplish this. The energy consumption and message delivery delay in such a network paradigm depend on the number and velocity of mobile sinks as well as the cluster size. In Ref. 2, energy consumption and message delivery delay are analyzed according to changes in number and velocity of mobile sinks and cluster size. For clustering of nodes, a novel concept called "characteristic distance clustering" is used. This suggests that if the distance between source and sink is less than character distance then single-hop communication is used, or else multi-hop communication is used. The approach is found to perform better than least-hop based clustering.

11.2.3. *Theoretical results for mobile GW movement problem*

In static networks, the problem of finding an optimal fixed location for the gateway node (base station) has been studied, with the objective of maximizing network lifetime. A similar formulation has been proposed to identify the optimal set of stopping points for the GW node. The mobile GW takes a particular path to cover the network so as to collect data from the sensor nodes, with minimum data loss. The energy expenditure of each node, in the given scenario, is hence proportional to the shortest distance to the mobile sink's path. If the sink moves in the sensor network in an optimal path such that the first instance of node failure is prolonged, then the network lifetime is also increased.

Theoretical results for the optimal movement of a mobile base station have been provided in Ref. 1. The problem is formulated so as to optimize the base station location and the corresponding flow routing. The flow routing problem is modeled using non-linear programming constructs, where the energy spent by every node is less than the total energy stored in the node. The total energy spent is found by integrating the energy spent in communication at each node over the network lifetime. The non-linear programming formulation thus obtained is NP-hard. In Ref. 1, the authors present the transformation of the joint mobile base station movement and flow routing problem from the time domain to space domain. This is done with regard to the fact that the specific time instances the base station is present at a point is not important as long as the total time duration the base station spends at the given location is fixed. Thus the problem is now formulated as a linear programming model which can be solved in polynomial time.

An approximate algorithm for joint mobile base station movement and flow routing protocol, which gives a network lifetime that is $(1 - \epsilon)$ times maximum network lifetime, is also suggested in Ref. 1. In the above method, an approximation is done by equating the unconstrained movement of the base station to a virtual constrained movement consisting of "fictitious cost points". Here the sensor network is divided into different sub-areas. The concept of fictitious cost points is introduced to represent each of these sub-areas. Each fictitious cost point is an N-tuple vector that contains the upper bound to the distance from all nodes in the sensor network to any point in the given sub-area. Then the problem is solved for the constrained mobile base station (C-MB) movement and a $(1-\epsilon)$ optimal solution to the unconstrained mobile base station movement is constructed from the C-MB problem's solution.

11.2.4. *Dealing with uncertainty in mobile WSN*

WSNs are more challenging due to the uncertainties in communication links, nodes' power levels, nodes' being alive, network topology, etc. In Ref. 8, data dissemination techniques based on information theoretical concepts are considered for a mobile sink based WSN. The paper presents a mechanism called WEDAS (Weighted Entropy Data dissemination) that combines the uncertainty in the position of the mobile sink with the uncertainty of the energy of sensor nodes. A probabilistic model is proposed where the disseminators are selected from the most appropriate ones. This is a hybrid approach where both the residual energy of the nodes and their respective distance from the sink is jointly considered for deciding the mobile sink path. The objective of increasing network lifetime is achieved by reducing the battery usage at each node. The data disseminators are selected not just by considering their estimated residual energy, but also their distance from the mobile sink. The selection of data disseminator is formulated as an optimization problem where the search space of coordinates is pruned and the mobile sink is assumed to follow the Random WayPoint (RWP) model. In RWP model the mobile sink randomly selects a point in the sensor field as its destination, known as waypoint, and moves towards it with a random constant speed. After reaching the destination the sink stays there for a particular amount of time and randomly selects the next waypoint.

A weighted entropy value is calculated based on Shannon's entropy to guide the selection of data disseminators. Entropy, as defined in the laws of thermodynamics, increases as the system spends more energy; a system with high entropy enters a disordered state. Hence, we see a direct correlation between the concept of entropy and energy efficient data dissemination being discussed. The residual energy and the position of the sink are calculated from the previously known accurate values. After calculating these, the values are used to find a weighted entropy value that decides the forwarding node selected for data dissemination to the mobile sink (probability estimation). The paper shows using analytical models that a sensor node selects the forwarding nodes based on both remaining power levels and position of the sink. Unfortunately, the paper does not present any simulation based analysis to help understand the probability of data dissemination and the increase in network lifetime.

11.2.5. *GW node mobility models*

So far, we have seen how the design tries to optimize the stopping points for the mobile GW or sink node; equally challenging is the task of designing the specific path taken by the mobile sink. Many mobility models have been suggested in the literature. Some examples are random walks, fixed walks and autonomous walks. In Ref. 5, a systematic comparison of system performance using four different sink mobility models is presented. The models studied include: (i) random walk, in which the mobile sink moves randomly in all possible directions; (ii) a modification to random walk called partial random walk with limited multi-hop data propagation that employs an approach in which the sink moves in a set of predefined areas using a random transition between each; (iii) biased random walk, in which some of the predefined areas are visited more often by the sink in a biased manner. This could alleviate the data collection in WSN and can overcome problems that arise from network topology; and (iv) deterministic walk with multi-hop data propagation where the mobile sink moves along a predetermined path. A detailed performance study is conducted for a 300 node network. The metrics studied include success rate, energy dissipation and average delay, for varying network speeds and topologies. Both linear and circular trajectories are also considered as the sink moves between locations. The results show that it is possible to get nearly 140% improvement in terms of packet success rates compared to the directed diffusion model[28] and nearly 40% reduction in energy consumption. For networks with higher delay tolerance, the random walk and biased random walk approaches are more suitable; with the second approach, the delay can be reduced at the cost of higher energy multi-hop paths; the deterministic walk is suitable for networks with limited mobility capabilities and higher loss tolerance.

In Refs. 6 and 7, different strategies for mobile sink movement and their respective impact on network lifetime are considered. In particular, an autonomous moving strategy that can work without prior knowledge of the network topology and energy distribution of sensor nodes is presented. In this method, data-gathering stage consists of reporting periods, each of which has 3 phases. In the first phase, each node updates its one-hop neighbors. In the second phase, the node sending data checks whether the mobile sink is within one-hop distance in which case the data is sent to the sink; else, the data is sent to a forwarding node according to the routing protocol. The third phase is characterized by the moving decision the sink makes according to the data packets it receives. Two moving schemes are proposed for the mobile sinks in a WSN. In the one-step moving scheme, the sink moves to the node with the highest residual energy. The sink does not move frequently; it moves only if it finds a node that has energy at least a threshold value above the energy of the current node. In the multi-step moving scheme, the sink decides its direction of motion according to the destination node's (one with highest residual energy) position and the energy information of neighboring nodes in two-hop distance. The simulation results show that both schemes prolong the network lifetime and are adaptive to different network topologies.

11.2.6. *Dealing with cost of mobile sink location updates*

In the previous approaches, the "data push" has been considered where the sensor nodes send the collected data to the mobile GW node. An alternate model for sink mobility is the "data pull" model related to the "directed diffusion" concept.[28] Here, there are several mobile nodes that move around the network. Note that the sinks in this context are not gateway nodes that have been described so far. A sink node in this context merely sends queries for data and expects suitable responses from sensor nodes that have answers to the query. The challenge involved here is in routing the query to all sensor nodes and routing the query response to the mobile sink, as it moves in the network. As mentioned earlier, the current position of the mobile sink should be known to the source sensor nodes so that they can send the query response to the sink, which might have moved after placing the query. However, ensuring that the information about sink position is consistently and continuously updated in the entire sensor field will lead to large overhead and reduced performance. In order to deal with this problem, a two-tier data dissemination (TTDD) architecture is proposed in Ref. 9. The technique has been developed specifically for large scale sensor networks. The TTDD architecture incorporates a grid structure that is constructed by each data source; the grid is used by the sink to receive data by only flooding the local cells. Propagating queries to update data forwarding information, the sink does not send query messages to all sensor nodes, instead only sensors located at the grid positions acquire the update information. The grids or the data dissemination nodes hence form the middle tier with local subset of nodes as lower tier. When a query is generated, it is sent to the sink's immediate dissemination node which in turn sends the query to the upstream dissemination node from which its receiving the corresponding data information. The upstream dissemination node then propagates the query in the same manner until the query reaches the source node that is generating the data. The reverse path is traversed by the data from the source node to reach the sink. Since the sink is moving and updates its immediate dissemination node, duplicate data might be received from the old and new dissemination node. This is checked by keeping a timer on the dissemination node that is initialized to the approximate time the sink stays near the node.

In Ref. 10, the use of locators to track and update the position of the mobile sinks is proposed. Locators are normal sensor nodes in the network, and do not require specialized hardware. Locators contain the updated sink location. The mobile sink sends updates on its location, periodically, to the immediate locators. When a source node needs to send data reports to the sink as a response to the latter's query, it queries the locators for sink's location. After acquiring the information on sink's location the data is forwarded using a geographic routing, like greedy forwarding.

A mobile sink updates it position to the nearest locators by sending a Location Update Packet (LUP). An LUP contains sink id, current sink location, time stamp of that location, expire time, destination location and id. Thus we see that updating location incurs an overhead. This can be minimized by sending the LUPs to only the nearest locators. Also the update rate should be selected appropriately. Too frequent updates could lead to excess energy consumption for packet forwarding, and too few updates could lead to an unsuccessful routing due to inaccurate destination location.

The results of simulations prove that this method has less delay and higher data delivery success as compared to the TTDD method.[9] The proposed method does not work well with high number of mobile sinks and fewer nodes, where as TTDD does not need additional mechanisms other than few more forwarding agents. This can be taken care of by adding more locators as the number of mobile sinks increases. The use of locators to update the position of mobile sinks work better than TTDD except for cases with high number of mobile sinks and fewer nodes.

This section has summarized some of the different methods that deal with moving the gateway node around the network. Additional research continues to be published on this topic, as in Refs. 13 to 16. However, an exhaustive summary of all such methods is out of scope of this chapter.

11.3. Mobile Data Collector Based Approach

In this section, the mobile data collector based approach is presented. In such systems, the GW or sink node is statically located at one or more points in the network. The network uses a set of mobile data collector nodes that travel the network and collect data from the sensor nodes at regular intervals, and finally deliver the data collected in a single network traversal to the GW node. The mobility of the data collector nodes is either random (also called opportunistic) or controlled by the network. If single-hop communication is used between the sensor nodes and collector node, this approach can significantly reduce the energy consumption compared to the traditional multi-hop transmission to the GW node. This approach is also especially effective in sparse networks. However, the disadvantage of this approach is the higher latency incurred due to waiting for the collector node and the delay in the collector node ultimately delivering the data to the GW node. Thus, this approach will be suitable for delay tolerant applications. The performance of the system will also depend to a major extent on the speed, buffer capacity and path of the mobile.

Some of the significant challenges in designing such mechanisms include:

- How many mobiles are needed to cover a given network, given delay and other constraints?
- What should be the path taken by each mobile node starting from the GW node, traversing the network and back to the GW node? Should it be random or computed by some entity?
- Should the path taken by the mobile depend upon network conditions (e.g. hotspots generating more traffic) and/or traffic priority?
- Can the mobile's path change dynamically during network traversal, due to some demanding network conditions?
- What should the network architecture be? Either a flat network model, where the mobile node visits each sensor node, or a hierarchical model, where sensor nodes

send data to a cluster head (also called rendezvous point in some articles), with the mobile node collecting data from the clusterheads?
- How do buffer constraints at the sensor nodes and the mobile node affect system performance?
- How does the velocity of the mobile node impact system performance?
- Should the mobile nodes meet at specific points (also called "rendezvous" in some articles) during network traversal and exchange/fuse data collected up to the meeting time?

In the following sections, we describe some of the mechanisms studied for the mobile data collector model. Note that the mobile nodes are called mobile node, mobile element and mobile data collectors (MDC) in different research articles. We use these terms interchangeably in the rest of this section.

The concept of mobile data collection was first proposed in Ref. 17 and later revised in Ref. 18. Around the same time, a similar concept called sensor networks with mobile agents (SENMA) was proposed in Ref. 19. The main difference is that the work of Refs. 17 and 18 considered terrestrial mobile nodes (humans, animals, robots, etc.) while Ref. 20 considered aerial mobile vehicles. Here, we primarily focus on the approach of Refs. 17 and 18. A three-tier approach is used in this approach: (i) the top tier consists of access points or gateway nodes; (ii) the middle tier consists of the mobile data nodes, called data MULEs, that execute a random walk, and (iii) the lower tier consists of the sensor nodes. The data MULE node traverses the network, collects the data from sensor nodes using single-hop communication and reports the data to the GW node. It is assumed the sensor node has limited buffer capacity (which will overflow if the data MULE does not communicate with the node in time), while the data MULE has no capacity limitations. The use of MDCs for data collection in a sensor network has its benefits and limitations. The benefits, as reported in Ref. 17, are:

- Energy efficiency: Since the sensor communicates to the MDCs in short range rather than to the base station located far, the energy consumption for communication is less.
- No routing overhead: Since MDCs deliver the data collected from the sensors directly to the base station, the sensor nodes do not have to maintain routing tables, virtually eliminating the need for routing overhead.
- Robustness: Failure of an MDC will only increase the latency in communication whereas the network is still connected as the other MDCs carry on the communication. But in the case of multi-hop sensor data transmission, a node failure can lead to disrupted communication and a disconnected network.
- Scalablility: Deployment of more MDCs does not incur network reconfiguration. Adding more nodes to an existing network can be done without increasing the number of hops traversed or increased medium access level contentions.

- Simplicity: The design of the sensor can be simplified since the sensor have less traffic forwarding requirements.

In Ref. 18, a detailed analytical model of system performance and a simulation-based study are presented. The results show that if the rate of arrival of the MULE to the static sensor node increases, then the buffer occupancy at the sensor node and the rate of successfully delivering sensor data increases. In comparison to a regular multi-hop sensor network, this architecture was shown to reduce energy by a factor of more than 100 for sparse networks and by a factor of 10 for dense networks. Similar results are observed when the ratio of maximum energy used by a sensor node in the multi-hop network and the data MULE network. However, the main limitation of using this approach is, as observed earlier, the higher latency that primarily depends upon the field size and the speed of the mobile. A similar approach for mobile node based data collection using opportunistic means was presented in Ref. 42.

In Ref. 32, an experimental mobile data collection testbed using Mica2 motes deployed in a large parking lot is presented. The performance was measured in terms of: (i) contact time, defined as the time interval during which the mobile node is in contact with a sensor node and packet deliver rate is at least 85%; and (ii) the number of packets successfully transferred by the sensor nodes. The mobile node moves along a predetermined path in the network. For a low mobility scenario (1 to 2 m/s), it was seen that the impact of perpendicular distance between a sensor node and the mobile's path was significant in terms of contact time. For instance, when this distance was increased from 15 to 35 m, the average number of packets delivered was reduced by nearly 58% in the reported experiments. When the speed of the mobile node was increased to 2 m/s from 1 m/s, the average number of packets delivered was reduced by nearly 49% in the reported experiments. Note that there is sojourn time in these studies for the mobile nodes, i.e. it keeps moving in the network and does not slow down or stop when it approaches a sensor node. For a MULE (on a bus or car) moving at 20 km/h, the packet drop rate was less than 15% only when the distance between the mobile node and sensor node was less than 10 m; the contact time between the two nodes was measured to be 2 seconds. This in turn reduced the number of packets that are collected by the mobile node.

Allowing the mobile node to process the sensor data before delivering to the GW node can enhance the data collection model. This can include data aggregation, data fusion, and other data processing tasks. This can also help in reducing the amount of code that has to be executed at the sensor node, simplifying its design and the process of updating software/firmware at the sensor node.

11.3.1. *Computing ME routes*

One of the important constraints in mobile data collection based approaches is the limited buffer capacity of sensor nodes. The mobile node's path computation should be designed such that there is no sensor data loss even at low speeds.

This problem, referred to as the Mobile Element Scheduling problem (MES) is shown to be NP-complete in Ref. 20. Heuristic algorithms based on Earliest Deadline First (EDF) and Minimum Weighted Sum First (MWSF) are presented. The MWSF algorithm is shown to be better than the EDF based algorithms. In Ref. 21, another heuristic to solve the MES problem, called the Partitioning Based Scheduling (PBS) algorithm, is presented. The PBS algorithm computes paths based on knowledge of the data generation rate of sensors, buffer size and their locations. The algorithm uses partitioning of nodes based on data generation rate and location, followed by scheduling the mobile node's movements between the nodes in each partition. It is possible to have the mobile node visit a given sensor node multiple times in order to meet buffer constraints. In this approach, all nodes are divided into different bins according to their buffer overflow time values such that ones in the same bin have similar values of buffer overflow time. Then each bin is divided into sub-bins such that each sub-bin contains nodes that are geographically close to each other. Next, a TSP (Traveling Salesman Problem) solution is determined for each sub-bin. The TSP path for each sub-bin is joined into a super-cycle that forms the schedule for the mobile node's movement. Compared to the MWSF algorithm, PBS algorithm is shown to have higher performance in terms of decreasing loss rate, reducing the minimum required speed, and providing high predictability.

In Ref. 37, a data mule based approach is presented, where a single mule moves in a straight line through the network, as shown in Figure 11.5. Sensor nodes that can transmit data to the mule in a single hop do so when the mule approaches them. Other nodes use a multi-hop path to reach the sensor nodes that lie along the mule's path. The paper describes how the speed of the mobile is controlled in order to determine suitable sojourn times at each stopping point along the path. This concept has been extended in Ref. 31 to consider multiple data mules traveling along their respective straight-line paths through the network. The important task

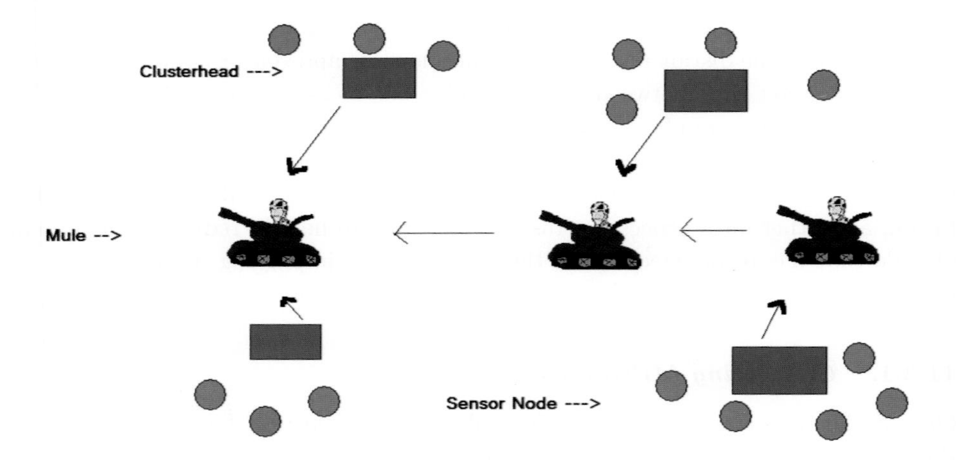

Fig. 11.5. Single data mule, traveling in a straight line across the network.

of load balancing, i.e. the number of sensor nodes serviced by a data mule should be uniformly spread among the data mules.

In most of the above approaches, the mobile node's path is either random or a straight line. In Ref. 38, a Mobile Data Gathering (DGR) algorithm that tries to minimize the mobile node's overall travel time and packet delay. The mobile node will follow a prescribed route with predetermined stop-and-gather locations on the route. At each such location, it will simultaneously gather data from sensor nodes that are in communication range of the mobile node. The novelty of this work is the consideration of a full-fledged ultra-wideband (UWB) transceiver at the mobile node and a UWB transmitter-only at the sensor node. UWB communications has the capability to support very high bandwidth (close to Gbps) at short distances (few meters). Thus, the data collected at the sensor node can be transferred in very short time to the mobile node. The goal of the DGR is to find a set of minimal points which the mobile node will visit to gather data. The problem of planning the path of a mobile node is similar to robot motion planning, with additional constraints such as node lifetime and transmission range. The DGR algorithm uses the network's Voronoi diagrams for finding the minimal set of stopping points, as shown in Figure 11.6. The sojourn time at each stop depends upon the number

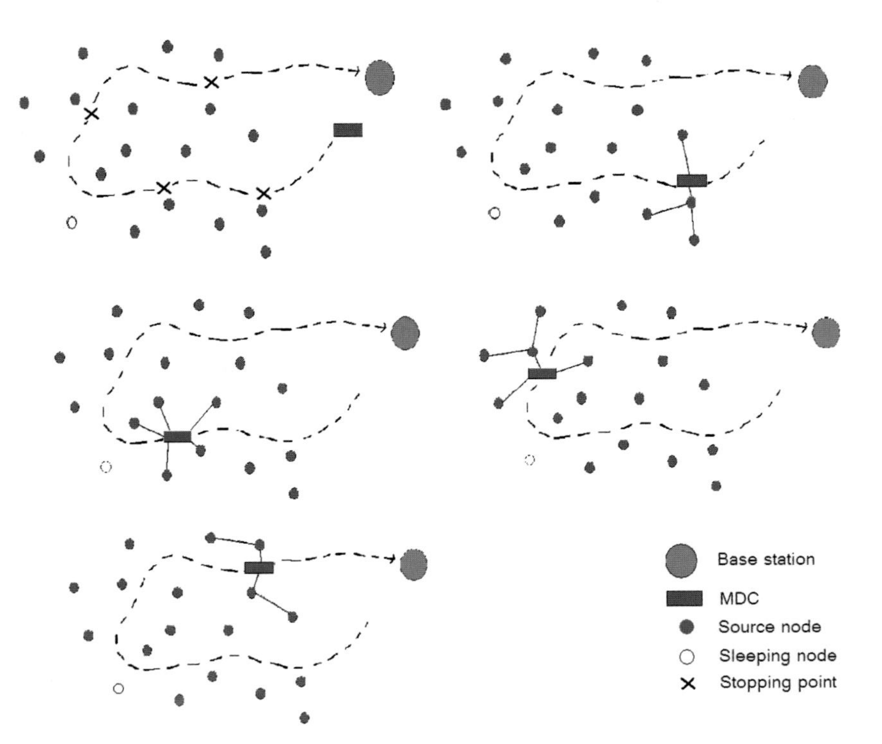

Fig. 11.6. Mobile data collection using a mobile node whose path is dependent upon sensor locations.

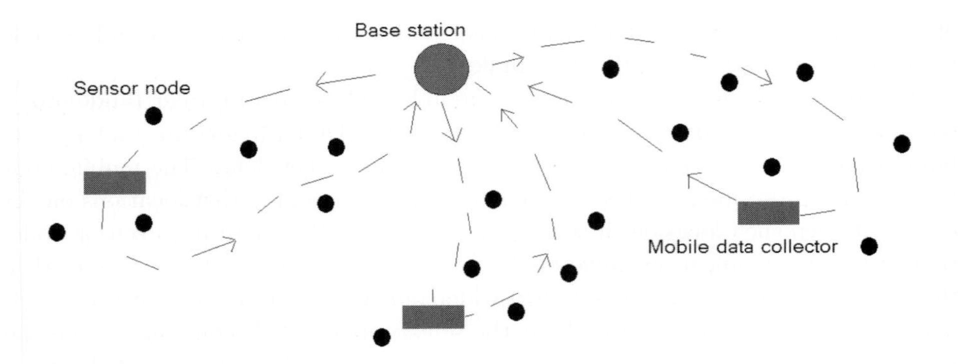

Fig. 11.7. Multiple mobile data collector based approach.

of sensor nodes that will communicate (via single-hop) at the stopping point. The performance metrics studied are the time taken for each tour of the mobile node and the network coverage provided during each tour of the mobile node. The performance of DGR is compared to the performance of random mobility pattern, a predetermined circular pattern, a "visit each node" approach and a "visit random nodes" approach. The results show a significant savings using DGR algorithm in terms of time required to collect data from the network nodes.

In Ref. 39, the above problem has been considered in the context of multiple mobile data collection nodes. In this paper, a hierarchical network architecture that uses RF and Ultra-wideband (UWB) communication is presented. Each cluster of sensor nodes (and the corresponding cluster head) will be working as an independent unit, with mobile agents such as unmanned aerial vehicles (UAV) or mobile robots used to gather the data from the cluster heads. Intra-cluster data gathering will be done using traditional RF communication. Cluster heads will also be equipped with UWB transmitters to transfer data with high rate to mobile nodes. The paper describes a multiple mobile agent based data gathering algorithm (MMDG) that takes into account the priority of data and life time of the cluster heads. The algorithm uses Solomon's heuristic[40] and Tabu search[41] to determine the mobile node's paths. The simulation results show that algorithm meets the needs of using minimal number of vehicles for data gathering, with equitable load distribution on these vehicles and use of priority based model for attending cluster heads with critical data or limited life time.

11.4. Conclusions

In this chapter, we presented a survey of WSNs that exploit the availability mobile nodes in addition to the traditional static sensor networks. Mobility can be used in the form of a mobile gateway node (or sink node) that uses mobility to balance energy consumption among the static sensor nodes, thereby increasing network lifetime. This form of mobility can be termed "slow mobility" where the mobile gateway node stays in a given location for a long period of time, before moving to a new location. An alternative approach is to use mobile data collector nodes that move around the network collecting data from the sensor nodes when they are in vicinity of the sensor nodes. This basic approach can also be extended to include clustering among sensor nodes, with the mobile node collecting data from the clusterheads. The potential benefits of this approach are lower energy for data transmission and reduced number of hops per data packet. It is suitable for networks that are sparse and also that can tolerate higher packet delay. Mobility can also be reached to improve the network connectivity in the form of mobile relay nodes; for improved location accuracy and for secure localization. Future research in this work can address efficient techniques for handling mobility in challenging terrain, for efficient scheduling of mobile data collection nodes, for improving the delay performance using techniques such as rendezvous based data exchange.

Acknowledgments

Part of the research was supported by a grant from Air Force Office of Scientific Research (AFOSR) grant No. FA9550-06-1-0103.

Problems

1. What are the common types of mobility proposed for wireless sensor networks?
2. What are the main advantages of using a mobile base station (or sink node) in a wireless sensor network?
3. What are the main disadvantages of using a mobile base station (or sink node) in a wireless sensor network?
4. What are the main challenges involved in designing a mobile base station (or sink node) based approach?
5. For a circular deployment area, what is the optimum mobility and routing strategy? Why do you think that this would be optimal?
6. How does the dual-sink architecture, consisting of a mobile sink and a static sink, help in improving system performance? How can this be extended to handle multiple mobile and static nodes?
7. What are the main advantages, disadvantages and challenges of an adaptive mobile base station based strategy that reactively moves the base station based on network events?
8. Summarize the main theoretical results for the base station mobility problem, as presented in Shi and Hou,[1] paper.
9. Explain the main concepts behind the Weighted Entropy Data dissemination model and how it deals with uncertainty while selecting forwarding nodes.

10. Discuss the impact of various mobility models on the performance a WSN using mobile base stations.
11. Discuss the key challenges of designing a mobile data collector based WSN.
12. Explain the main ideas behind the Data Mule architecture. Under what conditions would this architecture be preferred over a mobile base station based WSN?
13. What are the main advantages of mobile data collector based WSN?
14. In the mobile data collector approach, should the mobile node stop at regular intervals to collect data from neighboring nodes or should it continuously move without stopping anywhere? Discuss the merits and demerits of each approach.
15. How does the buffer size at the sensor node and at the data collector node impact the design of the mobile data collector approach?
16. Explain how the multiple mobile data collector approach works.
17. In the multiple mobile data collector model, consider an approach where the mobile nodes rendezvous with each other in the middle of their current collection round and exchange some of the data collected so far. What would be the advantages of this approach? Under what conditions would this approach be useful?

Bibliography

1. Y. Shi and Y. T. Hou, Theoretical results on base station movement problem for sensor networks, *Proceedings of IEEE INFOCOM*, Phoenix, AZ (2008).
2. C. Chen, J. Ma and K. Yu, Designing energy-efficient wireless sensor networks with mobile sinks, *Proceedings of First ACM Workshop on World-Sensor-Web* (in conjunction with ACM SenSys), Boulder, CO (2006).
3. Z. M. Wang, S. Basagni, E. Melachrinoudis and C. Petrioli, Exploiting sink mobility for maximizing sensor networks lifetime, *Proceedings of the 38th Annual Hawaii International Conference on System Sciences (HICSS)*, Big Island, Hawaii (2005).
4. J. Luo and J.-P. Hubaux, Joint mobility and routing for lifetime elongation in wireless sensor networks, *Proceedings of IEEE INFOCOM*, Miami, FL (2005) 1735–1746.
5. I. Chatzigiannakis, A. Kinalis and S. Nikoletseas, Sink mobility protocols for data collection in wireless sensor networks, *Proceedings of ACM International Workshop on Mobility Management and Wireless Access (MobiWac)*, Terromolinos, Spain (2006) 52–59.
6. L. Sun, Y. Bi and J. Ma, A moving strategy for mobile sinks in wireless sensor networks, *Proceedings of IEEE WiMesh Workshop*, Reston, VA (2006) 151–153.
7. Y. Bi, J. Niu, L. Sun, W. Huangfu and Y. Sun, Moving schemes for mobile sinks in wireless sensor networks, *Proceedings of IEEE International Conference on Performance, Computing, and Communications Conference (IPCCC)* (2007) 101–108.
8. H. M. Ammari and S. K. Das, Data dissemination to mobile sinks in wireless sensor networks: an information theoretic approach, *IEEE International Conference on Mobile Ad hoc and Sensor Systems (MASS)* (2005).
9. H. Luo, F. Ye, J. Cheng, S. Lu and L. Zhang, TTDD: two-tier data dissemination in large-scale wireless sensor networks, *Wireless Networks* **11** (2005) 161–175.

10. G. Shim and D. Park, Locators of mobile sinks for wireless sensor networks, *Proceedings of International Workshop on Wireless and Sensor Networks* (in conjunction with ICPP), Columbus, OH (2006) 159–164.
11. X. Wu and G. Chen, Dual-sink: using mobile and static sinks for lifetime improvement in wireless sensor networks, *IEEE ICCCN 2007*, Honolulu, Hawaii (2007).
12. A. Kinalis and S. Nikoletseas, Scalable data collection protocols for wireless sensor networks with multiple mobile sinks, *Proceedings of the 40th Annual Simulation Symposium*, Norfolk, VA (2007) 60–72.
13. D. Vass and A. Vidacs, Positioning mobile base station to prolong wireless sensor network lifetime, *Proceedings of ACM CoNEXT*, Toulouse, France (2005) 300–301.
14. S. A. Munir and W. B. Yu and J. Ma, Efficient minimum cost area localization for wireless sensor network with a mobile sink, *Proceedings of the 21st International Conference on Advanced Networking and Applications (AINA)* (2007) 533–538.
15. A. P. Azad and A. Chockalingam, Mobile base stations placement and energy aware routing in wireless sensor networks, *Proceedings of IEEE WCNC* (2006) 264–269.
16. Z. Vincze, D. Vass, R. Vida, A. Vidacs and A. Telcs, Adaptive sink mobility in event-driven multi-hop wireless sensor networks, *Proceedings of the 1st International Conference on Integrated Internet Ad hoc and Sensor Networks*, Nice, France (2006).
17. R. C. Shah, S. Roy, S. Jain and W. Brunette, Data MULEs: modeling a three-tier architecture for sparse sensor networks, *Proceedings of IEEE Workshop on Sensor Network Protocols and Applications (SNPA)*, Anchorage, Alaska (2003) 30–41.
18. S. Jain, R. Shah, W. Brunette, G. Borriello and S. Roy, Exploiting mobility for energy efficient data collection in wireless sensor networks, *ACM/Springer Mobile Networks and Applications* **11**(3) (2006) 327–339.
19. L. Tong, Q. Zhao and S. Adireddy, Sensor networks with mobile agents, *Proceedings of IEEE MILCOM*, Boston, MA (2003).
20. A. Somasundara, A. Ramamoorthy and M. Srivastava, Mobile element scheduling for efficient data collection in wireless sensor networks with dynamic deadlines, *Proceedings of IEEE International Real-Time Systems Symposium (RTSS)*, Lisbon, Portugal (2004) 296–305.
21. Y. Gu, D. Bozdag, E. Ekici, F. Ozguner and C.-G. Lee, Partitioning based mobile element scheduling in wireless sensor networks, *Proceedings of IEEE SECON 2005* (2005) 386–395.
22. H. Qi, S. S. Iyengar and K. Chakrabarty, Multi-resolution data integration using mobile agents in distributed sensor network, *IEEE Transactions on Systems, Man, and Cybernetics, Part C: applications and Review* **31**(3) (2001) 383–391.
23. Q. Wu, N. S. V. Rao, J. Barhen, S. S. Iyengar, V. K. Vaishnavi, H. Qi and K. Chakrabarty, On computing mobile agent routes for data fusion in distributed sensor networks, *IEEE Transactions on Knowledge and Data Engineering* **16**(6) (2004) 740–753.
24. M. Chen, T. Kwon, Y. Yuan, Y. Choi and V. C. Leung, Mobile agent-based directed diffusion in wireless sensor networks, *EURASIP Journal of Applied Signal Processing.* (Article ID 36871) (2007) p. 13.
25. S. R. Gandham, M. Dawande, R. Prakash and S. Venkatesan, Energy efficient schemes for wireless sensor networks with multiple mobile base stations, *Proceedings of IEEE Global Telecommunications Conference (GLOBECOM)* (2003) 377–381.
26. E. Ekici, Y. Gu and D. Bozdag, Mobility-based communication in wireless sensor networks, *IEEE Communications Magazine* **44**(7) (2006) 56–62.
27. A. Kansal, M. Rahimi, D. Estrin, W. J. Kaiser, G. J. Pottie and M. B. Srivastava, Controlled mobility for sustainable wireless sensor networks, *Proceedings of IEEE SECON 2004* (2004) 1–6.

28. C. Intanagonwiwat, R. Govindan and D. Estrin, Directed diffusion: a scalable and robust communication paradigm for sensor networks, *Proceedings of ACM Mobicom*, Boston, MA (2000).
29. K. Akkaya, M. Younis and M. Bangad, Sink repositioning for enhanced performance in wireless sensor networks, *Computer Networks* **49**(4) (2005) 512–534.
30. J. Luo, J. Panchard, M. Piorkowski, M. Grossglauser and J.-P. Hubaux, MobiRoute: routing towards a mobile sink for improving lifetime in sensor networks, *Proceedings of IEEE/ACM DCOSS*, San Francisco, CA (2006).
31. D. Jea, A. Somasundara and M. Srivastava, Multiple controlled mobile elements (data mules) for data collection in sensor networks, *Proceedings of IEEE Distributed Computing in Sensor Systems*, Marina Del Ray, CA (2005) 244–257.
32. G. Anastasi, M. Conti, E. Gregori, C. Spagoni and G. Valente, Motes sensor networks in dynamic scenarios: an experimental study for pervasive applications in urban environments, *International Journal of Ubiquitous Computing and Intelligence* **1**(1) (2006).
33. M. Laibowitz and J. A. Paradiso, Parasitic mobility for pervasive sensor networks, *Proceedings of 3rd International Conference on Pervasive Computing (PERVASIVE)*, Munich, Germany (2005) 255–278.
34. K. Dantu, M. Rahimi, H. Shah, S. Babel, A. Dhariwal and G. S. Sukhatme, Robomote: enabling mobility in sensor networks, *Proceedings of IEEE Information Processing in Sensor Networks (IPSN)* (2005) 404–409.
35. http://youtube.com/watch?v=D0Rgf-cKpV8&feature=related.
36. http://youtube.com/watch?v=Iv7xt6IOFF8&feature=related.
37. A. Kansal, A. Somasundara, D. Jea, M. Srivastava and D. Estrin, Intelligent fluid infrastructure for embedded networks, *ACM International Conference on Mobile Systems, Applications and Services (MobiSys)*, Boston, MA (2004) 111–124.
38. D. Bote, K. M. Sivalingam and P. Agrawal, Data gathering in ultra wide band based wireless sensor networks using a mobile node, *Proceedings of IEEE Communications Society/CREATENET International Conference on Broadband Communications and Networks*, Raleigh, NC (2007).
39. P. Shah, K. M. Sivalingam and P. Agrawal, Efficient data gathering in distributed hybrid sensor networks using multiple mobile agents, *Proceedings of IEEE Communications Society/CREATENET COMSWARE*, Bangalore, India (2008).
40. M. M. Solomon, Algorithms for the vehicle routing and scheduling problems with time window constraints, *Operations Research* **35**(2) (1987) 254–265.
41. F. S. Y. Rochat, A Tabu search approach for delivering pet food and flour in Switzerland, *The Journal of the Operational Research Society* **45**(11) (1994) 1233–1246.
42. A. Chakrabarti, A. Sabharwal and B. Aazhang. Using predictable observer mobility for power efficient design of sensor networks, *Proceedings of IEEE International Workshop on Information Processing in Sensor Networks (IPSN)* (2003).
43. C. S. Raghavendra, K. M. Sivalingam and T. Znati, *Wireless Sensor Networks* (Kluwer (now Springer) Academic Publishers, 2004).
44. H. Karl and A. Willig, *Protocols and Architectures for Wireless Sensor Networks* (Wiley, 2005).

Chapter 12

Delay-Tolerant Mobile Sensor Networks

Yu Wang* and Hongyi Wu†

Center for Advanced Computer Studies
University of Louisiana at Lafayette
P.O. Box 44330, Lafayette, LA 70504
**yxw1516@cacs.louisiana.edu*
†wu@cacs.louisiana.edu

The main-stream approach of wireless sensor networks is to densely deploy a large number of sensor nodes, forming a well-connected mesh network for data delivery. This approach, however, cannot be applied to the scenarios with extremely low and intermittent connectivity due to sparse network density and sensor node mobility. In such networks, the traditional routing protocols counting on end-to-end connections simply fail. In order to address this problem, the Delay-Tolerant Mobile Sensor Network (DTMSN) has been recently proposed. A typical DTMSN consists of two types of nodes, the wearable sensor nodes and the high-end sink nodes. The former are attached to mobile objects, gathering target information and forming a loosely connected mobile sensor network for data delivery; the latter collects data from the mobile sensor nodes and forward the data to the end user through a backbone network. This chapter focuses on a survey of current research on DTMSN. We first give an overview of the typical architecture of DTMSN. Then, several representative DTMSN systems are discussed in detail. We expect that this chapter will provide a deep understanding of the characteristics of DTMSN, leading to useful insights for future protocol design.

Keywords: Delay-tolerant mobile sensor network; delivery delay; delivery probability; DTMSN; pervasive information gathering; replication; transmission overhead.

12.1. Introduction

The continued scaling of microelectronics has enabled the modern tiny sensors, with sensing, data processing, and communication components integrated on a single chip. Given the limited resources of individual sensor nodes, massive information gathering usually relies on a distributed, embedded sensor system, where a large number of small and inexpensive sensor nodes with low power, short range radio are densely deployed to form a well connected wireless mesh network. The sensors in the network are self-organized and collaborate together to acquire the target data and transmit them to the sink nodes.[1]

The wireless sensor network has been extensively studied in the past several years, with numerous approaches proposed for routing, medium access control, data aggregation, topology control, power management, etc. While these mainstream approaches in the literature are well suitable for many sensor applications, they cannot be applied to the scenarios with extremely low and intermittent connectivity due to sparse network density and sensor node mobility. Two typical examples are pervasive air quality monitoring, where the goal is to track the average toxic gas inbreathed by human beings everyday, and flu virus tracking, which aims to collect data of flu virus (or any epidemic disease in general) in order to monitor and prevent the outbreak of devastating flu. Note that, while samples can be collected at strategic locations for flu virus tracking or air quality monitoring, the most accurate and effective measurement shall be taken at the people, naturally calling for an approach of deploying wearable sensors that closely adapt to human activities. As a result, the connectivity among the mobile sensors is poor, and thus it is difficult to form a well connected mesh network for transmitting data through end-to-end connections from the sensors to the sinks.

In order to address this problem, the Delay-Tolerant Network (DTN) technology has been recently introduced into such sensor networks. DTN is an occasionally connected network that may suffer from frequent partitions and that may be composed of more than one divergent set of protocol families.[2,3] Although DTN originally aimed to provide communication for the Interplanetary Internet, it has been also applied to many other network domains with extreme environments lacking continuous connectivity, such as sensor networks and ad hoc networks. In the remaining of this chapter we will first discuss the background of DTN-based sensor networks. Then the typical DTMSN model is presented, with its unique characteristics being outlined. Finally data delivery in DTMSN is discussed in detail, with several representative DTMSN systems being studied. We expect that this chapter will present a deep understanding of the characteristics of DTMSN, and stimulate further research in this emerging area.

12.2. Background

DTN technology has been recently introduced into wireless sensor networks. Its pertinent work on sensor networks can be classified into the following three categories, according to their differences in nodal mobility.

• **Network with Static Sensors**
The first type of DTN-based sensor networks are static. Due to a limited transmission range and battery power, the sensors are loosely connected to each other and may be isolated from the network frequently. For example, the Ad hoc Seismic Array developed at the Center for Embedded Networked Sensing (CENS) employs seismic stations (i.e. sensors) with large storage space and enables store and forward of bundles between intermediate hops.[4] In Ref. 5, wireless sensor networks are deployed for habitat monitoring, where the sensor network is accessible and controllable by the users through the Internet. The SeNDT (Sensor Networking with Delay Tolerance) project[6] targets at developing a proof-of-concept sensor network for lake water quality monitoring, where the radio connecting sensors are mostly turned off to save power, thus forming a loosely connected DTN network. Reference 7 proposes to employ the DTN architecture to mitigate communication interruptions and provide reliable data communication across heterogeneous, failure-prone networks.

• **Network with Managed Mobile Nodes**
In the second category, mobility is introduced to a few special nodes to improve network connectivity, while most other sensor nodes are still static. A typical example is the Data Mule system,[8] where a mobile entity called data mule receives data from the nearby sensors, temporarily store them, and drops off the data to the access points. This approach can substantially save the energy consumption of the sensors as they only transmit over a short range, and at the same time enhance the serving range of the sensor network.

• **Network with Mobile Sensors**
In the last category, sensor nodes can move around with various mobility patterns, and thus the network is called delay-tolerant mobile sensor network (DTMSN). This type of sensor network has recently attracted the interest of many researchers because it has a wide range of applicable scenarios in various areas, such as wildlife tracking, social network analysis, ubiquitous computing, future healthcare, etc. Meanwhile, due to the sparse deployment and nodal mobility, the connectivity of this type of networks is poor, and sometimes unpredictable, making it extremely difficult to design an efficient data delivery protocol for DTMSN. The most representative examples of this category are ZebraNet,[9] SWIM,[10,11] and DFT-MSN.[12]

The remaining of this chapter will focus on the study of recent research on the delay-tolerant mobile sensor network (DTMSN).

12.3. Typical System Model of Delay-Tolerant Mobile Sensor Networks

A typical DTMSN has a two-layer hierarchical architecture. The higher layer is a backbone network, which can be either a wireless infrastructure network or a wired network with wireless access interface. The lower layer consists of two types of nodes,

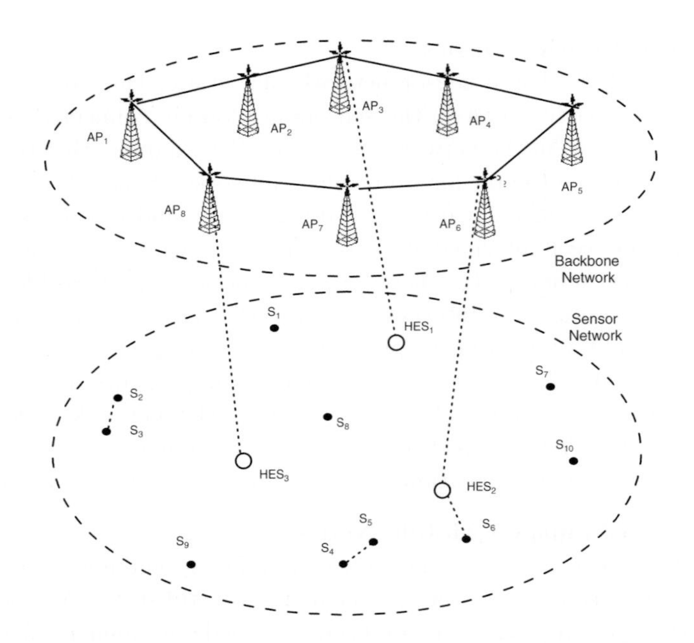

Fig. 12.1. An overview of DTMSN architecture. S_1-S_{10}: sensors; HES_1-HES_3: high end sensors (sinks); AP_1-AP_8: access points of backbone network.

the wearable sensor nodes and the high-end sink nodes. The former are attached to mobile objects, gathering target information and forming a loosely connected mobile sensor network for information delivery (see Figure 12.1 for mobile sensors S_1 to S_{10} scattered in the field, where only S_2 and S_3, S_4 and S_5, and S_6 and HES_2 can communicate with each other at this moment). Since the transmission range of a sensor is usually short, it cannot deliver the collected data to the destination (e.g. a data server) directly. As a result, a number of high-end nodes, (e.g. mobile phones or personal digital assistants with sensor interfaces), are deployed at strategic locations with high visiting probability or carried by a subset of people, serving as the *sinks* to receive data from wearable sensors (see HES_1, HES_2 and HES_3 in Figure 12.1). The high end sink nodes are assumed to have sufficient power compared with the sensor nodes, and thus capable of performing further processing on the collected data, such as aggregation, fusing, filtering, etc. Meanwhile, the high end nodes are equipped with more powerful wireless transceivers so that they can forward the data to the end user through the backbone network. With its self-organizing ability, DTMSN is established on an ad hoc basis without preconfiguration.

Although it is with similar hardware components, DTMSN distinguishes itself from conventional sensor networks by the following unique characteristics:

- **Nodal mobility**: The sensors and the sinks are attached to people with various types of mobility. Thus the network topology is dynamic (similar to that of a mobile ad hoc network but with very frequent partitions). Nodal mobility incurs continuous change on the network structure, which can heavily deteriorate the quality of the established wireless links, but at the same time also brings opportunities for establishing new, high-quality connections.

- **Sparse connectivity**: The connectivity of DTMSN is very low, forming a sparse sensor network where a sensor is connected to other sensors only occasionally.
- **Delay tolerability**: Data delivery delay in DTMSN is high, due to the loose connectivity among sensors. Such delay, however, is usually tolerable by the applications that aim at pervasive information gathering from a statistic perspective.
- **Fault tolerability**: Redundancy (e.g. multiple copies of a data message) may exist in DTMSN during data acquisition and delivery. Thus, a data message may be dropped without degrading the performance of information gathering.
- **Limited buffer**: Similar to other sensor networks, DTMSN consists of sensor nodes with limited buffer space. This constraint, however, has a higher impact on DTMSN, because the sensor needs to store data messages in its queue for a much longer time before sending them to another sensor or the sink, exhibiting challenges in queue management.

In addition, DTMSN shares characteristics of other sensor networks such as a short radio transmission range, low computing capability, and limited battery power.

Due to the above characteristics of DTMSN, there are usually no end-to-end connections between the sensor nodes and the sink nodes at any specific moment, and the topology of the network can be changing consistently. This makes the mainstream approaches used for data delivery in traditional wireless sensor networks ineffective and inefficient in DTMSN, because most of them have a strong assumption on end-to-end connectivity. Therefore, designing an effective data delivery scheme, which accommodates the unique features of DTMSN, becomes challenging.

12.4. Data Delivery in DTMSN

DTMSN is fundamentally an opportunistic network, where the communication links exist with certain probabilities and become the scarcest resource. Given their very low nodal density, the mobile sensors are intermittently connected, thus calling for the needs of making the utmost use of the temporarily available communication links. As a result, replication is usually necessary for data delivery in order to achieve certain success ratio. Clearly, replication also increases the transmission overhead. Thus it is a key issue for a DTMSN data delivery scheme to properly deal with the tradeoff between data delivery ratio/delay and overhead.

In order to illustrate these design issues, we start our discussion with two simplest data delivery schemes, i.e. direct transmission and flooding. Then we introduce several representative DTMSN systems proposed in the literature.

12.4.1. *Study of two basic approaches*

In this subsection we study two simple data delivery schemes, which are not efficient but provide a useful insight into the data delivery in DTMSN.

• *Simple Direct Transmission*
One of the simplest approaches is direct transmission, where a sensor transmits directly to the sink nodes only. More specifically, assume that the generated data message is inserted into a data queue at the corresponding sensor node. Whenever a sensor meets a sink, it transmits the data messages in its queue to the sink. A sensor does not receive or transmit any data messages of other sensors. Since there is only one transmission for each individual message, the overhead is minimal. However, due to the limited buffer space of a sensor node, some messages may have to be dropped before the sensor meets any sink node, resulting in low delivery ratio and long delivery delay.

• *Flooding Scheme*
The second approach is flooding. In the simple flooding scheme, a sensor always broadcasts the data messages in its queue to nearby sensors, which receive the data messages, keep them in queue, and rebroadcast them. Compared with the direct transmission approach, flooding achieves a lower data delivery delay at the cost of more traffic overhead and energy consumption. At the same time, flooding produces excessive message copies, leading to frequent buffer overflow. As a result, even more message copies may be dropped, resulting in low delivery ratio.

As we can see, the direct transmission approach minimizes transmission overhead (i.e. the number of message copies) and energy consumption, at the expense of a long message delivery delay (with large buffer space) or a low message delivery ratio due to a high message dropping rate (with small buffer space). In contrast, the flooding approach catches every possible opportunity for data transmission and thus minimizes the message delivery delay. At the same time however, it results in very high transmission overhead and energy consumption. Clearly, an efficient DTMSN data delivery scheme will take into consideration the tradeoff between delivery delay/ratio and transmission overhead/energy. In particular, the following three key issues need to be addressed.

• When to transmit data messages?
When a sensor moves into the communication range of another sensor, it needs to decide whether to transmit its data messages or not, in order to achieve a high message delivery ratio, and at the same time, minimize transmission overhead.

• Which messages to transmit?
The data messages generated by the sensor itself or received from other sensors are put into the sensor's data queue. After deciding to initiate data transmission, the sensor needs to determine which messages to transmit if there are multiple messages with different degrees of importance in its queue.

• Which messages to drop?
A data queue has a limited size. When it becomes full (or due to other reasons as to be discussed later), some messages have to be dropped. The sensor needs to decide which messages to drop according to their importance in order to minimize data transmission failure.

With the above issues taken into consideration, we will study several DTMSN systems in the following subsections.

12.4.2. *ZebraNet*

ZebraNet[9] employs the mobile sensors to support wildlife tracking for biology research. It targets at building a position-aware and power-aware wireless communication system. More specifically, a special wireless sensor node, namely *collar*, is attached to each selected zebra. A collar continuously collects and buffers location data of the corresponding zebra, and delivers the data to the base stations whenever possible. ZebraNet proposes a history-based protocol that intelligently selects neighbors for data forwarding. In this history-based scheme, each node maintains a hierarchy level, which is the likelihood of this node being in the transmission range of the base station, based on its success ratio of transmitting data directly to the base station in the past. When a node meets other sensor nodes with higher hierarchy level, it sends its collected data to the neighboring node with the highest hierarchy level, with ties randomly broken. As an example shown in Figure 12.2(a),

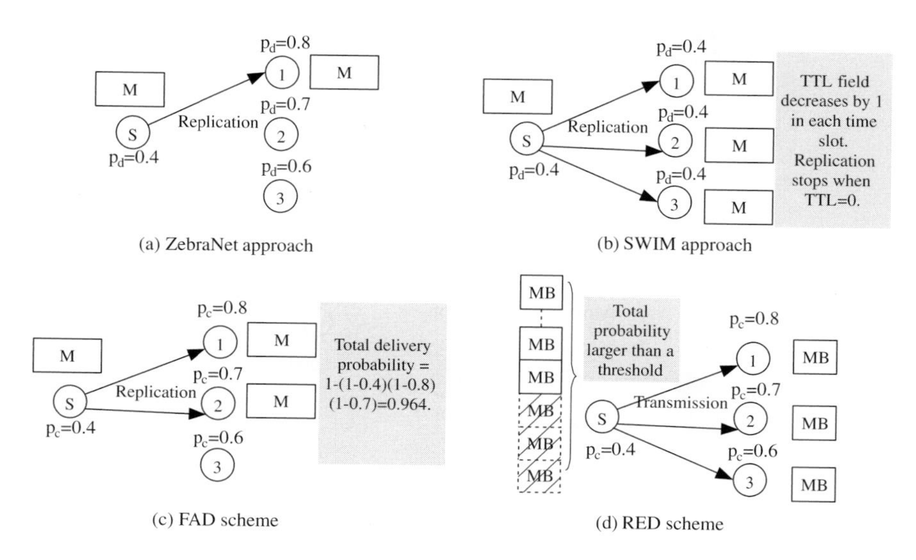

Fig. 12.2. Comparison of data transmission in different approaches. S: source node; $1, 2, 3$: qualified receivers; p_d: direct sink contact probability; p_c: cascaded nodal delivery probability; M: message copy; MB: message block generated by erasure coding.

when node S meets several other neighbors (i.e. nodes $1-3$), it only replicates the message to node 1, which has the highest hierarchy level.

This approach simply assumes that each node can reach the base station directly with certain probability. However, it is possible that some nodes may never move close to the base station, resulting in inefficient forwarding or even failures. In the scenario shown in Figure 12.3(a), node i always has connection to the sink node, while nodes j and k never move into the range of the sink, rendering a hierarchy level of 0. Since nodes k and j both have hierarchy level of 0, they have no idea on the correct direction of message transmission between them, resulting in random and unpredictable transmission. Worse yet, in the scenario shown in Figure 12.3(b) wireless links always exist between node i and the sink, and between node i and node j. However, node j never moves close to the sink node by itself, and thus has a hierarchy level of 0. On the contrary, node k is located far away from the sink, but with a small and positive hierarchy level because it occasionally moves close to the sink. Thus, data may flow incorrectly from node j to node k when they meet together. Another problem with this protocol is its lack of proper overhead control. More specifically, at each time a node meets other neighbors, it

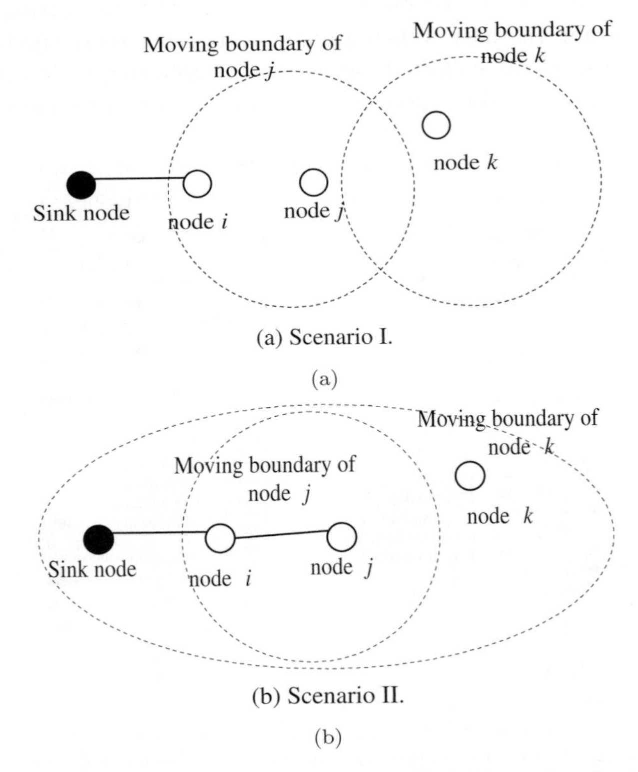

(a)

(b)

Fig. 12.3. Examples of inefficiency in ZebraNet's history-based scheme.

transmits data to one of them with the highest hierarchy level. If a node moves very dynamically and meets many different nodes, a lot of copies of the same message may be generated and transmitted in the network simultaneously, resulting in high energy consumption and low performance. In order to address these issues, some better approaches are needed to present the network hierarchy, and to balance between energy consumption and message delivery probability.

12.4.3. *The shared wireless info-station (SWIM) system*

The Shared Wireless Info-Station (SWIM) system is proposed in Refs. 10 and 11 for gathering biological information of radio-tagged whales. More specifically, whales are attached with a special equipment with sensor interface, RF component, and a large amount of memory. Data collected on a particular whale is first stored locally, and then may be transmitted to other whales within the transmission range. The stored data can be offloaded, when the whales move close to some special nodes called SWIM stations, which can then transmit the data through traditional network infrastructure. With this *store-carry-and-forward* communication paradigm, the message delivery delay depends on how frequently a whale meets other whales, as well as other resources such as bandwidth, storage space, and energy.

In Ref. 11, the tradeoffs of the energy-delay and the storage-delay in SWIM are discussed. More specifically, a data packet has a TTL (Time-To-Live) field indicating how long the packet can exist. When a packet is generated, it is associated with an initial TTL, which is determined for creating a minimum number of copies of the data packet so as to reach the desired data delivery probability. When a node meets a neighbor, its data packets are replicated to the neighbor. The TTL field of each message copy decreases as time goes on until it reaches 0, which means the packet is very likely to be offloaded to the sink node already by some other nodes. Thus the copy with TTL of 0 can be safely dropped to save resource. As can be seen, SWIM is a limited flooding scheme. An example of SWIM is shown in Figure 12.2(b). The source node S replicates message M to all the neighbors. Each neighbor continues this replication process, until the estimated number of copies reaches the expected value (i.e. TTL $= 0$), in order to guarantee the desired delivery probability with minimum overhead.

It is assumed in SWIM that the sensor nodes move randomly and thus every node has the same chance to meet the sink. In many practical applications, however, different nodes may have different probabilities to reach the sink, and thus SWIM may not work efficiently. Worst yet, some nodes may never meet the sink, resulting in failure of data delivery in SWIM.

12.4.4. *Delay/fault-tolerant mobile sensor network (DFT-MSN)*

The Delay/Fault-Tolerant Mobile Sensor Network (DFT-MSN) is introduced in Ref. 12 for pervasive information gathering. Two schemes have been proposed for efficient data delivery in DFT-MSN, namely the Fault-tolerance-based Adaptive Delivery (FAD) scheme and the Replication-based Efficient Delivery (RED) scheme. In the former, a message is replicated dynamically according to its fault tolerance by both the source and the intermediate nodes.[12] In the latter scheme, the replication is done by the source node via erasure coding.[13]

- *Fault-tolerance-based Adaptive Delivery (FAD)*

The proposed Fault-tolerance-based Adaptive Delivery (FAD) scheme consists of two key components for data transmission and queue management, respectively.

Data transmission decision is made based on the cascaded nodal delivery probability, which is the estimated total probability that the node transmits messages successfully to the sink node in the past, either directly or indirectly through other sensor nodes. Let ξ_i denote the cascaded delivery probability of a sensor i. ξ_i is initialized with zero and updated upon an event of either message transmission or timer expiration. More specifically, the sensor maintains a timer. If there is no message transmission within an interval of Δ, the timer expires, generating a timeout event. Timer expiration indicates that the sensor could not transmit any data messages during Δ, and thus its delivery probability is reduced to be $\xi_i = (1-\alpha)[\xi_i]$, where $[\xi_i]$ is the delivery probability of sensor i before it is updated and α $(0 \leq \alpha \leq 1)$ is a constant employed to keep the partial memory of historic status. Whenever sensor i transmits a data message to another node k, ξ_i should be updated to reflect its current ability in delivering data messages to the sinks. Since end-to-end acknowledgement is not employed in DFT-MSN due to its low connectivity, sensor i does not know whether the message transmitted to node k actually reaches the sink or not. Therefore, it estimates the probability of delivering the message to the sink by the delivery probability of node k, i.e. ξ_k (usually $\xi_k > \xi_i$). More specifically, ξ_i is updated as $(1-\alpha)[\xi_i] + \alpha\xi_k$.

Each sensor has a data queue that contains data messages ready for transmission. The messages in the queue are sorted with an increasing order of their fault tolerance, which is defined to be the probability that at least one copy of the message is delivered to the sink by other sensors in the network. The fault tolerance of a message is initialized to be zero when the message is generated and is updated every time when the message is transmitted. Let's consider a sensor i which is multicasting a data message j to Z nearby sensors, denoted by $\Xi = \{\psi_z \,|\, 1 \leq z \leq Z\}$. This multicast transmission essentially creates totally $Z+1$ copies. An appropriate fault tolerance needs to be assigned to each of them. More specifically, the message transmitted to sensor ψz is associated with a fault tolerance of $\Gamma^j_{\Psi_z} = 1 - (1 - [\Gamma^j_i])(1-\xi_i)\prod_{m=1,m\neq z}^{Z}(1-\xi_{\Psi_z})$, and the fault tolerance of the message at sensor i is updated as $\Gamma^j_i = 1 - (1-[\Gamma^j_i])\prod_{m=1}^{Z}(1-\xi_{\Psi m})$, where $[\Gamma^j_i]$ is the fault tolerance of message j at sensor i before multicasting. The above process repeats at each time when message j is transmitted to another sensor node. In general, the more times a message has been forwarded, the more copies of the message are created, thus increasing its delivery probability. As a result, it is then associated with a larger fault tolerance.

Message with the smallest fault tolerance is always at the top of the queue and transmitted first. A message is dropped at the following two occasions. First, if the queue is full, the message with the largest fault tolerance (lest important) is dropped. Second, if the fault tolerance of a message is larger than a threshold, it is dropped, even if the queue is not full, in order to reduce transmission overhead, given that the message will be delivered to the sinks with a high probability by other sensors.

When a sensor i has a message j at the top of its data queue ready for transmission and is moving into the communication range of a set of other sensors, it first learns the neighbors' delivery probabilities and available buffer spaces via simple handshaking messages. Then, it sends the data message to a subset of the neighbors (denoted by Φ) with higher delivery probabilities, and at the same time, controls the total delivery probability of the message (i.e. $1-(1-\Gamma_i^j)\prod_{m\in\Phi}(1-\xi_m)$) just enough to reach a predefined threshold γ in order to reduce unnecessary transmission overhead. As an example shown in Figure 12.2(c), node S meets three nodes with higher delivery probabilities. It finds out that if it replicates message M to node 1 and node 2, the total delivery probability of this message will be 0.964, which is higher than the predefined threshold (e.g. set to be 0.95). Therefore, no extra copy is replicated to node 3, in order to minimize overhead.

In this approach, the number of copies of each message is dynamically controlled by both source node and intermediate relaying nodes. Thus the network performance can be significantly improved, achieving a high message delivery ratio with acceptable delay and transmission overhead.

- *Replication-based Efficient Delivery (RED)*

In this scheme, an erasure-coding approach tailored for DFT-MSN is proposed to efficiently address the tradeoff between delivery ratio and replication overhead.

In the erasure coding approach, a message is first split into b blocks with equal size. Erasure coding is then applied to these b blocks, producing $S \times b$ small messages (which are referred as block messages), where S is the replication overhead. The gain of erasure coding stems from its ability of recovering the original message based on any b block messages.[13] The objective of this scheme is then to determine the optimal erasure coding parameters (i.e. b and S) with given inputs (i.e. the cascaded nodal delivery probability as discussed for FAD scheme in previous subsection) in order to meet the desired message delivery probability while minimizing the transmission overhead. This can be achieved by standard approach.[14] More specifically, each node maintains a nodal delivery probability as we have discussed in the FED scheme. When a node generates a message, it first calculates the optimal values for parameter b and S according to its current nodal delivery probability, and then generates $S \times b$ block messages by using erasure coding. Finally these block messages are simply forwarded towards the sink nodes from the nodes with lower delivery probability to neighboring nodes with higher nodal delivery probabilities.

The advantage of RED is simplified message manipulation and queue management at intermediate nodes, since all computation is done by the source node. At the same time, however, the optimization of erasure coding parameters is usually inaccurate because they are calculated according to the current data delivery probability of source node, especially when the source is very far away from the sinks. In addition, propagating many small messages in the network may incur further processing overhead and inefficiency of bandwidth utilization.

Figure 12.2(d) illustrates the data transmission in RED. Node S first applies erasure coding to a new message M, generating optimal number of small message blocks to achieve the desired success ratio. Then, these message blocks are simply transmitted to neighbors with higher nodal delivery probabilities, until they reach the sink nodes.

The summary and comparison of ZebraNet, SWIM, FAD, and RED are given in Table 12.1.

12.4.5. *Performance evaluation*

Extensive simulation has been carried out to evaluate the performance of the aforementioned data delivery schemes in a community-based scenario, where the sensors

Table 12.1 Characteristics comparison of different data delivery schemes.

	How to calculate nodal delivery probability	Where to replicate message	How to identify receivers	How to control overhead
ZebraNet	Direct sink node contact probability	Source and intermediate nodes	The neighbor with the highest delivery probability	No
SWIM	Direct sink node contact probability (same for all nodes)	Source and intermediate nodes	All neighbors	Predetermined by source node (with a proper TTL value) in order to achieve certain delivery probability
FAD	Cascaded probability to deliver either directly or indirectly (via intermediate nodes) to the sink	Source and intermediate nodes	A subset of neighbors with higher delivery probabilities	Dynamically controlled by using message fault tolerance at both source and intermediate nodes
RED	Cascaded probability to deliver either directly or indirectly (via intermediate nodes) to the sink	Source	Any neighbor with higher delivery probability	Predetermined by source node as an erasure coding parameter in order to achieve certain delivery probability

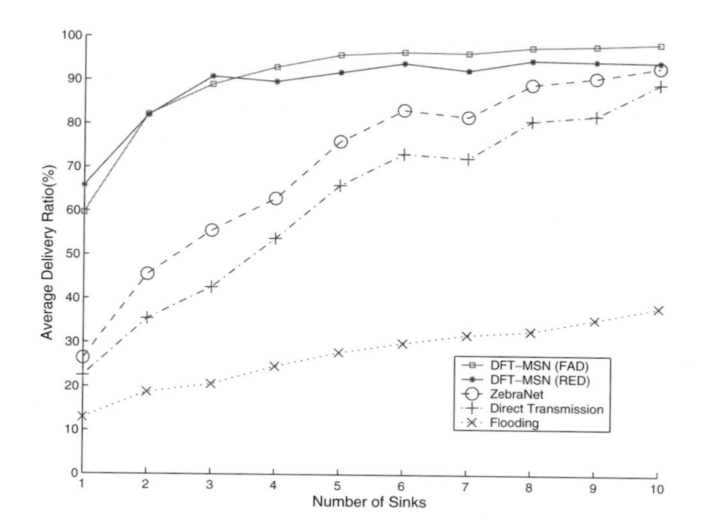

Fig. 12.4. Performance comparison.

Table 12.2 Performance comparison.

	Delivery ratio	Delivery delay	Communication overhead	Fairness	Scalability	Complexity
DFT-MSN	High	Moderate	Low	Good	Good	Moderate
ZebraNet	Moderate	Moderate	Moderate	Unbalanced	Moderate	Low
Direct transmission	Low	High	Lowest	Highly unbalanced	Bad	Very low
Simple flooding	Lowest	Lowest	Highest	Good	Bad	Very low

have different mobility patterns and thus resulting in different probabilities to reach the sink nodes.[12] Figure 12.4 compares the performance of the direct transmission approach, the simple flooding approach, the ZebraNet approach, and the DFT-MSN approach (with the FAD and RED schemes), by varying the number of sink nodes. Note that, the SWIM approach is not compared in our simulation because it is based on the assumption that each node has the same probability to reach the sink node, and thus cannot be applied here. The simulation results show that the DFT-MSN approach always has a higher delivery ratio than other approaches, especially when a small number of sinks are deployed. With a large number of sinks, all the approaches, except simple flooding, yield similar results. This is reasonable because the sensors then have high probabilities to reach the sink nodes, thus achieving high delivery ratio close to 100%. As expected, flooding has a much lower delivery ratio than other approaches because it generates too many message copies, which leads to excessive buffer overflow and message dropping. Other simulations reveal various characteristics of the aforementioned approaches, as summarized in Table 12.2.

In general, the performance of ZebraNet is only marginally higher than that of directly transmission. DFT-MSN can achieve high delivery ratio, low overhead, and moderate delay with slightly complicated protocol design. In terms of scalability and fairness, the DFT-MSN approach also performs well, in sharp contrast to other approaches, which are either unscalable (i.e. their performance declines sharply with the increase of network size), or unfair (i.e. they can only deliver the messages generated by the nodes near the sink nodes). The advantages of DFT-MSN are mainly due to its novel design based on message fault tolerance (or special erasure coding scheme) and the cascaded nodal delivery probability, which make efficient use of available communication links and at the same time effectively control the transmission overhead.

12.5. Challenges and Open Research Issues

Based on the unique characteristics of DTMSN and the existing works, we have also identified several open research issues, as listed below.

- The performance requirement of a DTMSN may vary from application to application. Most of the existing approaches regard the overall delivery ratio as the most important criteria. Some applications, however, may have special needs, such as the delivery rate and delay of certain important messages. How to incorporate these requirements into the data delivery scheme is worth of studying.
- Nodal mobility can be utilized to enhance the network connectivity. However, selecting the proper nodes for data forwarding is difficult, especially when the network is highly dynamic. In order to address this problem, more adaptive approach needs to be considered.
- In DTMSN, communication links exist only with certain probabilities, and thus are the most crucial resource. At the same time, the sensor nodes in DTMSN have very limited battery power alike those in many other sensor networks. How to deal with the tradeoff between link utilization and energy efficiency is a challenging problem.
- Several preliminary analytic models have been proposed in Refs. 11 and 15. Although useful at revealing the characteristics of DTMSN, they employ several assumptions and simplifications, which may not hold in the real system. It is interesting yet challenging to develop more realistic analytic models.
- Several testbeds have been established for small-scale experiments of DTMSN.[16] In order to better evaluate the effectiveness of DTMSN in real world applications, more empirical studies should be carried out in large-scale testbed.

12.6. Conclusion

The technology used in Delay-Tolerant Network (DTN) has been recently introduced into sensor networks with extremely low and intermittent connectivity.

This article focuses on a survey of current research on delay-tolerant mobile sensor networks (DTMSNs). We have given an overview of the typical architecture of DTMSN and its unique characteristics, such as sensor mobility, loose connectivity, fault tolerability, delay tolerability, and buffer limit. Then, several representative DTMSN systems and their data delivery schemes have been discussed and compared. In addition, we have identified several open research issues in this emerging area. We expect that this article will lead to a deep understanding of the sparsely connected mobile sensor network and provide the useful insight for future protocol design of DTMSN.

Problem

(1) What are the unique characteristics of DTMSN?
(2) Why cannot the mainstream data delivery approaches of traditional sensor networks be applied directly to DTMSN?
(3) What are the pros and cons of the two simple data delivery approaches, i.e. the direct transmission and flooding?
(4) What are the major considerations when we design an efficient data delivery approach for DTMSN?
(5) In the ZebraNet approach, how does a message be delivered from a sensor node to the sink node? What's the major issue of this approach?
(6) What is the assumption on node delivery probability in the SWIM approach? How message is delivered?
(7) How is the nodal delivery probability calculated in DFT-MSN's data delivery approaches?
(8) How is the message delivery ratio and energy consumption balanced in DFT-MSN?
(9) What is the major difference between FAD and RED data delivery (in terms of where and when the message is manipulated)?
(10) What is the principle of Erasure Coding?
(11) Compare the ZebraNet, SWIM, and DFT-MSN data delivery approaches.

Bibliography

1. I. Akyildiz, W. Su and Y. Sankarasubramaniam, A survey on sensor networks, *IEEE Communications Magazine* **40**(8) (2002) 102–114.
2. V. Cerf, S. Burleigh, A. Hooke, L. Torgerson, R. Durst, K. Scott, K. Fall and H. Weiss, *Delay Tolerant Network Architecture*, draft-irtf-dtnrg-arch-02.txt (2004).
3. K. Fall, A delay-tolerant network architecture for challenged internets, *Proceedings of the 2003 Conference on Applications, Technologies, Architectures, and Protocols for Computer Communications (SIGCOMM)* (2003) 27–34.
4. http://www.cens.ucla.edu/.

5. A. Mainwaring, J. Polastre, R. Szewczyk, D. Culler and J. Anderson, Wireless, sensor networks for habittat monitoring, *Proceedings of ACM International Workshop on Wireless Sensor Networks and Applications (WSNA)* (2002) 88–97.
6. http://down.dsg.cs.tcd.ie/sendt/.
7. M. Ho and K. Fall, Poster: delay tolerant networking for sensor networks, *Proceedings of IEEE Conference on Sensor and Ad Hoc Communications and Networks* (2004).
8. R. C. Shah, S. Roy, S. Jain and W. Brunette, Data MULEs: modeling a three-tier architecture for sparse sensor networks, *Proceedings of the 1st International Workshop on Sensor Network Protocols and Applications* (2003) 30–41.
9. P. Juang, H. Oki, Y. Wang, M. Martonosi, L. S. Peh and D. Rubenstein, Energy-efficient computing for wildlife tracking: design tradeoffs and early experiences with ZebraNet, *Proceedings of the 10th International Conference on Architectural Support for Programming Languages and Operating Systems* (2002) 96–107.
10. T. Small and Z. J. Haas, The shared wireless infostation model — a new ad hoc networking paradigm (or Where there is a Whale, there is a Way), *Proceedings of ACM MOBIHOC'03* (2003) 233–244.
11. T. Small and Z. J. Haas, Resource and performance tradeoffs in delay-tolerant wireless networks, *Proceedings of ACM SIGCOMM05 Workshop on Delay Tolerant Networking and Related Topics (WDTN-05)* (2005).
12. Y. Wang and H. Wu, DFT-MSN: the delay fault tolerant mobile sensor network for pervasive information gathering, *Proceedings of IEEE INFOCOM'06* (2006).
13. W. K. Lin, D. M. Chiu and Y. B. Lee, Erasure code replication revisited, *Proceedings of the 4th International Conference on Peer-to-Peer Computing* (2006).
14. Y. Wang and H. Wu, Replication-based efficient data delivery scheme (RED) for delay/fault-tolerant mobile sensor network (DFT-MSN), *Proceedings of the 1st International Workshop on Ubiquitous & Pervasive Healthcare (UbiCare'06)*, in conjunction with PerCom'06 (2006).
15. Y. Wang and H. Wu, Analytic study of delay/fault-tolerant mobile sensor networks (DFT-MSN's). Technical Report, CACS, University of Louisiana at Lafayette (2006).
16. F. Lin, Y. Wang and H. Wu, testbed implementation of delay/fault-tolerant mobile sensor network (DFT-MSN), *Proceedings of the 2nd International Workshop on Sensor Networks and Systems for Pervasive Computing (PerSeNS'06)*, in conjunction with PerCom'06 (2006) 321–327.

Chapter 13

Integration of RFID and Wireless Sensor Networks

Hai Liu,[*,‡] Miodrag Bolic,[*,§] Amiya Nayak[*,¶] and Ivan Stojmenovi[*,†,||]
[*]*SITE, University of Ottawa, Ottawa, K1N 6N5, Canada*
[†]*EECE, The University of Birmingham*
Birmingham, B15 2TT, United Kingdom
[‡]*hailiu@site.uottawa.ca*
[§]*mbolic@site.uottawa.ca*
[¶]*anayak@site.uottawa.ca*
[||]*ivan@site.uottawa.ca*

Radio frequency identification (RFID) and wireless sensor networks (WSN) are two important wireless technologies that have a wide variety of applications and provide limitless future potentials. RFID facilitates detection and identification of objects that are not easily detectable or distinguishable by using current sensor technologies. However, it does not provide information about the condition of the objects it detects. Sensors, on the other hand, provide information about the condition of the objects as well as the environment. Hence, integration of these technologies will expand their overall functionality and capacity. This chapter first presents a brief introduction on RFID and then investigates recent research works, new patents, academic products and applications that integrate RFID with sensor networks. Four types of integration are discussed: (1) integrating tags with sensors; (2) integrating tags with wireless sensor nodes and wireless devices; (3) integrating readers with wireless sensor nodes and wireless devices; and (4) mix of RFID and wireless sensor networks. New challenges and future works are discussed at the end.

Keywords: RFID; wireless sensor networks; integration; survey; applications; products; patents.

13.1. Introduction

RFID is a means of storing and retrieving data through electromagnetic transmission using an RF compatible integrated circuit.[7] It is usually used to label and track items in supermarkets and manufactories. For example, Wal-Mart, Procter and Gamble, and the US Department of Defense have deployed RFID systems with their supply chains.[20] However, the potential of RFID is much more than this. Today, RFID has been widely applied in supply chain tracking, retail stock management, library books tracking, parking access control, marathon races, airline luggage tracking, electronic security keys, toll collection, theft prevention, and healthcare.[39] Current trends and forecasts indicate that the market will grow fast in the next 10 years. In 2006 alone, 1.02 billion RFID tags were sold.[13] The total value of the market, including hardware, systems and services, is expected to grow from €500 million to €7 billion by 2016.[11]

Briefly, RFID systems consist of two main components: tags and readers. A tag has an identification (ID) number and a memory that stores additional data such as manufacturer name, product type, and environmental factors such as temperature, humidity, etc. The reader is able to read and/or write data to tags via wireless transmissions. There are basically two types of communications between tags and readers. One is inductive coupling, which is done by antenna structures forming an integral feature in both tags and readers. The other is propagation coupling, which is done by propagating electromagnetic waves.[7] In a typical RFID application, tags are attached or embedded in objects that need to be identified or tracked. By reading tag IDs in the neighborhood and then consulting a background database that provides mapping between IDs and objects, the reader is able to monitor the existence of the corresponding objects.[37]

A sensor network is composed of a large number of sensor nodes that can be deployed on the ground, in the air, in vehicles, inside buildings or even on bodies. Sensor networks are widely employed in environment monitoring, biomedical observation, surveillance, security, etc.[4] Since sensors' energy cannot support long range communication to reach a sink, which is generally far away from the data source, multi-hop wireless connectivity is required to forward data to the remote sink.

WSNs are different from RFID networks. WSNs are usually employed to monitor objects in interest areas or to sense environments while RFID systems are used to detect presence and location of objects which have RFID tags. In typical applications, relay nodes are deployed to forward data from sensor nodes to remote sinks in WSNs. It forms a multi-hop network while traditional RFID is only single-hop and consists of a batch of tags and several readers. Sensor nodes are more intelligent than RFID tags. Sensor nodes' firmware can be easily reprogrammed which is not the case for RFID tags. RFID readers can be parameterized, but they are rarely user-programmed. Hence, RFID networks and WSNs represent two complementary technologies and there exist a number of advantages in merging these two technologies. Some of them include: adding ad hoc capabilities to RFID network, adding sensing capabilities to RFID tags and adding tracking capabilities to RFID tagged objects that are difficult to detect otherwise.

The chapter investigates recent research works, new products/patents and applications that integrate RFID with wireless sensor networks. We classify current works into four categories according to manner of integration. They are (1) integrating RFID tags with sensors; (2) integrating RFID tags with wireless sensor nodes; (3) integrating RFID readers with wireless sensor nodes or wireless devices; and (4) mixing RFID systems and wireless sensor networks. The difference between class 1 and class 2 is that the tags with integrated sensors are traditional RFID tags which communicate with only readers, while the tags in class 2 are able to communicate with each other and form a multi-hop network. The investigation starts with a brief introduction on RFID. Since WSNs have already been introduced in the book, we skip its introduction and focus only on RFID in rest of this section.

13.1.1. *Previous work*

Surveys and classifications on the integration of RFID and wireless sensor network technologies are attempted in several publications but they either are outdated or are missing a comprehensive study. In Ref. 41, three types of integrations are suggested. The first one is heterogeneous network architecture with a mix of RFID tags and WSN nodes and a smart station that will be used for collecting information from two networks. The second type of integration includes integration of the reader and a WSN node in one device. The third type is a smart active tag that merges functionality of a WSN node and an active tag. These three classes are similar to our classes 4, 3 and 2, respectively. However, in this chapter we will perform a comprehensive study of many current and potential solutions that correspond to these classes and not only point out specific solutions as described in Ref. 41. Besides, there are some serious flaws in the proposed architectures of Ref. 41. For example, a smart station is based on a single reader with multiple antennas. However, the cable between the reader and the antennas have to be short in practical implementations so that such a solution with a single reader is not feasible when there is a need to cover relatively large areas where both RFID and WSN units operate.

In Ref. 26, several different classifications are presented. This paper describes possible ways of integrating WSNs into the existing RFID network based on the standards defined by EPCGlobal (www.epcglobal.org). The paper is different from this chapter because: (1) it considers only integration of WSNs into RFID frameworks, and (2) on the RFID side, it considers only the EPCGlobal-based RFID technologies. However, a number of useful classifications and architectures are defined. RFID tags with added sensors are classified as fixed and variable function sensor tags. Fixed function sensor tags are black boxes that have pre-defined functionality. Variable function sensor tags can change their function by changing, for example, the air protocol. At the level of integration, four reference models for integration are proposed. They include (1) mix of RFID and WSN with the integration at the application level, (2) mix of RFID and WSN with the integration at the filter and collection level (this level in the EPC framework is intended for providing useful non-redundant data to upper levels), (3) hardware integration in which the RFID reader is responsible for collecting data from both tags and WSNs, and (4) logical integration at the EPCIS level that allows for a mix of RFID and WSNs.

In Ref. 6, the taxonomy of wireless sensor network devices that includes RFID devices is proposed. Attributes for characterization of wireless devices include (1) communication, (2) power, (3) memory, (4) sensors, and (5) other features. The paper does not consider integration of devices but it gives a uniform classification that can be applied to integrated WSN and RFID devices. Each attribute is further divided into a set of sub-attributes so that fine classification of wireless devices is possible. Communication attributes are divided into communication protocols/standards, communication modes, communication modules and supports for mobility. Communication modes are "device talk first", "beacon", "ad hoc" and "human controlled". Power is further classified into three categories: "storage", "energy harvesting mechanisms" and "transfer". Storage uses batteries or passive devices such as capacitors. Energy transfer is the way by which passive RF devices are powered. The energy transfer mechanisms are inductive coupling, capacitive coupling, and passive backscattering. Memory can be classified into read-only, write-once and read-write.

13.1.2. *RFID technology*

13.1.2.1. *Types of RFID tags*

Based on power source, RFID tags can be classified into three major categories: *active* tags, *passive* tags, and*semi-passive* (*semi-active*) tags. An active tag contains both a radio transceiver and a battery that is used to power the transceiver. A passive tag reflects the RF signal transmitted to it from a reader or a transceiver and adds information by modulating the reflected signal. The passive tag does not use any battery to boost the energy of the reflected signal.[7] Every passive tag contains an antenna needed to collect electromagnetic energy in order to wake up the tag and to reflect (backscatter) the portion of the energy back to the reader. In addition, tags have transmitter/receiver circuits, power generating circuits and the state machine logic. The state machine logic is needed in order to follow the RFID protocol and support communication between the reader and tags. Similar to passive tags, semi-passive tags use the radio waves of senders as an energy source for their transmissions. However, a semi-passive tag may be equipped with batteries to maintain memory in the tags or power some additional functions. Active tags are more powerful than passive tags/semi-passive tags. For example, they have larger range/memory and more functions. However, they are also more expensive than passive/semi-passive tags.

Based on memory type, RFID tags can also be classified into two categories: tags with read/write memory, and tags with read-only memory. As their names imply, the tags with read/write memory allow both read and write operations on the memory while data in the tags with read-only memory cannot be modified after the manufacturing process. The tags with read/write memory are more expensive than the tags with read-only memory.

13.1.2.2. *Radio frequency*

RFID tags operate in three frequency ranges: low frequency (LF, 30–500 kHz), high frequency (HF, 10–15 MHz), and ultra high frequency (UHF, 850–950 MHz, 2.4–2.5 GHz, 5.8 GHz).[20] LF tags are less affected by the presence of fluids or metals when compared to the higher frequency tags. They are fast enough for most applications and are also cheaper than any of the higher frequency tags. However, low frequency tags have shorter reading ranges and low reading speeds. Typical applications of LF tags are access control, animal identification and inventory control. The most common frequencies used for LF tags are 125–134.2 kHz and 140–148.5 kHz.

HF tags have medium transmission rates and ranges but are more expensive than LF tags. Typical applications of HF tags are access control and smart cards. RFID smart cards, working at 13.56 MHz, are the most common members of this group.

UHF tags have the highest transmission rates and ranges among all tags. They range from 3 to 6 m for passive tags and more than 30 m for active tags. The high transmission rates of UHF tags allow the reading of a single tag in a very short time. This feature is important in the application where tagged objects move very fast and remain within a reader's range only for a short time. However, UHF tags are severely affected by fluids and metals. These properties make the UHF tags most appropriate in automated toll collection systems and railroad car monitoring systems. UHF tags are more expensive than any other tag. The typical frequency of UHF tags are 868 MHz (Europe), 915 MHz (USA), 950 MHz (Japan), and 2.45 GHz. Frequencies of LF and HF tags are license exempt and can be used worldwide while frequencies of UHF tags require a permit and differ from country to country.

13.1.3. *Motivation of integration of RFID and WSNs*

13.1.3.1. *Applications of RFID and WSNs*

The major application of RFID networks is to detect the presence of tagged objects and/or people. Another important application of RFID systems is to provide the location of the objects. We distinguish among several different approaches for providing the location: detecting position of the object with the mobile reader based on detection of tags placed at fixed known locations and detection of position of the object with tags based on the position of fixed readers. In case of long-range RFID systems, the estimation of the position of RFID tags can be further improved by using triangulation and/or signal processing techniques.

WSNs are mainly used for sensing the environment, positioning and identifying objects/people. WSNs are used for sensing temperature, humidity, pressure, vibration intensity, sound intensity, power-line voltage, chemical concentrations, pollutant levels, etc. WSNs can be used for sensing the environment, or sensing phenomena related to objects or people when sensors are attached to them.

By combining the properties of RFID (identifying and positioning) and WSNs (sensing, identifying and positioning) we can define four different application scenarios for combining RFID and WSNs. Some of them are used much more in industry and academia than others. In addition, WSN nodes can be independent or attached to objects/people. Some examples of integration at the application level are provided next:

(1) Integration in which RFID is used for identifying and WSN for sensing.

   (a) WSN and RFID are attached to the same object.

   In the type, RFID tags and integrated sensors are attached to the same object to perform both detecting and sensing task. The tags are used to detect presence of the objects and the integrated sensors are used to sense the objects' temperature,[22] PH value,[8] vibration,[5] angular tilt,[31] blood pressure and heartbeat rate,[14] etc. In some applications, integrating RFID readers with wireless devices and wireless sensor nodes, the integrated sensors provide both sensing and communication functionalities, such as Refs. 19 and 27.

   (b) RFID tag is attached to the object and WSN is used for sensing the object.

   An example of this type of application is using RFID tags to provide information about the objects that were photographed in museums. Here, the reader is integrated to the camera and tags are attached to the objects in the museum. After the object is sensed, RFID can provide additional information about the object.

   (c) RFID tag is attached to the objects and WSN is used for sensing the environment.

   Typical applications include mobile robots that rely on RFID for detecting non-sensible objects and on WSN for collecting information about temperature, humidity and other environment-related information.

(2) Integration in which RFID is used for identifying objects/people together with sensors.

An interesting solution is from Broadcom (http://www.broadcom.com), in which RFID information can be read from the tag only after the fingerprint scan matches the one that is already stored in the chip. In this way sensor and RFID technology can be combined to enhance security.

(3) Integration in which RFID is used for identifying objects/people and WSN is used for providing location.

Table 13.1 Difference between RFID and WSN.

	Wireless sensor networks	RFID systems
Purpose	Sense interested parameters in environments and attached objects	Detect presence and location of tagged objects
Component	Sensor nodes, relay nodes, sinks	Tags, readers
Protocols	Zigbee, Wi-Fi	RFID standard
Communication	Multi-hop	Single-hop
Mobility	Sensor nodes are usually static	Tags move with attached objects
Programmability	Programmable	Usually closed systems
Price	Sensor node — medium	Reader-expensive Tag-cheap
Deployment	Random or fixed	Fixed

In Ref. 10, RFID tags are used for identifying people in the museum tour and WSN is used to locate the leader of the group. After the locations of the leaders of a group are known, the lost members are guided in finding their group leaders.

(4) Integration in which RFID is used to assist positioning that is identified using sensors.

In Ref. 25, a positioning system for first responders is described in which different sensors are used to estimate the position of first responders and RFID tags placed at known positions are used to correct the estimated position.

13.1.3.2. *Difference between RFID and WSN technologies*

In this section, we will point out the major differences between RFID and WNS technologies.

The major components of WSNs are sensor nodes. Besides sensor nodes, the network can contain relays, sinks and some other nodes. Communication among the nodes is multi-hop. On the other side, classical RFID systems are composed of RFID tags and readers. Since the price and power consumption of the tags is a very important issue, most of the complexity is transferred to the reader side. Communication between the reader and the tags is single-hop.

Standardization efforts in RFID networks are significant. There is a large interest and large investments in RFID in industry and hence a number of standards. Major standardization organizations are EPCGlobal and ISO. Both organizations define a set of standards that can be divided into data and interface standards. Data related standards define the format of the identification number. Interface standards define protocols between the different levels of the stack including both reader-to-tag communications as well as the interface among different software layers. This is very different from the existing trends in WSNs where there is much less industrial involvement. WSNs usually rely on existing standards borrowed from other areas.

Existing protocols such as Zigbee are accepted in general — however there are many solutions that are proprietary or based on different standards.

Deployment of WSN nodes can be both random and fixed. On the other hand, if we consider long range RFID readers, the position of RFID antennas must be computed carefully in order to cover all the tags in the range as well as to prevent interference. Several types of interferences exist in RFID networks. They include tag-to-tag, reader-to-tag and reader-to-reader interference. Tag-to-tag interference occurs if several tags try to communicate with the same reader at the same time. Reader-to-tag interference occurs if the tag receives signals from two readers. Tags are not frequency selective and they can get confused in that case. Reader-to-reader interference means that the signal from the neighboring readers will be stronger than the backscattered signals from the tags. There are several standards that deal with tag-to-tag interference. The EPC Class 1 Generation 2 UHF standard[16] is based on slotted Aloha reader-to-tag communication protocol. The collision occurs if several tags select the same time slot for communication. The reader is responsible for running the optimization algorithm that will minimize the query time (throughput). The same standard[16] suggests solutions for reader-to-reader and reader-to-tag interference based on separating reader and tag transmissions spectrally or temporarily. Hence, RFID readers are deployed at fixed positions.

Functionality of RFID tags is usually fixed. Since the main goals in designing RFID tags are to lower power consumption and lower cost, tags are usually implemented in hardware. RFID readers are usually black boxes without possibility of firmware modification. On the other side, in many WSN nodes microcontrollers are reprogrammable. This enables easier modifications and facilitates research on WSNs.

Since the difference between WSNs and RFID technologies are significant, their integration would enable the combination of both their merits. WSNs offer a number of advantages over traditional RFID implementations such as multi-hop communication, sensing capabilities and programmable sensor nodes. On the other hand, WSNs are also required to be integrated with RFID. First, RFID tags are much cheaper than wireless sensor nodes. In some WSN applications it is an economical solution to utilize RFID tags instead of wireless sensor nodes. This is true in objects in which we care only for their presence and locations. Second, integration with RFID equips sensor nodes with tag IDs. One may say that the MAC address of a sensor node can be treated and utilized as an ID. Note that RFID is a mature technology of storing and retrieving data, and has been widely used in manufactures and retailers. Use of tag IDs instead of MAC addresses is an efficient solution for wireless sensor nodes. Moreover, although RFID technology has limitations such as low tolerance to fluid or metal environments, it can extend the ability of a sensor network by providing sensible properties to otherwise unsensible objects. For instance, sensor nodes may fail to work in harsh environments or some special applications and RFID could be an alternative solution.

These two technologies have almost converged in long-range active RFID tag solutions. They can be implemented using RFID protocols or some other widely accepted standards such as WLAN. They have the unique ID numbers like RFID tags and they are usually attached to one or more sensors and as such they represent a sensor node. In some solutions, MAC addresses are used as an RFID tag.

13.2. Integrating RFID Tags with Sensors

Integration of this class is used to equip RFID tags with sensors which provide sensing capabilities for RFID tags. The RFID tags with sensors (sensor-tags) use the same RFID protocols and mechanisms for reading tag IDs as well as for collecting sensed data. For example, the Class 1 Generation 2 UHF protocol from EPCGlobal[16] allows the specification of which part of the tag's memory is used for reading. Thus, sensed data can be selected and read by EPC Generation 2 compliant readers by using properly configured commands. Since integrated sensors inside RFID tags are used for only sensing purposes, current protocols of these RFID tags rely on single-hop communication. That is, integration of this class is typical in RFID systems with additional sensing abilities for integrated tags.

A distributed architecture for ubiquitous RFID sensing networks is studied in Refs. 30 and 35. RFID tagged objects communicate an EPC code to identify themselves as a unique entity. Integrated sensors in the tags are used to sense the objects and environments. The Analog signal of the sensors is converted by the A/D module and the resulting data is forwarded by readers to the base station which provides service layer functionalities. Readers are able to detect certain events or query objects with certain RFID labels to obtain event data. The application system then responds to these events and corresponding actions are processed. The system architecture is shown in Figure 13.1.

Current RFID sensing applications include monitoring physical parameters, automatically detecting product tampers, detecting harmful agents, and non-invasive monitoring.[38] A large portion of applications are used to monitor the temperature of tagged objects and environments. There are several commercial RFID tags with integrated sensors. Based on the way the tags are powered up, they can be classified into passive tags, semi-passive tags and active tags. Three representative classes are introduced below.

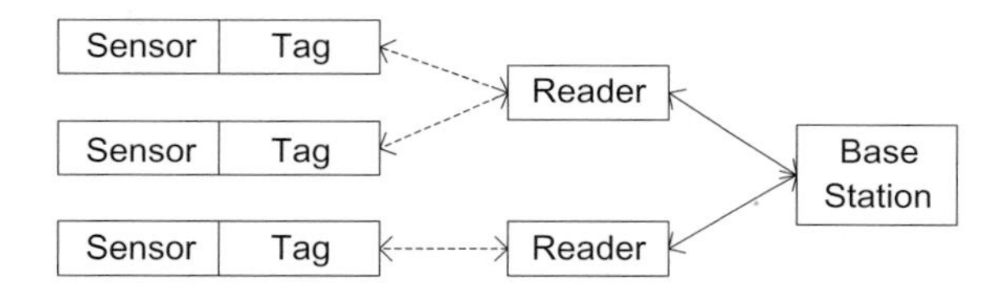

Fig. 13.1. System architecture.

13.2.1. *Passive tags with integrated sensors*

Since integrated sensors are usually powered by additional batteries, most tags with integrated sensors are semi-passive or active tags which are battery powered. However, there exist some passive tags with integrated sensors which operate without a battery and instead gather power from the RF signal of readers. Passive tags with integrated sensors are used in the following applications: temperature sensing and monitoring,[12,23] PH value detection,[8] and photo detection.[12] We describe several solutions below.

Instrumental passive tags developed by Instrumentel Ltd. (http://www.instrumentel.com) are capable of powering sensors and actuating switches as well as holding data. The tags capture enough power from reader signals to drive integrated sensors. Unlike the sensors in an active tag, the sensor in an Instrumentel tag monitors the environment only when it is interrogated by a reader. It operates in 13.56 MHz and is able to provide a reading range of up to 200 millimeters (8 inches), which is longer than the range of many other 13.56 MHz passive RFID tags currently available. The size of an Instrumentel tag is about 20 mm by 10 mm. Its price is expected to be below 5 pounds each.[8] In one of its applications, an Instrumentel tag with a PH sensor is placed into dentures and is used to monitor the level of acidity or alkalinity of food in the mouth of test patients. Using its technology's ability to actuate switches, an Instrumentel tag can also be integrated with a locking mechanism, so that it can be applied to containers to secure goods throughout the supply chain. In the application, a smart container includes an Instrumentel tag which is able to store data and lock/unlock the container using a reader signal. The technology is suited to a range of applications, including securing police evidence and medical specimens.

Besides industrial products, there are some academic solutions. An RF-powered RFID tag with temperature and photo sensors is studied and designed in Ref. 12. It consists of a supply voltage generator, a temperature-compensated ring oscillator, a synchronizer, a temperature sensor and a photo sensor. The tag gathers power from external ISM (860–960 MHz) band RF signals and senses ambient temperature and light. It operates in three states: ready, interrogating and active. The tag enters ready state when it receives an energizing RF signal. In ready state, only the internal clock generator is activated. The tag enters interrogating state when the base station sends a request to it. In interrogating state, the demodulator and decoder are activated to enable one of two sensors. When the selected functional block is activated, the tag enters active state and requested data is transmitted to the base station. It automatically switches to ready state after active state. The tag is fabricated in a 0.25-μm CMOS process. The chip size is 0.6×0.7 mm excluding pads and the total power consumption is 5.14 μW when in the active state.

A solution for long range passive RFID tags is proposed in Ref. 23. It consists of a divided micro-strip antenna and a rectifying circuit which is used to boost the DC voltage. Two passive tags are implemented. The tag working at 860–950 MHz achieves a 30 m read range. The tag working at 2.45 GHz is integrated with a temperature sensor. Its range is longer than 9 m. The tags have sizes of $90 \times 60 \times 4$ mm and $60 \times 25 \times 4$ mm for 900 MHz and 2.45 GHz, respectively.

13.2.2. *Semi-passive tags with integrated sensors*

Semi-passive tags with integrated sensors are used in the following applications: temperature sensing and monitoring,[22,32] location recording,[32] vehicle-asset tracking,[1] and access control. Next, we describe industrial solutions.

KSW — TempSens developed by KSW Microtec AG (http://www.ksw-microtec.de/) is able to periodically measure temperature in a configurable measurement interval. It is a semi-passive tag and can be attached to/embedded in any product that could spoil during transport due to temperature fluctuations. For example, it can be attached to frozen chickens that have a risk of salmonella contamination if the temperature of the environment becomes too high for a long time. The main features of KSW — TempSens are shown in Table 13.2.[22]

ThermAssureRF developed by Syntax Commerce (http://www.syntaxcommerce.com) is a semi-passive tag which is able to monitor cold chain and temperature fluctuations over time during shipment and storage of temperature sensitive goods. It is credit card size and is able to record both temperature and location information. ThermAssureRF is also capable of receiving programmable thresholds through the software application. It can be used for tracking inventory throughout the entire facility. The main features of ThermAssureRF are shown in Table 13.3.[22]

Table 13.2 Main features of KSW — TempSens.

Operating model	Semi-passive
Operating frequency	13.56 MHz (HF)
Operating temperature	−15°C to +50°C
Operating voltage	3V
Security	6 Byte Code
Memory	2 Kbit SRAM, 4 Blocks of 256 Byte each, configurable
Storable values	64 time and temperature values storable
Memory access	Read/write
Sampling interval	10 seconds to 16 hours (configurable)

Table 13.3 Main features of ThermAssureRF.

Operating model	Semi-passive
Operating frequency	13.56 MHz (HF)
Operating temperature	−40°C to +50°C
Accuracy	+/ − 0.2°C
Memory	4000 readings
Setting	Programmable alarms, programmable reading intervals

The Alien Technology Battery Power Backscatter 2450 MHz system developed by Alien Technology (http://www.alientechnology.com/) is an extendable solution to fill the gap between short-range passive systems and high-cost active systems. The tag of the system is easily extendable so that many different types of sensors can be connected to it through a serial I2C bus. The tag provides reliable read/write capabilities with a range of up to 30 m. It can be applied in long range identification, sensor monitoring, vehicle-asset tracking, etc. Equipped with a small battery, the tag is able to store 4 Kbytes of data which includes locally acquired sensor data. It works in 2450 MHz frequency range and there is an onboard temperature sensor. The tags can be attached to temperature-sensitive products at production or shipment time, and the temperature history of the product can be downloaded wirelessly at the final destination or at any point along the way. The low power backscatter technology is employed and therefore a small battery can provide several years of operation.[1]

Another interesting solution that allows for integration with external sensors is Enterprise Dot from Axcess Inc. (http://www.axcessinc.com/). Dot technology incorporates a battery powered and software definable wireless transceiver that is compatible with multiple global regulations, including EPC Class 1 and Generation 2 standards. It has three radios on the same chip and 3 antennas: UHF 315/433 MHz range in which it acts as an active tag, UHF 900 MHz range for EPC based applications in which it acts as a passive tag and LF 100–150 kHz range. The chip also has an external connection through an I2C bus. Dot supports a range of applications including manufacturing, the enterprise, oil and gas, utilities, education, government and the military. It is a good solution for access control badges, passive RFID product tags, active RFID asset tags, real time location systems (RTLS) and distributed sensor transmitters.

13.2.3. *Active tags with integrated sensors*

Active tags with integrated sensors are used in the following applications: temperature sensing and monitoring,[3] vibration detection,[5] blood pressure and heartbeat rate monitoring, etc.[14] Next we describe industrial and academic solutions.

Log-ic Temperature Tracker developed by American Thermal Instruments (www.americanthermal.com/) is an active tag. It acts as a watchdog in temperature-regulated or temperature-sensitive environments. The Log-ic Temperature Tracking system consists of three components: Log-ic Temperature Tracking Tags, a CertiScan RFID Reader and CertiScan Log-ic Software which runs on a standard windows based PC or laptop. The Log-ic Temperature Tracking tag is capable of receiving programmable temperature thresholds via 2-way RFID communication. The LED indicator on the tag can blink with warning signals if temperature exceeds the thresholds. The main features of the Log-ic Temperature Tracker are shown in Table 13.4.[3]

Table 13.4 Main features of Log-ic temperature tracker.

Operating model	Active
Operating frequency	13.56 MHz (HF)
Operating temperature	−40°C to +65°C
Accuracy	+/ − 1.0°C
Memory	Up to 64,000 readings

Table 13.5 Main features of 2.4 GHz vibration sensor tag.

Operating model	Active
Operating frequency	2.45 GHz (UHF)
RFID sensors	vibration
Min sensitivity	4 V/g
Read range	0–100 m
Data rate	1 Mbps
Power	12∼18 μA, 3V
Battery life	4 years

Bisa Technologies (http://www.bisatech.com/) is an active RFID products provider. It provides many active sensor tags which operate at 2.4 GHz–2.5 GHz. For example, the 2.4 GHz Temperature Sensor Tag (model No. 24TAG02T) is capable of collecting real-time temperatures of tagged items as well as identifying and locating them. It also can set off an alarm if the detected temperature is beyond the reasonable temperature. The size of the tag is 90 × 31 × 11 mm. The reading range is up to 100 m and the reading rate is 100 tags per second. The lifetime is estimated to be 4 years with a 3V battery. Typical applications of the tag include cold logistics and medicine transportation. Another active sensor tag from the company is 2.4 GHz Vibration Sensor Tag (model No. 24TAG02V). It is able to detect and record tagged items' either continuous or impulsive vibrations. So, it is applicable to various alarm systems. The main features of the tag are shown in Table 13.5.[5]

Besides the above industrial products, there are solutions presented by academia, such as in Ref. 14. Two different architectures for sensor-embedded RFID (SE-RFID) systems are proposed. The first architecture is illustrated in Figure 13.2. In the architecture, multiple sensors with different functions can be embedded in a tag and these sensors are controlled by a programmable timer. The sensors sample external data independently and periodically. The obtained raw data are preliminarily processed by the microprocessor before sending to a more powerful database via the reader. The database integrates the data from the reader and from other sources such as user interventions and the Internet. The sampling of sensing data can be turned on/off and be programmed through the reader. Since the tags in the system need to be periodically turned on to sample data, they need to be battery-powered.

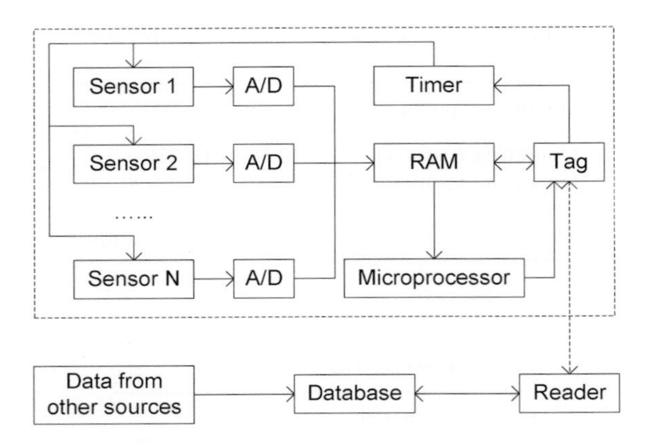

Fig. 13.2. Architecture I of SE-RFID systems.

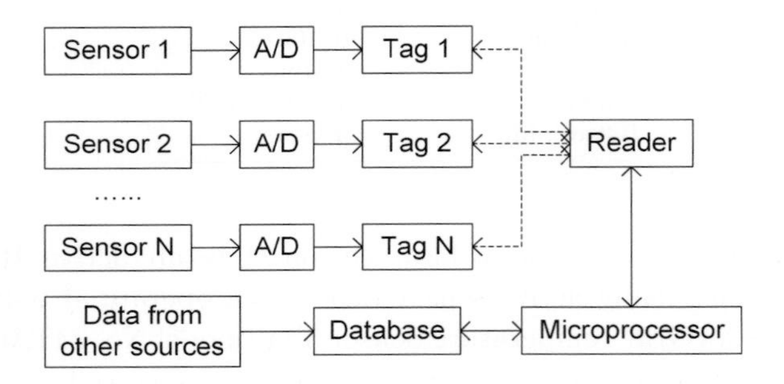

Fig. 13.3. Architecture II of SE-RFID systems.

Once the remaining energy of the tags is less than a particular unworkable level, the tags are able to automatically switch to passive mode. This means that the sensing function operates only when the tag is in the interrogative zone of the reader.

In the second architecture shown in Figure 13.3, each tag has only one integrated sensor. Similar to the first architecture, sensors in this architecture sample environment data periodically and independently and the sensing data is transmitted to the reader. Since the tag contains only one sensor and the microprocessor is embedded inside the reader, the tag in the second architecture consumes much less energy compared with that in the first architecture. Moreover, it allows different sensing sources to be located at different geographical positions. If different sensor sources are located at the same position, the first architecture is a better solution.

To validate the proposed architectures, a real-time health monitoring system is further developed in Ref. 14. The SE-RFID system is able to continually monitor, re-evaluate and diagnose the medical condition of chronic patients. The sensor tag is installed within a convenient device, such as a watch, worn by the patient. Three different sensors for measuring temperature, blood pressure and heartbeat rate are adopted. The integrated sensors periodically sample the patient's parameters and then analyze the preliminary data. The sensing data are further processed by readers which are installed in the patient's vehicle, office or home. The data will be transmitted to a doctor through the Internet. The sensor tag is also able to contact the reader to send an emergency signal directly to the doctor in emergency cases.

Since the sensing sources, i.e. temperature, blood pressure and heartbeat rate, can be detected at the same position of the patient's body, such as the patient's wrist, the first architecture is a good solution for the real-time health monitoring system. The system operates at UHF 915 MHz. The SE-RFID reader is directly connected to a PC computer via a USB connection. The PC computer is further connected to the Internet via wireless LAN using Wi-Fi standard. This allows critical data to be stored in a remote database for future diagnoses.

13.3. Integrating RFID Tags with Wireless Sensor Nodes and Wireless Devices

RFID tags integrated with sensors have limited communication capabilities. In high-end applications, it is possible to integrate RFID tags with wireless sensor nodes and wireless devices, such that the integrated tags are able to communicate with many wireless devices which are not limited to readers. The main difference between the tags of this class and the tags discussed in Section 2 is that the tags with integrated sensors are traditional RFID tags which communicate only with readers, while the tags in this class are able to communicate with other wireless devices, including tags themselves. Therefore, the tags in this class are able to communicate with each other and form a multi-hop network. These new tags may be compliant with existing RFID standards or they can have proprietary protocols.

CoBIs RFID tags[9] are designed to monitor the ambient conditions around them and provide alerts when detected conditions break predetermined business rules. Each CoBIs tag carries an accelerometer (movement) sensor, a wireless transceiver, up to 10 kilobytes of memory and other computing components for storing and processing business rules. The tags are able to communicate with each other via a proprietary peer-to-peer protocol. Each node transmits not only its unique ID number but also details of its sensed data to all other nodes within a 3-m range. These communications enable CoBIs tags to be tagged to chemical containers to monitor ambient parameters and total volume of stored chemicals. It will show alarm messages and take corresponding actions if the detected ambient parameters or total volume of chemicals exceed some predetermined threshold. The application requires communications not only between tags and readers, but also among tags for cooperative control on the total volume of chemicals. Moreover, it also helps ensure that potentially reactive chemicals have not been stored close to each other.

AeroScout T3 Tags developed by AeroScout (http://www.aeroscout.com/) are Wi-Fi-based active RFID tags which utilize standard Wi-Fi networks to track high-value assets and people in real time in both indoor and outdoor applications. There is a built-in motion sensor and an optional built-in temperature sensor. The tag is tamper proof and can periodically retrieve valuable data such as fuel gauge, mileage, or pressure measurements. Since it follows Wi-Fi standard, the generated data can be received and processed by any wireless access points (802.11b/g) from various networking vendors. This enables customers to have multiple reader options. The tag's small size allows it to be attached to very small or oddly-shaped assets. AeroScout T3 offers multiple mounting options, such as a trailer mount, an ID badge clip and a vehicle hanger. Expanded LED functionality adds the option for up to three different colored LED lights on the tag when unique visual identification is needed in large inventory scenarios. The tag is equipped with a powerful and replaceable battery which provides up to 4 years operation. The battery level is periodically reported, so that replacement of the battery can be done efficiently.

In Asia, multi-hop RFID tags had been developed by NTT labs (http://www.nttcom.co.jp) to prevent monkeys/animals from messing up farms. These sensor tags are battery-powered and operate in 429 MHz band. The communication range is less than 1 km. The sensor tags not only can send out information, but also can read/relay information to other tags. Monkeys are first attached with tags that act as transmitters. The tags which are installed in the environment act as receivers and detect monkeys as they approach farms. The information will eventually be gathered at the RFID reader and be reported to residents. Since there is no wireless/network connectivity in mountains, some tags act as relay devices and are installed at key points.

Actually, the tags in this class can be treated as wireless sensor nodes with RFID ability in some degree. A wireless smart sensor platform is studied in Ref. 31. It is capable of "plug-and-play" capability and uses RF links, such as Wi-Fi, Bluetooth, and RFID, for communications in a point-to-point topology. The platform consists of a collection of sensors, and actuators which communicate with the central control unit through standard RF links. Each sensor or actuator is equipped with a smart sensor interface (SSI). The interface extracts data from sensors, sends commands to the actuators, and provides a data communication interface to the central control unit. The sensor/actuator coupled with SSI is the smart sensor node. The architecture of the smart sensor node is shown in Figure 13.4.

The wireless smart sensor platform is further implemented to demonstrate its non-deterministic real-time performance. To achieve near real-time performance, the smart sensor node tracks the traffic of wireless channels and uses a simple TCP-like congestion control scheme to regulate the traffic. That is, it increases packets linearly in case of low load of the system and drops packets exponentially in case of heavy load. Once there is congestion, high traffic, or connection loss, the node turns to safe mode, and waits for reconnection of the central control unit or degradation of collision signals.

The implemented system is a proportional gyro motor-encoder system. It consists of two smart sensor nodes and a laptop which acts as the central control unit.

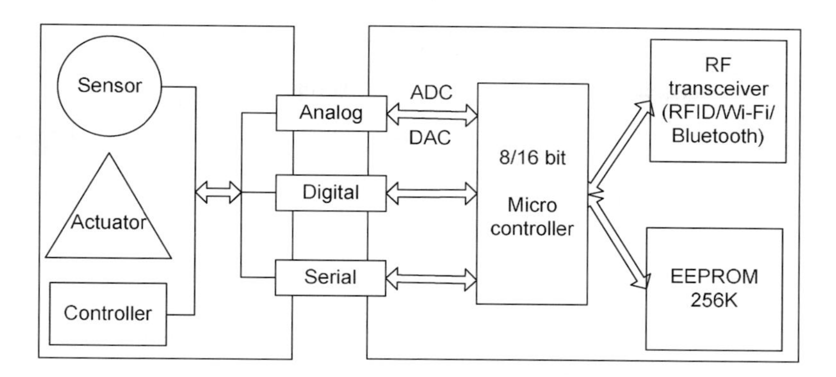

Fig. 13.4. Architecture of the smart sensor node.

Each sensor/actuator pair is connected to a SSI and uses Bluetooth to communicate with the central control unit. One smart sensor node combines a Bluetooth radio, a rate sensor and an RFID tag into a smart sensor board. The other smart sensor node integrates an actuator and a Wi-Fi radio into its smart sensor board. The gyro senses the angular tilt and communicates with the laptop which in turn sends appropriate commands to the motor. The encoder, attached to the motor, tracks the motor's position. The safe mode of the system is used to bring the motor to a halt in this implementation. The smart sensor nodes monitor the status of a machine and store the health information in the RFID tag. The tag is used as a plain wireless non-line-of-sight data storage. One can retrieve the required health information of the machine by querying the tag with a handle RFID reader even when the central computer (laptop) has been switched off. In the implementation, ISO 15693 (13.56 MHz) tags are utilized to store data. The memory of the tags ranges from 256 bytes to 2 KB. A handheld RFID reader connected to a PDA is used to read data from these tags.

Another academic solution is presented in Ref. 29. An RFID-impulse technique is proposed to eliminate node idle listening and save energy which is the critical resource in wireless sensor nodes. In wireless sensor networks, communication protocols usually adopt a periodical sleep-active schedule on sensor nodes to reduce energy consumption. However, the sleep-active schedule often results in idle listening and high end-to-end delay. It will degrade performance of the networks. The RFID-impulse technique provides an on-demand wake-up capability for wireless sensor nodes. Each sensor node is integrated with an RFID tag and is also provided with RFID reader capability. There are two sets of radios in each sensor node. One is an RF sensor radio for communications with sensor nodes and the other is an RFID radio, i.e. wake-up radio. A component of the sensor nodes is illustrated in Figure 13.5. The integrated tag listens to the RFID reader radio of neighboring nodes. If some channel activity is detected, the tag awakes the sensor to listen to the channel and then receives data through the RF sensor radio. Otherwise, the

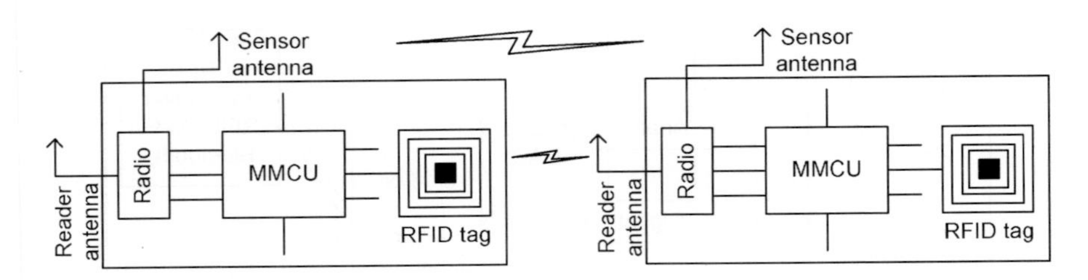

Fig. 13.5. Component of integrated sensor nodes.

sensor node can stay in sleep status. Since RFID radio uses much less energy than RF sensor radio, the RFID-impulse technique is able to significantly reduce energy consumption while providing short end-to-end delay.

It is possible to integrate or to add the tag's features to sensor motes, such as Mica motes, so that tag sensors can cooperate with each other to form an ad hoc network. The tag sensors are able to decide by themselves where and when data should be transmitted/received as long as they are equipped with microcontrollers. However, the cost of Mica motes is too high for commercial applications. It is possible to substitute Mica motes for RFID tags when the Mica motes become much cheaper in the future. As the cost of devices is low and data flow in the networks is little, ZigBee/IEEE 802.15.4 standard[40] is applicable to the sensor tag networks to achieve end-to-end system security.

13.4. Integrating Readers with Wireless Sensor Nodes and Wireless Devices

Another type of integration of RFID and sensors is the combination of RFID readers with wireless sensor nodes and wireless devices. The integration enables new functionalities and opens the door to a number of new applications. The integrated readers are able to sense environmental conditions, communicate with each other in wireless fashion, read identification numbers from tagged objects and effectively transmit this information to the host. Based on functions of integrated sensor nodes, current solutions can be classified into three categories. Both industry and academic solutions are presented next.

In the first category, RFID readers are integrated with wireless devices which are used for wireless communications in Wi-Fi standard. One industry solution is ALR-9770 series multi-protocol RFID reader that was developed by Alien Technology. The devices support all current EPC protocols and are upgradeable to EPC Class 1 Generation 2. The reader is equipped with up to 4 antenna sets for reliable tag reading and is able to communicate via 802.11 b/g standard. Product specifications are listed in Table 13.6.[2]

There are also academic solutions. For example, the integration of RFID technologies into an ad hoc network such that information can be easily collected from

Table 13.6 Product specifications of ALR-9770 series RFID readers.

Protocols support	EPC class 1 Gen 1 EPC class 1 Gen 2 EPC class 0 Rewritable class 0
Operating frequency	902–928 MHz
Memory	64 Mbytes DRAM, 16 Mbytes Flash
Antenna ports	ALR-9774-bdl 4-port bundle ALR-9772-bdl 2-port bundle
Network	10.100 Base-T Ethernet Optional WLAN 802.11 b/g
Protocols	TCP/IP, UDP, DHCP, HTTP, NTP
Power	12 VDC, 2A
Dimensions	25.4 × 25.4 × 3.8 cm, 3 lb 10 oz

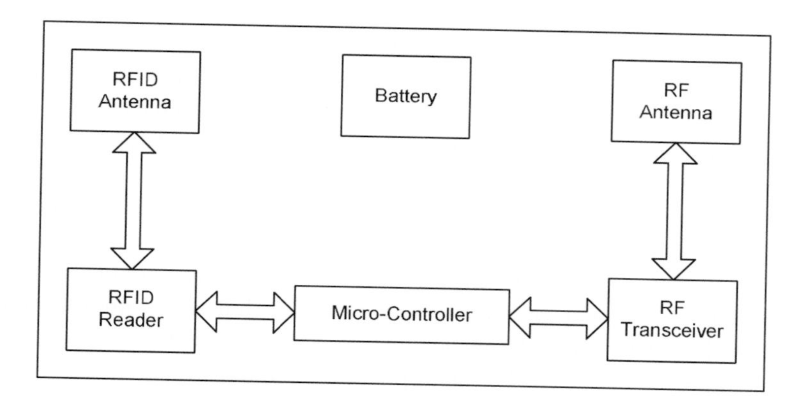

Fig. 13.6. Structure of the integrated reader.

multiple RFID tags spread over a large area has been studied in Ref. 17. The basic idea of integration is to connect the RFID reader to an RF transceiver which has routing functions and can forward information to and from other readers. Users are able to read tags from a distance that is well beyond that of the normal range of readers through hop-by-hop communication of the readers. The integrated node consists of an RFID reader, an RF transceiver and a micro-controller that coordinates different components in the node. The micro-controller is also used to control the RFID reader and other components that go into sleep mode when they are not busy. The structure of the node is shown in Figure 13.6. The Philips ICODE standard was adopted for the RFID system. The ICODE standard is the industry standard for HF RFID solutions. It has the ability of reading from and writing to the tag, and can read up to 200 tags per second.[28] There are a number of embedded platforms available for RF transceivers, such as Berkeley mote, Mica, Mica2, Mica2Dot, and MCS Cricket manufactured by CrossBow Technology (http://www.xbow.com). The Mica2 platform and event driven operating system TinyOS were adopted to construct the network in Ref. 17. Moreover, the lifetime of the battery is shown to be up to 1.04 years based on the assumptions on the transmission rate, the amount of data between the reader and micro-controller per second and the response delay of the reader, etc.

Another prototype in Ref. 24 combines wireless sensor nodes and RFID readers to provide a distributed application for automated asset tracking and inventory management. The prototype contains one wireless sensor node connected to the host device, called the host node, and one wireless sensor node connected to a RFID reader, called the reader node. The host device can be an ordinary PC that has an inventory of tagged items in a database. A user on the PC is able to use the computer to execute a query on the database. The command from the user is transmitted to the host node and is relayed to the reader node via the wireless sensor network. The reader node passes the command to the RFID reader to get the desired data. The communication is bi-directional. Using the same interface, data can be transmitted from the reader to reach the host device.

In the second category, the integrated sensor nodes provide both sensing and communication functionalities. One industry solution is Smart Rack from HP Labs.[27] It uses thermal sensors and HF RFID readers to identify and monitor the temperature of servers which are in large metal server cabinets. A 13.56 MHz tag is attached to each server and each server cabinet is mounted with a reader that is connected to several thermal sensors. These sensors are used to monitor the temperature of the servers sitting in the cabinet. All readers are connected to form a network. The prototype can be applied to companies that maintain a large number of sensors in different areas. If the network is connected with portal readers and contactless employee badge cards, the application could also indicate when a sensor is removed from the server area and by which employee.

For academic solutions, a prototype on RFID and sensor networks for elder healthcare was studied in Ref. 19. The system consists of seven components: three motes, a HF RFID reader, a UHF RFID reader, a weight scale, and a base station. The system component configuration is shown in Figure 13.7. The HF RFID tags are attached to each medicine bottle for identification. The HF RFID reader is used to track all medicine bottles within the range of the reader. The system is able to determine when and which bottle is removed or replaced by patients by reading all tags at a predetermined interval. The weight scale is used to monitor the amount of medicine in the bottle. By combining information from the weight scale and HF tags, the system is able to determine how much medicine is taken from which medicine bottle when the patients take their pills.

A UHF RFID system includes a reader and several tags that are used to track the elder patient who needs the medicines. A UHF tag attached to a patient can be detected by the associated RFID reader within 3–6 m. The system is able to determine if the patient is in the neighborhood, and reminds the patient to take the required medicines via a beep sound or a blinking light. All motes are used

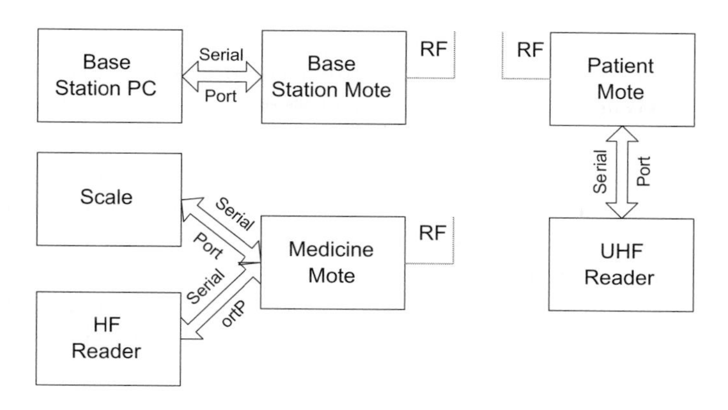

Fig. 13.7. Component of the integrated RFID and WSN system.

to communicate with the control system (Base Station PC). The Medicine Mote communicates with the HF RFID reader and weight scale to monitor HF tags and medicine weight. The Patient Mote communicates with the UHF RFID reader to monitor patients when they are close to the room. All data from the Medicine Mote and the Patient Mote is relayed by the Base Station Mote to the Base Station PC. The energy consumption model of the system was further studied in Ref. 41. It shows that the lifetime of battery can be 1.7 years if the amount of data between the reader and the microcontroller is 60 bytes per read instruction and the microcontroller works with a duty-cycle of 1%.

Another type of architecture is the 3-tier hierarchical architecture.[21] The lowest tier is the RFID tag tier. The second tier is the RFID readers which are embedded in wireless sensor nodes in the third tier. The lowest two tiers are normal RFID systems and standard RFID protocols can be adopted for communications. The sensor tier is connected to the base station and the Internet. Sensor nodes provide sensing abilities as well as delivery functions for both tags and sensors to the base station. There are several applications in a variety of different fields for this architecture. For example, it can be used in ecosystem and wildlife habitat monitoring systems. A large-scale RFID-embedded sensor network can be deployed in the field to monitor and capture the information about migration patterns, population count and other environmental data. By tagging a habitat's occupants with RFID tags, required information can be delivered to the base station via a multi-hop network. The base station can disseminate the information over Internet or store the information in the database for future use.

In the third category, RFID readers and sensors are combined with multi-functional devices, such as PDAs and cell phones. The RFID Based Sensor Networks patent[18] from Gentag (http://www.gentag.com/) provides a way to add sensor networks to RFID readers in wireless devices such as cell phones, PDAs and laptops. It describes the system and a method for the real-time concurrent detection of

13.56 MHz RFID and 8.2 MHz EAS (Electronic Article Surveillance) identification tags using a single stimulus signal. The patent comprises an RFID reader and an EAS step-listen receiver that may be placed in a single housing. Both the RFID reader and the EAS step-listen receiver have respective antennae. In operation, the RFID reader emits the stimulus on frequency 13.56 MHz. A nearby RFID tag that is tuned to that RFID frequency emits a response that is detectable by the RFID reader. If an EAS tag is also present in the neighborhood and is tuned to an EAS frequency, e.g. 8.2 MHz, the tag emits a response "ring down" signal which is detectable by the EAS step-listen receiver. Notice that the stimulus and the "ring-down" signal are nearly concurrent in time whereas the RFID response occurs later in time.[36]

The basic idea of the patent is that by combining RFID cell phones and RFID sensors in cellular networks or the Internet, the consumers will be able to read any RFID sensor tag in almost any application. Information of RFID tags can be also downloaded to a cell phone from a remote database for some applications. For example, consumers can pay their bill using their cell phones once credit card information is embedded in the cell phone. Nokia has already developed the cell phone that is embedded with HF RFID readers. Koreans are going to release the UHF cell phone. However, the ability of reading information with the cell phone also brings up security or privacy concerns.

13.5. Mix of RFID and Sensors

Different from previous cases, RFID tags/readers and sensors in this class are physically separated. An RFID system and a wireless sensor network both exist in the application and they work independently. However, there is an integration of RFID and WSN at the software layer when data from both RFID tags and sensor nodes are forwarded to the common control center. In such scenarios, successful operation of either RFID system or WSN may require assistance from the other. For example, the RFID system provides identification for the WSN to find specific objects, and the WSN provides additional information, such as locations and environmental conditions, for the RFID system. The advantage of the mix of RFID and sensors is that there is no need to design new integrated nodes and all operations and collaborations of RFID and WSN can be done at the software layer. However, since RFID tags/readers and sensors are physically separated and they work in the same system, it may cause some communication interference issues. It results in overhead of scheduling on communications to avoid interference.

A representative application of the class can be found in Ref. 10. It introduces a framework for group tour guiding services based on techniques of RFID and wireless sensor networks. It assumes that the sensing field is mixed with multiple independent tourist groups. Each group has a group leader and several members. Each member may follow the moving path of its leader, or occasionally roam around randomly based on its interest. The group tour guide system provides the following services: (1) tracking location of leaders and maintaining the guiding paths to each leader; (2) guiding lost members to their leaders; (3) helping leaders to call their members.

The system design is as follows. Sensor nodes are distributed and installed in the application area. Each sensor is connected to a direction board for displaying simple guiding direction. Some sensor nodes in the WSN serve as help centers which are connected to a laptop and a RFID reader. Each group leader carries a badge which is composed of a buzzer, a switch circuit, a control module, control buttons, and a power supply. The badge periodically broadcasts audio signals on a 4 KHz band to allow the WSN to track its location. Each group member carries a passive RFID tag which contains a group ID.

To track the location of leaders and maintain the guiding paths to each leader, each leader's badge will periodically broadcast signals, so that sensor nodes can cooperate with each other to track the locations of leaders and maintain the guiding paths from each sensor to each leader. To guide lost members to their leaders, the lost member can go to any help center and let the RFID reader read his/her tag. Note that the tag contains the group ID and WSN maintains the guiding paths from each sensor to each leader. The guiding direction will be shown on the screen of the help center and the direction boards of those sensors which are on the guiding path from the lost member to the leader. To help leaders call their members, a group leader just needs to push a button on the badge. A broadcast message will be flooded to the network. All direction boards of sensors will show the guiding directions to the sensor which is tracking the leader. Wherever the group members are, they can follow these directions to find their leaders.

The system architecture of mix RFID tags and sensor nodes was studied in Ref. 41. The system consists of three classes of devices. The wireless devices in the first class are called smart stations which have no serious power constraints. A smart station consists of an RFID reader, a microprocessor for data processing and a network interface. It can be treated as a wired device but uses wireless connections to backbone the network. The purpose of using wireless connection here is for more convenient deployment. The devices in the second and the third class are normal tags and sensor nodes, respectively. Smart stations gather information from tags and sensor nodes and then transmit that information to local host PC or remote LAN. The information coming from RFID and WSN can be further integrated into the base station for some specific application. For example, detection of sensed data values that exceed some threshold may trigger RFID readers to read data from tags in some area.

The traditional Internet protocol architecture can be employed in the smart stations since there is no limitation on resources and power. This means that there is a multi-layer networking stack implemented in each smart station. Such implementation allows not only data processing, but also data routing and even reliable

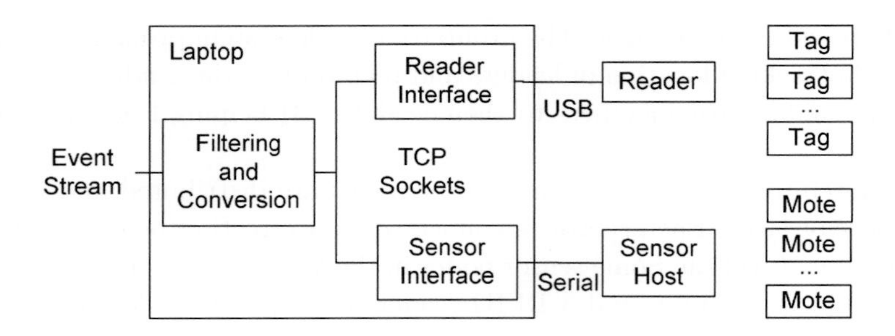

Fig. 13.8. System architecture of the implementation.

transport protocols such as TCP. The 802.11b/Wi-Fi platform can be adopted for such heterogeneous networks. The 802.11b/Wi-Fi platform uses the unlicensed 2.4 GHz band and DSSS technique in physical layer, and uses Carrier Sense Multiple Access with Collision Avoidance (CSMA/CA) scheme in MAC layer. The maximum data rate can reach 11 Mbps.[41]

Another architecture used for querying streaming RFID and sensor data in a supply chain was studied in Ref. 15. The architecture is decomposed into four layers: physical layer, data layer, filtering layer and application layer. An implementation of track and trace and cold-chain model was further presented. The implementation consists of a laptop, a server, readers and sensors. The system architecture is shown in Figure 13.8.

The reader adopts the 1356-MINI reader from TagSense (http://www.tagsense.com). It operates at 13.56 MHz and its reading range is about 1 cm. The reader is connected to the laptop via a USB cable. The laptop runs a Windows application written in C# called Reader Interface which listens on the COM port. The Tag's ID is forwarded by the reader to the COM port. The sensor board mote adopts the MTS300CA from Crossbow. It is powered by two AA batteries and is able to detect sound, light, temperature, etc. TinyOS runs in these motes. A host sensor is connected to the laptop via a serial port. It is the gateway to the laptop for a set of wireless motes. The host sensor distributes TinyDB queries over the motes. TinyDB is a program which is implemented with a SQL-style query and is running on TinyOS. The Sensor Interface is implemented by a Java program. It works with the host sensor to broadcast TinyDB queries over the sensor network and gather query results. The Filtering and Conversion is implemented by C# on a Windows platform. It receives data from the Reader Interface and the Sensor Interface via TCP sockets.

The EPC sensor network for the RFID and WSN integration infrastructure was studied in Ref. 34. The EPC sensor network provides an infrastructure to link RFID systems and sensor networks, so that useful data can be retrieved from heterogeneous sources according to different application requirements. A typical application is to build a cultural property management system. The system consists of a large number of nodes with various sensors which are utilized to monitor property assets.

RFID is adopted to track the movement of tourists and prevent them from going to prohibited area. The challenge is to combine RFID and WSN, which are totally different networks. In the RFID system, control is limited to reading, writing, and a few security options. In contrast, WSN requires more complicated algorithms/protocols for routing, data dissemination, data aggregation, and data processing.

To integrate distinct technologies in one system, the basic idea in Ref. 34 is to introduce an additional component, reader management, which extends current reader management to include WSN by adopting UPnP and SNMP. RFID data and WSN data are configured by the reader management component to provide a proper and uniform interface, such that the WSN data can be delivered to upper layers as the RFID data. Therefore, upper layers do not need to distinguish the data sources of RFID and WSN. That is, the EPC sensor network uses the concept of readers instead of the base stations which are used to gather sensing data in traditional WSNs.

Currently, some software platforms for integration of RFID and sensors are available. For example, RFID middleware software — RFID Anywhere, developed by Sybase iAnywhere (http://www.ianywhere.com), is a flexible software platform that integrates business logic and processes with a variety of automatic data collection and sensor technologies, including RFID, barcodes, mobile devices, locating systems, environmental sensors, and feedback mechanisms.[33] Another example is the WinRFID and ReWINS technologies developed by the WINMEC RFID lab (http://www.winmec.ucla.edu/rfid/). WinRFID is the middleware which supports a variety of readers/tags from different hardware vendors and provide intelligent data capturing, smoothing, filtering, routing and aggregation. ReWINS is the solution to wireless monitoring and control. It consists of two components — a wireless interface for a remote data collection unit and a control architecture with central control unit for smart data processing. ReWINS is able to support integration with RFID network via the WinRFID middleware.

13.6. Conclusions and Future Challenges

Although RFID has received more and more attention from industries and academics, more effort is needed in the study of integration of RFID and WSNs in the future. In this section, we discuss new applications as well as the need for modifications of existing application or development of new standards and integrated simulators.

At this point in time, the largest number of industrial solutions exists for sensor-tags described as class 1 in this chapter. RFID technology is seen as a cheap way to add wireless capabilities to sensors. Passive sensor-tags are a completely new area and we expect a number of new solutions to appear in the near future. The major challenges here are the need for extremely low power consumption and the limited accuracy of very low-power sensors. Passive and in some applications semi-passive RFID sensor-tags can be used in applications where, due to power/area/reliability constraints, most of the system complexity is moved to the reader side. An example of implantable RFID technology is a solution from Verichip (http://www.verichipcorp.com) based on passive chips that can carry electronic health records. The next step is the addition of sensors to these RFID chips.

Multi-hop RFID tags are a new area as well. At this point in time, ZigBee and WLAN networks are used for multi-hop communication. Multi-hop is not envisioned as a part of RFID standards right now. This is an open area of research and a lot of novelties can be expected here. We see the opportunity in using multi-hop networks to extend the range of passive RFID networks.

In class 3 and 4, significant challenges will come with ubiquitous deployment of RFID and WSNs. Current RFID deployments have not yet included hundreds of readers in a single environment. Having such a large number of readers in a single place will require detailed analysis of several related issues in order to minimize the interference, cost of RFID equipment and query time. These issues include the selection of the appropriate RFID components, positions of the readers, network topology, and synchronization of the readers in time and frequency. The situation will be even more complicated when the network is composed of a large number of RFID and WSN components. This problem requires development an RFID and WSN deployment simulator to tackle these problems. The simulator should include modeling, analysis, optimization, and performance evaluation of RFID and WSN networks. At this point in time, there are more matured simulators in the area of WSN than in RFID networks. Existing RFID simulators are missing more detailed RF simulations which are necessary when deployment is considered. To the best of our knowledge there are no simulators that combine WSN and RFID technology. We expect that significant research efforts will be put in that direction.

We envision that integration of RFID and WSN will open a large number of applications in which it is important to sense environmental conditions and to obtain additional information about the surrounding objects. One possible application is in robotics (for example, rescue missions based on robots) where robots are equipped with RFID readers and WSN. These robots will be able to acquire environmental conditions from sensors and to better understand their environment after reading IDs from the tagged objects that surround them. The information about the environment is very important and can be used for navigation of the robots or for making time-critical decisions. For example, stationary objects can be tagged in the known environment which can be used for the navigation of robots.

Actually, any application where there is a need for collecting information about the environment can be a candidate for the mix of RFID and WSN. Since RFID readers have relatively low ranges and are quite expensive, we envision that the first applications will not have RFID readers deployed ubiquitously. The applications which allow mobile readers to be attached to a person's hands, cars or robots will be good candidates. Moreover, since both sensor nodes and RFID readers are expensive, integration of WSN and RFID is a good solution for large networks, where a number of RFID tags are deployed and only necessary sensors and readers are adopted.

The use of semi-passive or active RFID technology in combination with WSN also has a promising future as the reading range of RFID systems becomes larger. In addition, the appearance of new RFID chipsets from several companies (e.g. Intel) will reduce the price of RFID readers in UHF range in future. Using cheaper readers will allow fixed stationary deployment of the readers in a similar way to how WSNs can be deployed. Similar multi-hop communication from sensor networks can be used to extract information from the readers.

There have been several attempts to incorporate WSN into existing RFID standards. In Ref. 30, it is suggested that the integration be performed between RFID tags and sensors and that the whole framework of EPCGlobal network be used. The sensor data will be stored in the tags' memory and accessed using protocols defined in Ref. 16. As described in the previous work, a whole architecture that allows for connecting the two networks at different software levels is described in Ref. 26. Significant work has to be put towards integration of RFID standards and standards that are used in sensor technologies such as IEEE 802.15.4.

Problems

1. What is RFID?
2. How many types of tags are there in current applications?
3. What is the advantage and disadvantage of each type of tag?
4. What are the frequencies used by RFID?
5. What is the advantage and disadvantage of each frequency?
6. What is WSN?
7. What is the difference between RFID and WSN?
8. How many types of integration are there in integrating RFID with WSN?
9. What are the obstacles for random deployment of RFID systems?
10. What are the main characteristics of passive, semi-passive and active sensor-tags regarding power sources, frequency ranges, using integrated or external sensors, implementation of digital logic, programmability and memory size?

Bibliography

1. Alien Technology, *Alien Technology Battery Powered Backscatter 2450 MHz System*, http://www.rfidsolutionsonline.com/Content/ProductShowcase/product.asp?DocID=ec528315-5e82-497d-8a0d-df687b314e83.
2. Alien Technology, *ALR-9770 Series Multi-protocol RFID Readers*, http://store.trustrfid.com/ProductsURL/alien/alr9774.pdf.

3. American Thermal Instruments, *LOG-IC Temperature Tracker*, http://www.americanthermal.com/ati-log-ic-ecm.pdf.
4. I. F. Akyildiz, W. Su, Y. Sankarasubramaniam and E. Cayirci, A survey on sensor networks, *IEEE Communications Magazine* (2002).
5. 2.45 GHz Vibration Sensor Tag (24TAG02V), *Bisa Technologies*, http://www.bisatech.com/product.asp?pid=2&zid=3&do=view&id=50.
6. S. Cheekiralla and D.W. Engels, A functional taxonomy of wireless sensor network devices, *Proceedings of the 2nd International Conference on Broadband Network* **2**(3–7) (2005) 949–956.
7. M. Chiesa *et al.*, RFID: a week long survey on the technology and its potential, *Technical Report on Hamessing Technology Project* (2002).
8. J. Collins, *RFID Journal*, http://www.rfidjournal.com/article/articleview/1520/1/1/ (2005).
9. J. Collins, *BP Tests RFID Sensor Network at UK Plant*, http://www.rfidjournal.com/article/articleview/2443/1/1/ (21 June, 2006).
10. P. Y. Chen, W. T. Chen, C. H. Wu, Y. C. Tseng and C. F. Huang, A group tour guide system with RFIDs and wireless sensor networks, *IPSN'07* (2007) 561–562.
11. CORDIS, *Communication Lays Out RFID Research Priorities*, http://www.feast.org/?articles&ID=568, CORDIS: News (16 March, 2007).
12. N. Cho, S.-J. Song, J.-Y. Lee, S. Kim and H.-J. Yoo, A (5).1-uW UHF RFID tag chip integrated with sensors for wireless environmental monitoring, *European Solid-State Circuits Conference (ESSCIRC)*, Grenoble (2005).
13. R. Das, RFID forecasts, players & opportunities 2007–2017, *Enterprise Networks & Servers* (March 2007).
14. H. Deng, M. Varanasi, K. Swigger, O. Garcia, R. Ogan and E. Kougianos, Design of sensor-embedded radio frequency identification (SE-RFID) systems, *Proceedings of IEEE International Conference on Mechatronics and Automation* (June 2006) 792–796.
15. K. Emery, *Distributed Eventing Architecture: RFID and Sensors in a Supply Chain*, Master thesis, MIT http://db.lcs.mit.edu/madden/html/theses/emery.pdf.
16. EPCglobal, *Class 1 Generation 2 UHF Air Interface Protocol Standard*, Version 1.0.9 (2005).
17. C. Englund and H. Wallin, *RFID in Wireless Sensor Network*, Technical Report, Department of Signals and Systems, Chalmers University of Technology, Sweden (2004).
18. R. B. Ferguson, *Gentag Patent Adds RFID Sensor Network Feature to Mobile Device*, http://www.eweek.com/article2/0,1895,2078070,00.asp (29 December, 2006).
19. L. Ho, M. Moh, Z. Walker, T. Hamada and C.-F. Su, A prototype on RFID and sensor networks for elder healthcare, Process Report, *SIGCOMM'05Workshops*, Philadelphia, PA, USA (22–26 August, 2005).
20. C. Jechlitschek, *A Survey Paper on Radio Frequency Identification (RFID) Trends*, http://www.cse.wustl.edu/~jain/cse574-06/rfid.htm (2006).
21. K. C. Ko, Z. H. Mir and Y. B. Ko, *An Energy Conservation Method for Ubiquitous Sensor Networks Integrated with RFID Readers*, submitted to WWIC (2007).
22. KSW microtec AG, *KSW — TempSens*, http://www.ksw-microtec.de/www/doc/overview_tempsens_1124436343_en.pdf.
23. H. Kitayoshi, and K. Sawaya, Long range passive RFID-tag for sensor networks, *Proceedings of the 62nd IEEE Vehicular Technology Conference* (2005).

24. M. L. Mckelvin, M. L. Williams and N. M. Berry, *Integrated Radio Frequency Identification and Wireless Sensor Network Architecture for Automated Inventory Management and Tracking Applications*, *TAPIA'05*, Albuquerque, New Mexico, USA (19–22 October, 2005).
25. L. E. Miller, *Indoor Navigation for First Responders: a Feasibility Study*, Technical Report, National Institute of Standards and Technology (2006).
26. J. Mitsugi, T. Inaba, B. Patkai, L. Theodorou *et al.*, *Architecture Development for Sensor Integration in the EPCglobal Network*, Auto-ID Labs white paper (2007).
27. M. C. O'Connor, *HP Kicks Off US RFID Demo Center*, http://www.rfidjournal.com/article/articleview/1211/1/50/ (2004).
28. Philips Semiconductors, ICODE, http://www.nxp.com/acrobat_download/datasheets/SL1ICS3001_2.pdf.
29. A. G. Ruzzeli, R. Jurdak and G. M. P. O'Hare, On the RFID wake-up impulse for multi-hop sensor networks, *Proceedings of 1st ACM Workshop on Convergence of RFID and Wireless Sensor Networks and Their Applications* (2007).
30. D. C. Ranasinghe, K. S. Leong, M. L. Ng, D.W. Engels and P. H. Cole, A distributed architecture for a ubiquitous RFID sensing network, *Proceedings of the International Conference on Intelligent Sensors, Sensor Networks and Information Processing* (2005) 7–12.
31. H. Ramamurthy, B. S. Prabhu, R. Gadh and A. M. Madni, Wireless industrial monitoring and control using a smart sensor platform, *IEEE Sensor Journal* **7**(5) (2007) 611–618.
32. RFID Temperature Tracking, *Syntax Commerce*, http://www.syntaxcommerce.com/files/24784374.pdf.
33. Sybase iAnywhere, *RFID Anywhere Review*, http://www.ianywhere.com/downloads/whitepapers/rfidanywhereoverview.pdf.
34. J. Sung, T. S. Lopez and D. Kim, The EPC sensor network for RFID and WSN integration infrastructure, *Proceedings of the 5th Annual IEEE International Conference on Pervasive Computing and Communication Workshops (PerComW'07)* (2007).
35. X. Su, B. S. Prabhu and R. Gadh, *RFID Based General Wireless Sensor Inferface*, Technical Report, UCLA (2003).
36. N. Shah, J. Paranzino and R. Salesky, *System and Method for Detecting EAS/RFID Tags Using Step Listen*, US Patent 7148804 (2005).
37. N. Vaidya and S. R. Das, *RFID-Based Networks — Exploiting Diversity and Redundancy*, Technical Report (June, 2006).
38. R. Want, Enabling ubiquitous sensing with RFID, *Computer* **37**(4) (2004) 84–86.
39. S. W. Wang, W. H. Chen, C. S. Ong, L. Liu and Y. W. Chuang, RFID application in hospitals: a case study on a demonstration RFID project in a Taiwan hospital, system sciences, HICSS'06, *Proceedings of the 39th Annual Hawaii International Conference* **8** (2006).
40. ZigBee, http://www.palowireless.com/zigbee/.
41. L. Zhang and Z. Wang, Integration of RFID into wireless sensor networks: architectures, opportunities and challenging problems, *Proceedings of the 5th International Conference on Grid and Cooperative Computing Workshops (GCCW'06)* (2006).

Chapter 14

Integrating Sensor Networks with the Semantic Web

Yong Pei and Bin Wang

Department of Computer Science and Engineering
Wright State University, Dayton, OH 45435, USA

Semantic Web and Sensor Networks are two exciting areas of ongoing research and development. While much of the existing research in sensor networks has focused on the design of sensors, communications, and networking issues, the needs and opportunities arising from the rapidly growing capabilities and scale of dynamically networked sensing devices open up demands for efficient data management and convenient programming models. Semantic web represents a spectrum of effective technologies that support complex, cross-jurisdictional, heterogeneous, dynamic and large scale information systems. Recently, growing research efforts on integrating sensor networks with semantic web technologies have led to a new frontier in networking and data management research. The goal of the chapter is to develop an understanding of the semantic web technologies, including sensor web, ontologies, and semantic sensor web services, which can contribute to the growth, application and deployment of large-scale sensor networks, leading to a broad interdisciplinary scope, such as ontology-based sensor networks, semantic sensor networks, cognitive radio networks, and so on.

Keywords: Sensor network; sensor web; semantic sensor web; ontology; metadata; OWL, XML, RDF.

14.1. Introduction

Recent technological advances in the design of sensors, communications, and networking have led towards successful sensor network deployments for various applications, such as environmental and wildlife monitoring and conservation, health-care and tele-medicine, and home/commercial automation and so on. For example, the mature miniaturized design techniques in RF device make the low cost sensors, such as RF tags, possible. Newly developed wireless communication techniques such as MIMO, OFDM, UWB and cognitive radio have made high data rate communication

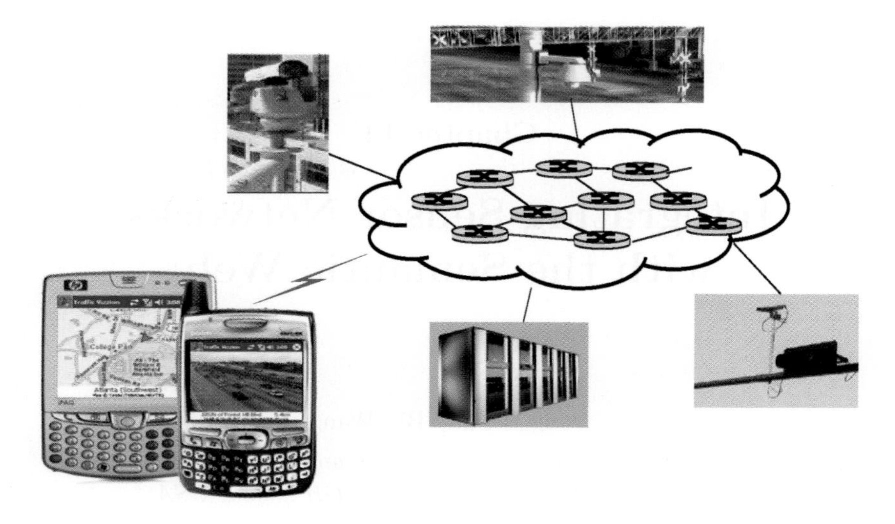

Fig. 14.1. A sample traffic information system using Sensor Web.

between sensors a reality, thus enabling the connectivity of sensors and sensor networks to the Internet infrastructure. It is envisioned[1–3] that the emerging wide deployments of sensor networks will substantially augment the Internet to form a global scale information system. Just as other information sources such as databases, sensor networks will supply the World Wide Web with real-time sensor readings. The advance in sensor web will most certainly enhance the information exchanges between people and their surrounding as well as remote environments, thus making our lives more convenient and safer, bringing us closer to the nature and helping us understand and protect our environment.

Imagine a traffic information system, as illustrated in Figure 14.1, with database services and a web of sensors such as cameras, motion detectors, road condition monitors, connected together by Internet web services. A traffic manager in a control center or a driver in a car can use a computer or a mobile device such as a cell phone to browse the most up-to-date traffic scene pictures through the sensor web that connects the traffic cameras mounted along the major highways, streets and traffic lights with the Internet. With additional information from database services, e.g. Google Earth, that provide information about routes, streets, route construction sites, detours and events or emergency hot spots, the user may assess the traffic conditions along the route.

14.2. Sensor Web Technologies

A Sensor Web[1] refers to web accessible sensor networks and archived sensor data that can be discovered and accessed using standard protocols and application program interfaces (APIs). It represents an emerging trend of making various types of web-attached sensors, instruments, image devices, and repositories of sensor data, discoverable, accessible, and controllable via the World Wide Web.[1,2]

However, to support the web service expansion to include sensors and sensor networks, there exist several challenges because, traditionally, sensor applications are hard to develop and deploy, and are not designed for reuse by other applications. So there is a lack of global resource discovery and sharing mechanism. Based on the survey in Ref. 3, most existing sensor systems are customized for specific applications and use highly customized protocols and data formats. Their primary design goal is directly collecting and using sensor level data from isolated sensor networks. In such cases, specific sensor readings are posted directly to the users; and the users are burdened with the task of interpreting the context and meaning of sensor data in order to infer high-level phenomena. However, for many emerging sensor applications, to obtain such high-level information, sensor data need to be filtered, aggregated, correlated, and translated from potentially a large number of heterogeneous and dispersed sensors and sensor networks.[4]

Different characteristics of various sensor networks such as data content, formats, modality and quality pose a challenge to make use of diverse data beyond the individual sensor network level.[4] On the other hand, different users or applications may be interested in different high-level information which can be inferred from potentially overlapped sensing capabilities of different sensor networks. For example, one application may only be interested in the health condition reports of wild animals based on specific audio, temperature, and location sensor readings collected; while another application may be interested in tracking the animal mirgration patterns based on the same audio, temperature and location sensor readings collected.

Similar to the other web services, it is expected that new sensors and sensor network services will be developed and attached to the web service at any time.[1] In order to support the progressive nature of the sensor augmentation to the web, common standards must be defined which will serve as a spanning layer to handle the heterogeneity of existing and emerging sensor systems in order to provide a seamless integration to the worldwide web services. The idea of designing such a spanning layer is similar to the Internet design strategy which uses IP (Internet Protocol) to operate over various physical communication networks, e.g. Ethernet, WiFi, optical, satellite and etc. For sensor system designers, the standards serve as guidelines for integrating their products into the web. For users and application designers, the use of such a layered design is also particularly important because it provides common interfaces for accessing various sensor systems and eases the technical demand for domain specific knowledge and also makes the high-level constructs reusable.[1,2]

14.2.1. *Sensor Web Enablement (SWE)*

As a major effort to integrate sensors and sensor networks as new data sources to augment its web services, the Open Geospatial Consortium (OGC) started the

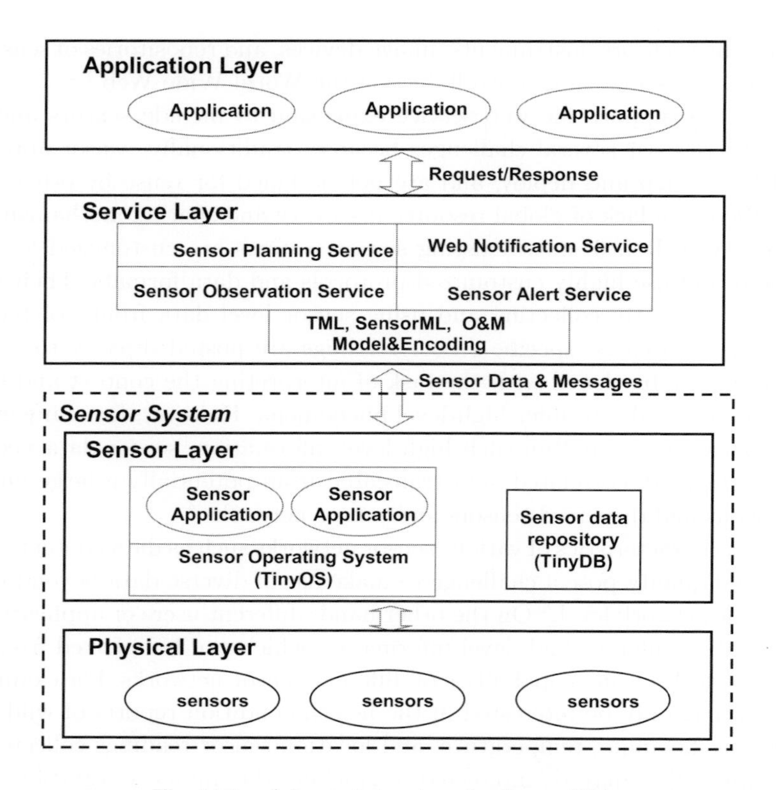

Fig. 14.2. A layered structure for Sensor Web.

Sensor Web Enablement (SWE) initiative. The primary goal of SWE is to enhance the web-accessibility of sensors by specifying interoperability interfaces and metadata encodings that enable real-time integration of heterogeneous sensor webs into the information infrastructure.[1]

The OCG SWE initiative has adopted a layered and service-oriented architecture, which can be illustrated in Figure 14.2.[2,6] The service layer serves as a spanning layer for the upper applications to integrate various sensor systems together. The major design goals are:

- Enable sensor resource discovery and sharing over Internet.
- Define standards for representing, accessing and controlling sensor and sensor data processing resources.
- Provide reliable and easy-to-use services to end-users.
- Provide infrastructure and middleware support for product designers.

Thus, some of the major functionalities associated with the service layer include[1]:

- Standard description of a sensor's properties and capabilities and quality of measurements.

- Standard data representation of real-time or time-series observations and coverage.
- Registry of individual or multiple sensor systems, observations, and observation processes which can be used to support one or more applications.
- Compose sensors for a given task to acquire observations of interest.
- Subscribe to and publishing of alerts to be issued by sensors or sensor services based upon certain criteria.

The OGC SWE[1] has defined four Web service protocols and three modeling and encoding protocols for data readings and sensors, respectively. It provides a service-based infrastructure that follows the publish-find-bind paradigm known from the service oriented architecture (SOA).[5] It makes use of a message communication protocol using eXtensible Markup Language (XML) and SOAP (Simple Object Access Protocol). It also allows fusing multiple data models and formats by using a common data model and representation, and provides an open, standards based architecture that allows sensor and data resources to be published and accessed in a standard way by simply implementing public interface and encoding specifications. Now, we give a brief review of these protocols.

The data models and encodings[1] include:

(1) **Observations & Measurements (O&M)** — SWE defines standard models and XML Schema for representing sensor data: observations and measurements, both archived and real-time. A sensor measurement will have a value and unit. An observation is an event with a result that has a value describing some phenomenon. The observation is modeled as a feature within the context of the General Feature Model. And an observation feature binds a result to a feature of interest, upon which the observation was made.
(2) **Sensor Model Language (SensorML)** — It defines standard models and XML encoding for describing sensors and processes, including the process of measurement by sensors and instructions for deriving higher-level information from observations. Processes described in SensorML can also be discovered and executed. Each process defines its input, output, parameters, and method, as well as provides relevant metadata for discovery. This metadata may include identifiers, classifiers, constraints (time, legal, and security), capabilities, characteristics, contacts, and references, in addition to inputs, outputs, parameters, and system location. As a matter of fact, SensorML models sensors as processes that convert real phenomena to data. It also support for tasking, observation, and alert services.
(3) **Transducer Markup Language (TransducerML or TML)** — It mainly concerns with data transfer in the sensor web. The conceptual model and XML Schema for describing transducers and supporting real-time streaming of data

to and from sensor systems. Here a transducer is a superset of sensors and actuators. TML provides a mechanism to efficiently and effectively capture, transport and archive transducer data, in a common form, regardless of the original source.

The Web services[1] include:

(1) **Sensor Observations Service (SOS)** — It provides the means to insert and retrieve sensor observations from sensor networks by defining standard web service interfaces for requesting, filtering, and retrieving observations and sensor system information. It is an interface between a client and an observation repository or real-time sensor networks.
(2) **Sensor Planning Service (SPS)** — It defines a standard web service interface for requesting user-driven acquisitions and observations in sensor tasking. By using SPS, a client can identify, use and manage sensors or sensor platforms; and sensors can be reprogrammed or calibrated, sensor missions can be started or changed, simulation models can be executed and controlled. It may also serve as a coordinator to evaluate the feasibility of a tasking request and provide alternatives as needed. Current SPS implementations cover a wide range of application scenarios, as well as many sensor systems ranging from simple web cams to sophisticated satellites.
(3) **Sensor Alert Service (SAS)** — It defines standard web service interfaces for publishing and subscribing to alerts from sensors. Through SAS, a sensor system can produce alerts that can be advertised, published, and subscribed to. A consumer who wants to subscribe to an alert will need to send a subscription-request to the SAS based on observation from one or more sensor system.
(4) **Web Notification Services (WNS)** — Standard web service interface for asynchronous delivery of messages or alerts from SAS and SPS web services and other elements of service workflows. It supports two different kinds of notifications: the one-way-communication, providing the user with information without expecting a response, e.g. notifying a user and/or service that a specific event occurred; and the two-way-communication, providing the user with information and expects asynchronous response.

Developers will use these specifications in creating applications, sensor platforms, and products involving Web-connected devices such as flood gauges, air quality monitors, traffic cameras as well as space and airborne earth imaging devices. Figure 14.3 illustrates the interactions among these services to accomplish a sensor web task requested by a client.[2,6]

For example, a driver may be interested in getting an alert if the visibility condition in certain area is not good enough for safe driving. With the sensor web service available, he can send an SAS request to be notified when the camera data available from that area indicate the visibility below a desirable threshold. He can specify to the SAS the geographic area and range of visibility values he is interested

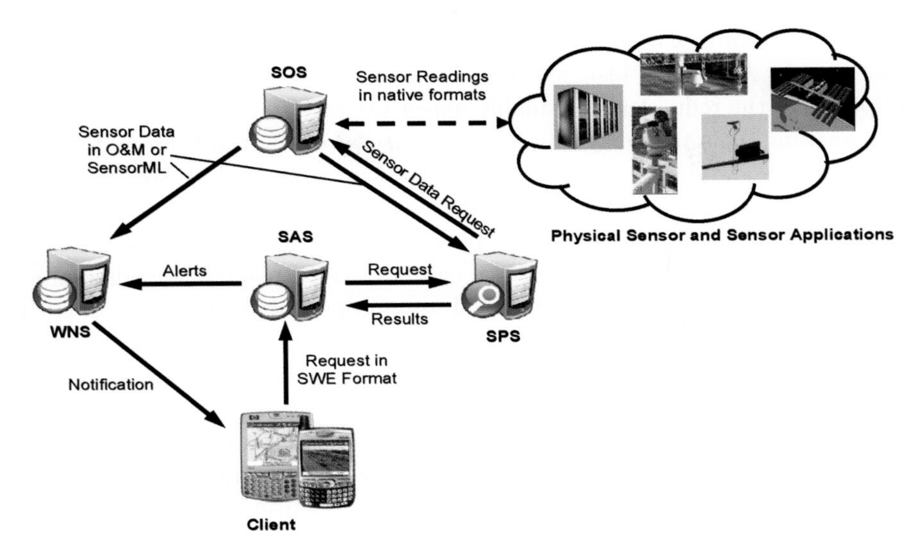

Fig. 14.3. Service models and their interactions in SWE.

in, along with other alert thresholds such as the number of cameras containing data values of interest.

Through a middleware agent, the client invokes a Sensor Planning Service (SPS) to do the required analysis. The SPS, in turn, invokes a Sensor Observation Service (SOS) to request the needed data, in this case, images captured by cameras, and a SensorML description of the cameras in the area. The SOS reads each dataset in its native format, determines images where cameras are located within the user's region of interest and intersect with roads. Depending on the area that the user requests, there may be multiple cameras for the region of interest. The SOS returns a set of geo-temporally located data points to the SPS, using the appropriate Observation and Measurement (O&M) schema. The data points will be structured as a series of arrays, each corresponding to an image. The SPS examines every image, determining whether or not the visibility value falls under the user-specified threshold. For each image, the SPS records its temporal and spatial endpoints, calculates and returns the visibility indicators to the SAS. The SAS monitors the results, and when the visibility is detected as low as determined by user-supplied thresholds, the SAS sends an alert to the Web Notification Services (WNS), which will in turn send an email alert to the user with the relevant statistics and temporal and spatial information.

14.2.2. *Limitation of Sensor Web*

Although OGC SWE sensor web solution represents an important step towards supporting sensor web services, there are some major gaps when it comes to successful integration of large scale sensor networks into the web and support various application demands.[7–10]

Ideally, a sensor web user can initiate a sensing task by sending a query or request to the system. A sensing task will then be constructed on the fly based on the descriptions of the request and available sensor network resources, by using a planning algorithm. This task may require real-time data collected from many heterogeneous sensors and dispersed sensor networks being activated, as well as recorded data available in databases. Then, these data can be processed and fused by certain processing agents to generate high-level phenomena information. If needed, the produced high-level phenomena information can be verified before the end results are delivered back to the user as a response. Internal operations of the sensor system, e.g. potential conflict handling in the use of shared sensors and sensor networks among multiple simultaneous user tasks, can be kept hidden from or made visible to end users depending on the user authority.

Clearly, it is infeasible for a human user to sort out, from thousands or millions of sensor data sources and data processing agents, which ones are appropriate for his needs, and the correct and efficient ways to process, fuse and verify them. From these observations, it is clear that: (i) there is a need to introduce automation to the sensor networks to support dynamic binding, such that sensor resources can be planned and organized for sensing tasks requested by the users, and resulting high-level phenomena can be presented to the application; (ii) there is a need to hide the sensor network internals from users for sharing sensor networks resources; and (iii) there is a need for interoperability cross various sensor networks. This is especially true in large-scale heterogeneous sensor network deployments for monitoring and observing a broad spectrum of emergent phenomena. Such needs and opportunities, arising from the rapidly growing sensor network capabilities and scales of dynamically networked sensing devices, open up demands for efficient data management and convenient programming models.

However, the services and encodings of SWE focus on providing standard encoding schemas, and are not grounded in formal ontologies. Recent research[7–10] identified several critical limitations of the sensor web based on the OGC SWE suite of protocols.

On the sensor side, SensorML provides models and XML encoding to describe sensor's service capability, and allows the user to discover and access the services. But, it does not provide a semantic description of the sensor nor the phenomenon that it measures. For example, although SensorML provides annotations for simple concepts such as spatial coordinate and time stamp, it lacks the capability of expressing more abstract concepts such as spatial region and temporal interval, or any domain-specific thematic entity.[10] As a result, it is difficult to automatically discover related sensor sources for dynamic task composition and integrating data from sensors that are described in SensorML.

On the data side, O&M lacks semantics and cannot effectively support automatically fusing data with different granularities of time, space and measured phenomena.[7,8] Suppose that a highway surveillance camera captures a surveillance image at a rate of 1 frame per second and adjacently another acoustic sensor produces measurements at a rate of 2000 samples per second. As the data has no explicit semantics there is no means to discover that both measurements are related. There is no way of automatically fusing this data unless there was software explicitly configured with the relations and necessary conversions to cater for this. The spatial and temporal conversions are possible, but there are no explicit temporal (e.g. before, after, during) and spatial operators (e.g. within, outside, adjacent) that are often required for dynamic data fusion.

As a result, due to the lack of semantics, the SWE framework mostly addresses data acquisition but neglects data filtering resulting in information overload.[7] This may significantly impact the user's ability to perceive the right information at the right time and in the right format. In SWE, data is currently pulled from passive services, rather than being pushed from active services in terms of a publish-subscribe paradigm. Massive observation data from sensors are directly exposed to the user through the service providers via predefined interfaces. In most cases, the data is presented to the user by predefined specific views of the data.

When users require alternative views, not provided by existing service providers they are forced to interact directly with sensors and determine and implement the necessary processing steps to provide these views. Thus even though services can be dynamically discovered and accessed in SWE, their offerings to service consumers are static and not dynamically integratable.

Finally, the SWE framework lacks effective support for deploying, discovering and accessing Sensor Web applications. Service providers hide complex application logic behind OGC services.[7] Users may be aware of individual instances of OGC services, but it will not be clear on exactly what applications each service, or combination of services supports. As end-user applications are not specifically addressed in the SWE framework, they will be developed on an individual basis and will often require manual upgrade when new services appear. This will eventually result in a duplication of efforts in terms of building new services, when parts of existing service could be reused as well as duplication in developing end-user applications for the Sensor Web.

14.3. Semantic Sensor Web Technologies

Semantics-based approaches[7–11] have been proposed to address the above challenge in sensor network design and provide a viable solution for integrating sensor systems with the new generation semantic web service over the Internet. Semantic web represents a spectrum of effective technologies to support such complex, cross-jurisdictional, heterogeneous, dynamic and large scale information systems. Latest research efforts on integrating sensor networks with the semantic web technologies have led to a new frontier in networking and data management research. It is expected that, through the formal semantic descriptions of sensor sources and processing agents and application domain-specific ontologies, the planning algorithm is able to compose various related sensors and sensor networks in legitimate and efficient ways so that they collectively produce meaningful end results.

14.3.1. *Semantic Web*

In her introductory article "How Semantic Web Works",[12] Tracy V. Wilson illustrated the limitation of current web service by examining the actions an online shopper has to take to place an order for a "Star Wars Trilogy" boxed DVD set with her specific preferences, e.g. widescreen, not full-screen, with the extra disc of bonus materials, lowest available price, a new set, low-cost shipping and handling, and/or short delivery time. With current web services, the shopper likely will have to look and compare at different retailers' web pages, or, if available, at one site that will compare prices and shipping options from several retailers all at once. In either situation, the shopper have to do most of the comparison and tradeoff by herself based on the data supplied by the web, then make the buying decision and place the order herself.

Why can't we simply place a request to the computer with our shopping item and preferences, and then the order will be made by the computer automatically? Fundamentally, the current World Wide Web service is designed for the human to consume the information not the computers who carry them.[13–15] The sites we visit every day use natural language, images and page layout to present information in a way that is easy for us to understand, but not for the computers. Even though they are central to creating and maintaining the Web, the computers themselves really cannot make sense of all this information. They cannot read, see relationships or make decisions like we can. As a result, there is lack of interaction between computers at the data layer, and thus current web lacks service automations.

The Semantic Web is designed with a database approach.[12] A mesh of information need to be linked up in such a way that the computers are able to effectively understand and process them, but not necessarily the same way as human does. The idea is actually very simple — provide computers with certain description about the data, i.e. metadata, such that computers can search the metadata to find, exchange and, to a limited extent, interpret information carried by the data. It is a functional extension of, not a replacement for, the World Wide Web.

For the shopping example,[12] with semantic web services in place, the shopper could enter her shopping item with preferences using a request to a semantic web service provider, which would automatically search the Web, find the best option based on her preference inputs, and place the order. Moreover, the service agent could then open personal finance software on her computer and record the amount she just spent, and it could also mark the date her DVDs should arrive on her digital calendar with a reminding message sent as the date approached.

The Semantic Web concept was proposed by Tim Berners-Lee, the same inventor of the WWW himself.[12] The World Wide Web consortium (W3C)[16] has been leading the coordinated effort to improve, extend and standardize the semantic web, and several semantic languages, tools and so on have already been developed. The Semantic Web is generally built with the following supports:

- Uniform resource identifiers (URIs) is used to direct the computer to a document or object that represents the resource.
- The metadata, associated with the document or object, use machine-readable languages: eXtensible Markup Language (XML) and Resource Description Framework (RDF). The RDF triples and XML tags are used to describe its attributes.
- Common ontologies must be used as vocabulary by the RDF and XML for constructing metadata that describe all objects and their attributes.
- Computerized applications or agents would read all the metadata found at different sites. The applications could also compare information, verifying that the sources were accurate and trustworthy.

While some websites are already using Semantic Web concepts, a lot of the necessary tools are still under development. In this section, we review some of the concepts and tools developed for semantic web which can also be applied for semantic sensor web design, including metadata, ontologies, RDF, OWL, sensor ontologies and semantic web architecture, designs and applications.

14.3.2. *Semantic Sensor Web*

The Semantic Sensor Web[7–10] is envisioned as an infrastructure that allows end users to automatically access, extract and use appropriate information from multiple sensor sources as well as existing databases, all attached to the Internet. There are three key technical challenges to realizing this vision.[7] The first challenge is to create a publicly accessible open distributed computing infrastructure where heterogeneous sensor resources and complex end-user applications can be deployed, and automatically discovered and accessed. The second challenge is computer assisted fusion of data from different sensor sources that have different temporal and spatial resolutions and different data models and formats. By fusing data from different sensors a higher spatial coverage and higher temporal resolution is achieved. The third challenge is to perform context based information extraction. The technical skill and time required to extract appropriate information from sensor data impose a barrier to a potentially large end user community who could benefit from this data. Users do not want to deal with the complexity and scale of sensor data, but would prefer a more intuitive view of the data that only exposes information that can aid them in their application. Depending on their needs (context), users would be interested in different aspects of the sensor data. Thus the same data could be used for a variety of applications. This section provides a review about some of the major concepts and architecture proposed for building semantic sensor web.

14.3.2.1. *Metadata*

Because of the opaque nature of observed phenomena encodings, metadata play an essential role in managing sensor data.[10] Metadata are defined as data about data.[13] In information system, semantics, referred to meaning and use of data, can be described using semantic-rich metadata. In sensor networks, such metadata may represent the properties of and relationships between sensors and useful derived properties of sensor data which may or may not capture information content of actual underlying data.

Semantic metadata is different from a declarative interface to access sensor networks.[4] Specifically, in the case of a programming interface, such as the Web Service Description Language (WSDL) used in sensor web service, it does not include any semantics of services, parameters, or conditions. As a result, applications are burdened with the task of interpreting the context and meaning of sensor data. In contrast, semantic-rich metadata have the advantage of enabling automation between sensors and sensor networks and to provide a common ground such that different sensor sources and processing agents from various sensor networks can be shared and reused among applications.

Firstly, with diverse and heterogeneous sensors, various forms and contents of sensed data, and large numbers of processing agents, it is a grand challenge to dynamically compose sensing tasks to meet the need of an application request.[8,9] By using the formal semantic descriptions of sensors, sensor data and the processing agents, automatic planning techniques can be developed to compose them in legitimate and efficient ways so that they collectively produce meaningful end results.

Secondly, without a common ground for the description of sources and processing agents, one party's sources and processing agents cannot be reused by others.[8] A universal and formal semantic description makes it possible to share sensor sources and processing agents among different applications. By tapping into the growing reservoir of data sources and processing agents, increasingly powerful applications can be built progressively. More specifically, metadata will specify the characteristics of service providers, e.g. sensor and processing agents. A service registry uses these characteristics to categorize service providers, and a service planner uses them to discover matched sensing and processing resources to its requirements. Metadata will also specify non-functional characteristics which may be used to help a service planner find an appropriate service provider, e.g. availability or reliability measures (e.g. mean time to fail, mean time to repair); description of needed resources (CPU, network bandwidth, memory) for the component to work properly; or trust-based information on other components.[17] Finally, metadata describes the interfaces used to access its service. The interface description includes its signature, allowed operations, data typing, and access protocols. A service planner can then use this information to bind to the service provider and invoke its service using the published interfaces.

14.3.2.2. *Ontologies*

Building a system capable of semantic brokering and dynamic task composition requires the use of universal language to construct the metadata.[13] This is achieved by composing metadata using a standard ontology to describe the sensing and processing resources in the network. Figure 14.4 illustrated the relationships between ontology, metadata and data.[18] An ontology in semantic web technologies is referred to a formal method for describing the terms, definitions and inter-relationships relevant to a certain domain of interest. It specifies a representational vocabulary used by semantic metadata to capture the information content of underlying data repositories, or service providers. The general sensor ontology[8] includes major properties such as sensor location and sensing mechanism. Each major property can be specified with low-level properties. For example, sensing mechanism is a property associated with a sensor. Every sensor has to choose a value for its sensing mechanism to describe itself, e.g. ElectricMagnetic, mechanic, or optical.

The ontologies can be classified into three levels, i.e. top-level or upper ontologies, domain and generic task ontologies, and application ontologies.[7] Developers of Semantic Sensor Web applications will typically use domain and domain task ontologies to specify new application ontologies.

14.3.2.3. *Web Ontology Language (OWL)*

Ontologies can be described in a formal ontology language, e.g. DAML (DARPA Agent Markup Language), or OWL (Web Ontology Language). OWL[19] is a standard

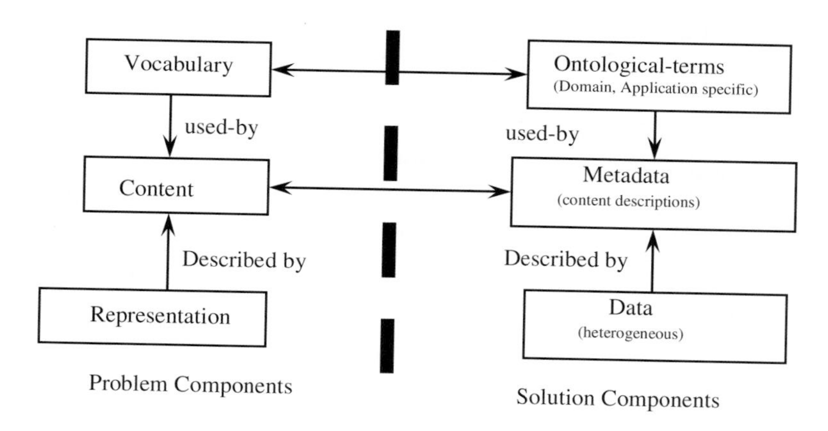

Fig. 14.4. Relationships between ontology, metadata and data.

representation language in the Semantic Web. It describes concepts (or classes), properties and individuals (or instances) relevant to a domain of interest. Descriptions of data sources, processing agents and users' queries will use the terms and relations defined in the OWL ontologies. Data sources are described by the semantics of typical data objects they produce; processing agents are described by the semantics of data objects they consume and produce; queries express the semantics of end results in a way users desire.

Sensor ontologies can be created using ontology languages, e.g. OWL or DAML, to abstractly describe the capabilities of sensors and/or sensor networks. In general, sensor ontologies may include major properties such as location, sensing mechanism, and output measures. One example of applying sensor ontologies in constructing semantic metadata was provided in Ref. 8 as demonstrated in Figure 14.5. In this example, Sensor, Location, and Multimedia Data are considered as concepts. A traffic camera is an individual that belongs to concept Sensor, and BroadwayAt42nd is individual that belongs to concept Location. Properties are used to describe the relationship between concepts. For example, Sensor and Location can be related by using the property "atLocation". Concepts can also have hierarchical relation via "subclassOf" relation. For example, SoundSensor, MotionDetector, and Camera are subclasses of Sensor, and Fixed PositionTrafficCamera is a subclass of camera.

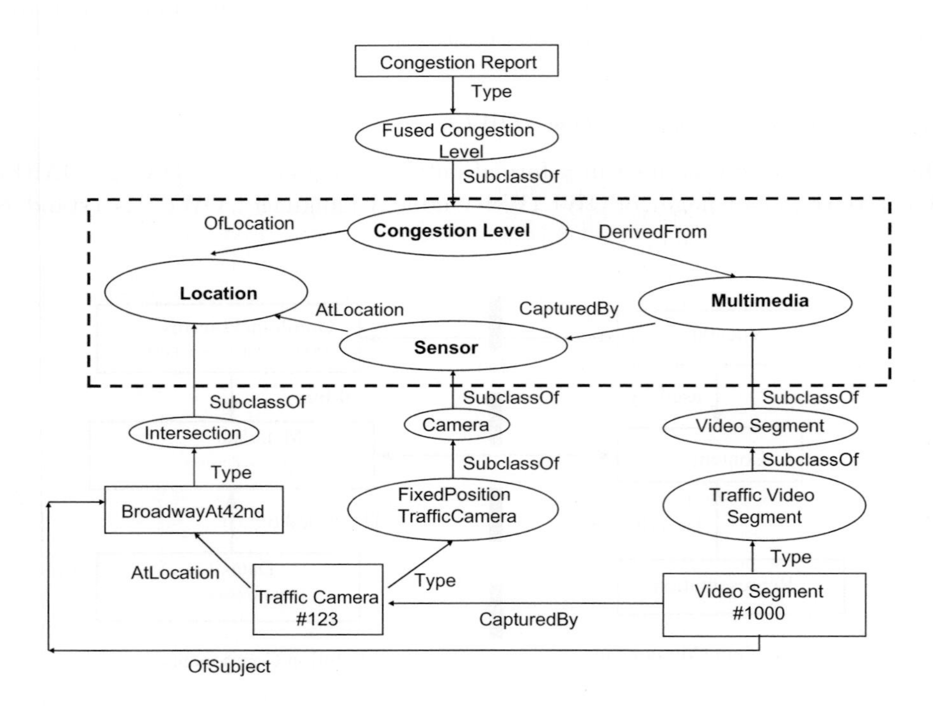

Fig. 14.5. Sample sensor and data ontologies in RDF.

14.3.2.4. *Resource Description Framework (RDF)*

In semantic sensor web, sensors are marked up with ontologies using tools such as RDF Instance Creator (RIC).[9] As a result, these sensor instances can be dynamically discovered by a search within the metadata during service brokering. The basic format for semantic descriptions is an RDF triple. An RDF triple consists of three components: a subject, a predicate and an object. The subject and object can be concepts or individuals, and the predicate is the property that associates them. RDF triples can describe OWL axioms and OWL facts, i.e. the semantic information about concepts and individuals, respectively.[8] For example, in Figure 14.5 (Camera subClassOf Sensor) is an OWL axiom; and (TrafficCamera#123 atLocation BroadwayAt42nd) is an OWL fact. A set of RDF triples can be represented as a graph, called RDF graph. The nodes in the graph are subjects and objects in the triples, and the edges are predicates (properties) between them.

Similarly, sensor data can also be described using ontologies and RDF graphs in order to support computer assisted data fusion. For example, Video Segment #1000 in Figure 14.5, belongs to the concept of TrafficVideoSegment and is captured by a camera TrafficCamera#123 atLocation BroadwayAt42nd, which is a FixedPositionTrafficCamera and located at BroadwayAt42nd, the intersection of Broadway and the 42nd Street. The semantics may also specify the time interval of the video segment.

14.3.2.5. *Sensor ontologies*

The development of upper ontologies and sensor ontologies is an ongoing process in building the semantic sensor web.[7] Currently, semantic sensor web use OWL as the web ontology language. There are several ongoing efforts on developing sensor ontologies. The OntoSensor ontology[20] is designed to provide an OWL representation of Sensor ML and serves as a linking mechanism to bridge the gap between the XML-based metadata standards of the OGC SWE and the RDF/OWL-based metadata standards of the semantic Web.

A separate effort, the NASA SWEET ontologies (Semantic Web for Earth and Environmental Terminology)[21,22] provide an upper-level ontology for Earth system science. As an upper ontology it also defines ontologies in support of the sensor systems, which can be used as an alternative to the OntoSensor ontology. SWEET ontologies include several thousand of terms, spanning a broad extent of Earth system science and related concepts (such as data characteristics) using the OWL language. Other than sensor, it includes ontologies, for example, earth, space, time, data, units, biosphere, human activity, and etc. The ontologies can be downloaded from http://sweet.jpl.nasa.gov/sweet. Compared to OntoSensor,[20] currently the SWEET ontologies[21] provide more detailed concepts that are useful for building semantic web, particularly the specialized domain and application ontologies, e.g. for earth, space and human activity.

Domain and domain task ontologies specify terms, relationships between terms and generic tasks and activities that are relevant in a particular domain.[7] Domain concepts provide further specialization of concepts from the upper ontology. Moreover, sensor data must be described in terms of the phenomenon being measured, the time and space over which it is measured and in terms of the larger data set to which this data belongs. For example, the SWEET ontologies provide support for representing time and space.

14.3.3. *Layered Semantic Sensor Web architecture*

The semantic sensor web likely will adopt a layered architecture.[7–9,23–25] For example, Figure 14.6 illustrates one of such layered structure separating Physical Sensor Layer, Sensor Service Layer, Knowledge Layer and Application Layer.[18] The Sensor Service Layer provides an abstract representation of the service offered by individual sensor or sensor networks, and deals with components working directly on physical or virtual sensors in a tight or loose coupling. The knowledge layer actually realizes the organization of and coordination between complex processing chains based on expert knowledge that was explicitly expressed in the ontology. It contains reusable data processing agents. These agents can be dynamically assembled by workflow agents to support a variety of end-user applications. The application layer contains the user interface that allows human users or other client computers to interact with the sensor web in the client's choice of data formats and transport protocols.

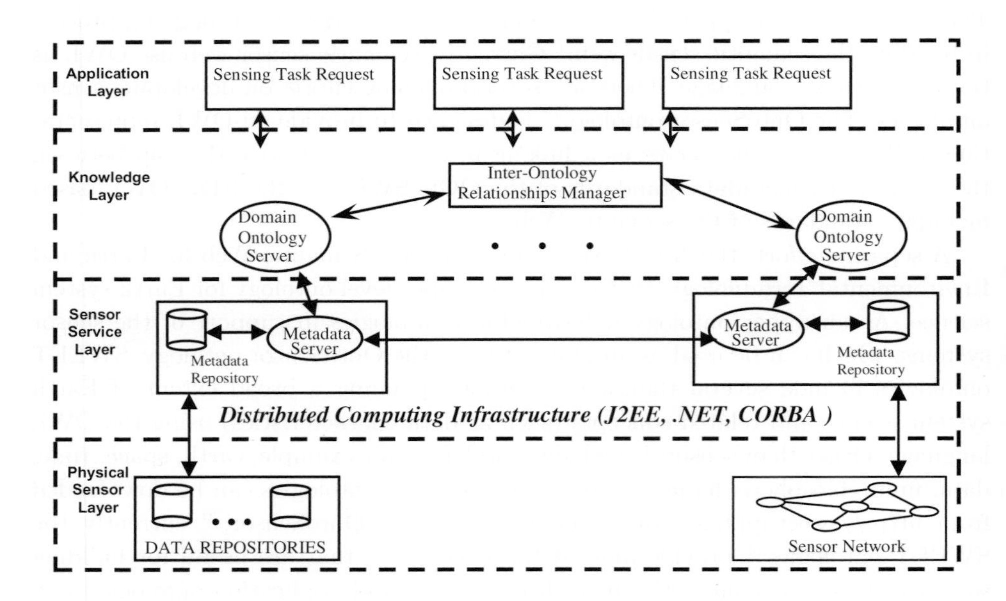

Fig. 14.6. A layered structure for Semantic Sensor Web.

14.3.3.1. *Sensor service layer*

Sensor service layer will operate directly above physical sensors or make use of intermediary services like those specified by the OGC SWE, e.g. Sensor Observation Service, Sensor Planning Service, Sensor Alert Service, Web Notification Service, etc.[7] Downwardly, sensor services belonging to this layer can use any specific and internal approaches to communicate with their corresponding physical sensors. Such services can be implemented using sensor agent which is dependent on particular sensors or sensor networks. Upwardly, through messages constructed using formally specified structured-content that is described in ontologies, sensor web services can be dynamically interfaced and invoked by components from the knowledge layer. And the service agent are able to automatically generate and interpret these messages and provides an automated mapping from conceptual intent represented using ontologies to the actual interactions required with physical sensors to fulfill this intent.

14.3.3.2. *Knowledge layer*

Representation of tasks is a crucial component in the system as they describe actions and processes in the system.[7] Workflow composition agents are capable of specifying task chains of processing agents automatically. In semantic web, OWL-S is being used for representing processes in the system. OWL-S is an extension of OWL, which provides support for modeling processes and can be integrated with the web services framework. Furthermore there are OWL-S tools available such as the OWL-S Editor. As an augmentation to the semantic web, semantic sensor web can also use OWL-S to compose sensing tasks.

The core components of the knowledge layer in semantic sensor web are workflow composition agents and processing agents.[7–9] Workflow composition agents capture and store expert knowledge in the system. Based on a workflow description language, such as OWL-S, the workflow composition agent combines a set of data processing agents to handle a specific sensing task. All agents in the knowledge layer are sensor independent. Based on a common protocol of exchange formats and general concepts specified in multiple ontologies, the workflow composition agents designate the sensor data flows from the sensor agents and invokes data processing agents to run predefined processing steps and collect additional data for further analysis. The processed data will be passed back to the workflow composition agent in order to be passed on to the next service or forwarded to the application agent in case the workflow has reached its final step.

Processing agents usually represent well defined data processing steps based on deterministic or stochastic modeling and analysis. Whereas deterministic agents run predefined processes that usually do not require further data other than the one provided or referenced by the workflow agent, stochastic processing agents provide complex processing capacities that might take some time to get accomplished or even fail in producing a proper outcome.

Stochastic processing agents may request additional data other than the one provided by the workflow composition agent and usually receive additional processing instructions on request. Furthermore, stochastic processing agents will be able to provide scenario based information (what if-scenarios) and provide multiple, concurrent outcomes on a single request.

14.3.3.2.1. Semantic composition of service workflows

The semantic descriptions of sensor and processing agents enable automatic composition of workflows using an AI agent — intelligent workflow composition agents.[7,8] A workflow composition agent has two basic components: a composer and a dynamic binding agent. The composer is the application interface that handles the interaction between the application and the dynamic binding agent. Simply speaking, the dynamic binding agent is an ontology reasoner that searches and matches properties of application requests with the related sensors and services provided by processing agents.

In the semantic sensor web, there are two types of services[9]: sensor services provided by sensor and data processing services through processing agents. While sensor services provide sensor data, processing services are reusable, shareable and linkable computing services that process sensor data. The composition of semantic services in the semantic sensor web can be accomplished with machine assisted automatic composition.

A processing agent can first be associated with a semantic web service based on the match between its semantic metadata and the semantic description of the service through a semantic based search process. This service description contains a link to the description of the processing agent that is marked up in ontology language. Then, these identified semantic descriptions of processing agent are loaded to the dynamic binding agent to browse their detailed information, and filter the services by additional constraints on the non-functional attributes of the service such as the accuracy of its outputs. During the composition process, the composer may also provide a user interface to the user to view the available services discovered, and allow the user to participate in a composition of services by presenting the available choices at each step. When a processing agent is selected from the list, a query is sent to the processing agent to retrieve the information about the inputs of the service if any exist. For each of the inputs, a new query is run to get the list of possible sensor services that can supply the data for this input. The composer may also show the user the different types of services available in the system and filters the results based on the constraints defined by the user on the attributes of a service. Finally, service workflow is constructed by chaining them together.

14.3.3.3. *Application layer*

The application layer provides the end-user interface that supports effective and convenient interactions between the user and the semantic sensor web systems.

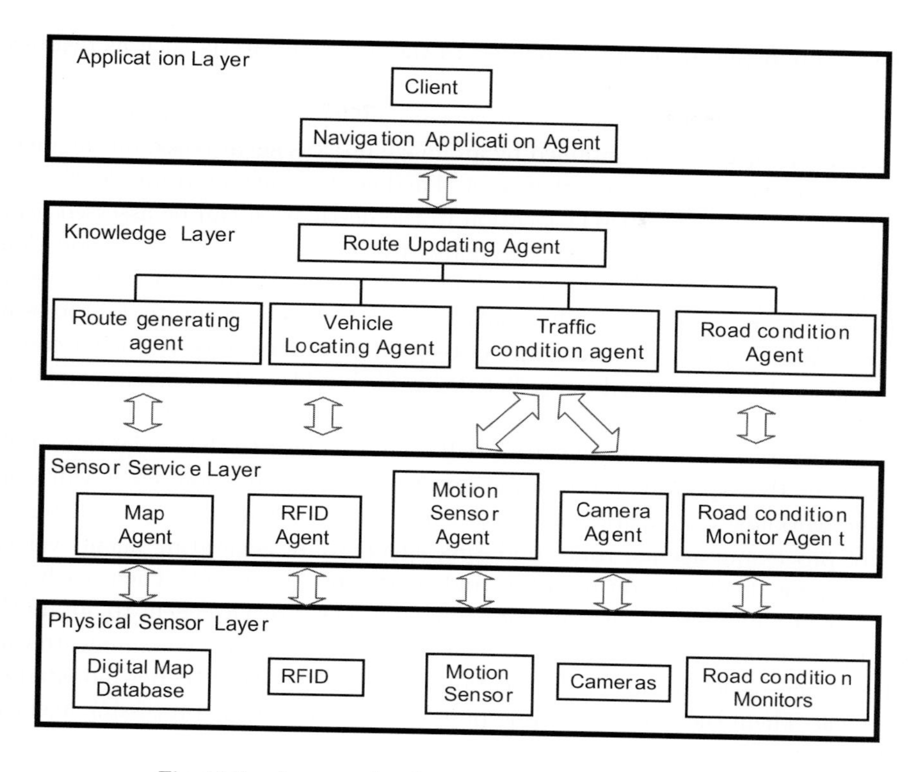

Fig. 14.7. An example of semantic Sensor Web application.

Among such interactions via the interface, users can select or construct the combination of services that meet their needs, and will be able to create a custom view of related results to realize specific tasks by integrating the output of one or more workflow agents. The application layer must also assume the responsibility of updating the user with new applications and capabilities available in the sensor web system on the fly and the interface should be easily reconfigured to reflect the changing requirements of the user.[7]

14.3.3.4. *Semantic Sensor Web applications*

Consider again a traffic information system application that provides a vehicle with navigation service from its current location to its destination based on the analysis of digital map database and real-time data obtained from sensor web including sensors such as cameras, motion detector and road condition monitors.

The objective is to provide the driver with the optimal driving directions based on distance, traffic, road conditions from his current location to the destination. This navigation task requires the computer-assisted decision making based on information collection and fusion from a broad spectrum of data sources: database, cameras, RFIDs, road condition monitors and motion sensors. And the various data collected have to be properly processed by different processing procedures to extract the necessary supporting information related to the decision criteria such as distance, traffic and road condition.

For example, current traffic level at certain key locations such as intersections, exits and tollbooths can be obtained by an image process agent based on the image input from a traffic camera on that spot. Further, the current traffic level along the entire route from the current location to the destination can be assessed using certain combining technique that aggregates the traffic level of all key locations along the path. For instance, an approach may simply use the highest traffic level of any location along that route as its traffic level. To be more accurate, a processing agent may also consider the time gap between different locations. In such a case, a prediction model may be used to predict the traffic level based on current condition, time gap, traffic pattern (e.g. rush hour), weather information, and so on.

In a semantic sensor web, the client will send a request to the navigation application agent that specifies the destination. Next, the application agent will work with the workflow composition agent to construct a workflow, which will bind a series of processing agents: route updating agent, route generating agent, vehicle locating agent, traffic condition agent and road condition agent. Then, each processing agent will interact with the corresponding sensor service agent to obtain the current sensor data. For example, the vehicle locating agent will communicate with the RFID sensor service agent for location readings from the RF tag mounted on the vehicle. Then, the readings will be processed by the vehicle processing agent to extract the current location of the vehicle.

Coupled with the destination input from the user request, the route generating agent will then generate certain number of candidate routes for consideration by the route updating agent. Before the navigating direction can be made and sent to the application, the route updating agent will use the traffic and road condition agents to assess the traffic and road conditions of each candidate route. Once the optimal route is chosen, the route updating agent will notify the navigation application agent with the driving direction to the next location along the route. The navigation application agent can then notify the client through a manner that client prefers, e.g. voice from the car's speaker, text or graphical display on an LED screen. The navigation application agent can also control the frequency of the driving direction updates, e.g. at each intersection.

As multiple clients may use the navigation services simultaneously, the workflow agent also has to handle the coordinated accesses to the same sensor and processing agents between different clients. Such coordination can be achieved through standard tools in distributed system design. Some sensor readings and processed results can be shared by multiple clients through the workflow agents.

14.4. Conclusion

By augmenting the Internet with a real-time dimension and with continuous information supplies, it is expected that a full-fledged sensor web will greatly enhance the information collection, processing and exchanges over Internet and the World Wide Web services. Meanwhile, with a rapid expanding network of sensors, semantic web technologies will likely revolutionize the operatability of sensor network and lift it to the scale of the Internet. Although semantic sensor web technologies are still very much in their infancies, the maturity of semantic web technologies and growing number of semantic web services will likely stimulate increasing research interest towards this new frontier in networking and data management research, particularly on sensor and application related ontology and tool developments. The future of the semantic sensor web has never been brighter.

Problems

(1) Please give a list of data sources in WWW.
(2) Describe how the sensor network and the World Wide Web may benefit from each other.
(3) Why are common standards required for building sensor web?
(4) Two or more applications may request access to the same sensor or sensor server in the sensor web. Give arguments for and against allowing the client requests to be executed concurrently. In the case that they are executed concurrently, give an example of possible "interference" that can occur between the operations of different clients. Suggest how such interference may be prevented.
(5) Layered architecture is used in sensor web. Please provide arguments for this design.
(6) In OGC SWE architecture, how to discover a sensor service that's attached to the web?
(7) The need to store metadata requires extra storage space for a document. What's the benefit from storing metadata in semantic web?
(8) XML and RDF are the official languages of the Semantic Web, but by themselves they're not enough to make the entire Web semantically accessible to a computer. Why?
(9) Please describe the difference and relations between upper ontologies and domain ontologies.
(10) What factors may affect the responsiveness of an application in semantic sensor web? Describe remedies that are available and discuss their usefulness.

Bibliography

1. OGC White Paper, *Sensor Web Enablement: overview and High Level Architecture* (2006).
2. X. Chu, T. Kobialka, B. Durnota and R. Buyya, Open sensor web architecture: core services, *Proceedings of the 4th International Conference on Intelligent Sensing and Information Processing*, Bangalore, India (15–18 December, 2006) 98–103.

3. I. Akyildiz, W. Su, Y. Sankarasubramanian and E. Cayircl, A survey on sensor networks, *IEEE Communication Magazine* (August 2002).
4. A. Wun, M. Petrovi and H. Jacobsen, A system for semantic data fusion in sensor networks, *Proceedings of the 2007 Inaugural International Conference on Distributed Event-Based Systems*, Toronto, Ontario, Canada (20–22 June, 2007) 75–79.
5. E. Newcomer and G. Lomow, *Understanding SOA with Web Services* (Addison Wesley, 2005), ISBN 0-321-18086-0.
6. X. Chu and R. Buyya, *Service Oriented Sensor Web, Sensor Network and Configuration: Fundamentals, Standards, Platforms, and Applications*, ed. N. P. Mahalik, Springer-Verlag (Springer, Berlin, Germany, 2007) 51–74; ISBN: 978-3-540-37364-3.
7. D. Moodley and I. Simonis, A new architecture for the sensor web: the SWAP framework, *Proceedings of the Semantic Sensor Networks Workshop, Supplemental Proceedings of the 5th International Semantic Web Conference*, Athens, Georgia (5–9 November 2006).
8. E. Bouillet, M. Feblowitz, Z. Liu, A. Ranganathan, A. Riabov and F. Ye, A semantics-based middleware for utilizing heterogeneous sensor networks, *IEEE DCOSS '07* (*Distributed Computing in Sensor Systems*), Santa Fe, HM (18–20 June, 2007) 174–188.
9. G. Jiang, W. Chung and G. Cybenko. Semantic agent technologies for tactical sensor networks, ed. E. M. Carapezza, *Proceedings of the SPIE Conference on Unattended Ground Sensor Technologies and Applications V, SPIE*, Orlando, FL (2003) 311–320.
10. C. Henson, A. Sheth, P. Jain, J. Pschorr and T. Rapoch, Video on the semantic sensor web, *W3C Video on the Web Workshop*, San Jose, CA and Brussels, Belgium (12–13 December, 2007).
11. M. Imai, Y. Hirota, S. Satake and H. Kawashima, Semantic connection between everyday objects and a sensor network, *Proceedings of the Semantic Sensor Networks Workshop, Supplemental Proceedings of the 5th International Semantic Web Conference*, Athens, Georgia (5–9 November, 2006).
12. T. V. Wilson, *How Semantic Web Works*?, http://computer. howstuffworks. com/semantic-web.htm.
13. J. Cardoso, *Semantic Web Services: Theory, Tools and Applications, Idea Group*, ISBN 978-1-59904-045-5 (2007).
14. J. Cardoso and S. Amit, *Semantic Web Services, Processes and Applications* (Springer), ISBN 0-38730239-5 (2006).
15. M. C. Daconta, L. J. Obrst and K. T. Smith, *The Semantic Web: a Guide to the Future of XML, Web Services, and Knowledge Management* (John Wiley & Sons), ISBN 0-471-43257-1 (2003).
16. WC3, http://www.w3.org/.
17. G. Serugendo, J. Fitzgerald, A. Romanovsky and N. Guelfi, A metadata-based architectural model for dynamically resilient systems, *Proceedings of the 2007 ACM Symposium on Applied computing*, Seoul, Korea (2007) 566–572.
18. V. Kashyap, Enabling the semantic web: the role of metadata, semantics and domain ontologies, Colloquium Talk, CSEE Department, UMBC (3 October, 2003).
19. D. McGuinness and F. van Harmelen, Owl web ontology language overview, W3C Recommendation (2004).
20. D. J. Russomanno, C. Kothari and O. Thomas, Building a sensor ontology: a practical approach leveraging ISO and OGC models, *The 2005 International Conference on Artificial Intelligence*, Las Vegas, NV (2005) 637–643.
21. NASA, *Semantic Web for Earth and Environmental Terminology* (SWEET), http://sweet.jpl.nasa.gov/ontology/.

22. R. Raskin, Semantic Web for Earth and Environmental Terminology (SWEET), *NASA 4th Annual Earth Science Technology Conference (ESTC)*, Palo Alto, CA (22–24 June, —).
23. M. Goodman *et al.*, The Sensor Management for Applied Research Technologies (SMART) Project, *The 2007 NASA Science Technology Conference (NSTC2007)*, Baltimore, MD (19–21 June, 2007).
24. C. Goodwin and D. J. Russomanno, An ontology-based sensor network prototype environment, *5th International Conference on Information Processing in Sensor Networks*, Nashville, TN, 1–2.
25. J. Liu and F. Zhao, Towards semantic services for sensor-rich information systems, *2nd IEEE/CreateNet International Workshop on Broadband Advanced Sensor Networks (Basenets 2005)*, Boston, MA (2005).

Chapter 15

Effective Multi-User Broadcast Authentication in Wireless Sensor Networks

Kui Ren
Illinois Institute of Technology, Chicago, IL 60616, USA
kren@ece.iit.edu

Wenjing Lou
Worcester Polytechnic Institute, Worcester, MA 01609, USA
wjlou@ece.wpi.edu

Yanchao Zhang
New Jersey Institute of Technology, Newark, NJ 07102, USA
yczhang@njit.edu

Broadcast authentication is a critical security service in wireless sensor networks (WSNs), as it allows the mobile users of WSNs to broadcast messages to multiple sensor nodes in a secure way. Previous solutions on broadcast authentication are mostly symmetric-key-based solutions such as μTESLA and multilevel μTESLA. These schemes are usually efficient; however, they all suffer from severe energy-depletion attacks resulted from the nature of delayed message authentication. Being aware of the security vulnerability inherent to existing solutions, we present several efficient public-key-based schemes in this chapter to achieve immediate broadcast authentication with significantly improved security strength. Our schemes are built upon the unique integration of several cryptographic techniques, including the Bloom filter, the partial message recovery signature scheme and the Merkle hash tree. We prove the effectiveness and efficiency of the proposed schemes by a comprehensive quantitative analysis of their energy consumption regarding both computation and communication.

Keywords: Security; wireless sensor networks; broadcast authentication; multi-user.

15.1. Introduction

Wireless Sensor Networks (WSNs) have enabled data gathering from a vast geographical region and present unprecedented opportunities for a wide range of tracking and monitoring applications from both civilian and military domains.[2,3,31] In these applications, WSNs are expected to process, store, and provide the sensed data to the network users upon their demands.[20] As the most common communication paradigm, the network users are expected to issue the queries to the network in order to obtain the information of their interest. Furthermore, in wireless sensor and actuator networks,[3] the network users may need to issue their commands to the network (probably based on the information they received from the network). In both cases, there could be a large number of users in the WSNs, which might be either mobile or static; and the users may use their mobile clients to query or command the sensor nodes from anywhere in the WSN. Obviously, broadcast/multicast[a] operations are fundamental to the realization of these network functions. Hence, it is also highly important to ensure broadcast authentication for the security purpose.

Broadcast authentication in WSNs was first addressed by μTESLA.[27] In μTESLA, users of WSNs are assumed to be one or a few fixed sinks, which are always assumed to be trustworthy. The scheme adopts a one-way hash function $h(\ )$ and uses the hash preimages as keys in a message authentication code (MAC) algorithm. Initially, sensor nodes are preloaded with $K_0 = h^n(x)$, where x is the secret held by the sink. Then, $K_1 = h^{n-1}(x)$ is used to generate MACs for all the broadcast messages sent within time interval I_1. During time interval I_2, the sink broadcasts K_1, and sensor nodes verify $h(K_1) = K_0$. The authenticity of messages received during time interval I_1 are then verified using K_1. This delayed disclosure technique is used for the entire hash chain and thus demands loosely synchronized clocks between the sink and sensor nodes. μTESLA is later enhanced in Ref. 17 to overcome the length limit of the hash chain. Most recently, μTESLA is also extended in Ref. 18 to support multi-user scenario but the scheme assumes that each sensor node only interacts with a very limited number of users.

It is generally held that μTESLA-like schemes have the following shortcomings when applied to multi-hop large scale WSNs even in the single-user scenario: (1) all the receivers have to buffer all the messages received within one time interval; (2) they are subject to Wormhole attacks,[12] where messages could be forged due to the propagation delay of the disclosed keys. However, here we point out a much more serious vulnerability of μTESLA-like schemes when they are applied in multi-hop WSNs. Since sensor nodes buffer all the messages received within one time interval, an adversary can hence food the whole network arbitrarily. All the adversary has to do is to claim that the flooding messages belong to the current time interval which should be buffered for authentication until the next time interval.

---

[a]For our purpose, we do not distinguish multicast from broadcast in this chapter.

Since wireless transmission is very expensive in WSNs, and WSNs are extremely energy constrained, the ability to food the network arbitrarily could cause devastating Denial of Service (DoS) attacks. Moreover, this type of energy-depletion DoS attacks become more devastating in multi-user scenario as the adversary now can have more targets and hence more chances to generate bogus messages without being detected. Obviously, all these attacks are due to delayed authentication of the broadcast messages. In Ref. 12, TIK is proposed to achieve immediate key disclosure and hence immediate message authentication based on precise time synchronization between the sink and receiving nodes. However, this technique is not applicable in WSNs as pointed out by the authors. Therefore, multi-user broadcast authentication still remains a wide open problem in WSNs.

Observing that symmetric-key-based broadcast authentication schemes such as μTESLA are insufficient for WSNs, we resort to public key cryptography (PKC) for more effective solutions in this chapter. We address multi-user broadcast authentication problem in WSNs by designing PKC-based solutions with minimized computational and communication costs.

Objectives of the Chapter. We focus on providing multi-user broadcast authentication in WSNs, where the broadcast messages are initiated by a number of network users. Please note that the network users in this chapter refer to personnel or devices that use the WSN; they are not sensor nodes. On the one hand, we aim to achieve immediate message authentication and resist DoS attacks in the presence of both user revocation and node compromise. On the other hand, we want to optimize both computational and communication costs.

Overview of the Chapter. In this chapter, we propose four different public-key-based approaches and provide in-depth analysis of their advantages and disadvantages. In all the four approaches, the users are always authenticated through their public keys. We first propose a straightforward certificate-based approach and point out its high energy inefficiency with respect to both communication and computation costs. We then propose a direct storage based scheme, which has high efficiency but suffers from the scalability problem. A Bloom filter based scheme is further proposed to improve the memory efficiency over the direct storage based scheme. Further techniques are also developed to increase the security strength of the proposed scheme. Lastly, we propose a hybrid scheme to support a larger number of network users by employing the Merkle hash tree technique. We give an in-depth quantitative analysis of the proposed schemes and demonstrate their effectiveness and efficiency in WSNs in terms of energy consumption.

Organization of the Chapter. The remaining part of this chapter is as follows. In Section 2, we introduce the cryptographic mechanisms to be used. Section 3 presents the system assumption, adversary model, and security objectives. In Section 4, we introduce two basic schemes. We next propose two advanced schemes and detail the underlying design logic in Section 5. In Section 6, we analyze the performance and security strength of the proposed schemes. Section 7 then discusses further enhancements of the proposed schemes. Finally, Section 8 is the future work and Section 9 is the conclusion.

15.2. Background

15.2.1. *Digital signature*

A digital signature algorithm is a cryptographic tool for generating non-repudiation evidence, authenticating the integrity as well as the origin of a signed message. In a digital signature algorithm, a signer keeps a private key secret and publishes the corresponding public key. The private key is used by the signer to generate digital signatures on messages and the public key is used by anyone to verify signatures on messages. The digital signature algorithms mostly used are RSA[32] and DSA.[26] ECDSA is referred to Elliptic Curve Digital Signature Algorithm.[11] While RSA with 1024-bit keys (RSA-1024) provides the currently accepted security level, it is equivalent in security strength to ECC with 160-bit keys (ECC-160). Hence, for the same level of security strength, ECDSA uses a much short key size and hence has a short signature size (320-bit).

15.2.2. *The Bloom filter and counting Bloom filter*

A Bloom filter is a simple space-efficient randomized data structure for representing a set in order to support membership queries.[23] A Bloom filter for representing a set $S = S_1, S_2, \ldots, S_n$ of n elements is described by a vector v of m bits, initially all set to 0. A Bloom filter uses k independent hash functions $h_1, \ldots, h_k$ with range $0, \ldots, m-1$ which map each item in the universe to a random number uniform over $[0, \ldots, m-1]$. For each element $s \in S$, the bits $h_i(s)$ are set to 1 for $1 \leq i \leq k$. Note that a bit of v can be set to 1 multiple times. To check if an item x is in S, we check whether all bits $h_i(x)$ are set to 1. If not, x is not a member of S for certain, that is, no false negative error. If yes, x is assumed to be in S. A Bloom filter may yield a false positive. It may suggest that an element x is in S even though it is not. The probability of a false positive for an element not in the set can be calculated as follows. After all the elements of S are hashed into the Bloom filter, the probability that a specific bit is still 0 is $(1-1/m)^{kn} \approx e^{-kn/m}$. The probability of a false positive f is then $(1-(1-1/m)^{kn})^k \approx (1-e^{-kn/m})^k$.We let $f = (1-p)^k$. From now on, for convenience, we use the asymptotic approximations p and f to represent, respectively, the probability that a bit in the Bloom filter is 0 and the probability of a false positive. Let $p = e^{-kn/m}$.

The counting Bloom filter is a variation of the Bloom filter, which allows member deletion. In the counting Bloom filter, each entry in the Bloom filter is not a single bit but a small counter that tracks the number of elements that have hashed to that location.[10] When an element is deleted, the corresponding counters are decremented. To avoid overflow, counters must be chosen large enough.[10]

15.2.3. *The Merkle hash tree*

A Merkle Tree is a construction introduced by Merkle in 1979 to build secure authentication schemes from hash functions.[22] It is a tree of hashes where the leaves in the tree are hashes of the authentic data values $n_1, n_2, \ldots, n_w$. Nodes further up in the tree are the hashes of their respective children. For instance, assuming that $w = 4$ in Figure 15.1, the values of the four leaf nodes are the hashes of the data values, $h(n_i)$, $i = 1, 2, 3, 4$, respectively, under a one-way hash function $h(\ )$ (e.g. SHA-1[25]). The value of an internal node A is $h_a = h(h(n_1) \,\|\, h(n_2))$, and the value of the root node is $h_r = h(h_a \,\|\, h_b)$. h_r is used to commit to the entire tree to authenticate any subset of the data values n_1, n_2, n_3, and n_4 in conjunction with a small amount of auxiliary authentication information AAI (i.e. $\log_2 N$ hash values where N is the number of leaf nodes). For example, a receiver with the authentic h_r requests for n_3 and requires the authentication of the received n_3. The source sends the AAI : $\langle h_a, h(n_4)\rangle$ to the receiver. The receiver can then verify n_3 by first computing $h(n_3), h_b = h(h(n_3) \,\|\, h(n_4))$ and $h_r = h(h_a \,\|\, h_b)$, and then checking if the calculated h_r is the same as the authentic root value h_r. Only if this check is positive, the user accepts n_3. The Merkle hash tree can prevent an adversary from sending bogus data to deceive the client. In the earlier example, an adversary impersonating cannot send a bogus n_3 to the client without being detected. This is because he cannot find h_a and $h(n_4)$ such that $h(h_a \,\|\, h(h(n_3) \,\|\, h(n_4))) = h_r$ as $h(\ )$ is one-way.

15.3. System Model, Adversary Model, and Design Goals

System Model. In this chapter, we consider a large spatially distributed WSN, consisting of a fixed sink(s) and a large number of sensor nodes. The sensor nodes are

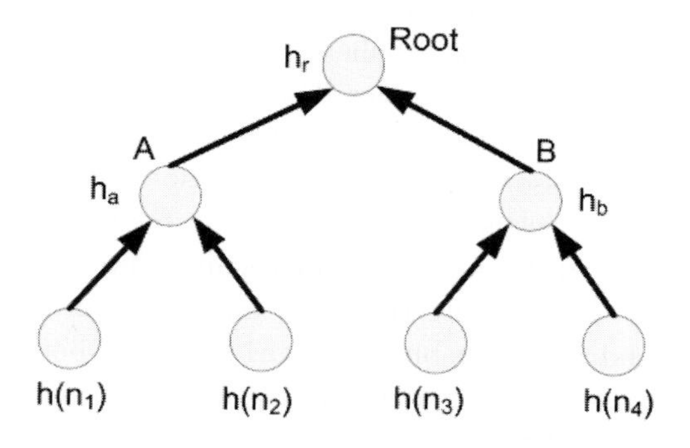

Fig. 15.1. An example of Merkle hash tree.

usually resource-constrained with respect to memory space, computation capability, bandwidth, and power supply. The WSN is aimed to offer information services to many network users that roam in the network, in addition to the fixed sink(s).[20] The network users may include mobile sinks, vehicles, and people with mobile clients, and they are assumed to be more powerful than sensor nodes in terms of computation and communication abilities. For example, the network users could consist of a number of doctors, nurses, medical equipment (acting as actuators) and so on, in the case of CodeBlue,[19] where the WSN is used for emergency medical response. These network users broadcast queries/commands through sensor nodes in the vicinity, and expect the replies that reflect the latest network information. The network users can also communicate with the sink or the backend server directly without going through the WSN if necessary. We assume that the sink is always trustworthy but the sensor nodes are subject to compromise. At the same time, the users of the WSN may be dynamically revoked due to either membership changes or compromise, and the revocation pattern is not restricted. We also assume that the WSN is loosely synchronized.

Adversary Model. In this chapter, we assume that the adversary's goal is to inject bogus messages into the network, attempt to deceive sensor nodes, and obtain the information of his interest. Additionally, Denial of Service (DoS) attacks such as bogus message flooding, aiming at exhausting constrained network resources, is another important focus of the chapter. We assume that the adversary is able to compromise both network users and the sensor nodes. The adversary hence could exploit the compromised users/nodes for such attacks. However, we do assume that adversary cannot compromise an unlimited number of sensor nodes.

Design Motivation. When μTESLA was proposed, sensor nodes were assumed to be extremely resource-constrained, especially with respect to computation capability, bandwidth availability, and energy supply.[27] Therefore, PKC was thought to be too computationally expensive for WSNs, though it could provide much simpler solutions with much stronger security resilience. At the same time, the computationally efficient onetime signature schemes are also considered unsuitable for WSNs, as they usually involve intense communications.[27] However, recent studies[9,29,35] showed that, contrary to widely held beliefs, PKC with even software implementations only is very viable on sensor nodes. For example, in Ref. 35, Elliptic Curve Cryptography (ECC) signature verification takes 1.61 seconds with 160-bit keys on ATmega128 8 MHz, a processor used in current Crossbow motes platform.[8] Furthermore, the computational cost is expected to fall faster than the cost to transmit and receive. For example, ultra-low-power microcontrollers such as the 16-bit Texas Instruments MSP430[34] can execute the same number of instructions at less than half the power required by the 8-bit ATmega128L. The benefits of transmitting shorter ECC keys and hence shorter messages/signatures generation sensor nodes are expected to combine ultra-low power circuitry with so-called power scavengers such as Heliomote,[15] which allow continuous energy supply to the nodes. At least

8–20 μW of power can be generated using MEMS-based power scavengers.[4] Other solar-based systems are even able to deliver power up to 100 mW for the MICA Motes.[15,16] These results indicate that, with the advance of fast growing technology, PKC is no longer impractical for WSNs, though still expensive for the current generation sensor nodes, and its wide acceptance is expected in the near future.[9]

Design Goals. Our security goal is straightforward: all messages broadcasted by the network users of the WSN should be authenticated so that the bogus ones inserted by the illegitimate users and/or compromised sensor nodes can be efficiently rejected/filtered. We also focus on minimizing the overheads of the security design. Especially, energy efficiency (with respect to both communication and computation) and storage overhead are given priority to cope with the resource-constrained nature of WSNs.

15.4. The Basic Schemes

We explore the PKC domain for the possible solutions to multi-user broadcast authentication in WSNs. The PKC-based solutions realize immediate message authentication and thus can overcome the delayed message authentication problem present in μTESLA-like schemes.

15.4.1. *The certificate-based authentication scheme (CAS)*

CAS works as follows. Each user (not a sensor) of the WSN is equipped with a public/private key pair (PK/SK), and signs every message he broadcasts with his SK using a digital signature scheme such as ECDSA.[11] Note that in all our designs, we do not require sensors to have public/private key pairs for themselves. To prove the user's ownership over his public key, the sink[b] is also equipped with a public/private key pair and serves as the certification authority (CA). The sink issues each user a public key certificate, which, to its simplest form, consists of the following content $Cert_{U_{ID}} = U_{ID}, PK_{U_{ID}}, ExpT, SIG_{SK_{Sink}}\{h(U_{ID}\|ExpT\|PK_{U_{ID}})\}$, where U_{ID} denotes the user's ID, $PK_{U_{ID}}$ denotes its public key, $ExpT$ denotes certificate expiration time, and SIG is a signature over $h(U_{ID}\|ExpT\|PK_{U_{ID}})\}$ with SK_{Sink}. Hence, a broadcast message is now of the form as follows:

$$\langle M, tt, SIG_{SK_{U_{ID}}}\|\{h(U_{ID}\|tt\|M)\}, Cert_{U_{ID}}\rangle. \tag{15.1}$$

Here, M denotes the broadcast message and tt denotes the current time. For the purpose of message authentication, sensor nodes are preloaded with PK_{Sink} before the network deployment; and message verification contains two steps: the user certificate verification and the message signature verification.

[b]We assume that the sink represents the network planner.

CAS suffers from two main drawbacks. First and foremost, it is not efficient in communication, as the certificate has to be transmitted along with the message across every hop as the message propagates in the WSN. A large per message overhead will result in more energy consumption on every single sensor node. In CAS, the per message overhead is as high as $|tt| + |SIG_{SK_{U_{ID}}}\{h(U_{ID}\|M)\}| + |Cert_{U_{ID}}| = 128$ bytes. As in Ref. 35, the user certificate is at least 86 bytes, when ECDSA-160[11] is used. Here, we assume that tt and U_{ID} are both two bytes, in which case the scheme supports up to 65,535 network users. Moreover,$|SIG_{SK_{U_{ID}}}\{h(U_{ID}\|M)\}| = 40$ bytes, when ECDSA-160[11] is assumed. Second, to authenticate each message, it always takes two expensive signature verification operations. This is because the certificate should always be authenticated in the first place.

15.4.2. *The direct storage based authentication scheme (DAS)*

One way to reduce the per message overhead and the computational cost is to eliminate the existence of the certificate. A straightforward approach is then to let sensor nodes simply store all the current users' ID information and their corresponding public keys. In this way, a broadcast message now only contains the following contents:

$$\langle M, tt, SIG_{SK_{U_{ID}}}\{h(U_{ID}\|tt\|M)\}, U_{ID}, PK_{U_{ID}}\rangle. \tag{15.2}$$

Verifying the authenticity of a user public key is reduced to finding out whether or not the attached user/public key pair is contained in the local memory. Upon user revocation, the sink simply sends out ID information of the revoked user, and every sensor node deletes the corresponding user/public key pair in its memory.

The drawbacks of DAS are obvious. Given a storage limit of 5 KB, only 232 users can be supported at most; even with a memory space of 19.5 KB, DAS can only support up to 1000 users. At the same time, CAS can support up to 2560 users given the same storage limit 5 KB. The reason is that in CAS only the ID information of the revoked users is stored by the sensor nodes. Therefore, DAS is neither memory efficient nor scalable. However, the advantage of DAS is also significant as compared to CAS. It successfully reduces the per message overhead down to $|tt| + |SIG_{SK_{U_{ID}}}\{h(U_{ID}\|M)\}| + |U_{ID}| + |PK_{U_{ID}}| = 64$ bytes. The above analysis clearly shows that more advanced schemes are needed other than DAS and CAS. And the direction to seek is to improve storage efficiency while retaining or further reducing the per message overhead.

15.5. The Advanced Schemes

In this section, advanced schemes are proposed to achieve both storage efficiency and communication efficiency simultaneously. The proposed schemes significantly outperform the previous basic schemes through a novel integration of several cryptographic techniques.

15.5.1. *The Bloom filter based authentication scheme (BAS)*

System Preparation. The sink generates the public keys for all network users, and constructs the set:

$$S = \{\langle U_{ID_1}, PK_{U_{ID_1}}\rangle, \langle U_{ID_2}, PK_{U_{ID_2}}\rangle, \ldots\},$$

where $\#\{S\} = N$, and $\#\{\}$denotes the cardinality of the set. Using the Bloom filter, the sink can apply k system-wide hash functions (cf. Section 15.2.2) to map the elements of S (each with $L + 2$ bytes, that is, $|U_{ID}| = 2$ bytes, and $|PK_{U_{ID}}| = L$ bytes to an m-bit vector v with $v = v_0 v_1 \cdots v_{m-1}$ where we have $m < N(L = 2)$ to reduce the filter size and $m > kN$ to retain a small probability of a false positive. These k hash functions are known by every node and the sink. For each v_i, $i \in [0, m-1]$, we have

$$vi = \begin{cases} 1, & \text{if } \exists l \in [1,k], j \in [1,N], \\ & \text{s.t. } h_l(U_{ID_j} \| PK - U_{ID_j}) = i. \\ 0, & \text{otherwise} \end{cases}$$

Additionally, the sink constructs a counting Bloom filter v of $m * c$ bits with $\bar{v} = \bar{v}_0 \bar{v}_1 \cdots \bar{v}_{m-1}$, where each $\bar{v}_i$, $i \in [0, m-1]$ is a c-bit counter, i.e. $|\bar{v}_i| = c$ bits. The value of $\bar{v}_i$ is determined as follows:

$$v_i = \#\{(ID_j, PK_{U_{ID_j}}) | h_l(U_{ID_j} \| PK_{U_{ID_j}}) = i, \text{ for } \exists l \in [1,k], j \in [1,N]\}.$$

$c = \lceil \log_2(\max(\bar{v}_i, i \in [0, m-1])) \rceil$ bits, which is usually of 4 bits for most applications.[10] The above operations are illustrated in Figure 15.2 The sink finally preloads each sensor node with v (not including $\bar{v}$), as well as the sink's public key and the common domain parameters of the ECDSA signature scheme.

Message Signing and Authentication. Let $PK_{U_{ID}} = sG$, be the public key of user U_{ID}, where s is the private key of the signer, and G is the generator of a subgroup of an elliptic curve group of order r. Let $S_K(.)$ be a symmetric key cipher such as AES. To broadcast a message $M(|M| \geq 10$ bytes), U_{ID} takes the steps below,[24] a variant of ECDSA with the partial message recovery property:

- Concatenate $\langle M \| tt \| U_{ID}\rangle$, and break it into two parts M_1 and M_2, where $|M| \leq 10$ bytes.
- Generate a random key pair $\{u, V\}$, where $u \in [1, r-1]$, $V = uG = (x_1, y_1)$, and $(x_1 \bmod r) \neq 0$.
- Encode-and-hash V into an integer I.[24]
- Form F_1 from M_1 by adding the proper redundancy.[1]
- Compute$c = (I + F_1) \bmod r$, and make sure that $c \neq 0$ or repeat the above steps otherwise.

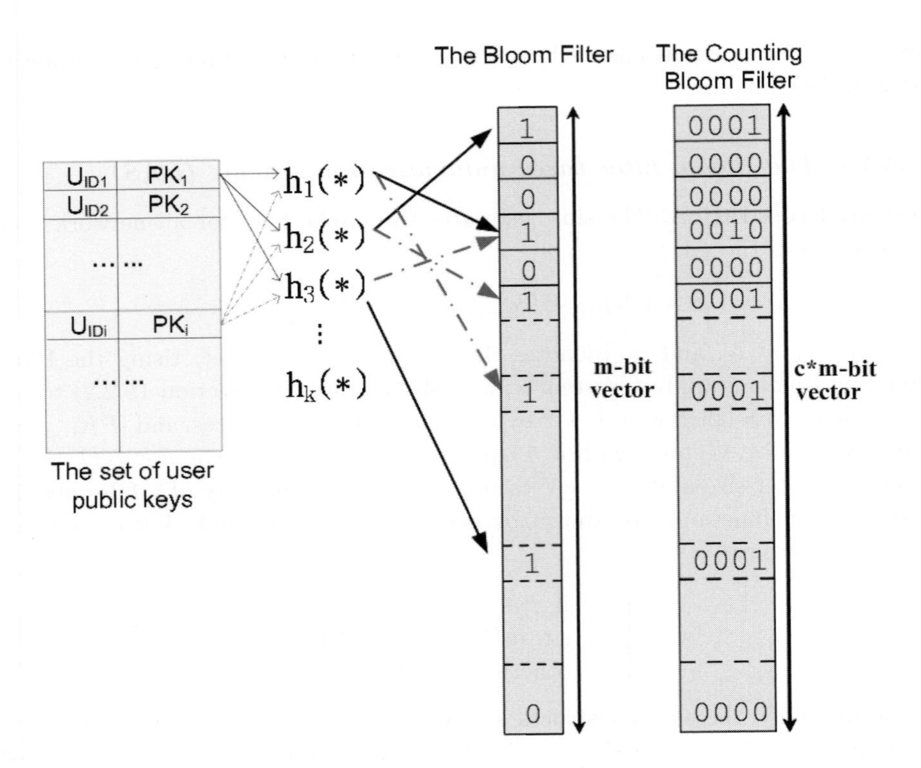

Fig. 15.2. An example of the Bloom filter and Counting Bloom Filter.

- Compute $F_2 = h(M_2)$, and $D = u^{-1}(F_2 + sC) \bmod r$.
- Repeat all the above steps if $D = 0$; output the signature as $\langle C, D\rangle$ otherwise.

Then, U_{ID} broadcasts

$$\langle M_2, C, D, PK_{U_{ID}}\rangle, \tag{15.3}$$

where tt and U_{ID} are parts of M_2. And this is the known simplest message format that can be achieved using PKC.[c] Now, upon receiving a broadcast message (not from the sink), a sensor node checks the authenticity of the message in two steps. First, it checks the authenticity of the corresponding public key by verifying its membership in S. To do so, the sensor node checks whether $v[h_l(U_{ID}\|PK_{U_{ID}})] = 1$, $l \in [1, k]$, and a negative result will lead to the discarding of the message. We note that here a false positive may happen due to the probabilistic nature of the

[c] The claim is true only when ID-based cryptography[33] is excluded from consideration, in which case the user's ID is also his public key. Furthermore, the shortest signature size possibly obtained from pairing is around 22 bytes,[7] which is shorter than 40 bytes obtained from ECDSA. However, to apply a pairing-based scheme (i.e. an ID-based signature or short signature) on sensor nodes, the known reachable signature size has to be 84 bytes, even when a 32 bit microprocessor can be used.[36] And the energy cost is also multiple times higher than that of an ECDSA-160 signature.

Bloom filter, but only with a very small (negligible) probability when appropriate parameters are chosen as we will analyze later. Second, it verifies the attached signature as follows:

- Discard the message if $C \notin [1, r-1]$ or $D \notin [1, r-1]$.
- Compute $F_2 = h(M_2)$, $H = D^{-1} \bmod r$, and $H_1 = F_2 H \bmod r$.
- Compute $H_2 = CH \bmod r$, and $P = H_1 G + H_2 PK_{U_{ID}}$.
- Discard the message if $P = O$.
- Encode-and-hash P into an integer I[24] and compute $F_1 = C - I \bmod r$.
- Discard the message if the redundancy of F_1 is incorrect.
- Otherwise accept M_1 (obtained from F_1) and the signature and reconstruct $M \| tt \| U_{ID} = M_1 \| M_2$.

User Revocation/Addition. To revoke a user, say U_{ID_j}, the sink follows the steps below:

- First, it hashes $h_l(U_{ID_j}) \| PK_{U_{ID_j}}) = i$ and decreases $\bar{v}_i$ by 1. It repeats this operation for all h_l, $l \in [1, k]$.
- From the updated counting Bloom filter $\bar{v}$, the sink obtains the corresponding updated Bloom filter v' with $v' = v'_0 v'_1 \cdots v'_{m-1}$. Here, $v'_i = 1$ only when $\bar{v}_i \geq 1$, and $v'_i = 0$ otherwise.
- The sink further calculates $v_\Delta = v' \oplus v$ and deletes v afterwards. Here $\oplus$ denotes bitwise exclusive OR operation. Obviously, v_Δ is an m-bit vector with at most k bits set to 1. Hence v_Δ can be simply represented by enumerating its 1-valued bits, requiring $\overline{k} \lceil \log_2 m \rceil$ bits for indexing ($\overline{k} \leq k$). This representation is efficient for a small $\overline{k}$ as will be analyzed in Section 6.2.
- The sink finally broadcasts v_Δ after signing it. The message format follows (3) but with the sink's public key omitted, as every sensor already has it.
- Upon receiving and successfully authenticating the broadcast message, every sensor node updates its own Bloom filter accordingly, that is, if $v_{\Delta,i} = 1$ then $v_i = 0$, $i \in [0, m-1]$.

BAS also supports simultaneous multi-user revocation. Suppose that N_{rev} users are revoked simultaneously. The sink follows the same manner to construct v_Δ with $\overline{k}$ bits set 1. Now we have $\overline{k} \leq kN_{rev}$. Furthermore, the compressed message for representing v_Δ now could achieve $mH(p)$ bits theoretically, where $H(p) = -p \log_2 p - (1-p) \log_2 (1-p)$ is the entropy function and $p = (1 - 1/m)^{\overline{k}}$ is the probability of each bit being 0 in v_Δ. As pointed out in Ref. 23, using arithmetic coding technique can efficiently approach this lower bound.

BAS supports dynamic user addition in two ways. First, it enables a later binding of network users and their (ID, public key) pairs. In this approach, the sink may generate more (ID, public key) pairs than needed during system preparation. When a new network user joins the WSN, it will be assigned an unused ID and public key pair by the sink. Second, BAS could add new network users after the revocation of old members. This approach, however, could only add the same number of new users as that of the revoked. This requirement ensures that the probability of a false positive never increases in BAS. To do so, the sink updates its counting Bloom filter by hashing the new user's information into the current Bloom filter. The sink then obtains a v_Δ in the same way as in the revocation case, and broadcasts it after compression. This time, if $v_{\Delta,i} = 1$, sensor nodes will set $v_i = 1, i \in [0, m-1]$ to update their current Bloom filters.

15.5.2. *Minimize the probability of a false positive*

Since the Bloom filter provides probabilistic membership verification only, it is important to make sure that the probability of a false positive is as small as possible.

Theorem 1. *Given the number of network users N and the storage space m bits for a single Bloom filter, the minimum probability of a false positive f that can be achieved is 2^{-k} with $k = \frac{m}{N}\ln 2$, that is,*

$$f = (0.6185)^{\frac{m}{N}}.$$

Proof. since $f = (1 - (1 - \frac{1}{m})^{kN})^k \approx (1 - e^{-kN/m})^k$, we then have $f = e^{k\ln(1-e^{-kN/m})}$. Let $g = k\ln(1 - e^{-kN/m})$. Hence, minimizing f is equivalent to minimizing g with respect to k. We find

$$\frac{dg}{dk} = \ln(1 - e^{-kN/m}) + \frac{kN}{m}\frac{e^{-kN/m}}{1 - e^{-kN/m}}.$$

It is easy to check that the derivative is 0 when $k = \frac{m}{N}\ln 2$. And it is not hard to show that this is a global minimum.[23] Note that in practice, k must be an integer. □

Figure 15.3 shows the probability of a false positive f as a function of $\frac{m}{N}$ i.e. bits per element. We see that f decreases sharply as $\frac{m}{N}$ increases. When $\frac{m}{N}$ increases from 8 to 96 bits, f decreases from $2.1 * 10^{-2}$ to $9.3 * 10^{-21}$. Obviously, f determines the security strength of our design. For example, when $\frac{m}{N} = 92$ bits, the adversary has to generate around $2^{63.8}$ public/private key pairs on average before finding a valid one to pass the Bloom filter. This is almost computationally infeasible, at least within the lifetime of the WSN (usually at most several years). However, when $m/N = 64$ bits, the adversary is now expected to generate around $2^{44.4}$ public/private key pairs before finding a valid pair. The analysis below shows the time and cost of the attack. To generate a public/private key pair in ECDSA-160, a point multiplication operation has to be performed, for which the fastest known implementation speed is 0.21 ms through a specialized FPGA design.[14] Suppose the adversary could afford 100,000 such FPGAs, which would cost no less than one million dollars. Then, by executing 100,000 FPGAs simultaneously, to generate one valid key pair still takes 13.2 hours roughly. With the above analysis, we suggest to select the value of f

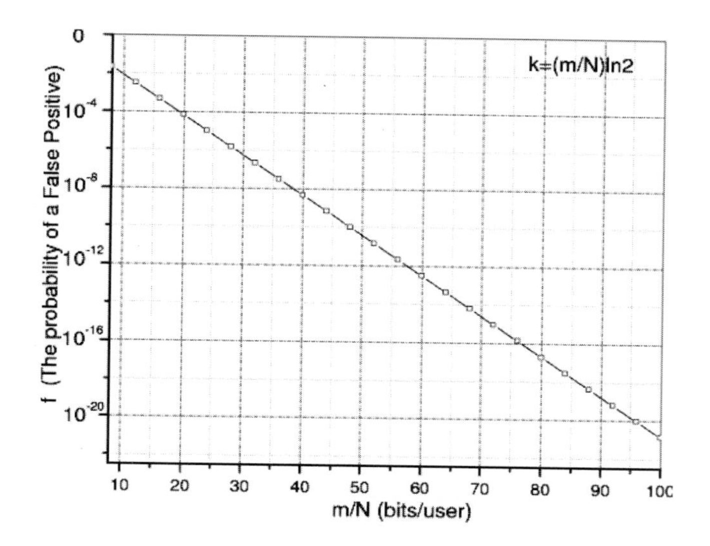

Fig. 15.3. The minimum probability of a false positive regarding $\frac{m}{N}$.

carefully according to the security requirements of the different types of applications. Given a highly security sensitive military application, we suggest that f should be no larger than $6.36 * 10^{-20}$, i.e. $m/N \geq 92$ bits. On the other hand, when the targeted applications are less security sensitive as in the civilian scenario, we can tolerate a larger f. This is because the adversary is now generally much less resourceful as compared to the former case.

15.5.3. *Maximum number of network users supported*

It is important to know how many network users can be supported in BAS so that the WSN can be well planned. The following theorem provides the answer.

Theorem 2. *Given the storage space m bits for a single Bloom filter and the required probability of a false positive $f_{req}(f_{req} \in (0,1))$, the maximum number of network users that can be supported is $\frac{-m(\ln 2)^2}{\ln f_{req}}$, that is,*

$$N \leq \frac{-0.4805m}{\ln f_{req}}.$$

Proof. Since the minimal probability of a false positive $f = 2^{-k}$ is achieved with $k = \frac{m}{N} \ln 2$, we have $f_{req} = 2^{-\frac{m}{N} \ln 2}$. Then, we can easily get $N = \frac{-m(\ln 2)^2}{\ln f_{req}}$ in this case; and this is the maximum number of users that can be supported given f_{req} and m. □

Figure 15.4 illustrates the maximum supported number of network users as a function of the storage limit. Figure 15.4 shows that BAS supports up to 1250 users

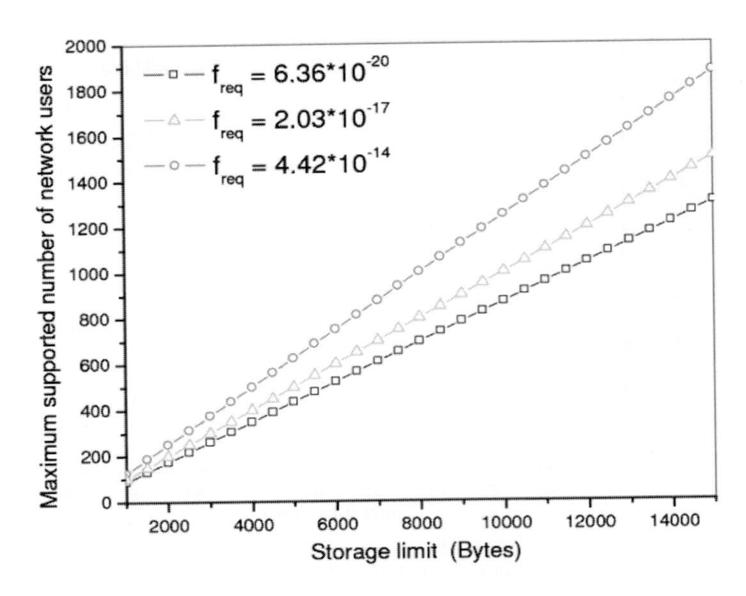

Fig. 15.4. Maximum supported number of network users with respect to storage limit.

when $f_{req} = 4.42 * 10^{-14}$, 1000 users when $f_{req} = 2.03 * 10^{-17}$ and 869 users when $f_{req} = 6.36*10^{-20}$, for a storage space of 9.8 KB. Obviously, BAS also allows tradeoff between the maximum supported number of network users and the probability of a false positive given a fixed storage limit.

15.5.4. *Supporting more users using the Merkle hash tree: the hybrid authentication scheme (HAS)*

Through the above analysis, we know that the maximum supported number of network users is usually limited given the storage limit and the probability of a false positive. For example, if $f_{req} = 6.36 * 10^{-20}$ and the storage limit is 4.9 KB, the maximum number of users supported by BAS is 434. Therefore, an additional mechanism has to be employed to support more users when necessary. HAS achieves this goal by employing the Merkle hash tree technique, which trades the message length for the storage space. That is, by increasing the per message overhead, HAS can support more network users. Specifically, HAS works as follows.

The sink first calculates the maximum number of users supported in case of BAS according to the given storage limit and the desired probability of a false positive. It then collects all the public keys of the current network users and constructs a Merkel hash tree. In fact, the sink constructs N leaves with each leaf corresponding to a current user of the WSN. For our problem, each leaf node contains the binding between the corresponding user ID and his public key, that is, $h(U_{ID}, PK_{U_{ID}})$. The values of the internal nodes are determined by the method introduced in Section 2.3. The sink further prunes the Merkle hash tree into a set of equal-sized smaller trees.

We denote the value of the root node of a small hash tree as h_r^i, $i = 1, \ldots, |S|$, where $|S|$ equals the maximum number of supported users the sink calculates in BAS.

Next, the sink constructs a Bloom filter v following the same way as described in the last section. The difference is that now the member set $S = \{h_r^1, h_r^2, \ldots, h_r^{|S|}\}$. Then, the sink preloads each sensor node with v. At the same time, each user should obtain its AAI according to his corresponding leaf node's location in the smaller Merkle hash tree. Let T denote all the nodes along the path from a leaf node to the root (not including the root), and A be the set of nodes corresponding to the siblings of the nodes in T. Then, AAI further corresponds to the values associated with the nodes in A. Obviously, AAI is of size $(\bar{L} * \log_2 \frac{N}{|S|})$ bytes, where $\bar{L}$ is the length of the hash values. Upon user revocation, the sink simply updates all the sensor nodes with the ID information of the revoked users. And each node directly stores the revoked IDs as described earlier. Now a message sent by a user U_{ID} is of form

$$\langle M_2, C, D, PK_{U_{ID}}, AAI_{U_{ID}} \rangle. \tag{15.4}$$

Each node verifies the authenticity of a user public key in two steps. First, it calculates the corresponding root node value h_r^i using $AAI_{U_{ID}}$ attached in the message. Second, it checks whether or not the calculated h_r^i is a member of v stored by itself. By checking Message (4), we can easily find that HAS doubles the maximum supported number of users as compared to BAS at the cost of 20 more bytes per message overhead, assuming SHA-1 is used.[25] And the number can be further doubled with 40 more bytes per message overhead.

15.6. Performance Analysis

In this section, we analyze the performance of BAS and HAS with respect to communicational and computational overheads (in terms of energy consumption), and security strength. We give a quantitative analysis of the schemes and compare them with the other two basic schemes.

15.6.1. *Communication overhead*

We study how the message size affects the energy consumption in communication in a WSN. We investigate the energy consumption as the function of the size of the WSN (denoted as W). We denote by E_{tr} the hop-wise energy consumption for transmitting and receiving 1 byte. As reported in Ref. 35, a Chipcon CC1000 radio used in Crossbow MICA2DOT motes consumes 28.6 and 59.2 μJ to receive and transmit 1 byte, respectively, at an effective data rate of 12.4 Kb/s. Furthermore, we assume a packet size of 41 bytes, 32 bytes for the payload and 9 bytes for the header.[35] The header, ensuring an 8-byte preamble, consists of source, destination, length, packet ID, CRC, and a control byte.[35] We also assume that $|M| = 20$ bytes.

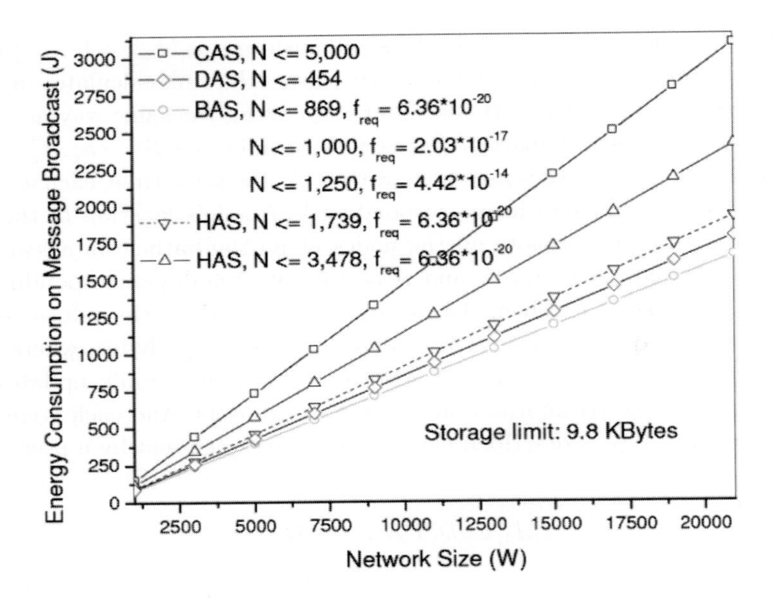

Fig. 15.5. Energy consumption in communication regarding different schemes.

Then, for BAS, the signature size is still the same as that of ECDSA, but only part of the message now has to be transmitted, with the saving of up to 10 bytes. Therefore, the per message overhead of BAS is 54 bytes, which is 10 bytes less than that of DAS. As Message (3) is 74 bytes, there should be 3 packets in total, among which two of them are 41 bytes, and one is 19 bytes. Therefore, there should be $41 * 2 + 19 * 1 + 8 * 3 = 125$ bytes for transmission (including 8-byte preamble per packet). Hence, the hop-wise energy consumption of message transmission is $125 * 59.2\,\mu\text{J} = 7.40\,\text{mJ}$; and the energy consumption of message reception is $125 * 28.6\,\mu\text{J} = 3.58\,\text{mJ}$. For each message broadcast, every sensor node should retransmit the message once and receive ω' times of the same message assuming the blind flooding is *used*[d] Here, ω' denotes node density in terms of the total number of sensor nodes within one unit disc, where a unit disc is a circle area with radius equal to the transmission range of sensor nodes.[e] Hence, the total energy consumption in communication will be $W * (7.4 + 3.58 * \omega')$ mJ.

Figure 15.5 illustrates the energy consumption in communication as a function of W with $\omega' = 20$. Clearly, BAS consumes a much lower energy as compared to others. For example, when $W = 15{,}000$, CAS always costs 2.20 KJ, while BAS costs only 1.18 KJ. The energy saving for a single broadcast can be more than 1000 J

[d] In an idealized lossless network, blind flooding, i.e. every node always retransmits exactly once every unique message it receives, is wasteful, as individual nodes are likely to receive the same broadcast multiple times. In practice, however, blind flooding is a commonly used technique, as its inherent redundancy provides some protection from unreliable (lossy) wireless networks.[21]

[e] We assume a uniform transmission range for all sensor nodes.

between BAS and CAS. Note that although DAS also consumes much less energy than CAS, DAS only supports up to $1000/22 \approx 454$ users. At the same time, BAS can handle 869 users even when $f_{req} = 6.36 * 10^{-20}$ CAS handles more users than BAS and DAS, however, at the cost of much higher energy consumption. Moreover, HAS can handle a large number of users but with a much lower energy consumption when compared to CAS. In summary, BAS demonstrates the highest communication efficiency, as well as desirable storage efficiency. From Figure 15.5, we also find that the energy consumption in communication is the critical cost for WSNs, as a single broadcast of a message of only 20 bytes in length could cost energy on the order of KJ. This also exposes the severe vulnerability of the μTESLA-like schemes, as they allow the adversary to flood the WSN arbitrarily.

15.6.2. *User revocation/addition traffic overhead*

Another important performance metric for the broadcast authentication schemes is the overhead of the user revocation/addition traffic. As analyzed in Section 5.1. BAS requires the sink to broadcast v_Δ upon user revocation/addition. We have shown that in the single user case, v_Δ can be efficiently represented by simply enumerating all its 1-valued bits, the length of which is bounded by $\overline{k}\lceil\log_2 m\rceil$ bits. That is, the per user revocation traffic overhead is upper bounded by $\overline{k}\lceil\log_2 m\rceil$ bits. And the theoretical lower bound obtained from the entropy function is $mH(p)$ bits with $H(p) = -p\log_2 p - (1-p)\log_2(1-p)$ and $p = \left(1 - \frac{1}{m}\right)^k$. It is not hard to see that the expectation value of $\overline{k}$ is around $k/2$, where $k = \frac{m}{N}\ln 2$. Our simulation shows that $\overline{k}$ is always around $k/2$. Hence, for a given $f_{req} = 6.36 * 10^{-20}$, we will have $\overline{k}\lceil\log_2 m\rceil = 68$ bytes, and $mH(p) \approx 52$ bytes, for $N = 1000$. This implies that the per user revocation traffic v_Δ only ranges from 52 to 68 bytes on average for $N = 1000$, depending on the used coding method.[f] And for $N \leq 11{,}000$, v_Δ is at most 80 bytes on average. This overhead is much lower as comparable to that of the μTESLA-like scheme proposed for supporting multiple users.[18] In Ref. 18, the per user revocation traffic (i.e. a revocation certificate) is no less than $1 + \lceil\log_2 N\rceil$ hash values, which is 220 bytes for $N = 1000$, and 300 bytes for $N = 11,000$, assuming the same hash length of 20 bytes. We further note that in contrast to μTESLA-like schemes, BAS does not require periodic key chain update (for running out of available keys) among users and sensor nodes. This is the advantage inherent to the PKC-based schemes.

15.6.3. *Computational overhead*

It was previously widely held that PKC is not suitable in WSNs, as sensor nodes are extremely computation constrained. However, recent studies[9,35] showed that PKC with only software implementations is very viable on sensor nodes. For example, in

[f] We assume that the number of simultaneous network users is always around N.

Ref. 35, an ECC signature verification takes 1.61 seconds with 160-bit keys on ATmega128 8 MHz processor used in a Crossbow mote. We analyze the computation cost of the proposed schemes to further justify the suitability of PKC-based schemes in WSNs. In all our proposed schemes, the major computational cost is due to the signature verification operation. In the following analysis we omit the cost of other operations such as hash operations and table lookup, as they are negligible as compared to the signature verification operation.[35]

In CAS, two ECDSA signature verifications are needed for each broadcast message. In BAS, to verify a message takes $k = \frac{m}{N} \ln 2$ hash operations and one ECDSA signature verification. It was reported in Ref. 35 that an ECDSA-160 signature verification operation costs 45.09 mJ on a 8-bit ATmega128L processor running at 4 MHz. If we assume that the sensor CPU is a low-power high-performance 32-bit Intel PXA255 processor, the energy cost can be further minimized. Note that the PXA255 has been widely used in many sensor products such as Sensoria WINS 3.0 and Crossbow Stargate running at 400 MHz. According to Ref. 13, the typical power consumption of PXA255 in active and idle modes are 411 and 121 mW, respectively. It was reported in Ref. 5 that it takes 92.4 ms to verify an ECDSA-160 signature with the similar parameters on a 32-bit ARM microprocessor at 80 MHz. Therefore, the same computation on PXA255 roughly needs $80/400 \times 92.4 \approx 18.48$ ms, and the energy cost is hence around 7.6 mJ. Therefore, we can obtain the computational costs of the proposed CAS and BAS schemes on different sensor platforms.[g] The results are summarized below.

Scheme	ATmega 128L	PXA 255
CAS	90.18 mJ	15.4 mJ
BAS	45.09 mJ	7.6 mJ

BAS is obviously also more computationally efficient than CAS. Furthermore, when we compare the computational cost with the communication cost on hop-wise message transmission, we can find that both are on the same order, which justifies the suitability of PKC-based schemes in WSNs.

15.6.4. *Security strength*

The Bloom filter based public key verification ensures the security strength of the proposed scheme by enabling immediate message authentication. That is, there is no authentication delay on messages being broadcast. Therefore, it is very hard for the adversary to perform network wide flooding in the WSN. As we analyzed

[g] DAS and HAS consume similar amount of energy as BAS does, as they both require one signature verification.

above, by appropriately choosing a suitable value of f_{req}, such as $6.36 * 10^{-20}$ in military applications, it is infeasible to forge a valid public/private key pair during the lifetime of the WSN. Furthermore, by embedding a time stamp into the message, the message replay attack is also effectively prevented, as WSN is assumed to be loosely synchronized.[28] Therefore, the immediate message authentication capability provided by the proposed schemes can effectively protect the WSN from network wide flooding attacks. This is the most significant security strength over the μTESLA-like schemes, in which network wide flooding attacks are always possible.

Moreover, since the public key operation is expensive, it is also important that sensor nodes can be resistant to the local jamming attacks. Under such attacks, the adversary may simply broadcast random bit strings to the sensor nodes within his transmission range. If these neighbor sensors have to perform the expensive signature verification operation for all received messages, it will be a heavy burden on them. CAS obviously suffers from this type of attacks, as the signature verification operation has to be performed for every received message. However, in both BAS and HAS, such an attack can be effectively mitigated. This is because in both schemes, a sensor node first verifies the authenticity of the attached user public key through hash operations, so it performs signature verification operation for a bogus public key only with a negligible probability (e.g. $6.36 * 10^{-20}$). As reported in Ref. 35, the energy cost of SHA-1 is only $5.9\,\mu$J/byte on a 8-bit ATmega128L processor, while ECDSA-160 could consume 45.09 mJ on signature verification. An adversary may also flood the sensor nodes with forged messages but containing valid user public keys, which can be obtained by eavesdropping the network traffic. In this case, the forged messages can only be discarded after signature verification, and sensor nodes that are physically close to the adversary can thus be abused. We note that this type of attacks is always possible for PKC-based security mechanisms. However, this attack can still be mitigated in BAS by implementing an alert report mechanism. If a sensor node fails to authenticate the received messages multiple times in a row, it will derive that an attack is going on and alert the sink about the attack. The sink further carries out field investigations or other means to detect the adversary and take corresponding remedy actions that are outside the scope of this chapter.

15.7. Further Enhancements

15.7.1. *Dealing with long messages*

The messages broadcast in WSNs are usually short, due to the application specific nature of WSNs. The query or command messages can be less than 100 bytes. However, there are few cases that long messages may be required to be broadcast in WSNs. For example, the sink may broadcast code images to the sensor nodes for the purpose of retasking WSNs.[37] The size of such code images can be on the order of KB. In this case, it is not desirable to apply the proposed BAS or HAS scheme directly by signing the whole message (i.e. the message hash) only once or signing on every single packet otherwise. This is because of two reasons. First, if we sign the whole message once, then each sensor node can authenticate a message only after it obtains the entire message. That is, the sensor nodes have to buffer a large number of received packets before it can authenticate them. This obviously introduces a severe vulnerability that could result in message flooding attacks. Second, if we sign every packet belonging to the same message, the scheme overheads will increase significantly with respect to both computation and communication. This is because now every packet is attached with a signature, which is 40 bytes in our setting.

Fortunately, several solutions were proposed to solve this problem in the context of code update in WSNs.[38,39] The first solution is suitable for lossless network environments, which employs off-line hash chain technique to amortize the cost of a single digital signature over multiple packets and allow for incremental message authentication and packet pipelining.[38,40] The second solution is aimed at tolerating packet losses. This solution makes use of a signed hash tree technique and trades message overhead for potential packet losses.[39] Both solutions can be directly superimposed with BAS and HAS in dealing with long messages. We omit the details of these solutions for space interest.

15.7.2. *Reducing the probability of a false positive*

In Ref. 41, a method is introduced to use two families of k hash functions, instead of using one. And an element is in the set if either family gives back all 1s from the filter. The trick is to choose one of the two families of the hash functions adaptively: choose which family of hash functions to use for each element of your set in such a way to keep the number of 1s in the filter as small as possible. In such a way, a smaller false positive probability in the same space can be achieved at the cost of more hashing. This method can reduce the probability of a false positive to the half under certain conditions using the same storage space. This technique can be exploited by BAS so that we achieve a desirable probability of a false positive with a smaller storage space.

15.7.3. *Optimization on constructing the Merkle hash tree*

Different types of network users may have different broadcast frequencies in practice. This fact can be exploited by HAS, when supporting a vast number of network users is a must. Instead of pruning the user Merkle hash tree into a set of equal-sized smaller trees, now the tree can be trimmed into the same number but different-sized smaller ones based on user broadcast frequency. The higher the frequency is, the smaller hash tree the user is grouped in. In such a way, the energy efficiency can be improved in the overall sense, as more messages being broadcast containing only smaller AAI sizes. This is similar to the idea introduced in Ref. 9.

15.7.4. *Using "fast forward, slow react" to prevent public key forgery attacks*

Since a false positive is still possible though small, the adversary can forge different user public/private key pairs and seek to pass the membership test. In this way, the adversary could possibly pretend to be a valid network user and has its messages authenticated. The sensor nodes, however, will not be able to distinguish the bogus public keys, once a successful guess is made.

On the other hand, a bogus public key can always be deterministically detected by the sink, as the sink keeps the copies of all the user public keys. Furthermore, as the sink is always connected to the WSN as assumed, it also receives the broadcast messages sent by the users. Thus, the sink can always detect a bogus public key that cannot be detected by the sensor nodes through analyzing the received messages. The following "fast forward, slow react" policy is designed to leverage this fact and does not affect message propagation efficiency. In "fast forward, slow react" policy, each sensor node estimates the round trip time between itself and the sink. We denote this time as ΔU_{ID}. The estimation of ΔU_{ID} can be obtained from either direct location-based calculation or the previous interactions between the node and the sink. If there are multiple sinks, ΔU_{ID} is estimated for the closest one. Then, upon successfully authenticating a received message, each sensor node waits for ΔU_{ID} additional time to take the further reactions ("slow react") except for forwarding the message ("fast forward"). Suppose the message is a query. If node U_{ID} has the (partial) answer to this query, it delay replying the user with the answer ΔU_{ID} time. When ΔU_{ID} is timeout and there's no warning received from the sink, U_{ID} now sends out the answer. And if the sink does broadcast a revocation message, U_{ID} will be able obtain it before any further action is taken. Note that no matter whether U_{ID} is an intended recipient, it always forwards the message without delay, as long as the message is successfully authenticated by itself. Therefore, a public key forgery attack will only result in one successful message broadcast, given the "fast forward, slow react" policy enforced. And we note that the "fast forward, slow react" policy can be implemented in an on-demand manner. That is, only when the WSN is under attack will sensor nodes implement this policy. The sink can be used to control the starting and ending time.

15.8. Challenges for Future Research

There are many challenges regarding multi-user broadcast authentication in WSNs. One of the foremost important issues is to further reduce the computational overhead of PKCs when applied on resource constrained sensor nodes. Research along this direction can be two-fold: one is to adapt and test more efficient state-of-the-art PKC algorithms on sensor nodes, which also includes design new approaches to allow trade-offs between security strength and computational complexity. The other is to use hardware-software co-design for speeding up PKC computations on sensors. As sensor nodes are usually specially purposed according to the desired application, its functionality can be predetermined, which allows specific hardware design optimized both for the application and security related algorithms. Other challenges include better protocol design according to different user query patterns and data storage mechanisms. It is also useful to make use of the potential heterogeneity among the sensor nodes to design more resilient and efficient mechanisms for broadcast authentication.

15.9. Conclusion

In this chapter, we studied the problem of multi-user broadcast authentication in WSNs. We pointed out that symmetric-key-based solutions such as μTESLA are insufficient for this problem by identifying a serious security vulnerability inherent to these schemes: the delayed authentication of the messages can easily lead to severe energy-depletion DoS attacks. We then came up with several effective PKC-based schemes to address the problem. Both computational and communication costs of the schemes are minimized through a novel integration of several cryptographic techniques. A quantitative energy consumption analysis, as well security strength analysis were further given in detail, demonstrating the effectiveness and efficiency of the proposed schemes.

Problems

1. What is a wireless sensor network?
2. What is a digital signature?
3. What is a Bloom filter?
4. What is a Merkle hash tree?
5. How is the False Positive rate of a Bloom filter defined?
6. How are broadcast messages authenticated according to the solutions proposed in this chapter?
7. What is the security issue of μTESLA scheme when applied in large scale multi-hop wireless sensor networks?
8. How to trade off between storage overhead and the false positive rate of the Bloom filter in this chapter?
9. What is a partial message recovery digital signature?
10. How does partial message recovery signature technique help improve communication efficiency in this chapter?
11. How is user revocation operation performed in this chapter?

Bibliography

1. IEEE P1363a Standard, Standard specifications for public key cryptography, http://grouper.ieee.org/groups/1363/index.html (2000).
2. I. Akyildiz, W. Su, Y. Sankarasubramaniam and E. Cayirci, A survey on sensor networks, *IEEE Communications Magazine* **40**(8) (2002) 102–116.
3. I. Akyildiz and I. Kasimoglu, Wireless sensor and actor networks: research challenges, *Ad Hoc Networks* **2**(4) (2004) 351–367.
4. R. Amirtharajah and A. Chandrakasan, Self-powered signal processing using vibration-based power generation, *IEEE Journal of Solid-State Circuits* **33** (1998) 687–695.
5. M. Aydos, T. Yanik and C. Koc. An high-speed ECC-based wireless authentication protocol on an ARM microprocessor, *Proceedings of ACSAC*, New Orleans, LA (2000) 401–409.
6. J. Baek, J. Newmarch, R. Naini and W. Susilo, A survey of identity-based cryptography, *AUUG 2004* (2004) 95–102.
7. D. Boneh, H. Shacham and B. Lynn, Short signatures from the Weil pairing, *Journal of Cryptology* **17**(4) (2004) 297–319.
8. Crossbow Technology Inc., Wireless sensor networks, http://www.xbow.com/ (2004).
9. W. Du, R. Wang and P. Ning, An efficient scheme for authenticating public keys in sensor networks, *Proceedings of MobiHoc'05* (25–28 May, 2005) 58–67.
10. L. Fan, P. Cao, J. Almeida and A. Z. Broder, Summary cache: a scalable wide-area web cache sharing protocol, *IEEE/ACM Transactions on Networking* **8**(3) (2000) 281–293.
11. D. Hankerson, A. Menezes and S. Vanstone, *Guide to Elliptic Curve Cryptography*, ISBN 0-387-95273-X (2004).
12. Y. Hu, A. Perrig and D. Johnson, Packet leashes: a defense against wormhole attacks in wireless ad hoc networks, *Proceedings of INFOCOM* (2003).
13. Intel PXA255 Processor Electrical, Mechanical, and Thermal Specification, http://www.intel.com/design/pca/applications processors/manuals/278780.htm.
14. T. Itoh and S. Tsujii, A fast algorithm for computing multiplicative inverse in GF (2^m) using normal bases, *Information and Communication* **78** (1988) 171–177.
15. A. Kansal, D. Potter and M. Srivastava, Performance aware tasking for environmentally powered sensor networks, *Proceedings of ACM SIGMETRICS'04* (2004).
16. A. Kansal and M. Srivastava, An environmental energy harvesting framework for sensor networks, *ACM/IEEE ISLPED* (2003).
17. D. Liu and P. Ning, Multi-level mTESLA: broadcast authentication for distributed sensor networks, *ACM Transactions in Embedded Computing Systems (TECS)* **3**(4) (2004).
18. D. Liu, P. Ning, S. Zhu and S. Jajodia, Practical broadcast authentication in sensor networks, *Proceedings of MobiQuitous 2005* (July 2005).
19. K. Lorincz, D. Malan, T. Fulford-Jones, A. Nawoj, A. Clavel, V. Shnayder, G. Mainland, S. Moulton and M. Welsh, Sensor networks for emergency response: challenges and opportunities, *IEEE Pervasive Computing* (2004).
20. C. Lu, G. Xing, O. Chipara, C. Fok and S. Bhattacharya, A spatiotemporal query service for mobile users in sensor networks, *Proceedings of ICDCS*, Columbus (2005).
21. J. McCune, E. Shi, A. Perrig and M. Reiter, Detection of denial-of-message attacks on sensor network broadcasts, *IEEE Symposium on Security and Privacy* (2005) 64–78.
22. R. Merkle, Protocols for public key cryptosystems, *Proceedings of the IEEE Symposium on Research in Security and Privacy* (1980).

23. M. Mitzenmacher, Compressed Bloom filters, *IEEE/ACM Transactions on Networks* **10**(5) (2002) 613–620.
24. D. Naccache and J. Stern, Signing on a postcard, *Proceedings of Financial Cryptography'00, Lecture Notes in Computer Science*, Vol. 1962 (2000) 121–135.
25. NIST, Digital hash standard, *Federal Information Processing Standards Publication* **180**(1) (1995).
26. NIST, *Proposed Federal Information Processing Standard for Digital Signature Standard (DSS), Federal Register* **56**(169) (1991) 42980–42982.
27. A. Perrig, R. Szewczyk, V. Wen, D. Culler and D. Tygar, SPINS: security protocols for sensor networks, *Proceedings of MobiCom'01* (2001).
28. A. Perrig, R. Canetti, J. Tygar and D. Song, The TESLA broadcast authentication protocol, *RSA CryptoBytes* **5** (2002).
29. K. Ren, K. Zeng, W. Lou and P. Moran, On broadcast authentication in wireless sensor networks, *IEEE Transactions on Wireless Communications* **6**(11) (2007) 4136–4144.
30. K. Ren, W. Lou and Y. Zhang, Multi-user broadcast authentication in wireless sensor networks, *Proceedings of IEEE SECON 2007*, San Diego (2007).
31. K. Ren, W. Lou and Y. Zhang, LEDS: providing location-aware end-to-end data security in wireless sensor networks, *Proceedings of IEEE INFOCOM*, Barcelona, Spain (23–29 April, 2006).
32. R. Rivest, A. Shamir and L. Adleman, A method for obtaining digital signatures and public-key cryptosystems, *Communication ACM* **21**(2) (1978) 120–126.
33. A. Shamir, Identity based cryptosystems and signature schemes, *Proceedings of CRYPTO'84, Lecture Notes in Computer Science*, Vol. 196 (1984) 4753.
34. Texas Instruments Inc., *MSP430 Family of Ultra-lowpower 16-bit RISC Processors*, http://www.ti.com.
35. A. Wander, N. Gura, H. Eberle, V. Gupta and S. Shantz. Energy analysis of public-key cryptography on small wireless devices, *IEEE PerCom* (2005).
36. Y. Zhang, W. Liu, W. Lou and Y. Fang, Location based security mechanisms in wireless sensor networks, *IEEE JSAC, Special Issue on Security in Wireless Ad Hoc Networks* **24**(2) (2006) 247–260.
37. J. Hui and D. Culler, The dynamic behavior of a data dissemination protocol for network programming at scale, *Proceedings of ACM SenSys'04*, Baltimore (2004).
38. P. E. Lanigan, R. Gandhi and P. Narasimhan, Secure dissemination of code updates in sensor networks, *Proceedings of ACM SenSys'05* (2005).
39. J. Deng, R. Han and S. Mishra, Secure code distribution in dynamically programmable wireless sensor networks, *Proceedings of ACM/IEEE IPSN* (2006) 292–300.
40. R. Gennaro and P. Rohatgi, How to sign digital streams, *Information and Computation* **165**(1) (2001) 100–116.
41. S. Lumetta and M. Mitzenmacher, *Using the Power of Two Choices to Improve Bloom Filters*, preprint.

Chapter 16

Security Attacks and Challenges in Wireless Sensor Networks

Al-Sakib Khan Pathan* and Choong Seon Hong†

Department of Computer Engineering, Kyung Hee University
1 Seocheon, Giheung, Yongin, 449701 Gyeonggi, South Korea
**spathan@networking.khu.ac.kr*
†cshong@khu.ac.kr

With the advancements of networking technologies and miniaturization of electronic devices, wireless sensor networks (WSN) have become an emerging area of research in academic, industrial, and defense sectors. Sensors combined with low power processors and wireless radios will see widespread adoption in the new future for a variety of applications including battlefield, hazardous area, and structural health monitoring. However, many issues need to be solved before the full-scale implementations are practical. Among the research issues in WSN, security is one of the most challenging. Securing WSN is challenging because of the limited resources of the sensors participating in the network. Moreover, the reliance on wireless communication technology opens the door for various types of security threats and attacks. Considering the special features of this type of network, in this chapter we address the critical security issues in wireless sensor networks. We discuss cryptography, steganography, and other basics of network security and their applicability to WSN. We explore various types of threats and attacks against wireless sensor networks, possible countermeasures, and notable WSN security research. We also introduce the holistic view of security and future trends for research in wireless sensor network security.

Briefly, in this chapter we will present the following topics:

- Basics of security in wireless sensor networks.
- Feasibility of applying various security approaches in WSN.
- Threats and attacks against wireless sensor networks.
- Key management issues.
- Secure routing in WSN.
- Holistic view of security in WSN.
- Future research issues and challenges.

Keywords: Attacks; constrain; energy; key; security; sensor; threat.

16.1. Introduction

Wireless Sensor Networks (WSNs) offer a unique way of extracting data from hazardous geographical regions where human intervention is extremely difficult, the network is often unattended, and where a specified level of security has to be maintained for each step of the network's operation. Among all varieties of wireless networks, WSNs are the type of networks that demand high-level security as one of their core features. In practical terms, a WSN is considered a class of ad hoc networks which can form whenever needed, sometimes without a fixed infra-structure. We define a sensor network as a network consisting of a set of small sensor devices that are deployed in an ad hoc fashion to cooperate with each other for sensing certain physical phenomenon. Typically a WSN has one or more base stations (sometimes called as *sink*) and a large number of sensing devices.

Various issues in WSNs are still under investigation and many of them have yet to reach desired standards. Over the past few decades, with the advancements of ad hoc networking technologies, the research on WSNs has also been benefited. However, because of the differences in the nature of the works and the constrained resources of WSNs, many solutions that are devised for traditional ad hoc networks will not work for WSNs.

Security in wireless sensor network has a great number of challenges, ranging from the nature of wireless communications, constrained resources of the sensors, unknown topologies of the deployed networks, unattended environment where sensors might be susceptible to physical attacks, dense and large networks, etc.[1,2] In fact, each of these issues leads to different research direction. Whenever we think about any feasible security scheme for WSNs, we focus on a specific aspect and often ignore the other associated threats. It is in reality impossible to deal with all the security threats with a single mechanism. Hence, the approach is often to choose the most appropriate mechanism, based on the situation at hand and the settings of the network.

From the high-level point of view, we consider the following six principles while considering security for any system. These are collectively known as the *philosophy of mistrust*:

- Don't talk to any one you don't know.
- Accept nothing without a guarantee.
- Take everyone as an enemy until proved otherwise.
- Don't trust your friend for long.
- Use well-tried solutions.
- Watch the ground you are standing on for cracks.

Maintaining all these principles at the same time requires a lot of computational, memory, and energy resources which are not available in wireless sensor networks. Many security solutions that are well-established for other wireless networks are often not fit for direct use in WSNs and any security solution that needs periodic renewal of any security component (e.g. secret key, secret hash value, session key, etc.) might not at all be viable because of the energy constraints. Considering these factors, devising efficient security mechanisms for WSNs is a challenging issue.

In this chapter, we present a detailed review of security in wireless sensor network considering all the challenges, prospects, and the futuristic views. As we will mainly focus on the issue of security in WSN, other aspects of sensor networks will not be discussed. Interested readers are advised to review[3] for a good survey on the basics of wireless sensor networks.

We have started this chapter with a brief introduction of wireless sensor network, its characteristics, and the major challenges that it faces to develop efficient security solutions. In the rest of this chapter, we will first discuss the key aspects to consider for WSN security, various security approaches, and their applicability to WSN, major threats and attacks against WSNs, their detection, prevention, and attack countermeasures, key management issues, and secure routing issues. Finally, we will present a holistic view of security and what we expect in the future for research on WSN security.

16.2. Background

Before an in-depth investigation of the security threats and attacks in wireless sensor networks, let us first look at the major aspects that make the issue of maintaining security difficult for wireless sensor networks.

16.2.1. *Key aspects to consider for WSN security*

16.2.1.1. *Constrained resources of sensors*

The sensors that build up the network are usually of inadequate memory, processing, and communication capabilities and cannot support the execution of a large amount of code. Their energy sources are also very limited. As an example, Crossbow MICA2 mote[4] is a well-known sensor node with an ATmega128L 8-bit processor at 8 MHz, 128 KB program memory (flash), 512 KB additional data flash memory, 433, 868/916, or 310 MHz multi-channel radio transceiver, 38.4 kpbs radio, 500–1000 feet outdoor range (depending on versions) with a size of only $58 \times 32 \times 7$ mm. Usually it is run by TinyOS operating system and powered by 2 AA sized batteries. Clearly a device with this configuration cannot support security mechanisms that require executing a large amount of instructions. In addition, a sensor network usually contains a large number of sensor nodes. The number of sensors in the network might directly affect the use of memory space of nodes participating in the network, because often they store pre-distributed secret keys, keying information, or the codes to calculate pairwise secret keys between nodes in the network. Node failure is another problem that could also affect the network severely. If a node is busy relatively longer than other nodes in the network (e.g. performing huge calculations related to security), it might lose its energy rapidly and can fail much sooner than the other less active nodes.

16.2.1.2. *Nature of work of WSN*

Many applications of wireless sensor networks require deployment of sensors in remote, unattended, hostile, or hazardous areas. The sensors are often exposed to various types of adversaries and could be attacked physically. Even if they are deployed over a field, a passing vehicle can run over and physically damage them. An adversary can physically search and destroy the nodes.[5] Environmental conditions might also affect the performance of the sensors or can cause physical damage. All these unintentional or intentional events that can cause physical damage to a sensor are considered physical security issues. Sometimes physical attacks (like the capture or destruction of nodes) can cause several types of logical security attacks. A good deployment or management policy, tamper-proofing mechanisms of the physical package of the sensors, camouflaging, protective shields, or other available techniques[6] could be used for dealing with physical security threats in wireless sensor network. More discussions on these issues will be provided later in this chapter.

16.2.1.3. *Use of wireless communications*

Wireless technology is used for communications in a wireless sensor network. As with any other wireless network, it is also prone to various types of threats related to the unreliable nature of wireless links like: undelivered packets, collisions of packets, latency, etc. Because of the broadcast nature of wireless channels, any adversary can even eavesdrop or passively listen to the transmissions of any legitimate node. In case of wired communication, the guided media would be well-protected by using various means and usually the end devices come with sufficient protective mechanisms. On the contrary, in wireless communication, because of its unguided medium and open nature, many types of attacks could be launched. In fact, many of the security threats in WSN exist because of the use of wireless technology for communications among the nodes.

16.2.2. *Feasibility of different security approaches in WSN*

Security is a broadly used term encompassing the characteristics of authentication, integrity, privacy, non-repudiation, and anti-playback.[7] Over the past few decades, the more the dependency on network-provided information has increased, the more the risk has increased for secure transmission of information over the networks. To ensure various aspects of security (i.e. authenticity, integrity, privacy, etc.), diverse approaches like cryptography, steganography, physical layer security, etc. are used. In this section, we will examine which of the major security approaches can be viable for wireless sensor networks.

16.2.2.1. *Cryptography*

Most of the encryption-decryption techniques devised for traditional wired networks are not fit for direct use in wireless networks. As mentioned previously, WSNs consist of tiny low-cost devices which possess scarce processing, memory, and battery power resources. Applying any kind of encryption scheme requires transmission of extra bits, and thus requires extra processing, memory, and battery power which can impact the network's longevity. Encryption and decryption operations can also increase delay, jitter, and packet loss in wireless sensor networks. Moreover, critical questions arise when applying an encryption-decryption scheme to WSN like: How the keys should be generated? How the keys should be disseminated? How the keys should be managed? What is the procedure to revoke the keys? How the keys could be assigned to a newly added sensor? As minimal (or no) human interaction is one of the fundamental features of WSN, it is also a crucial point to decide how the keys could be modified/refreshed from time to time for encryption. Adoption of pre-loaded keys or embedded keys might always not be the best solution. Overall, schemes that are based on cryptographic techniques must be lightweight so sensors can support them along with other programs, which are running and sharing the same resources.

16.2.2.2. *Steganography*

While cryptography aims at hiding the content of a message, steganography[8,9] aims at hiding the existence of the message. Steganography is the art of covert communication by embedding a message into the multi-media data (image, sound, video, etc.).[10] The main objective of steganography is to modify the carrier in a way so that it is not perceptible and hence, looks ordinary. It hides the existence of the covert channel, and furthermore, if we want to send a secret data without sender information or want to distribute secret data publicly, it is very useful. However, securing wireless sensor networks is not directly related to steganography. Processing multi-media data (like audio, video) with the inadequate resources of the sensors is difficult. Steganography in WSNs remains as an open research issue that will not be solved until the sensors acquire enough capabilities to support extensive computations associated with it.

16.2.2.3. *Physical layer secure access*

Physical layer secure access in wireless sensor networks could be provided by using frequency hopping. A dynamic combination of the parameters like hopping set (available frequencies for hopping), dwell time (time interval per hop), and hopping pattern (the sequence in which the frequencies from the available hopping set is used) could be used with a little expense of memory, processing, and energy resources. Important point in physical layer secure access is the efficient design so that the hopping sequence is modified in less time than is required to discover it. One drawback for employing this is that both the sender and receiver should maintain a synchronized clock, therefore time synchronization in WSN[11] is another important research issue.

Considering all the basic security approaches, lightweight cryptography, logical or algorithmic schemes could be the best choice for WSN security. We must keep in mind that the higher the level of security of a WSN, the higher the amount of resources needed to support it.

16.3. Security Issues in Wireless Sensor Networks

Let us now investigate the security threats and attacks in wireless sensor networks. We can consider several factors for categorizing the attacks like — the approach of attack, target of the attack, position of attacker, role of attacker, etc. Overall, we can classify all of the known attacks into three basic types:

Type I

Attacks on the Basic Mechanism (e.g. attacks against routing in the network). *Attacks on the Security Mechanisms* (e.g. against cryptographic scheme or against key management scheme).

Type II

Passive Attack — It typically means eavesdropping of data. In this case, the attacker passively listens to the transmitted data in the network and can use the collected information later for launching other types of attacks.

Active Attack — It means any type of direct attack caused by an adversary. The attacker actively participates in the collection, modification, and fabrication of data. Sometimes, the information collected by passive attacks can be used for active attacks.

Type III

External Attack — In an external attack, an outsider is involved. These attacks can cause denial of service (DoS) situation, congestion, propagation of wrong routing information, etc. Typically external attacks can be resisted using firewalls, encryption mechanisms, good security management policy, and other techniques.

Internal Attack — An internal attack sometimes could be very harmful for the network as any node within the network works as an attacker in this case. Internal attack is performed by compromising node(s) in the network. Compromising a node means convincing a legitimate node to help the attacker or persuading a node in the network to work on behalf of the attacking entity. Often it is difficult to detect an internal attacker within the network that has a legitimate identity. Various kinds of authentication schemes, intrusion detection schemes, or membership verification schemes can be used for preventing internal attacks.

Other than these basic categories of attacks, several attacks are given formal names. Here, we will discuss all the known attacks in WSN with their major features and possible defense mechanisms.

16.3.1. *Denial of service (DoS) attack*

Strictly speaking, we consider any kind of attempt of an adversary to disrupt, subvert, or destroy the network as a Denial of Service (DoS) attack. In reality, any kind of incident that diminishes, eliminates, or hinders the normal activities of the network can cause a DoS situation. Some examples include hardware failures, software bugs, resource exhaustion, environmental conditions, or any type of complicated interaction of these factors. Note that, *DoS* (Denial of Service) is basically a given formal name of a particular condition of the network but when it occurs as a result of an intentional attempt of an adversary, it is called *DoS attack*. In general, "*Denial of Service* (*DoS*)" is an umbrella term that can indicate many kinds of events in the network in which legitimate nodes are deprived of getting of expected services for some reasons (intentional attempts or unintentional incidents).

DoS attacks can mainly be categorized into three types:

(1) Consumption of scarce, limited, or non-renewable resources.
(2) Destruction or alteration of configuration information.
(3) Physical destruction or alteration of network resources.

Among these types of DoS attacks, the first one is the most significant for wireless sensor networks as the sensors in the network suffer from the lack of resources. Other than these basic types, categorization of DoS attacks can be done according to the layers of the network structure.[12] An attacker can choose different targets at different layers to stop proper functioning of legitimate nodes so that they cannot get the services they are entitled to. Though it is quite difficult to know whether any particular DoS situation is caused intentionally or unintentionally, there are some common prevention and detection methods for each of the DoS attacks.

Let us now have a look at the DoS attacks in wireless sensor networks by layer:

16.3.1.1. *DoS attacks in physical layer*

Jamming — Jamming means the deliberate interference with radio reception to deny a target's use of a communication channel. For single-frequency networks, it is simple and effective, causing the jammed node unable to communicate or coordinate with others in the network. Due to their very nature, wireless sensor networks are probably the category of wireless networks most vulnerable to "radio channel jamming"-based Denial of Service (DoS) attacks.[13] Mainly two types of jamming could be possible; constant and sporadic. In case of constant jamming, attacker interferes with the signals of a legitimate node continuously for a certain period of time while in case of sporadic jamming, the attacker intermittently causes jamming. Sporadic jamming in the network is often more difficult to detect than detecting constant jamming. Some solutions to deal with jamming in WSN are proposed in Refs. 13, 14 and 15.

Tampering — Due to the unattended feature of wireless sensor networks, an attacker can physically damage/replace sensors, parts of computational and sensitive hardware, even can extract cryptographic keys to gain unrestricted access to higher communication layers. Tampering is actually any type of physical attack on sensors in the network. Success in tampering depends on:

- how accurately and efficiently the designer considered the potential threats at design time,
- resources available for design, construction, and test,
- attacker's cleverness and determination.

16.3.1.2. *DoS attacks in link layer*

Collision — Adversaries may only need to induce a collision in one octet of a transmission to disrupt even a relatively longer packet. As the resources of the sensors are scarce, such loss could be significant in many cases. Also it is a great hurdle for acquiring timely and accurate data from the sensors. Unfortunately, in wireless networks, detection of a collision with a node's own transmission is difficult. Standard collision avoidance mechanisms also cannot help as they are cooperative by nature. An attacker simply can ignore the avoidance protocol and transmit at the same time as the victim. One possible solution could be the use of error correction codes (ECC) but with the use of ECC, more processing and communication overheads are incurred.

Exhaustion — Battery exhaustion attack could be launched with repeated requests for using the channel. A naive link layer implementation could be a target for this type of attack. Feasible defense mechanisms against battery exhaustion caused by repeated transmissions could be the use of time division multiple access (TDMA) or rate limitation. Additional logic could also be developed to help these mechanisms.

Unfairness — Unfairness is a weaker form of DoS attack. This threat may not entirely prevent legitimate access to the channel, but could degrade service for real time MAC protocols. In fact, ensuring fairness in WSN is often viewed as a separate research issue. Use of small frames might be helpful in this case. However this would also incur some framing overhead.

16.3.1.3. *DoS attacks in network layer*

Neglect and Greed — If a node drops packets or denies transmitting legitimate packets or if a node is very greedy to give undue priority to its own messages, these could be considered as "*neglect and greed*". Dynamic Source Routing (DSR) protocol or the protocols that are based on DSR are especially vulnerable to this type of attack. Use of multi-path routing or redundant message transmission could be the solutions for handling such attacks. However, for WSNs these solutions might not be feasible. Instead, use of some other routing mechanisms could help.

Homing — Sometimes in wireless sensor networks, some nodes are given some special responsibilities like managing cryptographic keys, making use of acquired data, maintaining a local group, etc. Often the adversaries are attracted to these leader nodes and try to eavesdrop on their activities. In case of homing attack, the adversaries try to hamper the normal functioning of such types of leader nodes within a WSN. Homing attack is especially dangerous for the location-aware routing protocols which rely on geographic information. Different types of cryptographic schemes, algorithms, hiding management messages, etc. could be used for preventing homing attack.

Misdirection — Misdirection means simply directing the legitimate packets to the wrong path. A malicious insider can cause misdirection of traffic. Egress filtering (in hierarchical routing protocols), authorization and monitoring, or any kind of intrusion detection scheme (IDS)[16] could be used to prevent this type of DoS attack.

Blackhole — Blackhole (or Sinkhole) attack itself is one of the major attacks in WSN. We will discuss this attack in detail later in this chapter. However, when this attack causes any sort of *denial of service* in the network, it is considered as a DoS attack in network layer.

16.3.1.4. *Transport layer DoS attacks*

Flooding — Protocols which must keep the states of both end-nodes are particularly vulnerable to this attack. It aims at memory exhaustion of the nodes by flooding of a great number of packets. Client puzzles or traceback mechanisms could be used to deal with such type of DoS attack.

Desynchronization — This attack means forging of packets during transmission. Existing connection between two endpoints could be effectively disrupted by desynchronization. Any kind of authentication mechanism for the packets could be used to handle desynchronization attack.

Other than these attacks, many other individually considered attacks like wormhole attack, hello flood attack, Sybil attack etc. can also cause *denial of service* situation in the network. In fact, many of the methods of attacking and targets of attacks overlap with each other, but considering different circumstances, are given different tags and names. It should be clear that, any sort of intentional attempt that causes any sort of *denial of service* situation in the network is considered as *DoS attack*. As we will examine all other attacks in the rest of the chapter, here we conclude this section with the names of the major types of DoS attacks only.

16.3.2. *Attacks on information in transit*

The basic task of a sensor is to monitor the changes of some specific parameters (like temperature, sound, magnetism, light level, etc.) and to report those to the base station. The readings from the sensors could be transmitted using various methods. While in transit, the packets may be altered, spoofed, or vanished on the way (this type of attack could also be considered as network layer DoS attack when it resists a valid node from getting its expected service). As wireless communication is susceptible to eavesdropping, any attacker can monitor the traffic flow and get into action to interrupt, intercept, modify, or fabricate packets. If the routing method does not have proper security measures, wrong information even can reach up to the base station and thus can influence the decision taken by the central authority. Such an event might be extremely dangerous for a military reconnaissance scenario which could lead to taking disastrous military decisions. As sensor nodes typically have short range of transmission and scarce resources, an attacker with adequate processing power and larger communication range can attack several sensors at the same time to modify the actual information during transmission. Among several works, a good approach to tackle this and to filter out falsely injected data in sensor networks is presented in Ref. 88.

16.3.3. *Sybil attack*

Sometimes the sensors in a wireless sensor network might need to work together to accomplish a task, hence the management policy of the network can use distribution of subtasks and redundancy of information. In such a situation, a node can pretend to be more than one node at the same time using the identities of other legitimate nodes. This type of attack is called a *Sybil attack*.[17] The malicious device's additional identities are called the *Sybil nodes*. Sybil attack tries to degrade the integrity of data, level of security, and resource utilization that a distributed algorithm targets to achieve. This type of attack can be performed for downgrading the performances of distributed storage, routing mechanism, data aggregation, voting, fair resource allocation, and misbehavior detection mechanisms. A conceptual view of Sybil attack is shown in Figure 16.1. Basically, any peer-to-peer network (any kind of wireless ad hoc network) is vulnerable to Sybil attack. Newsome *et al.*[18] presented a taxonomy of Sybil attacks in WSN based on three orthogonal dimensions.

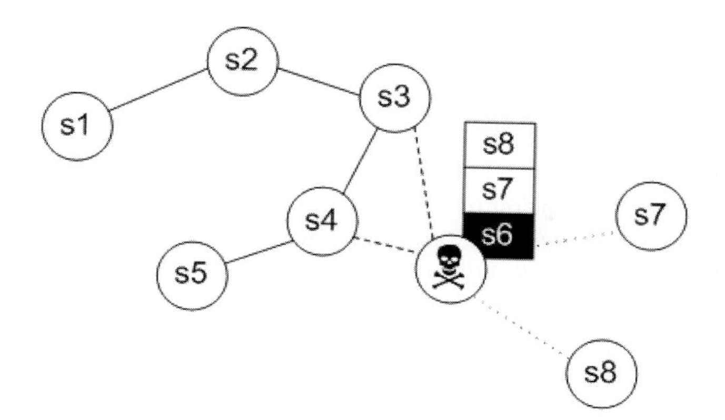

Fig. 16.1. Conceptual view of a Sybil attack. The node with id s6 is pretending to be three nodes at the same time (s6, s7, and s8), the nodes s3 and s4 do not have direct contacts with s7 and s8, so s6 can pretend to them as it is s7 or s8. Here, additional ids of s6 are called the "Sybil nodes" (s7 and s8).

16.3.3.1. *Dimension I*

Direct Communication — In this case, Sybil nodes directly communicate with the legitimate nodes. When a legitimate node sends message to a Sybil node, malicious device listens to the message. In the same way, messages sent from the Sybil nodes are actually sent from the malicious device.

Indirect Communication — In this case, the legitimate nodes cannot directly communicate with the Sybil nodes rather a malicious device convinces them that it can reach to the Sybil nodes. Any message sent by a legitimate node to a Sybil node is routed through the malicious node which can do anything (modification, fabrication, dropping, etc.) with the received messages.

16.3.3.2. *Dimension II*

Identities used for the Sybil nodes could be obtained in one of two ways:

Fabricated Identities — Attacker can simply generate a fake identity supported by the network and perform Sybil attack.

Stolen Identities — Attacker in this case steals the identities of the legitimate nodes and uses those for launching attacks.

16.3.3.3. *Dimension III*

The identities of the Sybil nodes could be used in two ways:

Simultaneous — The malicious node or the attacker can pretend to have multiple identities at the same time (as shown in Figure 16.1).

Non-simultaneous — The attacker can somehow obtain a large number of valid identities but, instead of using all the identities at the same time, it can use those one after another in different time slots.

One advantage for WSN to face Sybil attack is that, it can have some sort of centralized entity (base station or cluster head) in the network. Hence, this attack could be prevented using efficient protocols. Douceur[17] showed that, without a logically centralized authority, Sybil attacks are always possible except under extreme and unrealistic assumptions of resource parity and coordination among entities. However, detection of Sybil nodes in a network is not so easy. Some of the recently proposed detection and prevention mechanisms could be found in Refs. 19 to 22, and 90.

16.3.4. *Blackhole/Sinkhole attack*

In this attack, a malicious node acts as a blackhole[23] to attract all the traffic in the network. Especially in a flooding based protocol, the attacker listens to the route request and then replies to the target node saying that it has a high quality or shortest path to the base station. A victim node is thus lured to select it as a forwarder of its packets. Once the malicious device is able to insert itself between the communicating entities (between the base station and sensor node), it is able to do whatever it wishes with the packets that pass through it. The blackhole (i.e. malicious node or the attacker) can drop the packets, selectively forward those to the base station or to the next node, or even can change the content of the packets. This type of attack could be very harmful for those nodes that are considerably far from the base station. We should keep in mind that blackhole attack and sinkhole attack are basically the same attack but these two terms are often used interchangeably. As mentioned earlier, this attack can cause DoS in the network and thus could be considered as one type of DoS attack. Figure 16.2 shows a conceptual view of a blackhole/sinkhole attack. Some recent works addressing this attack and possible solutions to deal with it are in Refs. 24 to 31.

16.3.5. *Hello flood attack*

Hello flood attack was first detected and introduced by Karlof and Wagnor in Ref. 32. This attack uses HELLO packets as a weapon to convince the sensors in the network. Many protocols require broadcasting of HELLO packets for neighbor discovery. In this case, a node receiving such a packet may assume that it is within (normal) radio range of the sender node. This assumption could be exploited by an attacker. An attacker with a large radio transmission range (termed as a laptop-class attacker in Ref. 32) and enough processing power can send HELLO packets to a large number of sensors in the network. Thus the sensors could be persuaded that the adversary is their neighbor.

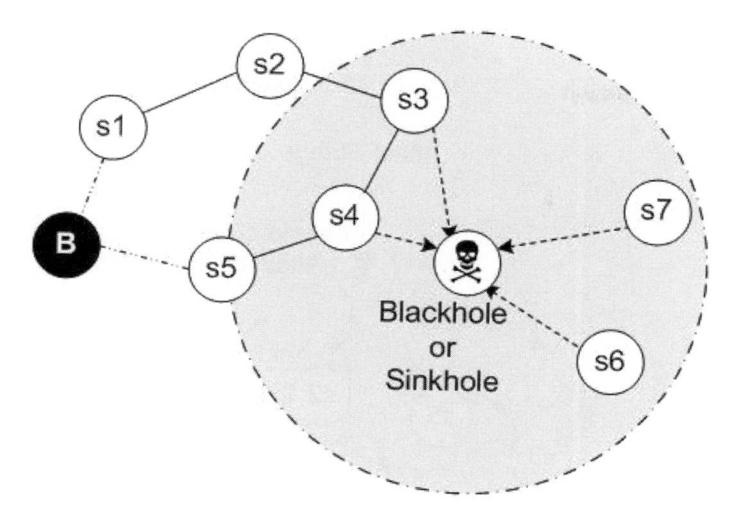

Fig. 16.2. Conceptual view of a Blackhole/Sinkhole attack. The attacker advertises high quality link through it which tempts s3, s4, s6, and s7 to select itself as a forwarding node for their packets. In the figure, B is the base station and the large gray circle is the attacker's radio range.

As a consequence, while sending the information to the base station, the victim nodes try to go through the attacker as they know it as their neighbor. In this way the attacker cheats the victims. A conceptual picture of hello flood attack is presented in Figure 16.3. Possible countermeasures to handle hello flood attack could be the use of bi-directional verification of links before using them, multi-path routing, use of multiple base stations,[33] or any kind of lightweight packet authentication scheme.

16.3.6. *Wormhole attack*

Wormhole attack is a very critical attack in which the attacker records the packets (or bits) at one location in the network and tunnels those to another location.[34] The tunneling or retransmission of bits could be done selectively. Wormhole attack is a significant threat to wireless sensor networks because this is possible even if the attacker has not compromised any node, and even if all communications provide authenticity and confidentiality. It could be performed even at the initial phase when the sensors start discovering the neighborhood information. In a nutshell, attacker's goal in wormhole attack is to disrupt routing information by creating shortcuts in the network.

Figure 16.4 shows a graphical representation of wormhole attack. In the figure, two adversaries are communicating with each other through a direct and dedicated channel by using wired link or additional RF (radio-frequency) transceivers with longer transmission range. The route via the wormhole looks like an attractive path

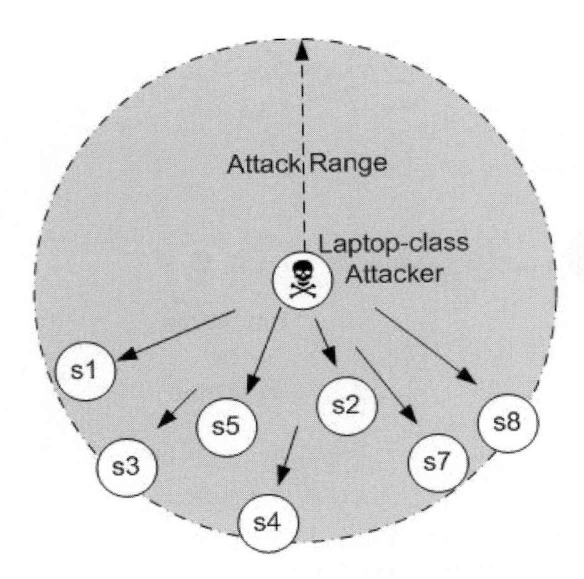

Fig. 16.3. Hello flood attack. A laptop-class attacker (attacker with large radio range) is transmitting the HELLO packets and pretending to be a neighbor of all other legitimate nodes within its radio range.

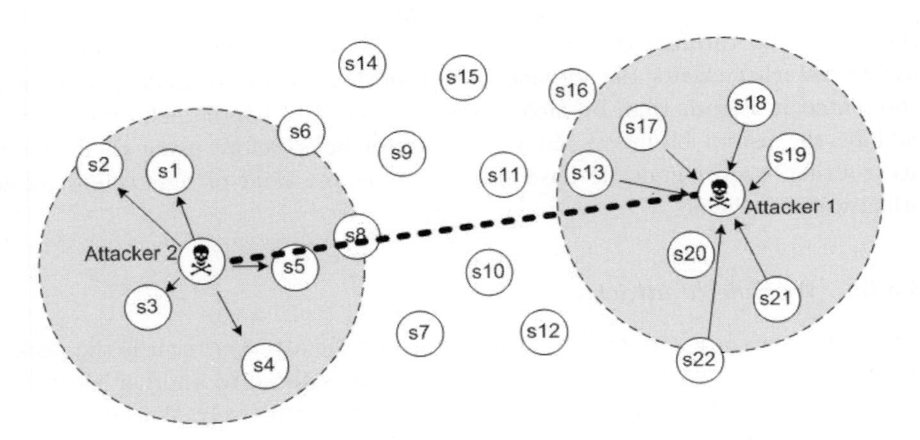

Fig. 16.4. Wormhole attack. Two attackers have created a dedicated tunnel between them and are attracting traffic.

to the legitimate sensor nodes because it generally offers less number of hops and less delay than other normal routing paths. While relaying packets, the adversaries can arbitrarily drop the packets. Therefore data communications through the wormhole suffer from severe performance degradation. In a recently published work, Sharif and Leckie propose three new variants of wormhole attacks namely Energy Depleting Wormhole Attack (EDWA), Indirect Blackhole Attack (IBA), and Targeted Energy

Depleting Wormhole Attack (TEDWA). Interested readers are suggested to read more in Ref. 35.

Several works tried to defend against this attack by detection of intruder nodes in the network. Some of them are in Refs. 24, 29, 36 to 43. Other than these works, Ref. 44 proposes an approach to deal with wormhole attacks using directional antennas, which is often not feasible for sensor networks.

So far, we have talked about various security threats and attacks in wireless sensor networks. Most of these attacks can be tackled by using proper cryptographic mechanisms. If the node authentication method is robust and messages in the network are made illegible to the outside entities, many security problems are eventually resolved or just need a little add-on with the defense mechanism. For utilizing any kind of cryptographic operation in the network, key management is a fundamental issue to deal with. Given the constrained resources of the sensors and the special characteristics of wireless sensor networks, key management in WSN is considered to be a very challenging topic and a hot research issue. Efficient mechanisms and management policies are needed to determine how the keys in such a network would be generated, stored, used, manipulated, renewed, or revoked. In the next section, we will try to get some insights on these issues.

16.3.7. *Key management issues in WSN*

Primary goal of key management is to set up secure links among the neighboring nodes in the network at the formation phase. Some of the major challenges any kind of key management mechanism faces are:

(i) **Unknown scalability of the network.** It means that if there are n number of nodes initially in the network, n' more nodes could be added to it later. The key management scheme must consider the tactics to handle the addition of nodes in the network.

(ii) **Unknown topological distribution of sensors in the network.** As the topological information of the sensors are often very difficult to obtain and not known in prior in most of the cases, the key management scheme must distribute keys or keying information in such a way that the neighbor nodes could communicate securely with each other.

(iii) **Limited available resources of the sensors.** Like any other mechanism, this is a great hurdle that the key management scheme must confront with.

(iv) **What if the nodes in the network are captured by adversaries?** The key revocation mechanism should ensure that the captured keys cannot be used further in the network and still the network should be able to keep functioning with proper level of security.

(v) **Re-keying.** If there is any re-keying mechanism in the management scheme, how to generate or distribute the new keys among the already deployed sensors in the network?

There are mainly three kinds of approaches for key management in wireless sensor network:

- Key pre-distribution.
- Key management based on public key.
- Key management based on online server.

16.3.7.1. *Key pre-distribution*

In case of key pre-distribution schemes, keys or the keying materials are delivered to all sensor nodes prior to their deployment. Keying materials are partial information of the keys that could be used by the nodes to derive keys for node-to-node secure communications. Among all the key management approaches, key pre-distribution seems to be the most feasible solution. This is because; most of the operations in this approach can be done prior to the deployment of the network.

For key pre-distribution, we mainly consider two phases of operations: initialization phase and network formation phase. In the initialization phase, most of the planning and computations are done so that the sensors could get relief of the heavy computational burdens. In the formation phase, the sensors establish secure links among themselves based on the pre-stored information in their memories.

There are mainly three approaches of key pre-distribution:

System key pre-distribution — Same key k is stored in each sensor. k could also be used for deriving other keys for secure communications among the sensors. The advantage of this approach is the use of little memory to store the key. The drawbacks are little resilience and weak authentication.

Trivial key pre-distribution — Distinct pairwise keys $k_{i,j}$ are stored for each pair of nodes s_i and s_j. The two nodes contact with each other to derive the pairwise key for further secure communications. The advantage of this approach is greater resilience and strength of authentication. However, this approach is not scalable and in this case, it is hard to handle the addition of new sensors in the network.

Random key pre-distribution — In this approach, a number of random keys (say w keys) from a key pool is stored in the sensors. Any two nodes in the network may share a key with probability p. The advantage of this type of scheme is the resiliency and support for addition of new sensors in the network. On the other hand, the drawbacks are the lose node authentication and possibility of not finding a common key even among the neighboring nodes. One of the legendary works on random key pre-distribution, known as the *basic scheme* was proposed by Eschenauer and Gligor.[45] The *basic scheme* is one of the early works which opened the door for further research on various aspects of key management in this type of network.

16.3.7.2. *Key management based on public-key*

Public key based schemes use asymmetric keys for encryption and decryption operations. There are some well-established public-key based schemes like Diffie-Hellman, Digital Signature Standard, ElGamal, Elliptic Curve Cryptography (ECC), RSA, etc.[46] But the reality is, public key cryptography (PKC) based schemes are often not directly applicable for wireless sensor networks. As mentioned earlier, the limitation of resources of the sensors is the major hurdle for using these mechanisms. Also the need for a certificate authority or a trusted middle-man, unknown topology of the network, and random deployment of sensors often make their use more difficult. In spite of the existence of these barriers, the existing PKC schemes could somehow be modified for making them suitable for use in the sensors. Often the number of operations is reduced to make the PKC schemes a bit lightweight. Though in the early days, the researchers thought that the PKC schemes are in all the ways inappropriate for WSN, some recent works have shown that some lightweight versions of these schemes might be very effective for high-security demanding applications. The works like Refs. 47 to 58 have presented some success stories and gains regarding using public key based security mechanisms and key management in WSN.

16.3.7.3. *Key management based on online server*

In this approach, an online server provides the necessary keys to the sensors for communications among themselves. The key could be provided by the base station or by the group leaders (sometimes called as cluster heads) in the network. However, this approach is not as efficient as the key pre-distribution approach as in this case, the special nodes must have relatively more memory, processing power, and energy than those of the ordinary sensors in the network. Also, the special nodes should be well-dispersed in the network so that they can cover the whole network for providing the keys with minimum effort. Maintaining security during the transmission of keys also requires some other supporting mechanisms or some trust-based approach. Overall, most of the researchers agree that this approach in most of the scenarios, not a good solution for managing keys in this type of network.

Some of the recent and notable works on key management in sensor networks are in Refs. 59 to 75, 87 and 89. Readers are encouraged to go through these for gaining in-depth knowledge on key management in wireless sensor networks. Other than these works, a recent survey on the key management schemes in WSN is presented in Ref. 76.

16.3.8. ***Secure routing in WSN***

Basically, secure routing is not a separate issue than that we have discussed so far. If we have an efficient key management scheme with a supporting security infrastructure, this issue is easily solved. In that case, the whole thing reduces to the task of verifying who is communicating with whom and through whom. A number of routing protocols are proposed for wireless sensor networks (for further reading, Refs. 77 and 78 are suggested to the interested readers). However, the key point is that most of the routing protocols have overlooked the issue of security at their design phase. Sometimes it is quite impossible to fit a good security mechanism with a good routing protocol.

If the operational method of a routing protocol does not support a particular security mechanism, we need to choose any other suitable security approach for that one. In such a case, often the suitable security solution might not be the best solution or might not at all help for secure routing using that particular protocol. A routing protocol may focus on saving energy resources of the sensors, but if a security mechanism is added to it, it might not hold its major point of advantage or could even turn into an energy-consuming routing protocol. Therefore, it is better to consider the security issues at the design phase of any routing protocol. If the structural design and communication methods of the routing protocol allow the security solutions to run side-by-side or on top of it, then it could be beneficial for secure routing as well as for handling almost all types of threats and attacks in WSN. Nonetheless, it should be noted that a single solution cannot solve all the problems at the same time. Instead, based on the application requirements and network settings, the strategy of routing and security should be set. Often we need to consider some trade offs among some parameters like security, QoS (Quality of Service), latency, packet loss, etc.

In one of the prominent works on secure routing in wireless sensor networks, Karlof and Wagner[32] noted that:

"*One aspect of sensor networks that complicates the design of a secure routing protocol is in-network aggregation. In more conventional networks, a secure routing protocol is typically only required to guarantee message availability. Message integrity, authenticity, and confidentiality are handled at a higher layer by an end-to-end security mechanism such as SSH or SSL. End-to-end security is possible in more conventional networks because it is neither necessary nor desirable for intermediate routers to have access to the content of messages. However, in sensor networks, in-network processing makes end-to-end security mechanisms harder to deploy because intermediate nodes need direct access to the content of the messages. Link layer security mechanisms can help mediate some of the resulting vulnerabilities, but it is not enough: we will now require much more from our routing protocols, and they must be designed with this in mind.*"

In general, for secure routing in wireless sensor networks, the following points could be considered:

- Multi-path routing can help for introducing some sort of security.
- Use of symmetric key cryptography can reduce the processing overhead.
- The routing protocols should be intrusion tolerant and should be able to keep on functioning at least up to a certain level so that the overall network operations are not hampered in case of the presence of intruders.

- As involvement of security mechanisms can increase the overheads of the protocol, the overall design should be kept as simple as possible.
- Any broken routing path should not hamper the functions of the associated security mechanisms. The working method of the protocol should allow finding an alternate path to the destination within a minimum interval.

16.3.9. *Physical security issues*

Earlier we have introduced the types of physical attacks in WSN in brief. In this section, we will have a closer look at the physical security issues in wireless sensor networks. We know that the sensors in the network could be physically reached by adversaries because of the network's unattended nature. There are several ways to protect a sensor network from the physical attacks.

- The most suitable way to tackle this is the concept of *self-destruction*. In this case, a sensor detects a physical attack and quickly deletes all of its hidden information to become non-functional. For a large-scale sensor network, this could be a feasible solution as their might be several backups of the sensors' data, cryptographic keys, codes, and other secret information. Also if a part of the network is attacked, the sensors in other parts can be ready to destroy themselves before getting captured. Though this sort of *self-destruction* mechanism is expensive to incorporate with the sensor's physical package, it is not impossible.
- An alternate solution could be using a mechanism where each sensor monitors the status of its neighboring sensors. Any suspicious behavior or lack of response of a neighbor for a certain period of time might trigger a warning. Consequently, the other neighbors can get ready for hiding all of their secret information.
- Analyzing the deployment policy and detailed mapping of the network could also be effective for reducing the probability of physical attacks. However, in many applications, such kind of thorough study of the deployment area might not be possible.
- Camouflaging of sensors could be efficient in some deployment scenarios. Say, for example, a wireless sensor network is to be deployed over a rocky hilly area. In that case, the sensors could be colored like rocks or could be given the shapes of rocks (with some outer coverings!), which can make the task of physically locating them more difficult.
- Sensors might have some sort of protective shields that can save the internal hardware from external pressure or from other environmental conditions.

However, applying any of these approaches depends on the deployment budget and requirements of the application. Some of the recent works on physical security issues in wireless sensor networks can be found in Refs. 5, 6, 79 and 80.

16.4. Challenges for Future Research

With the sophistication of various communication protocols and rapid advancements of Micro-Electro-Mechanical Systems (MEMS) technologies,[81] sensors are gaining more resources and capabilities with which many barriers of security could be surmounted. In spite of the previous advancements and those that are coming in the near future, some issues regarding security in WSN could still pose great challenges. In this section, we will talk about those issues and will try to visualize the future so that the research works on security in WSN may get a proper direction towards devising realistic solutions.

16.4.1. *Holistic approach to security in WSN*

A holistic approach aims at improving the performance of wireless sensor networks with respect to security, longevity, and connectivity under changing environmental conditions. This approach of security concerns about involving all the operational layers for ensuring total security in the network. When talking about layering concepts, it should be mentioned that the security in network layer is mainly concerned about authentication, availability of routing information, and integrity of information, the data link layer is concerned mainly about data confidentiality and data freshness, and the physical layer is concerned about tamper-resistance. Holistic approach tries to lead to a single architecture so that different security mechanisms can work in tandem for different layers. Some key principles of holistic approach are:

- In a given network, the cost for ensuring security should not surpass the assessed security risk at a specific time.
- If there is no physical security ensured for the sensors, the security measures must be able to exhibit a graceful degradation, if some of the sensors in the network are compromised, out of order, or captured by the enemy.
- The security measures should be developed to work in a decentralized fashion.

Considering all types of security threats and attacks in WSN, we can understand that for this type of network, a single security solution for a single layer cannot be considered as a reasonable solution. It is better to employ a holistic approach so that all facets of the network could be made secure at the same time. As an example, if a WSN has very good security solutions for almost all the layers but physically the network is vulnerable, we cannot guarantee that the total security of the network is ensured. In such a case, any adversary can go and pick up the sensors from the field, extract the cryptographic keys, can use jamming for causing physical layer DoS attacks, destroy the sensors, and so on. Though physical security is often not possible to ensure for WSNs, at least the overall system must allow a graceful degradation of the network's operation when it is attacked. However, designing and developing such type of efficient security architecture and management policy

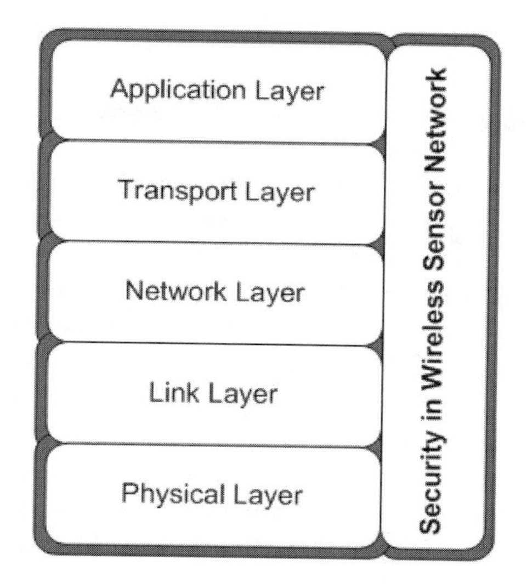

Fig. 16.5. Holistic view of security in WSN.

remain as an open challenge. At least we can hope that with the advancements of technological capabilities of sensors, this task will become a bit easier in the future.

16.4.2. *What to expect next?*

Today, the limitation of resources of the sensors is considered as the primary obstacle for applying robust security mechanisms. In future, this barrier might totally be vanished or might be reduced by significant extent. We might see sensors capable of handling even the heavy computations associated with public key cryptography schemes (like RSA, SHA-1, etc.) without any reduced operation. Say, for example, one of the latest advanced wireless sensor platforms, Imote2[82] is built with the low power PXA271 XScale processor at 13-416 MHz and it integrates an 802.15.4 radio (CC2420) with a built-in 2.4 GHz antenna. Imote2 has 256 kB SRAM, 32 MB FLASH, and 32 MB SDRAM. It is a modular stackable platform and can be expanded with extension boards to customize the system to a specific application. Through the extension board connectors, sensor boards can provide specific analog or digital interfaces. A battery board is provided to supply system power, or even it can be powered via the integrated USB interface. All these features make it a very powerful sensor node compared to its predecessors. The rechargeable feature of the sensor's battery opens the door to overcome the problem of constrained and non-renewable energy.

Considering today's achievements, it is reasonable to assume that some years later we could even see sensors with much higher configurations with the same tiny size! If it becomes true, some interesting questions may arise. What will be the case if these tiny devices get the capabilities like high configuration computers? Will we be able to run classic security schemes that require heavy computations? If so, will all the works done so far be meaningless? The answer to all of these questions is; "No work will be thrown away even if the sensors achieve very high configurations". Basically the researchers have been working on the fact that, given such low-configuration devices, how best level of security can be provided for the network. Yes, in future the sensors might get more capabilities keeping even today's physical size, but even then the devices with current specifications could remain as low-cost alternatives. Also, some other tiny devices might have such limited resources. It is also reasonable to think that the sensors with current specifications might become much smaller in physical size. If it becomes true, in that case, reduction of physical size would ultimately increase the level of physical security of these devices. In fact, reduced size of sensors would make them more *physically secure* in the hostile deployment areas as a relatively smaller object is harder to notice! Hence, the major point is, no matter how much capabilities a sensor node attains in future, the research works done with today's given limitations (like MICA2's specifications) will still be useful for use for the devices with such capabilities. As a whole, the research area will still remain challenging.

In future we might also see wide-spread use of wireless multi-media sensor networks[83,91] for various security applications like; distributed vision, tracking, and monitoring applications. At that time, processing multi-media data might become a little bit easier. However, when issues like QoS (Quality of Service) and latency are involved with this, the challenge is likely to remain for finding efficient solutions. In fact, ensuring a good level of QoS and a good level of security at the same time is always very difficult and often *contradictory*! Not only for sensor networks but also for other types of networks this statement is true. This is because, any sort of security operation requires some processing time. If the level of security is increased, the processing delay also increases causing degradation of quality of service. For real-time multi-media applications (if at all possible using WSNs or if at all required!), this challenge will remain for a long time.

Some of the recent works show that, in future some applications might need to handle multiple types of data within the same network.[84] The development of sensors like ExScal motes.[85,86] has already opened the door for further research on heterogeneous applications using homogenous multi-purpose nodes. The heterogeneous data generated from such multi-purpose nodes might have different levels of security based on their priorities. Handling these heterogeneous data with different security levels could also be an interesting topic for research in the near future.

16.5. Conclusions

In this chapter, we have discussed security in wireless sensor networks considering five major aspects; (i) security approaches for sensor networks, (ii) threats and attacks against sensor networks, (iii) key management issues, (iv) secure routing issues, and (v) future possibilities and challenges. We have seen that most of the attacks against security mechanisms in wireless sensor network are launched by injecting false information either by compromised nodes residing in the network or by attackers. Most attacks can be resisted by employing efficient schemes to detect the attackers or the compromised nodes. Distributed detection and prevention mechanisms can really help to resist lots of attacks. Since it is not desirable to give sensors extra tasks, developing distributed detection and prevention schemes still remains a challenging research issue.

Considering current direction of research and advancements of the methodologies, we can expect that in the coming days, holistic security in wireless sensor network will become a major research issue. Many of today's proposed security mechanisms are based on specific applications, network models, or application specific assumptions. Till today there is a lack of proper effort to make the security mechanisms operable with one another. If the efficient and lightweight security mechanisms for different layers could be made compatible for providing holistic security, it would be a great achievement for this research area.

With the increase of innovative applications of wireless sensor networks, the security issue is expected to get more emphasis. There is a well-known adage, "*An ounce of prevention is worth a pound of cure.*" If the solutions of other issues are developed keeping security in mind or at least they are made operable with the available security mechanisms, we can expect wide-spread use of wireless sensor networks for many security-demanding data extraction applications in the coming days.

Problems

(1) What is "*Philosophy of Mistrust*"?
(2) What are the key aspects that we should consider for security in WSN?
(3) Why the classic cryptographic schemes cannot be applied directly for WSN?
(4) What is steganography? Can it be applied in WSN?
(5) What are the major types of attacks in WSN?
(6) What is DoS? Explain the terms "DoS" and "DoS attack".
(7) What are the DoS attacks in physical layer?
(8) What is a Sybil node?
(9) What is the difference between blackhole and sinkhole attack?
(10) What type of protocol is vulnerable to HELLO Flood attack?
(11) Why wormhole attack is difficult to deal with?
(12) What are the major challenges that the key management in WSN faces?
(13) What is a "node compromise"?
(14) Mention some of the ways to tackle physical attacks in WSN.
(15) Why ensuring a good level of security and QoS (Quality of Service) at the same time is difficult?

Terminologies

DoS — Denial of Service.
ECC — Error correction codes/Elliptic Curve Cryptography.
EDWA — Energy Depleting Wormhole Attack.
IBA — Indirect Blackhole Attack.
IDS — Intrusion detection scheme.
MEMS — Micro-Electro-Mechanical Systems.
QoS — Quality of Service.
TDMA — Time Division Multiple Access.
TEDWA — Targeted Energy Depleting Wormhole Attack.
WSN — Wireless Sensor Network

Bibliography

1. A.-S. K. Pathan, H.-W. Lee and C. S. Hong, Security in ireless sensor networks: issues and challenges, *Proceedings of the 8th IEEE ICACT*, Vol. II, Phoenix Park, Korea (2006) 1043–1048.
2. P. Hämäläinen, M. Kuorilehto, T. Alho, M. Hännikäinen and T. D. Hämäläinen, Security in wireless sensor networks: considerations and experiments, *SAMOS 2006*, Lecture Notes in Mathematics, Vol. 4017 (Springer-Verlag, 2006) 167–177.
3. I. F. Akyildiz, W. Su, Y. Sankarasubramaniam and E. Cayirci, Wireless sensor networks: a survey, *Computer Communications* **38** (2002) 393–422.
4. http://www.xbow.com/Products/productsdetails.aspx?sid=72.
5. W. Gu, X. Wang, S. Chellappan, D. Xuan and T. H. Lai, Defending against search-based physical attacks in sensor networks, *IEEE International Mobile Adhoc and Sensor Systems Conference* (2005).
6. A. Becher, Z. Benenson and M. Dornseif, Tampering with motes: real-world physical attacks on wireless sensor networks, *SPC 2006*, Lecture Notes in Computer Science, Vol. 3934 (Springer-Verlag, 2006) 104–118.
7. J. Undercoffer, S. Avancha, A. Joshi and J. Pinkston, Security for sensor networks, *CADIP Research Symposium*, http://www.cs.sfu.ca/~angiez/personal/paper/sensor-ids.pdf (2002).
8. C. Kurak and J. McHugh, A cautionary note on image downgrading in computer security applications, *Proceedings of the 8th Computer Security Applications Conference*, San Antonio (1992) 153–159.
9. I. S. Mokowitz, G. E. Longdon and L. Chang, A new paradigm hidden in steganography, *Proceedings of the 2000 Workshop on New Security Paradigms*, Ballycotton, County Cork, Ireland (2001) 41–50.
10. C. H. Kim, S. Lee, W. I. Yang and H. W. Lee, Steganalysis on BPCS steganography, *Pacific Rim Workshop on Digital Steganography (STEG'03)*, Japan (2003).
11. K. Römer, P. Blum and L. Meier, Time synchronization and calibration in wireless sensor networks, *Handbook of Sensor Networks: Algorithms and Architectures* ed. I. Stojmenovic (John Wiley & Sons, 2005) 199–237, ISBN 0-471-68472-4.
12. A. D. Wood and J. A. Stankovic, A taxonomy for denial-of-service attacks in wireless sensor networks, *Handbook of Sensor Networks: Compact Wireless and Wired Sensing Systems* eds. M. Ilyas and I. Mahgoub (CRC Press, 2004).

13. M. Čagalj, S. Čapkun and J.-P. Hubaux, Wormhole-based antijamming techniques in sensor networks, *IEEE Transactions on Mobile Computing* **6**(1) (2007) 100–114.
14. A. D. Wood, J. A. Stankovic and S. H. Son, Jam: a jammed-area mapping service for sensor networks, *Proceedings of the 24th IEEE Real-time Systems Symposium* (2003) 54–62.
15. G. Alnifie and R. Simon, A multi-channel defense against jamming attacks in wireless sensor networks, *Proceedings of ACM Q2SWinet'07*, Crete Island, Greece (2007) 95–104.
16. H. Chen, P. Han, X. Zhou and C. Gao, Lightweight anomaly intrusion detection in wireless sensor networks, *PAISI 2007*, Lecture Notes in Computer Science, Vol. 4430 (Springer-Verlag, 2007) 105–116.
17. J. R. Douceur, The Sybil attack, *IPTPS 2002*, Lecture Notes in Computer Science, Vol. 2429 (Springer-Verlag, 2002) 251–260.
18. J. Newsome, E. Shi, D. Song and A. Perrig, The Sybil attack in sensor networks: analysis & defense, *Proceedings of ACM IPSN'04*, California, USA (2004) 259–268.
19. Q. Zhang, P. Wang, D. S. Reeves and P. Ning, Defending against Sybil attacks in sensor networks, *Proceedings of the 25th IEEE International Conference on Distributed Computing Systems Workshops* (2005) 185–191.
20. D. Mukhopadhyay and I. Saha, Location verification based defense against Sybil attack in sensor networks, *ICDCN 2006*, Lecture Notes in Computer Science, Vol. 4308 (Springer-Verlag, 2006) 509–521.
21. H. Yu, M. Kaminsky, P. B. Gibbons and A. Flaxman, SybilGuard: defending against Sybil attacks via social networks, *Proceedings of ACM SIGCOMM*, Pisa, Italy (2006) 267–278.
22. W. Jiangtao, Y. Geng, S. Yuan and C. Shengshou, Sybil attack detection based on RSSI for wireless sensor network, *Proceedings of the International Conference on Wireless Communications, Networking and Mobile Computing (WiCom 2007)* (2007) 2684–2687.
23. N. Ahmed, S. Kanhere and S. Jha, The holes problem in wireless sensor networks: a survey, *ACM SIGMOBILE Mobile Computing and Communications Review (MC2R)* **9**(2) (2005) 4–18.
24. A. A. Pirzada and C. McDonald, Circumventing sinkholes and wormholes in wireless sensor networks, *Proceedings of International Workshop on Wireless Ad Hoc Networks*, London, England, Kings College, London (2005).
25. Z. Karakehayov, Using REWARD to detect team black-hole attacks in wireless sensor networks, *Proceedings of the Workshop on Real-World Wireless Sensor Networks (REALWSN'05)*, Stockholm, Sweden (2005).
26. J. Yin and S. K. Madria, A hierarchical secure routing protocol against black hole attacks in sensor networks, *Proceedings of IEEE International Conference on Sensor Networks, Ubiquitous, and Trustworthy Computing* **1** (2006) 376–383.
27. S. S. Ramaswami and S. Upadhyaya, Smart handling of colluding black hole attacks in MANETs and wireless sensor networks using multipath routing, *Proceedings of the 2006 IEEE Workshop on Information Assurance*, New York, USA (2006) 253–260.
28. E. C. H. Ngai, J. Liu and M. R. Lyu, An efficient intruder detection algorithm against sinkhole attacks in wireless sensor networks, *Computer Communications* **30** (2007) 2353–2364.
29. H. A. Nahas, J. S. Deogun and E. D. Manley, Proactive mitigation of impact of wormholes and sinkholes on routing security in energy-efficient wireless sensor networks, *Wireless Networks* (Springer, Netherlands), DOI 10.1007/s11276-007-0060-7 (2007).

30. M. Demirbas and Y. Song, An RSSI-based scheme for Sybil attack detection in wireless sensor networks, *Proceedings of the 2006 International Symposium on a World of Wireless, Mobile and Multimedia Networks (IEEE WoWMoM 06)* (2006) 564–570.
31. I. Krontiris, T. Dimitriou, T. Giannetsos and M. Mpasoukos, Intrusion detection of sinkhole attacks in wireless sensor networks, *3rd International Workshop on Algorithmic Aspects of Wireless Sensor Networks (AlgoSensors'07)*, Wroclaw, Poland (2007).
32. C. Karlof and D. Wagner, Secure routing in wireless sensor networks: attacks and countermeasures, *Ad Hoc Networks* **1** (2003) 293–315.
33. M. A. Hamid, M. Mamun-Or-Rashid and C. S. Hong, Routing security in sensor network: HELLO flood attack and defense, *Proceedings of IEEE ICNEWS*, Dhaka, Bangladesh (2006) 77–81.
34. Y. C. Hu, A. Perrig and D. B. Johnson, Wormhole attacks in wireless networks, *IEEE Journal on Selected Areas in Communications* **24**(2) (2006) 370–380.
35. W. Sharif and C. Leckie, New variants of wormhole attacks for sensor networks, *Proceedings of the Australian Telecommunication Networks and Applications Conference*, Melbourne, Australia (2006) 26–30.
36. Y.-C. Hu, A. Perrig and D. B. Johnson, Packet leashes: a defense against wormhole attacks in wireless ad hoc networks, *Proceedings of the 22nd Annual Joint Conference of the IEEE Computer and Communications Societies (INFOCOM 2003)*, San Francisco, CA **3** (2003) 1976–1986.
37. L. Buttyán, L. Dóra and I. Vajda, Statistical wormhole detection in sensor networks, *ESAS 2005*, Lecture Notes in Computer Science, Vol. 3813 (Springer-Verlag, 2005) 128–141.
38. H. Alzaid, S. Abanmi, S. Kanhere and C. T. Chou, Detecting wormhole attacks in wireless sensor networks, Technical Report, Computer Science and Engineering School, The Network Research Laboratory, University of New South Wales (2006).
39. R. Maheshwari, J. Gao and S. R. Das, Detecting wormhole attacks in wireless sensor networks using connectivity information, *Proceedings of the 26th Annual IEEE Conference on Computer Communications (INFOCOM'07)* (2007) 107–115.
40. I. Khalil, S. Bagchi and N. B. Shroff, MOBIWORP: mitigation of the wormhole attack in mobile multihop wireless networks, *Ad Hoc Networks* **6**(3) (2008) 344–362.
41. J.-H. Yun, I.-H. Kim, J.-H. Lim and S.-W. Seo, WODEM: wormhole attack defense mechanism in wireless sensor networks, *ICUCT 2006*, Lecture Notes in Computer Science, Vol. 4412 (Springer-Verlag, 2006) 200–209.
42. R. Poovendran and L. Lazos, A graph theoretic framework for preventing the wormhole attack in wireless ad hoc networks, *Wireless Network* (Springer) **13** (2007) 27–59.
43. Y. Xu, G. Chen, J. Ford and F. S. Makedon, Distributed wormhole detection in wireless sensor networks, *Critical Infrastructure Protection*, eds. E. Goetz and S. Shenoi (Springer, Boston, 2008).
44. L. Hu and D. Evans, Using directional antennas to prevent wormhole attacks, *Proceedings of the 11th Network and Distributed System Security Symposium* (2003) 131–141.
45. L. Eschenauer and V. D. Gligor, A key-management scheme for distributed sensor networks, *Proceedings of the 9th ACM Conference on Computer and Communications*, Washington, DC, USA (2002) 41–47.
46. M. Y. Rhee, *Internet Security: Cryptographic Principles, Algorithms and Protocols* (Wiley, 2003).
47. D. J. Malan, M. Welsh and M. D. Smith, A public-key infrastructure for key distribution in tinyOS based on elliptic curve cryptography, *Proceedings of the First Annual IEEE Communications Society Conference on Sensor and Ad Hoc Communications and Networks (SECON'04)* (2004) 71–80.

48. R. Watro, D. Kong, S.-F. Cuti, C. Gardiner, C. Lynn and P. Kruus, TinyPK: securing sensor networks with public key technology, *Proceedings of ACM SASN'04*, Washington, DC, USA (2004) 59–64.
49. W. Du, R. Wang and P. Ning, An efficient scheme for authenticating public keys in sensor networks, *Proceedings of ACM MobiHoc'05*, Illinois, USA (2005) 58–67.
50. G. Gaubatz, J. Kaps and B. Sunar, Public keys cryptography in sensor networks — revisited, *ESAS 2004*, Lecture Notes in Computer Science, Vol. 3313 (Springer-Verlag, 2005) 2–18.
51. G. Gaubatz, J.-P. Kaps, E. Öztürk and B. Sunar, State of the art in ultra-low power public key cryptography for wireless sensor networks, *Proceedings of the Third IEEE International Conference on Pervasive Computing and Communications Workshops (PerCom 2005)* (2005) 146–150.
52. E.-O. Blaß and M. Zitterbart, Towards acceptable public-key encryption in sensor networks, *Proceedings of ACM 2nd International Workshop on Ubiquitous Computing* (INSTICC Press, Miami, USA, 2005) 88–93.
53. Q. Jing, J. Hu and Z. Chen, C4W: an energy efficient public key cryptosystem for large-scale wireless sensor networks, *Proceedings of the IEEE International Conference on Mobile Adhoc and Sensor Systems (MASS)* (2006) 827–832.
54. D. Nyang and A. Mohaisen, Cooperative public key authentication protocol in wireless sensor network, *UIC 2006*, Lecture Notes in Computer Science, Vol. 4159 (Springer-Verlag, 2006) 864–873.
55. E. Mykletun, J. Girao and D. Westhoff, Public key based cryptoschemes for data concealment in wireless sensor networks, *Proceedings of IEEE International Conference on Communications (ICC'06)* **5** (2006) 2288–2295.
56. O. Arazi, I. Elhanany, D. Rose, H. Qi and B. Arazi, Self-certified public key generation on the intel mote 2 sensor network platform, *Proceedings of the 2nd IEEE Workshop on Wireless Mesh Networks (WiMesh 2006)* (2006) 118–120.
57. O. Arazi, H. Qi and D. Rose, Public key cryptographic method for denial of service mitigation in wireless sensor networks, *Proceedings of the 4th Annual IEEE Communications Society Conference on Sensor, Mesh and Ad Hoc Communications and Networks (SECON'07)* (2007) 51–59.
58. A.-S. K. Pathan, J. H. Ryu, M. M. Haque and C. S. Hong, Security management in wireless sensor networks with a public key based scheme, *APNOMS 2007*, Lecture Notes in Computer Science, Vol. 4773 (Springer-Verlag, 2007) 503–506.
59. A.-S. K. Pathan, T. T. Dai and C. S. Hong, A key management scheme with encoding and improved security for wireless sensor networks, *ICDCIT 2006*, Lecture Notes in Computer Science, Vol. 4317 (Springer-Verlag, 2006) 102–115.
60. G. Jolly, M. C. Kuşçu, P. Kokate and M. Younis, A low-energy key management protocol for wireless sensor networks, *Proceedings of the 8th IEEE International Symposium on Computers and Communication (ISCC 2003)* (2003) 335–340.
61. D. Huang, M. Mehta, D. Medhi and L. Harn, Location-aware key management scheme for wireless sensor networks, *Proceedings of ACM SASN'04*, Washington, DC, USA (2004) 29–42.
62. B. Dutertre, S. Cheung and J. Levy, Lightweight key management in wireless sensor networks by leveraging initial trust, SDL Technical Report SRI-SDL-04-02, *SRI International* (2004).
63. Y.-H. Lee, V. Phadke, A. Deshmukh and J. W. Lee, Key management in wireless sensor networks, *ESAS 2004*, Lecture Notes in Computer Science, Vol. 3313 (Springer-Verlag, 2005) 190–204.
64. D. Liu, P. Ning and R. Li, Establishing pairwise keys in distributed sensor networks, *ACM Transactions on Information and System Security* **8**(1) (2005) 41–77.

65. F. An, X. Cheng, J. M. Rivera, J. Li and Z. Cheng, PKM: a pairwise key management scheme for wireless sensor networks, *ICCNMC 2005*, Lecture Notes in Computer Science, Vol. 3619 (Springer-Verlag, 2005) 992–1001.
66. W. Du, J. Deng, Y. S. Han, P. K. Varshney, J. Katz and A. Khalili, A pairwise key predistribution scheme for wireless sensor networks, *ACM Transactions on Information and System Security* **8**(2) (2005) 228–258.
67. W. Du, J. Deng, Y. S. Han and P. K. Varshney, A key predistribution scheme for sensor networks using deployment knowledge, *IEEE Transactions on Dependable and Secure Computing* **3**(2) (2006) 62–77.
68. C. Yang, J. Zhou, W. Zhang and J. Wong, Pairwise key establishment for large-scale sensor networks: from identifier-based to location-based, *Proceedings of the First International Conference on Scalable Information Systems*, Hong Kong (2006).
69. T. T. Dai, A.-S. K. Pathan and C. S. Hong, A resource-optimal key pre-distribution scheme with improved security for wireless sensor networks, *APNOMS 2006*, Lecture Notes in Computer Science, Vol. 4238 (Springer-Verlag, 2006) 546–549.
70. S. A. Çamtepe and B. Yener, Combinatorial design of key distribution mechanisms for wireless sensor networks, *IEEE/ACM Transactions on Networking* **15**(2) (2007) 346–358.
71. M. Chorzempa, J.-M. Park and M. Eltoweissy, Key management for long-lived sensor networks in hostile environments, *Computer Communications* **30** (2007) 1964–1979.
72. J. Großschädl, A. Szekely and S. Tillich, The energy cost of cryptographic key establishment in wireless sensor networks, *Proceedings of the 2nd ACM Symposium on Information, Computer and Communications Security (ASIACCS'07)* (2007) 380–382.
73. D. Huang, M. Mehta, A. V. D. Liefvoort and D. Medhi, Modeling pairwise key establishment for random key predistribution in large-scale sensor networks, *IEEE/ACM Transactions on Networking* **15**(5) (2007) 1204–1215.
74. D. Huang and D. Medhi, Secure pairwise key establishment in large-scale sensor networks: an area partitioning and multigroup key predistribution approach, *ACM Transactions on Sensor Networks* **3**(3) (2007) Article 16.
75. W. Zhang, M. Tran, S. Zhu and G. Cao, A random perturbation-based scheme for pairwise key establishment in sensor networks, *Proceedings of ACM MobiHoc'07*, Montreal, Canada (2007) 90–99.
76. Y. Xiao, V. K. Rayi, B. Sun, X. Du, F. Hu and M. Galloway, A survey of key management schemes in wireless sensor networks, *Computer Communications* **30** (2007) 2314–2341.
77. M. Younis and K. Akkaya, A survey on routing protocols for wireless sensor networks, *Ad Hoc Networks* **3** (2005) 325–349.
78. H. Karl and A. Willig, *Protocols and Architectures for Wireless Sensor Networks* (Wiley, 2006).
79. X. Wang, W. Gu, S. Chellappan, K. Schosek and D. Xuan, Lifetime optimization of sensor networks under physical attacks, *Proceedings of 2005 IEEE International Conference on Commmunications (ICC 2005)* **5** (2005) 3295–3301.
80. X. Wang, W. Gu, K. Schosek, S. Chellappan and D. Xuan, Sensor network configuration under physical attacks, *ICCNMC 2005*, Lecture Notes in Computer Science, Vol. 3619 (Springer-Verlag, 2005) 23–32.
81. B. A. Warneke and K. S. J. Pister, MEMS for distributed wireless sensor networks, *Proceedings of the 9th IEEE International Conference on Electronics, Circuits and Systems*, Vol. 1, Dubrovnik, Croatia (2002) 291–294.
82. http://www.xbow.com/Products/Product_pdf_files/Wireless_pdf/Imote2_Datasheet.pdf.

83. I. F. Akyildiz, T. Melodia and K. R. Chowdhury, A survey on wireless multimedia sensor networks, *Computer Networks* **51** (2007) 921–260.
84. A.-S. K. Pathan, G. Heo and C. S. Hong, A secure lightweight approach of node membership verification in dense HDSN, *Proceedings of the IEEE Military Communications Conference (IEEE MILCOM 2007)*, Orlando, FL, USA (2007).
85. L. Gu, D. Jia, P. Vicaire, T. Yan, L. Luo, A. Tirumala, Q. Cao, T. He, J. A. Stankovic, T. Abdelzaher and B. H. Krogh, Lightweight detection and classification for wireless sensor networks in realistic environments, *Proceedings of ACM SenSys 2005*, San Diego, CA, USA (2005) 205–217.
86. P. Dutta, M. Grimmer, A. Arora, S. Bibyk and D. Culler, Design of a wireless sensor network platform for detecting rare, random, and ephemeral events, *Proceedings of the 3rd Symposium on Information Processing in Sensor Networks (IPSN'05)*, Los Angeles, CA (2005) 497–502.
87. S.-P. Chan, R. Poovendran and M.-T. Sun, A key management scheme in distributed sensor networks using attacks probabilities, *Proceedings of IEEE GLOBECOM 2005* **2** (2005).
88. F. Ye, H. Luo, S. Lu and L. Zhang, Statistical en-route filtering of injected false data in sensor networks, *IEEE Journal on Selected Areas in Communications* **23**(4) (2005) 839–850.
89. H. Chan, A. Perrig and D. Song, Random key predistribution schemes for sensor networks, *Proceedings of Security and Privacy Symposium 2003* (2003) 197–213.
90. S. Tanachaiwiwat and A. Helmy, Correlation analysis for alleviating effects of inserted data in wireless sensor networks, *Proceedings of MobiQuitous'05* (2005) 97–108.
91. E. Gurses and O. B. Akan, Multimedia communication in wireless sensor networks, *Annals of Telecommunications* **60**(7–8) (2005) 799–827.

Chapter 17

Information Security in Wireless Sensor Networks

Abdelraouf Ouadjaout and Miloud Bagaa

Houari Boumediene University of Science and Technology Informatics Systems Lab. Algiers, Algeria

Abdelmalik Bachir

Grenoble Institute of Technology Informatics Lab. Grenoble, France

Yacine Challal

Université de Technologies de Compiègne, UMR CNRS 6599 Heudiasyc, Compiègne, France

Nouredine Lasla

National Institute of Informatics, Algiers, Algeria

Lyes Khelladi

Scientific and Technical Information Research Center Basic Software Lab. Algiers, Algeria

In this chapter, we provide a comprehensive survey of security issues in wireless sensor networks. We show that the main features of WSNs, namely their limited resources, wireless communications, and close physical coupling with environment, are the main causes of the their security vulnerabilities. We discuss the main attacks stemming from these vulnerabilities, along with the solutions proposed in the literature to cope with them. The security solutions are analyzed with respect to the different layers of the network protocol stack and cover the following issues: key management, secure data dissemination, secure data aggregation, secure channel access and secure node compromise.

Keywords: Key management; data dissemination security; data aggregation security; denial of service countermeasures; energy exhaustion attacks.

17.1. Introduction

The main security issues in WSN stem from the properties that make them efficient and attractive, which are:

(1) Resource Limitations: Energy is perhaps the greatest constraint to sensor node capabilities. The energy reserve of each node should be conserved to extend its lifetime and thus that of the entire network. Most of time, the sensed information is redundant due to geographically collocated sensors. Most of this energy can be saved through data aggregation. This requires a particular attention to detect faulty injected or modified data. In addition, energy scarcity opens new opportunities for DoS (Denial of Service) attacks aiming to exhaust nodes' batteries.
(2) Wireless and Multi-hop Communication: In addition to providing easy deployment, wireless communication has the advantage of offering access to hardly accessible locations such as disastrous and hostile terrains. Unfortunately, the radio communication range of sensor nodes is limited due to energy considerations. Multi-hop communication is thus indispensable for data dissemination in WSN. This introduces many security vulnerabilities at two different levels: attacking route construction and maintenance, and attacking data payload through packet injection, modification or suppression. Moreover, the wireless communication introduces other vulnerabilities at the link layer opening the door to jamming-style DoS and energy exhaustion attacks.
(3) Close physical coupling with environment: Most of WSN applications require a close deployment of nodes inside or near the phenomena to be monitored. This close physical coupling with environment leads to frequent intentional or accidental compromising of sensor nodes. As the success of WSN applications also depends on their low cost, nodes cannot afford expensive elaborated tamper protection. Therefore, a sufficiently capable adversary can manage to extract cryptographic information from nodes. As WSN missions are typically unattended, the potential for tamper attacks is significant.

As illustrated in Figure 17.1, we distinguish five main building blocks to secure WSN: (i) key management for WSN, (ii) securing radio channel access, (iii) securing data dissemination, (iv) securing data aggregation, and (v) securing the network against sensor nodes compromise. For each building block, we explain the security vulnerabilities and survey work in the area. Also, we briefly discuss advantages and drawbacks of existing solutions.

17.2. Background: Information Security and Cryptography

In this section, we review some common information security properties and present an overview of the cryptographic mechanisms used to achieve

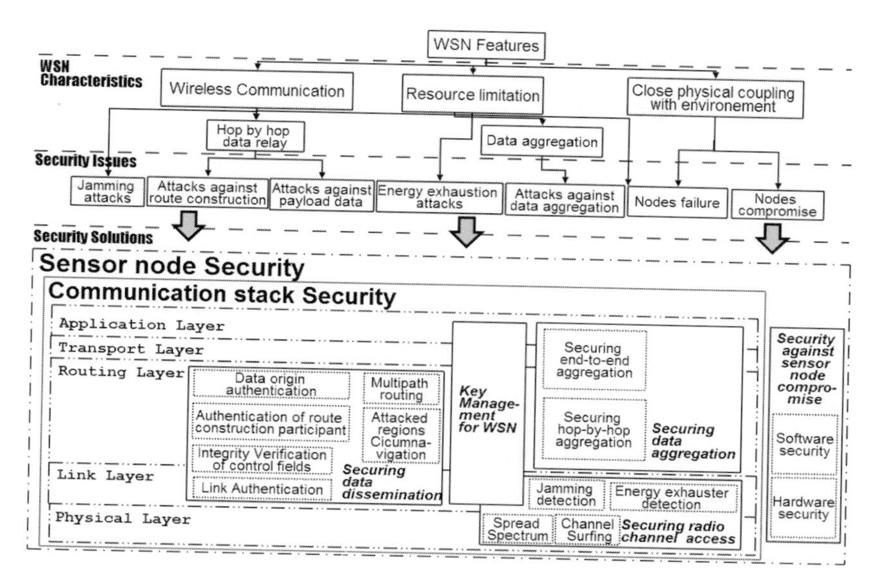

Fig. 17.1. Taxonomy of WSN security solutions.

them.[a] We consider a sender that sends a message to a receiver. For each security service, we present the cryptographic mechanism that the sender and the receiver should apply to guarantee the security service.

17.2.1. *Data integrity*

Data integrity is the property that data has not been changed, destroyed, or lost in unauthorized or accidental manner.[20] Cryptographic hash functions are typically used to assure data integrity.[21] Let us consider the following definition and then present a typical usage of cryptographic hash functions to assure data integrity:

A hash function is a computationally efficient function that maps binary strings of arbitrary length to binary strings of some fixed length, called hash-values.[21] *We denote the hash-value of a message m by* $h(m)$. *Cryptographic hash functions have the following properties* [20–23]: (i) *Given m, it is easy to compute* $h(m)$. (ii) *Given* $h(m)$, *it is hard to compute m such that* $h(m) = h$. (iii) *Given m, it is hard to find another message,* m', *such that* $h(m) = h(m')$.

Example[b]: Suppose that you want to save a large digital document (a program or a database) from alterations that may be caused by viruses or accidental misuses. A straightforward solution would be to keep a copy of the digital document on some

[a] The given measurements are for cryptographic systems implemented on the Atmel ATMega128L (found in the Mica family: Mica2, MicaZ, Mica2Dot, ... etc.).

[b] This example is cited by many authors such as Kaufman *et al.* in Ref. 23 and Menezes *et al.* in Ref. 21

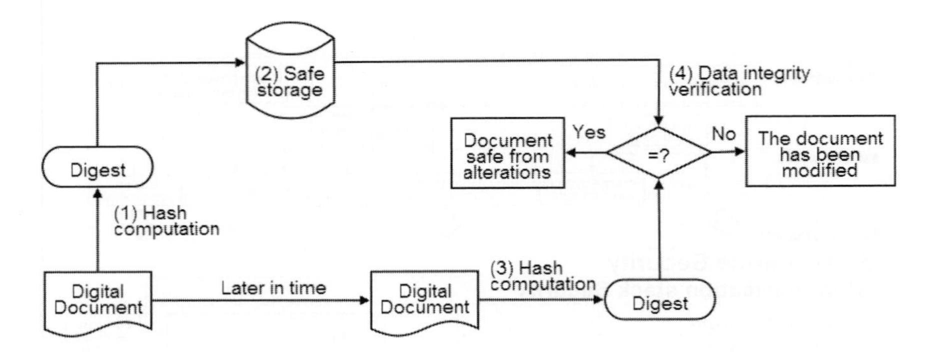

Fig. 17.2. Assuring data integrity using message digests.

Table 17.1 Measurements of some hash algorithms.

Hash algorithm	Digest size (bits)	Computation speed (bytes/ms)
MD5	128	23.81
SHA-1	160	7.93

tamper-proof backing store and periodically compare it to the active version. With a cryptographic hash function, you can save storage: you simply save the message digest of the document on the tamper-proof backing store (because the hash is small could be a piece of paper or a floppy disk) (see Figure 17.2, steps 1 and 2).

Then, periodically, you recalculate the message digest of the document (see Figure 17.2, step 3) and compare it to the original message digest (see Figure 17.2, step 4). If the message digest has not changed, you can be confident none of the data has.

Examples of hash functions are: MD2 (Message Digest 2),[24] MD5,[25] SHA-1 (Secure Hash Algorithm 1)[26] Table 17.1 gives some measurements[74] for usually used hash algorithms.

17.2.2. *Data origin authentication*

Data origin authentication is the validation that the source of data received is as claimed.[20] Message Authentication Code (MAC) is a cryptographic mechanism that can be used to assure data origin authentication and data integrity at the same time.

A Message Authentication Code (MAC) algorithm is a family of functions h_k parameterized by a secret key k, with the following properties: (i) *Given a key k and an input m, $h_k(m)$ is easy to compute.* (ii) *h_k maps an input m of an arbitrary finite bit length to an output $h_k(m)$ of fixed bit length. Furthermore, given a description of the function family h, for every fixed allowable value of k (unknown to an adversary), the following property holds:* (iii) *Given zero or more pairs $(m_i, h_k(m_i))$, it is computationally infeasible to compute any pair $(m, h_k(m))$ for any new input m.*[21]

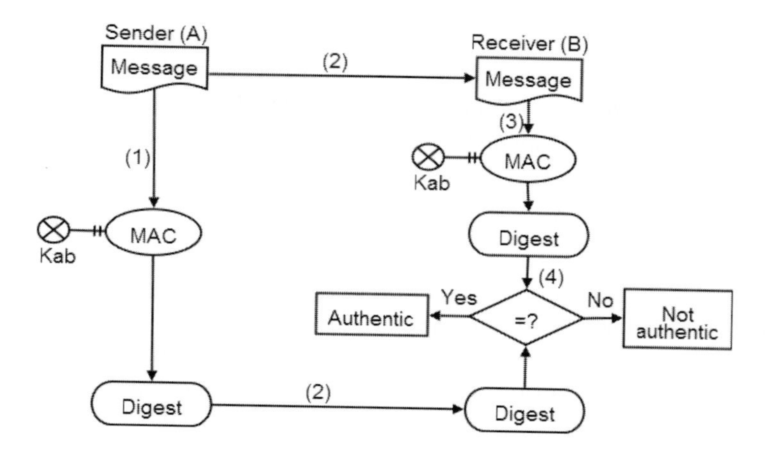

Fig. 17.3. Assuring data origin authentication using MACs.

The point of a MAC is to send something that only someone knowing the secret key can compute and verify. For example, a *MAC* can be constructed by concatenating a shared secret K_{AB} with the message m, and use $H(m|K_{AB})$ as the *MAC* (where H is a hash function).[23] Then, to assure data origin authentication, a sender (A) and a receiver (B) have to share a secret key K_{AB}. Then the sender computes the digest $(MAC(K_{AB}, m))$ corresponding to the message (m), to be sent, using the secret key (K_{AB}) (see Figure 17.3, step 1). Upon receiving the message as well as the digest, the receiver verifies the origin of the received message as follows: it recalculates the digest of the received message using the secret key K_{AB} (Figure 17.3, step 3) and compares it to the received digest (Figure 17.3, step 4). If the two digests are equal, the message is said to be authentic (has not been altered) and originates from the sender (A) since only (A) and (B) know the secret K_{AB}. Otherwise, the received message has been altered or fabricated by a sender who is not (A). An example of MAC is HMAC.[27]

17.2.3. *Data confidentiality*

Data confidentiality is the property that information is not made available or disclosed to unauthorized individuals, entities, or processes.[20] Confidentiality is guaranteed using encryption.

Encryption is a cryptographic transformation of data (called "plaintext") into a form (called "ciphertext") that conceals the data's original meaning to prevent it from being known or used. If the transformation is reversible, the corresponding reversal process is called decryption, which is a transformation that restores encrypted data to its original state.[20]

With most modern cryptography, the ability to keep encrypted information secret is based not on the cryptographic encryption algorithm, which is widely known, but on a piece of information called a key that must be used with the algorithm to produce an encrypted result or to decrypt previously encrypted information. Depending on whether the same or different keys are used to encrypt and to decrypt the information, we distinguish between two types of encryption systems used to assure confidentiality:

17.2.3.1. *Symmetric-key encryption*

In a symmetric-key encryption system, a secret key is shared between the sender and the receiver and it is used to encrypt the message by the sender and to decrypt it by the receiver. The encryption of the message produces a non-intelligible piece of information and the decryption reproduces the original message (see Figure 17.4).

Examples of symmetric-key encryption systems are: DES,[28] AES,[29] IDEA.[23] Table 17.2 gives some measurements for usually used symmetric encryption algorithms.

17.2.3.2. *Public-key encryption*

Public-key encryption (also called asymmetric encryption) involves a pair of keys (a public key and a private key) associated with the sender. Each public key is published, and the corresponding private key is kept secret by the sender. Data encrypted with the sender's public key can be decrypted only with the sender's private key (see Figure 17.5).

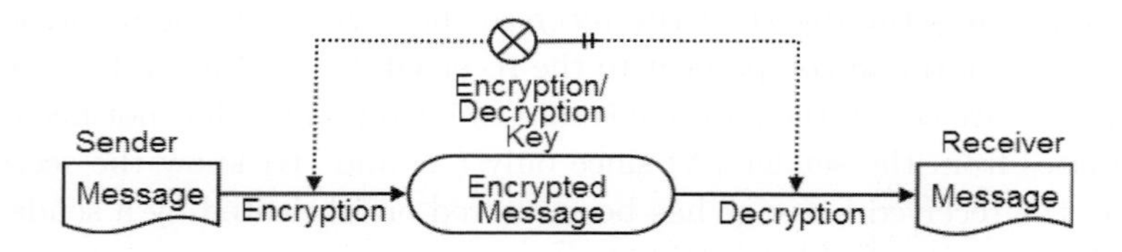

Fig. 17.4. Assuring confidentiality using symmetric-key encryption.

Table 17.2 Computation speed of some encryption algorithms.

Encryption algorithm	Encryption speed (Bytes/milli-second)
IDEA[75]	6.74
AES[74]	0.613

Fig. 17.5. Assuring confidentiality using asymmetric-key encryption.

In general, to send encrypted data to someone, the sender encrypts the data with that receiver's public key, and the receiver of the encrypted data decrypts it with the corresponding private key. Compared with symmetric-key encryption, public-key encryption requires more computation and is therefore not always appropriate for large amounts of data. However, it is possible to use public-key encryption to send a symmetric key, which can then be used to encrypt additional data. An example of asymmetric encryption systems is: RSA.[30]

17.2.4. *Non-repudiation with proof of origin*

Non-repudiation with proof of origin provides the recipient of data with evidence that proves the origin of the data, and thus protects the recipient against an attempt by the originator to falsely deny sending the data.[20]

Asymmetric cryptography is the basic answer for non-repudiation. With asymmetric cryptography, the piece of information sent with the message as a proof of integrity and data origin is computed using a private key held only by the sender and is verified by the receiver using the public key that corresponds to the private key. Hence, since only the sender can compute the piece of information, the latter can be used as a proof of origin to a third party and hence non-repudiation is assured. This cryptographic mechanism is called Digital Signature.

To sign a message, a sender generates a pair of private/public keys using some asymmetric cryptographic system. The sender keeps the private key secret and publishes the public key. Then the sender calculates the digest of the message to be sent using any hash function (see Figure 17.6, step 1). The digest is then cryptographically transformed using the private key (Figure 17.6, step 2). The result of this

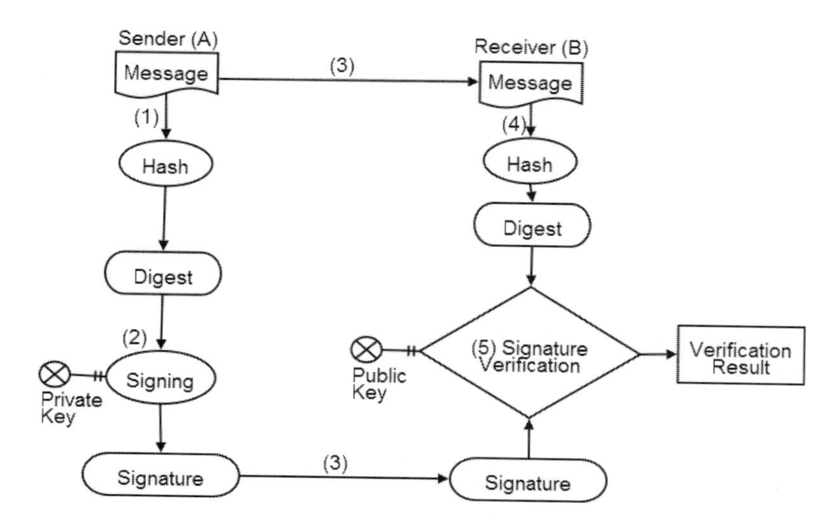

Fig. 17.6. Assuring non-repudiation using Digital Signatures.

Table 17.3 Measurements of some digital signature systems.

Digital signature	Signing speed (s)	Verification speed (s)	Public key size (bits)
RSA-1024	22.03	0.86	1024
ECC-160	1.65	3.27	160

transformation is called: the digital signature of the message. Upon receiving the message and the signature, the receiver verifies the signature using the public key as follows: first, the receiver recalculates the digest of the received message (Figure 17.6, step 4). Then, the receiver verifies the received signature using the public key (Figure 17.6, step 5). If the signature is valid then the message as well as its origin is authentic and non-repudiation is guaranteed. Otherwise the message is rejected.

Examples of digital signature schemes are: RSA,[30] DSA.[31] Table 17.3 gives some measurements[76] for usually used digital signature systems.

17.3. Key Management for WSN

Key management functions are the keystone of any communication-based system's security. In traditional networks such as the Internet, asymmetric cryptography is one of the most used key management schemes. Although asymmetric cryptography is very attractive as it provides reliable mechanisms for authentication and key distribution, its computation complexity, large memory footprints requirement and high computational overhead make it unsuitable for use in WSN. Recent researches[36,37] show, however, that it is possible to apply public key solutions to sensor networks by selecting the right algorithms and the appropriate parameters. In Refs. 34, 35 and 37, the energy cost of public key cryptography algorithms is quantified and it is shown that Elliptic Curve Cryptography (ECC)[72] has a significant advantage over RSA[30]: not only it reduces computation time but also it reduces the amount of transmitted and stored data. Some implementations of ECC for WSN can be found in Refs. 38 to 40.

17.3.1. *Literature overview*

Figure 17.7 shows taxonomy of a set of pre-distribution based key-management methods. In this taxonomy, protocols are classified into several categories according to the way neighboring nodes share common keys (probabilistic, deterministic) and to the network organization (hierarchical, flat).

17.3.1.1. *Probabilistic key management*

In a *pure probabilistic* key pre-distribution scheme,[41] each node is preloaded, prior to deployment, with a subset of keys taken from a very large *key-pool*. The main idea of this scheme is that two neighboring nodes have a certain probability of sharing

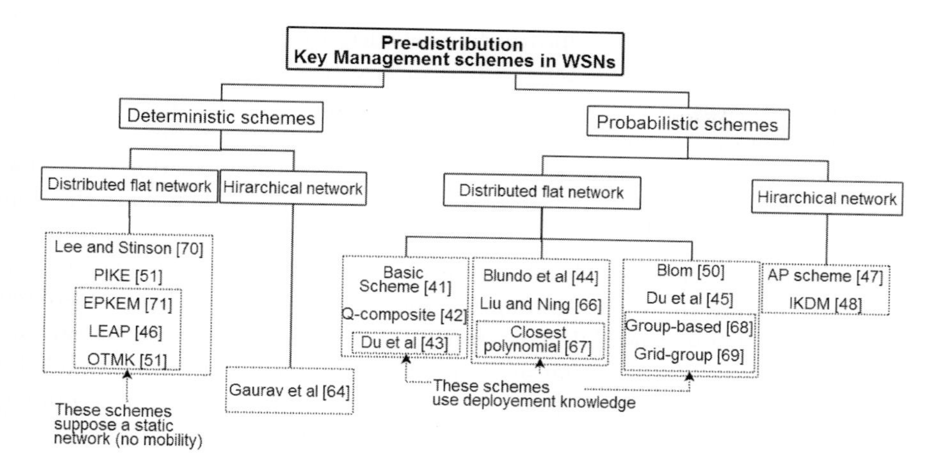

Fig. 17.7. Taxonomy of pre-distribution key management schemes in WSNs.

a common key that belongs to both key subsets of those neighbors. Although this scheme is simple and completely distributed, it has two main drawbacks. First, it requires large memory footprints to contain large subsets of preloaded keys so that the probability of neighboring nodes sharing common keys increases. Second, it cannot still provide sufficient security when the number of compromised nodes increases. To increase the resiliency of pure random key pre-distribution against compromised nodes, the *Q-composite* approach proposed in Ref. 42 allows two neighboring nodes to establish a secure communication link only if they share Q ($Q \geq 2$) common keys. Another work[43] takes advantage of deployment knowledge to improve both memory overhead and resiliency to node compromise. In this protocol, it is assumed that the deployment leads to collocated groups of nodes. Thus a sub-pool of keys is assigned to each group of collocated sensors. Each node in the same group selects a set of keys from its group sub-pool. Authors showed that two nodes in the same group will have a high probability to share a common key.

In Ref. 44, the authors use a symmetric bivariate λ-degree polynomial $\mathrm{P}(x, y)$ (such as $\mathrm{P}(x, y) = \mathrm{P}(y, x)$). Each sensor i with a unique ID, is preloaded with a polynomial share $\mathrm{P}_i(y) = \mathrm{P}(i, y)$. The common key between nodes i and j is $K_{ij} = \mathrm{P}_i(j) = \mathrm{P}_j(i)$. The solution is λ-secure, which means that keys are secure if no more than λ nodes are compromised. In order to make a solution more secure, a set F of λ-degree polynomials is generated. Each node i is preloaded with a subset F_i of the polynomial set F. Two nodes can establish a key, if they share a same polynomial in their subset of polynomials.

Du *et al.*,[45] adapted the idea of Blom's[50] symmetric key generation system in which each node i stores a column i and a row i of size $\lambda+1$ from matrices G (public matrix) and $(D * G)$ respectively, where $K = (D * G)^T * G$ is a symmetric matrix. A secret key between nodes i and j is generated as $K_{ij} = \mathrm{row}_i^*\ \mathrm{column}_j = \mathrm{row}_j^*\ \mathrm{column}_i$. The system is λ-secure, which means that keys are secure if no more than

λ nodes are compromised. To improve the resilience, the authors generate a space of w symmetric matrices $(D_1, D_2, \ldots, D_w)$ of size $\lambda + 1$ and select a random ζ distinct matrix spaces from the w matrix spaces for each node, this is similar to the assignment of random generated keys from a large key pool in Ref. 41. For each space selected by node j, the jth row is stored in the memory of this node. Two neighbouring nodes are able to calculate a common key only if they share the same matrix in their matrix spaces.

In networks with a hierarchical organization, probabilistic key management schemes take advantage of the assumption that some nodes have more resources than the others. Those powerful nodes usually elected to act as cluster heads, are preloaded with subsets of keys that are larger than the subsets of keys preloaded in the other ordinary nodes. These schemes reduce storage footprint at the ordinary nodes and achieve higher connectivity than those in flat networks.

17.3.1.2. *Deterministic key management*

In contrast to probabilistic schemes, deterministic schemes ensure that each node is able to establish a pair-wise key with its neighbors. To guarantee determinism, protocols such as LEAP[46] make use of a common transitory key that is preloaded into all nodes prior to deployment. The transitory key is used to generate session keys between neighboring nodes. To secure nodes against capture attacks, the transitory key is cleared from the memory of nodes after the generation of session keys.

In networks with a hierarchical organization, deterministic key management schemes centralize the responsibility for establishing secure links between the members of a cluster at the cluster-head. Although these schemes offer the attractive feature of being low energy, they are vulnerable to node capture attacks particularly when the cluster head is compromised.

17.3.2. *Description of some solutions*

17.3.2.1. *Random key pre-distribution*

The proposed protocol in Ref. 41 is the basis of probabilistic key pre-distribution schemes. It consists of three phases to establish a secure link between two neighboring nodes:

- Key pre-distribution phase: Each node, prior to deployment, is preloaded with a subset of keys of size K from a very large key pool of size $P \gg K$. A key identifier is associated to each key and stored into the sensor memory with the keys.
- Shared key discovery phase: After deployment, each node broadcasts the identifiers' list of its preloaded keys. Then, each node receiving this list discovers, eventually, at least a common key with its neighbor and establishes a secure link with it.

- Path key establishment phase: After the accomplishment of the above steps, two neighboring nodes may not share a key in their subsets of keys but are connected securely through one or more nodes. The latter are used to transmit securely a chosen key that will be used to secure a direct link between them.

Authors in Ref. 42 proposed a *q-composite* protocol that enhances the resilience to node capture of the basic scheme. In this protocol, two neighbouring nodes are able to establish a secure link only if they share at least q $(q > 1)$ common keys in their preloaded subset of keys. For example, a pair-wise key, K, between two nodes that share q' keys $(q' \geq q)$ is computed as: $K = \text{hash}\ (k1\|k2\|\ldots\|kq')$.

17.3.2.2. *LEAP*

LEAP[46] is a deterministic scheme that provides a mechanism to establish a pair-wise key between two neighbouring nodes. The protocol is based on the assumption that a sensor node, after its deployment, is secure during a time T_{min} and cannot be compromised during this period. The following steps illustrate how a newly deployed node establishes a pair-wise key with its neighbors.

- Key pre-distribution phase: The controller generates an initial key K_{IN} and preloads each node with this key. Each node u derives its master key from K_{IN}: $Ku = f_{KIN}(u)$.
- Neighbor discovery phase: After deployment, each node u broadcasts a HELLO message (1) that contains its identifier. When a node v receives the HELLO message it replies with an ACK message (2). The latter is authenticated using K_v that is calculated like this $K_v = f_{KIN}(v)$

$$u \quad * : u \ (1)$$

$$v \quad u \ : v, MAC(K_v, u \,|\, v) \ (2)$$

- Pair-wise key establishment phase: The node u calculates its pair-wise key with v: $K_{uv} = f_{Kv}(u)$. With the same formula, node v computes its pair-wise key with u.
- Key erasure phase: After the expiration of time T_{min}, the node u erases K_{IN} from its memory and hence forbids an intruder from getting this key.

After the above steps, the established pair-wise key between two nodes is used to secure communication between them.

17.3.3. *Analysis and discussion*

The main drawback of the probabilistic schemes is the high storage overhead that must be balanced against connectivity and resilience to node compromise. The deterministic schemes have fewer overheads but are more vulnerable to node compromise. For example, if the transitory initial key used in protocols such as LEAP is discovered then the entire network can be compromised. The hierarchical schemes offer better performances than the flat ones. However, they rely on the existence of powerful nodes heading clusters, which might not be the general case of WSN. In what follows we define some basic metrics that are used to analyze the previous presented schemes.

17.3.3.1. *Connectivity*

We define the connectivity as the probability that a node can establish secure links with all of its neighbors. In the deterministic schemes all nodes are preloaded with the same secret that is used to establish a secure link with any neighboring node. Therefore, the connectivity is 100%. In the probabilistic schemes two neighboring nodes can build a secure link only if they share Q common keys in their subset of keys of size m. The smaller is Q the better is the connectivity. Thus, The basic probabilistic scheme,[41] the polynomial-based scheme[44] and Blom's matrix-based scheme[45] with $Q = 1$ have a better connectivity than the Q-composite scheme[42] with $Q > 1$.

17.3.3.2. *Resilience to node capture*

When a sensor node is captured by an adversary, its entire secret and the entire established links with its neighbors are compromised. The effect of such attack can be beyond the neighboring nodes of the compromised sensor. We define the resilience against node capture as the probability of not compromising any other link in the network when x sensor nodes are compromised.

In the deterministic scheme[46] all nodes are preloaded with the same initial key, which will be erased after establishing a secure link with all neighbors. If an adversary can compromise a node before erasing the secret, the security of the entire network is compromised. Therefore, the resilience of this scheme is very poor from this point of view. Nevertheless, the authors showed that it is infeasible using actual means to break this initial key in a duration less than the fixed threshold before erasing the key. In the probabilistic schemes, because the preloaded keys in sensor nodes are taken from the same pool, all the probabilistic schemes are vulnerable to sensor node compromising. Furthermore, the resilience to sensor node capture changes from a scheme to another according to the number of keys needed to establish a secure link. The Q-composite scheme with a high Q presents a best resilience comparing to the other probabilistic schemes. Indeed, an attacker needs to break at least Q keys of a sensor's key pool before compromising it. On the other hand, the greater the number (m) of the preloaded keys in a sensor, the worse is the resilience. This is due to the fact that compromising a sensor leads to compromising its key ring (m keys) and hence compromising all the secure links based on those compromised keys.

17.3.3.3. *Resilience to node replication*

After compromising a sensor node from the network, the adversary may attempt to add one or more nodes to the network that use the same key information of the compromised node. The replicated nodes can be used to inject false data, modify or delete legitimate data of other sensor nodes.

The deterministic schemes, such as Refs. 46 and 51, use the same initial key in all nodes for authentication and the establishment of pair-wise key between nodes. Because the initial key is used to establish all pair-wise keys in the network, an adversary that obtains this key and duplicates it in other nodes, can successfully establish a pair-wise key with any other node in the network. In such a case, the network will be insecure.

All probabilistic schemes must provide the node-to-node authentication to prevent node replication attacks. However, the actual resilience against node replication attack is the same for all the probabilistic schemes, and depends on the probability that a replicated node can find a common key with other legitimate nodes. In Ref. 49, the authors studied the problem of node replication attack on probabilistic key management schemes. In their study they find that WSN with probabilistic key management schemes become almost 100% insecure in the presence of one replicated sensor, and the Q-composite scheme with large Q is more resilient against replication attack compared with the other schemes.

17.4. Securing Data Dissemination

In a multi-hop communication system, routing represents a fundamental service to guarantee the proper operation of the network. This predominant position makes this component an ideal target for an adversary who wants to disturb the normal operation of the network. In addition, the intrinsic properties of a WSN, such as the limited energy and short range communications, make such networks more vulnerable to security threats than traditional wireless networks.

Generally, routing mechanisms are mainly based on a hop-by-hop collaboration among sensors in order to gather the network's data and make it available for the end users. However, an uncontrolled collaboration, as found in most existing routing protocols, constitutes the major weakness that opens a variety of doors to an adversary in order to defeat the network security.

17.4.1. *Literature overview*

To build a secure dissemination module, two important components should be considered: *route construction* and *data relay.*

17.4.1.1. *Securing route construction*

The route management engine is responsible for the construction and maintenance of the communication backbone represented by the routing topology. Since the process of route establishment aims to build the backbone of the communication

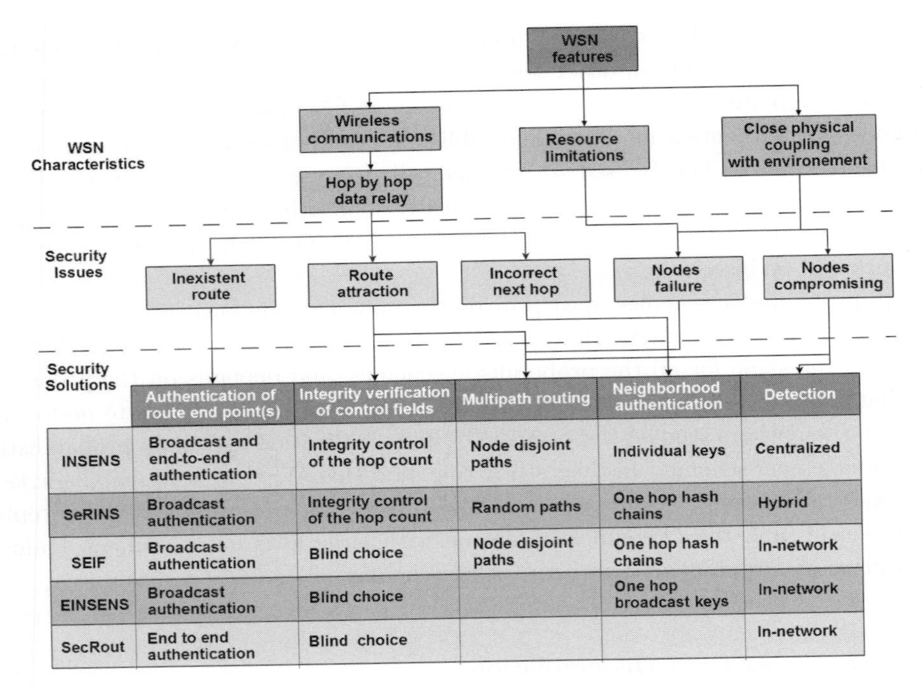

	Authentication of route end point(s)	Integrity verification of control fields	Multipath routing	Neighborhood authentication	Detection
INSENS	Broadcast and end-to-end authentication	Integrity control of the hop count	Node disjoint paths	Individual keys	Centralized
SeRINS	Broadcast authentication	Integrity control of the hop count	Random paths	One hop hash chains	Hybrid
SEIF	Broadcast authentication	Blind choice	Node disjoint paths	One hop hash chains	In-network
EINSENS	Broadcast authentication	Blind choice		One hop broadcast keys	In-network
SecRout	End to end authentication	Blind choice			In-network

Fig. 17.8. Taxonomy of secure route construction in WSN.

system, an intruder will certainly try to disturb the normal operation of this important phase. Figure 17.8 gives a general overview of the security requirements and existing approaches to guarantee the construction of a correct routing state.

Before describing the proposed solutions, we analyze the goals of an intruder and the impact of the attacks at the routing layer:

- The first goal is to *inhibit the routing functionality* of some nodes in the network. This can take two forms: physical and logical inhibition. Logical inhibition is achieved by providing victim nodes with false routing information in order to isolate them from communicating with the sink. Loop creation and false neighborhood relationships are two examples of such attacks.

  On the other hand, attacks based on physical inhibition try to benefit from sensors' constraints to make them physically disconnected from the rest of the network. Generally, these DoS attacks try to exhaust the energy of sensors by forcing them to perform useless calculations, which will have more impact if the protocol employs some cryptographic operations during the route discovery phase.
- The second goal is to take *control over the traffic* by attracting the maximum number of routes. An intruder can "pretend" to have the best routes toward the sink, which gives him an important role in the final routing topology.

To protect a WSN against these malicious behaviors, several defense mechanisms were proposed in the literature. They can be classified into three main axes: *avoidance*, *detection* and *tolerance*.

17.4.1.1.1. Avoidance of route compromising

One of the main goals of a secure routing protocol is to provide each communicating sensor with a correct path to the corresponding sink. In order to choose one path among the discovered ones, routing protocols use some metrics. However, this enables an intruder to modify the metric and to "pretend" having the best route to obtain a predominant position among the routers. To overcome this problem, two classes of solutions were proposed:

- Some existing solutions, like EINSENS,[52] SEIF[60] and SecRout,[53] employ a radical method by using a *blind path choice* that does not reflect any meaningful metric. This approach aims at reducing attack opportunities on the path choice mechanism. EINSENS[52] and SecRout[53] are both based on choosing the first neighbor advertising a correct route as the next hop. The protocol SEIF[60] uses a more resilient blind choice since it selects a random next hop among a set of alternative routes.
- A second family of protocols ensures an integrity control of the hop count metric, used for building short paths. There exist two main approaches to guarantee this integrity. The protocol INSENS[52] employs a totally centralized approach requiring excessive communication overhead, while the protocol SeRINS[54] uses a semi-distributed solution providing a more scalable scheme.

Another avoidance method is *to verify the link state and authenticate the next hop node*, in order to prevent incoherent neighborhood information. Since a WSN may contain powerful intruders, an attacker may use a high-powered transmitter to reach a large set of nodes, to make them believe that they are neighbors of him while they are not. To defend against this Hello flooding attack, each sensor should discover its *reachable neighborhood*, consisting of neighbors having a bidirectional link using a challenge-response mechanism. We can distinguish between three existing mechanisms. In INSENS,[52] sensors are preloaded with individual keys and will generate and exchange proofs of neighborhoods that will be forwarded to the sink node. The latter can verify the proofs and infer correct neighborhood relationships. A second approach, found in EINSENS,[52] is to use *one hop broadcast keys* to authenticate the relay of the RREQ message. Using a key management protocol such as LEAP,[46] each sensor should establish a secret key shared only with its reachable neighborhood. The third approach, used by SeRINS[54] and SEIF,[60] is based on the concept of *one way hash chains* for one hop authentication.

A one way hash chain (OHC) is used as a generator of one way sequence numbers (OWS), which provides an efficient manner for authenticating a sequence of messages in a one-to-many communication. It consists of a sequence of numbers

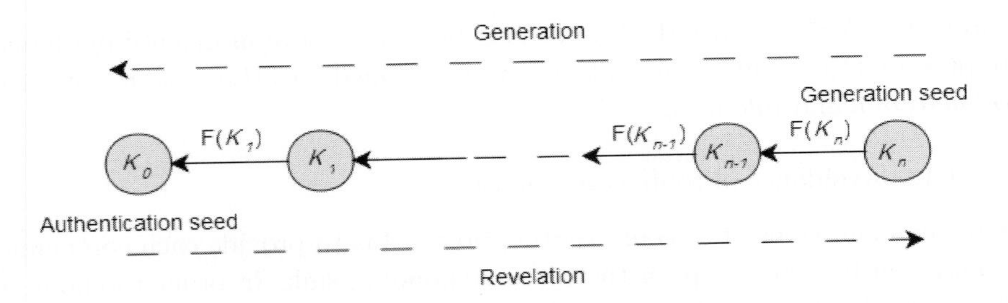

Fig. 17.9. A one way hash chain.

$(K_i)_{0 \leq i \leq n}$ *generated from a random seed* K_n *as follows:* $K_i = F(K_{i+1})$, *where* F *is a one way function (see* Figure 17.9). *This entire chain is stored in the source node, whereas the eventual recipients of future messages are loaded only with the first value* K_0 *(named chain verifier or authentication seed). For each sent message, the source node reveals a new value from the chain in the reverse order of its generation, i.e.* $K_1, K_2, \ldots, K_n$. *Since nodes are initialized with* K_0, *they can verify the belonging of any received value to the chain by using the following property:* $K_0 = F^i(K_i), \forall 0 < i \leq n$.

Whatever the used route compromising avoidance technique, to guarantee that a discovered route really exists, it is necessary to ensure *authentication between route end points*, with defense against replay attacks. In tree-based protocols, such as INSENS,[52] SeRINS,[54] SEIF[60] and EINSENS,[52] the construction is initiated by the sink node using a global diffusion. Thus, this family of protocols employs a broadcast authentication of the sink node using one way hash chains, comparable to the μ Tesla[73] mechanism. However, in reactive protocols, like SecRout,[53] sensors represent the starting points in route discovery. Therefore, communication is one-to-one, allowing an end-to-end authentication based on shared secret keys with the sink.

17.4.1.1.2. Detection of corrupt routes

The verification of route correctness and the correction of the eventual problems is the cornerstone of each secure routing protocol. There exist three classes of mechanisms:

- In centralized solutions, like INSENS,[52] all verifications are processed at the sink node. The latter should collect neighborhood proofs from all the sensors, producing a correct cartography of the whole network.
- In in-network detection schemes, like EINSENS[52] and SEIF,[60] the verification is done hop-by-hop. The sensors perform all security controls without referring to the sink.

A trade off between the above solutions, used in SeRINS,[53] consists in delegating only partial verifications to sensors. Due to this partial information, when a node detects a problem, it cannot make a decision without referring to the sink node. So, the role of the sink node is curative and intervenes only in presence of false routing information.

17.4.1.1.3. Fault tolerance of compromised routes

Since sensors can fail or become compromised, the discovered routes may expire and become unusable. To overcome the shortcomings of a periodical reconstruction, sensors should "anticipate" these situations by constructing alternate paths. This multi-paths solution significantly enhances the resiliency against node failure and presence of intruders. In the literature, the most important secure protocols for WSN providing multi-paths routing are INSENS,[52] SeRINS,[54] and SEIF.[60]

17.4.1.2. *Securing data relay*

Securing route construction is not sufficient to ensure secure data collection in a WSN: an intruder can be interested in falsifying the current readings, in order to make part of the network reflect an inconsistent state from the real sensed environment. This falsification can be conducted by direct injection of bogus data or by modification of forwarded packets. Another issue consists in injecting a large burst of data packets in order to drain the energy of relay nodes. This type of attack is known as a Path-based DoS attacks (PDOS). Figure 17.10 gives an in-depth view of the issues and solutions for the data relay problem.

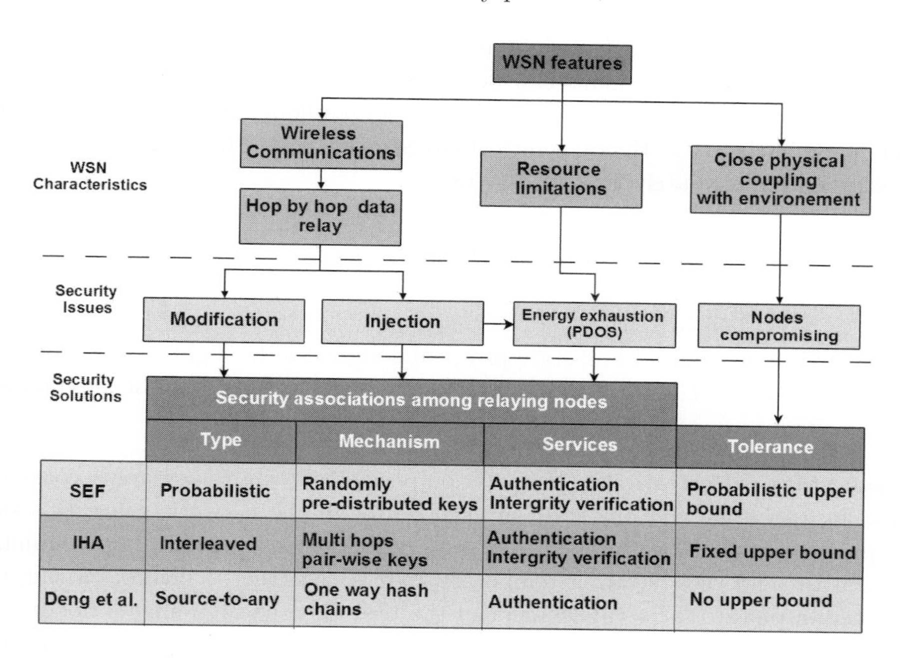

	Type	Mechanism	Services	Tolerance
SEF	Probabilistic	Randomly pre-distributed keys	Authentication Intergrity verification	Probabilistic upper bound
IHA	Interleaved	Multi hops pair-wise keys	Authentication Intergrity verification	Fixed upper bound
Deng et al.	Source-to-any	One way hash chains	Authentication	No upper bound

Fig. 17.10. Taxonomy of secure data relay in WSN.

To address these problems, it is important to embed a data origin authentication mechanism within the relay engine, which will allow an en-route filtering of bogus data. In existing solutions, the essential idea to implement this origin authentication is to establish *security associations among relay nodes.* One naïve approach is to establish for every path one secret key shared among forwarding nodes and the source of the data. Using this secret key, the source node can generate a MAC for each transmitted data, which allows relay nodes to verify the origin of the received data packets. However, this simple solution does not provide any resiliency in presence of malicious nodes inside the path. More sophisticated approaches were proposed to provide a stronger and more resilient data origin authentication during the relay phase:

- Instead of using one secret association among relay nodes, IHA[56] proposed to establish several associations using multi-hop pairwise keys between distant nodes in the path. Each sensor in the path discovers a multi-hop upstream and a downstream node, and establishes a secret key with each one of these associated nodes using an id-based key protocol.[44,59]
- SEF[57] tries to avoid the key establishment phase of IHA by using a random pre-distribution approach. Each sensor is preloaded with a ring of keys from a common pool. The source node sends its data with a MAC generated using a randomly chosen key. SEF relies on the fact that since keys are from the same pool, there is a certain probability that one relay sensor possesses the key used by the source node to generate the MAC.
- A more efficient solution for origin authentication is to use one way hash chains (Deng *et al.*[55]). These chains can be considered as an authentication sequence number that can be generated only by the source node, and can be verified by any node in the path. Hence, intermediate sensors can authenticate any received packet and immediately reject faulty ones.

17.4.2. *Description of some solutions*

17.4.2.1. *INSENS*

INSENS[52] is a centralized link-state protocol designed to be highly tolerant to failures and intruders. The main functional blocks of this solution can be summarized in two points:

- ***Link-state collection*:** The base station periodically starts new route construction by gathering the link-state of each sensor in the network. Since this step is prior to route establishment and sensors have to use a multi-hop communication to send their state, a temporary tree is constructed before starting the collection phase. The establishment of the tree is based on a simple flooding of a RREQ message initiated by the sink node. The latter includes in the RREQ a one way sequence number to enable a one-to-many and network-wide origin authentication. Each sensor chooses the sending node of the first RREQ message as its parent. After this choice, the node relays the RREQ message by adding its identifier, and its neighborhood proof consisting in a MAC generated with its own secret key shared with the base station. Therefore, this hop-by-hop relay of the RREQ message allows nodes to discover the proofs of the direct neighbors and in the same time construct the temporary tree. The collected link-state information, consisting of the list of neighbors and their respective proof, is sent to the base station using a RREP message. This message is relayed from parent to parent until reaching the base station.

- ***Routing table construction*:** After receiving RREP messages, the base station starts the construction of a correct connectivity graph by verifying the coherence of the link-states. This verification is performed by recomputing the proof for each sensor and comparing it with the one provided by every "claimed" neighbor. Afterwards, the base station can apply on the resulting graph any type of path finding algorithm to build the routing tables. INSENS builds a pair of 2-hops disjoint routes for every sensor, meaning that each node in the first path should be at least at two hops far from every node in the second path. The routing table of each sensor i will contain the source node and the next hop for every path passing through i. To communicate this information without flooding the entire network, a path should be used for every node. Since the previous temporary tree is not adequate for this sink-to-node communication, INSENS employs the following solution. The base station operates by rings and sends these tables to nodes at hop h before nodes at hop $h+1$. When a node receives a table sent to another node, it has to check if the target node is included in its routing table as a source node. In this case, the receiving node just relays the message to its neighbors.

17.4.2.2. *EINSENS*

ENINSENS[52] was proposed by the same authors of INSENS to provide a better scalability. However, this new version gives a lower tolerance since it uses a single-path routing. The authors proposed a solution for this shortcoming by "emulating" a multi-path behavior using several sinks with one path to each one of them.

The distributed nature of EINSNES imposed also another constraint. To support the in-network verification, sensors must establish some key materials before starting the protocol, leading to more processing and storage requirements.

EINSENS can be considered as a secure version of the TinyOS beaconing protocol. Periodically, the sink node constructs a new spanning tree by flooding a RREQ message having the following format:

$$i \rightarrow * : i, E(BK_i, ows \| i).$$

The one way sequence number ows and the id of the sending node are encrypted with the broadcast key BK_i to allow a one-hop authentication among reachable neighborhood. When a node j receives such message, it decrypts the secret part and checks if it indicates a valid new round. In this case, node j chooses i as its parent and relays the RREQ using the same encryption procedure. Consequently, each sensor will discover a valid parent resulting in a reliable communication tree rooted at the sink node.

17.4.2.3. *SEIF*

SEIF[60] is a secure multi-path protocol based on a totally in-network verification scheme. In contrast to other solutions, SEIF provides an efficient and secure method to build node disjoint paths in a totally distributed manner without any referring to the base station. Path construction in SEIF is based on the idea of branch-aware route discovery.[63] It consists in *tagging* the exchanged RREQ messages with the identifier of the first relaying node after the base station. We call these nodes *root nodes*, and their sub-trees *branches*. Using these tags, a sensor can easily decide whether two RREQ came from disjoint routes by comparing their branch id. Figure 17.11 shows an example of a branch-aware discovery process. Nodes a, b, and c represent the root nodes of the branches. When two neighboring nodes belong to two distinct branches, they discover a new disjoint path through each other. For instance, node e posses a main route through its main branch rooted at node a, and has also an alternative route through its neighbor h in the branch rooted at node b (and both routes are node disjoint).

To secure this simple but efficient mechanism, SEIF replaces the branch ids with one way hash chains in order to avoid fabrication of bogus branches. In fact, using plain text identifiers for root nodes allows an intruder to attract more routes by

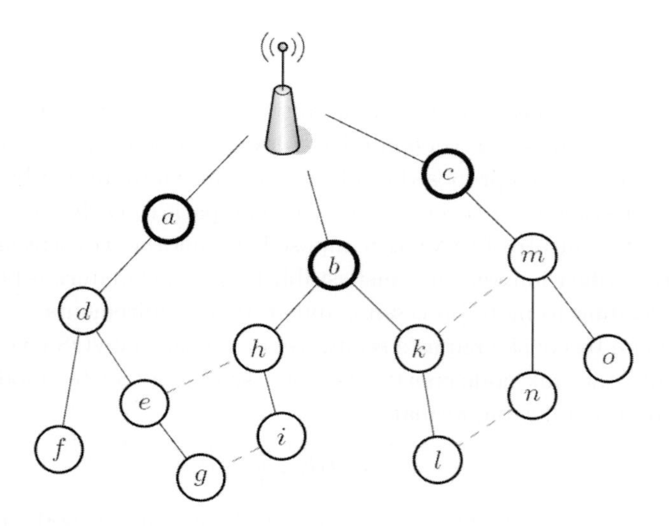

Fig. 17.11. Discovery of node disjoint paths using a branch-aware discovery.

injecting inexistent branches. The one way hash chains will prevent such malicious misbehavior by allowing legitimate sensors to authenticate the root nodes.

Before deployment, a set of hash chains are generated and stored in the base station. Each sensor i is preloaded with the first value of each chain j, which will constitute the initial value of the chain verifier $CV_{i,j}$. In addition, sensor i also maintains the position $P_{i,j}$ of that value within its corresponding chain. At each round, the base station discovers its neighborhood, maps each neighbor i to a hash chain n and sends to it the next value $V$and its position p. This message contains also a one way sequence number ows for round identification and the whole message will be encrypted using the secret key K_i shared between i and the base station:

$$BS \rightarrow i : E(K_i, n \,\|\, p \,\|\, V \,\|\, ows).$$

After verifying the one way sequence number ows, i authenticates the new hash value by verifying the belonging of V to the claimed chain as follows:

$$\begin{cases} p > P_{i,n} \\ CV_{i,n} = F^{p-P_{i,n}}(V) \end{cases}.$$

This authenticated V value will allow i to securely construct its branch since it will be its proof of being a valid root node, i.e. a real neighbor of the base station. Node i forwards the RREQ message having the following format:

$$i \rightarrow * : i, n, p, V, ows, oha.$$

The oha field represents the *one hop authentication* mechanism. It is the next value of a local one way hash chain stored in each sensor. Before deployment, each node i generates its own hash chain and will distribute its first value to the reachable neighborhood.

When a neighbor j receives the RREQ, it will verify the three chain values: V, ows and oha. Especially for the round verifier, node j should accept messages with the same ows in order to discover alternate paths. If these three values belong to their respective chains, j accepts i as an alternative parent since it provides a valid disjoint path form a new branch. After a random period of alternative route discovery, j relays the RREQ by selecting a random parent and relaying the RREQ with the corresponding branch authentication credentials.

SEIF also proposed an extension to find more alternative paths and increase the tolerance of the topology without requiring additional messages. Instead of tagging routes with the roots' ids, the tagging responsibility will be assigned to the neighbors of root nodes, i.e. 2-hops neighbors of the base station. By adding this second level tagging, SEIF allows root nodes as the solely intersection points between routes. Neighboring nodes of roots can become sub-roots and thereby construct their own sub-branches. A sensor will accept paths within the same branch only if they come from different sub-branches, which will increase the number of alternative routes since the basic version presented earlier discards any additional route from an already discovered branch.

17.4.2.4. *SERINS*

SeRINS[53] falls into the category of secure and tolerant routing protocols based on multi-path topologies. One major contribution of this work is its semi-distributed protection of the hop count metric using a set of one way hash chains. Three types of chains are present in SeRINS:

- As for previous protocols, one chain is used for sink authentication and round identification. At each round, the sink node reveals a new value of the chain in the reverse order of its generation.
- An "on the fly" chain is constructed during the relay of the RREQ message to prevent an intruder from decreasing its hop count. This chain is implemented by adding a field in the RREQ message named nI I that will represent the hop count as successive calls to a one way function F on a random seed value.
- The last chain is similar to the previous one and uses a field named nII that aims to detect a malicious node trying to relay the nI of its parent without applying the one way function.

Initially, the BS broadcasts a RREQ with a null hop count and a random nI:

$$BS \rightarrow * : BS, \quad h = 0, \quad nI = r, \quad nII = 0, ows.$$

When a sensor detects a valid new round, it starts an initial phase for discovering its *main parent* before relaying the RREQ. This node is selected as being the neighbor presenting the lowest hop count.[c] When the main parent is selected, the node relays the RREQ as follows:

$$i \rightarrow * : i, \quad h_p + 1, \quad F\left(nI_p\right), \quad F\left(nI_p \| i\right), ows$$

where h_p, nI_p and nII_p represent the received values from the chosen main parent.

Just after relaying the message, node i starts discovering the eventual alternate paths. When i receives a sub-sequent RREQ message $\langle j, h_j, nI_j, nII_j, ows\rangle$ indicating the same round, it has to perform two tests in order to accept the sending node as an alternative parent:

(a) If $h_j \leq h_p$, node i checks whether $F^{h_p - h_j}(nI_j) = nI_p$. This test will be successful only if both $nI_j$and nI_p were generated from the same seed, since $F^{h_p}(r) = nI_p$ and $F^{h_j}(r) = nI_j$.

(b) If $h_j = h_p + 1$, nodes i and j are at the same distance to the base station. To verify that j did not relay the nI of its parent without applying the function F on it, node i uses the nII field. Since nI of the parent of j should be equal to nI_p, i can verify if $nII_j = F(nI_p \| j)$. This means that the parent of node j is at h_p hops and that j relayed correctly the hop count metric of its parent.

[c] One main drawback of SeRINS is that this choice is performed without any security check, which represents a considerable vulnerability.

If one of the previous tests fails, node i sends an alert message to the base station. To guarantee that the attacker will not block this important message, SeRINS sends it using a global network flood. After authenticating this alert, the base station interrogates the neighbors of the suspect nodes in order to collect more information about it. If the collected hop counts are consistent, the suspect node is declared as compromised; otherwise the base station deduces that the source node of the first alert sent a false alarm and is considered as compromised.

17.4.2.5. *SecROUT*

SecRout[53] is a hierarchical routing protocol based on a reactive and sensor-initiated route discovery. Nodes are organized into clusters forming a two-level communication architecture. Cluster-heads collect data readings from their members, and aggregate them using some chosen functions (such as SUM, MAX, AVG, ... etc.). If the cluster-head does not posses a fresh route to send this aggregated report to the base station, it should start a new route discovery process. The node starts by broadcasting a RREQ message containing a MAC generated using the secret key of the cluster-head shared only with the base station. The RREQ message is relayed using a simple flooding technique until reaching the base station. During this relay, the message keeps track of the ids of the last two forwarding nodes. Using this information, each relaying node discovers the two next hops to reach the source cluster-head.

When the RREQ message reaches the base station, the latter authenticates the source node by verifying the provided MAC. If this authentication succeeds, the base station replies with a RREP message, which will travel on the reverse path of the received RREQ message. The RREP message contains also a MAC generated with the secret key of the source cluster-head. Similarly to the RREQ, the RREP message will also keep track of the last two relaying nodes, which will help to trace the path toward the base station. When this RREP reaches the source cluster-head, it verifies the MAC field using its secret key. If the MAC is correct, the node refreshes its routing table and sends the aggregated report to the newly discovered next node. The data message will be relayed hop-by-hop along the path traced by the forwarding of the RREP message.

17.4.2.6. *IHA*

IHA[56] was proposed as a solution for the false data injection problem. The protocol uses the concept of multi-hop pairwise keys to achieve remote security associations among forwarding nodes, which is necessary to ensure origin authentication during the relay (see Figure 17.12). IHA assumes that the network is organized into clusters that should contain at least t members. To establish the security associations, each node should discover its t-hops neighbors along the path in both directions: upstream and downstream. For that, the base station broadcasts a Hello message relayed recursively, which will keep track of the last $t+1$ relay nodes. Therefore, each

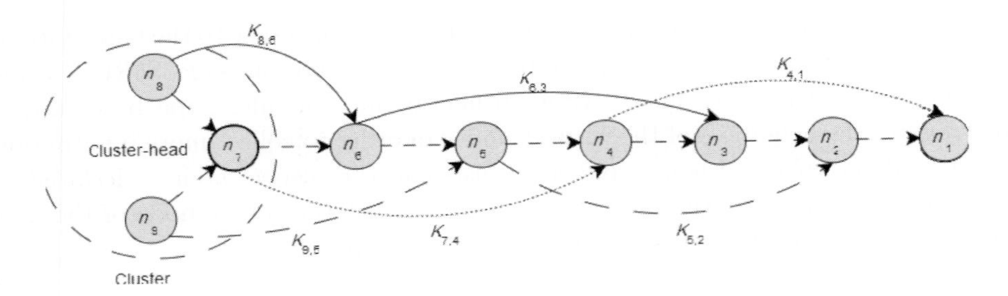

Fig. 17.12. The interleaved hop-by-hop authentication mechanism.

sensor knows the id of its upstream association node, which will be the first node in the received list in the Hello message. When a cluster-head receives the Hello message, it assigns each of the $t + 1$ ids in the message to one of its cluster nodes (including itself). The cluster-head replies to the base station by sending an ACK message using the reverse path. This message is similar to the Hello message and aims to discover the associated downstream nodes in the path. When a sensor discovers a new associated node, it can use an id-based key establishment scheme to generate a pairwise key without additional messages.

IHA is based on a two-level data gathering scheme. When an event is detected in a cluster, at least t members should respond to the event by sending to the cluster-head two MACs: an *individual MAC* computed using the secret key of the node and a *pairewise MAC* computed using the pairwise key shared with the upper association node. The cluster-head collects the t individual and pairwise MACs with the ids of the members to construct the final report. To reduce the size of the packet, the individual reports can be compressed by XORing them into one MAC. When a forwarding node u receives the report, it checks the pairwise MAC of its downstream association node. If the MAC is correct and u is more than $t+1$ hops from the base station (i.e. u has an upstream association node), u replaces this MAC with its own pairwise MAC generated using the key shared with the upstream association node. When the base station receives the report, it can check the compressed individual MACs by recomputing them using the secret keys of the source nodes. The resulting MACs are XORed and compared to the received value.

17.4.2.7. *SEF*

Instead of establishing associations with multi-hop keys, SEF[57] employs a pre-deployment key assignment. Before network installation, a pool of N keys is generated and stored in the base station. This global pool is divided into n non-overlapping partitions of m keys. Keys are also assigned a unique identifier known by every node. Each deployed sensor is preloaded with k random keys from a partition chosen randomly.

When a group of sensors detect an event, they elect a center of stimulus (CoS), which can be similar to the clustering approach of IHA. Each detecting sensor

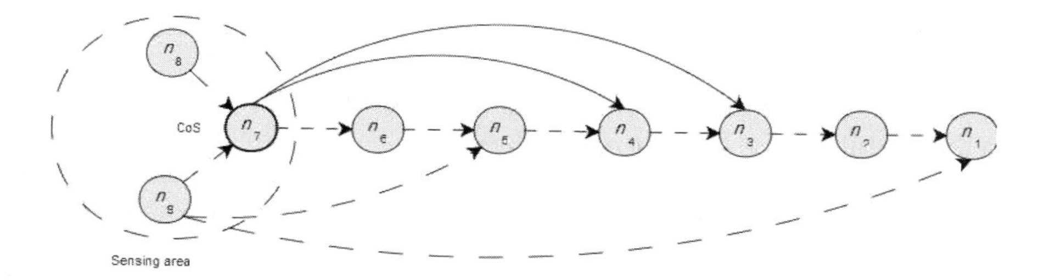

Fig. 17.13. Probabilistic security association in SEF.

chooses randomly one of its k keys and computes a MAC on the event value. The MAC and the id of the used key are sent to the CoS, which reorganizes the received MACs by the partition of their keys. The authors of SEF name the MACs generated by keys from the same partition as one category. The CoS selects T categories and picks one random MAC from each category. These MACs with the ids of the keys and the event value will form the final report. When a relaying sensor receives this report, it will check if it has one of the T keys used in the report. If there is no common key or the node possesses the right key and the MAC is correct, the packet is forwarded to the next hop.

17.4.3. *Analysis and discussion*

17.4.3.1. *Detection overhead*

The impact of a detected attack represents an important criterion for comparing existing solutions. In INSENS, forged routing information will have no consequence on sensor nodes, because nodes are no longer overburden with security checks, which are all performed at the sink level. In distributed solutions, incorrect RREQ messages will also be without negative effects, since any attack is immediately stopped by sensors. Sensors rely only on local information to successfully detect forged routing messages. Therefore, any intrusion attempt is instantly detected without additional delay. In contrast, SeRINS is a hybrid protocol in which sensors can perform only partial verifications that limit the ability of sensors to make local decisions in presence of suspect messages. Indeed, when a sensor detects a suspicious packet, it alarms the sink which must collect more information on the suspect node from its neighbors. This process is achieved via successive broadcasts, which is too expensive in large networks causing additional delay and overhead to detect the intruder.

17.4.3.2. *Tolerance*

In many situations, sensors have to be deployed in hostile or inaccessible environments, making very difficult the manual control and the individual monitoring of sensors. To maximize the network lifetime, the routing topology should tolerate as much as possible the presence of failures by preserving the connectivity of the remaining active nodes. A disconnection occurs when the graph corresponding to the network topology contains more than one connected component. In this case, we will have some sensors without a valid route toward the base station.

When classifying routing algorithm depending on the provided tolerance, we can distinguish between three categories. *Single-path protocols*, like EINSENS and SecRout, provide the poorest resiliency due to the lack of redundancy in their topologies. *Multi-path routing approaches* provide better performance but with different levels. SeRINS builds alternative paths without considering their intersection. This uncontrolled route selection will have a negative impact on the probability of network disconnection. In fact, when a node belongs to more than one path for a given sensor, its failure will cause the disconnection of all these alternative paths. On the other hand, node-disjoint routes do not suffer from such problem since a node failure will cause at most one path to fail for every sensor in the network. Therefore, the impact of a failure on the routing topology decreases, meaning that the system will tolerate more failures, which is the case of SEIF protocol.

17.4.3.3. *Scalability*

To guarantee large scale deployments of WSN, it is important to study the scalability of the proposed solutions and the required number of control messages. Protocols requiring feedbacks from sensor nodes, such as INSENS and SecRout, suffer from a poor scalability. This is due to the one-to-one dialogue between each sensor and the base station, which leads to an excessive communication overhead when dealing with a large number of sensors. By removing the feedback phase, tree-based protocols (like EINSENS, SEIF and SeRINS) improves significantly the overall scalability by requiring only one message per sensor in order to establish the communication topology.

17.5. Secure Aggregation in WSN

Data aggregation is a collection of data readings that represents a collaborative view of a set of nodes. Generally, it is applied in a very specific area limited by the sink node. The latter sends a final aggregated data to the base station using a routing protocol. The number of source nodes sending information to the sink may be large and the information carried within the packets they transmit may be redundant. Data aggregation aims at increasing the energy saving by reducing the forwarding load of intermediate nodes. When data aggregation is used, intermediate nodes merge multiple packets into one to reduce the amount of transmitted packets. By reducing the number of transmitted messages, data aggregation also contributes to reducing the number of collisions especially in dense networks. The aims of securing data aggregation can be summarized in the following points.

- Ensuring authentication and data integrity of aggregation results: Data aggregation requires that each intermediate node be able to read and modify the data transmitted in packets. Under this condition, ensuring data integrity becomes challenging. As using a shared key between each node and the sink to secure the data contents denies the use of aggregation mechanisms, nodes use pair-wise keying instead. A pair-wise key is a common key shared between a node and its upstream node. It allows the upstream node to manipulate the received data without any control. Thereby, securing data aggregation should allow detecting corrupt aggregation operations, and/or injection of faulty data in the aggregation process.
- Preserving the benefits of data aggregation in terms of energy consumption: The main objective of data aggregation is to reduce energy consumption through minimizing data transmission. Thus, securing data aggregation should not thwart this objective through introducing supplementary computations and transmissions.

17.5.1. *Literature overview*

When an intruder node has no access to the key material of a legitimate node, all the proposed protocols can prevent data aggregation corruption. However, when an intruder obtains the key material of a legitimate node, it can carry out attacks by injecting faulty data or by falsifying the aggregation content. To avoid these problems, many protocols have been proposed in the literature.

Data aggregation protocols may be divided into two categories: *End-to-end* encrypted data, and *hop-by-hop* encrypted data, as shown in Figure 17.14.

17.5.1.1. *Protocols based on end-to-end encrypted data*

Protocols of this category make use of a shared key between each node and the sink to guarantee the integrity of the transmitted data. As the data content is encrypted, intermediate nodes use a particular encryption transformation called Privacy Homomorphism (**PH**)[8] to be able to perform aggregation without disclosing the content of data. The **PH** encryption verifies the following property:

$$E_{K1}(x_1) \oplus E_{K2}(x_2) = E_{K1 \oplus K2}(x_1 \oplus x_2),$$

where k_i are keys and x_i are data. The single point of verification in this type of protocols (such as DEPH,[1] CMT,[2] and ECEG[3]) is the sink node that holds all the keys used to encrypt data in the network. This idea has been used in many protocols with different cryptographic techniques such as modular addition[8] that uses a temporary symmetric key, or Elliptic Curve Cryptography.[72]

17.5.1.2. *Protocols based on hop-by-hop encrypted data*

In contrast to protocols based on end-to-end encrypted data, protocols based on hop-by-hop encryption make use of other mechanisms to guarantee integrity while

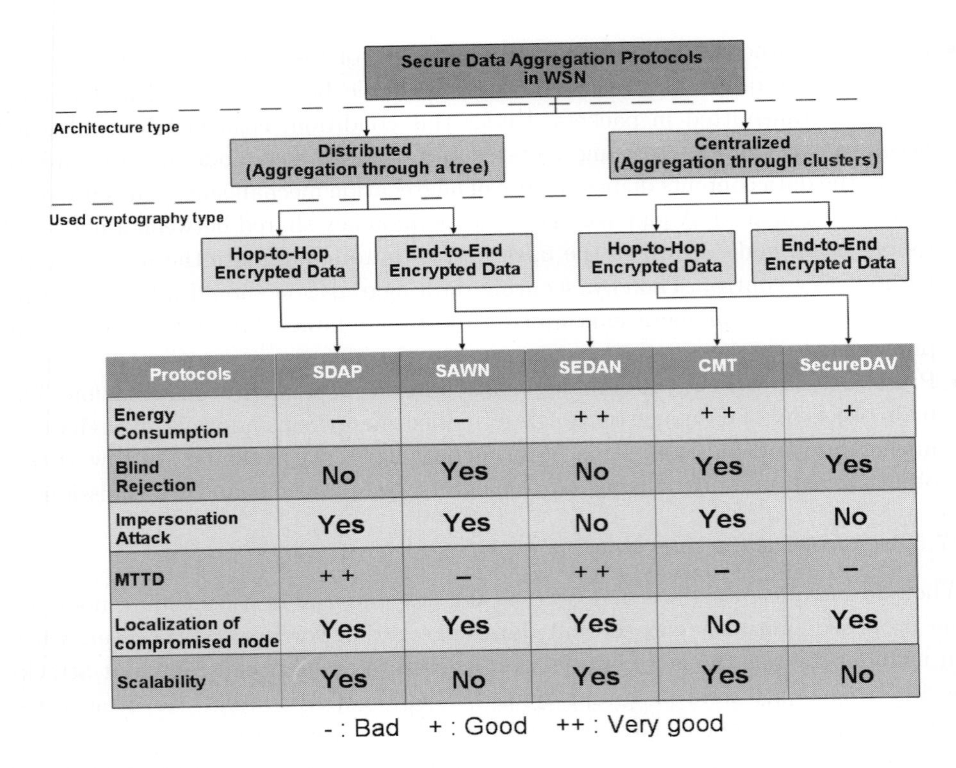

Protocols	SDAP	SAWN	SEDAN	CMT	SecureDAV
Energy Consumption	–	–	+ +	+ +	+
Blind Rejection	No	Yes	No	Yes	Yes
Impersonation Attack	Yes	Yes	No	Yes	No
MTTD	+ +	–	+ +	–	–
Localization of compromised node	Yes	Yes	Yes	No	Yes
Scalability	Yes	No	Yes	Yes	No

- : Bad + : Good ++ : Very good

Fig. 17.14. Taxonomy of secure data aggregation protocols.

allowing data aggregation in plain text. To ensure the integrity of data transmitted between nodes, each protocol uses a different verification mechanism. SAWN[4] proposes a two hop verification mechanism to prevent data modification by the next hop during the aggregation process. Each node generates a MAC of its data to its grandfather in the aggregation tree using a temporary key. This MAC allows the grandfather to verify aggregation value calculated by its child. The verification of MACs is delayed until the sink node discloses the series of temporary keys used to calculate the MACs. Based on the same mechanism of SAWN for verification, Bagaa *et al.* proposed SEDAN.[5] The protocol improves the energy consumption by using a new type of keys between each node and its grandfather. This type of keys eliminates the broadcasting of series of temporary keys. Both SAWN and SEDAN rely on the assumption that the network does not contain two consecutive compromised nodes.

Kui *et al.*,[6] proposed Secure Data Aggregation without Persistent cryptographic operations in WSNs (SDAP). The protocol is based on a clique topology and uses a watch dog mechanism; each node in the same clique can listen to each other, then verify the aggregation value calculated by their parent. Similarly, SecureDAV[7] is a mechanism that uses a collaborative participation of some nodes chosen randomly from the members of a cluster to generate a signature over the average value of the cluster. This signature will be verified by the sink node.

17.5.2. *Description of some solutions*

17.5.2.1. *CMT*

The CMT protocol is based on the assumption that each node shares a symmetric key with the sink node, this key must be renewed at each aggregation process. In this protocol each node encrypts its sensed data by adding the shared key to the data, using a modular addition function. Using the same modular addition function, each aggregator node aggregates all received encrypted data. When the sink node receives the final aggregation value, it decrypts the received value through subtracting the added nodes' keys.

17.5.2.2. *SAWN*

SAWN is a secure hop-by-hop aggregation protocol based on delaying both, the aggregation and the aggregation verification processes. Instead of calculating the aggregation value at the parent level, it is delayed to the grandparent level. Moreover, the verification is delayed until receiving the final aggregation value and revealing all needed temporary keys by the sink node. This protocol is the first which introduces the two hop verification mechanism: first, each node uses its own temporary shared secret key with the sink node to calculate a Child-MAC over the data, and sends both the data and the calculated MAC to its parent. Second, the parent node sends all the received data and MACs to the grandparent node and also a Parent-MAC for the calculated aggregation value over the received data using its own temporary shared secret key. Third, in order to verify the calculated aggregation value, the grandparent node stocks all the received data, Child-MACs and Parent-MACs sent by the parent node. Finally, after receiving all child and grandchild keys, the grandparent node can verify the correctness of the calculated aggregation value.

Example: Figure 17.15 illustrates the different steps described above: nodes A and B send their sensed data, in addition to a MAC calculated over that data using a secret key shared with the sink, to their parent C. E and F do the same with their parent G. The parents C and G calculate the aggregation value using a function f and authenticate the aggregation value with a MAC. Then, they forward the received child data in addition to the authenticated aggregation value to their parent D which repeats the process with its parent. When the sink disclosed the secret keys, the grandparent D will be able to verify the correctness of the received aggregated value. Indeed, node D recalculates the aggregation values and the MACs using the appropriate disclosed keys and compares the results with the MACs received from nodes C and G.

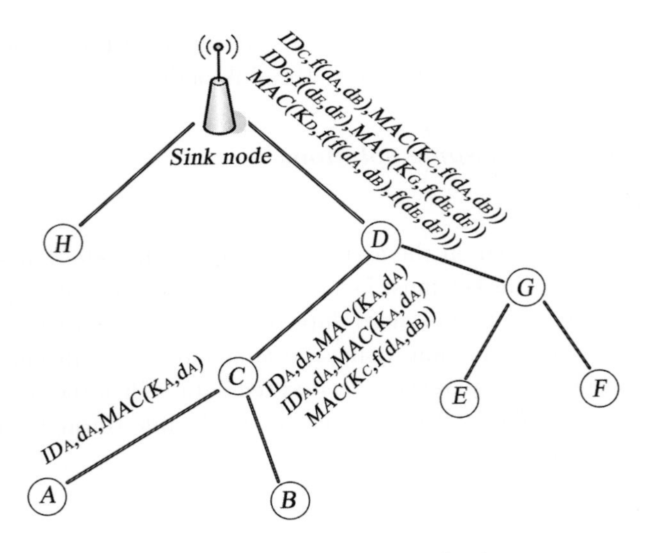

Fig. 17.15. SAWN aggregation and verification processes.

17.5.2.3. *SEDAN*

SEDAN uses the same two hop verification mechanism as SAWN. SEDAN enhances this mechanism by eliminating the use of temporary keys and hence avoids the overhead due to key broadcasting. Indeed, SEDAN introduces a new type of keys shared between each node and its grandparent. The use of this new type of keys allows a better scalability, enhances the time of detecting a malicious node and removes the blind rejection problem. Moreover, SEDAN uses a symmetric key shared between any node and its parent to ensure the origin authentication and eliminate the impersonation attack. The establishment of new type of keys in SEDAN is inspired from LEAP.[46] Each node establishes a pair-wise key shared with its one hop and two hops neighborhood using a Transitory Initial Key (TIK) preloaded into each node prior to deployment. The TIK is erased from the memory of each node after establishing the needed keys to prevent node compromising attack. Figure 17.16 illustrates the aggregation process in SEDAN.

17.5.2.4. *SecureDAV*

SecureDAV is based on a hierarchical topology. It ensures authentication and data integrity using elliptic curve cryptography. Each cluster member uses a secure link to send its data to the cluster head. After receiving all members' data by the cluster head, the latter calculates and broadcasts the average data to the cluster members. Each member compares its data with the received average data, if the difference does not exceed a defined threshold, the cluster member sends to the cluster head the average received data signed with a partial secret key. After that, the cluster head sends the average data and the combined received partial signatures to the SINK

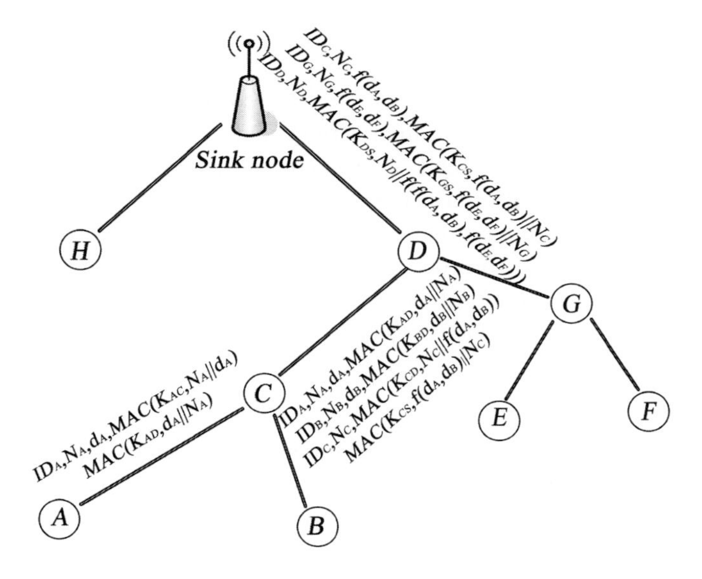

Fig. 17.16. SEDAN aggregation and authentication steps.

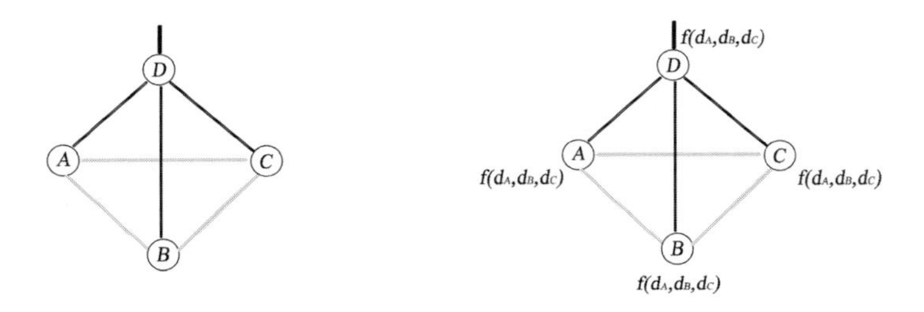

Fig. 17.17. Aggregation and watch dog mechanism in SDAP.

node. To verify integrity, the base station can query repeatedly the cluster-heads on individual readings.

17.5.2.5. *SDAP*

SDAP uses a clique topology, where each clique is composed of a set of neighbouring nodes that elect one parent node. The parent calculates the aggregation value over all the received members' data and sends it to its parent. Because each member in the same clique can listen to the others, every node uses a watch dog mechanism to calculate the aggregation value in the purpose of controlling the correctness of the data sent by their parent (cf. Figure 17.17).

17.5.3. *Analysis and discussion*

17.5.3.1. *Blind rejection*

Blind rejection is an important parameter to evaluate secure aggregation protocols. A protocol suffering from this kind of problem cannot prevent a bogus data from infecting the global aggregation. Both SEDAN and SDAP, by stopping immediately invalid data during the aggregation process, overcome the blind rejection of the final aggregation value. However, protocols proposed in Refs. 4, 2 and 7 suffer from this phenomenon. Indeed, in these protocols, the verification is done at the sink level, and hence the final aggregation value is rejected after it has been relayed up to the sink.

17.5.3.2. *Impersonation attack*

Impersonation attack is the possibility of launching an attack by injecting false data carrying the source address of another node. If this attack happens, the pretended source will be considered as an intruder and then, will be revoked from the network.

In SAWN, when a node detects an invalid MAC, it must exclude the two downward nodes (child and grandchild) from the sensor network. However, there is no mechanism in SAWN that enables to verify the origin of a packet. This enables an intruder to launch an impersonation attack to remove legitimate nodes from the network. In SEDAN and SecureDAV, the use of the pair-wise key between a node and its upstream allows data origin authentication, and rejects any message coming from unauthenticated nodes. CMT and all end-to-end encryption protocols ruin to a local authentication mechanism, allowing the intruder to execute an impersonation attack. The SDAP protocol is vulnerable to the impersonation attack. Indeed, a malicious node can send a faulty data to one selected parent, using the identity of one chosen member node belonging to the same clique of the selected parent. To launch such an attack, the malicious node must be positioned in the neighbourhood of the parent. In this case, the parent calculates a fault aggregation value and hence will be considered as a malicious node by its child members.

17.5.3.3. *Scalability*

Depending on the mechanism used to secure data aggregation, protocols[2,4,5,7] react differently when the network size increases. The revelation of keys in SAWN is based on the assumption that the SINK node can reach all sensor nodes in the network, using only one hop broadcast. When the network size increases, this assumption will be hardly verified. Therefore, SAWN does not scale well with large networks. SecureDAV, which assumes that cluster heads send the aggregation value through only one hop to the SINK node, does not satisfy also scalability requirements. Protocols SEDAN and SDAP, however, are scalable because they rely on a distributed verification mechanism and do not make any reference to the SINK node. Furthermore, CMT that is similar to a simple aggregation process offers a best scalability too.

17.5.3.4. *Localization*

When detecting faulty data aggregation values, it is important to drop them, and also localize the compromised node to revoke it from the network. The protocols based on end-to-end encrypted data suffer from the lack of localization of the intruder, since the sink receives and verifies only the final result. However, the localization of malicious nodes in the protocols based on hop-by-hop encryption is possible because intermediate nodes have access to payload data and thus can detect the malicious nodes that falsify the aggregation.

17.5.3.5. *Resilience against aggregator node capture*

In any aggregation protocol, it is very important to verify the behaviour of aggregator nodes. A compromised aggregator node can falsify the aggregation value by rejecting the received data value from its children or simply modifying it. In the first case, all protocols, except CMT, prevent such attacks by using a simple watchdog mechanism. In the second case, SEDAN and SAWN under the assumption that two consecutive nodes cannot be compromised and by employing the two hops verification mechanism, detect any modification tentative at the parent node level. SDAP by using a watchdog mechanism in the same clique can detect this kind of attacks if the number of compromised nodes is smaller than $n/2$, with n being the number of nodes in a given clique. In SecureDAV, the aggregation values sent by each cluster head are verified by the sink node. Each cluster head sends, in addition to the mean of the received data, the mean data signatures of some of its cluster members. However, detecting the cluster head compromise cannot be guaranteed if some of the cluster members are also compromised. CMT uses the PH encryption in the purpose of detecting any faulty aggregation value. However, authors in Ref. 8 show that it is possible for an attacker to alter the encrypted aggregation value without knowledge of the plaintext, which forbids the detection of an existing compromised aggregator node.

17.6. Securing Channel Access

The wireless communication medium and the limited resources of sensor networks generate a set of vulnerabilities that make them prone to various attacks both at the physical and link layers. The wireless communication medium opens the door to jamming-style DoS attacks and the limited resources to energy exhaustion ones.

17.6.1. *Jamming attacks*

The main goal of a jammer is to break down communication links to prevent sensor nodes from exchanging information. A jammer can impede communication either by preventing nodes from accessing the channel or by letting them access the channel but corrupting the transmitted messages so that they cannot be successfully received.

Depending on its knowledge on the network, the jammer may operate more or less efficiently. If the jammer does not know the technique used in the physical transmission, it can only jam by continuously transmitting strong noise in the suspected communication band to create noise or to cause interference. The presence of noise blocks nodes using CSMA-based channel access methods from accessing the channel as it continuously appears busy to them. With a continuous noise transmission, even TDMA-based channel access methods suffers from communication jamming as the transmitted frames suffer from interferences and thus end up in collisions.

When the jammer has more information on the network it can carry out more efficient jamming. The key idea behind such a clever jamming is to achieve targeted attacks at critical instants, such as causing deliberate collisions, while running at a low duty cycle to save energy. The paper[1] describes a number of energy-efficient jamming techniques that can be used with a large set of well-known link protocols for sensor networks.

The main counter-measures proposed to combat jamming attacks can be classified in two approaches: (i) avoidance or (ii) detection and evasion as shown in Figure 17.18.

17.6.1.1. *Jamming avoidance*

The jamming avoidance approach proposes to use robust communication techniques to compete against both intentional jamming and accidental noise.

Among these schemes, we can cite, for example, the use of Spread Spectrum with a pseudo-random sequence such as Time Hopping that is envisaged for the promising low power UWB transceivers. The use of Spread Spectrum communication can be efficient only if the pseudo-random sequence is kept secret and is long enough to be difficultly detectable. In addition to Spread Spectrum communications, sensor nodes can use directional antennas to minimize omni-directional noise and reduce the jamming-to-signal ratio at the receiver. The main issue with omni-directional antennas is that upper communication layers should adapt to maintain network connectivity. Other techniques such as power control, error correcting codes and lowering the information rate can also be used to increase the rate of successfully received messages at the receivers.

17.6.1.2. *Jamming detection and evasion*

The second counter-measure to jamming contains two parts: detection and evasion. While detection is common to many protocols and naturally executed at the lower layers (i.e. physical and link), evasion may take place even at the upper layers such as the routing layer. At the physical layer, nodes can execute a channel surfing procedure[13] upon the detection of jamming in the communication channel.

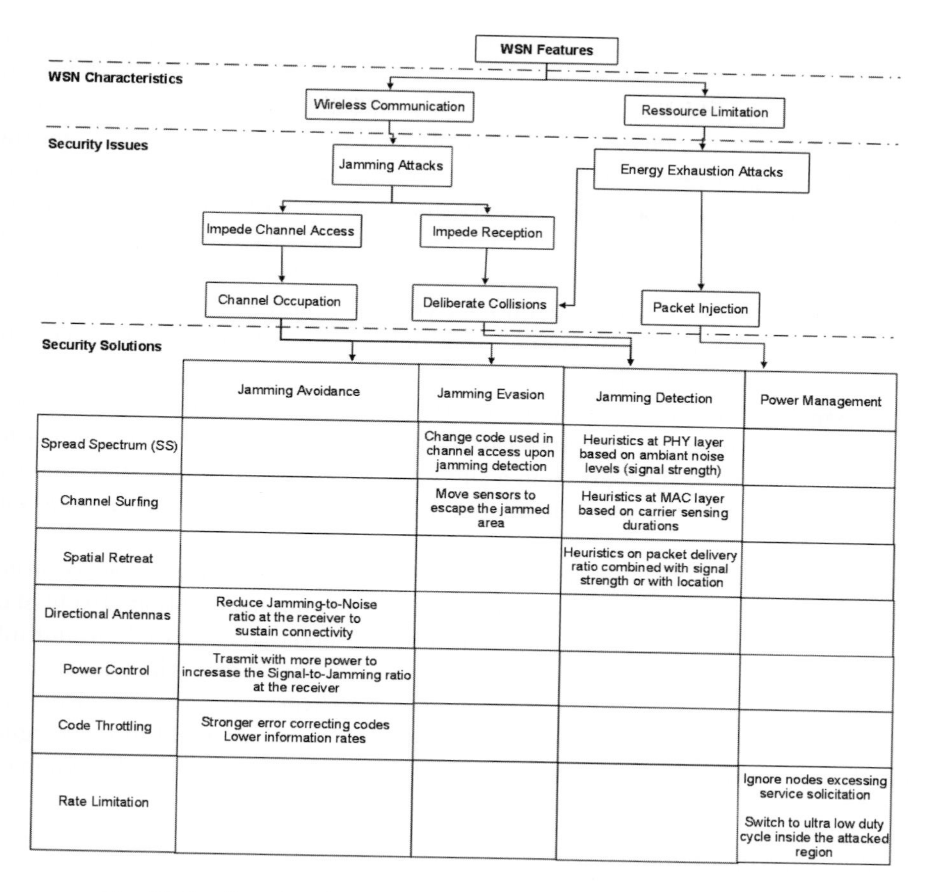

Fig. 17.18. Securing the access to radio channels in WSN: A taxonomy.

In channel surfing, nodes switch to another orthogonal channel to escape jamming. At the upper layers, nodes aim at circumnavigating the jammed region. They can, for example, map the jammed region[14] and route messages around it or physically move to escape jamming.[d]

Jamming detection techniques are based on heuristics. To detect a jammed channel, a node gathers information about channel state in normal situations and uses it as reference to detect jamming situations. The most common detection techniques are those based on signal strength levels and carrier sense durations. For signal strength, a node builds a statistical model describing normal energy levels in the network by gathering measurements on signal strength level during a given time interval then it uses the established model to determine whether a situation is jamming or not. Another way to detect jamming situations is to use carrier sense

[d]Note that the latter is particularity valid for sensor and actuator networks.

duration. By building a statistical model describing the distribution of carrier sense durations a node can determine whether the channel is jammed or not. Jamming can also be detected by building a statistical model for the PDR (Packet Delivery Ratio) and using it to detect pathological situations. Both of the receiver and the transmitter can measure the PDR. The receiver measures the PDR by calculating the ratio of the number of received packets that pass the CRC check to the number of all received packets. However, the transmitter determines the PDR by calculating the ratio of the number of received acknowledgments to the number of transmitted packets. While the PDR can be useful for receivers, it may be inefficient for transmitters to detect an intelligent jammer that only corrupts broadcast messages. To cope with this, SIS (Secure Implicit Sampling)[15] makes it possible for a broadcasting base station to probabilistically detect the failure to receive its broadcast, even if the attacker is intended to insert this attack to be undetectable. By soliciting authenticated acknowledgments from a subset of nodes per broadcast, the subset is unpredictable to the attacker and tunable to avoid acknowledgments implosion at the base station.

The main issue with these methods is that they can lead to false positive situations such as declaring a congestion situation to be a jamming situation. To enhance jamming detection, multimodal detection, in which detection is based on multiple criteria such as combining packet delivery ratio with signal strength or location, is used. For example, if a node detects a low packet delivery ratio then it suspects a jamming situation. The low packet delivery ratio should correspond to weak signal strength or a transmission from a far node to weaken the hypothesis of jamming.

17.6.2. *Energy exhaustion attacks*

In addition to communication jamming, the limited energy resources of sensor nodes make them prone to another type of attacks such as energy exhaustion. This type of attacks may have two scopes: network-wide or local. In contrast to network wide attacks, an attacker carrying out a localized attack aims at exhausting the energy of neighboring nodes. Although such an attack has only localized issues, it may be very dangerous and hard to cope with. This attack is particularly disastrous if the attacker is intending to exhaust critical nodes. Moreover, finding a counter measure for such an attack can be difficult as the node should know at the link layer whether the message it is receiving could be classified as an attack. The only existing solution to those attacks is rate limitation[16] in which a node decides to ignore messages from blacklisted nodes.

17.7. Security Against Sensor Node Compromise

Once deployed, WSN are expected to run without human attendance. This makes nodes highly vulnerable to compromising and physical tampering. If an adversary captures a node, it can easily undertake a large number of attacks exploiting possible hardware or software implementation shortcomings. Preventing such attacks mandates the use of tamper-resistance mechanisms addressing two principal security issues: hardware and software security.

17.7.1. *Hardware security*

In this category, attacks deal with nodes hardware by exploiting the physical-side of implementation flaws. Two types of attacks are distinguished: invasive and non-invasive attacks. Invasive attacks, like micro-probing techniques, require access to the chip-level components of a node in order to observe, manipulate or interfere with the system internals. In contrast, non-invasive attacks do not need to have physical access to a node. For example: Side-channel attacks may allow to break a cryptographic algorithm by simply observing the node's power consumption or its timing profile, which reveals some sensitive information about the executed cryptographic primitives, or the used secret keys. The study presented in Ref. 32 shows how a simple differential power analysis can achieve side-channel attacks on MAC.

Although both types of physical attacks represent a significant threat for sensor networks, few works have been proposed to tackle them. To cope with invasive attacks, we believe that the FIPS 140-2 requirements should be applied to WSN to increase the levels of nodes tamper-resistance. For example, the use of zeroization circuitry that immediately zeroize all plain text secrets and private keys located in memory, when a valid or invalid access is detected.

For non-invasive attacks such as side-channel attacks, future research should focus on mitigating the symptoms that allow the leak of the system's side-channel information like power dissipation, timing and electromagnetic radiations. Possible solutions include randomizing the clock signal or the instruction execution sequence, introducing dummy instructions, balancing Hamming weights of the internal data and bit splitting.

17.7.2. *Software security*

In this category, attacks aim at compromising the software running on nodes by exploiting known vulnerabilities. This kind of attacks is generally based on malicious software, like viruses and worms, and can be based on various techniques. In interception-based software attacks, the goal is to passively eavesdrop sensitive data. An attacker may make use of a logical analyzer to probe the inner lines of a node. Interruption-based attacks target destabilizing the software to disrupt the system availability. This kind of DoS attacks is facilitated due to nodes limited resources (energy, computation and memory). Finally, modification-based attacks, such as buffer overflow, modify the software code to compromise its integrity. Current WSN are considerably vulnerable to software attacks. For instance, TinyOS, a widely used operating system for sensor networks, does not provide any user/process access control to system resources. Moreover, its serial forwarder component allows an adversary to open a port to a node without any authentication mechanism. This greatly facilitates uploading malicious code or downloading sensitive information from the sensor.

We believe that research efforts should be conducted in order to adapt software security issues like secure OS bootstrapping, or secure software design and coding, to WSN software. Techniques such as software authentication and validation using remote software-based attestation,[33] the use of restricted environment such as Java Virtual Machine (JVM) to execute untrusted code, and the use of encryption wrappers to allow dynamic run-time software encryption/decryption should also be adapted for WSN.

17.8. Challenges for Future Research

The unsuitability of asymmetric cryptography has driven research in key management to symmetric cryptography and thus to key pre-distribution methods. We believe that this situation may change in the future and low overhead asymmetric cryptography methods based on Elliptic Curve Cryptography (ECC)[72] will aid to solve many challenges relating to key management and authentication.

The absence of a hardware protection to prevent tampering, expose sensor nodes to a physical compromising by an adversary. The latter can extract the key materials contained in a deployed node and launch multiple attacks. To relax the consequences of such vulnerability, authors in Ref. 62 define two properties that must be verified in the design of a tolerant key management scheme: (1) *the opaqueness property* — an adversary cannot deduce most of the keys being used in the network by compromising a small number of sensor nodes. (2) *The inoculation property* — an adversary cannot aid unauthorized sensor node to successfully join a network by compromising a small number of sensor nodes. Thus future key management schemes should tolerate the compromise of some sensors while maintaining a safe operation of the whole network.

So far, existing secure routing solutions have been focused mainly on ensuring a correct routing state for every sensor. By analyzing Figure 17.18, we remark clearly that even the best secure routing topology in terms of performance is based on the hop count metric. Nevertheless, WSN communication mechanisms should rather be based on more meaningful properties that address real problems. By choosing appropriate metrics, the resulting topology should be more "profitable" aiming at increasing the survivability and the usability of the network. To achieve this, future protocols should devise secure ways for building paths using sensor and environment oriented metrics, such as energy and link reliability.

For the data relay problem, there are still some important problems remaining unresolved by WSN security community. Existing solutions that provide countermeasures against the false reports generated by compromised source nodes are based on data comparison by a controller node (such as a cluster-head or a Center of Stimulation). Consequently, this method can only be applicable if the controller node can collect a certain number of samples from the same sensing region. However, sparse networks do not have such correlation between readings and collected data may vary dramatically. For instance, an event detection network with sparse nodes can observe an event only from one node, and data comparison method cannot be used in this scenario. Therefore, a compromised node can easily cheat in its reports with a low probability for being detected.

The use of efficient cryptographic techniques to guarantee integrity for aggregated data should be reinforced with lightweight statistical methods to detect faulty and corrupt data. Indeed, while cryptographic mechanisms guarantee that data is transmitted without modification, there is no mean to detect faulty data originating from a corrupt node using cryptographic techniques.

The most efficient way to cope with jammers in wireless systems is hiding the communication channel by using secret pseudo-random spread spectrum codes. Although this technique, largely used in military communication, is efficient, it may not apply to most of the wireless sensor networks. Current applications envisaged for sensor networks are built upon open standards to ensure inter-operability between different manufacturers product. For example, in Zigbee Alliance,[17] the wireless communication is based on the IEEE 802.15.4[18] open standard physical specifications. Therefore, it is very vulnerable to intelligent jamming at the physical layer. In addition, even if the physical and access layer specifications are not kept secret at the instant of deployment, they can be discovered by means of traffic analyzing or node capture. Therefore, securing channel access should include jamming detection techniques that should be able to efficiently map the jamming region. When the jamming region is detected, nodes need to either evade and/or compete with the jammers. In the evasion techniques such as channel surfing or spatial retreat, most challenges are traditional and related to distributed computing. Nodes should cooperate efficiently to provide low-latency, convergent and scalable solutions. For example, when nodes physically move to evade the jammed region, they should move efficiently to generate a new advantageous topology maintaining routes between communicating nodes. The same applies for channel surfing. In addition, in channel surfing, it should be determined whether all the nodes or only the subset of them that are affected by jammers should switch to another channel. In both case, coordination is needed to maintain connectivity throughout the network.

To realize their full potential, sensor networks require connectivity to the Internet. One benefit in connecting sensor networks to the Internet using IPv6, is to take advantages of the huge (128-bit) address space of IPv6.[61] Preparing sensor networks for IP communication and integrating them into the Internet, however, requires certain features and specification to work, for example, in the adaptation of the respective link technology, specification of ad hoc networking, handling the security issues, and auto-configuration to support ad hoc deployment. Security is one major concern in every part of the Internet, covering areas like encryption, detection of intrusion, access control, authentication, authorization, integrity protection, prevention of denial of service etc. In principle, in IP-enabled sensor networks standard security mechanisms based on IP could be applied. However, especially sensor networks are resource constraint concerning processing power and network bandwidth, putting limits on security. Therefore, new lightweight security mechanisms appropriate for sensor networks have to be used. On the other hand, adaptation of the existing solutions to take advantage of available IPv6 services with regard to security, auto-configuration and mobility management will be required.

Most of the proposed security mechanisms are designed for static networks with flat organization. However, several applications require using mobile sensors/actuators to save energy and to increase the connectivity of the network. Future security mechanisms should take into consideration the existence of mobile components in the network. This will have necessarily an impact on route construction and updates, authentication mechanisms of mobile actuators, and data dissemination which would be mostly delivered through one hop transmissions.

17.9. Conclusions

In this paper, we provided a comprehensive taxonomy of security attacks in sensor networks and their corresponding solutions. Through the survey of existing work, we noticed that research on security issues have matured over the years. However, there remain several open problems that need to be resolved to make WSN secure to the extent required by their applications. We noted that secure data dissemination is an important area in WSN security as it addresses a core service in sensor networks, namely routing protocols. The main lesson learned in this field, is that conventional solutions, early proposed for ad hoc networks are not applicable. Moreover, secure routing for specific classes of sensor networks, such as sensor-actor networks or underwater sensor networks, remains highly unexplored. Routing services in such networks differs in terms of node mobility, and traffic patterns. Hence, adequate schemes need to be developed in order to guarantee their security. Concerning data aggregation security, most of the proposed solutions are still exposed to one or more security threats. Proposing more complete solutions, based on a fully distributed approach, constitute an attractive research field. Finally, we remarked that few works have been devoted to address security issues in link and physical layers. Similarly, the domain of physical and software security of the sensor mote remains in its infancy, and an additional endeavour is mandatory to prevent sensor node compromise, which represents a significant threat for the whole mission of the network.

Problems

1. Why asymmetric cryptography can hardly be used in securing WSN? Is there a specific asymmetric cryptosystem that can be used to secure WSN? Which one, and why?
2. In probabilistic key pre-distribution scheme, each sensor is preloaded with a key ring composed of m keys randomly picked from a huge key pool. Explain why the resiliency of such a scheme is inversely proportional to the key ring size (m)?
3. Explain why SAWN and SecureDAV do not scale to very large networks, while SEDAN and SDAP scale better?
4. Explain how SEDAN and SecureDAV allow authenticating data origin? And explain why SAWN is vulnerable to impersonation attacks?
5. How can an intruder exclude/revoke a valid sensor from a network using SDAP for data aggregation?
6. Is it always possible to localize the node responsible for injecting faulty aggregation value in a WSN? Explain.
7. In SeRINS, is it possible that a legitimate node chooses an intruder as a main parent? If yes, explain what could be the impact on the security of the protocol.
8. EINSENS uses one hop broadcast keys to authenticate one hop communications during the relay of the RREQ. Explain why these types of keys are not suitable for such communications, and give an example of a possible attack.
9. In your opinion, what is the best layer for implementing an efficient solution against energy exhaustion attacks? Why most of the proposed solutions are inefficient?

Bibliography

1. J. Domingo-Ferrer, A provably secure additive and multiplicative privacy homomorphism, *ISC '02: Proceedings of the 5th International Conference on Information Security* (2002).
2. C. Castelluccia, E. Mykletun and G. Tsudik, Effcient aggregation of encrypted data, wireless sensor networks, *MobiQuitous, IEEE Computer Society* (2005) 109–117.
3. E. Mykletun, J. Girao and D. Westhoff, Public key based cryptoschemes for data concealment in wireless sensor networks, *IEEE International Conference on Communications (ICC2006)* (2006).
4. L. Hu and D. Evans, Secure aggregation for wireless networks, *Workshop on Security and Assurance in Ad Hoc Networks* (January 2003).
5. M. Bagaa, N. Lasla, A. Ouadjaout and Y. Challal, SEDAN: secure and efficient protocol for data aggregation in wireless sensor networks, *32nd IEEE Conference on Local Computer Networks (LCN 2007), Workshop on Network Security*, pp. 1053–1060.
6. K. Wua, D. Dreefa, B. Sunb and Y. Xiao, Secure data aggregation without persistent cryptographic operations in wireless sensor networks, *Ad Hoc Networks* **5**(1) (2006) 100–111.
7. A. Mahimkar and T. S. Rappaport, *SecureDAV: A Secure Data Aggregation and Verification Protocol for Sensor Networks* (2004).

8. S. Peter, K. Piotrowski and P. Langendoerfer, On concealed data aggregation for wireless sensor networks, *IEEE CCNC 2007*, Las Vegas Nevada, USA (January 2007).
9. H. Chan, A. Perrig and D. Song, Secure hierarchical in-network aggregation in sensor networks, *Proceedings of the ACM Conference on Computer and Communications Security*, New York, NY, USA (2006) 278–287.
10. H. O. Sanli, S. Ozdemir and H. Cam, SRDA: secure reference-based data aggregation protocol for wireless sensor networks, *Proceedings of the VTC Fall 2004 Conference*, Los Angeles, CA, USA (2004).
11. Y. Sang and H. Shen. Secure data aggregation in wireless sensor networks: a survey, *7th International Conference on Parallel and Distributed Computing, Applications and Technologies (PDCAT'06)* (2006) 315–320.
12. Y. Law, L. Van Hoesel, J. Doumen, P. Hartel and P. Havinga, Energy-efficient link-layer jamming attacks against three wireless sensor network MAC protocols, *Proceedings of ACM SASN*, Alexandria, Virginia (November 2005).
13. W. Xu, K. Ma, W. Trappe and Y. Zhang, Jamming sensor networks: attack and defense strategies, *IEEE Network* **20**(3) (2006) 41–47.
14. A. Wood, J. Stankovic and S. Son, *JAM: A Mapping Service for Jammed Regions in Sensor Networks*, RTSS, Cancun, Mexico (December 2003).
15. J. McCune, E. Shi, A. Perrig and M. Reiter, Detection of denial-of-message attacks on sensor network broadcasts, *Proceedings of the IEEE Symposium on Security and Privacy*, Oakland, CA (May 2005).
16. A. Wood and J. Stankovic, Denial of service in sensor networks, *IEEE Computer* (October 2002) 48–56.
17. www.zigbee.org.
18. IEEE 802.15.4, *Wireless Medium Access Control (MAC) and Physical Layer (PHY) Specifications for Low-Rate Wireless Personal Area Networks (LR-WPANs)* (2003).
19. W. Dai, *Comparison of Popular Cryptographic Algorithms*, http://www.eskimo.com/\ ∼weidai/benchmarks.html (2003).
20. R. Shirey, Internet security glossary, *RFC 2828* (May 2000).
21. A. J. Menezes, P. C. van Oorschot and S. A. Vanstone, *Hand book of Applied Cryptography* (CRC Press, 1996).
22. B. Schneier, *Applied Cryptography: Protocols, Algorithms, and Source Code in C* (1996).
23. C. Kaufman, R. Perlman and M. Speciner, *Network Security: Private Communication in a Public World*, Printice Hall Series in Computer Networking and Distributed Systems (2002).
24. B. Kaliski, The MD2 message-digest algorithm, *RFC 1319* (1992).
25. R. Rivest, The MD5 message-digest algorithm, *RFC 1321* (1992).
26. D. Eastlake and P. Jones, US secure hash algorithm 1 (SHA1), *RFC 3174* (2001).
27. H. Krawczyk, M. Bellare and R. Canetti, HMAC: keyed-hashing for message authentication, *RFC 2104* (1997).
28. Federal Information Processing Standards Publication, Data encryption standard (DES), *FIPS PUB 46* (1993).
29. Federal Information Processing Standards Publication, Advanced encryption standard (AES), *FIPS PUB 197* (2001).
30. R. L. Rivest, A. Shamir and L. M. Adelman, A method for obtaining digital signatures and public-key cryptosystems, *Communications of the ACM* **21**(2) (1978) 120–126.
31. Federal Information Processing Standards Publication, Digital signature standard (DSS), *FIPS PUB 186* (1994).

32. K. Okeya and T. Iwata, Side channel attacks on message authentication codes, *ESAS 2005 European Workshop on Security and Privacy in Ad Hoc and Sensor Networks* (2005).
33. M. Shanek, K. Mahadevan, V. Kher and Y. Kim, Remote software-based attestation for sensor networks, *European Workshop on Security and Privacy in Ad Hoc and Sensor Networks* (2005).
34. A. S. Wander *et al.*, Energy analysis of public-key cryptography for wireless sensor networks, *PerCom '05: Proceedings of the 3rd IEEE International Conference of Pervasive Computing and Communications* (2005).
35. N. Gura *et al.*, Comparing elliptic curve cryptography and RSA on 8-bit CPUs, *6th International Workshop on Cryptographic Hardware and Embedded Systems*, Boston, Massachusetts (2004).
36. G. Gaubatz, *et al.*, Public key cryptography in sensor networks-revisited, *ESAS '04: 1st European Workshop Security in Ad-Hoc and Sensor Networks* (2004).
37. K. Piotrowski *et al.*, How public key cryptography influences wireless sensor node lifetime, *SASN'06*, Alexandria, Virginia, USA (2006).
38. A. Liu and P. Ning, *TinyECC: Elliptic Curve Cryptography for Sensor Networks (Version 0.3)* (February 2007), available at http://discovery.csc.ncsu.edu/software/TinyECC/.
39. D. J. Malan, M. Welsh and M. D. Smith, A public-key infrastructure for key distribution in TinyOS based on elliptic curve cryptography, *Proceedings of the 1st IEEE International Conference of Sensor and Ad Hoc Communications and Networks*, Santa Clara, CA (2004).
40. H. Wang, B. Sheng and Q. Li, Elliptic curve cryptography-based access control in sensor networks, *International Journal of Security and Networks* **1**(3/4) (2006).
41. L. Eschenauer and V. D. Gligor, A key-management scheme for distributed sensor networks, *Proceedings of the 9th ACM Conference on Computer and Communications Security* (2002).
42. H. Chan, A. Perrig and D. Song, Random key predistribution schemes for sensor networks, *IEEE Symposium on Security and Privacy*, Berkeley, CA (11–14 May, 2003) 197–213.
43. W. Du, J. Deng, Y. S. Han, S. Chen and P. K. Varshney, A key management scheme for wireless sensor networks using deployment knowledge, *Proceedings of IEEE INFOCOM'04* (IEEE Press, Hong Kong, 2004) 586–597.
44. C. Blundo *et al.*, Perfectly-secure key distribution for dynamic conferences, *Proceedings of the 12th Annual International Cryptology Conference on Advances in Cryptology* (Spring-Verlag, Berlin, 1992) 471–486.
45. Z. Yu and Y. Guan, A robust group-based key management scheme for wireless sensor networks, *Proceedings of IEEE Wireless Communications and Networking Conference (WCNC 2005)*, New Orleans, LA (USA, IEEE Press, 2005) 13–17.
46. S. Zhu, S. Setia and S. Jajodia, LEAP: efficient security mechanisms for large-scale distributed sensor networks, *ACM Conference on Computer and Communications Security (CCS '03)* (2003).
47. X. Du *et al.*, An effective key management scheme for heterogeneous sensor networks, *Security Issues in Sensor and Ad Hoc Networks* (2007) 24–34.
48. X. Du *et al.*, An improved key distribution mechanism for large-scale hierarchical wireless sensor networks, *Security Issues in Sensor and Ad Hoc Networks* (2007) 35–48.
49. H. Fu *et al.*, Replication attack on random key pre-distribution schemes for wireless sensor networks, *IEEE Workshop on Information Assurance and Security*, United States Military Academy, West Point, NY (2005).

50. R. Blom, An optimal class of symmetric key generation systems, *Proceedings of the Eurocrypt 84 Workshop on Advances in Cryptology: Theory and Application of Cryptographic Techniques* (Springer Verlag, 1985) 335–338.
51. J. Deng, C. Hartung, R. Han and S. Mishra, A practical study of transitory master key establishment for wireless sensor networks, *Proceedings of the 1st IEEE International Conference of Security and Privacy for Emerging Areas in Communication Networks* (*SecureComm '05*) (2005).
52. J. Deng, R. Han and S. Mishra, Intrusion-tolerant routing for wireless sensor networks, *Computer Communications* **29**(2) (2006) 216–230.
53. J. Yin and S. Madria, Secrout: a secure routing protocol for sensor networks, *AINA 06: Proceedings of the 20th International Conference on Advanced Information Networking and Applications*, Washington, DC, USA, IEEE Computer Society **1** (2006) 393–398.
54. S. Lee and Y. Choi, A secure alternate path routing in sensor networks, *Computer Communications* **30**(1) (2006) 153–165.
55. J. Deng, R. Han and S. Mishra, Defending against path-based DoS attacks in wireless sensor networks, *3rd ACM Workshop on Security of Ad Hoc Sensor Networks*, New York, USA (2005) 89–96.
56. S. Zhu, S. Setia, S. Jajodia and P. Ning, Interleaved hop-by-hop authentication against false data injection attacks in sensor networks, *ACM Transactions on Sensor Networks* **3**(3) (2007) 14.
57. F. Ye, H. Luo, S. Lu and L. Zhang, Statistical en-route filtering of injected false data in sensor networks, *IEEE INFOCOM'04* **4** (2004) 2446–2457.
58. L. Lamport, Password authentication with insecure communication, *Communications of the ACM* **24**(11) (November 1981) 770–772.
59. D. Liu and P. Ning, Establishing pairwise keys in distributed sensor networks, *Proceedings of the 10th ACM Conference on Computer and Communications Security (CCS '03)*, Washington DC (2003).
60. A. Ouadjaout, Y. Challal, N. Lasla and M. Bagaa, SEIF: secure and efficient Intrusion-Fault tolerant routing protocol for wireless sensor networks, *IEEE-ARES* (2008).
61. K. Das, *IPv6 and Wireless Sensor Networks*, http://www.ipv6.com/articles/sensors/IPv6-Sensor-Networks.htm.
62. J. Deng, C. Hartung, R. Han and S. Mishra, A practical study of transitory master key establishment for wireless sensor networks, *Proceedings of the First IEEE International Conference of Security and Privacy for Emerging Areas in Communication Networks (SecureComm '05)* (2005).
63. W. Lou and Y. Kwon, H-spread: a hybrid multipath scheme for secure and reliable data collection in wireless sensor networks, *IEEE Transactions on Vehicular Technology* **55**(4) (2006) 1320–1330.
64. G. Jolly, M. C. Kuscu, P. Kokate, and M. Younis, A low-energy key management protocol for wireless sensor networks, *IEEE Symposium on Computers and Communications (ISCC'03)*, Kemer, Antalya, Turkey (30 June–3 July, 2003).
65. H. Chan and A. Perrig, PIKE: peer intermediaries for key establishment in sensor networks, *INFOCOM* (2005).
66. D. Liu and P. Ning, Establishing pairwise keys in distributed sensor networks, *Proceedings of 10th ACM Conference on Computer and Communications Security (CCS' 03)*, Washington DC (ACM Press, 2003) 41–47.
67. D. Liu and P. Ning, Improving key pre-distribution with deployment knowledge in static sensor networks, *ACM Transactions on Sensor Networks* **1**(2) (2005) 204–239.
68. Z. Yu and Y. Guan, A robust group-based key management scheme for wireless sensor networks, *Proceedings of IEEE Wireless Communications and Networking Conference (WCNC 2005)*, New Orleans, LA (USA, IEEE Press, 2005) 13–17.

69. D. Huang, M. Mehta, D. Medhi and L. Harn, Location-aware key management scheme for wireless sensor networks, *Proceedings of ACM Workshop on Security of Ad Hoc and Sensor Networks (SASN' 04)*, Washington DC (USA, ACM Press, 2004) 29–42.
70. J. Lee and D R. Stinson, Deterministic key predistribution schemes for distributed sensor networks, *Proceedings of ACM Symposium on Applied Computing 2004*, Waterloo, Canada, Lecture Notes in Computer Science, Vol. 3357 (Springer, 2004), 294–307.
71. Y. Cheng and D. P. Agrawal, Efficient pairwise key establishment and management in static wireless sensor networks, *Proceedings of the 2nd IEEE International Conference on Mobile Ad-Hoc and Sensor Systems (MASS'05)* (November 2005), 544–550.
72. I. F. Blake, G. Seroussi and N. P. Smart, Advances in elliptic curve cryptography, *London Mathematical Society Lecture Note Series*, No. 317 (2005).
73. A. Perrig, R. Szewczyk, J. D. Tygar, V. Wen and D. E. Culler, Spins: security protocols for sensor networks, *Wireless Networks* **8**(5) (2002) 521–534.
74. M. Passing and F. Dressler, Experimental performance evaluation of cryptographic algorithms on sensor nodes, *3rd IEEE International Conference on Mobile Ad Hoc and Sensor Systems (IEEE MASS 2006)* 882–887.
75. K. J. Choi and J-I. Song, Investigation of feasible cryptographic algorithms for wireless sensor network, *Proceedings of the 8th International Conference on Advanced Communication Technology (ICACT 2006)* (2006).
76. K. Piotrowski, P. Langendoerfer and S. Peter, How public key cryptography influences wireless sensor node lifetime, *Proceedings of the 4th ACM Workshop on Security of Ad Hoc and Sensor Networks, SASN'06*, ACM, New York, NY (2006).

PART 3
Wireless Mesh Networks

Chapter 18

Network Architecture and Flow Control in Multi-Hop Wireless Mesh Networks

Deepti Nandiraju*, Nagesh Nandiraju† and Dharma P. Agrawal‡

OBR Center for Distributed and Mobile Computing
Department of Computer Science, University of Cincinnati
Cincinnati, OH 45221-0030
* *nandirds@cs.uc.edu*
† *nandirns@cs.uc.edu*
‡ *dpa@cs.uc.edu*

Recently, Wireless Mesh Networks (WMNs) have drawn considerable attention in academia and industry due to their potential to supplement the wired backbone in a cost-effective manner. A WMN comprises of Mesh Routers (MRs) forming the backbone of WMN and serve Mesh Clients (MCs) in their neighborhood. A small set of these MRs provide Internet connectivity, and these MRs are referred to as Internet Gateways (IGWs). The remaining MRs form into network's backbone and employ multi-hop communication paradigm to provide relay services to MCs, which are the end users.

The key advantages of WMNs are their rapid deployment and ease of installation. WMNs are capable of providing attractive services in a wide range of application scenarios such as broadband home/enterprise/community networking, public safety surveillance systems and disaster management. However, unpredictable interference, excessive congestion, and half-duplex nature of radios at MRs may hinder deployment of WMNs. Traffic in WMNs is predominantly between IGWs and the MRs in contrast to Mobile *Ad hoc* Networks (MANETs) where traffic is among peer nodes. This focused traffic flow within WMNs towards and from IGWs places higher demand on certain paths connecting IGWs and MRs, unlike that of MANETs where the traffic is more or less uniformly distributed.

Ensuring fairness to multi-hop flows in WMNs is a critical issue that needs to be considered for their successful deployment. In this chapter, we provide an insight to these issues, analyze various approaches discussed in the literature, and evaluate their applicability to WMNs.

Keywords: Wireless Local Area Networks; Wireless Mesh Networks; Fairness; buffer management; multi-radio multi-channel architecture.

18.1. Introduction

Due to the growing demand for ubiquitous broadband Internet connectivity and a widespread use of applications such as multimedia streaming (VoIP services, video streaming etc), wireless networking technology has been growing tremendously in recent years[1] [Nandiraju *et al.* (2007)]. Some key advantages of WMNs include their self-organizing ability, self-healing capability, low-cost infrastructure, rapid deployment, scalability, and ease of installation. WMNs are capable of providing attractive services in a wide range of application scenarios such as broadband home/enterprise/community networking and disaster management.

The mesh-networking technology attracted both academia and industry stirring efforts for their real-world deployment in a variety of applications. MIT deployed WMN in one of its laboratories for studying the industrial control and sensing aspects. Several companies like Nortel Networks, Strix Systems, Tropos Networks, MeshDynamics are offering mesh networking solutions for applications such as building automation, small and large scale internet connectivity, etc., using customary products. Strix systems has deployed a city-wide Wi-Fi mesh network in Belgium spanning an area of 17.41 KM^2 to provide wireless Internet access to its residents, tourists, businesses, and municipal and public-safety applications and advertising systems around the city. Strix also deployed a wireless tracking system called *project kidwatch* that traces the real-time location of a child in a beach area or around a city.

Further commercial interests in WMNs have prompted immediate and increasing attention for integrating WMNs with the Internet. IEEE has setup a task group 802.11s for specifying the PHY and MAC standards for WMNs. The current draft of the 802.11s standard targets defining an Extended Service Set (ESS) that provides reliable connectivity, seamless security, and assure interoperability of devices. It also proposes the use of layer-2 routing, frame forwarding and increased security in data transmission. Industry giants such as Motorola Inc., Intel, Nokia, Firetide, etc., are actively participating in these standardization efforts. Two main proposals, one each from consortiums SEEMesh and WiMesh Alliance, were considered and successfully merged into a single draft version of the IEEE 802.11s standard in July 2007. The task group is refining the specifications and aiming to finalize the standards by the end of year 2009.

The reminder of the chapter is organized as follows. In Section 18.2, we describe the traditional Wireless Local Area Networks and architecture of Wireless Mesh Networks. Section 18.3 discusses some of the challenges at layers that need to be considered while designing protocols for WMNs. As WMNs are based on a multi-hop communication paradigm, an understanding of various problems in a multi-hop communication is necessary. Section 18.4 delves into the issue of unfairness prevalent in a multi-hop wireless network. Various approaches in the wired networks to address the unfairness problem are analyzed in Section 18.5. This is followed by Section 18.6, which gives a detailed description of three recent approaches to ensure fair treatment of all flows in a multi-hop WMN. Finally, concluding remarks are included in Section 18.7.

18.2. Background

18.2.1. *Traditional Wireless Local Area Networks (WLANs)*

Traditional Wireless Local Area Networks (WLANs) are broadly characterized into two types[7,10]:

1. Infrastructure WLANs
2. Ad hoc WLANs, also called as Mobile Ad hoc Networks (MANETs)

This classification is based on whether or not there is a central controller providing Internet connectivity. Infrastructure WLANs are structured networks consisting of Access Points (APs) and the client-stations, or the subscriber units. APs are typically installed at fixed locations and are connected to a wired network also known as Distribution System (DS), and relay data between wireless and wired devices. The clients that could be either stationary or mobile, communicate with each other through APs. These client nodes are connected to the APs through wireless links. In other words, all the information exchange among the clients in the network occurs via an AP and the AP is also responsible for providing Internet connectivity to the clients registered with it. Multiple APs can be interconnected to form a large network which allows the clients registered with them to switch between the APs.

The other WLAN architecture, MANET, is characterized by the absence of any infrastructure in terms of AP, and the client devices communicate directly with other close by devices and relay each other's traffic. MANETs are easier to install and to configure due to the absence of any needed infrastructure, but have limited connectivity options for other devices and weaker security mechanism.

The IEEE 802.11 family of protocols standardizes WLAN technology and includes the three well known standards: 802.11a, 802.11b, and 802.11g. These standards operate in unlicensed Industrial Scientific Medical (ISM) bands. IEEE 802.11a operates at a frequency of 5.8 GHz, while 802.11b and 802.11g operate at 2.4 GHz. The maximum data rate supported by 802.11a and 802.11g is 54 Mbps and that supported by 802.11b is 11 Mbps. However, in case of any losses or errors on the data links, 802.11b reduces the data rate to 5.5 Mbps or 2 Mbps or 1 Mbps depending on the loss rate of the links. This method, called automatic fallback, is used in order to operate over extended range of communication and in areas with high levels of interference. Also, Wi-Fi alliance was created to enable compatibility and interoperability between products produced by different vendors in the industry.

These WLAN standards do not provide a significant improvement in achievable bandwidth for applications that span long distances such as mining industry.

For instance, with 802.11b, the data rate of the wireless links drops off as the distance or the number of hops increases. The 802.11g standard intends to provide higher bandwidth in a confined space such as inside a building, so that it can be used as a replacement for wired networks. 802.11b and 802.11g both operate in the same frequency band and use identical signal propagation. 802.11g aims to achieve performance improvement by using an encoding scheme Orthogonal Frequency Division Multiplexing (OFDM) that incorporates detailed information into the signal. A receiver requires higher power to decode the signal encoded using OFDM. When the signal is transmitted over large distances, the Signal to Noise Ratio (SNR) parameter measured at the receiver decreases. As a result, signals encoded using higher modulation techniques cannot be decoded at the receiver. Further, with increasing error rates in the medium, the radio employing 802.11g reverts back to 802.11b encoding scheme and its data rates. Also, with ever increasing wireless devices in the market operating in the same frequency band, interference from other sources cannot be avoided. Thus, the theoretical data rates specified in the standard are typically not achievable in a practical scenario.

A big leap in terms of achieved throughput of about 600Mbps and higher range is provided by 802.11g is promised by the emerging standard called 802.11n.[2,3] This standard offers improvement in many aspects such as throughput, range, channel reliability, and transmission efficiency. It can operate in either 2.4GHz or 5GHz frequency bands and use Multiple Input Multiple Output (MIMO) antennas for data transfer. A single transmission stream can be split into multiple (4 in 802.11n) substreams and sent over the available antennae. Further, certain improvements at the physical layer, and increased channel band achieve such escalation in the throughput of 802.11n.

18.2.2. *Architecture of Wireless Mesh Networks (WMNs)*

The architecture of Wireless Mesh Networks (WMNs) is derived largely as a combination of Infrastructure WLANs and MANETs described in previous section. WMNs encompass Mesh Routers (MRs) and Mesh Clients (MCs). Thus these WMNs can be organized into a three-tier hierarchical architecture, as shown in Figure 18.1. The first (or the top) tier includes a subset of MRs, called Internet Gateways (IGWs), which are connected to the wired network and these IGWs act as a bridge between the wireless mesh backbone and the wired network. IGWs also have an interface solely to communicate with the wired network. The second (or the middle) tier is the mesh backbone, consisting of relatively large number of wireless MRs which communicate with IGW and with each other using a multi-hop communication paradigm, thus forming a multi-hop wireless network. The MRs organize autonomously and are self-healing, facilitating the addition and deletion of nodes in the network on a dynamic basis. This backbone network of MRs is responsible for

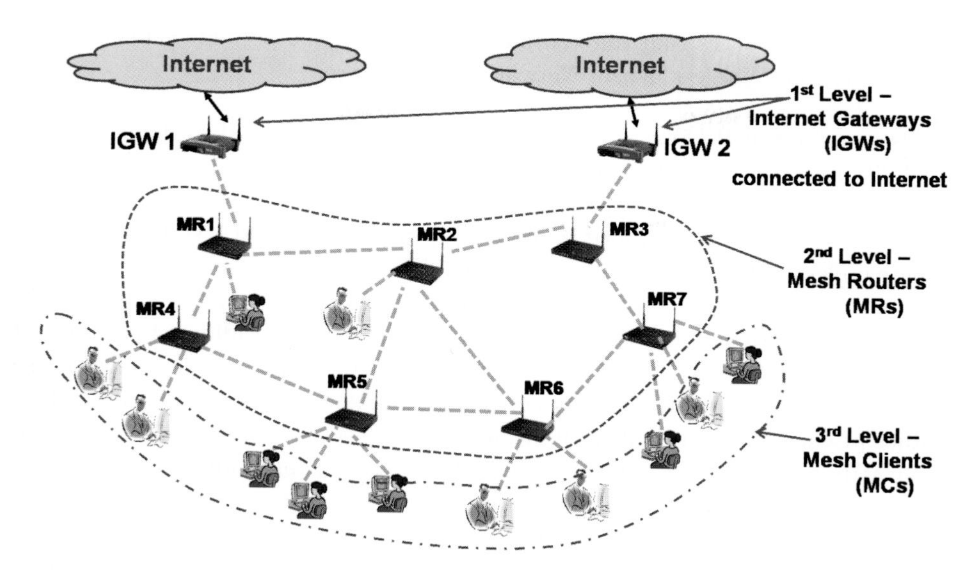

Fig. 18.1. Hierarchical architecture of wireless mesh networks.

providing services to the MCs by transporting traffic either to/from IGWs by cooperatively relaying each others' traffic and facilitating interconnectivity. The third (or the bottom) tier includes the end users or the MCs, which use the network to access the Internet and other services such as Internet Protocol (IP) telephony, etc. In WMNs, MRs are mostly static and MCs are typically mobile and get registered with different MRs at different points of time. It has to be noted that MRs and IGWs are similar in design, with the only one exception that an IGW is directly connected to a wired network, while MR is not. The links in a WMN can be either wired/wireless. In a WMN, only a subset of APs needs to be connected to the wired network in contrast to a traditional Wi-Fi network where each AP has to be connected to the wired network.

WMNs require relatively minimal planning, marginal infrastructure support and are easily scalable. Specifically, WMNs can be deployed in places where either infrastructure is unavailable or where it is difficult to plant the APs. Also, WMNs can be deployed with few IGWs and numerous wireless MRs requiring low infrastructure for setting them up. WMNs provide a cost-effective alternative to other types of networks, requiring meticulous planning and indulge in huge expenses. Further, these networks are scalable, meaning they can be extended to thousands of MRs by just deploying new MRs which self-configure themselves in a dynamic manner. Large number of MRs in the mesh backbone of a WMN provides high connectivity, facilitating availability of multiple routes between any two MRs. This feature can be used to increase reliability of the data transmission, allowing adequate fault tolerance.

Some characteristics that differentiate WMNs from the popular networks such as MANETs and Wireless Sensor Networks (WSNs) are:

- *MRs are static in WMNs* — MRs in WMNs are either immobile or have minimal mobility as compared to that of ad hoc networks, thereby minimizing any mobility related concerns of the network.
- *MRs are not power constrained in WMNs* — In contrast to MANETs and WSNs where nodes are typically power constrained, WMNs have MRs directly connected to a power outlet on a consistent basis and thus overcome this power issue.
- *MRs are equipped with multiple radios in WMNs* — Costs of radios got plummeted in recent years, thus making it practical to equip MRs with multiple radios. These radios can be used for facilitating concurrent transmissions in the network. Few radios at each MR can be dedicated for mesh backbone communication between peer MRs and others for relaying the traffic of its registered MCs. This way, the WMNs achieve increased capacity gains.
- *Traffic model is non-uniform in WMNs* — In WMNs, traffic from the clients is aggregated at the MRs to which they register. This traffic is predominantly directed between MRs and the IGW in contrast to MANETs where the traffic is among the peer nodes. This focused traffic flow in WMNs towards and from an IGW places higher demand on certain paths connecting IGWs and MRs, unlike that of MANETs where the traffic is more or less uniformly distributed. Also, in contrast to distribution of users in MANETs, the density of MCs is unevenly distributed in WMNs. For instance, a community mesh network for a city would have a high concentration of users and traffic in office buildings compared to roads or parks.

A generic system architecture depicting the functional component structure of WMN technology is shown in Figure 18.2 [Mesh network architecture] as described by Aoki *et al.* The main modules and their functionalities in the system architecture are:

- *Mesh Topology Learning, Routing and Forwarding* — Typically, the topology learning component is responsible for discovering neighbors of a node and obtaining radio metrics from the wireless links which has information about the links' quality. The routing component consists of a routing protocol which determines routes to transmit packets to their respective destinations based on MAC addresses. The forwarding module relays a packet from one node to another towards its destination. In order to make efficient use of radio resources, the routing protocol should make use of radio metrics and multiple frequency channels available in accordance with existing radio conditions.
- *Channel State Monitor* — This module calculates radio metrics used by the routing protocol and also measures the existing radio conditions in the network to select the appropriate frequency channel.

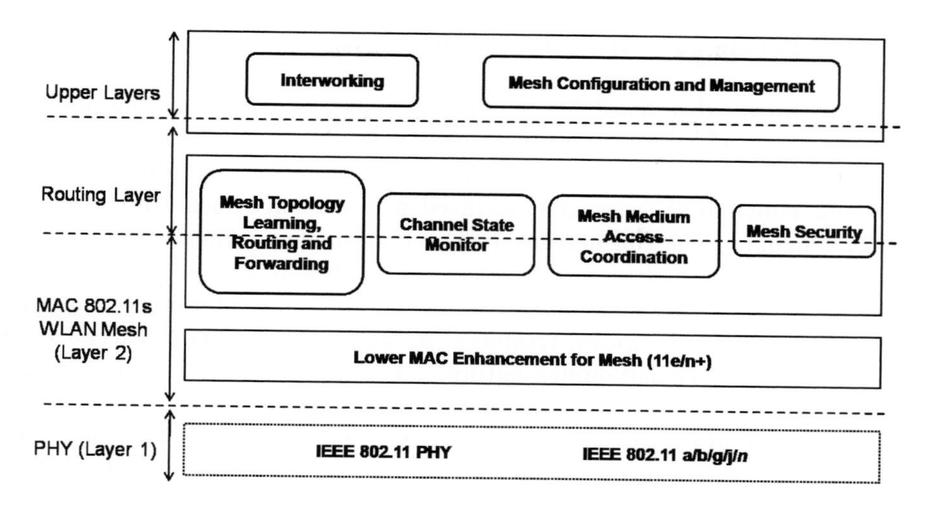

Fig. 18.2. WMN system architecture — functional components.

- *Mesh Medium Access Coordination* — This component contains two modules — one to prevent the performance degradation caused by the hidden and/or exposed terminal problems prevalent in wireless networks; and the other to perform priority control, congestion control, and admission control for traffic in the network.
- *Mesh Security* — This functional component addresses the security aspect in WMNs. Typically, this module carries functions to protect data-frames between source-destination pairs, and management-frames that are used as control packets, such as route discovery messages sent by a routing protocol. Currently, this component assumes the use of WLAN security schemes defined in the IEEE 802.11i standard.
- *Interworking* — As per the IEEE 802 standard, a WMN must conform to the standard's network architecture. This means that, for a WMN to be able to connect to other networks, it should implement a transparent bridge module in the Internet Gateway (IGW) placed at the network boundary. Also, in order to deliver forwarded packets to all the terminals connected to LANs, the WMNs should operate in a broadcast mode.
- *Mesh Configuration and Management* — This component includes a WLAN interface used for automatic setting of MR's Radio Frequency (RF) parameters such as frequency channel selection, transmit power, for QoS policy management, etc.

Some of these modules are between the MAC and the Routing layers as these functionalities require cross-layer coordination for measurements, etc. For instance, Channel State Monitor estimates certain parameters such as Expected Transmission Time (ETT) for a packet over the medium which is used in making routing decisions by the Routing layer.

18.3. Challenges for Future Research

WMNs have proven their feasibility to provide ubiquitous broadband Internet access at low costs and to support a large number of users. Though feasible, their performance is still considered to be far below the needed limits for practical applications. And so, unfortunately the companies involved in WMN deployments often face challenges in designing, deploying and ensuring their optimal performance due to certain inherent problems of multi-hop networks. The multi-hop wireless communication is beset with several problems such as unpredictable/high interference, increased collisions due to hidden/exposed terminals [Nandiraju *et al.* (2007)], excessive congestion and its typical half-duplex nature of radios. In addition to these, effects of fading and shadowing lead to unreliable link connectivity. This results in poor performance of WMNs with low end-to-end throughput and high latencies, which are undesirable in the perceived applications of WMNs. For instance, EarthLink[1] made efforts to deploy WMNs to provide community networking services and was compelled to back off from such projects due to aforementioned issues.

Though envisioned applications of WMNs seem luring, considerable research is still needed at all network communication layers before wide scale deployment of WMNs becomes practical. Following sections outline some major challenges that are present at various layers of network protocol stack and need attention while designing solutions for WMNs.

18.3.1. *Physical layer*

At the physical layer, the hardware of the node defines the maximal capacity of a link, i.e., the maximum achievable bandwidth for transmitting a packet on the link. With recent developments in VLSI and digital communication technologies, the maximal raw data rates that can be supported at the physical layer are rapidly increasing. For instance, the theoretical raw data rates at the physical layer increased from 1 Mbps (few years back) to about 600 Mbps (currently) using IEEE 802.11n.[2] However, though this *theoretical* possible bandwidth is high, the practical *achievable* bandwidth is still considerably low due to unpredictable interference and multipath fading.

In WMNs, the quality of the underlying wireless links often fluctuate mostly due to reasons such as multi-path fading effects, weather conditions, and external interference. Several devices operate in the unlicensed open frequency band of 2.4/5 GHz range and these could cause unpredictable interference with external devices of WMNs and vice-versa. Typically, the link quality information is an important metric of WMNs which determines the probability of reliable packet delivery, error or loss rate on the link, the link's overall status parameter, etc. This link quality information at a node plays a key role in determining whether or not the node can proceed with the reliable data transmission to its the next hop. Additionally, such data is also used by the higher layers to optimize capacity, make routing decisions, detect handoff imminence, etc.

The physical layer is responsible for collecting such link-quality information and informing its upper layers about the nature of the underlying link or the channel. The physical layer protocols need to be designed in a way so as to estimate the behavior of the link using parameters that best reflect its behavior and propagate the same to the higher layers efficiently.

In typical wireless networks, nodes use omni-directional antennas for transmission to help enable communication in all directions with the nodes in the neighborhood. Though using omni-directional antennas at a node is cost-effective and has several advantages, it is vulnerable to interference which results in severe degradation of achievable network capacity. Added to this, typically there is a constraint on the minimal physical distance among the antennae at a node, which depends on the gain-metric of antennae, for proper functioning with non-interference. For instance, antennae with 5dBi gain on a node should be placed at least 3 feet apart so that non-interfering channels can be properly utilized.[2]

One alternative to overcome these issues is the use of directional antennas which would minimize interference between simultaneous communications in a section of the network. Recent developments in multi-antenna technologies (like MIMO) and smart antennas allow the nodes to decode received signals even with low Signal to Noise Ratio (SNR), thus resulting in an increased obtainable capacity. However, the drawback with these antennas is — they make the cross-layer design more complex.

18.3.2. *MAC layer*

Typically, multi-hop wireless networks suffer from the performance degradation and unfairness. When the traffic is forwarded over multiple hops, such a flow receives only a fraction of the throughput compared to a single-hop flow because of the random access nature of the MAC protocol. A flow that traverses multiple hops has to compete multiple times for the medium at the intermediate MRs to reach the destination. With existing 802.11 protocols, each such competition at MRs is treated with the same weight as that of a single hop flow; due to multiple contentions en-route the probability of a multi-hop flow packet reaching the destination is significantly lower than that of a packet belonging to a single-hop flow. This also results in higher inter-arrival time for a multi-hop packet due to the fact that, at each hop when it loses the contention, it is queued for a period of time before getting retransmitted. This issue has to be considered while designing MAC protocols for WMNs.

In a single radio based multi-hop network that use only a single channel, as shown in Figure 18.3(a), effective throughput decreases drastically with increasing number of hops because of spatial contention.[9] Li *et al.*[18] have demonstrated in their assumed network models that the achieved throughput of IEEE 802.11 in a multi-hop network is only one-seventh of the effective or theoretical bandwidth of the channel. When a MR is equipped with only a single radio, this radio needs to switch its mode back and forth — for transmitting the backhaul traffic within the mesh backbone, and for communication with its registered MCs. And, this back and forth switching results in significant latencies. Furthermore due to the half-duplex nature of the radio, a MR cannot send and receive traffic simultaneously. Thus, equipping a MR with more than one radio allows concurrent communication with minimal or no interference.

Incorporating a single radio multi-channel transceiver at a MR that uses multiple non-interfering channels to communicate with its clients reduces the issue of spatial contention. However, it still requires complex MAC protocols and has high overhead in terms of channel switching delays. Wu *et al.*[37] propose a MAC layer strategy, Dynamic Channel Assignment (DCA), which employs two transceivers — one for control packets transmission, and the other for switching among different channels for data transmission with different receivers. So *et al.*[36] propose multi-channel MAC protocols using single radio for ad hoc networks and show the performance gains. However, such a methodology incurs considerable delays while initiating a communication session. Also, maintaining a dedicated control channel by a MR can be expensive and results in wastage of bandwidth when the total number of available channels is limited. Usage of a time multiplexed control channel[35] could address the limitations of dedicated control channel architecture, but it still suffers from synchronization problems.

One approach to overcome these limitations is to increase the number of radios at each MR and balance the resource allocation for the needed backbone communication and for relaying the traffic of its registered clients. MRs can use multiple radios tuned to orthogonal channels for simultaneously communicating with their neighbors. Figure 18.3(b) shows the use of dual-radio WMN. In a dual-radio model, each MR is equipped with two radios — one dedicated to the clients' access and the other used for backbone communication. However in this model, typically the radio used for backbone communication still results in considerable amount of channel switching due to sharing of bandwidth with several MRs in the mesh backbone, and simultaneous communication could still be a problem. And, the performance improvement achieved compared to that of single-radio architecture, is still only marginal. Further performance enhancement can be achieved by increasing the number of radios per MR to more than two, so that majority of radios can be dedicated for the backbone communication and the remaining ones for the client access.

Figure 18.3(c) shows the use of a multi-channel approach using multiple radios that successfully overcomes these typical problems encountered in the other architectures. The uplink, downlink backhaul radios and the service radio are all operated at non-overlapping channels, eliminating potential co-channel interference. Pathamasuntharam *et al.*[31] and Kyasanur *et al.*[20] propose a multi-interface architecture which employs three half-duplex interfaces, one of each dedicated for transmitting, receiving and broadcasting. They present an interface switching strategy in which

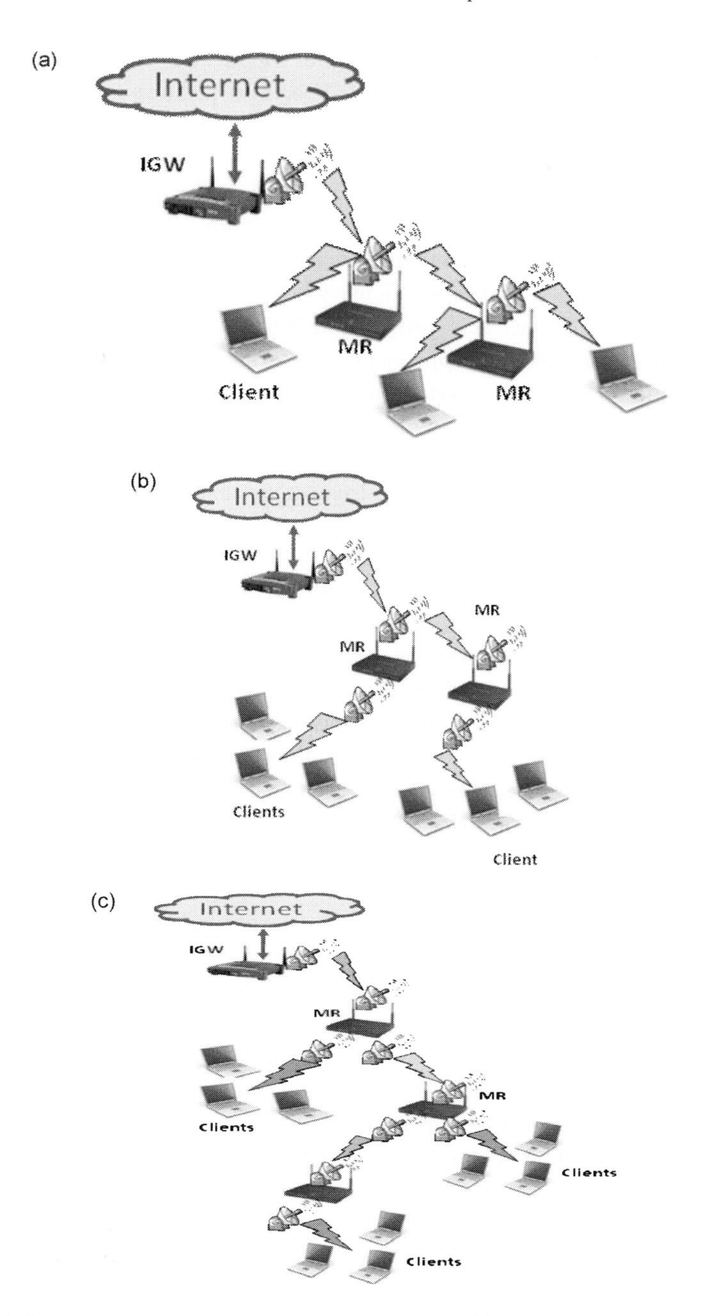

Fig. 18.3. (a) Single Radio, Single Channel Mesh Network; (b) Dual Radio, Single Channel Mesh Network; (c) Multi-Radio, Multi-Channel Mesh Network.

receiver interface is fixed on a specific channel for reception of data, and transmit interface gets tuned according to the needed receiver's channel.

With multiple interfaces for communication at a MR in a WMN, proper channel assignment among these interfaces is a primary challenge at the MAC layer. As each MR can be equipped with multiple radios, fixed channel assignment to these radios is a viable solution. However, efficient and intelligent channel assignment schemes have to be designed as the number of channels is not infinite or may not be sufficient in a high node density scenario. Several channel assignment mechanisms[6,20,33,34] address this problem. Elegant assignment schemes can also be derived from graph coloring approaches.[15] Ramachandran *et al.* investigate the channel assignment problem based on an interference-estimation technique. Adya *et al.* perform the channel assignment using a measurement-based approach which is dependent on the channel quality for selecting the appropriate channel. Deafness is another key issue that needs to be addressed in such an environment.

Many commercial MRs that are currently in the market use multiple radios with multi-channel capability for improving the channel capacity. As all these vendors use their own proprietary MAC and Routing protocols for their products, interoperability cannot be guaranteed. Multi-radio unification protocol, a virtual MAC proposed by Adya *et al.* is an interesting approach to manage multiple radios. But, concrete and robust native layer-2 protocols can be more effective and efficient than such virtual MAC solutions. With multiple radios communicating on non-interfering channels, a key aspect to focus is an efficient flow scheduling scheme in a neighborhood to maximize resource usage.

Each radio at a MR has multiple channels to which the radio can be tuned onto when communication needs to take place. The MAC layer should be aware of the non-interfering channels on which concurrent communication can happen and hence the protocols designed at MAC layer should perform proper channel assignment and maintain coordination among concurrent transmissions at the MRs, with minimal channel switching delay.

Another key issue that needs consideration is the QoS provisioning at the MAC layer. The envisioned scenario of WMNs is expected to support applications like broadband Internet access, real time applications such as video streaming and voice conferencing. QoS provisioning for such applications is an essential requirement and is a key challenge. The performance of IEEE 802.11e which is proposed for QoS provisioning in wireless LANs is yet to be investigated over multi-hop networks. Also, the suitability of IEEE 802.11 MAC in a WMN is debatable. Thus, existing MAC protocols have to be redesigned for efficient operation over WMNs.

18.3.3. *Routing layer*

In a multi-hop WMN, the primary function of the networking/routing layer is to transfer packets from the source MR to the destination MR using multiple hops.

WMNs aim to provide high bandwidth broadband connections to a large community and should be able to accommodate a large number of MCs to access the Internet. In a WMN, the expected traffic volumes are high; so scalability and load balanced routing become important issues to be considered while designing the protocols.

For several reasons, traditional routing solutions of MANETs are not directly useful for WMNs. Most of them are usually designed around single-path routing which can result in an unbalanced network load, with some links being highly utilized while others seldom used. Such links are referred as "hot-paths" and often lead to packet loss resulting in sub-optimal performance of the network. The scenario leading to hot-paths is depicted in Figure 18.4. Also in single path routing, if a link in the chosen path fails, applications will be interrupted and rediscovering an alternate path could result in delays. To increase the reliability, extensions to single-path routing protocols have been designed which typically use backup paths to route the traffic when primary path fails.[16,35] However, due to path switching, even most of these models result in higher latencies.

Traffic in WMNs is predominantly between IGWs and the MRs which places higher demand on certain paths, connecting IGWs and MRs. The advantage of WMNs is the high connectivity of the mesh backbone, which facilitates availability of multiple routes between any two MRs. This benefit of availability of multiple

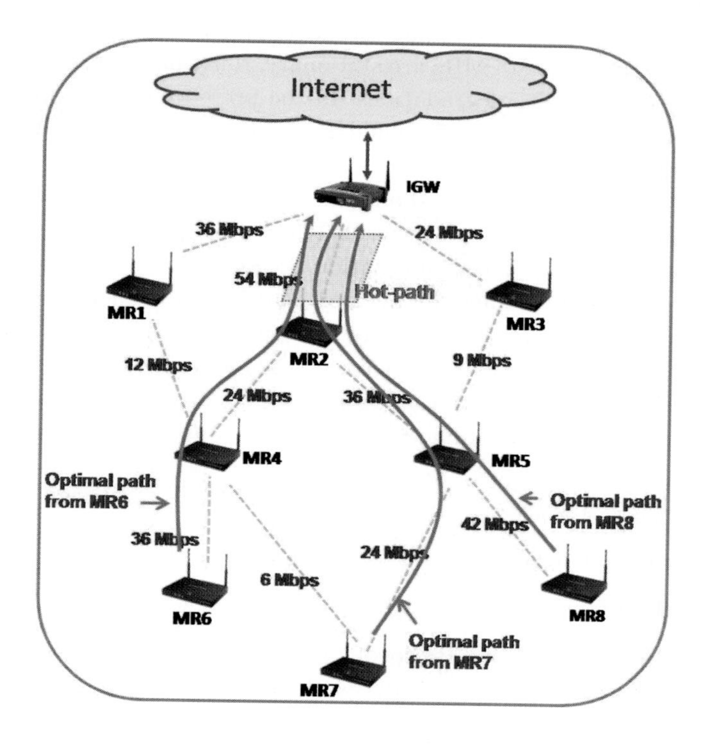

Fig. 18.4. Hot-paths.

routes needs to be further exploited to increase reliability of data transmission and to allow adequate fault tolerance.

Existing multi-path routing protocols advocate the use of disjoint paths and do not consider the delays (such as queuing delays) and congestion experienced over the links, once the paths are readily selected. Authors in Ref. 24 reveal that the multiple paths need not be disjoint and in fact, use of disjoint paths could be counter-productive. Use of multiple paths offers an opportunity of error resilience, and hence traffic load distribution as the spatial diversity and data redundancy could be exploited.

In addition to maintaining multiple paths, optimally distributing the traffic load onto those routes is important to achieve load balancing, thereby enhancing the reliability. It also aids in achieving better throughput as it allows bandwidth aggregation in the network. Typically, applications that are supported over WMNs such as Video on Demand (VoD), Voice over IP (VoIP) require certain QoS provisioning, which need a routing solution with reliable routes. Also, certain applications require high bandwidth which may not be supported by an underlying link at a node. When multiple routes are used and the traffic is judiciously partitioned onto those paths, considering the loss ratio of the routes and choosing a reliable route in most circumstances, this procedure results in increased aggregate bandwidth and better reliability. Thus, proper traffic distribution policy is an important aspect that needs to be considered while designing a routing protocol for WMNs.

Further in WMNs, though MRs are stationary, route fluctuations can still occur due to frequent link quality degradations. In the presence of any failed nodes and broken links, the routing protocol should re-route traffic around such nodes and redistribute orphaned clients to other active IGWs. With changing network topologies of WMNs, the routing protocol should be flexible and adapt to such changes and function by incurring less routing overhead. The routing protocol should make routing decisions based on less global information because relying on global information requires frequent updates along the network and incurs tremendous overhead. In WMNs, it is common to support MCs which require the routing protocol to be able to handle handoffs in a dynamic manner to support seamless connectivity for such MCs. Also, in WMNs, with clients requiring services for different applications, the routing protocol should route traffic depending on the priority in providing different QoS levels.

18.3.4. *Transport layer*

18.3.4.1. *TCP unfairness*

Existing Transmission Control Protocol (TCP) implementations that are widely used at the transport layer do not perform well in multi-hop wireless networks. Typically, the TCP congestion control mechanism at a sender responds to any packet loss recorded in the network by reducing the packet transmission rate. It resumes its packet transmission rate to the default value only after confirming that the packets which are sent are received without any errors. Such a congestion control mechanism faces several challenges in WMNs[19]:

- Transmission errors & loss unfairness — In wireless networks, transmission errors over the wireless medium are frequent and often result in packet loss. Also, due to varying channel conditions over different timescales, there can be variations in the observed Round Trip Time (RTT) of packets. TCP, unaware of the wireless nature of the medium and the resulting transmission errors, assumes that congestion in the network is the cause of the packet loss and slows down its packet sending rate. Also, random packet loss in the network slows down the traffic flows unevenly (for instance, Web access by the users is usually slowed).
- Multi-path packet reordering — In WMNs, multiple paths can be exploited to route packets from a source MR to the destination MR. However, packets traveling over multiple paths typically arrive at the destination in out-of-order sequence. The re-ordered packets need to be handled properly at the destination; otherwise this situation may trigger spurious re-transmissions and cause confusions in TCP congestion control.
- Distance unfairness or the spatial bias — In multi-hop wireless networks, for flows sharing the same wireless link, the flow which traverses less number of hops tends to acquire relatively higher share of the link bandwidth compared to other flows. Thus this scenario leads to the widely known spatial bias problem where flows emanating from long-distance users get slower service compared to other users in the network. In other words, for WMNs, this problem is the smaller share of bandwidth to nodes that are located farther away from IGWs.

Added to this, the traditional TCP congestion control mechanism has two major drawbacks — the multi-stream and persistence phenomena.[5] Some applications can exploit these two phenomena and result in a scenario where some flows could capture a much larger share of the network bandwidth at the expense of others.

Typically, TCP follows fairness provisioning at flow-level granularity and it assigns equal share of bandwidth to all the flows existing in the system. For instance, if an application at a node A starts 10 flows and an application at node B emanates only one flow, the TCP tries to give 10 times more bandwidth to A compared to that of B. This phenomenon is called multi-stream or multi-flow fairness which is often exploited by certain applications leading to unfairness. In other words, applications such as peer-to-peer (P2P) can capture several times more bandwidth than a traditional single-stream application under a congested Internet link. Since networks have bottleneck links in some areas of the network, a small percentage of Internet users utilizing P2P applications can hog the vast majority of network resources at the expense of other users. Figure 18.5 illustrates the multi-stream exploit in action where Node/User A hogs more and more bandwidth compared to Node/User B by opening more TCP streams. In this illustration, let the large blue pipe represent a congested network link with finite capacity.

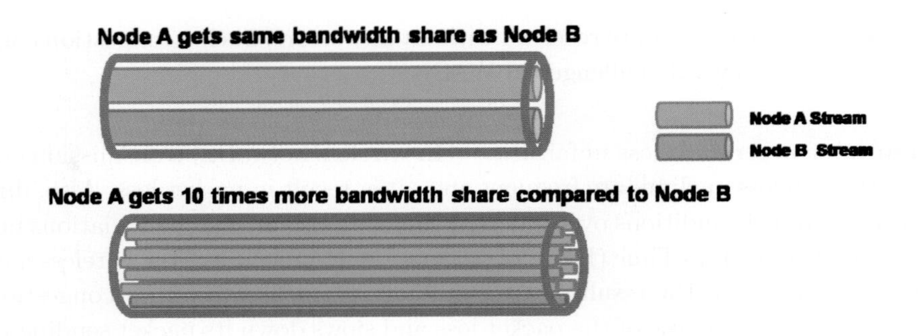

Fig. 18.5. Multi-flow unfairness.

One solution to mitigate multi-flow unfairness is to control each flow in the network such that the total traffic to a destination node is equally and fairly distributed regardless of the number of flows they generate. This way, a fair distribution of Internet capacity can be accomplished in a network.

Another problem with conventional TCP is that if an application consistently connects to the network and uses the network's bandwidth, it captures higher bandwidth share than other applications connecting to the network and requiring service on an ad hoc or need basis. This phenomenon is called persistence and applications such as P2P continuously use the network and obtain vast share of network's resources as compared to others. Thus future design and enhancements to TCP should adapt and mitigate scenarios such as large delay variations, path asymmetries, varying channel conditions, and multi-stream fairness.

18.4. Unfairness in Multi-hop Wireless Networks

Fairness in networks refers to the optimal allocation of available network resources such as channel access and bandwidth, to the flows originating from various nodes based on a pre-determined and balanced criterion. Users in conventional single hop networks such as a cellular network typically get fair access to resources and this process is managed by its Base-Station or a central controller. However, in a multi-hop network like a WMN, an IGW cannot be assigned nor can perform the role of a centralized coordinator, as MRs are connected in a multi-hop fashion to the IGW. In such a scenario, MRs solely depend on their peer MRs to relay their traffic. Thus, though multi-hop communication facilitates increased coverage, low deployment cost and other such advantages, it suffers from drawbacks such as spatial bias, collisions, hidden/exposed terminal problem, which are explained under in detail.

In a multi-hop network, flows spanning multiple hops experience dismal throughput performance as compared to flows traversing fewer hops, leading to a spatial bias. Apart from hidden and exposed terminal problems that substantially contribute to the spatial bias, the link layer buffer/queue management scheme at the intermediate MRs[14] is another factor that plays a major role. Most of the existing

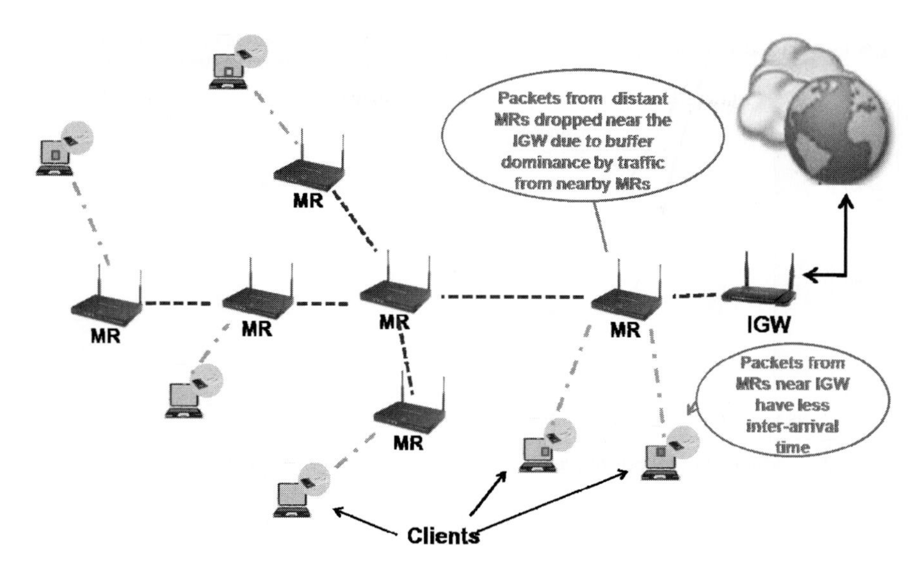

Fig. 18.6. Spatial bias — unfair queue management.

queuing mechanisms do not consider the parameter — *number of hops a packet has traversed* — in their queuing logic and drop packets when there is no space in its Interface Queue (IFQ), independent of the number of hops they have already traversed. An IFQ is a queue maintained at a node to keep track of packets that are later transmitted over the medium one at a time. The packets in the queue comprise of those generated at the node as well as those arriving from other nodes in the network which need to be forwarded by this node.

The problem of spatial bias, shown in Figure 18.6, affects the network's performance in two ways. Firstly, it results in wastage of valuable network resources, and secondly, clients of a MR far away from IGW will obtain very low throughput and undergo starvation as compared to the clients connected to a MR that is near to an IGW. Generally, the link layer at a MR uses a drop-tail queue management scheme, where the drop-tail mechanism is one which discards a packet at the end (tail) of the Interface Queue (IFQ). If the IFQ at the link layer of a MR is full, any newly arriving packet is dropped, regardless of the number of hops it has already traversed. The link layer IFQ of an upstream MR near IGW is filled up faster with packets belonging to shorter hop flows originating from nearby MRs and those from the MR itself. On the other hand, packets belonging to flows traversing multiple hops, have to contend for the medium at each hop in the route. Typically, the arrival rate of relay traffic in a wireless network is controlled by the underlying channel access mechanism like a Carrier Sense Multiple Access with Collision Avoidance based Medium Access Control (CSMA/CA) based MAC. In a multi-hop scenario, as the relay nodes repeatedly contend for the channel at every hop, the channel access delay increases and the arrival rate of distant multi-hop flows at an upstream MR near

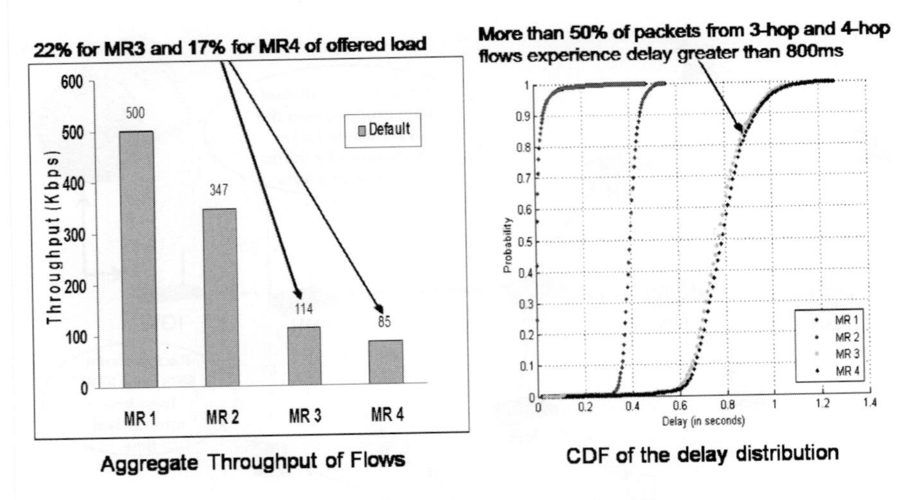

Fig. 18.7. Illustration of buffer dominance at MRs near an IGW.

the IGW is considerably reduced. Thus, shorter hop length flows experience lesser drop probability and have higher performance while the longer hop length flows suffer inordinately as their packets are often dropped at the intermediate MRs due to lack of IFQ space.

The intermediate MRs contend for the common wireless channel in order to send their own traffic and the relayed traffic. A typical scenario to illustrate the spatial bias causing unfairness can be explained as follows. If a MR near the IGW has huge amount of traffic (both forwarded and its own), it repeatedly contends for the wireless channel leading to starvation of MRs present much lower in the hierarchically structured WMN. Additionally, a potential collision in the congested link near the IGW further degrades the network performance. A closed loop feedback based protocols such as TCP suffers from much severe loss as it adapts to the congestion and packet losses by decreasing its congestion window size. Thus, the envisioned goal of WMN to provide broadband connectivity to a large community of end users is hampered by these issues. It is critical to eliminate this problem of spatial bias and provide an impartial service to all flows, irrespective of the number of hops it has already traversed. Figure 18.7 illustrates the impact of unfairness in a typical linear scenario of a WMN.

18.5. Queue Management Approaches in Wired Networks

Before discussing queue management approaches in *wireless* networks, we briefly outline various approaches prevalent in wired networks. Existing approaches for traffic regulation and congestion mitigation in multi-hop wired networks can be broadly classified into two categories — queue-based mechanisms and flow-based mechanisms. The main goal of a queue-based mechanism is to effectively manage the available buffer (queue) at a given node, while the objective of a flow-based mechanism is to ensure fair treatment for traffic flows. Flow-based mechanisms discourage the aggressive flows (high-rate flows) from monopolizing the available network resources and thus fairly accommodate the non-aggressive flows (low-rate flows) in the network.

Several Active Queue Management (AQM) mechanisms are proposed in literature which dynamically adapt to congestion in the network and guarantee a fair treatment for all flows. These mechanisms differ primarily in the parameters used to indicate congestion and the policy which is used to detect and respond to the congestion. Such mechanisms can be broadly categorized based on the method used for the notification of congestion — Explicit notification, and Implicit notification. An explicit notification sends an Explicit Congestion Notification (ECN) during congestion and the ECN signals the source node about the prevalent congestion so that the source can adjust its traffic rate accordingly. Implicit notification mechanisms can be further classified as *proactive* or *reactive* notification. While implicit reactive buffer management mechanisms are exercised only upon the detection of congestion, their proactive counterparts are administered beforehand (a-priori) to check for any impending congestion.

Further, there are two prevalent models used by these queue management mechanisms to control congestion — *heuristic* and *predictive* models. Typically, heuristic-based models compute a pre-defined metric to estimate the level of congestion and then act accordingly to cope up with the congestion. On the other hand, prediction-based models take subsequent measures for alleviating congestion by predicting the impendent congestion in the network on a proactive basis. Random Early Detection (RED)[13] and its variants fall under the category of heuristic based models while BLUE 0 and PURPLE[32] fall under the category of prediction-based models. RED maintains a moving average of the queue length and drop probability at each node. The drop probability at a node determines whether a packet should be dropped or admitted in the link layer IFQ. In RED, if the average instantaneous queue length of the queue at a node lies between a minimum and a maximum threshold, the subsequent packets are dropped with a random probability; and if it exceeds the maximum threshold all subsequent packets are dropped at the IFQ. Several variants of basic RED scheme such as Adaptive-RED,[13] FRED[8,22] have been proposed, especially to control misbehaving and unresponsive flows in the network. Dynamic RED is another approach that uses a feedback method to randomly and periodically discard packets regardless of the network traffic loads. It pre-computes an optimal queue length and maintains the instantaneous queue length at a node at this level at all times. Stabilized RED[29] extends RED by improving the scheme for misbehaving sources. It employs a zombie list containing the most recently seen flows and replaces a random packet from the list probabilistically with a new arriving packet. It promises a stabilized buffer occupation independent of the traffic load levels by statistically estimating the number of active flows. BLUE tunes the drop probability of a node by observing the link utilization and the number of packets dropped due to buffer overflow. PURPLE estimates the congestion in a network by monitoring the ECN bits embedded in the packets.

CHOKe[30] offers a simpler solution for penalizing the aggressive misbehaving flows. It identifies the packets of an aggressive flow by comparing it with a packet, chosen randomly from the First In First Out (FIFO) buffer. If the packet of the aggressive flow and the one from the buffer match, the algorithm drops those packets. As packets of an aggressive flow are more likely to be present in the buffer, they are pre-dominantly dropped. RED-PD[23] ensures fair bandwidth to flows by identifying the aggressive flows at an intermediate node. It maintains a history of dropped packets which is used to assign a drop probability for each monitored flow. Thus, aggressive flows are assigned high drop probability. In contrast, Stochastic Fair Blue (SFB)[11] protects TCP flows from non-responsive flows by using a *Bloom filter*. All incoming packets are mapped to multiple bins at different levels by using independent hash functions corresponding to each level. Each bin is accompanied with a drop probability which is increased or decreased depending on the number of packets in the bin (if it is greater than a certain threshold). SFB effectively minimizes the per-flow state information required to identify the high bandwidth flows.

However, applicability of aforementioned schemes to multi-hop WMNs is debatable due to the basic policy employed by them. For example, randomly dropping packets at an intermediate MR does not guarantee a fair treatment for distant multi-hop flows in WMNs. In a WMN, it is better to give preferential treatment to packets arriving from distant sources, as they have already consumed fair amount of network's bandwidth. The channel bandwidth is a very scarce resource in a wireless network. Thus, dropping packets that have already traversed multiple hops and consumed bandwidth will result in severe performance degradation. Schemes such as RED cannot effectively control these types of traffic flows.

18.6. Queue Management Approaches in Multi-hop Wireless Mesh Networks

Although, considerable research has been done to improve the performance of multi-hop flows in a wired network, they are not directly applicable to a WMN. This section mainly focuses on three schemes that address the unfairness problem in a multi-hop WMN: QMMN,[26] CBTR [Nandiraju et al. (2007)], and DQSD [Nandiraju et al. (2007)]. Achieving fairness in wireless networks can be broadly measured in terms of — per-node fairness, and per-flow fairness. Per-node fairness involves guaranteeing a fair access of the communication medium and/or ensuring a fair degree of performance for every node in the network. In per-flow fairness, individual flows are monitored and their traffic rates are regulated so that every flow in the network obtains a fair access to the network resources. As will be discussed in next sections, while QMMN ensures per-node fairness, CBTR provides a more versatile solution by regulating the traffic at a flow level granularity. Similar to CBTR, DQSD also works at the flow-level granularity and is concerned with achieving better performance to a group of flows rather than individual flows.

18.6.1. *Queue Management in Multi-hop Networks (QMMN)*

Queue Management in Multi-hop Networks (QMMN) is a novel link layer queue management scheme for IEEE 802.11s based WMN that targets to improve the performance of multi-hop flows. The main goal of this scheme is to guarantee a fair buffer share at each intermediate MR, for all the flows traversing this MR irrespective of their hop length. QMMN achieves this by fairly sharing the available buffer at each intermediate (relay) MR among all active source MRs. Every sender that is currently forwarding one or more flows through a given intermediate MR, is referred to as an active source at this MR. For each active source MR, the intermediate MR allocates a fair share of the buffer by equally dividing among all the MRs whose flows are being forwarded. A packet arriving at any MR is either admitted into the queue or dropped depending on its source MR's share of allocated buffer space. If the source has fewer packets buffered than its maximum allowable limit, the packet will be admitted; otherwise the packet will be dropped. Thus, aggressive MRs (with a high traffic rate) can not overwhelm other peer MRs by merely increasing their traffic generation rate. The QMMN effectively limits the allocated buffer shares, protecting the flows originating from distant MRs that are several hops away from those generated at the intermediate MRs relatively near the IGW.

The algorithm makes use of several data structures like the maximum share of the buffer that any node's traffic can occupy. Initially, fair share of an active node at each MR is equal to the maximum share it can occupy. However, due to the higher inter-arrival rates and/or different traffic generation rates, packets from certain source nodes may not occupy their entire share of the buffer. In such a case, it is better to exploit the residual share. In order to make this possible, QMMN scheme calculates the fair share allocation of the buffer according to the moving average of the service time and the arrival rate. The average arrival rate and service time are estimated using moving averages, as follows:

$$arrv\_rate = \alpha^* arrv\_rate + (1-\alpha) * curr\_arrv\_rate \tag{18.1}$$

$$serv\_time = \alpha^* serv\_time + (1-\alpha) * curr\_serv\_time \tag{18.2}$$

The *new_fair_share* for a given source is:

$$new\_fair\_share = \alpha * old\_fair\_share + (1-\alpha) * \left(\frac{serv\_time}{arrv\_rate}\right) \tag{18.3}$$

And the *fair_share* is computed as:

$$fair\_share = \min(\max\_share, new\_fair\_share) \tag{18.4}$$

After updating the fair shares, residual share of the buffer is computed as

$$residual\_share = \sum (max\_share_i - fair\_share_i) \quad \forall i \in sources \tag{18.5}$$

At any time, a MR can buffer more packets than its estimated *fair_share* only if there is residual space available. In this way, the *residual_share* can be temporarily utilized by the sources that generate higher traffic loads thereby preventing any underutilization of the buffer.

18.6.2. *Cache Based Traffic Regulator (CBTR)*

QMMN based buffer sharing scheme alleviates improper buffer utilization at intermediate nodes, and does not focus on ensuring fair sharing of the buffer between multiple flows originating from the same MR. Usually, packets belonging to aggressive flows capture a larger share of the buffer at intermediate MRs, thus causing the packets belonging to longer hop length flows and/or the slower traffic rate flows to be dropped. Thus, a protocol needs to be designed to prevent aggressive flows from overwhelming the intermediate buffers and protect non-aggressive and/or longer hop flows to obtain a fair buffer share. Cache Based Traffic Regulator (CBTR) is an attempt in this direction which protects the non-aggressive flows from being dominated by the aggressive flows. The objective of the CBTR scheme is two-fold: to identify the aggressive flows passing through a relay MR; and to ensure a fair share of the buffer at the intermediate MRs. CBTR achieves the first objective by employing a "*most frequently seen*" cache discipline. The packets from nearby sources arrive more frequently at an intermediate MR than the packets belonging to distant sources. Hence, the packets from a near-by source are more likely to be present in the cache with a large count. If the generation rate of a near-by flow is high, it is reasonable enough to treat such a flow as aggressive as they would fill up the buffer quickly, resulting in buffer overflows for subsequent packets belonging to longer hops. Thus, a *most frequently seen* cache discipline identifies aggressive flows using the flow-specific entries in the cache.

Upon identifying the aggressive flows, the rate regulator module is responsible for controlling the traffic from these aggressive flows. The main idea of the rate regulator is to drop packets from aggressive sources in order to service flows that have traversed multiple hops. To accomplish this objective, the scheme makes use of a node-wise dynamically tunable parameter called *drop_probability*. Packets from selective aggressive flows are dropped depending on the value of the *drop_probability*. In other words, a packet belonging to an aggressive flow upon arriving at a node is admitted into the buffer with a probability of (1-*drop_probability*). The use of this parameter is introduced only when a packet belonging to a non-aggressive flow is dropped due to lack of the buffer space. In this scheme, tuning of the parameter follows a Multiplicative Increase and Additive Decrease (MIAD) approach to properly capture the degree of regulation. Thus, after the first non-aggressive flow's packet is dropped, the *drop_probability* at this node is multiplicatively increased,eventually causing the frequently arriving aggressive flows' packets to be dropped. As CBTR employs a multiplicative increase for the *drop_probability*, it quickly controls the rate of aggressive flow. But, if a multiplicative decrease method is folllowed, the *drop_probability* may decrease quickly and would let the aggressive flows to fill up the buffer again. So, a slow linear decrease is advocated for the *drop_probability* so that the aggressive flows remain under control for an appropriate amount of time. The linear decrease of *drop_probability* also helps in mitigating any ping pong effect in the control of aggressive flow.

18.6.3. *Dual Queue Service Differentiation (DQSD)*

In a multi-hop WMN, the proximity of client's corresponding MR to the IGW plays a significant role in the performance obtained. Often the clients attached to MRs that are closer to the IGW receive greater throughput and experience lesser end-to-end delays when compared to the clients attached to MRs far away from the IGW. In other words, the longer hop length flows receive extremely low throughputs and experience high end-to-end delays. The envisioned goal of WMNs to replace the wired backbone implies an implicit requirement of unbiased treatment to all flows regardless of their spatial origin.

Dual-Queue Service Differentiation (DQSD) algorithm at a MR ensures fairness to the multi-hop traffic from the traffic originating from local neighborhood of the MR. Broadly, this algorithm works by maintaining two queues at each MR to separately host the locally generated traffic at the MR and the multi-hop traffic traversing through this MR. The scheduling of the packets from either of these queues is based upon a service rate defined at each MR, typically giving more priority to the forwarded traffic as compared to locally generated traffic.

The DQSD algorithm uses a Queue Management (QM) module for the IEEE 802.11s based mesh networks to ensure proportional level of service to multi-hop traffic compared to the local traffic at each MR. Local traffic at a MR is the traffic generated from all the clients that are being served by the MR and can be maintained in the Local Queue (LQ). Traffic arriving from far away MRs which has to be relayed is stored in a separate queue, called the Multi-hop Queue (MQ), thus shielding from the local traffic. The algorithm works by elegantly segregating and exclusively reserving queues for either of the traffic. In other words, while one of the queues buffers self-originated packets at a MR, the local traffic; the other queue exclusively stores the multi-hop traffic; i.e., traffic traversing multiple hops.

The QM module governs the service schedule sequence of both the queues at each MR. Whenever a packet arrives at the link layer, the MR checks whether there is an entry for the corresponding *source* in the Flow table. If there is no entry for that *source*, a new entry is created and the Flow table is updated with necessary parameters. Then the packet is buffered into one of the two queues depending on whether it belongs to a flow that is relayed through this MR or generated by the clients under this MR. Whenever a packet has to be dequed, the service rate is computed for each of the queues. More clearly, at each MR, the QM computes the service rate of the LQ, denoted by *Own_Service_Rate*; for this source based on the number of flows originating from this source divided by the total number of flows currently occupied in the buffer space in the Flow table. The service rate of flows in the other queue, MQ, denoted by *Others_Service_Rate*, is recalculated for other sources in the table.

The DQSD algorithm identifies and distinguishes the local flows from the multi-hop flows and correspondingly ensures proportional service fairness. DQSD guarantees a fair buffer share at each intermediate MR, for all flows traversing through the MR, irrespective of their hop length. The proportional service schedule of the LQ and MQ can be computed using the total number of flows currently being serviced at each node and the own flows at each node. The average service rate of own flows and forwarded flows are estimated as follows:

$$Own\_Service\_Rate = \left(\frac{1}{total\ no.\ flows}\right) * no.\ own\ flows \tag{18.6}$$

$$Others\_Service\_Rate = (1 - Own\_Service\_Rate) \tag{18.7}$$

18.7. Conclusion

In this chapter, we introduce WMNs, an upcoming technology that aims to provide broadband wireless Internet to a wide range of MCs. For WMNs, factors such as load balancing, avoiding congested routes, and interference patterns directly affect their performance. And, Achieving good end-to-end throughput for longer hop flows is one of the major issues hampering the acceptability of this technology. We try to lay emphasis on several aspects that need to be focused for designing efficient protocols for WMNs. We also describe and analyze the importance of flow control and buffer management schemes in such networks, and present three recent approaches — QMMN, CBTR, and DQSD — to ensure fairness of multi-hop flows in WMNs.

Questions

(1) Differentiate between Wireless Local Area Networks and Wireless Mesh Networks. List any two real-world deployment examples of WMNs.

(2) Describe the various functional components in a typical Wireless Mesh Network architecture.
(3) What is the purpose of IEEE 802.11s standard?
(4) Enlist two challenges, each encountered in the design of protocols at physical, MAC, and routing layers.
(5) Illustrate the multi-flow unfairness problems prevalent in wireless networks and state a solution or approach to alleviate this problem.
(6) Discuss the limitations of existing queue management approaches for wired networks.
(7) State two queue management approaches for WMNs.
(8) What do you mean by per-flow fairness and per-user fairness in wireless networks? Illustrate them with examples.
(9) Explain in detail the differences in single-radio single channel, single-radio multi-channel and multi-radio multi-channel networks with the help of a diagram.
(10) Mention any two advantages of equipping a node with multiple interfaces in a network.

Bibliography

1. http://www.earthlink.net/
2. http://www.wirevolution.com/2007/09/07/how-does-80211n-get-to-600mbps/
3. http://i.i.com.com/cnwk.1d/html/itp/burton_80211nbeyon.pdf
4. http://www.nttdocomo.co.jp/english/binary/pdf/corporate/technology/rd/tech/main/mesh_network/vol8_2_13en.pdf
5. http://blogs.zdnet.com/Ou/?p=1078&page=1
6. A. Adya, P. Bahl, J. Padhye, A. Wolman and L. Zhou (2004) A multi-radio unification protocol for IEEE 802.11 wireless networks, In *BroadNets*
7. D. P. Agrawal and Q-A Zeng (2003). Introduction to Wireless and Mobile Systems, (*Brooks/Cole Publishing*).
8. J. Aweya, M. Ouellette and D. Y. Montuno (2001). A control theoretic approach to active queue management, *In the IEEE Transaction on Computer and Networking*, **36**(2–3).
9. K. Chowdhury, N. Nandiraju, D. Cavalcanti and D. P. Agrawal (2006). C-MAC — A Multi-Channel Energy Efficient MAC for Wireless Sensor Networks, in *Proc. IEEE WCNC*.
10. C. Cordeiro and D. P. Agrawal (2006). Ad hoc and Sensor Networks — Theory and Applications (*World Scientific Publishing*).
11. W. Feng, D. Kandlur, D. Saha and K. Shin (2002). The Blue Queue Management Algorithms, *In the IEEE/ACM Transactions on Networking* **10**(4).
12. S. Floyd and V. Jacobson (1993). Random early detection for congestion avoidance, *In the IEEE/ACM Transactions on Networking* **1**(4), 397–413.
13. S. Floyd, R. Gummadi and S. Shenker (2001). Adaptive RED: An algorithm for increasing the robustness of RED's active queue management.
14. V. Gambiroza, B. Sadeghi and E. W. Knightly (2004). End to end performance and fairness in multihop wireless backhaul networks, in *Proc. ACM Mobicomm*.

15. T. R. Jensen and B. Toft (1995). Graph Coloring Problems, *Wiley Interscience*, New York.
16. S.-J. Lee and M. Gerla (2001). AODV-BR: Backup routing in ad hoc networks, in *Proc. WCNC*.
17. S.-J. Lee and M. Gerla (2001). Split multipath routing with maximally disjoint paths in ad hoc networks, in *Proc. ICC*.
18. J. Li, C. Blake, D. S. De Couto, H. I. Lee and R. Morris (2001). Capacity of ad hoc wireless networks, in *Proc. ACM MOBICOM*.
19. Liu Chunlei, Shen Fangyang and Sun, Min-Te (2007). A unified TCP enhancement for wireless mesh networks, *Parallel Processing Workshops*.
20. P. Kyasanur and N. H. Vaidya (2005). Routing and interface assignment in multi-channel multi-interface wireless networks, in *Proc. WCNC*.
21. S. Liese, D. Wu and P. Mohapatra. Experimental characterization of an 802.11b wireless mesh network, *UC Davis Computer Science Department Technical Report*.
22. D. Lin and R. Morris (1997). Dynamics of random early detection, *In the Proc. SIGCOMM Symposium on Communications Architectures and Protocols*.
23. R. Mahajan, S. Floyd and D. Wetherall (2001). Controlling high-bandwidth flows at the congested router, *In the Proc. of ICNP*.
24. M. Mosko and J. J. Garcia-Luna-Aceves (2005). Multipath routing in wireless mesh networks, *WIMESH*.
25. D. Nandiraju, N. S. Nandiraju and D.P. Agrawal (2007). Service differentiation in IEEE 802.11s mesh networks: A dual queue strategy, in *proc. IEEE MILCOM*.
26. N. S. Nandiraju, D. Nandiraju, D. Cavalcanti and D. P. Agrawal (2007). A novel queue management mechanism for improving performance of multihop flows in IEEE 802.11s based mesh networks, in *Proc. IPCCC*.
27. S. Nandiraju, D. L. Santhanam, B. He, J. F. Wang and D. P. Agrawal (2006). Wireless mesh networks: Current challenges and future directions of web-in-the-sky, in *Proc. IEEE Wireless Communications Magazine*.
28. N. S. Nandiraju, D. Nandiraju, L. Santhanam and D. P. Agrawal (2007). A cache based traffic regulator for improving performance in IEEE 802.11s based mesh networks, in *Proc. IEEE RWS*.
29. T. J. Ott, T. V. Lakshman and L. H. Wong (1999). SRED: stabilized RED, in *Proc. INFOCOM*.
30. R. Pan, B. Prabhakar and K. Psounis (2000). CHOKe — A stateless active queue management scheme for approximating fair bandwidth allocation, in *Proc. INFOCOM*.
31. J. S. Pathmasuntharam, A. Das and A. K. Gupta (2004). Primary channel assignment based MAC (PCAM) — A Multi-channel MAC protocol for multi-hop wireless networks, in *Proc. IEEE WCNC*.
32. R. Pletka, M. Waldvogel and S. Mannal (2003). PURPLE: predictive active queue management utilizing congestion information, In *Proc. the LCN*.
33. K. Ramachandran, E. Belding, K. Almeroth and M. Buddhikot (2006). Interference-aware channel assignment in multi-radio wireless mesh networks, *In Infocom*.
34. A. Raniwala and T. Chiueh (2005). Architecture and algorithms for an IEEE 802.11-based multi-channel wireless mesh network, in *Proc. IEEE Infocom*.
35. P. A. Sambasivam Murthy and E. Belding-Royer (2004). Dynamically adaptive multipath routing based on AODV, in *Proc. MedHocNet*.
36. J. So and N. H. Vaidya (2004). Multi-channel MAC for ad hoc networks: Handling multi-channel hidden terminals using a single transceiver, in *Proc. ACM MobiHoc*.
37. S.-L. Wu, C.-Y. Lin, Y.-C. Tseng and J.-P. Sheu (2000). A new multi-channel MAC protocol with on-demand channel assignment for multi-hop mobile ad hoc networks, in *Proc. ISPAN*.

Chapter 19

Multi-hop MAC: IEEE 802.11s Wireless Mesh Networks

Ricardo C. Carrano,* Débora C. Muchaluat Saade,*
Miguel Elias M. Campista,† Igor M. Moraes,†
Célio Vinicius N. de Albuquerque,‡ Luiz Claudio S. Magalhães,*
Marcelo G. Rubinstein,§ Luís Henrique M. K. Costa†
and Otto Carlos M. B. Duarte†

** TET/UFF, Brazil*
carrano@midiacom.uff.br; debora@midiacom.uff.br; schara@midiacom.uff.br

† GTA/COPPE/POLI/UFRJ, Brazil
miguel@gta.ufrj.br; igor@gta.ufrj.br; luish@gta.ufrj.br; otto@gta.ufrj.br

‡ IC/UFF, Brazil
celio@ic.uff.br

§ PEL/DETEL/FEN/UERJ, Brazil
[i]rubi@uerj.br

This chapter presents IEEE 802.11s, an emerging standard for wireless mesh networks (WMNs). IEEE 802.11s proposes multi-hop forwarding at the MAC level, which is a new approach for building WMNs. Traditional solutions for WMNs use network-level routing protocols to allow multi-hop forwarding among wireless mesh nodes. IEEE 802.11s specifies multi-hop MAC functions for mesh nodes using a mandatory path selection mechanism named HWMP (Hybrid Wireless Mesh Protocol) and also provides a path selection framework for alternative mechanisms and future extensions. This chapter discusses the emerging standard details and compares this new solution for WMNs to traditional ones.

Keywords: Wireless mesh networks; multi-hop; IEEE 802.11s; HWMP; path selection; routing protocols; routing metrics.

19.1. Introduction

Wireless local area networks (WLANs) are well-known for being easy to deploy and support for user mobility. Although IEEE 802.11a, b and g standards are extremely popular and can be found in most laptops, PDAs and all sort of untethered equipment, wireless technology still faces some challenges and many research fields related to it are open to active development.

One of the main evolving fields is multi-hop ad hoc wireless networks that are based on, or extend, current wireless standards and technologies. This new trend is relevant since infrastructured wireless networks, though providing a number of advantages, can be highly empowered if nodes are able to forward traffic sourced by other nodes in an ad hoc self-configuring fashion. Multi-hop forwarding can, for instance, extend the coverage of wireless access points without the need of additional infrastructure.

Inexpensive IEEE 802.11 routers are also currently used in the deployment of low cost wireless backbones. Networks where the placement of each router forming a wireless backbone is chosen in order to create radio coverage for network access in a certain area, or to interconnect distant wired networks, are called Wireless Mesh Networks, or WMNs. Therefore, by this definition, WMNs would not be true ad hoc networks because they are planned (or engineered) but would nevertheless benefit from the wireless technology. Examples of WMN pilots can be found in Refs. 1, 3, 4, 7, 9 and 13.

In contrast to a WMN, a Mobile ad hoc Network (MANET) is a self-configuring network where there are no fixed routers. In a MANET, routers are free to move and network topology can change quickly and dramatically. Traffic routing functions are carried on by some or all of the participating nodes. Being ad hoc or engineered, or somewhere in between these two paradigms, wireless multi-hop networks share a common challenge: the development of routing protocols capable of coping with specific challenges posed by wireless networks, such as node mobility, fast-changing characteristics of the radio environment and medium access contention. After some decades of research in routing algorithms and routing metrics, there is a natural tendency that routing protocols shall be based, in varying degrees, on pre-existent routing mechanisms.

The traditional approach to multi-hop forwarding has been the implementation of routing protocols at the network level, which brings the obvious advantage of being link-layer independent. After all, internetworking has been the realm and main goal of routing protocols.

A more recent proposal in WMN design addresses the implementation of multi-hop forwarding techniques at the link level, as an extension of WLAN functionalities.[8,19] This type of solution can widely spread the use of WMNs since it is going to be provided by end-user equipment. Additionally, quality-aware metrics can be easily implemented at the MAC level, allowing a better utilization of the wireless mesh network.

The IEEE Task Group 802.11s[24] is currently developing an emerging standard for mesh networking at the MAC level. This new approach for building a WMN makes it appear as a LAN for layer-three protocols. IEEE 802.11s specifies multi-hop MAC functions for mesh nodes using a mandatory path selection mechanism, named HWMP (Hybrid Wireless Mesh Protocol), and provides a path selection framework for alternative mechanisms and future extensions.

The main goal of this chapter is to present the IEEE 802.11s emerging standard proposals, focusing on path selection mechanisms, and discuss and compare both-layer-two and layer-three-approaches for building WMNs. Section 19.2 describes the most relevant routing metrics and protocols used in current wireless mesh networks that employ the traditional layer-three approach. Section 19.3 on the other hand, is devoted to a detailed description of layer-two multi-hop techniques proposed by IEEE 802.11s. Final remarks are provided in Section 19.4.

19.2. Traditional Network-Level WMNs

In wireless mesh networks backbone routers communicate through multiple hops similar to that of ad hoc networks (Figure 19.1). On the other hand, users carrying devices such as laptops or PDAs are often not responsible for routing, and connect to the backbone via mesh routers playing the role of access points. Thus, as previously mentioned, in mesh networks some nodes are dedicated to provide a backbone, unlike the ad hoc case. Wireless mesh networks have other two peculiarities. First, routers are typically stationary. As a consequence, the WMN routing metrics should

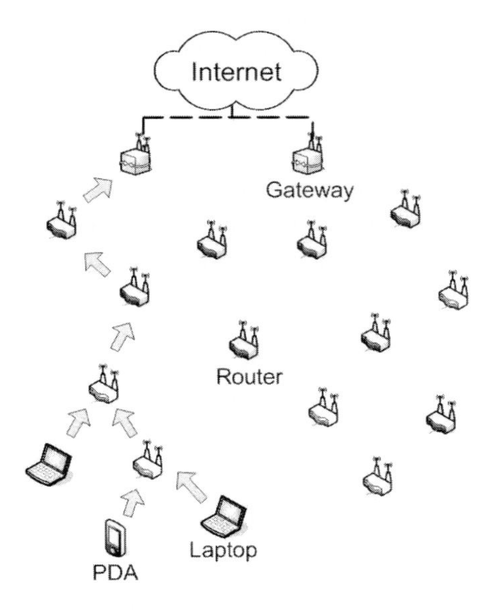

Fig. 19.1. A typical wireless mesh network.

measure wireless link quality instead of number of hops. Second, in WMNs, most of the network traffic flows towards the gateways, assuming that the common-case application is Internet access. This particular traffic matrix tends to present traffic concentration on the links close to the gateways. This characteristic can be explored to optimize WMN routing protocols. This section presents some of the most representative metrics and layer-three routing protocols proposed in the literature.

19.2.1. *Routing metrics*

In ad hoc networks, the most used metric is hop count. In such networks, this metric is convenient because of the user mobility, which may incur in many link breakages. It is more important to quickly recover and have a route to the destination than to have a high-quality route. On the other hand, in WMNs, routers are usually stationary. Hence, routing metrics are more concerned with link-quality variations than specifically with link breakages. We refer to these as quality-aware metrics.[25]

WMN routing metrics are continuously evolving. One of the first metrics specifically proposed to be used in WMNs is called Expected Transmission Count (ETX).[13,17] ETX represents the expected number of transmissions a node needs to successfully transmit a packet to a neighbor. To compute ETX, each node periodically broadcasts small probe packets containing the number of received probes from each neighbor. The number of received probes is calculated during the last T time interval using a sliding window. A node A computes the ETX of the link to a node B by using the delivery ratio of probes sent on the forward (df) and reverse (dr) directions. These delivery ratios correspond, respectively, to the fraction of successfully received probes from A as announced by B, and the fraction of successfully received probes from B, during the same interval, T. Thus, the ETX of link AB is given by:

$$ETX = \frac{1}{df \times dr}. \tag{19.1}$$

The ETX computation takes into account both forward and reverse directions to consider data-frame and ACK-frame transmissions. ETX is an additive metric, and the best route is the one with the lowest sum of ETXs along the route to the destination. It is worth noting that the number of broadcast probes in a network with n nodes is $O(n)$.

The Minimum Loss (ML) metric[31] is also based on probing to compute the delivery ratio. Rather than calculating ETX, the ML metric finds the route with the lowest end-to-end loss probability, as follows:

$$ML = \prod_{i=1}^{l} df_i \times dr_i, \tag{19.2}$$

where l is the number of links in a path. Thus, ML is not cumulative, differently from ETX. Instead, it multiplies the delivery ratio of the links in the reverse and forward directions to find the best path. The authors of ML argue that the use of multiplication reduces the number of route changes, improving network performance. Iannone *et al.*,[21] use another simple end-to-end metric, in which the idea is to avoid bottlenecks. Their metric is denoted here by I_p and is defined as follows:

$$I_p = \max_{1 \leq i \leq l} \left(\frac{1}{R_i} \right), \tag{19.3}$$

where R_i is the PHY rate of the link i in a path p composed of l links. The lower the I_p value, the better the route.

The implementation of ETX has revealed two shortcomings: broadcasts are usually performed at the network basic rate and probes are smaller than typical data packets. Thus, unless the network is operating at low rates, the performance of ETX is low because it neither distinguishes links with different bandwidths nor it considers data-packet sizes. To cope with these issues, the Expected Transmission Time metric (ETT)[2,18] was proposed. ETT represents the time a data packet needs to be successfully transmitted to each neighbor. Basically, ETT adjusts ETX to different PHY rates and data-packet sizes.

The ETT computation is a matter of implementation choice. Currently, there are two main approaches. For Draves *et al.*,[18] ETT is the product between ETX and the average time a single data packet requires to be delivered, as seen in Equation (19.4).

$$ETT = ETX \times t. \tag{19.4}$$

To calculate the time t, the authors divide a fixed data-packet size (S) by the estimated bandwidth (B) of each link ($t = S/B$). The authors prefer to periodically estimate the bandwidth rather than use the transmission rate as informed by the firmware. This is because the IEEE 802.11 standard does not define an algorithm to select the transmission rate, it only defines physical transmission rates according to modulation schemes. Therefore, manufacturers are free to implement algorithms to automatically select the best transmission rate. This feature is called auto-rate, and usually does not give information about bandwidth, which is required to compute ETT. The packet-pair technique is then used to calculate the bandwidth (B) per link. This technique is well-known from wired networks, and consists of transmitting a sequence of two back-to-back packets to estimate bottleneck bandwidth. In the implementation of Draves *et al.*, two packets are unicast in sequence, a small one followed by a large one, to estimate the link bandwidth to each neighbor. Each neighbor measures the inter-arrival period between the two packets and reports it back to the sender. The computed bandwidth is the size of the large packet of the sequence divided by the minimum delay received for that link. In a n-node network where each node has v adjacencies, estimating the bandwidth is $O(n.v)$. Another approach to compute ETT is considered by Aguayo *et al.*[2] The authors estimate the loss probability by considering that IEEE 802.11 uses data and ACK frames. The idea is to periodically compute the loss rate of data and ACK frames to each neighbor. The former is estimated by broadcasting a number of packets of the same size as data frames, one packet for each data rate as defined in IEEE 802.11. The latter is estimated by broadcasting small packets of the same size as ACK frames transmitted at the basic rate, which is the rate used for ACKs. Note that broadcasting packets at higher data rates may require firmware modifications. According to Aguayo *et al.*, ETT considers the best throughput achievable (r_t) and the delivery probability of ACK packets in the reverse direction (p_{ACK}). Thus, it is defined as:

$$ETT = \frac{1}{r_t \times p_{ACK}}. \tag{19.5}$$

Computing ETT in a n-node network is $O(n.m)$, where m is the number of possible data rates. Similarly to ETX, the chosen route is the one with the lowest sum of ETT values for each link to the destination. Cross-layer approaches are receiving special attention in WMNs.[3] Among the available techniques, the use of multiple channels is commonplace. Through multiple channels it is possible to improve network throughput by using, at the same time, the available non-overlapping channels defined by IEEE 802.11. This technique, however, needs to deal with two issues to become effective, namely, intra-flow and inter-flow interference. The intra-flow interference occurs when different nodes transmitting packets from the same flow interfere with each other. Maximizing the number of channels is not trivial, considering that nodes must maintain connectivity. The inter-flow interference, on the other hand, is the interference between concurrent flows. The Weighted Cumulative ETT (WCETT)[18] changes ETT to also consider intra-flow interference. This metric is the sum of two components: end-to-end delay and channel diversity. Let p denote a path composed of l links, and c the maximum number of channels in the wireless system, the first component (Γ_1) of WCETT is given by:

$$\Gamma_1 = \sum_{i=1}^{l} ETT_i, \tag{19.6}$$

where ETT_i is the ETT of link i. This first component computes the end-to-end ETT. The second component (Γ_2), on the other hand, computes the ETT of links on the same channel along the path to estimate channel diversity. Then, the second component is given by:

$$\Gamma_2 = \max_{1 \le j \le c}(X_j), \quad \text{where } X_j = \sum_{\text{link } i \text{ is on channel } j} ETT_i. \tag{19.7}$$

The higher the channel diversity, the more balanced is the sum of ETTs among the different channels along the path. Hence, computing Γ_2 gives an idea of the traffic balance among the different channels, or if there is a predominant channel being used. Furthermore, a tunable parameter (β) is used to combine both components or prioritize one of them. Equation 19.8 shows the WCETT metric.

$$WCETT = (1-\beta) \times \Gamma_1 + \beta \times \Gamma_2 = (1-\beta) \times \sum_{i=1}^{l} ETT_i + \beta \times \max_{1 \le j \le c}(X_j). \tag{19.8}$$

Unlike ETX and ETT, WCETT is an end-to-end metric. Thus, its outcome is the total cost of the route. This metric computes end-to-end values because it must consider all channels used along the path in order to avoid intra-flow interference.

The Normalized Bottleneck Link Capacity (NBLC)[26] metric also uses multiple radios and channels to improve network throughput. Differently from WCETT, NBLC considers the traffic load on each link and thus can perform load balancing among links. NBLC is an estimate of the residual bandwidth of a path. To compute NBLC, each node periodically measures the percentage of busy air time on each radio (tuned to a certain channel) and then obtains the percentage of residual air time. Each node periodically broadcasts this information to its k-hop neighbors using a dedicated control channel. For a specific channel used by a node, the actual residual channel capacity is approximated by the lowest residual channel capacity reported by interfering neighbors of the node or observed by the node itself. Based on this calculation, each node can determine the percentage of free-to-use channel air time on each outgoing link, called the Residual Link Capacity (RLC). Additionally, intra-flow interference is considered. For a link on a path, the actual air time consumed for the transmission of one packet along the path includes not only the air time spent in forwarding the packet on the link, but also the air time spent in keeping away from interference with the transmissions on links operating on the same channel on the same path. This amount of consumed air time, called Cumulative Expected Busy Time (CEBT), for a certain link on a path is obtained by aggregating the ETT values for the links of the path that operate on the same channel and interfere with this link. For a path p composed of l links, the NBLC metric is given by:

$$NBLC_p = \min_{link\ i \in p} \left(\frac{RLC_i}{CEBT_{i,p}} \right) \times \gamma^l, \tag{19.9}$$

where γ is the probability of a packet being dropped by an intermediate node. The chosen route is the one with the higher NBLC value.

Metrics such as WCETT do not guarantee shortest paths and do not avoid inter-flow interference.[38] Link-state routing protocols need minimum-cost routes to be loop-free. Moreover, not avoiding inter-flow interference may lead WCETT to choose routes in congested areas. The Metric of Interference and Channel-switching (MIC) addresses these issues.[38] MIC is also an end-to-end metric and its computation is based on two components: Interference-aware Resource Usage (IRU) and Channel Switching Cost (CSC). The first one is defined as:

$$IRU_i = ETT_i \times N_i, \tag{19.10}$$

where N_i is the set of neighbors affected by a transmission on link i, and ETT_i is the ETT of the same link. This first component takes into account the number of interfering nodes in the neighborhood to estimate inter-flow interference. The CSC computed by a node n of a path p is defined as:

$$CSC_n = \begin{cases} w_1, & \text{if } c_{n-1} \neq c_n \\ w_2, & \text{if } c_{n-1} = c_n, \end{cases} \tag{19.11}$$

where $0 \leq w_1 < w_2$, c_n is the channel used by node n, and c_{n-1} is the channel used by the previous node on the same path. CSC values are summed for all nodes of a path. As w_2 is always greater than w_1, when the same channel is repeated on consecutive links, the sum of CSC values gets higher and consequently the metric avoids intra-flow interference. In addition, MIC uses virtual nodes to guarantee minimum-cost route computation. The MIC metric is defined as:

$$MIC = \frac{1}{N_t \times \min(ETT)} \times \sum_{i=1}^{l} IRU_i + \sum_{n=1}^{N_p} CSC_n, \tag{19.12}$$

where N_t is the total number of nodes in the network, $\min(ETT)$ is the minimum known ETT, l is the number of links in path p, and N_p is the number of nodes in path p.

One critical problem of wireless networks is the fast link-quality variation. Metrics based on average values computed on a time-window interval, such as ETX, may not follow the link-quality variations or may produce prohibitive control overhead. Indoor environments make this problem even more difficult due to obstacles, interfering wireless devices, walking people etc. To cope with this, modified ETX (mETX) and Effective Number of Transmissions (ENT) were proposed.[25] These metrics consider the variance in addition to link-quality average values. Thus, the main goal is to reflect physical-layer variations onto routing metrics.

The mETX metric is also calculated by broadcasting probes. The difference between mETX and ETX is that rather than considering probe losses, mETX works at the bit level. The mETX metric computes the bit error probability using the position of the corrupted bit in the probe and the inter-dependence of bit errors throughout successive transmissions. This is possible because probes are composed by a previously known sequence of bits. The metric mETX is defined in Equation (19.13), where μ_Σ and σ_Σ^2 denote the average and the variance of the packet error probability summed over the duration of one packet, respectively.

$$mETX = \exp\left(\mu_\Sigma + \frac{\sigma_\Sigma^2}{2}\right). \tag{19.13}$$

ENT is an alternative approach that measures the number of successive retransmissions per link considering the variance. ENT also broadcasts probes and limits route computation to links that show an acceptable number of retransmissions according to upper-layer requirements. If a link shows a number of expected transmissions higher than the maximum tolerated by an upper-layer protocol (e.g. TCP), ENT excludes this link from routing computation assigning an infinity-value metric to it. ENT slightly modifies the mETX computation to consider the probability that the number of successive retransmissions per link exceeds a given threshold. ENT is defined as:

$$ENT = exp\left(\mu_\Sigma + 2\delta\sigma_\Sigma^2\right), \tag{19.14}$$

where δ denotes the strictness of the loss rate requirement. Both mETX and ENT are aware of the probe size, therefore the inclusion of the data rate is trivial with the two metrics.

The Distribution-Based Expected Transmission Count (DBETX)[14] metric also takes into account medium instability by considering fading of wireless channels. To compute DBETX, first each node must estimate the probability density function (pdf) of the SINR (Signal to Interference and Noise Ratio). Then, for a given modulation, it is possible to calculate the expected Bit Error Rate (BER) and the expected Packet Error Rate (PER). The success probability of a link (P_{Suc}) is $1 - PER$. Differently from other metrics, DBETX also takes into account the maximum number of MAC sublayer retransmissions (MaxRetry) to select a route. This cross-layer technique is similar to the one used by ENT. Nevertheless, DBETX considers lower-layer retransmissions instead of upper-layer requirements. DBETX is based on the Average Number of Transmissions (ANT) and on the MAC sublayer outage probability ($P_{out_{MAC}}$). ANT represents the expected number of retransmissions on a link considering the MaxRetry limit, as follows:

$$ANT = \begin{cases} \dfrac{1}{P_{Suc}}, & \text{if } P_{Suc} > P_{lim} \\ \dfrac{1}{P_{lim}}, & \text{otherwise,} \end{cases} \tag{19.15}$$

where

$$P_{lim} = \frac{1}{MaxRetry}. \tag{19.16}$$

The MAC sublayer outage is the condition that arrives when the current success probability of a link results in an expected number of retransmissions higher than MaxRetry. Thus, DBETX is given by:

$$DBETX = E[ANT] \times \frac{1}{1 - P_{out_{MAC}}}. \tag{19.17}$$

Another metric that also considers link-quality variation is iAWARE.[36] This metric uses SNR (Signal to Noise Ratio) and SINR to continuously reproduce neighboring interference variations into routing metrics. The iAWARE metric estimates the average time the medium is busy because of transmissions from each interfering neighbor. The higher the interference, the higher the iAWARE value. Thus, unlike mETX, ENT, and DBETX, iAWARE considers intra-flow and inter-flow interference, medium instability, and data-transmission time. Let i denote a link between nodes u and v. To compute iAWARE, a node u measures the Interference Ratio ($IR_i(u)$) for a node u on link i. $IR_i(u)$ is defined as follows:

$$IR_i(u) = \frac{SINR_i(u)}{SNR_i(u)}. \tag{19.18}$$

In Equation (19.18), the Signal to Interference and Noise Ratio at node u on link i ($SINR_i(u)$) is given by:

$$SINR_i(u) = \frac{P_u(v)}{N_t + \sum_{n \in N_u - v} \tau(v) P_u(n)}, \tag{19.19}$$

where $P_u(v)$ is the signal strength of a packet from node v at node u, N_t is the total number of nodes in the network, N_u is the set of nodes that interfere on node u, and $\tau(v)$ is the normalized weight at which node n produces traffic averaged over a period of time. The Signal to Noise Ratio at node u on link i ($SNR_i(u)$) is defined as:

$$SNR_i(u) = \frac{P_u(v)}{N_t}. \tag{19.20}$$

Equation (19.21) shows the iAWARE metric of a link i, where IR_i is the minimum Interference Ratio ($\min(IR_i(u), IR_i(v))$) to consider both data-transmission and ACK-transmission directions on link i.

$$iAWARE_i = \frac{ETT_i}{IR_i}. \tag{19.21}$$

If there is no interference on link i, $SINR_i(u) = SNR_i(u)$ and then $IR_i = 1$. In this case, iAWARE depends only on the ETT_i of the link. Similarly to WCETT, the iAWARE metric also avoids intra-flow interference. To reach this goal, the iAWARE metric becomes a sum of two components tuned by a parameter β. In a wireless system composed of c different channels, iAWARE defines X_j as follows:

$$X_j = \sum_{conflicting\ link\ i\ is\ on\ channel\ j} iAWARE_i, \tag{19.22}$$

where $1 \leq j \leq c$, and the conflicting links are those which interfere on link i. The iAWARE metric of a path p is defined as in Equation (19.23).

$$iAWARE_p = (1 - \beta) \times \sum_{i=1}^{l} iAWARE_i + \beta \times \max_{1 \leq j \leq c} (X_j), \tag{19.23}$$

where l denotes the number of links on p.

Although there is an increasing number of routing metrics, no consensus has been reached. Up to now, most routing protocol implementations prefer metrics with simpler designs such as ETX or ETT. A summary of the main characteristics of WMN routing metrics can be found in Ref. 9.

19.2.2. *Routing protocols*

Ad hoc routing protocols usually use one of three strategies, namely reactive, proactive, or a combination of them. In the reactive strategy, as soon as a node has a data packet to send, it requests a route to the intended destination. If a node does not have data packets to send to a particular destination, the node will never request a route to it. The proactive strategy operates similar to that of classic routing on wired networks. Routers have at least one valid route always up-to-date to any destination in the network.

Many routing protocols of wireless mesh networks still use similar strategies to compute routes. Nevertheless, they are adapted to the peculiarities of WMNs and use one of the quality-aware routing metrics (Section 19.2.1). We use a taxonomy for the main WMN routing protocols according to Ref. 9. We divide them into four classes, namely, ad hoc-based, controlled-flooding, traffic-aware, and opportunistic. These protocol classes mainly differ on route discovery and maintenance procedures. In WMNs, most routing protocols assume that the network is composed by the set of wireless backbone nodes, and user nodes do not participate in the routing process. If a user device temporarily works as a backbone node, it must run the same routing protocol.

19.2.2.1. *Ad hoc-based protocols*

WMN ad hoc-based protocols adapt ad hoc routing protocols to deal with link-quality variations. Routers keep track of the quality of other links to make route computation more accurate. Therefore, they continuously update their metrics and disseminate them to other routers. The Link Quality Source Routing (LQSR) protocol[18] is ad hoc based. LQSR combines link-state proactive routing with the reactive strategy from ad hoc networks. It is fundamentally a link-state routing protocol and uses a complete view of the network topology to perform shortest-path computation. Nevertheless, LQSR uses route discovery procedures similar to those of reactive protocols to reduce routing overhead, which may become high because of medium instability and node mobility. During route discovery, LQSR obtains current link-state information of the traversed links, reducing the periodicity of link-state updates.

The SrcRR protocol[12] is another example of ad hoc-based protocol. It uses a discovery procedure similar to reactive protocols only to update the routing information of the traversed links. SrcRR further reduces control overhead, but computes routes from a reduced view of the network. Both LQSR and SrcRR implement route discovery procedures based on the Dynamic Source Routing (DSR)[16] using source routing, and use ETX. The Mesh Distance Vector (MeshDV) protocol[20] deals with user mobility and unlike LQSR and SrcRR, MeshDV considers not only backbone nodes but also user devices. Each node of the backbone maintains one table with the IP addresses of directly connected users and another table with the IP addresses of users connected to other backbone nodes. This scheme allows routers to be aware of users current location. MeshDV runs the Destination-Sequenced Distance Vector (DSDV)[32] routing protocol in the backbone and can use two metrics: hop count or I_p, as seen in Section 19.2.1.

One open research issue in wireless networks is the deployment of physical layer techniques to improve the overall efficiency of routing protocols. The Multi Radio LQSR (MR-LQSR)[18] is one example of such protocol. It adapts LQSR to operate over multiple channels and multiple interfaces and uses the metric WCETT. Although the use of WCETT does not guarantee minimum-cost paths, MR-LQSR is loop-free because it uses source routing. Another routing protocol that exploits physical-layer techniques to improve network performance is DOLSR (Directional Optimized Link-State Routing), which employs directional antennas.[15] The DOLSR protocol can use metrics such as number of hops, residual bandwidth, or ETX.

19.2.2.2. *Controlled-flooding protocols*

Flooding the network with routing updates may produce scalability issues, especially if frequent changes on medium conditions are considered. Controlled-flooding protocols implement algorithms to reduce control overhead. One possible approach assumes that flooding the network is not efficient because the majority of traffic in wireless networks is between nodes close to each other. Therefore, there is no need to send control packets to nodes that are farther away as frequently as to nearby ones. Another way to reduce routing overhead is to limit the number of nodes responsible for flooding the network, avoiding redundancies. Protocols that adopt the second approach run algorithms to find the minimum set of nodes needed to forward routing information to all destinations in the network.

The Localized On-demand Link State (LOLS)[28,37] attributes a long-term cost and a short-term cost to links. Long-term and short-term costs represent, respectively, the usual (historical) and the current cost of a link. In order to reduce the control overhead, short-term costs are frequently sent to neighbors, whereas long-term costs are sent at higher periods of time. LOLS computes routes using ETX or ETT.

Another typical example of a controlled-flooding protocol is the Mobile Mesh Routing Protocol (MMRP) developed by MITRE Corporation.[27] MMRP assigns an age to its routing messages in the same way as OSPF protocol. Thus, whenever a node sends a routing message, it subtracts the age of the message from the estimated time needed to forward it. When age value reaches zero, the respective message is dropped, preventing its retransmission. MMRP does not specify a metric to be used with.

The Optimized Link State Routing (OLSR)[10,11] is yet another example of a controlled-flooding protocol. Original OLSR uses hop count as a metric, but OLSR was adapted to use ETX in WMNs. It uses the fraction of HELLO messages lost in a given interval of time to calculate ETX. OLSR could also be considered ad hoc-based; however, it uses Multi Point Relays (MPRs) to control flooding. OLSR limits the number of nodes in charge of disseminating control packets to avoid redundancies. Therefore, each node selects its MPR set, which is composed by nodes responsible for forwarding routing information from the selector node. Each node fills its MPR set with the minimum number of one-hop neighbors needed to reach every two-hop neighbors. There are also additional implementations of OLSR that use ML[31] and ETT[9] metrics.

19.2.2.3. *Traffic-aware protocols*

Traffic-aware or tree-based routing protocols explore the traffic matrix typical of WMNs. They assume that backhaul access is the common-case application and, therefore, consider a network topology similar to a tree. One example is the Ad Hoc On-demand Distance Vector-Spanning Tree (AODV-ST).[34] This protocol is an adaptation of the AODV reactive protocol from ad hoc networks. In AODV-ST, the gateway periodically requests routes to every node in the network to initiate the creation of spanning trees in order to maintain its routing table updated. Thus, AODV-ST maintains a tree where the gateway is the root. Communications that do not include the gateway work as in the original AODV. AODV-ST supports ETX and ETT metrics.

Raniwala and Chiueh[35] propose a routing algorithm based on the spanning tree used in wired networks. Route maintenance is done with join and leave requests. They use the hop count and other metrics for load-balancing not specific to WMNs.

19.2.2.4. *Opportunistic protocols*

Opportunistic protocols improve classical routing by exploring cooperative diversity schemes. Classical routing protocols compute a sequence of hops to the destination before sending a data packet, either using hop-by-hop or source routing. In case of link failure, successive link-layer retransmissions are performed until successful reception at the next-hop neighbor or until the maximum number of link-layer retransmissions is reached. This approach may incur in high delay and poor performance because wireless links require time to recover from transient failures. Cooperative diversity schemes, on the other hand, exploit the broadcast nature of radio-frequency transmissions to use multiple paths towards a destination. Each destination requires suitable transceivers to choose one of the relayed signals or to use a combination of them. Opportunistic protocols adapt cooperative diversity to standard IEEE 802.11 transceivers. Therefore, only one node forwards each packet. For example, opportunistic protocols choose, on-the-fly, which hop offers the best throughput. These protocols guarantee that data is always forwarded whenever there is at least one next hop. Besides, the chosen route likely uses the best quality links, considering short-term variations.

The ExOR protocol combines routing with MAC sublayer functionality.[6] Routers send broadcast packets in batches, with no previous route computation. Packets are transmitted in batches to reduce protocol overhead, which may lead to underutilization of network resources. In addition, broadcasting data packets improves reliability because only one intermediate router is required to overhear a transmission. Nevertheless, it does not guarantee that the packets are received, because they are not acknowledged. Thus, an additional mechanism is needed to indicate the correct reception of data. Among the intermediate routers that have heard the transmission, only one retransmits at a time. The source router defines a forwarding list and adds it to the header of the data packets. This list contains the addresses of neighbors ordered by forwarding priority. Routers are classified in the forwarding list according to their proximity to the destination, computed with a metric similar to ETX. The metric used by ExOR only considers the loss rate in the forward direction because there are no acknowledgments. Upon reception of a data packet, the intermediate router checks the forwarding list. If its address is listed, it waits for the reception of the whole batch of packets. It is possible, however, that a router does not receive the entire batch. To avoid this problem, ExOR operates as follows. The highest-priority router that has received packets forwards them and indicates to the lower-priority routers the packets that were transmitted. The lower-priority routers therefore transmit the remaining packets, avoiding duplicates. The transmissions are performed until the destination indicates the reception.

The Resilient Opportunistic MEsh Routing protocol (ROMER)[39] focuses on resilience and high throughput by using multi-path forwarding. ROMER combines long-term best routes, shortest-path or minimum-latency, and on-the-fly opportunistic gain to provide resilient routes and to deal with short-term variations on medium quality. ROMER computes long-term routes and opportunistically expands or shrinks them at runtime to fully exploit short-term higher-quality links. These long-term routes are computed using the minimum number of hops or the minimum average delay. Unlike ExOR, ROMER performs transmissions on a packet basis to enable faster reaction to medium variations. The highest-throughput route is chosen according to the maximum PHY rate as indicated by the MAC sublayer.

In order to improve the efficiency of WMNs, proposals of quality-aware metrics have become a recent trend. However, the pratical implementation of such metrics demands access to lower layer information thus cross-layer techniques have been proposed. Following this trend it is expected that WMNs could greatly benefit from multi-hop forwarding at the MAC layer.

19.3. Multi-hop MAC: IEEE 802.11s

In September 2003, IEEE started a study group to investigate adding wireless mesh networks as an amendment for its IEEE 802.11 standard. One year later, the study group became the Task Group "s" (TGs), which issued its first draft later in March 2006. By the time of this writing, IEEE 802.11s is still a draft (currently in version 1.08),[24] therefore some degree of change should be expected before IEEE 802.11s becomes a standard. In fact, many improvements have been made in the current draft, considering previous versions of the document, and the reader should always keep in mind that this is still a work in progress. Nevertheless, commercial implementations of this draft are already available in some wireless devices.[29,30]

The recent emergence of handheld communication devices, constrained in many ways (power, processing, memory), demands a solution that may be easily embedded in network interface cards (NIC) and in systems-on-chip (SoC), and a MAC layer solution, being lightweight in contrast to a full implementation of ad hoc routing, fits that purpose.

In order to support multi-hop forwarding at the MAC layer, the current draft introduces changes in MAC frame formats, and an optional medium access method as well as many other optimizations to improve performance and security of wireless mesh networks. In this section, we focus on path selection mechanisms and new frame formats, since these aspects are the most closely related to multi-hop forwarding at the MAC level. Additional features are briefly discussed in Subsection 19.3.5.

Originally, two path selection mechanisms were proposed in the draft. RA-OLSR (Radio-Aware Optimized Link State Routing),[40] which is a proactive controlled-flooding protocol based on OLSR[11] but adapted to work at layer-two instead of three, and a hybrid traffic-aware protocol, named HWMP (Hybrid Wireless Mesh Protocol),[5] based on AODV,[33] which is actually the mandatory protocol and the only one remaining on the current proposal (version 1.08). RA-OLSR was removed in favor of an extensible path selection framework that enables alternative implementations of path selection protocols and metrics within the mesh framework.

Before going into the path selection mechanisms though, we must briefly discuss the mesh creation mechanisms and describe the architecture proposed by the emerging standard.

19.3.1. *Multi-hop MAC mesh network architecture*

According to the IEEE 802.11s draft, nodes in a mesh network fall into one of the four categories as illustrated in Figure 19.2:

- Client or Station (STA) is a node that requests services but does not forward frames, nor participates in path discovery mechanisms.
- Mesh Point (MP) is a node that participates in the formation and operation of the mesh cloud.
- Mesh Access Point (MAP) is an MP who has an attached access point (AP) to provide services for clients (STA).
- Mesh Portal Point (MPP) is an MP with the additional functionality of acting as a bridge or gateway between the mesh cloud and external networks.

Figure 19.2 illustrates a possible ad hoc topology for this architecture. The doted lines represent the mesh network itself (mesh cloud) in which other non-802.11s nodes may participate indirectly (solid lines) connecting to mesh nodes extended with access point functionalities (MAPs).

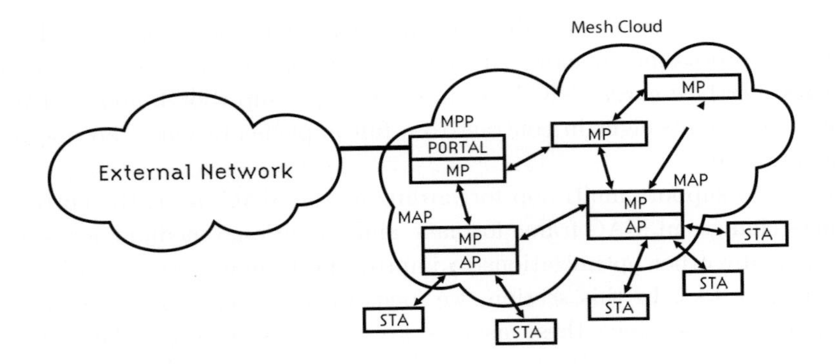

Fig. 19.2. IEEE 802.11s wireless mesh network.

In this topology example, there is only one MPP (PORTAL/MP), but nothing prevents a mesh network from having many. In that case each node must dynamically choose one of them for sending traffic outside the mesh network bounds.

Figure 19.2 must be understood as a snapshot for a dynamic topology, where nodes may move in unpredictable and diverse ways, and links are formed or disrupted not only because of mobility, but also due to the changing conditions of the wireless medium. In that sense, the role of MPP is opportunistic and the network should provide the means (protocols and mechanisms) for announcing throughout the mesh cloud the set of nodes that are able to work as MPPs. These announcement mechanisms will be described in the following sections.

19.3.2. *Mesh creation*

In infrastructured wireless networks, a Service Set Identifier (SSID) is used to distinguish the set of access points, which maintains a certain functional correlation and belong to the same local area network.

In a mesh network the same need for an identity exists, but instead of overloading the definition and function of the SSID, the draft proposes a Mesh identifier or Mesh ID. Similarly to 802.11, beacon frames are used to announce a Mesh ID, which should never be confused with the standard SSID employed by regular infrastructured wireless network. To avoid misleading a non-mesh station when trying to associate to a mesh network, Mesh Points (MPs) broadcast beacons with the SSID set to a wildcard value.

The Mesh ID is one of the three elements that characterize a mesh network. The other two are a path selection protocol and a path selection metric. Together these three elements define a profile. A Mesh Point may support different profiles, but all nodes in a mesh cloud, at a given moment, must share the same profile.

The 802.11s mandatory profile defines HWMP as the path discovery mechanism and the Airtime Link metric as the path selection metric, as it will be described in the following sections. The draft does not prevent other protocols or metrics from being used in a mesh cloud and even defines frameworks for those alternative mechanisms, but it advises that a mesh network shall not use more than one profile at the same time. This recommendation may be interpreted as an attempt to avoid complexity of profile renegotiation that may be too expensive for a simple device to handle. If a mesh cloud is formed with non-mandatory elements (protocol and metric), it is not obliged to fall back in order to accommodate a new mesh member that only supports the mandatory profile.

A mesh network is formed as MPs find neighbors that share the same profile. The neighbor discovery mechanism is similar to what is currently proposed by the 802.11 standard — active or passive scanning. In order to achieve this, regular (802.11) beacon frames and probe response frames are extended to include mesh related fields. As it will be discussed in the following sections, the draft does not only introduce new frames but also extends pre-existent ones.

To conclude our analysis on the mesh creation procedures we should comment on the establishment of the peer links — edges of a mesh graph. A Mesh Point shall create and maintain peer links to its neighbors that share its active profile (an MP may keep many profiles, but only one is active at a given moment). Once a neighbor candidate is found, through active or passive scanning, an MP uses the Mesh Peer Link Management protocol[24] to open a mesh peer link.

A mesh peer link is univocally identified by the MAC addresses of both participants and a pair of link identifiers, generated by each of the MPs in order to minimize reuse in short time intervals.

To establish a peer link, both MPs exchange `Peer Link Open` and `Peer Link Confirm` frames as depicted in Figure 19.3. Whenever an MP wants to close a peer link it should send a `Peer Link Close` frame to the peer MP.

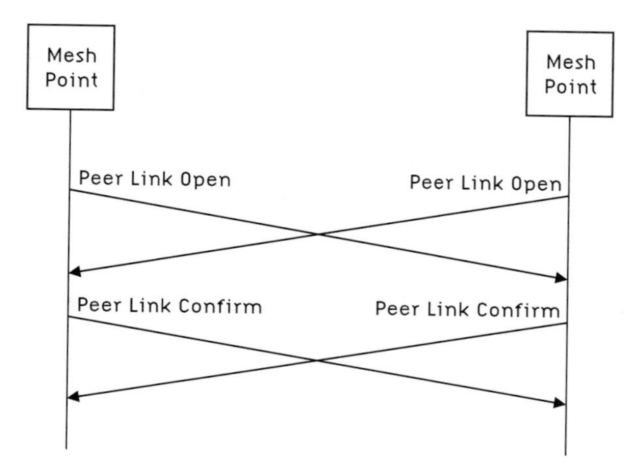

Fig. 19.3. IEEE 802.11s mesh peer-link creation.

19.3.3. *Path selection mechanisms*

IEEE 802.11s proposes a mandatory path selection protocol: a hybrid (proactive/reactive) traffic-aware protocol named HWMP — Hybrid Wireless Mesh Protocol. Although the standard assures compatibility between devices of different vendors by dictating mandatory mechanisms (HWMP and the Airtime Link Metric), it also includes an extensible framework that may be used to support specific application needs.

In order to exchange these configuration parameters, a `Mesh Configuration` element is transported by beacon frames, `Peer Link Open` frames and `Peer Link Confirm` frames. The `Mesh Configuration` element contains, among other subfields, an Active Path Selection Protocol Identifier and an Active Path Selection Metric Identifier.

19.3.3.1. *HWMP and airtime link metric*

As a hybrid protocol, HWMP aims at merging advantages of both proactive and reactive approaches. It is inspired on the Ad Hoc On Demand Distance Vector (AODV) protocol[33] and on its extension AODV-ST.[34]

HWMP can be configured to operate in two modes: on-demand reactive mode and tree-based proactive mode. On-demand mode is appropriate to establish a path between MPs in a peer-to-peer basis, while in proactive mode, a tree-based topology is calculated once an MP announces itself as a root MP. The tree-based approach can improve path selection efficiency when there is a tendency of forwarding significant portions of network traffic to some specific nodes, for instance, to a Mesh Portal Point (MPP).

What makes the protocol truly hybrid is the fact that both modes may be used concurrently. The main advantage of this approach is that, in certain circumstances, although readily available, the tree-based path may not be optimal and an on-demand path discovery may be employed to determine a more appropriate path. One example of such a circumstance is the case where two non-root nodes are able to exchange data through a low cost path (even directly by a single mesh link), but instead they are forced to send their frames to a distant root node up and down the tree.

In 802.11s the mandatory metric is the Airtime Link metric. This metric accounts for the amount of time consumed to transmit a test frame and its value takes into account the bit rate at which the frame can be transmitted, the overhead posed by the PHY implementation in use and also the probability of retransmission, which relates to the error rate in a link. The draft does not specify how to calculate the frame loss probability, leaving this choice to the implementation. Nodes transmitting at low data rates may use all the bandwidth in a network with their long transmissions the same way a high error rate link can occupy the medium for a long time. The Airtime Link metric is designed to avoid both. According to the standard, the Airtime Link metric is calculated as:

$$c_a = \left[O + \frac{B_t}{r}\right] \frac{1}{1 - e_f}, \tag{19.24}$$

where O is a constant overhead latency that varies according to the PHY layer implementation, B_t is the test frame size (1024 bytes), r is the data rate in Mb/s at which the MP would transmit a test frame and e_f is the measured test frame error rate.

During path discovery, each node in the path contributes to the metric calculation by using management frames for exchanging routing information. Independently of the operating mode (proactive or reactive), HWMP functions are carried on by management frames with the following set of information elements:

- `Path Request` (PREQ) elements are broadcast by a source Mesh Point that wants to discover a path to a destination Mesh Point.
- `Path Reply` (PREP) elements are sent from the destination Mesh Point back to the source Mesh Point, in response to a PREQ.
- `Path Error` (PERR) elements are used to notify that a path is not available anymore.
- `Root Announcement` (RANN) elements are flooded into the network in one of the proactive operation modes (there are two proactive modes in HWMP as it will be described later).

The above-listed frames are employed in all of the three mechanisms HWMP provides. The mechanisms are summarized in Figure 19.4. The first one, which is reactive, is called on-demand path selection. The other two are proactive and are named PREQ and RANN mechanisms.

Figure 19.5 displays a topology example for the on-demand path discovery mechanism. The Source Mesh Point (S-MP) needs to find a path to the Destination Mesh Point (D-MP) and in order to do so S-MP needs the cooperation of Intermediate Mesh Points (I-MPs).

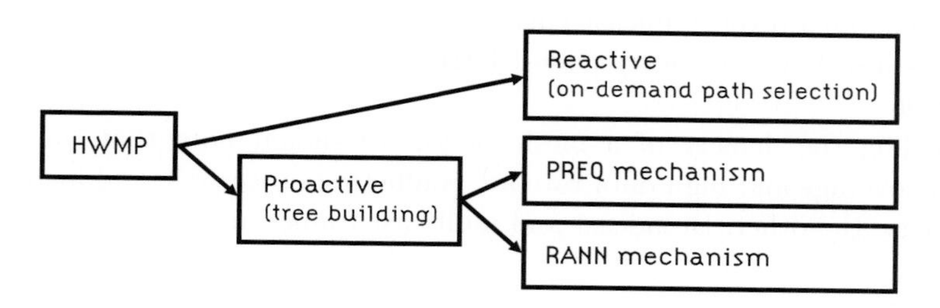

Fig. 19.4. IEEE 802.11s HWMP mechanisms.

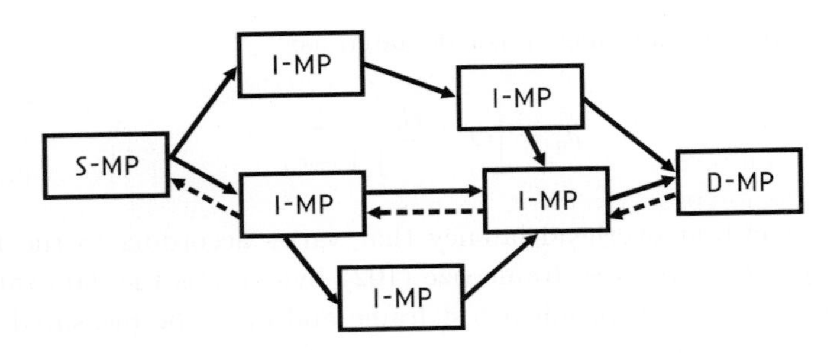

Fig. 19.5. 802.11s on-demand path discovery example.

The mechanism works as follows. First, S-MP broadcasts a PREQ frame.[a] Whenever an I-MP node receives a PREQ, it checks to see if it already knows a path to D-MP. If this is the case, this I-MP node issues a PREP frame back to S-MP. S-MP can prevent intermediate nodes from answering PREQs by setting a DO (Destination Only) flag in the PREQ frame. In that case, only D-MP is allowed to respond with a PREP frame. Therefore, when receiving a PREQ with DO set to "1", any I-MP node may broadcast the PREQ frame again and the process repeats until the request eventually gets to D-MP. Only if the DO flag is not set, an I-MP node (that knows a path to D-MP) may answer PREQ with a PREP frame. Solid-line arrows in Figure 19.5 represent PREQs while dotted-line arrows represent PREPs.

Another flag, RF (Reply and Forward) can also be used to control the behavior of intermediate nodes. If RF is set to 1, and DO is set to 0, an intermediate node may respond with a PREP frame but it must also broadcast the PREQ frame. Likewise, if both DO and RF flags are set to zero, an intermediate node responds but it does not broadcast the request farther. Hence, the RF flag can limit the quantity of PREPs received by S-MP.

Whenever an I-MP node receives a PREQ, it learns a path back to S-MP. This path is the reverse path and it may be used later (in case this I-MP node is in the selected path) to forward RREP frames to S-MP. Response frames can be unicasted using this reverse path.

Both PREQ and PREP frames carry a metric field and each I-MP node must increment this metric field accordingly. That is how the destination node (D-MP) is able to choose a reverse unicast path among many possibilities (in a dense mesh) and this is also how the source node (S-MP) chooses the forward path at the end of the cycle.

Regarding the density of a mesh cloud, we should note that, in a wireless medium, coverage and high data rate are conflicting objectives, and increasing one will decrease the other. Broadcast and multicast frames are usually transmitted at

[a]PREQ, PREP, PERR and RANN frames are management frames with the respective information element.

low rates in order to reach most nodes, since distant nodes will have a greater probability of receiving them. On the other hand, those frames will take a longer time propagating through the cloud, which may be problematic in a dense environment.

Besides the on-demand path discovery mechanism, HWMP provides two different mechanisms for proactively building a forwarding table, as previously stated. The first is based on the PREQ frames and called "proactive PREQ mechanism" and the second is based on the RANN frames, therefore named "proactive RANN mechanism".

In the proactive PREQ mechanism, when configured to work as a root MP, a node broadcasts a PREQ frame with DO and RF flags set to 1. This PREQ is sent periodically and every receiving MP updates the PREQ frame (decreasing the hop count and updating the path metric) and broadcasts the PREQ again, which eventually reaches all nodes in the mesh cloud.

Whether or not a node answers with a PREP frame upon receipt of a proactive PREQ depends in part on the setting of another flag, the "proactive PREP". If the root MP sets it on, all receiving nodes shall send a proactive PREP back to it. A node may send a PREP frame back if it has data to send to the root node and if it wants to establish a bidirectional link, even if the proactive PREP is not set.

The proactive PREQ mechanism is clearly chatty, particularly in its proactive PREP version. An alternative method is presented by the proactive RANN mechanism. Here, instead of sending PREQs out, a root node can flood the mesh with `Root Announcement` frames. Nodes willing to form a path to the root MP answer with a PREQ frame. This PREQ is sent in unicast mode to the root MP, through the node by which the RANN frame was received, and is processed by intermediate nodes with the same rules applied to PREQ broadcasts in the reactive mode.

The root node answers each of the received PREQs with a respective PREP, thus forming a forward path from each MP to the root MP. At the end, the RANN mechanism introduces one additional step and may be advantageous if compared to the PREQ mechanism only if a small subset of MPs wants to establish paths with the root node.

Finally, it is worth to comment about the role of PERR frames in the mechanisms previously described. Whenever a frame cannot be forwarded by an intermediate node, this fact should be informed to the previous nodes in the path, until it reaches the original sender that will then start a new path discovery cycle. Thus, PERR frames are used for announcing a broken link to all traffic sources that have an active path over this broken link.

19.3.3.2. *Internetworking with 802.11s*

The multi-hop capabilities of an 802.11s mesh network would be not very useful without the ability to connect the mesh cloud to other networks such as the wired Internet, as illustrated in Figure 19.6, which shows two examples of internetworking

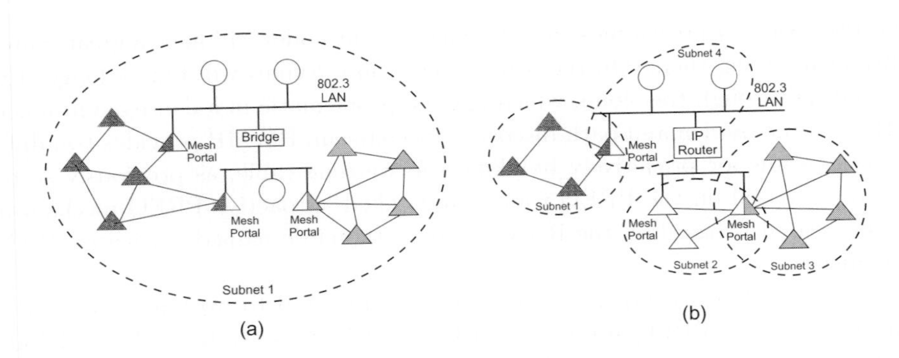

Fig. 19.6. 802.11s Internetworking examples. (a) Layer 2 bridging. (b) Layer 3 internetworking.

with mesh networks. As previously mentioned, the IEEE 802.11s draft names gateway nodes MPPs (Mesh Portal Points).

Figure 19.6(a) illustrates the use of MPPs to interconnect mesh clouds to other LAN networks, when they act like bridges and all nodes belong to the same subnet. Figure 19.6(b) depicts another scenario where MPPs act as gateways to different layer-three subnets. In a MANET where all nodes are potentially routers they are also potentially gateways to an infrastructured network.

An MPP basic characteristic is the fact that it is a mesh node (MP) that is also connected to another network, and this capability has to be announced for other MPs to benefit from its connectivity. Thus, once configured as an MPP, a node spreads the news sending a `Portal Announcement` (PANN) frame.

An MP that receives a PANN frame records the MPP MAC address and the associated path metric and then broadcasts the PANN frame again. Each mesh point in the cloud keeps a list of available MPPs and is able to choose among them when it needs to send traffic outside the mesh network limits.

A Mesh Portal Point may also interconnect mesh networks running different path selection protocols. It is also easy to design the interconnection of many wired 802.3 networks and mesh clouds in a big layer-two bridged network using protocols like 802.1D.[22]

19.3.4. *MAC frame structure and syntax*

In order to allow multi-hop functions at the MAC layer, the IEEE 802.11s emerging standard extends the original 802.11 frame format and syntax. The new frame format supports four or six MAC addresses and new frame subtypes are introduced as it will be described in the following subsections.

19.3.4.1. *IEEE 802.11s frame format*

The first two octets of an 802.11 frame contain the Frame Control field and the third and fourth bits of this field identify the frame type, as shown in Figure 19.7.

00 = management frames	01 = control frames
10 = data frames	11 = reserved

Fig. 19.7. 802.11s frame types.

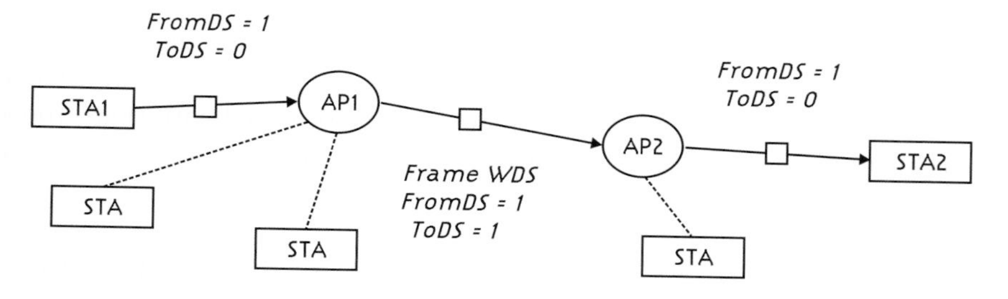

Fig. 19.8. 802.11s layer-2 wireless distribution system.

Besides those two bits, there are also four more bits devoted to a frame subtype. A beacon, for instance, is a management frame (0×0) of the beacon subtype (0×8), while an acknowledgement is a control frame (0×1) of subtype $0 \times \mathrm{D}$.

Since 802.11s is an amendment to 802.11, the frames it introduces must fall into the four pre-existing categories. Initially the reserved (0×3) type was considered for mesh traffic. Instead, it was decided to extend the data and management frames in the following ways:

- data exchanged between MPs are transported by `Mesh Data` frames, defined as data frames (type 0×2), where a mesh header is included in the frame body; and
- mesh-specific management frames, such as PREP or PREQ, belong to type 0×0 (management) and subtype $0 \times \mathrm{F}$, which was formerly reserved. This new subtype was named `Multi-hop Action` frames.

Another characteristic of the new frames is the use of the FromDS and ToDS flags. In 802.11, those bits marked frames as being originated from or destined to a distribution system, which is the infrastructure that interconnects access points. Figure 19.8 depicts a wireless distribution system that connects two access points (AP1 and AP2) and allows two stations (STA1 and STA2) to exchange frames without the intervention of layer-three protocols. In other words, the distribution system provides bridging for the extended service set.

In a wireless distribution system, or WDS, the backhaul connecting the access points is, as the name implies, wireless. A WDS frame is used to exchange frames between them and has both FromDS and ToDS frames activated. Its original role is to allow transmissions between stations connected to two different access points in the same wireless local area network. IEEE 802.11s also sets FromDS and ToDS flags in frames transmitted inside a mesh cloud.

The IEEE 802.11 standard defines frames where FromDS and ToDS flags are set to 1 as "data frames using the four-address format". This definition will be changed to "a data frame using the four-address MAC header format, including but not limited to mesh data frames" when the "s" amendment is approved. The fact that WDS implementations are vendor specific may potentially raise up issues of compatibility with the emerging standard.

As we described the use of the DS flags, we should notice that both are set to zero in ad hoc 802.11 frames. In an ad hoc network, peer-to-peer transfers can happen opportunistically in a way that should not be confused with that proposed by a mesh network, where frame forwarding, i.e. multi-hop forwarding, capabilities are present.

Figure 19.9 shows the general structure of an IEEE 802.11 frame extended by a Mesh Header (included in the frame body). The Mesh Header is represented in Figure 19.10 and contains four fields.

Currently, only the first two bits of the Mesh Flags field are defined. They inform the length of the last field in the header — the Mesh Address Extension field — and vary between zero and three, meaning the number of MAC addresses carried in the Mesh Address Extension field.

The Mesh TTL (Time To Live) field is decremented by each transmitting node, limiting the number of hops a frame is allowed to take in the mesh cloud and avoiding indefinite retransmissions in the case of a forwarding loop.

The three-octets-long Mesh Sequence Number identifies each frame and allows duplicate detection, preventing unnecessary retransmissions inside the mesh cloud. Finally, the aforementioned Mesh Address Extension field carries extra MAC addresses, since the mesh network might need up to six addresses as it will be discussed in the next section.

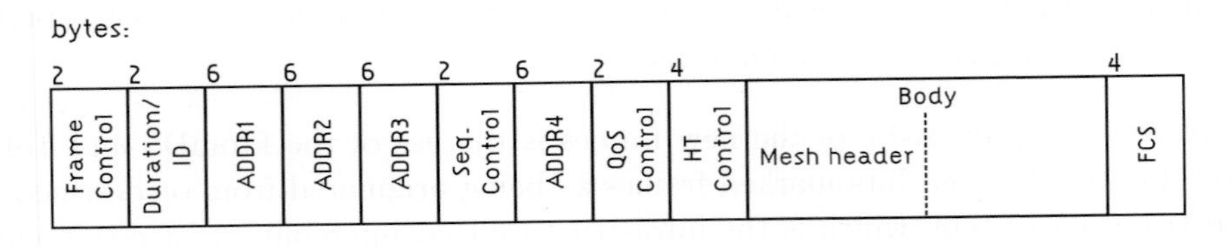

Fig. 19.9. 802.11s MAC frame structure.

bytes:

1	1	3	0,6,12 or 18
Mesh Flags	Mesh Time To Live (TTL)	Mesh Sequence Number	Mesh Address Extension (present in some configurations)

Fig. 19.10. 802.11s MAC frame Mesh Header.

19.3.4.2. *STAs connectivity and frame addressing*

According to IEEE 802.11s, non-mesh nodes (STAs) can participate in the mesh network through a Mesh Point with Access Point capabilities — MAPs in Figure 19.2. STAs communicating through the mesh cloud are proxied by their supporting MAPs and this scenario constitutes one example where the novel six-address frame format is employed.

We start our discussion on frame addressing by the more general four-address frame format, which may be used for both data or management frames. The four MAC addresses in this case are:

- SA (source address) is the MAC address of the frame source — the node that generated the frame.
- DA (destination address) is the MAC address of the node that is the final destination of the frame.
- TA (transmitter address) is the MAC address of the node that transmitted the frame. It can be the same as the source address, or the address of any MP that forwards the frame on behalf of the source (any intermediate node).
- RA (receiver address) is the MAC address of the node that receives the frame. It is the address of the next-hop node and, on the last hop to the destination, it becomes the same as DA.

In short, SA and DA are associated to the endpoints of a mesh path, while TA and RA are the endpoints of each single link. Four-address frames are originally supported by IEEE 802.11 for transmissions using a WDS (Wireless Distribution Systems) but, as mentioned before, they are not enough to implement all features described in the emerging standard.

As we exemplified above, if two non-mesh STAs are communicating through the mesh, two additional addresses will be necessary — the Mesh Source Address (Mesh SA) and the Mesh Destination Address (Mesh DA). In order to understand them, DA and SA entities are defined in a more general way:

- Mesh SA — In a six-address frame, the SA (source address) is the source communication endpoint, that is, the node outside the mesh cloud that originates the frame. Then, the Mesh SA is the node that introduces the frame in the mesh cloud (on behalf of the SA).
- Mesh DA — Likewise, the DA is defined as the final destination of the frame, while the Mesh DA must be understood as the address of the last node of the mesh cloud that handles the frame.

In Figure 19.11, we depict a scenario where STA1 wants to communicate to STA2, which is associated to another MAP in the mesh cloud. During this transmission, if we analyze a frame while it is being forwarded from node MP1 to node MP2, the six-address scheme will be in use (addresses are shown in the figure).

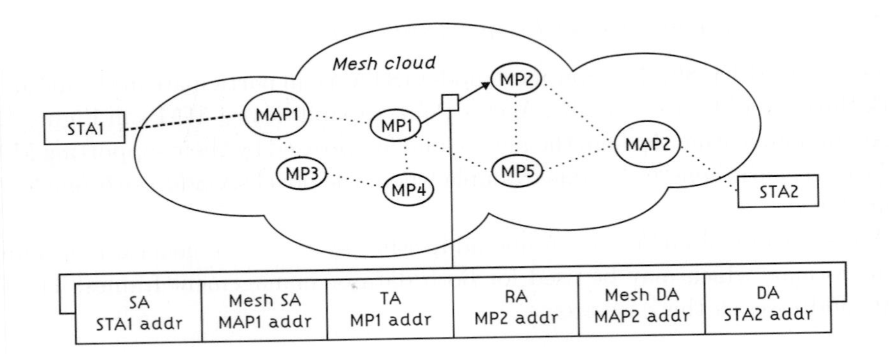

Fig. 19.11. 802.11s STA connectivity example through various MAPs.

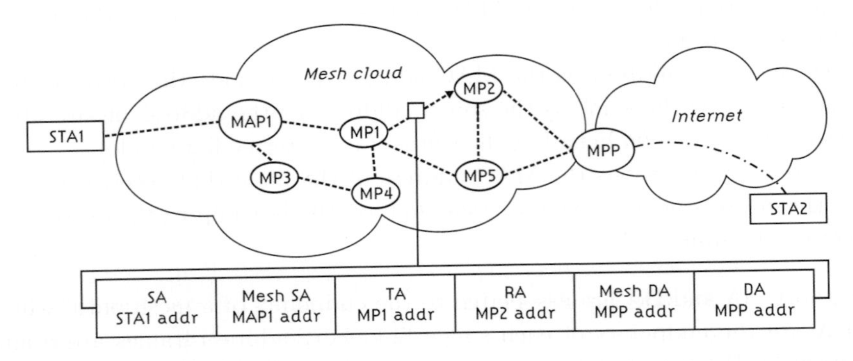

Fig. 19.12. 802.11s STA Internetwork connectivity example through an MPP.

Another case where the six address format is to be used comes from the HWMP tree-based mode, where two nodes can communicate through a root MP. In this scenario the complete path includes two sub-paths — one from the source MP to the root MP and another from the latter to the destination MP. Finally, mesh points can also communicate with the "outside world" through MPPs. In all those cases, more than four addresses are necessary.

In Figure 19.12, we present a scenario where MAP2 is substituted by an MPP node. In this case five different addresses are necessary. The six-address frame format is employed again and both the Mesh DA and DA are set to the MPP MAC address. It is responsibility of the MPP to act as a gateway and forward the traffic to an STA outside the mesh cloud, possibly using layer-three traditional routing.

19.3.5. *Additional features*

The IEEE 802.11s standard covers much more ground than we could address in a book chapter. We tried to cover the most important points that touch the operation of a mesh network, but there are still some interesting aspects.[8]

IEEE 802.11s introduces a medium access method called Mesh Deterministic Access (MDA), which helps reducing contention through the use of a new coordination function. This mechanism is optional and may be implemented by a subset of the MPs present in a mesh cloud. As a consequence, MDA-enabled Mesh Points must be able to interoperate with non-MDA MPs, potentially hurting the efficacy of the scheme.

The core idea of MDA is the introduction of periods of time, called MDAOPs (MDA Opportunities), during which MDA-capable nodes have the opportunity to access the medium with less contention (there may still be contention due to the presence of non-MDA nodes). MDA is implemented through five new action frames: `MDA Setup Request`, `MDA Setup Reply`, `MDAOP Advertisement Request`, `MDAOP Advertisements`, `MDAOP Set Teardown`.[24]

Congestion control is only quickly addressed in the standard proposal. A congestion control mechanism must be selected for the whole network and will be also advertised in the `Mesh Configuration` element, along with the path selection protocol and metric. The draft describes the format of the `Congestion Control Notification` frame to be sent by an MP to its peer MP (or MPs) in order to indicate its congestion status. However, details on how congestion is detected or what triggers the announcement of congestion are considered beyond the scope of the future standard.

Power savings, on the other hand, received more attention in the draft. The main idea is that some capable nodes, named Power Save Supporting MPs, will buffer frames to other nodes, called Power Saving MPs, and transmit them only at negotiated times. It is a service similar to the one access points may provide to its associated nodes in IEEE 802.11 networks.

In terms of security, IEEE 802.11s describes mechanisms to provide both authentication and privacy. Security is based on Mesh Security Association (MSA) services that provide link security between two MPs and may operate even if there is no central authenticator, i.e. it also supports distributed authentication.

Once configured to enable security, an MP shall establish only secure peer links and renegotiate pre-existing unsecure links. The establishment of a secure peer link involves the exchange of extra frames (a four-way handshake) that will start immediately after the initial exchange of `Peer Link Open` and `Peer Link Confirm` frames.

The IEEE 802.1X[24] standard is in the MSA core, but pre-shared keys (PSK) may also be employed, what seems viable only for centrally administered mesh networks.

19.4. Conclusions and Future Work

Handheld communication devices, such as PDAs, regular cell phones, smartphones, media players and many other gadgets yet to come, are not only near ubiquity, but converged into wireless devices that may account for most of the data traffic in a near future.

Although TCP/IP stacks are also incredibly universal, not all communication devices are IP devices or need to be one. Also, despite considerable advances in recent years, mobile devices are constrained in many ways if compared to their fixed equivalents. The price of mobility comprises reduced processing power, storage and memory capacity, due to weight and power-conservation issues.

In face of these factors, a layer-two network that is self sufficient in terms of data forwarding, i.e. a multi-hop layer-two network, seems to be a very appealing asset. It may support non-IP devices and it may be implemented in less capable ones. It may be less demanding, for instance, to port layer-two mesh functionalities to a network interface card, than to include an IP stack (on top of layer-two mechanisms).

An even bigger advantage though, might come from the fact that radio awareness is a natural capability of the link layer. And radio awareness is definitely important in a wireless network, where the communication medium is much more challenging than wired medium. Metrics based on the ever changing link conditions may be of vital importance in such deployments in order to find best multi-hop paths. One difficulty of layer-three routing protocols for wireless networks is measuring precise link quality based on layer-three packets instead of layer-two, as MAC retransmissions are usually not taken into account in those measures.

As the IEEE 802.11s draft indirectly implies in many instances, a mesh network need not to be homogeneous. More capable devices may have additional services or even support less capable devices in some activities. A Mesh Access Point is an example of such a capable device.

By not dictating every aspect of a mesh network, the future standard may extend its longevity. Its main goal should be to provide interoperability in a diverse and rapidly evolving environment. In that sense, an extensible framework for path selection or congestion control mechanisms might be a valid compromise between conciseness and completeness.

The first implementation of the IEEE 802.11s draft was brought to life by One Laptop Per Child[29] in its innovative low-cost educational laptop, called XO. OLPC layer-two mesh network is in many aspects a simplified version of the early versions of the draft, what in many ways helped it being consistent with the evolving proposal. In the OLPC implementation there are no proactive path selection mechanisms (either PREQ-based or RANN-based), but only a reactive on-demand version of HWMP was implemented. MPPs are not announced in advance but discovered by the willing MP, which actively searches for a special anycast MAC address.

The whole implementation lives in the Marvell 88W8388 SoC, which includes an ARM9 low consumption processor, volatile and ROM memory and an 88W8015 radio chip. The radio subsystem is connected to the CPU via an USB bus and may remain powered even when the main CPU is not. An XO can participate in a mesh cloud and forward frames on behalf of other nodes without the use of the main CPU or a TCP/IP stack.

OLPC has also developed standalone active antennas, which host basically the same independent radio subsystem of the XO. These devices help improving the coverage of the mesh cloud and are a good example of non-IP communication devices.

Another implementation initiative of the future standard is the Open802.11s project,[30] which is sponsored by a consortium of companies and has the final goal of creating a totally free implementation of the IEEE 802.11s standard.

It is expected that IEEE 802.11s will be a relevant amendment to the all successful IEEE 802.11 family of standards through its multi-hop forwarding capabilities.

Acknowledgments

The authors would like to thank Javier Cardona from Cozybit and Michalis Bletsas from OLPC for their valuable comments regarding IEEE 802.11s implementation. This work is supported in part by CNPq, CAPES, FAPERJ, FUNTTEL, and FUJB.

Problems

(1) What are the main differences between a WMN and a WLAN?
(2) Cite two examples of quality-aware routing metrics.
(3) Describe the main characteristics of ad hoc, controlled-flooding, traffic-aware and opportunistic routing protocols.
(4) Describe the main functionalities of MAPs and MPPs in the IEEE 802.11s mesh network architecture.
(5) What is the (current) mandatory path selection protocol for IEEE 802.11s and what are its main characteristics?
(6) What is the (current) mandatory metric for IEEE 802.11s and what are its main characteristics?
(7) May a vendor implement its own path selection protocol and metric and still be compliant to IEEE 802.11s?
(8) What advantages and disadvantages may the layer-two approach proposed by the IEEE 802.11s bring when compared to traditional layer-three solutions?
(9) Suppose there is a path between two nodes in a mesh network. Is it safe to assume that the reverse and forward paths include the same intermediate nodes?
(10) Describe the six-address format used in the IEEE 802.11s mesh frame structure.

Bibliography

1. D. Aguayo, J. Bicket, S. Biswas, G. Judd and R. Morris, Link-level measurements from an 802.11b mesh network, *ACM SIGCOMM* (2004) 121–132.
2. D. Aguayo, J. Bicket and R. Morris, *SrcRR: A High Throughput Routing Protocol for 802.11 Mesh Networks (DRAFT)*, Technical Report (MIT, 2005).
3. I. F. Akyildiz and X. Wang, A survey on wireless mesh networks, *IEEE Communications Magazine* **43**(9) (2005) S23–S30.
4. I. F. Akyildiz, X. Wang and W. Wang, Wireless mesh networks: a survey, *Computer Networks* **47**(4) (2005) 445–487.
5. M. Bahr, Proposed routing for IEEE 802.11s WLAN mesh networks, *WICON '06: Proceedings of the 2nd Annual International Workshop on Wireless Internet* (ACM, New York, NY, USA), ISBN 1-59593-510-X, p. 5, doi: http://doi.acm.org/10.1145/1234161.1234166 (2006).
6. S. Biswas and R. Morris, ExOR: opportunistic multi-hop routing for wireless networks, *ACM SIGCOMM* (2005) 133–144.
7. R. Bruno, M. Conti and E. Gregori, Mesh networks: commodity multi-hop ad hoc networks, *IEEE Communications Magazine* **43**(3) (2005) 123–131.
8. J. Camp and E. Knightly, The IEEE 802.11s extended service set mesh networking standard, *IEEE Communications Magazine* (2008).
9. M. E. M. Campista, P. M. Esposito, I. M. Moraes, L. H. M. K. Costa, O. C. M. B. Duarte, D. G. Passos, C. V. N. Albuquerque, D. C. Muchaluat-Saade and M. G. Rubinstein, Routing metrics and protocols for wireless mesh networks, *IEEE Network* **22**(1) (2008) 6–12.
10. T. Clausen and P. Jacquet, Optimized link state routing protocol (OLSR), IETF Network Working Group RFC 3626 (2003).
11. T. Clausen, P. Jacquet, A. Laouiti, P. Muhlethaler, A. Qayyum and L. Viennot, Optimized link state routing protocol, *IEEE International Multi-Topic Conference (INMIC)* (2001) 62–68.
12. D. S. J. D. Couto, *High-Throughput Routing for Multi-Hop Wireless Networks*, Ph.D. thesis (MIT, 2004).
13. D. S. J. D. Couto, D. Aguayo, J. Bicket and R. Morris, A high-throughput path metric for multi-hop wireless routing, *ACM International Conference on Mobile Computing and Networking (MobiCom)* (2003) 134–146.
14. D. O. Cunha, O. C. M. B. Duarte and G. Pujolle, An enhanced routing metric for fading wireless channels, *IEEE Wireless Communications and Networking Conference (WCNC)* (2008).
15. S. M. Das, H. Pucha, D. Koutsonikolas, Y. C. Hu and D. Peroulis, Dmesh: incorporating practical directional antennas in multi-channel wireless mesh networks, *IEEE Journal on Selected Areas in Communications* **24**(11) (2006) 2028–2039.
16. D. B. Johnson and J. Broch, DSR: the dynamic source routing protocol for multi-hop wireless ad hoc networks, *Ad Hoc Networking*, Chap. 5 (Addison-Wesley) (2001) 139–172.
17. R. Draves, J. Padhye and B. Zill, Comparison of routing metrics for static multi-hop wireless networks, *ACM SIGCOMM* (2004a) 133–144.
18. R. Draves, J. Padhye and B. Zill, Routing in multi-radio, multi-hop wireless mesh networks, *ACM International Conference on Mobile Computing and Networking (MobiCom)* (2004b) 114–128.
19. S. M. Faccin, C. Wijting, J. Kenckt and A. Damle, Mesh WLAN networks: concept and system design, *IEEE Wireless Communications Magazine* **13**(2) (2006) 10–17.

20. L. Iannone and S. Fdida, MeshDV: a distance vector mobility-tolerant routing protocol for wireless mesh networks, *IEEE ICPS Workshop on Multi-hop Ad hoc Networks: From Theory to Reality (REALMAN)* (2005) 103–110.
21. L. Iannone, K. Kabassanov and S. Fdida, Evaluation of cross-layer rate-aware routing in a wireless mesh network test bed, *EURASIP Journal on Wireless Communications and Networking* (2007) 1–10.
22. IEEE, IEEE 802.1d. media access control (MAC) bridges, Standard (1998).
23. IEEE, IEEE p802.1x.port-based network access control, Standard (2001).
24. IEEE, IEEE p802.11s/d1.08, draft amendment to standard IEEE 802.11: ESS mesh networking, Standard (2007).
25. C. E. Koksal and H. Balakrishnan, Quality-aware routing metrics for time-varying wireless mesh networks, *IEEE Journal on Selected Areas in Communications* **24**(11) (2006) 1984–1994.
26. T. Liu and W. Liao, On routing in multichannel wireless mesh networks: challenges and solutions, *IEEE Network* **22**(1) (2008) 13–18.
27. MITRE Corporation, http://wiki.uni.lu/secan-lab/MobileMesh.html (2006).
28. S. Nelakuditi, S. Lee, Y. Yu, J. Wang, Z. Zhong, G.-H. Lu and Z.-L. Zhang, Blacklist-aided forwarding in static multi-hop wireless networks, *IEEE Conference on Sensor and Ad hoc Communications and Networks (SECON'05)* (2005) 252–262.
29. OLPC, One laptop per child project, URL: http://laptop.org/ (2008).
30. Open802.11s, Open 802.11s project, URL: http://open80211s.org (2008).
31. D. Passos, D. V. Teixeira, D. C. Muchaluat-Saade, L. C. S. Magalhães and C. V. N. de Albuquerque, Mesh network performance measurements, *International Information and Telecommunicatios Technologies Symposium (I2TS)* (2006) 48–55.
32. C. Perkins and P. Bhagwat, Highly dynamic destination-sequenced distance-vector routing (DSDV) for mobile computers, *ACM SIGCOMM* (1994) 234–244.
33. C. E. Perkins and E. B. Royer, Ad hoc on-demand distance vector routing, *IEEE Workshop on Mobile Computing Systems and Applications* (1999) 90–100.
34. K. N. Ramachandran, M. M. Buddhikot, G. Chandranmenon, S. Miller, E. M. Belding-Royer and K. C. Almeroth, On the design and implementation of infrastructure mesh networks, *IEEE Workshop on Wireless Mesh Networks (WiMesh)* (2005).
35. A. Raniwala and T.-C. Chiueh, Architecture and algorithms for an IEEE 802.11-based multi-channel wireless mesh network, *IEEE Conference on Computer Communications (INFOCOM)* (2005) 2223–2234.
36. A. P. Subramanian, M. M. Buddhikot and S. C. Miller, Interference aware routing in multi-radio wireless mesh networks, *IEEE Workshop on Wireless Mesh Networks (WiMesh)* (2006) 55–63.
37. J. Wang, S. Lee, Z. Zhong and S. Nelakuditi, Localized on-demand link state routing for fixed wireless networks, *ACM SIGCOMM* (2005).
38. Y. Yang, J. Wang and R. Kravets, Designing routing metrics for mesh networks, *IEEE Workshop on Wireless Mesh Networks (WiMesh)* (2005).
39. Y. Yuan, H. Yang, S. Wong, S. Lu and W. Arbaugh, ROMER: resilient opportunistic mesh routing for wireless mesh networks, *IEEE Workshop on Wireless Mesh Networks (WiMesh)* (2005).
40. Y. Zhang, J. Luo and H. Hu, *Wireless Mesh Networking Architectures, Protocols and Standards*, Chap. 12, Wireless Networks and Mobile Communications Series (Auerbach Publications, Taylor & Francis Group) (2007) 391–423.

Chapter 20

Channel Assignment in Wireless Mesh Networks

Weihuang Fu,* Bin Xie,† and Dharma P. Agrawal‡

OBR Center for Distributed and Mobile Computing
Department of Computer Science
University of Cincinnati, Cincinnati, OH 45221-0030, USA
**fuwg@cs.uc.edu*
†*xieb@cs.uc.edu*
‡*dpa@cs.uc.edu*

This chapter discusses the problem of channel assignment in Wireless Mesh Networks (WMNs). Multiple channels can be employed by Mesh Routers (MRs) in such a network to increase the network capacity. Each MR is equipped with one or more radios (i.e. interfaces) each of which can operate on a selected channel. The purpose of channel assignment is to appropriately use radios on the available channels to satisfy specific objectives such as maximizing the network capacity. There are three major approaches for doing the assignment: static, dynamic, and hybrid approaches. In a static channel assignment, a link between a pair of neighboring MRs is assigned with a fixed channel and this binding does not change for a long duration. In a dynamic channel assignment, each MR is able to switch its radio to another channel in a dynamic fashion. Hybrid channel assignment, which integrates the two kinds of approaches, fixes some radios on a default channel while the rest of the radios are dynamically switched across channels. We present a detailed investigation of the current state-of-the-art protocols and algorithms in each category.

Keywords: Wireless mesh networks; channel assignment; multi-radio; multi-channel.

20.1. Introduction

Considering the potential of spatial reuse in radio spectrum, the achievable throughput in a mobile ad hoc network (MANET)[7] is expected to grow with an increase in the number of nodes to cover a large range. However, the average bandwidth available to each node is unexpectedly decreased. The reason behind this is that all nodes in the network share a common channel with a limited throughput and the communication between a pair of nodes depends on multi-hop transmission. Given an interference range, two nearby nodes cannot concurrently transmit data packets using the same channel due to co-channel interference.[2] The ad hoc node has to compete for the shared channel before accessing it for a packet transmission. On the other hand, with the help of intermediate nodes data packet is forwarded in a multi-hop manner before reaching the destination. When communication is established by a multi-hop fashion, the network throughput decreases with the increase of average path length (i.e. number of hops). Suppose n ad hoc nodes are randomly deployed in a unit disk area and use a common single wireless channel for transmission. If the one-hop link capacity is weighed as 1, the theoretical analysis of Gupta and Kumar[10] shows that the end-to-end throughput possible for an ad hoc node is:

$$\Theta\left(\frac{1}{\sqrt{n}}\right). \tag{20.1}$$

Equation (20.1) shows that the available network capacity per node in a MANET approaches to zero as the number of ad hoc nodes goes to infinity.

With the above critical limitation of a MANET in mind, a wireless mesh network (WMN) backbone is designed to form a multi-hop wireless backhaul network on ad hoc basis, with enhanced throughput for each node. It facilitates ubiquitous Internet access for Mesh Clients (MCs), e.g. mobile users, in a variety of applications such as voice, video, and data communication. Nodes consisting of the backbone of a WMN are referred as Mesh Routers (MRs). A WMN increases the throughput per MR in two ways:

- **Multiple orthogonal channels**: MR is equipped with multiple radios (i.e. interfaces) to utilize multiple orthogonal (i.e. non-overlapping) channels. Because no co-channel interference is present among orthogonal channels, MR is allowed to simultaneously transmit or receive data packets to/from its neighbors by using the radios on different channels. In this way, it increases the throughput per MR. In other words, distinct radios of a MR can operate on different frequent bands and the network capacity is therefore increased when the numbers of MR radios and the number of orthogonal channels in the network are increased. It should be noted that every MR provides wireless network accessibility to MCs such as laptops and PDAs. The channel used for the transmission between the MR and the associated MCs can be a separate channel or can be shared with the MR. The data packets from the MC is further forwarded toward the destination in a multi-hop fashion.
- **Deployment of gateway nodes**: Gateway nodes are placed in the network and connected to the Internet infrastructure network. When the gateway nodes are added into the WMN, the multi-hop transmission path length could be reduced since the gateway nodes perform as the ingress and egress of traffic and are

connected to the Internet. Every MR is able to access the gateway node and data packets can be forwarded to the destination via the Internet, instead of using long multi-hop paths. The Internet is assumed to have infinity bandwidth as compared to the wireless link, and thus the addition of the Internet gateway could reduce the number of hops while the traffic of MRs goes through the gateway node. Consequently the gateway nodes increase the throughput per MR. In a WMN, these gateway nodes are further referred as Internet Gateways (IGWs).

However, throughput augmentation by multiple channels cannot be achieved in a straightforward manner. Orthogonal channels have to be assigned carefully to the MR radios in a way to maintain the network connectivity and effectively handle the network traffic. The wireless link between two neighboring MRs can only be established when radios between these two neighboring MRs use a common channel. Otherwise, a network partition may occur in case neighboring MRs operate on different channels and do not have any common channel to establish the connection between them. In addition, if multiple orthogonal channels are available in the network, a fundamental issue is how to achieve high network capacity by deciding which channel to use for a pair of neighboring MR radios. To successfully receive a packet from a sending radio, Signal to Interference Ratio (SIR) at the receiving radio has to be higher than a given threshold. Otherwise, the receiving signals cannot be detected correctly and the packet is lost. Other MR radios within its transmission range using the same channel may increase the interference at the receiving radio seriously. Given available orthogonal channels, it is a critical issue to collaborate in determining channel optimally for the entire network. Such determination of orthogonal channel to be used by MR radios is called the *Channel Assignment*. As we can observe from the above discussions, the essential idea behind channel assignment is to effectively select multiple orthogonal channels for MR radios to maximally enhance the network performance while considering network resources such as the available channels, the number of radios, and the network traffic.

In the past few years, channel assignment for WMN has been investigated from different angles and in this chapter we present an overview of the progress in the literature. We discuss the issues of channel assignment and divide the current approaches into three categories: static, dynamic, and hybrid channel assignments. In the static approach, channels are statically assigned to different radios. The main concern remains the efficiency and guarantee of the network connectivity. In the dynamic approach, a radio is allowed to operate on multiple channels, implying that a radio can be switched from a channel to another channel, depending on the channel conditions such as the interference strength. The basic issues are the switching delay and the switching synchronization. In the hybrid approach, the radios are divided into two sets: one is fixed for certain channels and the other is switchable dynamically while deploying the channels. We further compare the benefits and limitations of these approaches in this chapter.

The rest of this chapter is organized as follows: Section 20.2 discusses the problem of channel assignment. Current channel assignment approaches are investigated in Section 20.3. Section 20.4 illustrates various static channel assignment approaches and then Section 20.5 investigates numerous dynamical channel assignment approaches. In Section 20.6, we discuss the hybrid channel assignment approaches. Comparison for all approaches covered in this chapter is given in Section 20.7 Finally, the chapter is concluded in Section 20.8.

20.2. Problem: Channel Assignment in a WMN

In a traditional MANET, each node is only equipped with one radio for communication, and all nodes utilize on the same channel (i.e. wireless medium). Despite recent advances in wireless physical layer technology, a wireless link still cannot offer the same level of reliable bandwidth as a wired counterpart. Due to significant Medium Access Control (MAC) layer overhead such as MAC contention, 802.11 headers, 802.11 ACK, packet errors, available data rate, etc. an IEEE 802.11 channel bandwidth is significantly less than the maximum physical speed (e.g. IEEE 802.11b at 11 Mbps, IEEE 802.11a/g at 54 Mbps, even IEEE 802.11n at 100 Mbps). On the other hand, if a channel is assigned to a link, this channel cannot be simultaneously used by any other link within its neighborhood due to co-channel interference. Thus, the node capacity (i.e. throughput) in a MANET is a major constraint in satisfying the required throughput. Recently, many researchers advocate the use of multiple radios of a node such that it can simultaneously deploy multiple radios over several orthogonal channels (i.e. non-overlapping channels), which helps in increasing the network capacity. As two orthogonal channels utilize different frequency bands, no interference between them occurs and they can be simultaneously used for packet transmission and reception by nodes. For example, the number of orthogonal channels can be 12 for IEEE 802.11a and 3 for IEEE 802.11b/g. Figure 20.1 shows an example of four nodes operating on three orthogonal channels. Suppose that node A has packets to send to node D via nodes B and C. If all nodes operate on the same channel, the transmission between nodes A and B would prohibit the transmissions at the rest of the links. In such a scenario, only one transmission over one of three nodes is allowed at a given instant. In contrast, if three links operate on three distinct channels (i.e. Ch_1, Ch_2, and Ch_3) and nodes B and C both have two radios (i.e. R_1 and R_2) as shown in Figure 20.1, the three links can work simultaneously, multiplying the throughput of the path by three time.

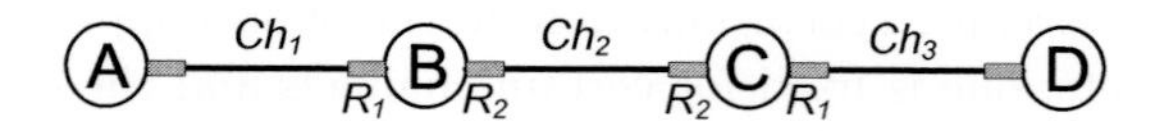

Fig. 20.1. Multi-channel and multi-radio in a linear topology.

In a traditional MANET, contention arises as all the nodes spatially and temporarily share a common channel for transmission. The radio signal of a node is propagated in all the directions, and thus, it may interfere with other nodes' signals. If two or more nodes within the interference range simultaneously access a common channel for transmission, it causes collision. Therefore, contention-based MAC protocols, which have the risk of collisions, have been developed for MANETs that reduce the probability of collision for accessing wireless medium. The contention-based MAC protocols can be further divided into two categories: random access and dynamical reservation/collision resolution. The former has been employed in ALOHA,[1] Slotted ALOHA,[13] CSMA (Carrier Sense Multiple Access).[12] The later includes MACA (Multiple Access with Collision Avoidance),[11] FAMA (Floor Acquisition Multiple Access),[9] and IEEE 802.11 DCF (Distributed Coordination Function).[15] These MAC protocols enable ad hoc nodes to collaboratively share wireless medium for their packet transmission.

However, these MAC protocols consider availability of only one channel for all the nodes and thus cannot be directly applied to multi-channel and multi-radio networks. Two radios at two adjacent MRs have to work on a common channel before trying to access the channel for transmission. When two radios operate on different channels, they are disconnected from each other. In other words, a radio or a node using a channel is isolated from the subset of nodes if this channel is not used by all of its neighboring MRs. In determining a channel for communication between a given pair of radios, the following issues should be taken into account, which are different from a traditional MANET:

- **Connectivity and Routing**: In a traditional MANET, since every node pervasively listens to a common channel, the connectivity and routing can be maintained if two nodes are within the transmission range of each other. On the contrary, MR cannot hear the radio signals from a neighboring MR if its radios are operating on distinct channels. In order for two neighboring MRs to communicate with each other, at least one of their radios in both the MRs should be assigned the same channel. Therefore, a multi-hop route can only be constructed if and only if every pair of MRs has a shared channel to maintain their connection. Otherwise, the connectivity from a source MR to a destination cannot be maintained. As a result, connectivity in maintaining a multi-hop route is one of critical constraints in channel assignment. In addition, routing in a WMN has to consider an optimal routing strategy that determines the load on each radio in the network and in turn affects the bandwidth requirements and the channel assignment of each radio.
- **Mobility and Traffic**: In a traditional MANET, all nodes are mobile, access a single shared channel, and are subjected to power constraints. The mobility also results in a dynamic network topology that should be addressed in its MAC design. In addition, the traffic in a MANET is randomly distributed and each node has the same probability to generate, forward and receive the network traffic.

In this case, each ad hoc node, on the average, has the same probability to access the channel for its use. On the other hand, MRs in the WMN are typically static and operate on multiple channels. For example, MRs in a community WMN are usually situated on the roof of buildings with external power supplies. In this case, the network topology is fixed and the MRs are not subjected to power constraints. The channel usage has to be considered from the point of view of multiple channels, e.g. selecting the least utilized channel or the least interference channel. In addition, the traffic at each MR may be significantly different to that in a MANET. For example, MRs closer to the IGW will have more traffic to forward since they act as the relaying nodes for the Internet access. Therefore, the channel assignment needs to consider the volume of traffic in the network.

- **Static and Dynamic**: In a traditional MANET, a shared channel is dynamically accessed by the nodes. On the contrary, multiple channels can be assigned statically or dynamically to the radios in a WMN. The dynamic channel assignment requires radios to be switched from a channel to another, which involves design issues such as fast channel switching and accurate synchronization.

Figure 20.2 shows the connectivity of a MANET and a WMN with six nodes (i.e. A–F). The network connectivity that all nodes have a single radio and operate on a common channel is shown in Figure 20.2(a). However, a WMN may have different connectivity, depending on the channel assigned on the MR radios. Suppose every node in Figure 20.2(b) and Figure 20.2(c) has two radios and it totally has six orthogonal channels. As shown in Figure 20.2(b), for example, one of the radios of nodes A and C is assigned with channel Ch_1 and thus the two nodes are connected. However, nodes C and D operate on different channels and therefore are disconnected. Due to this, the network in Figure 20.2(b) is partitioned into two isolated domains, i.e. $\{A, B, C\}$ and $\{D, E, F\}$. As a result, nodes A, B, or C cannot communicate with D, E, or F. Different from Figure 20.2(b), the network in Figure 20.2(c) is fully connected because of the existence of the link between nodes C and D. In this

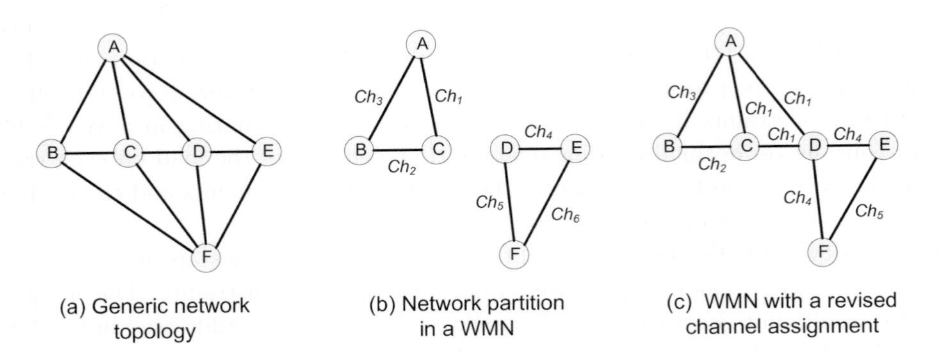

Fig. 20.2. Channel assignment and network connection of a MANET and a WMN.

case, channel Ch_1 is shared by the radios of nodes A, C and D. Any pair of nodes is allowed to communicate by using the indicated channel assignment.

20.3. Classification of Channel Assignment

The essential idea of channel assignment is to optimally assign channels to MRs in a way so as to maintain the required network connectivity, resulting in minimal interference for packet transmission and a higher network throughput. For this purpose, a number of channel assignment approaches have been recently proposed to enhance the MAC layer and support multi-channel and multi-hop networks. Raniwala *et al.*,[18] shows that the channel assignment is NP-hard and most of the existing approaches achieve a near-optimal solution. In a board sense, these approaches can be divided into three categories as follows:

- **Static Channel Assignment**: Static channel assignment allows a channel to be employed by a radio for a long period of time. The limitation of such an assignment is that it lacks flexibility in the channel usage.
- **Dynamic Channel Assignment**: Rather than using a fixed channel for each radio, dynamic channel assignment allows each node to possibly access all the channels by dynamically switching its radios among available channels. It therefore, improves the flexibility of dynamic channel usage in accordance with the channel availability, traffic information, and channel conditions.
- **Hybrid Channel Assignment**: In hybrid channel assignment, radios are divided into two sets: one is fixed for certain channels while the other is dynamically switchable in deploying channels. With the help of the fixed channel, hybrid channel assignment can coordinate the channel switching and facilitate the information exchange.

In the following sections, we consider existing channel assignment schemes according to the classification.

20.4. Static Channel Assignment Approach

In this section, we describe four different static channel assignment approaches: Identical Channel Assignment (ICA), Neighboring Partitioning Channel Assignment (NPCA), Load Aware Channel Assignment (LACA), and joint Routing, Channel assignment, and Link scheduling (RCL).

20.4.1. *Identical channel assignment (ICA)*

A simple way for implementing channel assignment is to sequentially assign the same set of channels to the radios of each node. Considering every MR in the network is equipped m radios, it assigns Ch_1 to the first radio, Ch_2 to the second radio, and so on for every MR in the network. Such a strategy is called as Identical Channel Assignment (ICA).[4] As discussed earlier, any two neighboring MRs can transmit data if they both have a radio that employs a common channel. In other words, a communicating link between two MRs can be established if two neighboring radios of MRs do have a common channel. In ICA, c channels are sequentially assigned to m radios on each MR. If the number of radios is greater than the number of channels, i.e. $m > c$, some radios will be in short of channels. On the other hand, if there are additional orthogonal channels, i.e. $m < c$, the channel from $m + 1$ to c will be rendered to be useless, resulting in inefficient channel utilization. Consequently, the network may be partitioned if some essential links are removed from the topology graph G formed by the MRs and their links. In addition, such an identical channel assignment may cause serious interference. An example is shown in Figure 20.3. In Figure 20.3(a), each dotted line denotes that two MRs are within the communication range and a transmission can be initiated if they have a common channel assigned to their radios. Suppose there are total of 6 channels and every MR has 3 radios. The rectangle attached on the MR in the figure denotes a radio and the associated number is the channel identification number. If MRs A, E, G, and F use Ch_4, Ch_5, and Ch_6 while MRs B, C, and D use Ch_1, Ch_2, and Ch_3, no MR can communicate with its neighbors because there is no common channel between them. Hence, the network is unconnected and every node becomes an isolated MR. If three radios of every node are allowed to operate on Ch_1, Ch_2, and Ch_3 (shown in Figure 20.3(b)), the network is connected but it still has some problems. Firstly, the channel resource is not fully utilized, for example, Ch_4, Ch_5, and Ch_6 are not being utilized in the network. At the same time, the radios suffer from extra interference. The transmission between any pair of nodes

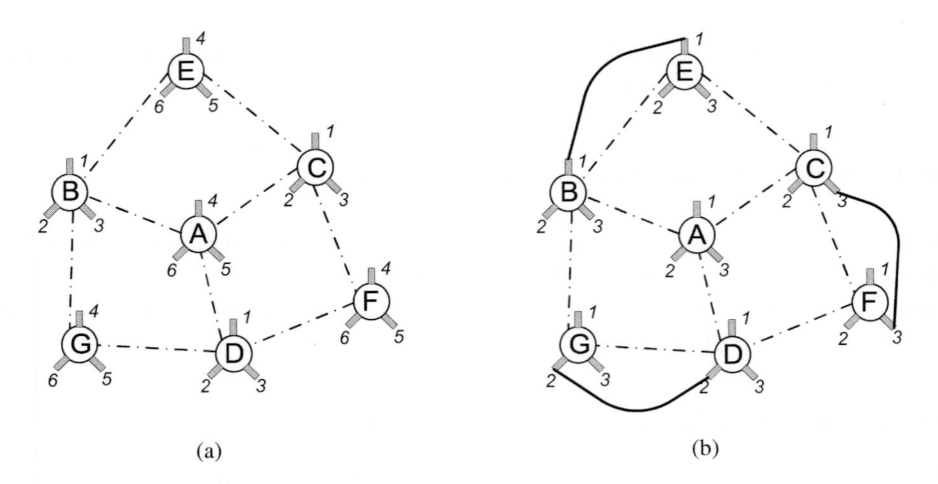

Fig. 20.3. Identical channel assignment.

in the network will prevent the transmission on other nodes on the same channel. The maximum number of simultaneous transmission is only three in the network. Therefore, it cannot effectively utilize the multi-radio multi-channel attribution in WMNs.

In order to overcome these limitations of ICA, two approaches for static channel assignment is proposed in Ref. 18. Neighboring Partitioning Channel Assignment (NPCA) and Load Aware Channel Assignment (LACA), which are considered next.

20.4.2. *Neighboring partitioning channel assignment (NPCA)*

In NPCA,[18] channel assignment procedure is divided into two steps:

- Neighbor-to-radio binding: Considering each MR has multiple radios, the first step is to determine which radio is to be used by two neighboring MRs to construct the communication link between them.
- Radio-channel assignment: This step considers which channel to be selected by the two neighboring radios to form the link.

In implementing these two steps, NPCA considers the following constraints:

- The total number of orthogonal channels is constrained by the number of radios on the MRs.
- The traffic demand over the ongoing link between two neighboring MRs is limited by the capacity of the channel.
- The total traffic load on the links within the interference range should be less than the channel capacity.

Graph edge coloring seems one way of assigning channels to the links. The objective is to assign a color to every edge in a graph such that the adjacent edges do not share a same color. Figure 20.4 shows an edge coloring case. The number shown on the edge denotes the color assigned to that edge. However, edge coloring has some limitations in doing channel assignment. The required number of colors depends on the topology of the graph. In Figure 20.4, it requires 5 colors in total to satisfy the coloring requirement. However, the total available number of channels

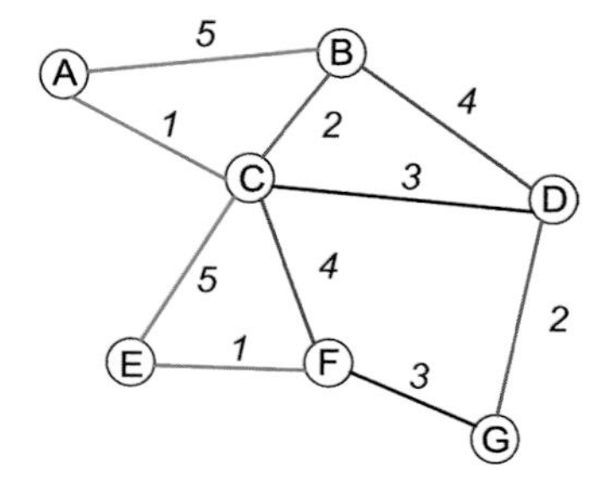

Fig. 20.4. Edge coloring.

may be less than this number. In addition, it does not consider the radio number of a MR in the network. For a static channel assignment, the maximum number of colors assigned to any MR cannot exceed the number of radios at that MR. The edge coloring does not take this factor into account. For example, the colors (e.g. channels) assigned to MR C is 5 in Figure 20.4. The problem is how to bind these channels to the radios if the number of radios is less than 5. Therefore, it has to find a way of satisfying the constraints on the channels and the radios.

NPCA starts from an arbitrary selected MR, partitions its neighbors into m groups (equal to the number of the radios) and assigns one group to each of its radios. Then, each of its neighbors, partitions its neighbors into m groups as well. The process is iterative until all MRs in the network have been partitioned. For assigning channel for a group, it uses the least used channel by the neighboring MRs. The basic process of NPCA can be described by the example shown in Figure 20.5. There are 7 MRs in the network. Each MR has 2 radios and the total number of channels is 4. Suppose the channel assignment proceeds alphabetically. MR A has two neighborhoods B and C. So, it divides them into two groups that group 1 contains MR B and group 2 contains MR C. We denote a group containing MRs u, v, ... by $\{u, v, \dots\}$ and a subscript x is used to denote the assigned channel for the group (e.g. $\{u, v, \dots\}_x$). Therefore, the groups of MR A is denoted by $\{B\}_1$ and $\{C\}_2$. MR A assigns Ch_1 to the first group and Ch_2 to the second group, shown in Figure 20.5. MR B also divides neighbors into two groups: $\{A, D\}_1$ and $\{C\}_2$. Because the link between MRs A and B is assigned Ch_1, the MRs in group 1 use Ch_1. The links for the nodes in the other group uses the least used channel which could select Ch_3. At MR C, the two divided groups could be $\{A, D, F\}$ and $\{B, E\}$. The two radios have to use the channels previously assigned by MRs A and B. MR D is also in a similar situation as MR C and uses Ch_1 and Ch_2. MR E has two groups $\{C\}$ and $\{F\}$. One of the radios is fixed on Ch_3 because of MR C. The other radio can pick up the least used channel: Ch_4. MRs F and G can use previously assigned channels. The channel assignment is accomplished when all MRs are assigned channels.

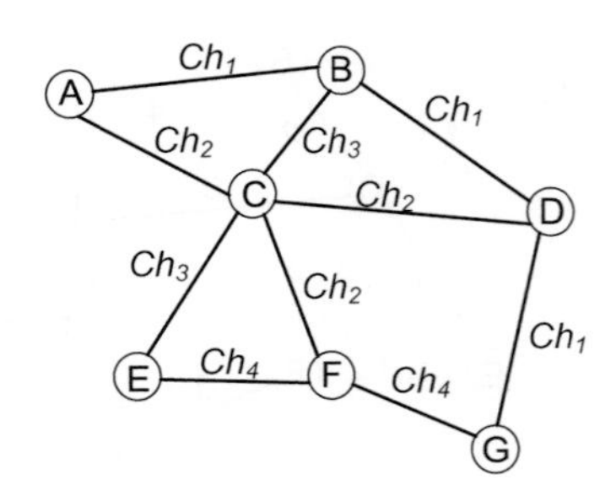

Fig. 20.5. NPCA.

In spite of the improvement, NPCA does not take the traffic load over links into account and all channels are uniformly assigned in the network. Such a schedule works well only when each of the link has the same traffic load, which may not be true in a realistic WMN. For example, the traffic in a WMN is predominantly directed toward or from the Internet, and thus, the link close to the IGW has more traffic than that in the periphery. Consequently, the MR with higher traffic load should be given a higher number of channels. Thus, a load aware channel assignment approach is required to consider the distribution of traffic while assigning channels to the radios.

20.4.3. *Load aware channel assignment (LACA)*

LACA[18] utilizes the local traffic load information to assign the channels to the radios of MRs. The results show that with 2 radios, it is possible to improve the network throughput by a factor of 6 to 7 as compared to a conventional single channel MANET. The assignment procedures include initial link load estimation, channel assignment, routing and link capacity estimation.

20.4.3.1. *Initial link load estimation*

LACA involves the calculation of the link load and the link capacity, which are two critical issues in doing the channel assignment. Because the routing and the channel assignment affect each other, the scheme starts with an initial estimation of the expected load on each link, and iterates the channel assignment and the routing until it converges to the expected load.

The algorithm starts with a rough estimated link load, which can be estimated by Equation (20.2). The initial capacity of link e, denoted by W_e', is assumed to equally share all channel bandwidth within the interference region and is expressed by:

$$W_e' = \frac{cW}{I_e}, \tag{20.2}$$

where c denotes the number of available channels, W denotes the capacity of each channel, and I_e denotes the number links within the interference range of link e. Then, given the traffic demand, network connectivity graph G, and the link capacity obtained from Equation (20.2), the routing algorithm can initially determine the routes and the initial link loads.

Suppose the traffic is perfectly balanced across all possible paths between source-destination MR pairs (s,d) in the initial link load estimation. Let $P(s,d)$ denote the number of possible paths between a pair of source-destination MRs, $P_e(s,d)$ denote the number of possible paths between the pair that passes through link e, and $D(s,d)$ denote the traffic demand between the pair. Then, the expected load on link e, denoted by f_e, is:

$$f_e = \sum_s \sum_d \frac{P_e(s,d)}{P(s,d)} D(s,d). \tag{20.3}$$

In the iteration, f_e is updated with a new calculated link capacity and the routing paths. If the expected link loads do not converge, the algorithm performs a new channel assignment to find other possible assignment which may have higher expected link loads.

20.4.3.2. *Channel assignment*

With the estimation of load, a greedy load-aware channel assignment algorithm can be developed by visiting the links in the non-increasing order of the expected load on a link. By visiting the links in such an order, it is possible to assign the link with heavier load to a channel with less interference so as to have higher capacity. When performing channel assignment for a link which has two incident MRs u and v, three cases are to be considered:

- Radios have not been assigned at MRs u and v. In this case, channel assignment can be done by following the least degree of interference channel. The degree of interference means the sum of expected load for the links in the interference region, which is $\sum_{e \in I} f_e$.
- MR u's radios have been assigned, but MR v has not been yet. Pick up one of the channels used by MR u and assign it to MR v.
- MR u and MR v both have been assigned. If they have common channel(s), channel assignment selects the one with the least degree of interference. Otherwise, it picks up the least degree of interfering channel for assignment.

After the channel assignment, the algorithm estimates the capacity for every link. At the same time, routing is determined based on the estimated link capacity.

20.4.3.3. *Routing and link capacity estimation*

The link capacity is determined by the number of links working on the same channel in the interference region and is represented by the portion of the channel bandwidth available to a link, which is expressed by:

$$W_e = \frac{f_e}{\sum_{l \in I(e)} f_l} cW, \tag{20.4}$$

where f_e denotes the flow on the link e, and $I(e)$ denotes the set of the links in the interference region of link e. The capacity of a link is proportioned to the expected traffic load. While the estimated link capacity can satisfy the traffic load on each link, the algorithm is terminated and provides the channel assignment for the radios on the MRs and the set of routing paths for every source-destination MR pair.

In sum, LACA starts with an initial case, and interactively adjusts the channel assignment and routing to approach a near-optimal solution. The algorithm considers the effect of traffic demand in the network and optimizes the network throughput by the channel assignment and the routing. It generates higher value than that in NPCA for non-uniformly distributed traffic demand.

20.4.4. *Joint routing, channel assignment, and link scheduling (RCL)*

Alicherry *et al.*,[3] proposed joint Routing, Channel assignment, and Link scheduling (RCL) which simultaneously considers channel assignment and routing. WMN may have several IGWs which have wired connections to the Internet. Due to rich Internet services, a large portion of the traffic will be directed toward or from the IGWs, which is called the Internet traffic. RCL is implemented with the focus on the Internet traffic, which is different from LACA that performs the channel assignment and the routing for MR pairs. RCL performs channel assignment by considering the traffic load as well as the routing, which affect each other. If the routing is changed, the distribution of traffic over channels in an interference region will also be changed and hence the channel assignment has to be adjusted so as to mitigate the interference. On the other hand, the routing is affected by the channel assignment. Some paths may not be available because two MRs within transmission distance cannot communicate with each other if they are not assigned a common channel. RCL considers this problem jointly with the objective of maximizing the throughput. In particular, RCL uses Linear Program (LP)[3] to find a channel assignment and scheduling that maximizes the throughput with two critical steps:

- **Routing based on linear programming**: The first step is to solve the LP modeling the problem of channel assignment and routing. The optimal solution is NP-hard, hence the algorithm relaxes the LP problem. A flow graph is generated from the results.
- **Channel assignment**: Due to the relaxation of the LP, the solution obtained from the first step may not be feasible. Channel assignment is to adjust the flows in the graph to ensure a feasible assignment. At the same time, the adjustment is to minimize the maximum interference over channels.

In the following subsections, we describe the steps.

20.4.4.1. *Routing based on linear programming*

Linear programming is an approach designed for optimization with a linear objective function that is subjected to linear equality or inequality constraints. In RCL, linear programming is firstly developed to model the problem of channel assignment with the routing constraint. The network is denoted by $G(V, E)$, where V denotes the set of MRs in the network and E denotes the set of links between MRs. If the optimization objective maximizes the total network throughput, the MR closer to the IGW will have higher throughput than other MRs. To avoid such a situation, the optimization objective in RCL is to maximize the proportion of traffic at every MR, denoted by λ. In other words, RCL achieves fairness that every MR has the same proportion traffic successfully delivered to the IGW. The optimization objective and constraints are given by LP1:

$$\text{Objective: } \max \lambda,$$

$$\text{Subject to: } \begin{cases} \lambda l(v) + \sum_{e(u,v)\in E} \sum_{i=1}^{c} f_e(i) = \sum_{e(v,u)\in E} \sum_{i=1}^{c} f_e(i), & \forall v \in V \\ f_e \le C_e, & \forall e \in E \\ \sum_{1\le i\le c} \left(\sum_{e(u,v)\in E} \frac{f_e(i)}{C_e} + \sum_{e(v,u)\in E} \frac{f_e(i)}{C_e} \right) \le m(v), & \forall v \in V \\ \frac{f_e(i)}{C_e} + \sum_{e'\in I(e)} \frac{f_{e'}(i)}{C_{e'}} \le c(q), & \forall e \in E,\ 1 \le i \le c, \end{cases} \tag{20.5}$$

where $l(v)$ denotes the proportion of the Internet traffic, $f_e(i)$ denotes the flow on link e over channel i, c denotes the total number of channels, $m(v)$ denotes the number of radios of MR v, and $c(q)$ is a constant that indicates the maximum number of simultaneous transmission in an interference region.

The LP1 considers the following linear constraints in the network:

- **MR-Traffic Balance**: The first expression illustrates the constraint of MR-traffic balance, which indicates the incoming traffic and the outgoing traffic at a MR is balanced with the assumption that no packet is dropped at the MR. This means that the traffic aggregated from the associated MCs and other MRs is equal to the traffic forwarding to the next hop MRs.
- **Link-Capacity Constraint**: The second expression indicates that the flow on link e is no more than the capacity of the link. The capacity of a link is limited by the capacity of the channel assigned to that link.
- **MR-Radio Constraint**: The third expression gives the MR-radio constraint, meaning that the maximum throughput of a MR is constrained by the number of radios of the MR.
- **Channel-Interference Constraint**: The last expression depicts the channel-interference constraint. The number of simultaneous traffic flows in an interference region is limited by $c(q)$.

Among all solutions that attain the optimal λ value of λ^*, it is useful to find the one for which the measurement of total interference is minimized. Consequently, LP2 is formed by fixing λ at λ^*, with the objective: $\min \sum_{1\le i\le c} \sum_{e(u,v)\in E} \frac{f_e(i)}{C_e}$ and the same linear constraints in Equation (20.5). The solution of LP2 provides the flows assigned to every link in graph G, and hence the flow graph H is generated from G.

20.4.4.2. *Channel assignment*

Channel assignment of RCL is performed on the flow graph $H(V_H, E_H)$ obtained from previous step and strives to minimize the interference for every channel. MRs and links in the network is denoted by $H(V_H, E_H)$, where V_H is the set of MRs and E_H is the set of flow links. During the channel assignment process, the algorithm firstly transforms the given network graph $H(V_H, E_H)$ to $H'(V'_H, E'_H)$ by creating approximately $r_u = \lfloor \frac{m(u)}{m} \rfloor$ copies for node u in H, where m denotes the minimum number of radios per MR and $m(u)$ denotes the number of radios of MR u. The copies in H' of node u ($u \in V_H$) are denoted by $u_1, u_2, \ldots, u_{r_u}$. The radios of MR u are distributed to each node copy in H' so that each copy has m radios except one of the copies has $m + (m(u) \mod m)$ radios. By this way, every node in the transformed network graph H' has m to $2m-1$ radios. Then, the channel assignment algorithm distributes the aggregate fractional flow obtained from the previous step to the nodes in H' so as to assign approximately the same fractional flow to all copies of the nodes in H. For a node $u_i \in V'_H$, the assigned fractional flow should be no more than the number of the assigned radios $m(u_i)$. After the processes above, the radios and the fraction flow of MR u are distributed to $\lfloor \frac{m(u)}{m} \rfloor$ corresponding nodes in the network graph H'.

At the next phase, the channel assignment algorithm assigns channels from Ch_1 to Ch_m to every node in H', with the objective that each connected component, whose links are connected in H' and assigned the same channel, has the minimum intra-component interference. To investigate the assignment, it needs to define the measurement for interference. The measurement of interference on link e over channel i is defined by:

$$Int(i, e) = \frac{f_e(i)}{C_e} + \sum_{e' \in I(e)} \frac{f_{e'}(i)}{C_{e'}}, \tag{20.6}$$

where $f_e(i)$ denotes the flow on link e over channel i and $I(e)$ denotes all edges within the interference region of link e. So, the interference measurement on link e over channel i is related to all the aggregate traffic in the interference region because these transmissions interfere with each other. Based on the measurement of links, the interference on channel i for a set of links A ($A \subseteq E$) could be defined by the maximum value of $Int(i, e)$, which is given by:

$$Int(i, A) = \max_{e \in A}\{Int(i, e)\}. \tag{20.7}$$

The total aggregate fractional flow on link e, denoted by $\mu(e)$, is expressed by:

$$\mu(e) = \sum_{i=1}^{c} \frac{f_e(i)}{C_e}. \tag{20.8}$$

The aggregate fractional flow of MR u is the summation of $\mu(e)$ toward/from MR u, which is expressed by:

$$\mu(u) = \sum_{e(u,v)\in E} \mu(e) + \sum_{e(v,u)\in E} \mu(e). \tag{20.9}$$

The assignment uses a greedy search by visiting every node in H' in non-increasing order of the fractional flow value. When visiting a node u, it considers the incident link that has not been considered by the algorithm in the non-increasing order of fractional flow value on the link. For each link, the algorithm distributes the traffic into channels from Ch_1 to Ch_m such that the interference in each channel is no more than $c(q)\frac{c}{m}$. To utilize all c available channels, the channel assignment algorithm needs to adjust the traffic distribution in the channels at the next phase.

While traffic is distributed at the channels from Ch_1 to Ch_m, the links assigned the same channel form the connected component(s) in the network. Now, the algorithm needs to shift some components to Ch_{m+1} to Ch_c to mitigate the interference in each channel. If there are at most c components generated from previous phase, each component can be assigned a distinct channel. Otherwise, these components have to be grouped into c groups. The grouping process is like this: the algorithm merges a pair of groups into a single group such that the merged group has the least interference increase over all other groups. By minimizing the maximum interference within each channel, this process continues until the number of groups is reduced to c. Then, all the links belonging to the same group are assigned a same channel and hence c groups are assigned c distinguished channels. The algorithm maps the channel assignment from the transformed network H' back to the original network H so that each MR knows the assigned channels for its radios. The routing of packets is also determined through the steps described above.

20.5. Dynamic Channel Assignment Approach

In a dynamic channel assignment, the radio is switched across available channels on demand, and thus, a radio can be used to transmit/receive packets to/from multiple neighboring MRs. In this case, channels are dynamically used by radios, resulting in high utilization of channels in the network. In addition, the ability of channel switching helps avoiding the network partition problem which may happen in a static channel assignment. On the other hand, dynamic channel assignment makes channel assignment complicated due to two issues: channel switching and switching synchronization. The channel switching is essential for two neighboring radios to operate on a common channel for data packet transmission by dynamic channel assignment. The switching of a radio from one channel to another costs time called switching delay. It is negligible in a static channel assignment because it is a very small fraction as compared to the long duration that a radio remains on a channel. In contrast, the switching delay is a main factor for dynamic channel assignment because a radio is frequently switched and remains on a channel for short time, which makes the delay comparatively a large fraction. Most of the current commercial radios need several milliseconds to hundred milliseconds[6,17] to switch the radio from one channel to another. It should be noted that the switching delay between two communication radios depends on the slower one. A fast switched radio has to wait for its correspondent radio to finish all operations related to channel switching.

Switching synchronization is another essential issue of dynamic channel assignment. It not only needs two transmitting MRs switch the radios to channel simultaneously but also requires the process to have a low overhead and high accuracy. There are two kinds of switching synchronization methods: out-of-band and in-band. Out-of-band method utilizes external information for each MR to acknowledge the time. For example, Global Positioning System (GPS) technology, which provides accurate time information, can be used for MRs to obtain the absolute time information from the satellite so that they can switch the radios to a channel at a desired time. However, it is still expensive to equip every MR in the network with a GPS receiver. In-band method seeks to synchronize the switching by information exchanging among MRs and is cost-effective. Beacon signal, for instant, broadcasted by every MR is considered as a in-band synchronization method. Every MR broadcasts own beacon signal with certain interval and listens to neighboring MRs' beacon signals to determine the relative time for channel switching. Compared to the method based on GPS technology, beacon signal suffers from many factors (such as propagation delay of beacon signal) that affect the accuracy.

In the following subsections, two featured dynamic channel assignment approaches: Multi-channel MAC (MMAC) and Slotted Seeded Channel Hopping (SSCH) are illustrated.

20.5.1. *Multi-channel MAC (MMAC)*

MMAC[19] is a dynamic channel assignment protocol that enables a single radio to utilize multi-channels. Because only a single radio is used, each MR needs to switch the radio on a common channel at certain time so that they can hear messages from each other. MMAC is proposed to divide the time into fixed time slots which are synchronized by beacon signals. All the beacon signals are sent on a common default channel, with a method similar to IEEE 802.11 Timing Synchronization Function (TSF).[15] At the beginning of a time slot, all MRs switch radios on a common default channel to indicate the volume of traffic and negotiate the channels for slot usage. Then, the radios are switched to the selected channel and perform the transmission like IEEE 802.11.

With the channel negotiation mechanism, neighboring MRs are able to operate on different channels while eliminating the interference and improving the throughput. MMAC is a distributed protocol and does not rely on any controlling infrastructure. It has good scalability and the overhead does not increase with the network size.

20.5.1.1. *ATIM window and channel negotiation*

In order for MRs to negotiate the channel, the beginning of a time slot called Ad hoc Traffic Indication Message (ATIM) window is used. During the ATIM window, all MRs switch their radios to a common default channel, a channel pre-assigned for the negotiation purpose, so that two MRs intending to communicate negotiate a channel for packet transmission.

The process of channel negotiation is shown in Figure 20.6. If a MR A has a packet to send to a MR B, it sends an ATIM message, including a Preferable Channel List (PCL) (discussed later). Upon receiving the packet, MR B selects a channel based on the channel usage information obtained from the ATIM message of MR A. Then, MR B sends an ATIM-ACK message back to MR A to notify the channel selection. When MR A receives the ATIM-ACK message, it decides if the channel specified in the ATIM-ACK message is available for use. If the channel is available, MR A sends an ATIM-RES message to MR B to confirm the selection. Then, MRs A and B use the negotiated channel for the rest of time during the slot for data transmission. If MR A cannot select the same channel chosen by MR B, it does not send any message back to MR B, but waits for the next ATIM window to negotiate for a channel. During this process, the MRs within the communication region of MR B such as a MR C in Figure 20.6, will overhear the ATIM-ACK message sent by MR B, so it knows the channel selection of MR B. Similarly, MRs within the communication region of MR A will overhear the ATIM-RES message if A sends the message, and remembers the channel selection. Those MRs avoid assigning the same channel for their transmission during that slot. For example, if MRs A and B select Ch_1 for their transmission in that slot, MR C will overhear the

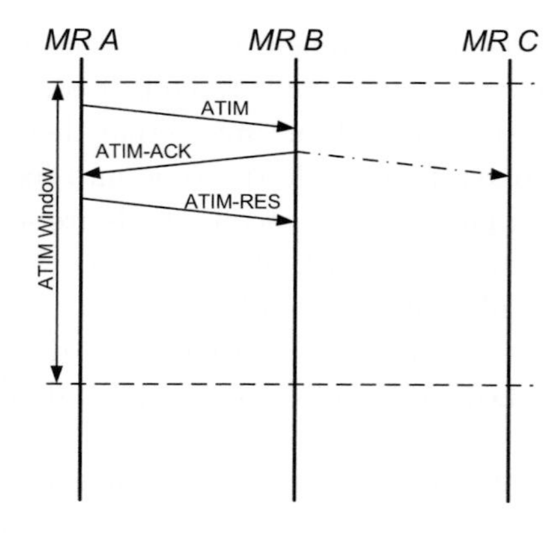

Fig. 20.6. Channel negotiation during an ATIM window.

selection and may choose another channel, e.g. Ch_2 for the transmission during that slot. Finally, multiple transmissions can exist simultaneously in the same time slot.

20.5.1.2. *Channel selection*

During an ATIM window, each MR has to make selection of the channel for data transmission. To choose an appropriate channel, each MR maintains a Preferable Channel List (PCL) based on the observation and the experience of the channel usage. According to the channel usage history, three classes of channels are divided: $HIGH$, MID, and *LOW*. When a channel has been selected by the MR for the usage of current slot, the state is $HIGH$, which means the channel is highly preferred. During each time slot, MR has at least one channel in the next state. MID means that a channel is a medium preference. A channel in this state is the channel that has not been taken for use in the interference region of the MR. If a $HIGH$ state channel is not available, the channel in this state will be taken. A low preference channel is denoted by *LOW*, which means a channel already has been taken by at least one of the neighboring MRs. In addition, a counter is used to record the usage times of a channel by the neighboring MRs in a time slot. If all the channels are in *LOW* state in the PCL, the MR selects the channel with the smallest count value.

Figure 20.7 illustrates the state machine of the channel selection. At the beginning of each time slot, all the channels in the PCL are reset to MID status. While two MRs exchange ATIM-ACK and ATIM-RES messages and agree on one specific channel, that channel state is set to $HIGH$. Overhearing of the channel usage by the neighboring MRs by an ATIM-ACK or ATIM-RES message also triggers the change of the channel state. If MR overhears a channel specified by an ATIM-ACK

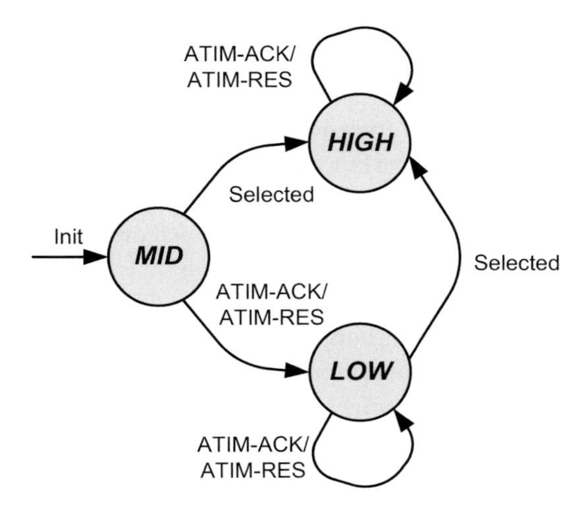

Fig. 20.7. State machine of a channel.

or ATIM-RES message from a neighboring MR, that channel state is set to *LOW* if the previous state is *MID*. At the same time, the associated counter is set to one. If the previous status is *HIGH*, it keeps that state without any change. If the previous state is *LOW*, the counter of that channel is increased by one.

MMAC prefers a channel with the least scheduled traffic and attempts to balance the channel load. During the ATIM window, each MR selects a channel for the transmission in that slot. While MR receives an ATIM message with PCL, it checks both the received PCL and its own PCL. If there is a *HIGH* channel state in its own PCL, the MR selects that channel. Else, if there is a *HIGH* channel state in received PCL, the MR selects that channel. Else, if there are *MID* state channels in both own and received PCL, the MR randomly selects one. If there are *MID* state channels only in one of the PCLs, it randomly selects one. If all the channels are in the *LOW* state, the channel with the least count value is selected by the MR.

After the MR selects the channel, it replies an ATIM-ACK message to notify the channel selection, as shown in Figure 20.7. The major overhead of MMAC depends on the length of the ATIM window, whose length should be long enough for all MRs within the transmission region to negotiate the channel usage.

20.5.2. *Slotted seeded channel hopping (SSCH)*

MMAC has to negotiate channel selection for every time slot, resulting in high overhead. In contrast, SSCH[5] reduces the channel negotiation times by implementing hopping schedule. The hopping schedule decides the channel on which a radio should be switched for the next time slot. Two MRs within the communication range having the same synchronized hopping schedule are able to communicate with each other. For two pairs of transmitting MRs within the interference region using two non-overlap hopping schedule, the transmissions can exist simultaneously without any interference. The details are described as follows.

20.5.2.1. *Hopping schedule*

Hopping schedule is essential to improve the performance of SSCH. The network requires enough number of non-overlapping hopping schedules within an interference region so as to support more simultaneous transmissions. Non-overlapping hopping schedules mean any two hopping schedules would not overlap their channels at any time slot, except for the parity slot (discussed later).

Given the multiple channels that identified by the Channel Identification number (CID) varied from 0 to $c-1$, the hopping schedule is designed like this: the next CID is determined by a linear function of current operating CID. Let (x_i, a_i) denote the ith hopping schedule pair (channel, seed), where x_i is an integer to denote the CID, and a_i is an integer number range from 1 to $c-1$. Then, the CID for the next time slot is decided by the function:

$$x_i \leftarrow (x_i + a_i) \mod c. \tag{20.10}$$

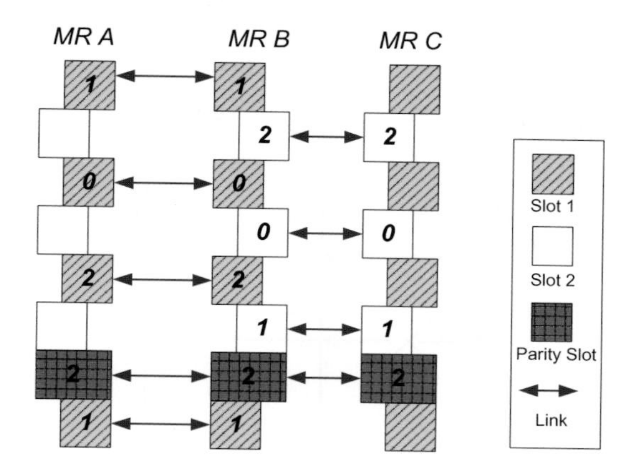

Fig. 20.8. Channel hopping schedule.

Because the MRs in the network may work on different hopping schedules, they need a time slot where all radios are switched to a common channel for information exchange, and this time slot is called the parity slot.

The example of Figure 20.8 illustrates the SSCH operations, with $c = 3$ channels. Each MR has two hopping schedules for two simultaneous transmissions. Suppose x_1 of MR B starts from 1 and a_1 is 2. Then, the sequence of hopping schedule for the channels is $\{1, 0, 2, \ldots\}$. The other hopping schedule of MR B starts channel from $x_2 = 2$ and uses $a_2 = 1$. The sequence for the channels is $\{2, 0, 1, \ldots\}$. Interleaving two hopping schedules, it is $\{1, 2, 0, 0, 2, 1, \ldots\}$. Then adding the parity slot, it is $\{1, 2, 0, 0, 2, 1, 2, \ldots\}$. MR A can use the same hopping schedule (x_1, a_1) for the time slot 1. So, MR A can communicate with MR B at the first time slot as shown in Figure 20.8. Similarly, MR C can use the (x_2, a_2) at the second time slot which is synchronized to the time slot 2 of MR B. So, MRs B and C can communicate with each other at the time slot 2 as shown in Figure 20.8. With such mechanism, SSCH can operate well in a multi-hop wireless network with a single radio.

To exchange the information of the hopping schedule, each MR in the network broadcasts its hopping schedule periodically within the cycle, so that the neighboring MRs can acknowledge the information. When MR intends to communicate with a neighboring MR, it adjusts one of its hopping schedules to be synchronized to the destination MR.

20.5.2.2. *Hopping synchronization*

Each MR needs to ensure that its slots start and stop at roughly the same time as other MRs. However, when MR decides to synchronize to the MR to which packets need to be sent, a problem may be present which is shown in Figure 20.9. Let MRs A, B, and C sequentially lie on a line. MR A has two hopping schedules represented by A_1 and A_2. Similarly, MR B has two hopping schedules represented by B_1 and

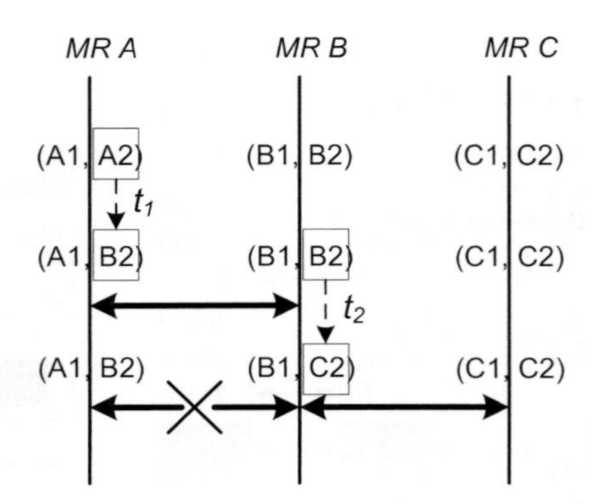

Fig. 20.9. Synchronization problem.

B_2; and MR C has two hopping schedules represented by C_1 and C_2. While MR A transmits packets to MR B, it changes one of the hopping schedules to synchronize to MR B at time t_1. For example, A_2 is changed to B_2 as shown in Figure 20.9. Then, while MR B transmits packets to MR C, it may change B_2 to C_2 at time t_2. Now, MRs A and B do not have a common hopping schedule and hence the transmission is interrupted.

To solve this problem, SSCH maintains a per-slot counter for each hopping schedule to count the number of packets received during a previous slot. If the value of the counter is greater than a defined threshold, the hopping schedule is labeled as a receiving schedule. When a MR needs to change hopping schedule, it selects a non-receiving schedule. If all schedules are receiving schedules, any one is allowed to be changed. Hence, it avoids frequent schedule change caused by the problem indicated.

20.6. Hybrid Channel Assignment Approach

Hybrid channel assignment allows some radios to work on a fixed channel while the rest of the radios are dynamically switched across channels. The fixed channel is called default channel that is usually used for exchanging control message. It is also used for data transmission in some schemes. This section illustrates two different hybrid channel assignment approaches: Interference-Aware Channel Assignment (IACA) and Multi-Channel Routing (MCR).

20.6.1. *Interference-aware channel assignment (IACA)*

Each MR in IACA[16] has a default radio operating on a default channel for information exchange and data transmission while the rest of the radios are dynamically switched among channels to reduce the interference and the packet delay. In addition to connect to the Internet, the IGW in the network manages all the MRs.

Each MR observes the real interference and traffic conditions to report to the IGW, which enables IGW to perform a global optimization of the channel assignment. A Channel Assignment Station (CAS) is located at the IGW and is executed periodically for the channel selection and the channel assignment. To optimally assign channels, the CAS at the IGW needs the information about interference and the delay at each MR. The estimation of interference and the delay can be performed at each MR and the default channel enables the MRs to send information to the IGW. After collecting information, the CAS performs the channel assignment.

20.6.1.1. *Interference and delay estimation*

MR evaluates the rank of the interference level by averaging the number of interfering radios and the channel utilization. The number of interfering radios and channel utilization can be estimated by the captured IEEE 802.11 management and the data frames. MR can configure one of the radios to capture the packets for a small duration on each channel. The number of interfering radios is identified by the MAC address of the captured packets. The traffic of an interfering radio is measured by counting the number and the size of packets associated with the MAC address. The channel utilization is further measured by counting the size of packets and the bit rate.

Link delay is measured by an Expected Transmission Time (ETT),[8] which is derived from the link's bandwidth and the loss rate:

$$ETT = ETX \times \frac{L}{W}, \tag{20.11}$$

where ETX denotes the expected number of transmission necessary to send a packet on the link, L denotes the length of the packet, and W denotes the bandwidth of the link.

20.6.1.2. *The default channel selection*

After the CAS collects the information, it first creates a Multi-radio Conflict Graph (MCG)[16] to perform the channel assignment. Considering a network graph G, with nodes corresponding to MRs and edges representing the wireless links. The MCG F is generated from G as follows: the nodes in F correspond to the radio links between pairs of MRs, rather than MRs, and the edges in F correspond to the interference relationship. For example, there is a topology shown in Figure 20.10(a). MRs A, B, and D only have a single radio, and MR C has two radios. Figure 20.10(b) is the corresponding conflict graph. The nodes in Figure 20.10(b) correspond to the edges in Figure 20.10(a). The edges in Figure 20.10(b) represent the interference between the links in Figure 20.10(a). Figure 20.10(c) is the corresponding MCG. Node AC_1 is to represent the link between the first radio of MR A and the first radio of MR B. Node AC_2 is to represent the link between the first radio of MR A and the second radio of MR C. Other labeled nodes represent a similar meaning.

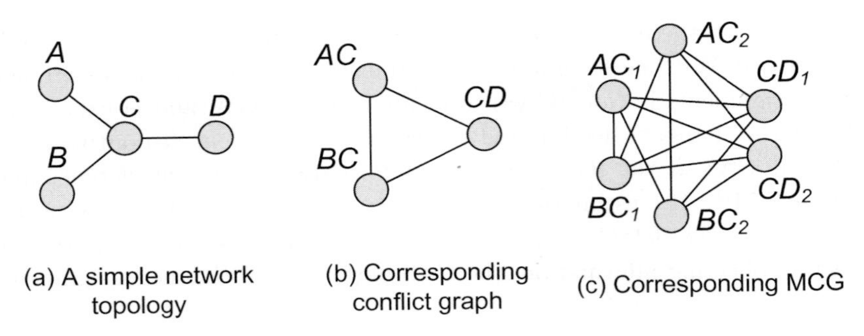

Fig. 20.10. MCG graph.

The edges between the nodes in Figure 20.10(c) mean the interference relationship between the edges in the original topology G.

To select a default channel, the CAS associates each node in MCG with a channel rank. The rank of a channel k is computed by:

$$R_k = \frac{\sum_{i=1}^{n} Rank_k^i}{n}, \tag{20.12}$$

where n denotes the number of MRs in the WMN and $Rank_k^i$ denotes the rank of the channel k at MR i. The channel with the least R_k value is selected as the default channel. It indicates that the default channel prefers the channel with the minimum interference.

20.6.1.3. *Channel assignment*

CAS constructs a MCG graph as discussed above and implements a Breadth First Search (BFS) to determine channel assignment. The search begins from the links adjacent to the IGW. The algorithm adds nodes of the MCG to a list V. All nodes with the same hop count from the IGW, are added into a priority queue Q. The queue sorts the nodes by non-decreasing delay values. Such a sorted list gives higher priority to better links with the shortest hop count to the IGW. The BFS algorithm visits each node in Q and assigns them the highest ranked channel that does not conflict with the channel assignments of its neighbors. If a non-conflicting channel is not available, a random channel is assigned to the node. After this, all nodes containing the radio from the assigned node are placed in a list M. All nodes from M are removed from the MCG so as to ensure only one channel is assigned to a radio. Then, nodes at the next level of the BFS are added to Q. Such procedure continues until all the nodes have been assigned. Because the traffic demand and the presence of MRs in the network may vary with time, the channel assignment needs certain mechanism to accommodate the changes. The CAS checks the MRs information and updates the assignment periodically. The update rate will be high if the number of the interfering MRs is large and the traffic changes frequently. Otherwise, the update rate could be low.

To assign non-default radios for MRs, the CAS sends messages to the MRs that need to be reconfigured. Once the MR receives the message, it invokes the link redirection process to avoid any interruption of the flows. Then, it initiates the switching process. After the radio is switched onto the assigned channel and obtains the confirmation from the neighboring MR that works on the same channel, the assignment process for the link is completed and the flow can be switched back from the default channel to the assigned channel.

Link redirection occurs when a flow intended for a MR's non-default radio is redirected to the MR's default radio. Link redirection can redirect packets when the intended receiver cannot receive the packets on the assigned channel. It redirects and sends the packets on the defaults channel to the receiver. This redirection is also used when the transmitter cannot work on the assigned channel. When MR cannot work on certain channel, it broadcasts a message periodically to inform other MRs that the radio is inactive.

20.6.2. *Multi-channel routing (MCR)*

MCR[14] has a cross layer design that utilizes a link layer protocol to manage the use of multiple radios and a routing protocol that interacts with the link layer protocol to select routes. In the link layer, the key issue is to select and manage the fixed radio. In the routing layer, cost of the link layer is computed for a routing decision. The cost of link layer is related to the switching delay and the traffic estimation on different channels. MRC incorporates these factors in the routing consideration.

20.6.2.1. *Fixed radio management*

In MCR, radios are divided into fixed and dynamic radios. The management of fixed radio includes two components: selecting the static channel and broadcasting the information to the neighbors. Each MR maintains two tables: NeighborTable and ChannelUsageTable. The NeighborTable contains the fixed channels being used by its neighbors. The ChannelUsageTable contains the information of channel usage in its two-hop neighborhood.

Initially, MR randomly chooses a channel for the fixed radio and broadcasts a Hello message, which includes current NeighborTable, periodically on every channel. When a neighboring MR receives the Hello message, it updates the NeighborTable with the fixed channel used by the sender and the ChannelUsageTable with the NeighborTable attached in the message. Such mechanism ensures that a MR acknowledges the channel usage information within the two-hop region. To keep information updated, any entry that has not been updated for a pre-defined time period will be removed from the table. Each MR checks its ChannelUsageTable to adjust its fixed channel. If the number of other MRs using the same fixed channel is large, MR changes its fixed channel to a less used channel with certain probability p. After the adjustment, the MR broadcasts the new message to its neighbors.

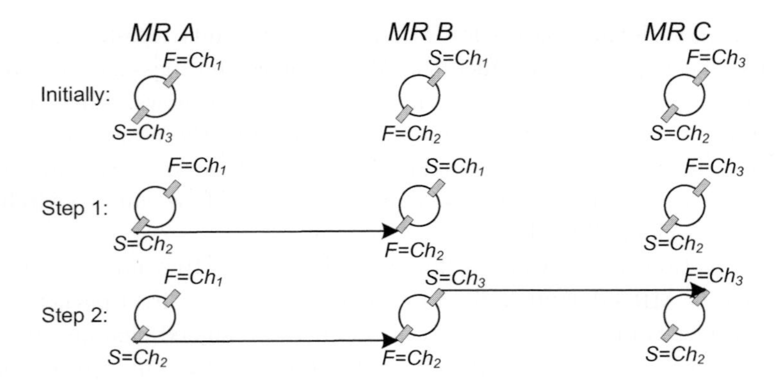

Fig. 20.11. Operation for 3 channels, 2 radios.

Figure 20.11 illustrates the protocol operation with 3 channels and 2 radios, where F denotes the fixed radio and S denotes the switchable radio. Assume that MR A has a packet to send to MR C via MR B. MRs A, B and C have their fixed radios on Ch_1, Ch_2, and Ch_3, and switchable radios on Ch_3, Ch_1, and Ch_2, respectively. At the first step, MR A switches its switchable radio from Ch_3 to Ch_2 before transmitting the packet. Since Ch_2 is the fixed channel of MR B, MR B can receive the packet immediately. At the next step, MR B switches its switchable radio to Ch_3, the fixed channel of MR C, and forwards the packet to MR C. Once the switchable radios are correctly set up during the flow initiation, there is no need to adjust the radios for subsequent packets on the same flow path unless the switchable radio has to be used for a different flow.

20.6.2.2. *Routing cost evaluation*

MCR integrates the multi-channel and the link condition factors for the routing consideration. The routing metric in MCR considers the following cost factors:

Radio switching cost: The radio switching delay is part of the packet delay. Let $U(j)$ denote the fraction of time a switchable radio transmitting on the channel j. The probability, denoted by $p_s(j)$, that the switchable radio will be on a different channel when a packet arrives on channel j as:

$$p_s(j) = \sum_{\forall i \neq j} U(i). \tag{20.13}$$

The switching cost of using channel j is then measured by:

$$SC(j) = p_s(j)\tau_d, \tag{20.14}$$

where τ_d is the radio switching latency, which depends on the hardware condition.

Individual link cost: The cost of a link is measured by the ETT (Expected Transmission Time) required to transmit a packet over the link. ETT has been mentioned in this chapter at 20.6.1.1.

Path metric: The path metric, named MCR metric, includes two components: the first one indicates the sum of ETT and switching cost values along the path; and the second one evaluates the cost of the bottleneck channel along the path.

Let ETT_i denote the ith hop ETT cost, and $SC(ch(i))$ denote the ith hop switching cost, where $ch(i)$ is the used channel at the ith hop. The total ETT cost of channel k, denoted by X_k, is given by:

$$X_k = \sum_{i=1}^{h} B(ch(i))ETT_i, \tag{20.15}$$

where h denotes the number of hops of the path, and $B(c(i))$ is equal to 1 if $c(i) = k$ and equal to 0, otherwise.

Then, MCR metric is given by:

$$MCR = (1 - \beta)\sum_{i=1}^{h}(ETT_i + SC(ch(i)) + \beta \max_{1\leq k\leq c}\{X_k\}, \tag{20.16}$$

where β denotes a weight between 0 and 1, and c denotes the total number of available channels. β is used to determine which component is more important for routing. From Equation (20.15), the value will be small if the channel diversity is used on the path. So, a channel path with that in diversity will be preferred during the route path selection. The defined route metric considering the channel assignment, and the routing protocol can be efficiently performed by the MCR.

20.7. Comparison

Table 20.1 compares the channel assignment approaches illustrated in this chapter. NPCA, LACA, and RCL are static channel assignment approaches that assign channels to radios for a long duration. The number of radios is important for static channel assignment to utilize multiple channels. Different to those static channel assignment approaches, dynamic channel assignment can be implemented on a single radio. For example, both MMAC and SSCH are capable of using a single radio and regularly switching the radio among multiple channels. In a dynamic channel assignment, the switching delay and the synchronization are two important issues that impact the network performance. No packet can be transmitted during the channel switching process, resulting in network overheads. Accurate synchronization is required to ensure two communicating MRs switch their radios at the same time. IACA and MCR are based on hybrid channel assignment approach, with one of the MR radios dedicated to a default channel which is fixed for a long duration while the rest of the radios operate dynamically.

Table 20.1 Comparison of different channel assignment schemes

	NPCA	LACA	RCL	MMAC	SSCH	IACA	MCR
Radio number	Multiple	Multiple	Multiple	Single	Single	Multiple	Multiple
Channel switching	Not often	Not often	Not often	Frequent, regular	Frequent, regular	Hybrid	Hybrid
Switching synchronization	no	no	no	Beacon signal	Beacon signal	Default channel	Default channel
Channel assignment	Centralized	Centralized	Centralized	Distributed	Distributed	Centralized	Distributed
Traffic consideration	no	yes	yes	no	no	yes	no
Routing consideration	no	yes	yes	no	no	yes	yes
IGW requirement	n/a	n/a	yes	no	no	yes	no

Channel assignment can be performed in a distributed or centralized manner. MMAC and SSCH are distributed approaches that achieve scalability. In contrast, RCL and IACA are implemented at the IGW to manage the assignment for all the MRs in the network. In the process of channel assignment, the traffic load should be considered so as to adapt to the change of traffic. NPCA fails to consider the traffic in the network, resulting in low channel utilization for non-uniformed traffic. LACA, RCL, and IACA would have better performance by considering the traffic on links. The impact of routing to channel assignment also should be investigated. LACA and RCL jointly consider routing and channel assignment problems and achieve better performance.

20.8. Conclusion

Multi-radio and multi-channel improve the network capacity by supporting simultaneous transmissions on every MR operating on different channels. The target of channel assignment is to improve the utilization of multi-channel in networks. It finds a proper mapping between the available channels and the radios at each MR in such a way that the network performance can be optimized. As we discussed in this chapter, there are many design issues such as interference, traffic, network topology, link delay, routing, etc. that should be considered for this purpose. We divide the existing approaches into three categories: static, dynamic, and hybrid channel assignments. For each category, we present two or three approaches developed in the last few years and identify the basic processes of channel assignment. Much more innovative approaches are needed to enhance the effectiveness of channel assignment in a WMN.

Problems

(1) What is the motivation behind doing channel assignment in a WMN?
(2) Assume that 81 MRs are deployed on 9×9 grids and the distance between two nearby MRs is 200. How many MRs are within the interference region of the MR at the center if the transmission distance is 250 and the interference distance is 500?
(3) Assume that IEEE 802.11 is implemented by the MRs in Problem (2). How many MRs will be prevented from transmitting if a transmission exists between the MR at the center and its neighboring MR?
(4) How many simultaneous transmissions are allowed in Problem (2)?
(5) Perform edge coloring for the topology graph shown below. Perform ICA and NPCA channel assignment methods for the topology graph with 2 radios and 3 channels.

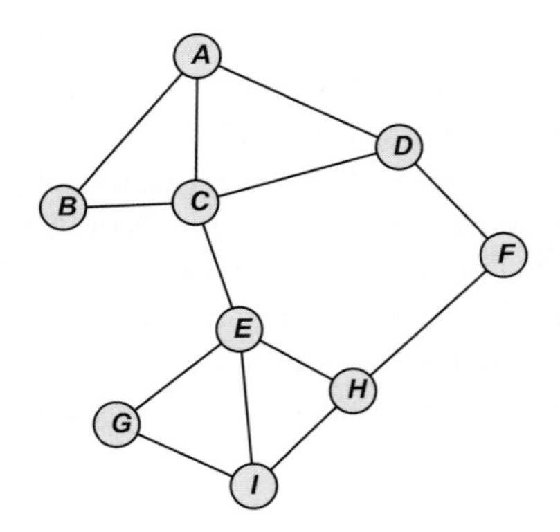

(6) Repeat Problem (5) if node E and the corresponding edges are removed from the topology.
(7) How many hopping schedules are needed if SSCH is used for the topology graph given in Problem (5)? Assume two simultaneous traffics exist on routes: B-C-E-I and A-D-F.
(8) What are the differences in the channel assignment for a WMN as compared to that in a cellular system?
(9) Compare the benefits and constraints of centralized to distributed channel assignment. What are the differences in their applications?
(10) Using a network simulator (e.g. ns-2), simulate one of the channel assignment approaches in this chapter or an approach developed by you. Compare the results with that of a single channel.

Bibliography

1. N. Abramson, The ALOHA system: another alternative for computer communications, *AFIPS Conference Proceedings FJCC* (1970).
2. D. P. Agrawal and Q. Zeng, *Introduction to Wireless and Mobile Systems* (Brooks/Cole Publishing, 2003).
3. M. Alicherry, R. Bhatia and L. E. Li, Joint channel assignment and routing for throughput optimization in multi-radio wireless mesh networks, *MobiCom* (2006).
4. V. Bahl, A. Adya, J. Padhye and A. Wolman, Reconsidering the wireless LAN platform with multiple radios, *Workshop on Future Directions in Network Architecture* (2003).
5. P. Bahl, R. Chandra and J. Dunagan, SSCH: slotted seeded channel hopping for capacity improvement in IEEE 802.11 ad hoc wireless networks, *MobiCom* (2004).
6. R. Chandra, P. Bahl and P. Bahl, Multinet: connecting to multiple IEEE 802.11 networks using a single wireless card, *InfoCom* (2004).
7. C. Cordeiro and D. P. Agrawal, *Ad Hoc and Sensor Networks: Theory and Applications* (World Scientific Publishing Company, 2006).
8. D. S. J. D. Couto, D. Aguayo, J. Bicket and R. Morris, A high-throughput path metric for multi-hop wireless routing, *MobiCom* (2003).
9. C. L. Fullmer and J. Garcia-Luna-Aceves, Solutions to hidden terminal problems in wireless networks, *Proceedings of ACM SIGCOMM* (1997).

10. P. Gupta and P. Kumar, The capacity of wireless networks, *IEEE Transactions on Information Theory* (2000).
11. P. Karn, MACA — a new channel access method for packet radio, *Proceedings of Amateur Radio 9th Computer Networking* (1990).
12. L. Kleinrock and F. A. Tobagi, Packet switching in radio channels: Part-I carrier sense multiple-access modes and their throughput-delay characteristics, *IEEE Transactions on Communications* (1975).
13. F. F. Kuo, The ALOHA system, *ACM Computer Communication Review* (1995).
14. P. Kyasanur and N. H. Vaidya, Routing and link-layer protocols for multi-channel multi-interface ad hoc wireless networks, *Mobile Computing and Communications Review* (2006).
15. IEEE STD. 802.11. Wireless LAN medium access control (MAC) and physical layer (PHY) specifications, *IEEE 802.11 Working Group* (1999).
16. K. N. Ramachandran, E. M. Belding, K. C. Almeroth and M. M. Budhikot, Interference-aware channel assignment in multi-radio wireless mesh networks, *InfoCom* (2006).
17. A. Raniwala and T. Chiueh, Architecture and algorithms for an IEEE 802.11-based multi-channel wireless mesh network, *InfoCom* (2005).
18. A. Raniwala, K. Gopalan and T. Chiueh, Centralized channel assignment and routing algorithms for multi-channel wireless mesh networks, *Mobile Computing and Communications Review* (2004).
19. J. So and N. Vaidya, Multi-channel MAC for ad hoc networks: handling multi-channel hidden terminals using a single transceiver, *MobiCom* (2004).

Chapter 21

Multi-hop, Multi-path and Load Balanced Routing in Wireless Mesh Networks

Sumita Mishra* and Nirmala Shenoy†

Department of Networking, Security & Systems Administration
Rochester Institute of Technology,
Rochester, NY, USA
**Sumita.mishra@rit.edu*
†nxsvks@rit.edu

Ad hoc networks are peer-to-peer multi-hop wireless networks where the member nodes cooperate with one another for successful routing of packets. Wireless Mesh Networks (WMN) can be considered as a special case of wireless ad hoc networks where the participating nodes have relatively fixed positions. The network backbone consists of these relatively stationary nodes (also known as "mesh routers"). Other network nodes (called "mesh clients"), which may or may not participate in the routing mechanism connect to these mesh routers for network and Internet access. Although traditional ad hoc routing protocols may be used for Wireless Mesh Networks, they do not provide the best results. This is because some of the assumptions made by these protocols are not valid for WMNs, thus leading to sub-optimal performance. Also, due to the relatively stationary nature of mesh routers, load-balanced routing is desired for WMNs to avoid network hot spots.

In this chapter, we provide a study of routing protocols for multi-hop wireless mesh networks. Besides an extensive literature search on the topic, we also provide the challenges and directions for future research.

Keywords: Wireless mesh networks; ad hoc networks; mesh routing protocols; routing metrics; multi-path routing; hierarchical routing; load balanced routing; multi-mesh tree; multi-radio routing; 802.11s.

21.1. Introduction

With the increasing popularity of mobile devices such as laptop computers, cellular phones and PDAs, there is a rise in demand for wireless access to the Internet. Wireless Local Area Networks (WLANs) provide an attractive solution for providing untethered wireless access. The IEEE 802.11 specification and all its subsequent variations were/are being developed to address the networking challenges of WLANs. There are two modes of operation specified in the 802.11 standard: the infrastructure mode and the ad hoc mode. In the infrastructure mode of operation, the access point provides the centralized point of control for wireless local area access. The access point is directly connected to the Internet via wired links. The other mobile nodes connect to the Internet through the access point via a one hop wireless link. In the ad hoc mode of operation, there is no centralized control. The nodes access the wireless medium in a random manner and communicate with each other in a peer-to-peer fashion. However when configured in this mode alone, they are not able to access the Internet. If all the mobile nodes are not within the radio transmission range of the central access point, they can group together and form a peer-to-peer "ad hoc network" to forward packets to the access point on behalf of their neighbors. Note that these ad hoc networks are very different from the ad hoc mode configuration of WLANs explained earlier. Ad hoc networks connected to Internet would be an attractive option to infrastructure WLANs and cellular networks as they provide low deployment and call costs like WLANs, but a much wider coverage with minimal infrastructure requirements. Unlike cellular networks that are designed primarily for point-to-point wireless communication, ad hoc networks are peer-to-peer multi-hop wireless networks, where the member nodes cooperate with one another for successful routing of packets. Besides sending and receiving messages, an ad hoc network node also functions as a router and relays messages for its neighbors. A packet of wireless data finds its way from the source to its destination, passing through intermediate nodes connected via wireless communication links. Hence efficient, robust and reliable routing in ad hoc networks is a major challenge.

Wireless Mesh Networks (WMN) can be considered as a variation of wireless ad hoc networks (note that sometimes ad hoc networks are considered as a variation of WMNs) where the participating nodes have relatively fixed positions. The network backbone consists of these relatively stationary nodes (also known as "mesh routers"). Other network nodes (called "mesh clients") as well as conventional clients, connect to these mesh routers for network and Internet access. Similar to ad hoc networks, each node operates as a host as well as a router and forwards packets on behalf of their neighbors that may not be within direct wireless transmission range of their destinations. The mesh clients may be stationary or mobile. Integration with other wireless technologies such as cellular, Wi-Fi and sensor technologies is achieved using the gateway functionality of mesh routers. Due to their self-organizing and self-configuration capabilities, they are very easily deployed and are robust to link failures (due to the existence of several paths from the source to the destination). Hence WMNs have a great potential to play a vital role as a last-mile broadband Internet access technology. Besides providing broadband home networking services, WMNs can also be deployed for neighborhood networking,[1] building automation, industrial monitoring, emergency response and security monitoring.[2–4] The popularity of mesh networking is evident from the fact that the 802.11 *Working Group* has a draft standard (802.11s) to define protocols specifically for mesh networks.[5] Note that there is another set of mesh networks based on Wireless Personal Area networks (WPANs and the 802.15 standard) and WiMAX (802.16 standard). Although many of the general characteristics covered in this work are applicable to these networks also, extensive coverage of the specifics of these mesh networks is beyond the scope of this chapter.

As we move towards the fourth generation of wireless networks (4G), IP has been proposed as the network layer *routed* protocol for WMNs as well as other interconnecting networks. However, the *routing* protocols in ad hoc networks and WMNs are very different from traditional protocols for wired networks. Traditional ad hoc network routing protocols like *dynamic source routing* (DSR) and *ad hoc on demand vector* (AODV) have been proposed for WMNs. For example, among several others, mesh networks developed at Microsoft Research[1] and MIT called "Roofnet"[6] are based on a modified version of DSR whereas several companies have developed mesh networks based on AODV.[7] Although traditional ad hoc routing protocols may be used for WMN, they do not provide the best results.[8] This is because some of the assumptions made by these protocols are not valid for WMNs, thus leading to sub-optimal performance.

Existing performance metrics for wireless ad hoc networks may not be suitable for evaluating the performance of WMNs. Hence, new performance metrics need to be defined and the WMN protocols have to be evaluated against them. Also, due to the relatively stationary nature of mesh routers, load-balanced routing is desired for WMNs to avoid network hot spots.[9]

In this chapter, we will provide a detailed discussion of routing for multi-hop WMN. A lot of work has been done in this area and it is not feasible to provide an in-depth coverage of all the routing protocols that have been developed; we provide details about a selected few. In the *Background* section we give details of these protocols, while in the reference section several others are mentioned. The remainder of the chapter is organized as follows. In Section 2, we provide the necessary background and details on the state of the art on the topic. Section 3 gives the challenges and directions for future research, followed by the conclusions in Section 4.

21.2. Background

Before getting into the discussion of existing routing protocols for WMNs, it is necessary to understand the WMN architecture and the desirable features of a routing protocol suitable for such networks.

21.2.1. *Architecture of wireless mesh networks*

As mentioned before, mesh routers are an integral part of WMNs as they form the backbone for providing the mesh connectivity to the client nodes. These routers

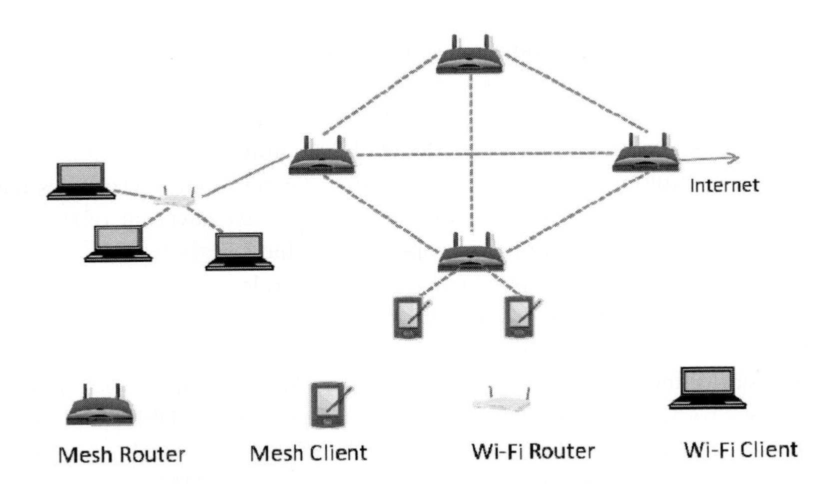

Fig. 21.1. A typical wireless mesh network.

are similar to conventional routers in terms of gateway and repeater functionality. The key difference is that these mesh routers are equipped with multiple wireless interfaces that are built on the same or different wireless access technologies.[10] The coverage range of these routers is similar to conventional routers but with much less transmission power requirements due to the multi-hop nature of communications. By integrating the *Medium Access Control* (MAC) and routing capabilities, better scalability can be achieved in the multi-hop mesh networking environment.[11] The mesh clients may also have the routing functionality besides host capabilities. However, typically they do not provide any gateway or bridging services to other network nodes. They normally have a single wireless interface and have much simpler hardware and software platforms compared to mesh routers.[10] An example of WMN scenario with mesh routers, mesh clients, a Wi-Fi router and Wi-Fi clients is shown in Figure 21.1.

The dashed lines represent the wireless links whereas the solid lines show wired connections. Although this figure shows a connection setup with integrated WMNs and Wi-Fi networks, similar integrations can be done for other existing wireless networks such as cellular, WiMAX and sensor networks, depending on the application requirements.

The mesh routers which form the backbone are connected to each other with wireless links. They can be interlinked using several radio technologies, ranging from the 802.11 suite to long range communication techniques with directional antennas.[1] The gateway functionality of these routers is used to connect to other wireless networks including cellular, WiMAX, sensor and Wi-Fi and their routing capabilities are exploited to provide connectivity with the clients within the WMN. "Neighborhood networks" mentioned in Ref. 1 are built using these backbone routers. On the other hand, the mesh clients can interlink with other clients, thus providing peer-to-peer networking, or can be connected to the mesh routers.

21.2.2. *Routing requirements of wireless mesh networks*

Although there are many similarities between wireless ad hoc networks and WMNs, there are some key differences that need to be taken into consideration while developing a routing protocol for WMNs. Most of the routing protocols for mobile ad hoc networks are developed while considering the dynamic nature of the topology due to node mobility. Some ad hoc networks (e.g. sensor networks) have severe power limitations. Thus there is a need for energy-aware routing protocols for such networks. For mesh networks, the backbone mesh routers are assumed to have minimal mobility and are typically not power-constrained, while the mesh clients require mobility support as well as energy-aware routing. Hence if the routing protocols developed for ad hoc networks are applied to the WMN scenario without any modifications, they may lead to sub-optimal performance. We also need to reevaluate the routing metrics for the WMN case keeping the new requirements in mind. Most ad hoc routing protocols are implemented at layer 3 along with the forwarding functions. Integration of routing and MAC functionalities for WMNs has been investigated and has shown very promising results.[11]

It is desirable that any routing protocol for WMNs should satisfy the following requirements.

(a) *Scalability*: The multi-hop communication paradigm in WMNs increases the coverage area of single-hop communication links without compromising on the channel capacity. It also ensures a non-line-of-sight connectivity between the nodes of the network. As more nodes join the already deployed WMN, the routing protocol should scale well and converge quickly to accommodate for the increase in the network size. The protocol should also adapt if the intermediate nodes become unavailable.

(b) *Robustness*: Since multiple paths exist between the nodes of a WMN, in the event of a link failure, the routing protocol should be able to select an alternative path with minimum delay.

(c) *Load Balancing*: Due to the existence of redundant paths between two network nodes, it is prudent to deploy a load balanced routing protocol that diverts the traffic from a congested region to alternative routes (if they exist). Note that load balancing does not always lead to optimal path selection (e.g. the number of hops might be more than the initial path) but it does improve the network performance drastically by minimizing congestion.[9]

(d) *Support for heterogeneous nodes*: As noted earlier, the mesh routers and mesh clients have different mobility and energy requirements. While a much simpler routing algorithm may be used for mesh routers, mesh clients would normally require all the functionalities of an ad hoc routing protocol.[10] Hence an ideal WMN routing protocol would cater to the needs of both types of network node requirements.

21.2.3. *Routing metrics for wireless mesh networks*

As discussed in the previous section, due to the difference in the characteristics of WMN, the routing metrics used by the routing protocols need to be evaluated for their suitability in the WMN scenario. The routing metrics should satisfy four requirements for ensuring the good performance of any routing protocol.[25]

(a) *Route Stability*: The routing metrics should be chosen such that frequent routing path changes are avoided. If paths change too often, they trigger too many routing information updates, resulting in the network becoming unstable while consuming costly bandwidth and battery power. This is not desirable in any scenario.
(b) *Good Performance for Minimum Weight Paths*: The chosen metrics should capture the mesh network features such that the paths with the least weight or cost as determined by the algorithm adopted by the protocol lead to the desired level of performance for the protocol in terms of throughput and delay.
(c) *Efficient Algorithms to Calculate Minimum Weight Paths*: Based on the routing metric, the algorithm used to calculate the minimum weight path (e.g. Dijkstra's algorithm is commonly used to calculate the shortest path) should be chosen such that they can calculate the optimal paths efficiently in polynomial time.
(d) *Loop free routing:* The routing metrics should be chosen such that the routing paths that are determined based on them lead to routes without any loops.

Based on the above requirements, the routing metrics should be dependent on the network topology and should capture the unique characteristics of mesh networks. There are several routing metrics that have been defined and investigated; we have chosen some of the most commonly used ones for discussion in this section. Several others will be introduced and discussed later under the appropriate routing protocols.

(i) Hop Count: Hop Count is the most common routing metric used by the routing protocols for ad hoc as well as mesh networks. AODV, DSR are some of the protocols that were developed for ad hoc networks and they all use the hop count between the source and the destination to determine the path length. However, the link quality, bandwidth, path interference and other route characteristics are not taken into account in this case. If the link exists between any two nodes, it is included in the hop count without considering any other factor. Even though minimum hop count paths lead to loop free transmission and the hop count metric is very simple to compute, this may not lead to the best performance due to the fact that it does not consider the changing link characteristics of wireless networks.[26]
(ii) Expected Transmission Count: The *Expected Transmission Count* (ETX) metric was introduced in Ref. 26 for finding the optimal paths in multi-hop wireless

networks. It is defined as the expected number of packet transmissions (including retransmissions) required to successfully deliver a packet along a wireless link. The total weight of a path is calculated by summing up the ETXs for all the links that form the path. Paths with the minimum ETX are expected to have the highest throughput. ETX captures the quality of the link (links with a high interference ratio would have a higher ETX value) as well as the path length (paths with more hops or links will have a higher total ETX value). However, ETX estimates the number of retransmissions needed to send regular data packets by measuring the loss rate of small broadcast packets between any two neighboring nodes of the network. These broadcast packets are sent at the lowest possible data rate and may not suffer the same amount of loss as regular packets sent at higher data rates. Also the link load and the data transmission rates are not taken into consideration. This might lead to misleading results in heavily loaded links with low loss scenario or two links with different data rates but same loss scenario.[27]

(iii) Per-hop Round Trip Time: Per-hop *Round Trip Time* (RTT) is calculated by measuring the roundtrip delay between a pair of neighboring nodes.[28] The total weight of a path is calculated by summing up the RTTs for all the links that form the path and the path with the minimum weight is chosen by the routing algorithm. This metric captures many aspects of the link quality including the queuing delay at the neighboring node, the channel contention delay due to several active nodes in the vicinity, retransmissions due to a lossy link with a high interference ratio etc. Hence, besides the loss factor of the link, this metric is able to take into account factors such as the link load and queuing delay which were not accounted for in the ETX calculations.

(iv) Expected Transmission Time: Another metric that is based on the delay factor of the link is called the *Expected Transmission Time* (ETT), which takes the link transmission rates into account.[25,29] ETT is defined as the expected time taken for a successful transmission of the packet from one node to the other neighboring node. The total weight of a path is calculated by summing up the ETTs for all the links along the path and the path with the minimum weight is chosen as the best path. ETT is sometimes defined as "bandwidth adjusted ETX"[29] and is calculated by multiplying the ETX by the packet size and dividing the product by the raw link data transmission rate. This gives an estimate of the time taken for the transmission of the packet. Hence the ETT metric takes into account the loss rate as well as the transmission rate for each link in the path. However it does not consider the impact of channel diversity.[29]

(v) Weighted Cumulative Expected Transmission Time: *Weighted Cumulative Expected Transmission Time* (WCETT) was defined in Ref. 29 to extend the ETT concept for the inclusion of intra-flow interference in the metric determination process. For example, without explicitly considering channel diversity, ETT might lead to a path selection where all the nodes are using the same channel for transmission, which would result in a lower throughput compared to a path with nodes transmitting on different channels. With WCETT, the paths that have more diversified channel assignments on their links are given a lower weight; hence when it comes to the optimal path selection, these paths are chosen over other paths that have less diversified channels. In other words, there is a reduction in the number of nodes transmitting on the same channel along a given path. Even though WCETT captures the intra-flow interference, it does not take inter-flow interference into account. Hence, the selected path may traverse regions of the network that are dense and hence prone to congestion. Also, there is no efficient algorithm that incorporates WCETT in the minimum weight path selection.[25] In Ref. 30, the authors propose a routing metric called *Weighted Cumulative Expected Transmission Time with Load Balancing* (WCETT-LB) for wireless mesh networks. WCETT-LB incorporates load balancing into WCETT with the load balancing feature implemented at the mesh routers. A congestion aware routing and traffic splitting mechanism is adopted to achieve global load balancing in the network.

(vi) Metric of Interference and Channel-switching: *Metric of Interference and Channel-switching* (MIC) is a modified version of WCETT that captures inter-flow interference.[25] Besides, using ETT (the minimum ETT for the network is used in the MIC calculations), two components are introduced in the metric. The *Interference-aware resource usage* (IRU) of a link l captures the total channel time of neighboring nodes that any transmission on $l$consumes. Hence inter-flow interference is taken into account as a dense region would have a higher IRU. The second component of MIC is the *Channel Switching Cost* (CSC). This factor takes the intra-flow interference into consideration. It assigns lower weights to links that have different channel assignment compared to the previous hop link along any given path. If consecutive links in a path have the same channel assignment, this path has a higher weight and hence is not chosen if a path with lower weight is available.

As mentioned at the beginning of this section, several routing metrics for WMNs have been proposed and the ones discussed above capture some of the characteristics of the links and the topology of WMNs. The performance of the routing protocol is highly dependent on the choice of the routing metric. Hence this requires careful consideration on the part of the protocol designers for any WMN application.

21.2.4. *Routing protocols for mesh networks*

Having discussed some of the desired features of a routing protocol for WMNs, in this section, we present some of the routing protocols that have been developed in the recent past to cater to the needs of WMNs.

21.2.4.1. *Hybrid wireless mesh protocol*

The *Hybrid Wireless Mesh Protocol* (HWMP)[13–16] has been defined as the routing protocol to be adopted for mesh networks in the 802.11s standard.[5] Although HWMP is the default routing protocol, the standard allows the vendors to use any alternative routing protocols that are suitable for these networks. The HWMP protocol uses on-demand (reactive) routing for dynamic topologies (with mobile nodes) and a tree-based proactive routing for the predominantly fixed backbone networks. Before we discuss the protocol details, it is necessary to discuss briefly the architectural definitions specified in the protocol.

There are two types of nodes in the network: *Mesh Points* (MP) that are mesh capable and stations (STA) that are not capable of meshing. Hence MPs are capable of establishing links with their MP neighbors (peer-to-peer links). The *Mesh Access Point* (MAP) has the combined functionality of the access point and a mesh point, thus providing communication support to the STAs. The *Mesh Portal* (MPP) provides the entry and exit point to the wired network. It supports transparent bridging as well as bridge-to-bridge communication using the spanning tree protocol for tree formation and loop resolution. The *root portal* (RMPP) is the MPP that is elected to be the root of the forwarding tree and is capable of building the topology. A mesh network example based on the 802.11s standard specifications is shown in Figure 21.2.

HWMP has two configuration options: without the root portal and with the root portal. If the root portal does not exist, the routing protocol that is used is an adaptation of AODV and is called Radio Metric AODV (RM-AODV). AODV is a reactive routing protocol and it is modified to include the information from layer 2

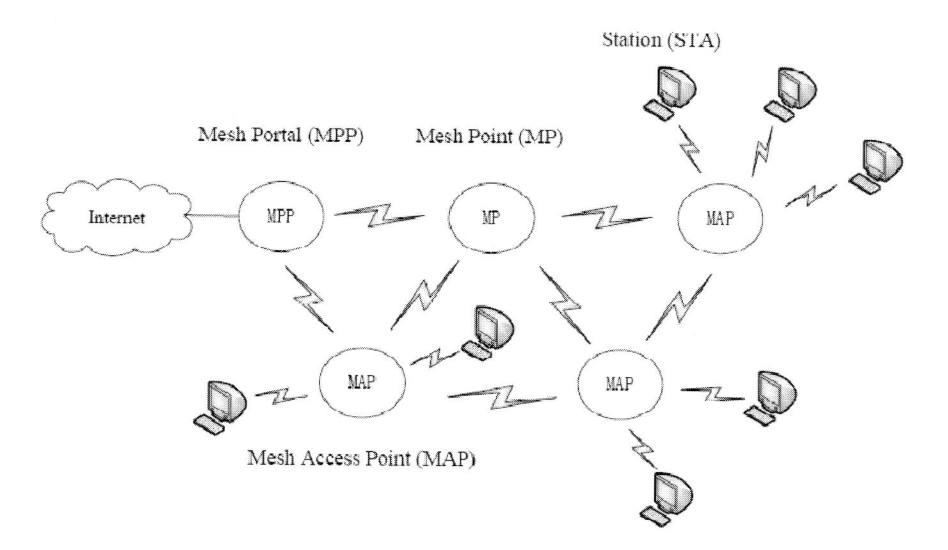

Fig. 21.2. Mesh architecture proposed in the 802.11s standard.

and hence the term "radio metric". The default radio aware link metric specified in the 802.11s standard is the *airtime link* metric.[14] It is calculated using the amount of channel resources used for transmitting a frame over a particular link. The path with minimum airtime link metric is chosen as the best routing path.

Although the airtime link metric is the default metric specified in the standard, the framework allows for any path selection metric based on QoS, load balancing, power-awareness, etc. The MPs use a route discovery process similar to AODV using Route Request (RREQ) and Route Reply (RREP) packets.[17] This process is used to determine the radio aware link metric information from the source to the destination. The MPs periodically send the RREQ packets for route maintenance. Since a radio aware link metric changes more frequently compared to a metric based on the hop count, it is preferred for the destination to answer the RREQ messages to ensure the correctness of the path metric.

With a root portal, a tree based proactive distance vector routing is used. The routes are maintained through the root. The RMPP broadcasts root announcements with a sequence number for each round. Every node that receives the announcement updates its metric and rebroadcasts the message. Based on the information received, the MPs chose a parent and also maintain the information about other potential parents. Route requests are sent periodically to the parents in order to maintain the path to the root. If three consecutive RREQs are sent without a reply from the parent, it is assumed that the path to the parent is lost. In this case, this particular MP informs its children and initiates the process of finding another parent. Once a new parent is found, the MP sends a RREP message to the root with the new information. This message is used by all the intermediate nodes to update the next hop information for this MP.

The HWMP protocol combines the flexibility of an on-demand route discovery with proactive routing to the root mesh portal. In dynamic environments, on-demand or reactive routing is more applicable and proactive tree based routing is efficient in the case of somewhat fixed mesh backbone networks. This combination makes the approach suitable for a variety of mesh networking applications.

21.2.4.2. *Radio aware optimized link-state routing (RA-OLSR) protocol*

The RA-OLSR is the optional routing protocol that is specified in the 802.11s standard[14] where there is a provision for using OLSR instead of AODV. It is a proactive link-state routing protocol based on the specifications of OLSR[18] and optional features from Fisheye State Routing (FSR).[19] Instead of IP addresses, MAC addresses are used and it can work with any link metric, including the airtime metric described in the previous section. The changes in the link state based on the link metric are communicated to the neighborhood nodes. The *multipoint relays* (MPRs) and the routing paths are selected based on the radio aware link metric. Recall that MPRs are a set of 1-hop neighbor nodes that cover the entire 2-hop network neighborhood.[18] These MPRs are responsible for propagating the topology information and retransmission of packets, thus reducing the overhead that would otherwise arise due to the flooding process. There is also an option to control the frequency of message exchanges using the mechanism in Ref. 19. The basic idea is that the nearer mesh nodes receive the topology updates more frequently than the distant nodes.

Even though the 802.11s working group has specified the routing protocols discussed in the sections above, it is clear that there is no mesh routing protocol that is optimal in every possible mesh networking scenario. Hence there is a provision of extensibility that allows interoperability between mesh nodes developed by different vendors.[14] Every mesh node announces the routing protocol and the metric that it uses. New nodes join the network only if they can support the announced protocol and metric. All IEEE 802.11s devices will be able to use the HWMP routing protocol and the airtime link metric. Other routing protocols and metrics that are more suitable for different application scenarios can be used besides the default protocol and the metric.

21.2.4.3. *Multi-path routing protocols*

As discussed in the previous section, the MPPs and the MAPs form the backbone network for mesh connectivity with the MPPs serving as gateways to the Internet. Hence the traffic is routed from various MPs and STAs towards these gateways. If the routing protocol selects a single best path based on the highest throughput and/or least delay, it might lead to an increased load on certain parts of the network, thus affecting the overall network performance. Also, if the route fails due to poor channel quality or mobility of the nodes, the route discovery process would need to be initiated, thus increasing the routing overhead. To address these issues, several research groups are working in the multi-path routing area to achieve load balancing and to provide fault tolerance.[31–35] Load balancing is achieved by distributing the traffic between a source and a destination via several paths. Due to the nature of the mesh creation, several paths exist between two nodes of the network and this feature is utilized in deploying multi-path routing protocols. In other words, several *good* paths are used to transmit the information between a given pair of nodes instead of using a single *best* path.[31]

Benefits of multi-path routing: Load balancing is one of the key benefits of any multi-path routing protocol. Network hot spots are avoided thus reducing congestion and traffic bottlenecks. The second advantage is the increase in the fault tolerance of the network in case of link failures. If one of the paths fails due to change in the network conditions or topology, the other paths can be used for sustained network connectivity. In case of applications with high *Quality of Service* (QoS) requirements, multiple paths may be used to send redundant data, thus ensuring the packet delivery. For applications with high bandwidth requirements, transmissions along multiple paths would lead to a better performance than those through single paths that might not be sufficient for providing the required bandwidth.[34] The availability of more bandwidth would result in decreasing the overall end-to-end delay. While designing multi-path routing protocols for mesh networks, link layer issues (e.g. channel access technologies) need to be taken into consideration to avoid interference due to transmissions from other nodes in the same path.

In Ref. 31, the authors explore the applicability of multi-path routing for WMNs. The argument is based on the fact that any ad hoc network (or mesh network) enjoys a rich connectivity between the participating nodes. Hence a good routing protocol for such networks should be able to determine many loop-free paths between a given pair of nodes and those paths that have poor link quality and prolonged delays due to a large number of hops should be eliminated from the list. They argue that as long as the protocol is able to determine several well-connected but loop free paths, disjoint paths are not necessary for improving the network reliability. Note that node disjoint paths are multi-hop paths that do not have any common nodes except for the source and the destination and link disjoint paths are paths that may have common nodes but no overlapping links. The authors use combinatorial analysis to show that the network reliability and the connectivity lifetime for a multi-path mesh connected network is better than that for a network with disjoint connections between the source and the destination.

In Ref. 32, a proactive multi-path routing scheme called Multi-path Mesh (MMESH) is introduced. It has mechanisms for discovering multiple paths as well as distributing the network traffic among these paths to achieve load balancing. Every MP in the network selects multiple routes to the gateway nodes and the information is shared with the intermediate nodes. The idea is derived from the traditional source routing mechanism where the source node calculates the route for the entire traffic and puts the path information in the header of every packet. Hence the intermediate nodes forward the packets based on the header information without processing the entire packet. In the initial network setup phase, all the nodes of the network discover multiple paths to the MPPs in the network. At startup, the MPPs broadcast their connectivity status to the Internet periodically. The neighboring MPs that are within the listening range of these MPPs, setup the paths to the MPPs based on the information received. They also rebroadcast the connection advertisements so that this information is propagated throughout the network. The protocol may be implemented with any chosen routing metric such as ETT, WCETT etc. Besides the route discovery phase, there is a route maintenance phase where all the MPs monitor all the active paths going through them. If a new route is discovered or a route becomes stale due to changes in network characteristics, node mobility or node failure, the information is updated in the local routing table and sent out to all the nodes in the MP's neighborhood. Subsequently, the neighboring MPs update their information and send the information out to other nodes so that the information is propagated in the entire network.

Faster failure recovery mechanisms are also proposed such that the intermediate nodes update their routing table information quickly. Based on a failed neighbor node, they suspend the routes that go through the failed node, and route the source routed packets via alternative routes. If the source routing specifies only one possible next hop for the path, that being the failed node, the intermediate MP can change the path based on the information in its routing table as long as the threshold time limit is observed. For load balancing among multiple paths, a simple round robin scheduling, where a node sends traffic to possible next hops in a sequential fashion, or a congestion aware routing mechanism may be used. Round robin scheduling may lead to out of order packet delivery.

21.2.4.4. *Load balancing routing protocols*

In Ref. 32, a load balancing routing scheme is introduced. The authors propose a centralized network management architecture called *Configurable Access Network* (CAN). The fact that the backbone network of WMNs is usually accessible through the wired infrastructure is used to define an external node called the *Network Operation Center* (NOC). Hence the network management and control functions are "outsourced" to the NOC, thus making the backbone routers responsible only for routing functions, and hence reducing their complexity. The algorithms presented by the authors are based on single path routing with bandwidth allocation which can achieve the near-optimal bandwidth allocation achieved by more complicated multi-path routing schemes. The approach fares better than the multi-path routing approach, particularly in a high node and user density test case. This scheme has a simpler traffic control mechanism, maintenance of the packet delivery order and works well with header compression mechanisms to achieve a much better throughput without increasing the capacity of the network. However it is not possible to use compression techniques with multi-path routing as the packets from the same traffic flow might traverse different gateways and arrive out of order. The proposed algorithms calculate the optimal multi-path routing solutions and then extract a single path solution from this set of solutions. A weight factor is allocated to every mesh router depending on the number of nodes connected to the router. The analysis is presented when nodes have the same and different weights and the wireless links have the same and different capacities. The approach claims to achieve high network utilization as well as long term fairness in the bandwidth allocation. One of the assumptions of the approach is that the WMN is accessible by the wired infrastructure and hence the NOC has knowledge of the network topology, the link capacities and the node weights and determines the routing and bandwidth allocation according to this information. In some WMN scenarios, it might not be feasible or desirable to have such centralized control since the self-configuring and decentralized nature of these networks is sometimes a desired feature.

Another load balanced routing scheme called *MAC-aware* and *Load Balanced* routing algorithm (MaLB) is proposed in Ref. 35 that takes MAC-layer interactions into account. Hence it has cross layer design as well as load balancing incorporated in the routing protocol. MaLB is shown to fare well particularly in scenarios when there is a load imbalance in the network due to MPP failures. Since MAC-layer interaction between the links is considered the routing protocol selects paths that have less congested links.

21.2.4.5. *Multi-mesh tree routing protocol*

For large multi-hop wireless ad hoc networks, a hybrid routing scheme that uses proactive and reactive approaches combined with hierarchy through a clustering approach was proposed by some researchers.[20,21] Hybrid routing protocols are useful as they compromise between high route discovery overheads and initial route discovery latency by using proactive schemes in a restricted area thus avoiding initial route discovery latency for communicating with nodes in the area, and reactive routing schemes that have higher route discovery overheads are used to communicate with distant nodes, assuming that such occurrences are infrequent. Hierarchical routing protocols are best suited to address scalability.[20–24] Most hierarchical routing schemes are based on the self-organizing feature of ad hoc networks wherein the network nodes group themselves into clusters based on some algorithm. Distributed clustering algorithms are highly desirable though difficult to implement. A cluster may have one or more cluster heads, depending on the adopted scheme. The nodes in a cluster can be one or more hops away from the cluster head. Inter-cluster communication is established using the border nodes which overlap between the clusters. Different routing mechanisms may be used for intra-cluster and inter-cluster communication. A combination of hybrid and hierarchical approach would normally adopt proactive routing from intra-cluster communications and reactive routing for inter-cluster communications.[20,21,23]

The Multi-Meshed Tree (MMT) routing protocol[11,20,21] combines the advantages of hybrid and hierarchical approaches. While most of the hybrid or hierarchical schemes use different algorithms for cluster formation, proactive and reactive routing, MMT is unique that is uses the same meshed tree principle for achieving all three operations. Once a cluster head is elected, as the cluster is formed, proactive routes are set up in the cluster. The proactive routes are established as a tree rooted at the cluster head. The tree branches are allowed to mesh without forming loops to enhance route robustness. The clusters overlap and the border nodes facilitate in inter-cluster reactive routing. The meshed tree principle is then extended to achieve a highly efficient route discovery and reactive route maintenance mechanism. MMT leverages the advantages of both a tree and a mesh; hence for packet forwarding the tree is used, while the meshing of the tree branches results in multiple routes which renders the scheme highly robust from the connectivity perspective.

The cluster head election can be based on any distributed scheme. However, when the cluster is formed due to the properties of the meshed tree approach the cluster head is able to control the size, number of hops of its client nodes and the usable bandwidth in the cluster among others, which can be ideal for load balancing

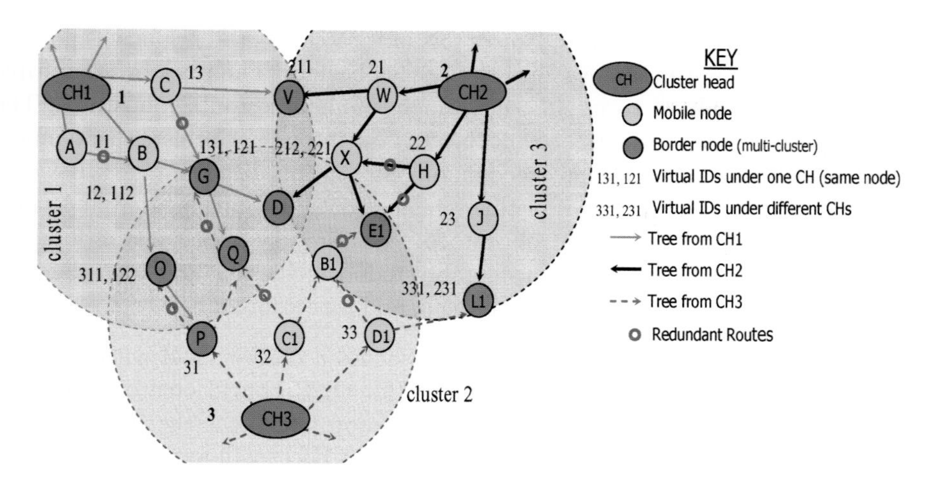

Fig. 21.3. Multiple meshed tree scheme.

in the network. MMT construction is based on a simple yet novel Virtual ID (VID) based addressing scheme that facilitates routing and forwarding at the same time. Hence in MMT, there is no need for nodes to store routing tables or states. The cluster head stores a map of its clients VIDs and their *Unique ID* (UID). The MMT based cluster can easily support several multi-hop mobile nodes.

We use Figure 21.3 to explain the operation of MMT. In this figure, CH1, CH2 and CH3 are elected cluster heads. For ease ion explanation assume that the cluster heads have VIDs 1, 2 and 3. Each cluster head announces its VID. The first hop mobile nodes listen to the announcement and send in a request to be connected i.e. a join request. The cluster head accepts the request and allocates VIDs to the children nodes. These VIDs are derivatives of the cluster head's VID; i.e. for this example case, mobile A gets VID 11, mobile B gets VID 12 and mobile C gets VID 13. The first hop nodes then register with the cluster heads along with their newly acquired VIDs and their UID which could be their MAC address or IP address or any other given ID. After registering, the first hop mobiles announce their VIDs and acquire second hop children that subsequently register with the gateway.

The branches of the tree thus grow using the VIDs. In this example, we can notice that B also became a child of A and acquires a VID from A i.e. 112. Any mobile can acquire several such backup VIDs, using an intelligent decision on suitable VIDs and hence routes to the cluster head. For example mobile O has VIDs 311 and 122 which it has acquired from mobiles P and B under two different trees which originate at two cluster heads. Hence when Q's moves and looses the route under one cluster head, it can fall back to the route under the other cluster head. Nodes do not maintain any routing tables or states as the routing information is carried in the VID. VIDs are acquired and updated dynamically and the scheme is proactive.

Though a node has multiple VIDs, it uses one of its VIDs as the primary (and uses one route only at any time) and stores the rest as backup. The trees from multiple cluster heads can be seen to mesh, which further enhances route robustness. The disadvantage of this scheme is that traffic has to flow through the cluster head unless the sending and receiving node have a common parent. This disadvantage is however offset by several other advantages of this scheme. MMT is implemented at layer 2 along with the medium access control protocol to aid in easy packet forwarding and makes it transparent to IP while considerably reducing the delays, processing and bit overheads that are incurred due to routing and forwarding at layer 3.

Packet forwarding in MMT: When a packet reaches a cluster head with the UID of a cluster client; e.g. CH1 receives a packet for O; CH1 will create a frame in which source address is VID of CH1 and destination address is the VID of O i.e. 122. When CH1 broadcasts this frame, only B with VID 12 will forward this frame, since it recognizes that VID 122 is its child. When B forwards the frame, node O is within its transmission range and will process the packet that has its VID as the destination. Before B forwards the packet, it will replace the source VID with its own VID i.e. 12. The source VID in the data frame thus serves as *implicit advertisement* for the VID of B. This substitution of source VID with self VID by forwarding nodes is done for all upstream and downstream packets and reduces traffic due to "hello" messages for self-advertisement. In MMT new branches are added only after a bidirectional exchange of messages — hence, when B forwards the frame, CH1 receives a copy, which serves as an *implicit acknowledgement*. Only destination nodes generate explicit acknowledgements.

Route failure/maintenance: When data frames are not available for forwarding for an interval of "neighbor_alive_time", a node sends an explicit advertisement frame. When neighboring nodes fail to hear any activity from its children for n*neighbor_alive_time, they will report the "missing VID" to the cluster head. The cluster head will remove that VID and its derivatives in its cache and use the backup VIDs for these cluster clients. As all VIDs including the backup VID are continually updated and there are no stale routes within the cluster.

In Figure 21.3, nodes O and Q are registered with CH1 and CH3, thus the cluster heads are aware that they are neighbors and can use nodes O and Q to provide connectivity and exchange information across these two clusters. Such nodes are the border nodes. There will be redundancy in cluster head to cluster head connectivity if nodes are allowed to select routes preferably under different cluster heads.

Inter-cluster routing: The border nodes help create paths for communications among cluster clients belonging to different clusters. For example, node B in one cluster wishes to communicate with H in another cluster (Figure 21.3). B sends a "route request" message to CH1 with UID of H. CH1 will forward this route request to CH3 via O (or Q) and CH2 via V (or D). As each cluster head keeps a local cache of the clients VIDs and their corresponding UIDs, CH2 will recognize the UID of node H and forward the route request to H. CH3 will drop the route request or may forward to its neighboring clusters if it has knowledge of such clusters. As the route request message is forwarded only the cluster head VIDs are recorded. The route discovery process though similar to DSR uses multicasting of the route query to border nodes of distinctly different clusters, which then forward to cluster heads other than the ones from which they received. A record of the past cluster heads will avoid looping the routes. The route request process explained helps avoid several problems that stem from query flooding; major one being redundant transmissions and possible collisions. All routes in MMT are concatenation of routes in a cluster, which were proactively setup, hence total route set up time could be low depending on the location of corresponding nodes. Using only the cluster heads recorded in the route discovery and route response helps provide highly stable routes inspite of node mobility as the dependency of the routes is only on a few cluster heads and not all the intermediate routes. As long as the cluster heads continue, the route will be valid.

The MMT routing scheme has the following features that are desirable for routing protocols for mesh networks.

(i) *Robustness*: Due to dynamic maintenance of multiple routes with fast route switchover times, the cluster-based MMT approach has a high fault-resilient routing structure. Alternative path routing and fast routing convergence are inherent to MMT.
(ii) *Easy Maintenance*: Due to the novel VID-based approach used for constructing trees and mesh without loops, the scheme does not require routing tables and states and best routes are acquired and maintained dynamically.
(iii) *Low Overheads and Processing Delays*: The MMT can be built most of the times using the header information in the packets that travel upstream and downstream. Hence, the scheme incurs very low overheads. Integration with the MAC also cuts down on the header overheads and layer 3 processing delays.
(iv) *Link Sensitive Routing and forwarding*: A cluster client in the MMT algorithm uses a decision block to decide on routes that it caches. Similarly a cluster head uses a decision block to decide whether to accept a cluster client or not. While forwarding a packet, quality of the route and application layer requirements can be used by MMT and prioritization if required can be handled by the MAC — both of which are at layer 2, and hence simple to achieve.

For networks with high node density, hierarchical routing protocols generally lead to a better performance than flat routing protocols due to shorter routing paths, faster setup and better manageability and scalability. However, the complexity of maintaining the hierarchy should be taken into account. For example, if the routing protocol requires the cluster head to have a significantly higher processing power and channel capacity, it might not be a feasible option for every WMN application.

MMT adopts dynamically changing the cluster head overtime. The cluster head change criteria can be varied.

21.2.4.6. *Multi-radio link quality source routing protocol*

In Ref. 29, the authors argue that WMNs have backbone nodes that are relatively stationary and not power-constrained. Hence the focus of the routing protocols for WMNs should be on improving the capacity and scalability of the network rather than on the energy concerns and mobility patterns of these backbone routers. One of the challenges encountered by WMNs is the decrease in the overall network capacity due to the interference between multiple transmissions at the same time. Hence nodes in a mesh network equipped with multiple radios may be preferred over single-radio nodes as the capacity of the network is increased without making changes to the MAC protocol. Multiple radios enable the node to receive and transmit at the same time thus increasing the relaying speed. Since multiple radios can be used to transmit on multiple channels simultaneously, the bandwidth can be used more efficiently. Also, multiple heterogeneous radios on a node will be desired for connecting networks based on different technologies such as 802.11b/g and 802.11a. In general, shortest path routing algorithms do not fare well in networks with nodes equipped with multiple radios. For example, other things being equal, 802.11b links cover a longer range than 802.11a links. Hence between a given pair of nodes, the 802.11b links will involve less number of hops and hence will be adopted. However the 802.11b links are much slower and hence will not lead to the optimal overall performance. To overcome these challenges, the authors propose a new routing protocol called Multi-Radio Link Quality Source Routing Protocol (MR-LQSR) which incorporates the new metric called WCETT discussed earlier in the routing metrics section of this chapter. MR-LQSR is a source routed link state routing protocol based on DSR. The neighbor discovery phase and the information propagation mechanism for MR-LQSR are similar to the ones in DSR. However the manner in which the weights are assigned to the links that connect a node to its neighbors and the algorithm used to determine the best path between the source and the destination using these link weights are different. In DSR, equal weights are assigned to all the links and the path metric is computed by summing up the link weights along a given path. Hence DSR is a shortest path routing protocol. Instead of this approach, the authors use the WCETT metric to assign the link weights in MR-LQSR. Hence the link quality as well as the hop count is taken into account while assigning the weights and both channel quality and channel diversity are considered at the same time. The protocol assumes that all the participating nodes are stationary, have multiple radios that are tuned to different non-overlapping channels (channel assignment is not handled in this work). Hence although it may be suitable for the WMN routers for the backbone network a different routing protocol needs to be implemented for the mesh clients that are mobile and are not equipped with multiple radios.

21.3. Challenges for Future Research

In the previous section, we have shown that the routing requirements for mesh networks are significantly different compared to those for other wireless networks like ad hoc networks. New routing metrics need to be defined to incorporate the characteristics of these networks and routing protocols based on these metrics need to be developed. Protocols that use two or more of these metrics in the path selection would lead to a better performance, and need to be investigated. However the complexity and efficiency of the algorithm should be considered while developing the protocols.

A hot research topic in the WMN area is the integration of multiple layers of the protocol suite (cross layer design) for improving the performance of routing protocols. Integrating the MAC and routing functionalities has been studied before[11,20,21,35] and one of our current projects at RIT involves a cross layered approach spanning the physical layer, MAC and routing functions integrated at layer 2, and application layer for cognitive airborne networking. When multi-radio and multi-channel mesh networking nodes are used, new routing protocols are needed for choosing the optimal radio transceiver, the best channel, as well as the best path connecting the source and the destination. In this case, the physical layer components (radio and the channel) are very closely linked to the routing layer component (routing path). Hence a routing protocol based on cross layer design would be desired.

Many WMN applications require multicasting capabilities (e.g. video distribution in a community network).[10] It would efficient to have the multicasting feature inherent in the proposed protocols. Scalability in routing protocols continues to pose challenges as the routing performance requirements vary depending on the applications, hence a routing protocol that scales well for a given applications may not delivery the required performance for certain other applications. Hierarchical routing protocols and hybrid routing protocols address scalability to some extent. In most of hierarchical schemes, clustering is used which requires a clusterhead with higher processing powers and capacity compared to other neighborhood nodes. A distributed clusterhead election process that is dynamically initiated would be ideal. New protocols that consider several aspects like different power constraints, mobility features and route robustness in WMN are still needed. A major challenge however is the variety of applications that can benefit by using the concept of WMN have varying requirements, because of which it is not possible to identify one routing protocol that will satisfy all WMN requirements.

21.4. Conclusions

WMN are very attractive due to their self-organizing capabilities and their unique features that provide a "last mile multi-hop connectivity to the Internet" solution for applications where it might not be feasible to provide direct single-hop connections. In fact many commercial wireless mesh solutions are available (some of them are mentioned in Refs. 39 to 45) and a lot of other companies are expected to develop products in this area as these networks become more and more popular. Routing in WMNs poses some unique challenges and the goal of this chapter was to discuss some of these requirements and present some of the multi-hop, multi-path, multi-radio and load balancing protocols that have been introduced by various research groups.

Problems

1. What is the primary difference between a conventional router and a mesh router?
2. What are some of the requirements that should be considered while determining the routing metrics for wireless mesh networks?
3. What is the ETX routing metric?
4. What is the difference between the ETT and WCETT routing metrics?
5. What are some of the desired features of a WMN routing protocol?
6. Which two routing protocols are described in the 802.11s standard?
7. How are the HWMP and RA-OLSR protocols different?
8. What are the advantages of multi-path routing?
9. What are some of the inherent features of the MMT protocol?
10. For what type of mesh nodes is the MR-LQSR protocol most suitable?

Bibliography

1. Microsoft Mesh Networks, http://research.microsoft.com/mesh.
2. N. Baker, ZigBee and Bluetooth strengths and weaknesses for industrial applications, *Computing & Control Engineering Journal* **16**(2) (2005) 22–25.
3. W. Kastner, G. Neugschwandtner, S. Soucek and H. M. Newman, Communication systems for building automation and control, *Proceedings of the IEEE* **93**(6) (2005) 1178–1203.
4. M. Mathiesen, G. Thonet and N. Aakvaag, Wireless ad hoc networks for industry automation: current trends and future prospects, *Proceedings of the IFAC World Congress*, Prague, Czech Republic (2005).
5. 802.11s standard, http://grouper.ieee.org/groups/802/11/Reports/tgs_update.htm (2008).
6. MIT Roofnet, http://pdos.csail.mit.edu/roofnet/.
7. Locust World, http://www.locustworld.com.
8. A. Lee and P. Ward, A study of routing algorithms in wireless mesh networks, *Proceedings of Australian Telecommunication Networks and Applications Conference* (2004).
9. Y. Yang, J. Wang and R. Kravets, Load-balanced routing for mesh networks, *ACM SIGMOBILE Mobile Computing and Communications Review* **10**(4) (2006).

10. I. F. Akyildiz, X. Wang and W. Wang, Wireless mesh networks: a survey, *Computer Networks* **47**(4) (2005) 445–487.
11. N. Shenoy, Y. Pan and V. Reddy, Quality of service in internet MANETs, *Proceedings of IEEE 16th International Symposium on Personal, Indoor & Mobile Radio Communications (PIMRC)* **2005**(3) (2005) 1823–1829.
12. J. Eriksson, M. Faloutsos and S. Krishnamurthy, Scalable ad hoc routing: the case for dynamic addressing, *Proceedings of INFOCOM* **2004**(2) (2004) 1108–1119.
13. M. Bahr, Proposed routing for IEEE 802.11s WLAN mesh networks, *Proceedings of the 2nd Annual International Workshop on Wireless Internet (WICON)* (2006).
14. Y. Zhang, J. Luo and H. Hu, *Wireless Mesh Networking: Architectures, Protocols and Standards* (CRC Press, 2006).
15. J. Camp and E. Knightly, The IEEE 802.11s Extended Service Set Mesh Networking Standard, *IEEE Communications Magazine* http://networks.rice.edu/papers/mesh80211s.pdf (2008).
16. M. Bahr, Update on the hybrid wireless mesh protocol of IEEE 802.11s, *Proceedings of the IEEE International Conference on Mobile Adhoc and Sensor Systems (MASS)* (2007) 1–6.
17. C. Perkins, Ad hoc on-demand distance vector (AODV) routing, *RFC 3561* (2003).
18. T. Clausen and P. Jacquet, Optimized link state routing protocol (OLSR), *RFC 3626* (2003).
19. G. Pei, M. Gerla and T. Chen, Fisheye state routing: a routing scheme for ad hoc wireless networks, *Proceedings of ICC* (2000).
20. N. Shenoy and P. Yin, Multi-meshed tree routing for Internet MANETs, *Proceedings of 2nd International Symposium on Wireless Communication Systems* (2005) 145–149.
21. N. Shenoy, Y. Pan, D. Narayan, D. Ross and C. Lutzer, Route robustness of a multi-meshed tree routing scheme for internet MANETs, *Proceedings of GLOBECOM* (2005).
22. E. M. Belding-Royer, Multi-level hierarchies for scalable ad hoc routing, *ACM/Kluwer Wireless Networks* **9**(5) (2003) 461–478.
23. S. Du, A. Khan, S. PalChaudhuri, A. Post, A. K. Saha, P. Druschel, D. B. Johnson and R. Reidi, Self-organizing hierarchical routing for scalable ad hoc networking, Technical Report TR04-433, Department of Computer Science, Rice University.
24. K. Xu, X. Hong and M. Gerla, Landmark routing in ad hoc networks with mobile backbones, *Journal of Parallel and Distributed Computing, Special Issue on Ad hoc Networks* **63**(2) (2002) 110–122.
25. Y. Yang, J. Wang and R. Kravets, Designing routing metrics for mesh networks, *Proceedings of IEEE Workshop on Wireless Mesh Networks (WiMesh)* (2005).
26. D. De Couto, D. Aguayo, J. Bicket and R. Morris, A high throughput path metric for multi-hop wireless routing, *Proceedings of ACM Annual International Conference on Mobile Computing & Networking (MOBICOM)* (2003) 134–146.
27. R. Draves, J. Padhye and B. Zill, Comparison of routing metrics for static multi-hop wireless networks, *Proceedings of ACM SIGCOMM* (2004) 133–144.
28. A. Adya, P. Bahl, J. Padhye, A. Wolman and L. Zhou, A multi-radio unification protocol for IEEE 802.11 wireless networks, *Proceedings of 1st International Conference on Broadband Networks (BroadNets)* (2004) 344–354.
29. R. Draves, J. Padhye and B. Zill, Routing in multi-radio multi-hop wireless mesh networks, *Proceedings of ACM Annual International Conference on Mobile Computing & Networking (MOBICOM)* (2004) 114–128.
30. L. Ma and M. K. Denko, A routing metric for load balancing in wireless mesh networks, *Proceedings of 21st International Conference on Advanced Information Networking and Applications Workshops (AINAW)* (2007) 409–414.

31. M. Mosko and J. J. G. L. Aceves, Multipath routing in wireless mesh networks, *Proceedings of IEEE Workshop on Wireless Mesh Networks (WiMesh)* (2005).
32. N. S. Nandiraju, D. S. Nandiraju and D. P. Agrawal, Multipath routing in wireless mesh networks, *Proceedings of 2nd IEEE International Workshop on Heterogeneous Multi-Hop Wireless and Mobile Networks (MHWMN)* (2006).
33. Y. Bejarno, S. J. Han and A. Kumar, Efficient load balancing routing for wireless mesh networks, *Computer Networks: The International Journal of Computer and Telecommunications Networking* **51**(10), (2007) 2450–2466.
34. S. Mueller, R. P. Tsang and D. Ghosal, Multipath routing in mobile ad hoc networks: issues and challenges, *Performance Tools and Applications to Networked Systems*, Revised Tutorial Lectures, Lecture Notes in Computer Science, Vol. 2965 (2004) 209–234.
35. V. Mhatre, F. Baccelli, H. Lundgren and C. Diot, Joint MAC-aware routing and load balancing in mesh networks, *Proceedings of ACM CONEXT* (2007).
36. J. Zheng and M. J. Lee, A resource-efficient and scalable wireless mesh routing protocol, *Ad hoc Networks* **5**(6) (2007) 704–718.
37. Motorola Mesh Networks, http://www.motorola.com/mesh.
38. Mitre Corporation (Mobile Mesh), http://www.mitre.org/work/tech_transfer/mobilemesh/.
39. Firetide Networks, http://www.firetide.com.
40. BelAir Networks, http://www.belairnetworks.com.
41. Mesh Dynamics, http://www.meshdynamics.com.
42. Strix Systems, http://www.strixsystems.com.
43. Meraki Networks, http://www.meraki.com.
44. Sonbuddy Inc, http://www.sonbuddy.com.
45. Radiant Networks, http://www.radiannetworks.com.

Chapter 22

Mobility Management in Wireless Mesh Networks

Paul Wu,* Bjorn Landfeldt† and Albert Y. Zomaya‡
School of Information Technologies
University of Sydney and National ICT Australia (NICTA)
**lpwu@it.usyd.edu.au*
†bjornl@it.usyd.edu.au
‡zomaya@it.usyd.edu.au

Over the past few years, wireless mesh networks (WMNs) have emerged to become a ubiquitous solution for providing wireless access at lower cost, greater flexibility, higher reliability and performance compared to conventional wireless local area network (WLANs). As the demand for delay sensitive real-time applications such as VOIP and streaming media grows, mobility management is becoming a critical function of WMN. Although Mobility management schemes designed for Mobile Ad Hoc Networks (MANETs) and traditional fixed networks might be useful for WMNs, mobility management in WMNs is still an active research issue because of the special properties of WMNs. This chapter focuses on the mobility management in wireless mesh networks, which takes into account the unreliability of wireless mesh links and other special properties of WMNs. The chapter began by discussing the special properties of WMNs and the general concepts of mobility management, and then gives an overview of the mobility management problems in wireless mesh networks (layer 2 handoff, vertical handoff and cross layer design, etc.) with some potential solutions.

Keywords: WMN; mobility management; handoff; cross layer design.

22.1. Introduction

Over the past few years, wireless mesh networks (WMNs) have emerged to become a ubiquitous solution for providing wireless access at low cost. These networks promise greater flexibility, higher reliability and performance compared to conventional wireless local area network (WLANs). There already exist well-established mesh equipment vendors in the market (Tropos, MeshDynamics etc.), which sell equipment as well as offering complete WMN integration solutions. Wireless mesh networks are also becoming an important technology for constructing 4G high speed networks and there is a broad ongoing standardization effort, and several standards, such as IEEE 802.11s,[3] IEEE 802.15.5[4] and IEEE 802.16a,[5] are in development for providing mesh support in different networking environments.

Wireless mesh networking and mobile ad hoc networking share the key concept of communication via wireless links between nodes over multi-hops on a meshed network graph. However, they are also very different in many respects. The primary difference between these two types of networks is the mobility of nodes. Mobile and ad hoc networks (MANET) focuses on ad hoc capability for devices with high mobility while WMN mainly focuses on reliable communication and practical deployment for static nodes. Compared to the dynamic topology of MANET which is changing over time, the topology of WMN is relatively static. Therefore, while on-demand routing protocols perform better in MANETs, table-driven or hierarchical routing protocols generally perform better in WMNs. Moreover, MANETs are infrastructure-less with no planning required at deployment, while WMNs usually have fixed infrastructure and require some planning at deployment. Finally, another important difference between MANET and WMN lies in energy constraints. Most nodes in WMNs are connected to a power source, thus removing the energy constraint of the mostly battery powered MANETs.

Mobility management is becoming a critical function of WMN as the demand for delay sensitive real-time applications such as VOIP and streaming media grows. Mobility management can be divided into two main functions, location management and handoff management.[1] Location management handles location registration and call delivery, while handoff management is responsible for handoff initiation, new connection generation, and data flow control for call handoff. In WMN, the mobility management protocols have to take into account the special properties of WMN such as the rather static graph topology and dynamic links.

Since WMNs share common features with mobile ad hoc networks (MANETs), mobility management protocols developed for MANETs such as the uniform quorum system[26] and group mobility[27] can be applied them if considering WMN as a special form of MANET where nodes velocity $v = 0$. IP mobility protocols such as Mobile IP,[8,9] Hierarchical Mobile IP[11] can also be applied to WMNs considering that most WMN nodes are fixed, in effect yielding infrastructure networks where routers are connected via wireless links. Despite the availability of mobility management schemes for MANETs and traditional infrastructure networks, the design of mobility management for WMNs is still an active research issue for several reasons[2]:

- Unlike MANET, most of the nodes in WMNs are either stationary or have minimum mobility. Hence the focus of mobility management schemes is on reducing the handoff delay for mobile clients instead of coping with the mobility of nodes.
- Unlike traditional networks, the links between nodes in WMNs are wireless and may suffer interference and are therefore not as stable as wired links.
- Unlike traditional networks which usually have tree hierarchies, the topology of WMNs is usually a meshed graph which does not have a well defined hierarchy.

Moreover, geographically neighboring mesh nodes could be very far from each other in a WMN connectivity graph.
- In WMNs, in order to benefit from routing optimizations on layer 2, cross-layer design becomes a necessity.

This chapter focuses on mobility management in wireless mesh networks, which takes into account the unreliability of wireless links and other special properties of WMNs. We begin by discussing the special properties of wireless mesh networks in Section 22.2. Section 22.3 introduces the general concepts of mobility management and Section 22.4 describes the mobility management schemes for wireless mesh networks.

22.2. Properties of Wireless Mesh Networks

WMNs comprise three kinds of mesh routers: *Mesh Point (MP), Mesh Portal Point (MPP)*, and *Mesh Access Point (MeshAP)*. MP is the basic entity of WMN that communicates with other MPs within its transmission range to deliver packets within the network; MPP is a MP collocated with a portal that acts as the gateway between a WMN and other types of networks like the Internet; MeshAP is a MP collocated with an access point to provide network access to mobile clients. Mesh clients equipped with wireless network interfaces connect directly to MeshAPs, which forward their data packets over multiple wireless hops in WMNs. Figure 22.1 depicts an example of such a WMN architecture.

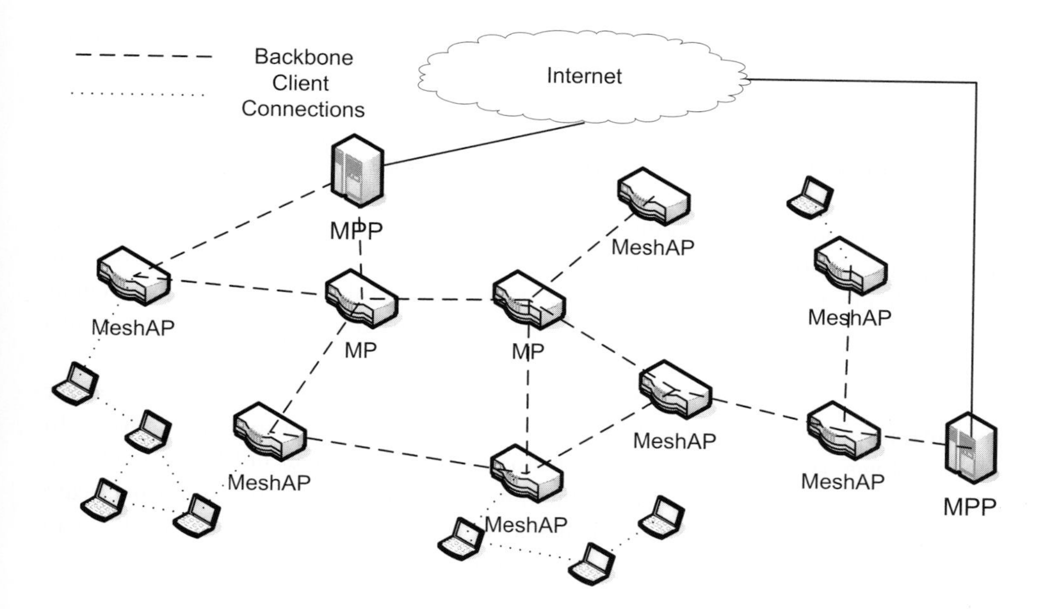

Fig. 22.1. An example of a WMN architecture.

According to Akyildiz *et al.*,[2] the architecture of WMNs can be classified into three main groups. *Infrastructure wireless mesh networks* consist of mesh routers and access points forming an infrastructure for mesh clients to connect to. The mesh client does not participate in the operation of the WMN. *Client mesh networks* constitute peer-to-peer networks among client devices such as PDAs, smart phones and laptops. The client nodes themselves constitute the actual network which routes data and carries out configuration functions. The last type of WMN is known as *Hybrid mesh networks*, which is a combination of infrastructure mesh network and client mesh network. Among these three types of networks, *infrastructure WMNs* are close to traditional infrastructure networks while *client WMNs* is closest to mobile ad hoc networks.

Reliability and network performance stability are important goals of WMNs, which is challenging due to the unreliable nature of the wireless channel. As Mesh Points are usually not mobile, they are considered to be stationary ($\nu = 0$). Moreover, Mesh points themselves are usually mounted on traffic light poles, lamb posts, building roofs etc. where there is power supply available. With these assumptions in mind, wireless ad hoc mobility management schemes should be optimized with respect to reliability and stability of WMNs instead of high mobility of nodes.

Another important goal of WMNs is extensibility. MPs form a meshed graph topology where new mesh nodes that come within range of existing MPs are simply incorporated into the network and the connectivity graph reformed. This way of forming networks is inherently scalable but it also creates problems for existing mobility management schemes. Using such network formation the network topology cannot be predetermined or planned and it is also a flat topology which is in stark contrast with traditional fixed hierarchical mobile networks. The traditional network mobility management schemes therefore needs to be altered to cater for these properties.

22.3. General Concepts of Mobility Management

Mobility management has been an interesting and challenging research topic for many years. Mobility management functions enable networks to locate roaming mobile clients for packets/call delivery and to maintain connections as a client moves to a new service area. Mobility management can therefore be divided into two main components: location management and handoff management.[1]

Location management is a two-stage process. It enables the network to discover the current point of attachment of the mobile client for packet and call delivery. In the first stage, the location registration stage, the mobile client notifies the network of its new point of attachment after mobility, allowing the network to authenticate the client and update/bind the client's location. The second stage is packet/call delivery. The network finds the current location of the client and forwards data packets or calls to the client.

Handoff management enables the network to maintain a client's connections as it continues to move and change its point of access to the network. This is a three-stage process: the initiation stage, new connection generation stage and data flow control stage. The first stage initiates the handoff management where the mobile client, the network, or changing of network conditions identifies the need for handoff. The second stage finds resources for a new connection and performs additional routing operations. For network-controlled or mobile-assisted handoff, like the ones in cellular data and voice systems, the network coordinates this stage. For mobile-controlled handoff as is done in 802.11-based systems, the mobile client finds the new resources and the network approves the connection. The final stage is the data flow control stage. At this stage, the network delivers data from the old connection to the new connection according to the data's Quality of Service (QoS) requirements.

Mobility management may also be classified into different scopes including intra-cell mobility, inter-domain mobility and intra-domain mobility. Intra-cell mobility is the mobility of clients within the coverage of the same access router between different access points (or base stations) and it usually happens at layer 2. Intra-cell mobility usually only involves handoff management. Inter-domain and intra-domain mobility are terms introduced by hierarchical mobility management protocols[11,12] and are both done at layer 3 involving both handoff and location management. Inter-domain mobility occurs when the mobile client moves to a new access router that is within the same network domain. A location update is only required to the anchor point of the domain. When the mobile client moves to a new network domain, intra-domain mobility management happens and a location update to the home network is required. For some mobility management protocols like in Refs. 8 and 9, only intra-domain mobility takes place as no anchor point exists and every router has its own domain.

As wireless mesh networking can use multiple types of radio access technologies including 802.11, 802.15, 802.16, UMTS[6] and CDMA2000,[7] there may exist both vertical and horizontal handoff.[1] Horizontal handoff refers to handoff within networks that uses the same radio access technology. Vertical handoff in this context refers to a mobile client changing the type of radio access technology used to access the Internet. For example, a laptop equipped with both 802.11 WLAN and UMTS devices might be able to use both high speed WLAN technology and W-CDMA technology for Internet access. Horizontal handoff happens when the mobile client does not change the type of radio technology it uses for communication (e.g. between UMTS base stations or between WLAN access points). When the mobile client moves from WLAN to UMTS or vice versa, a vertical handoff is required as the connection type changes.

Mobility management protocols track the location of mobile clients while maintaining their connections to the network for packet/call delivery. A good mobility management protocol should take the following issues into consideration: efficient packet processing; minimizing the signaling overhead; minimizing the handoff latency; optimizing the route for each connection; efficient bandwidth reassignment and refining the quality of service.

22.4. Mobility Management in Wireless Mesh Networks

This section will describe selected mobility management schemes that could be applied to wireless mesh networks as an illustration of the general concepts and the special properties. A comprehensive overview of all possible mobility management solutions is beyond the scope of this chapter.

22.4.1. *Layer 2 handoff*

A WMN consists of many mesh access points (MeshAP) connected to a few wireless gateways (MPP) through wireless links to provide network access to mobile clients. A mobile client may move freely within the range of a given MeshAP, but when it moves away from its current MeshAP and closer to another, it must at some point "handoff" to the new MeshAP in order to maintain connectivity. Ideally, the handoff should be completed transparently to end users without any noticeable interruption, loss of connection or degradation of quality of service for applications like VoIP.

WMNs may use both cellular and 802.11 technologies. In cellular data and voice systems, the handoff decision is typically coordinated by the network itself which is able to leverage considerable information about the network and client proximity. By contrast, 802.11 based networks dictate that the handoff should be managed autonomously and independently by mobile clients with no prior information about the network. Moreover, while cellular networks continuously monitor the signal quality between the mobile clients and all its neighboring base stations — handing off whenever a better signal quality is found for an alternative base station — 802.11 based clients only monitor the signal quality of their currently attached access points and only handoff after the signal quality degrades below a predefined threshold. Consequently, as a mobile client reaches the coverage limit of its current MeshAP, it must temporarily abandon its connection and actively probe the network to discover alternative MeshAPs and to find the one with the best signal quality to reconnect to. This approach minimizes the management overhead, but has a high handoff latency and even worse, can produce "gaps" in connectivity for a duration up to 1 second.[20] While this latency and "gap" are acceptable for some applications (e.g. file transfer, BitTorrent, email delivery, web browsing), it is too long for applications that are very sensitive to network delays like wireless Voice-over-IP and videoconferencing.

To make 802.11 based WMNs support telephony services, the network must support rapid disassociation from one MeshAP and re-association to the new MeshAP with a handoff delay less than 50 ms.[19] In order to achieve this, the "gap" in connectivity during handoff must be removed and thus the handoff latency reduced.[17]

The IEEE 802.11r working group[18] is drafting an amendment protocol that will facilitate fast handoff from one access point (MeshAP) to another managed in a seamless manner. IEEE 802.11r refines the transition process as a mobile client moves between access points. Mobile clients will be able to use the current AP as a conduit to other APs. Neighboring APs will continuously monitor and share the connectivity quality of the mobile clients to coordinate those that should serve the client. The protocol allows a mobile client to establish a QoS and security state at a new access point before making the transition, thus remove the "gap" in connectivity.

Layer 2 handoff constitutes several functions: scanning, re-association and re-authentication. Among the three factors, scanning overhead dominates the total handoff delay. Once a mobile client decides to attempt to handoff (usually initiated by a below threshold Signal-to-Noise ratio), it must identify the set of proximity candidate access points. Since 802.11 does not provide a shared control channel, the client must explicitly scan each channel (11 for 802.11b/g and 8 for 802.11a indoors) for potential APs. The scan can be completed passively or actively. Passive scanning is not feasible because the mobile client has to switch to every candidate channel and listen for periodic beacon packets from potential access points. Beacon intervals are typically 100 ms and as they are independent of each other clients must wait a full interval on each candidate channel causing the handoff latency using this approach to take up to 3 seconds. To reduce this delay, active scanning can be used where the client actively broadcasts a probe packet on each channel to force an AP to respond immediately. However, even with this approach, the scanning time usually exceeds the requirements of seamless handoff.[17,20] Preemptive scanning named Syncscan[21] was developed to overcome this problem. Synscan synchronizes the announcements of beaconing packets on each channel so a mobile client only has to switch to a particular channel at a predetermined synchronized time for all the beacon packets on that channel and there is therefore no waiting required. Moreover, instead of scanning only at handoff, the mobile client regularly scans the existence of other APs while still being associated with the current AP. Syncscan reports an order of magnitude smaller handoff latency than conventional approaches which is enough to support real-time applications including VoIP.

22.4.2. *Mobility management for ad hoc networks*

Ad hoc mobility management protocols can be directly applied to Client Mesh Networks. If we consider stable mesh points as mobile ad hoc nodes with velocity 0 and treat mesh clients as equal to mesh points, ad hoc mobility management schemes may also be applied to other types of WMNs. In this section, we will discuss some of the ad hoc mobility management protocols and their suitability to be applied in WMNs.

In Ref. 25, a distributed mobility management scheme is proposed using randomized database groups for ad hoc networks. In the proposed scheme, location databases are stored in the network nodes themselves, which form a virtual backbone within the flat network architecture. Upon a location update or call arrival, copies of the location information of a mobile node can be stored and retrieved in a distributed fashion through the randomized access of database groups that are themselves randomly distributed. In Ref. 26, a similar distributed mobility management system is proposed based on a uniform quorum system (UQS). Instead of randomized database groups, in this scheme, databases are dynamically organized into quorums, every two of which intersect at a constant number of databases. Wire and Read operations are done to all the databases of a quorum, chosen in a nondeterministic manner when a location update or call delivery is required. Although both solutions have demonstrated lag compared to conventional centralized schemes such as home location register (HLR), the schemes are more suitable for ad hoc networks in unstable connectivity and therefore unstable database access scenarios. However, both of the schemes focus on the location management part of the mobility management and are only applicable to Client Mesh Networks and not efficient for other types of networks.

Changes in network partitioning and topology can also be considered as mobility. Such changes are unexpected (as the mobility of users are unpredictable) and may disrupt ongoing connections. Reference 27 proposes a scheme to predict network partitioning and topology changes, and for applying the prediction to routing protocols to minimize disruptions, hence achieving seamless mobility. More specifically, it creates a group mobility model; a group-id based clustering algorithm and a partition prediction scheme. The mobility model is based on parameters such as distance, velocity and acceleration rate and is comparable to real user mobility patterns. The clustering algorithm forms mobility groups where the mobile nodes in each group have the same characteristics. With the grouping and information from the mobility model, the network changes and partitioning can be predicted. Thus, if a destination is unreachable, packets in the network can be rerouted to predicted locations, thereby minimizing the connection disruption and achieving seamless mobility management.

Because of the special properties of wireless mesh networks described in the previous section, mobility management schemes for ad hoc networks may not perform as efficiently in WMNs. Moreover, ad hoc mobility management schemes would not work well for infrastructure WMNs whose characteristics are closer to fixed networks. Thus, in order to be used in WMNs, ad hoc mobility management schemes have to be extended to take the properties of WMNs into consideration.

22.4.3. *Mobility management in fixed networks*

The most important wireless mesh network group is infrastructure WMN (including the infrastructure part of the hybrid WMN). Because the mesh points are stable, mobility management schemes designed for traditional fixed networks could be used in this type of WMN, e.g. Mobile IP, Hierarchical Mobile IP, Fast Handovers for Mobile IP, etc.

Mobile IP (MIP)[13] describes a global mobility solution that provides host mobility for a diverse array of applications and devices on the Internet. MIP can be directly applied in WMNs as the only difference between WMN and the fixed networks is that the communication channels are wireless. With MIP, each mobile client is identified by a static network address from its home network, regardless of its current point of attachment. When a mobile client moves into a foreign network, it obtains a temporary network care-of-address from its current point of attachment and updates its home agent with this temporary address. Thus, the home agent can intercept any packets destined for the mobile client and forward them to the mobile client's current location (most commonly using IP in IP tunneling).

The handoff latency in Mobile IP is primarily due to two procedures, the address resolution time and the (network) registration time. As the MIP handoff latency is too high for real-time applications, there have been numerous proposals for minimizing the handoff latency. These proposals can be broadly classified into two groups. The first group aims to reduce the address resolution time through address pre-configuration (Fast Handoff) while the second group aims to reduce the network registration time by using a hierarchical network management structure.

Fast-handoff Mobile IP[14] belongs to the first group and can also be directly applied to WMNs. It introduces seven additional message types for use between access routers and the mobile client to assist address pre-configuration. A fast-handoff is initiated on an indication from a wireless link-layer (L2) trigger which indicates that the mobile client will soon hand off. The mobile client sends a Router Solicitation for Proxy message to the current mesh point (oMP) containing the link layer information of the new mesh point (nMP). In response, oMP will send the mobile client a Proxy Router Advertisement message specifying the network prefix to help the client to form the new Care-of-Address using stateless address configuration.[15]

Hierarchical mobility management schemes[11,12] belongs to the second group. They separate mobility management into micro-mobility and macro-mobility, otherwise known as intra-domain and inter-domain mobility management. The central element of these schemes is the creation of a special conceptual entity called Mobility Anchor Points (MAP). The MAP is placed above several access routers, which constitutes its network domain. The domain usually consists of a set of routers and the MAP maintains a binding between itself and mobile clients currently entering its domain. When a mobile client moves to a new access router within a MAP domain (micro-mobility), it acquires a local care-of address (LCoA) from the new access router and registers this with the MAP by sending a regular MIPv6 Binding Update (BU). When a mobile client moves into a new MAP domain, it needs to acquire a new regional care-of address (RCoA) from the new MAP, which is usually the address of the new MAP, as well as a LCoA. After forming these addresses, the mobile client sends a BU to the MAP, which will bind its RCoA to the LCoA. Furthermore, the mobile client must also send another BU to its home agent that binds its home address to the new RCoA. Finally, the mobile client may also send a BU to its corresponding nodes (CN) specifying the binding between the home address and the newly acquired RCoA.

There have been a number of proposals investigating, extending and even combining the fast handoff and hierarchical mobility management schemes. Reference 22 proposes a "pre-handoff registration" scheme similar to fast-handoff which performs better that standard Mobile IPv4 handoff. Reference 23 shows that hierarchical together with fast-handoff schemes can greatly reduce the overall handoff latency. Reference 24 proposes the S-MIP architecture based on Fast Handover IPv6 and Hierarchical Mobile IPv6 that reduces the handoff latency to where the end devices can perceive a seamless connectivity at the network layer, namely, no packet loss.

Most Mobile IP based mobility management protocols can be directly used in WMNs, but not the hierarchical ones. The hierarchical mobility management protocols are not directly applicable on WMNs. In WMNs, the topology is a meshed graph and there is no obvious position to place Mobility Anchor Points (MAPs) and there is no easy way to form MAP domains. Furthermore, geographically close nodes in WMN can be very far from each other due to factors like fading and interference causing poor connectivity. In order to apply HMIP to WMNs, a scheme needs to be developed to find the correct locations in WMN for placing MAP(s) and forming network domains around these MAPs. In the other words, efficient virtual topology transformation algorithms or virtual head selection algorithms for WMNs needs to be developed.

22.4.4. *Vertical mobility management*

As a wireless mesh network may be built up from multiple wireless access technologies (e.g. a WMN using both 802.11 and 802.16 links), vertical handoffs may happen in addition to horizontal handoff. Thus, efficient mobility management protocols for handoff between networks with different radio access technologies are required for WMNs.

There has been a number of vertical mobility management protocols proposed for various number of wireless access technologies including WLAN, IMT-2000, GPRS, CDMA, etc. Most of the protocols can be applied to infrastructure WMNs as the basic concepts are the same but since the protocols are optimized for different network types performance will vary greatly.

Reference 28 proposes a mobility management system for vertical handoffs between wireless local area network (WLAN) such as 802.11 networks and wireless wide area networks such as GPRS and UMTS. The paper introduces the concept of a Connection Manager (CM) and a Virtual Connectivity Manager (VCM) to handle the handoffs in both directions. The CM intelligently detects the conditions of different types of networks and the availability of multiple networks. The VCM maintains the mobile client's connection to the network using an end-to-end argument when handoff occurs. Collaboration between the CM and VC accomplishes seamless handoff between WWAN and WLAN. More specifically, in the CM, MAC sensing,

Fast Fourier Transform detection and adaptive threshold configuration algorithms are proposed to significantly reduce the handoff rate and the so called ping-pong effect. This effect occurs when a mobile node ends up in a state so that it continuously hands off between two or more access points. In the VC, an end-to-end connection maintenance mechanism is used to make it independent of additional network infrastructure. Demonstrated in a prototype, it has been shown that seamless handoff can be achieved in the CM and VCM based system with much higher throughput than traditional schemes.

References 29 and 30 both propose schemes for vertical mobility management between 802.11 based WLAN and UMTS networks. Reference 29 creates a functional entity called a network inter-work unit (NIU) which is placed between SGSN (Serving Gateway Support Node) to hide the WLAN particularities in order to support Mobile IP based mobility management. The paper also presents an handover initiation model based on the criteria of received signal strength (RSS) and distance to reduce unnecessary handover probability. Reference 30 proposes a new method to facilitate seamless handover between cellular data and voice networks such as UMTS and 802.11 based WLAN using a modified version of Stream Control Transmission Protocol called Mobile SCTP (mSCTP). Extended with multi-homing capability and dynamic address configuration, the mSCTP are applied in a UMTS/WLAN overlay architecture to decrease handover delay and improve throughput performance.

In addition to the vertical handoff protocols proposed for various systems, there have also been several standards in development. IEEE 802.21[10] is a standard that defines algorithms enabling seamless handoff between networks of the same type as well as handover between different network types. The standard is expected to provide a framework to allow handing off to and from cellular, GSM, GPRS, 802.11, 802.15 and 802.16 networks through different handover mechanisms. The standard is also expected to allow roaming between 802.11 networks and 3G cellular systems. In comparison, Ref. 31 is a specification by 3GPP to support handoff to and from wireless local area networks.

Despite the large number of existing vertical handoff schemes and standards that can be applied to WMNs, vertical handoff is still an interesting research area for WMNs due the large number of access technologies and handoff types they support. For example, handoff from a Client Mesh Network to an Infrastructure Mesh Network does not only involve the change of network type, but also a change of network structure.

22.4.5. *Cross layer design*

In WMNs, mobility management is closely related to multiple protocol layers. Moreover, it is natural to attempt routing in Layer 2 in WMNs in order to get better access to information about network conditions. It is therefore natural to consider cross layer design in mobility management protocols.

An example of cross layer mobility management protocol design was presented in Ref. 16 where a multi-layer mobility management architecture was proposed to take care of both real-time and non-real-time traffic for both intra-domain and inter-domain mobility in ad hoc networks. The architecture consists of three components. Session Initiation Protocol (SIP) is used for real-time communication; Mobile IP with location register (MIP-LR) is used for non-real-time communication when the mobility happens between different domains; while a derivative of the Cellular IP family — Micro Mobility Management Protocol (MMP) takes care of inter-domain movement. SIP operates at the application layer while MIP-LR and MMP both operate at the network layer.

An interesting multi-layer mobility management protocols design is to follow the multi-layer design of routing protocols i.e. whenever routing takes place in layer 2, mobility management is performed in layer 2; and when routing takes place in layer 3, so does mobility management. As layer 2 routing may happen across multiple hops, a multi-hop layer 2 handoff protocol would need to be designed, while for layer 3 mobility management, the mobile IP family could be adopted.

22.5. Conclusion and Further Works

Recently, wireless mesh networking has attracted much attention from both industry and the research community alike. There are already many wireless mesh products on the market and many installations of wireless mesh networks in operation all around the world. Due to their great flexibility and robustness, wireless mesh networks looks to be an important part of the future 4G wireless access network architecture.

Although a large effort has been made in research and standards development, there are still many interesting problems left to be resolved. Among those open problems, an important one is seamless handoff support for delay sensitive applications such as VoIP in WMNs. Despite the huge amount of results from a decade of mobility management research in mobile ad hoc networks and fixed networks, the specific properties of WMNs require and allow optimizations of mobility management schemes in order to meet the performance goals for QoS sensitive traffic. The low mobility of mesh points require consideration of the efficiency of ad hoc mobility management schemes and the meshed graph topology requires additional transformation algorithms to take full advantage of hierarchical mobility management schemes. The existence of overlay networks in WMNs brings the need for vertical handoff protocols to handle frequent mobility between different network types. Location management is desired to enhance the performance of MAC and routing protocols and to help develop location-related applications. Last but not least, cross-layer mobility management protocol design is important in order to better utilize network condition information from the MAC and physical layers in routing decisions.

Although we have discussed a number of potential solutions to mobility management in wireless mesh networks, none of them can handle all mobility cases (e.g. currently none can handle mobility from client mesh network to infrastructure mesh network). Therefore, the development of a combination of efficient location management and handoff management schemes which handles all types of mobility in WMNs remains an unsolved issue.

Problems

(1) What are the differences between Wireless Mesh Networks and Mobile Ad Hoc Networks affecting mobility management?
(2) How many types of Wireless Mesh Networks are there and how do they differ?
(3) What are the components of Mobility Management?
(4) How does Vertical Handoff differ from Horizontal Handoff?
(5) Why is the 802.11 handoff process slow and how can it be improved?
(6) Describe one Ad Hoc Mobility Management protocol.
(7) What is Mobile IP? What are the schemes for improving the handoff latency of Mobile IP?
(8) IEEE 802.11r and IEEE 802.21 are two standards for mobility management in wireless networks, describe their main differences.

Bibliography

1. I. F. Akyildiz, J. McNair, J. S. M. Ho, H. Uzunalioglu and W. Wang, Mobility management in next generation wireless systems, *IEEE Proceedings* **87**(8) (1999) 1347–1385.
2. I. F. Akyildiz, X. Wang and W. Wang, Wireless mesh networks: a survey, *Computer Networks Journal, Elsevier* **47**(4) (2005) 445–487.
3. IEEE 802.11 Task Group s, IEEE 802.11 ESS Mesh Networking par (802.11s), Draft 1.06 (2007).
4. IEEE 802.15 WPAN Task Group 5, Mesh Enhancement for High Rate WPAN, Draft 0.01 (2006).
5. IEEE 802.16 Task Group a, *Medium Access Control Modifications and Additional Physical Layer Specifications for 2-11 GHz* (2003).
6. H. Kaaranen, S. Naghian, L. Laitinen, A. Ahtiainen and V. Niemi, *UMTS Networks: Architecture, Mobility and Services* (John Wiley & Sons, Inc., New York, NY, 2001).
7. V. Vanghi, A. Damnjanovic and B. Vojcic, *The cdma2000 System for Mobile Communications: 3G Wireless Evolution* (Prentice Hall PTR, Upper Saddle River, NJ, 2004).
8. D. Johnson and C. Perkins, *Mobility Support in IPv6, RFC 3775, IETF* (2004).
9. C. Perkins, *IP Mobility Support, RFC 2002, IETF* (1996).
10. IEEE 802.21 Standard Group website, available from "http://www.ieee802.org/21/".
11. H. Soliman, C. Castelluccia, K. El Malki and L. Bellier, *Hierarchical Mobile IPv6 Mobility Management, RFC4140, IETF* (2005).
12. A. G. Valko, Cellular IP: a new approach to Internet host mobility, *SIGCOMM Comput. Commun. Rev.* **29**(1) (1999) 50–65.

13. C. Perkins, *Mobile IP: Design Principles and Practices*, Addison-Wesley Wireless Communications Series (Addison Wesley, Reading, MA, 1998).
14. R. R. Koodli (ed), *Fast Handovers for Mobile IPv6, RFC4068, IETF* (2005).
15. S. Thomson and T. Narten, *Ipv6 Stateless Address Autoconfiguration, RFC2462, IETF* (1998).
16. A. Dutta, K. D. Wong, J. Burns, R. Jain, A. McAuley, K. Young and H. Schulzrinne, Realization of integrated mobility management protocol for ad hoc networks, *IEEE Military Communication Conference (MILCOM)* (2002) 448–454.
17. H. Velayos and G. Karlsson, Techniques to reduce IEEE 802.11b MAC layer handover time, Kungl. Tekniska Hogskolan, Stockholm, Sweden, Technical Report TRITA-IMIT-LCN R 03:02, ISSN 1651-7717, ISRN KTH/IMIT/LCN/R-03/02.SE (2003).
18. IEEE 802.11 Task Group r, Fast Roaming/Fast BSS Transition (2007).
19. Hearing, available from http://mysite. du.edu/~jcalvert/waves/hear.htm (2007).
20. A. Mishra, M. Shin and W. Arbaugh, An empirical analysis of the IEEE 802.11 MAC layer handoff process, *ACM Computer Communications Review* **33**(2) (2003).
21. I. Ramani and S. Savage, Syncscan: practical fast handoff for 802.11 infrastructure networks, *IEEE INFOCOM* (2005).
22. J. Kempf and J. Wood, Analysis and comparison of handoff algorithm for mobile IPv4, *Proceedings of the 52nd IETF Meeting* (2001).
23. R. Hsieh and A. Seneviratne, Performance analysis on hierarchical mobile IPv6 with fast-handoff over TCP, *Procedings of GLOBECOM* (2002).
24. R. Hsieh, Z. G. Zhou and A. Seneviratne, S-MIP: a seamless handoff architecture for mobile IP, *Proceedings of IEEE INFOCOM* (2003).
25. Z. J. Haas and B. Liang, Ad hoc mobility management with randomized database groups, *Proceedings of ICC'99* **3** (1999) 1756–1762.
26. Z. J. Haas and B. Liang, Ad hoc mobility management with uniform quorum systems, *IEEE/ACM Transactions on Networking (TON)* **7**(2) (1999) 228–240.
27. W.-T. Chen and P.-Y. Chen, Group mobility management in wireless ad hoc networks, *IEEE Vehicular Technology Conference* (2003) 2202–2206.
28. Q. Zhang, C. Guo, Z. Guo and W. Zhu, Efficient mobility management for vertical handoff between WWAN and WLAN, *IEEE Communications* **41**(11) (2003) 102–108.
29. H. Bing, C. He and L. Jiang, Performance analysis of vertical handover in a UMTS-WLAN integrated network, *IEEE PIMRC 2003* **1** (2003) 187–191.
30. L. Ma, F. Yu, V. Leung and T. Randhawa, A new method to support UMTS/WLAN vertical handover using SCTP, *IEEE Wireless Communications* **11**(4) (2004).
31. 3GPP TS 23.234, 3GPP system to Wireles Local Area Network (WLAN) interworking (2007).

Chapter 23

Selfishness and Security Schemes for Wireless Mesh Network

Lakshmi Santhanam*, Bin Xie† and Dharma P. Agrawal‡

OBR Center for Distributed and Mobile Computing
Department of ECECS, University of Cincinnati
Cincinnati, OH 45221-0030, USA
**santhal@ececs.uc.edu*
†xieb@ececs.uc.edu
‡dpa@ececs.uc.edu

Wireless Mesh Network (WMN) is a wired extension of multi-hop ad hoc network (MANET) which defines a new paradigm for broadband wireless Internet. A packet originating from a meshclient is relayed collaboratively in a multi-hop fashion by the intermediate mesh routers (MRs) towards the Internet Gateway (IGW). This is strictly true in a network managed by a single trusted authority. But, a WMN can be formed by a group of independent MRs operated by different service providers. It is a real challenge to establish *a priori* trust in a multi-operator WMN.

Unfortunately, the current thrust of research in WMNs, is primarily focused on developing multi-path routing protocols; and security is very much in its infancy. This book chapter provides a comprehensive coverage of various security issues pertinent to WMNs. We will systematically explore the vulnerabilities that can be exploited by attackers to conduct various attacks. We then provide a detailed description of some important security designs/proposals from industry and academia that will capture the current start-of-the-art solutions. We also cover key results from our research and other active researchers that have a great impact on the design of a secure WMN. Finally, we describe various open challenges which can catalyze new research efforts in this direction.

Keywords: Authentication; AAA server; denial of service attack; free rider; Internet gateway; mesh router; malicious; MANET; selfishness; spoofing; wireless mesh networks.

23.1. Introduction

The WMN promotes a tight integration of the concept in MANET (Mobile Ad hoc Networks) with infrastructure technology. Inspite of a decade of research effort on MANETs, their applicability domain is largely restricted to military arena. On the other hand, within a short span of time, WMNs have stirred considerable interest in the commercial and academic spheres. WMN envision a broad range of applications such as building automation, VoIP over wireless, video delivery, home networking etc. In order to ensure pragmatic deployment of MRs, it is critical to build a security framework for WMNs.

It is a common misconception that MRs are controlled by a single trusted entity.[1] *A priori* trust relationship exists only in some special application scenarios, like an enterprise or a military platoon. The MRs can be administered in different management styles: fully managed, semi-managed, and unmanaged. In a fully managed WMN, the web of interconnected MR is administered by one single trusted entity. Cisco's project covering various US cities is one such example. In applications of fully managed WMN, all MRs belong to a single authority (i.e. Internet Service Provider or ISP) and thus all MRs can be assumed to be trustworthy. In a semi-managed MN, while the core of the network is administered by a single trusted authority, majority of the MRs remain outside the jurisdiction of the trusted authority.[2] The "Freifunker" community mesh network deployed in the city of Berlin[3] is one such example of a semi-managed WMN. An unmanaged WMN are formed by independent entities and hence no single authority exists to administer such networks. These are prevalent in conference rooms, camping sites, trains etc. wherein some MRs in the network provide external internet connectivity through nearby available networks such as GSM, WLAN etc.

The existence of such a category of network has always been debated by researchers.[4] While, Bruno *et al.*[5] define WMN as an "opportunistic ad hoc networking", Kyasanur *et al.*[6] define WMN to be an infrastructure based networking. Salem *et al.*[1] classify the WMNs into two categories: single operator WMNs (owned by a single ISP) and multi-operator WMNs (MRs owned by independent operators/ networks).

Thus, unmanaged/semi-managed networks raise critical security issues. Each MR can freely decide on how they use their resources. A selfish MR can monopolize the resources for itself without providing any service to the community. They artificially declare a busy state to forward other's traffic. As WMNs operate in the unlicensed band, an adversary can gobble precious network bandwidth by generating excessive traffic along heavily congested paths. This in turn disrupts the normal network activity. This attack is well known in the literature as denial-of-service attack. Thus, it is important to design a defense mechanism for WMNs to combat selfishness and denial of service attacks.

The inherent characteristics of WMNs redefine the attack model and strategies of attacker. A WMN is organized in a hierarchical structure and consists of Internet

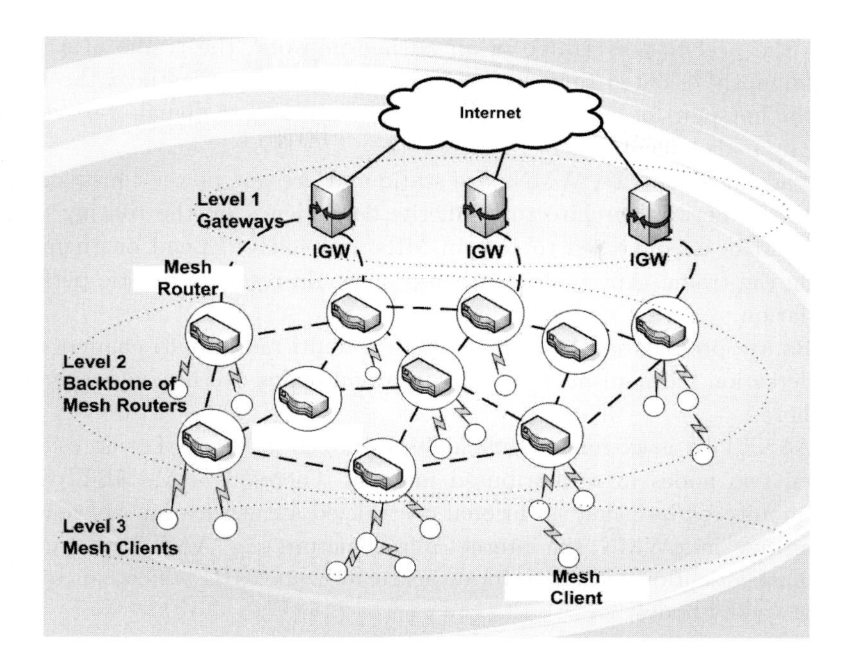

Fig. 23.1. Hierarchical architecture of WMN.

Gateway (IGW), Mesh Routers (MRs), and Meshclients (listed from top to bottom as shown in Figure 23.1). The MRs acts as an aggregation point of traffic originating from its meshclients and employs multi-hop forwarding to connect to the IGW. Thus, the MRs form a wireless backhaul network.

As MRs form the backbone infrastructure, highest level of security is required here for the successful operation of the network. A second level of security measure needs to be incorporated to prevent malicious users and provide end-to-end security for meshclients. The plug-and-play architecture of WMN, however paves way to malicious intruders.[1] An attacker if unnoticed can jeopardize the operation of the whole WMN. Hence, it is critical to prevent a malicious MR from joining and disrupting the network by suitable security mechanisms.

Though, the underlying paradigm of WMNs is somewhat similar to multi-hop ad hoc networks, they cannot be treated in the same way due to several discrepancies. Due to the unique properties of WMNs, the security threats are yet to be understood fully.

- The WMNs are non-power constrained when compared to MANETs and traditional wireless network and hence they have distinctly different motivations on exhibiting selfishness. A MR is motivated to act selfishly to maximize its throughput in the bandwidth-constrained links.

- Unlike the peer-to-peer traffic in an ad hoc network, the traffic in a WMN is predominantly oriented towards/from the IGW. The routers near the IGW are potential hotspots for congestion and can fall easy prey to denial-of-service (DoS) attack by malicious attackers.
- Unlike ad hoc networks, WMNs are static and are not power constrained. The fixed WMN network architecture inherits dependency on the routing paths, as a set of MRs is connected to certain MRs and solely depend on them for forwarding the traffic. Any malicious behavior would result in drastic performance degradation.
- As MRs are predominantly designed with a multi-radio multi-channel architecture, detection mechanisms that rely on promiscuous overhearing are not applicable here.
- In a MANET, it is extremely difficult to establish *a priory* of trust relationship between two nodes in a distributed manner. Therefore a MANET cannot be easily monitored by using traditional centralized authentication approaches. On the contrary, in a WMN, the internet infrastructure (e.g. AAA server) may serve as an authentication authority to authenticate some MRs which can be treated as trustworthy nodes.

In this chapter, we discuss the potential security threats in WMNs and analyze various current start-of-the-art solutions to solve them. The reminder of the chapter is organized as follows. Section 23.2 briefly focuses on the need for authentication mechanisms to establish a secure environment. But, authentication alone is an insufficient defense against attacks, like selfish node attack, Denial of Service (DoS) attack. Section 23.3 describes at length the problem of selfishness. In Section 23.4, we discuss about DoS attacks on the MRs. Finally, concluding remarks are included in Section 23.5.

23.2. Authentication of Mesh Routers

A wide variety of security threats exist in a WMN due to the multi-hop nature of communication and due to lack of any physical protection at MRs. The attacks in a WMN can be classified as impersonation attack, anti-integrity attack, and anti-confidentiality attack.[7]

An adversary can conduct impersonation attack by introducing a rogue MR that sends forged/replayed registration messages to entice meshclients. The mesh clients can accept this false connection and assume that it is connected to the Internet. The rogue MR can also send a fake registration message by masquerading as another node (Man-in-the-Middle attack) so that all the packets are tunneled to it. Thus, it can gain unauthorized access to the network information by a simple MAC address spoofing.

A rogue MR can conduct anti-integrity attack by poisoning the route tables. It can cause other MRs to redirect their traffic towards itself by advertising a higher

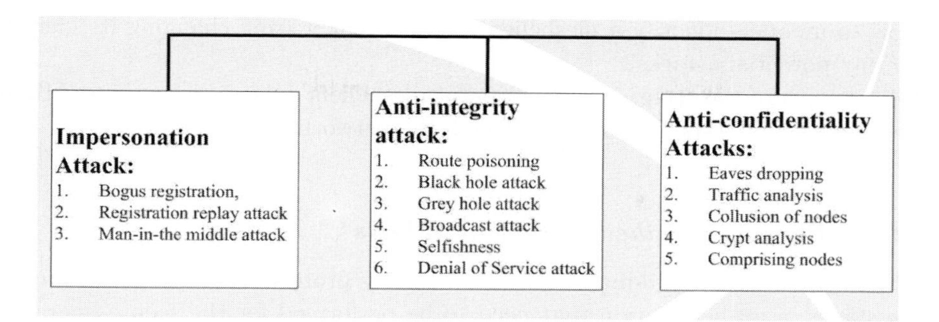

Fig. 23.2. Security attacks in WMNs.

rate link/less congested link to the IGW. It can also exploit the hidden vulnerabilities of the underlying routing protocol to conduct various attacks such as blackhole attack (advertise a shorter hop route with high sequence number so that all MRs route its traffic to the attacker, which drops the traffic fully), grey hole attack (a variant of blackhole attack with partial drop), broadcast attack (generating route request to unknown destination, generating false route error message to invalidate route), selfishness (drop forwarded traffic), Denial of Service attack (DoS) (generating flood of packets).

An attacker can conduct anti-confidentiality attack and reveal critical information (like keys) by eavesdropping, brute-force attack, or cryptanalysis. It can also collude with multiple nodes in the network and collectively compromise the secure communication. Figure 23.2 gives the summary of varies categories of attacks on WMNs.

23.2.1. *Security design challenges*

It is clear from the previous discussion that a security mechanism is very essential to ensure the safe operation of WMNs. The fundamental security primitives of authentication, integrity and confidentiality need to be implemented in WMNs.[1] Authentication refers to the verification of identities of the communicating entities, integrity refers to the validity of the original message so that it is not tampered by an adversary, and confidentiality refers to the establishment of a secure channel to transmit encrypted text so that it appears as a garble for an eavesdropper. This raises several issues such as:

- How to build a secure discovery protocol for a meshclient to establish a multi-hop communication with the IGW?
- How to enforce an authentication process for the mutual validation between a meshclient and the MR to which it wants to connect?
- How to establish trust relationships amongst multi-hop MRs?

- How to uniquely identify a meshclient and prevent it from changing its identity for any potential attacks?
- How to ensure the security association of meshclient is passed from one MR to another as the meshclient moves through the network?

23.2.2. *Phases of authentication in WMNs*

Authentication is the building block of a security protocol for WMNs. There are three stages of authentication that need to be performed for the secure operation of WMNs. The IGW is assumed to a trustable node in the network that lies under the jurisdiction of the ISP. Hence, the authentication mechanism requires a way to mutually authenticate the chain of communicating MRs and a way to mutually authenticate the meshclient establishing connection with the MR. We describe in the following the operation of each stage.

As the meshclients obtain Internet access through the MRs, they need to perform mutual authentication. It is important that a good meshclient does not join a rogue MR. And, it is equally important that a MR prevents a malicious meshclient from freely enjoying the network resources. In a WMN, the internet infrastructure (e.g. AAA server) may serve as an authentication authority to authenticate MRs. A simple authentication structure can employ certificates based on public/private keys. Hence, before a meshclient registers with a MR, it submits a request to the AAA server a registration request which contains the digital signature of the MR encrypted with the MR's private key. The registration request is encrypted by the secret key between meshclient and AAA server. The AAA server verifies the signature of the MR (MRs pre-register with the AAA server before being deployed and has a security association based on public/private key). It composes a registration message to the MR encrypted with the private key of the AAA server.[7] The MR decrypts the registration reply sent by the AAA sever using the public key of the AAA server and accordingly verifies the identity of the meshclient. Thus, the AAA server in the Internet acts as a bridge to authenticate MR and meshclient. If the AAA server trusts MR and meshclient, the MR and meshclient can mutually trust each other.

The second stage of verification involves the mutual authentication of the MRs and the IGW. As they are resource rich devices, they can afford expensive public key primitives. Authentication needs to be performed during the initialization phase as well as during the session establishment phase. As a communication session between a meshclient and an IGW involve a multi-hop path, a packet needs to be authenticated every time it is forwarded by an intermediate MR. This can be performed using pre-defined symmetric keys shared between neighboring MRs or between a MR and an IGW. More efficient techniques can be designed to employ secure polynomial based key shares to authenticate the chain of communicating MRs.

The third stage of an authentication mechanism is to ensure the integrity of messages (control packets and datagram) exchanged. This can be done using hash-based mechanisms.[9] Authenticated control messages prevent bogus registration messages, route poisoning, and node impersonations. It is also important to maintain the integrity of datagram. A malicious attacker can modify, inject or delete partially/fully a datagram. Thus, control messages need to be authenticated for the secure network operation; and datagram need to be authenticated to maintain their integrity. A cryptographic primitive for maintaining the integrity of messages can also be applied to encrypt the messages. Additionally, to handle mobile meshclients, a secure handoff has to be performed from one MR to another. The re-authentication of the meshclient at the new MR has to be performed within a short time to avoid loss of connectivity. Maccari *et al.*[10] present a secure handoff based on IEEE 802.1X security model.

The problem of secure authentication is well known one in Wireless Local Area Networks (WLANs) and Ad hoc networks. The growing interest in WLANs, prompted the IEEE Task Group to define 802.1X framework [IEEE 802.1X] that restricts network access to authorized client by three party authentication system. Thus, a simple way of securing the access to the MRs is by adapting 802.1X to WMNs. Before, we dwell into the specifics of the protocol, we introduce some basic terminology:

(1) Supplicant: The mobile client that wishes to join the network.
(2) Authenticator: The AP/MR that is directly connected to the mobile client seeking services.
(3) Authentication server (AS): The backend central server which acts as AAA (Authentication, Authorization and Accounting) server and maintains all user credentials like secret keys, public key certificates, and passwords.

The 802.1X framework offers port based network access control at Layer 2. When a mobile client attempts to connect to an AP/MR, its access is initially restricted to the unauthorized port only through which it sends authentication request messages. The authenticator acts as a proxy between the supplicant and the authentication server, and relays the request to the AS. The AS and the client mutual authenticate each other and generate a secret Pairwise Master Key (PMK). The IEEE 802.1X framework employs Extensible Authentication Protocol (EAP)[12] to execute the authentication using various cryptographic suites such as certificate based (EAP-TLS: EAP Transport Layer Security,[13] secret key based (EAP-TTLS: EAP Tunneled Transport Layer Security,[14] one-time passwords (EAP-MS-CHAP: EAP Microsoft Challenge-Handshake Authentication Protocol), smart-cards, etc. EAP offers great amount of flexibility by running any authentication method on top of 802.1X framework. But, so far only the client has been authenticated. The AS authenticates the AP using remote authentication dial in user service (RADIUS) and establishes an association security association between the two. The RADIUS[15] unlike EAP provides per-packet authentication and integrity. The AS delivers the previously generated PMK to the AP. Thus, the mobile client and the

AP mutually authenticate and derive a transitive trust through a common trusted friend which is the AS.[16] The mobile client trust AS and AS intum trusts AP. IEEE 802.11i standard used in conjunction with 802.1X further defines how PMK generated through EAP can be used to derive session keys for the communication between mobile client and the AP.[17]

23.2.3. *Design goals of authentication*

The design goals of an authentication mechanism for WMNs can be summarized as follows. Though an authentication authority (e.g. AAA server) can be conveniently deployed on IGW, the scalability of the approach should be considered so that the authentication overhead is not excessively high. As meshclients are battery operated devices, public key cryptography should be avoided. Maintaining a PKI for a large community of users is an expensive overhead.[18] The initialization phase of authenticating nodes should be followed by timely re-initialization so that a compromised host does not pave way for more adversaries. As meshclients can connect to MR in a single/multi-hop fashion, it is important to design a network layer authentication mechanism.

An authentication mechanism can secure the operation of a network to a certain degree only. Employing even a sophisticated authentication mechanism is an insufficient defense to block intruders in a public network like WMN. As MRs are typically deployed in public places like rooftops, buildings, poles, towers, traffic signals, even an authenticated node can be easily tampered by an adversary.

There are two kinds of security attacks on a WMN: active attacks and passive attacks. An active attack is conducted to intentionally disrupt the network activity (Flooding: Denial of Service attack, Suction of packet by luring nodes: Blackhole attack, Impersonation: Man-in-the-middle attack). A passive attack like a selfish node attack, eaves-dropping is conducted with a selfish motivation and does not cause any intentional damage to the network. In Section 23.3, we focus on the security threat posed by a selfish MR. And in Section 23.4; we discuss a more virulent attack called the DoS attack.

23.3. Selfishness of Mesh Routers

In this section, we discuss the problem of selfishness that silently disrupts network activity. A thin line of distinction exists between a selfish and a malicious MR. A malicious MR intentionally disrupts the network activity; while a selfish MR just greedily devours the network resources for itself and in the process affects the normal network operation. Selfishness is a common phenomenon rampant in any wireless network that relies upon a cooperative framework to operate. As a MR is the aggregation point of traffic originating from its meshclients, a group of selfish users might be motivated to introduce selfish MRs in the network to obtain maximum throughput for their applications. A selfish MR always favors the traffic originating from its meshclients; while fully or partly dropping the relayed traffic. We also call such a node "*free-rider*", as it freely enjoys network resources without providing any forwarding service to the community.

23.3.1. *Motivation for selfishness*

It should be, however noted that MRs exhibit distinct motivation for behaving selfishly when compared to the ad hoc nodes. An ad hoc node behaves selfishly to conserve its energy, but MRs being power rich nodes behaves selfishly for distinct motivation:

(1) Maximize Internet Throughput: The Internet capacity of a WMN is limited by the capacity of IGW since all the Internet traffic from/to each MR passes through it. The IGW forwards all the traffic aggregated by each MR to the Internet or routes the traffic from the Internet to the destined MRs. Thus, as the network grows the per-user Internet throughput decreases correspondingly. A selfish MR, is motivated to drop internet-oriented packets originating from other MRs so that packets from its MS get better opportunity to be forwarded at the IGW.
(2) Maximize Wireless Throughput: Though a WMN node is capable of employing multiple interfaces/radios and multiple channels to communicate, it is limited by the wireless link capacity which is still a limited resource. It has been shown that the network capacity suffers from losses by the factor of $\sqrt{n/m}$ (where n is the number of MRs and m is the number of IGWs) as the number of MR increases. Consequently, a MR behaves selfishly by dropping packets originating from other MRs so that packets originating from its meshclients have better chances of being forwarded at the upstream MRs near the IGW.
(3) Avoid Path Congestion: The paths leading to the IGW are hotspots for congestion due to the excessive traffic. Hence, a MR might act selfishly in an attempt to avoid congestion along the paths leading to the IGW.

23.3.2. *Negative impacts of selfishness*

Thus, selfishness in WMNs can lead to severe degradation of network performance. Due to the hierarchically structured WMN, closer the proximity of the selfish node higher is its level of impact.[19] If these MRs behave selfishly, it results in denial of service to flows originating from all the decadent MRs connected to them. In a sparse network, a selfish MR in an intermediate path can also result in network partition. As the mesh backbone is relatively static, the affected MRs are permanently disconnected from the network. Thus, the static topology of the MR, results in increased performance degradation until the selfish is detected and quarantined from the network. Even if a selfish MR is detected, the affected MRs would have to take remedial action to route their traffic through other alternate paths which could be circuitous leading to intolerable delays for real-time applications.

23.3.3. *Selfishness at various layers*

In order to achieve its goal of favoring its own traffic, a selfish MR implements its selfish behavior at various layers of the network as described below:

- Network Layer selfishness — Layer 3 selfishness
  A selfish MR fully or partially drops the relay packets oriented towards the IGW, in order to increase its own throughput. At a given instant at ith MR, say, the traffic demand for its own traffic is α_i, the traffic demand for the relay traffic is β_i, and the total available bandwidth is η_i. A selfish MR dynamically drops the relay packets, depending on its available bandwidth: $\eta_i = \alpha_i - \beta_i$. Whenever its residual bandwidth falls below a certain threshold, it drops relay packets in order to maximize its own throughput.

- Medium Access Layer (MAC) selfishness — Layer 2 selfishness
  This selfishness is exhibited at the MAC layer of the protocol stack, where in a selfish MR declares a continuous busy state in order to evade traffic request for sending relay traffic. This is called "busy radio" selfisness. For example, a selfish MR, can falsely claim that its receiver radios are busy to avoid packet forwarding duties. Thus, the neighbors of a selfish MR are forced to look for other alternatives to send their traffic or are forced to wait until the selfish MR has finished sending its own traffic. This results in denial of service for other MRs and the selfish MRs continues to obtain network services for sending its own traffic. In a multi-radio WMN architecture, two MRs dynamically negotiate the channel for communication. In such a scenario, a selfish MR might refuse to confer on a common free channel, thereby stalling the network traffic. This is called channel-riding. The selfish MR can also exploit the vulnerabilities in the IEEE 802.11 backoff mechanism (small value for itself), and NAV (Network Allocation Vector: assign large idle time for its neighbors). Each MR contends for the channel to send traffic. If two MRs concurrently contend for the channel, they adopt an exponential backoff mechanism to avoid collisiosn. A selfish MR chooses a small backoff time interval for sending its own traffic and a larger backoff time interval for the relaying traffic. This ensures that it obtains greater chance for sending its traffic in the network.

- Link Layer selfishness — Layer 2 selfishness
  Every MR has an Interface Queue (IFQ) to cache the network traffic traversing through it. However, this cache is of limited space and packets are dropped whenever the cache exceeds its capacity. A selfish MR can manipulate its IFQ such that it prioritizes the aggregate traffic originating from its meshclients, i.e. packets of its own traffic always find space in the IFQ.

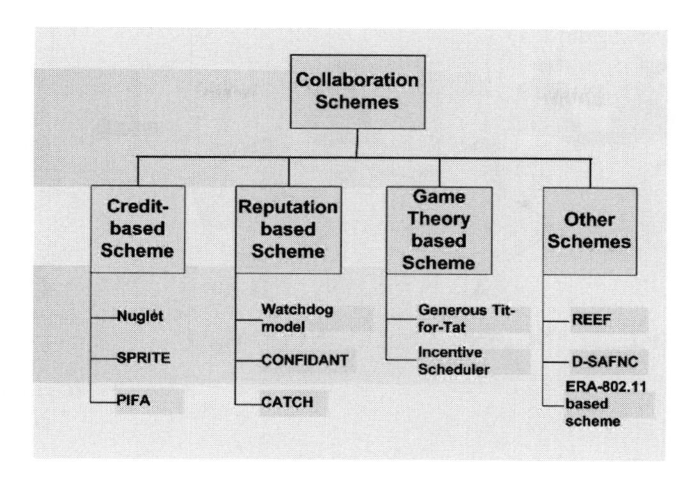

Fig. 23.3. Classification of collaboration schemes.

23.3.4. *Collaboration schemes*

Authentication is not a sufficient condition to prevent selfishness in WMNs. As there is no authority to monitor the bandwidth allocation at MR; each MR independently decides the distribution of the bandwidth for traffic originating from itself and the bandwidth for relay traffic originating from other MRs. Thus, we adopt schemes to enforce collaboration among MRs such that they abide by the ideology of "using" and "providing" network service and ensure that selfish MRs are punished severely.

Enforcing collaboration between nodes can be mainly achieved by four different approaches in ad hoc networks: credit based, reputation based, game-theory based, and others. Figure 23.3 shows the classification of various schemes in the literature.

23.3.4.1. *Credit based scheme*

In a credit based scheme, nodes are encouraged to forward packets in order to earn virtual currency. All nodes that forward other's traffic are rewarded for their services which act as an incentive to cooperate with each other. A source node using the forwarding service is charged and a rewarding authority allocates credits all intermediate nodes providing forwarding service. This indirectly imposes a limitation on a selfish node which cannot discretely drop other's traffic at the cost of sending its own traffic. A selfish node that refuses to cooperates would eventually run out of currency to pay other nodes and is thus forced to behave correctly to avoid being poor. The credit-based schemes primarily differ in their methodology adopted for the allocation of wealth.

In credit-based approaches, (like Nuglets,[20] SPRITE,[21] and PIFA,[22] each node earns virtual currency by forwarding other's packet as shown in Figure 23.4. Buttyan *et al.*[20] propose two models to reward forwarding nodes, PPM (Packet Purse Model)

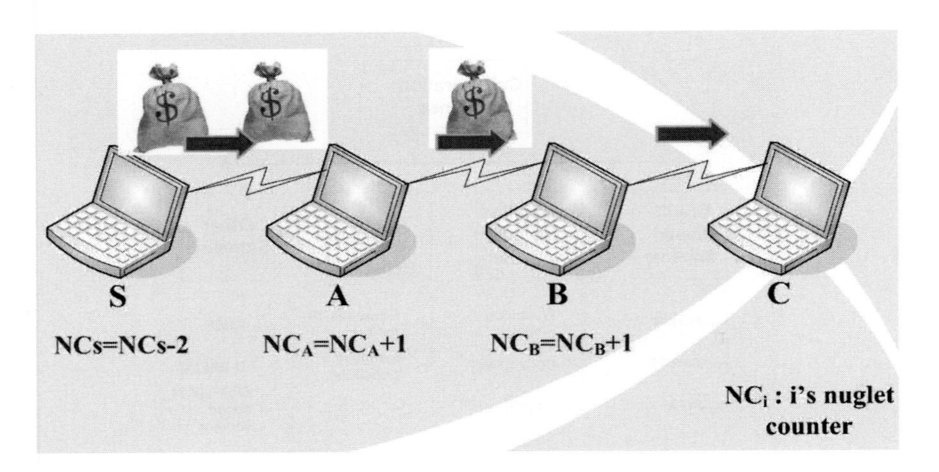

Fig. 23.4. Credit based scheme.

and PTM (Packet Trade Model). In PPM, a source node loads a packet with virtual currency called *Nuglets* to rewards all nodes that forwarded its packets. It loads as much currency as required to pay all intermediate nodes in the data forwarding path. Thus, *nuglets* are deducted at every hop by the intermediate nodes. This also restricts the applicability of the scheme to source routing protocol only. On the other hand, in PTM, each intermediate node conducts a local trading of *nuglets* with its neighbors. It buys packets from its previous hop node for the quoted price and sells it at a higher price to its next hop. PTM, however, is prone to Denial-of-Service (DoS) attack by a malicious source. A malicious source can send a flood of packet to exhaust the *nuglets* at all nodes. PPM and PTM, require tamper-resistant hardware to maintain the authenticity of currency, which is difficult to achieve in a public network.

SPRITE[21] ensures the authenticity of currency by employing a centralized authority called CCS (Credit Clearance Service). The CCS functions as centralized authority for allocation and distribution of wealth among nodes. Every node submits a receipt to the CCS on forwarding a packet to its neighbors. The CCS then determines the credit for every node in the transmission path from a game-theoretic perspective so that all nodes are motivated to report correctly, even when selfish nodes collude and submit false receipts. However, SPRITE suffers from scalability problem in a large-scale network, as nodes submit receipts to CCS for every packet. This also incurs considerable communication overhead. PIFA[22] works on the same principle as SPRITE and annuls the overhead by submitting periodic reports to a centralized credit manager (CM). The reports are aggregated information on a node's transaction with its neighbors like packets received, packets forwarded, packet originated. The CM applies necessary rules to validate the credibility of the reports (by comparing reports of adjacent nodes), and rewards the well-behaving nodes. This centralized scheme also suffers from the scalability problem. The security of this scheme is weak and is based on the assumption that the CCS is secure and safe from attack.

23.3.4.1.1. Applicability of credit-based scheme to WMNs

The static and hierarchical architecture of a WMN makes credit-based schemes largely unfit for it. The WMNs are relatively static due to which nodes in periphery may never earn any credit. This creates disparity in wealth among the leaf MRs and the central MRs. An egregious node can begin its selfish activity after accumulating enough credits, going against our goal of stimulating cooperation. A source node in a WMN might quickly run out of wealth due to the high-data rate of some applications such as multi-media.

23.3.4.2. *Reputation based scheme*

The goal of a reputation based scheme is to study the interactions of nodes by observing the traffic movements and build a database of trust index. Each node observes the behavior of other node's in its neighborhood by checking if they dutifully forward packets. It then derives a reputation index for each of its neighbors. This in turn reflects on whether a node cooperates or behaves selfishly. Figure 23.5 shows the central theme of reputation based scheme that builds trust relationships between users based on its prior interactions.

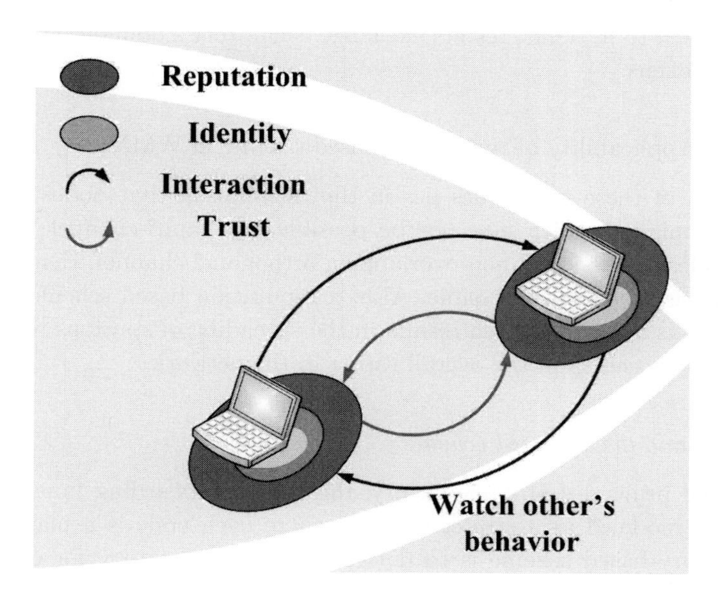

Fig. 23.5. Reputation based scheme.

In reputation-based approaches like Watch-dog and Path-rater model,[23] each node promiscuously eavesdrops on its neighboring nodes so as to detect any non-forwarding misbehavior. Accordingly, the source node selects the best path by avoiding selfish nodes. In Watch-dog model, a source node uses this observation and then selects a route with the highest rating. The scheme is prone to a replay attack since a node can pretend to forward by replaying an old cached packet. In CONFIDANT protocol,[24] each node employs a Watchdog technique to collect first-hand information on its neighbors and then compares this with the second-hand information sent by other nodes. The second-rating information prevents spurious rating and is guaranteed to detect inconsistencies in the two observations. However, the process of building reputation in both schemes can be distorted due to collision. They also face scalability problem due to centralized validation by the trust manager.

CATCH[25] combines Watchdog and anonymity approach. It consists of two sub-protocols called Anonymous Challenge Message (ACM) and Anonymous Neighbor Verification (ANV). A node whose connectivity is tested in called *testee.* Other nodes which participate in determining the trustworthiness of the *testee* are called *testers.* A tester broadcasts an ACM message at unpredictable intervals to all its neighbors, who upon receiving it are expected to re-broadcast it. As the identity of the sender is unknown, the *testee* is forced to re-broadcast every ACM message. The tester in the meantime keeps a watch on the *testee* to find if the *testee* had re-broadcasted the challenge message. If re-broadcasted, the *tester* considers *testee* to be normal. Next, in the ANV phase, a *tester* sends a cryptographic hash of a random token for re-broadcast. At the same time it also records hashes re-broadcasted by other nodes in the network. If a *tester* considers the *testee* to be normal, it releases the secret token to it. Thus, testers exchange each other's opinion by the release of their secret tokens.

23.3.4.2.1. Applicability of reputation based scheme in WMNs

The strength of these techniques lies in the assumption that nodes can overhear other's transmission which may not be possible in a multi-channel WMN, where adjacent nodes are assigned non-overlapping orthogonal channel. Hence, promiscuous listening is not always possible. Also, a reputation based scheme suffers from security attacks like lying which results in the spreading of spurious reputations. A set of bad nodes can spoil the overall rating in the network.

23.3.4.3. *Game theory based scheme*

Applying the principles of game theory, the packet forwarding functionality of a node can be modeled as a *strategic game*, where each node is a player. The goal of game theory-based scheme is to derive an optimal strategy for every rational node. It derives a *Nash Equilibrium* point for the forwarding rate of a node such that any deviation from the equilibrium results in lesser pay-off for the participant node. Assuming all players are rational; a selfish node cannot increase its pay-off by adopting a different strategy and is forced to cooperatively forward other's packets.

For the clear understanding of the principle of game theory, we describe the famous problem of prisoner's dilemma. Two prisoners, Dave and Harry, are interrogated by the police in two separate cell about a crime they could have possibly committed. The length of imprisonment depends on the following conditions:

(1) If both confess, Dave and Harry both will be sentenced for four years.
(2) If neither confesses, both will be sentenced for two years.
(3) If one of them confesses but the other doesn't, the confessor will go free while the other one goes to jail for five years.

Hence, the two prisoners are caught in dilemma to guess on the other's strategy and make the right move in revealing their crime. Hence, each prisoner adopts a strategy that yields maximum profit, which is to spend minimal years at jail.

Game theory approaches fix the forwarding rate of a node at certain Nash equilibrium for the network as in Generous Tit-for-Tat (GTFT).[26] GTFT equalizes the percentage of request served by a node j for others nodes (*Help-given*) against the percentage of request served by others for the node j (*Help-received*). The node then applies the following Equation (23.1) to determine whether or not to relay packet:

$$\text{Help-received} + \Delta > \text{Help-given}, \tag{23.1}$$

(where Δ is a small positive number to prevent initial system deadlock). If the above equation is true, a relayed packet is accepted and forwarded by this node. They prove theoretically that any deviation from this equilibrium results in loss for selfish nodes, but this is realistically infeasible to achieve. It also requires the knowledge on the utility of all nodes which is cumbersome and not always available.

Wei *et al.*[27] present an incentive scheduler that ensures a fair channel allocation for users in a WWAN/WLAN two-hop-relay system using game theory. Every user is assigned an incentive variable by the AP such that it is set to a constant value for a relay node and non-zero for others. The AP (or a base station) acts a central authority to fairly assign channel to users depending on the aggregate weightage of various factors such as instantaneous channel rate, average allocated rate, and the incentive variable. Hence, a user is motivated to cooperatively forward packets in the network to accrue incentive points.

23.3.4.3.1. Applicability of reputation based scheme in WMNs

The game theory approaches are based on several assumptions which are not necessarily true for WMNs. For example, a game theory based scheme like GTFT assumes all nodes are subject to energy constraint (i.e. limited node lifetime). But, MRs are power rich devices and hence such assumptions are invalid in WMNs. The incentive based scheduler assumes that the AP controls the radio resources for all the network entities in a WWAN/WLAN two-hop-relay system, which is not the case in WMN.

23.3.4.4. *Other schemes*

Conti *et al.*[31] propose REliable and Efficient Forwarding (REEF) scheme, in which a node uses the arrival of TCP acknowledgements to estimate the reliability of a path. Unlike conventional method of denying services to selfish users, it provides a differentiated quality of service by forwarding their traffic slowly. Though, REEF suggests building a cumulative acknowledgement mechanism for UDP kind of traffic, it is very cumbersome to build such a scheme. Hence, it is largely applicable to TCP traffic only. Santhanam *et al.*[29] propose a distributed detection scheme to identify selfish MRs. They employ specialized Sink Agents (SA), which are delegated the task of vigilantly policing the network to detect selfish MRs based on simple forwarding rules. These agents can be configured to operate in a static or mobile mode. The SA collects reports from MRs referred as member reports. A member report facilitates comprehensive representation of the statistical information on a MR's interaction with its neighbors. The scheme is independent of underlying architecture. Wang *et al.*[30] define a finite state machine model of AODV protocol and apply a clustering algorithm to partition the nodes as selfish or cooperative. They propose to study the problem of selfishness in WMN by observing the flow of protocol messages. They define a finite state machine model of AODV protocol action in order to gather statistics on a MR's interaction with its neighbors. It then proposes a clustering algorithm to partition the MRs into selfish and cooperative classes. However, a finite state machine modeling is dependent on the underlying routing protocol.

23.3.5. *Design goals of collaboration schemes*

Thus, selfishness is a rampant in any wireless network involving collaborative forwarding, and more so in a WMN. A scheme to stimulate cooperation in a WMN should be carefully redesigned such that it takes into consideration the unique characteristics of WMNs, like traffic pattern, multi-channel environment, relatively static topology, etc. For example, reputation scheme based on overhearing should be avoided. Instead, other indirect approaches should be employed to estimate the reliability of a node, for example, querying the two neighbor of a neighbor about a forwarded packet or by monitoring member reports. The approaches based on game theory should be carefully designed taking into consideration the traffic characteristics and the abundant power available to MRs. The collaboration schemes should perform distributed approach to reduce the latency in attack detection. Quicker an attack is detected, quicker is the implementation of remedial actions. Finally, a collaboration scheme should scale well and should be able to accommodate a large number of MRs.

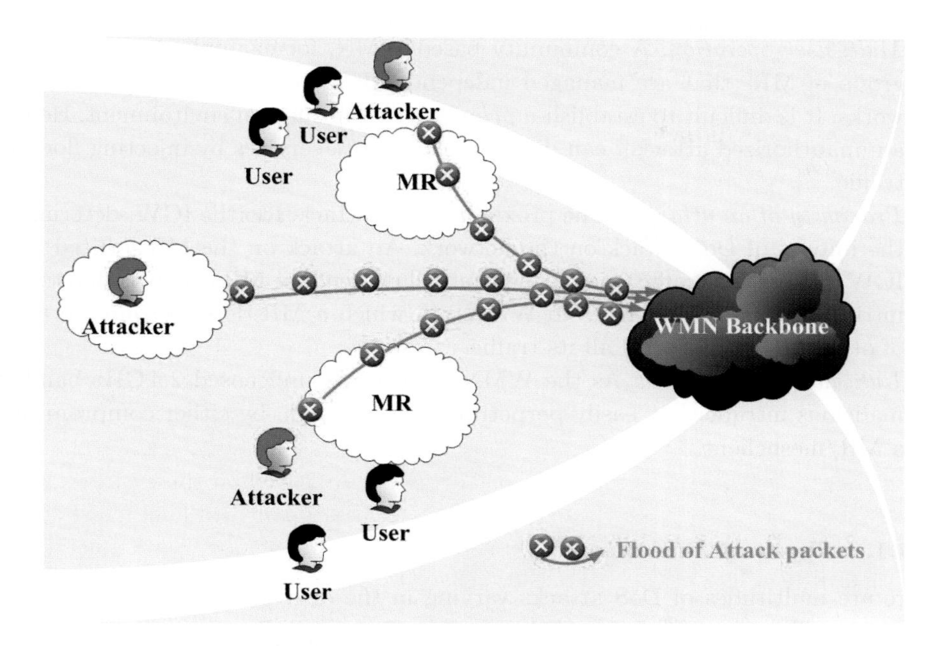

Fig. 23.6. Denial of service attack on WMN.

23.4. Detection and Prevention of Denial of Service Attack

A *Denial of Service* (DoS) attack is a deliberate attempt by a malicious intruder to render the system inaccessible to the legitimate users by incessantly pumping excessive traffic into the network. A DoS attack consumes the precious network bandwidth, blocks the buffers at the intermediate MRs, incurs excessive delays in the network, leads to the starvation of other innocent flows, and results in the wastage of the precious network resources. Figure 23.6 shows a distributed DoS attack scenario in WMNs.

A distributed DoS employs several attackers, that in unison target a single MR near the IGW and synchronously send a flood of traffic towards it. This results in monotonic increase in the congestion levels and leads to eventual collapse of the network. A flooding style DoS attack poses serious threat to WMNs as it can be easily launched by an attacker by generating spoofed source traffic. The victim is helpless in stopping the traffic unless action is taken at each and every intermediate route to throttle the attack flow.

The inherent characteristics of WMNs make them an easy prey for DoS attacks.

(1) *Traffic flow characteristics*: As the bulk of the Internet traffic in a WMN is predominantly directed towards the IGW, the paths leading to the IGW are heavily congested. Hence, an attacker can perpetrate a DoS attack easily.

(2) *Multi-user operation*: A community based WMN, for example, is formed by a group of MRs that are managed independently by different operators or networks. It is difficult to establish *a priori* of trust in such an environment. Hence, an unauthorized attacker can disrupt the network services by injecting flood of traffic.
(3) *Proximity of an attacker*: The proximity of an attacker to the IGW, determines the impact of DoS attack on the network. An attack on the MR located near IGW, affects the traffic originating from all descendant MRs. This is due to the intrinsic route dependencies in WMNs, in which a MR depends only on a set of other MRs to forward all its traffic.
(4) *Unlicensed traffic band*: As the WMNs run in the unlicensed 2.4 GHz band, a malicious intruder can easily perpetrate a DoS attack, by either compromising a MR/meshclient.

23.4.1. *Types of DoS attack*

There are multitudes of DoS attacks varying in the degree of their sophistication and the level of impact.[29] Some of the common form of DoS attacks discussed here are SYN attack, Teardrop attack, and Smurf attack.

The SYN attack misuses the TCP's 3-way handshake mechanism as shown in Figure 23.7. The TCP session is established by first sending a SYN message to the

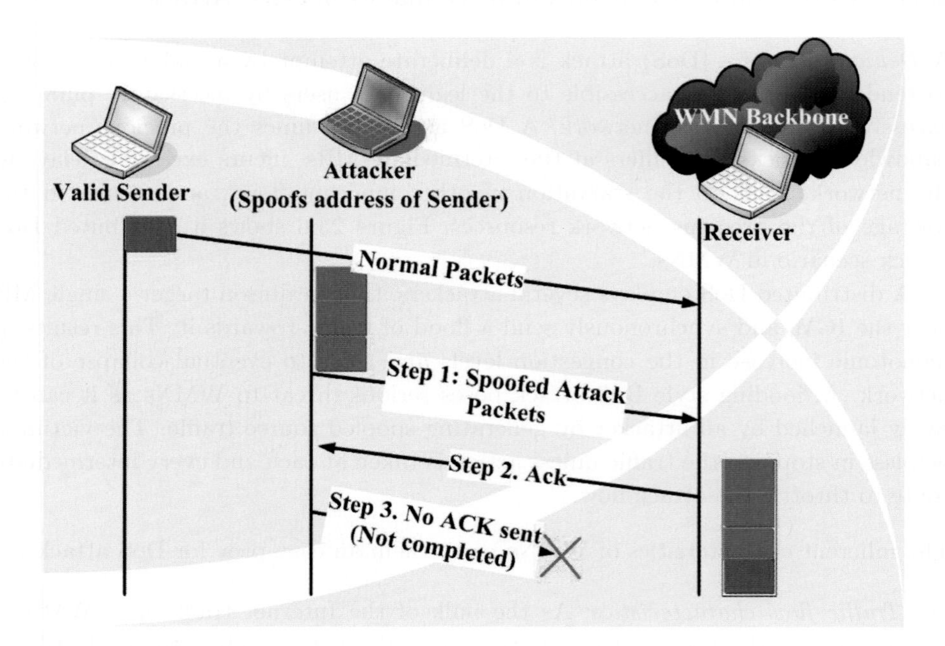

Fig. 23.7. SYN attack.

server, which then sends an acknowledgement (ACK) and as a final step the client that initiated this connection completes the handshake by sending an ACK. A malicious attacker opens multiple bogus connections with the server and refuses to send TCP ACK. This results in exhausting the resources of the server. In Teardrop attack, the attacker exploits the fragmentation process at a router. A router fragments a packet, if it cannot handle a large packet. An attacker disrupts the reassembly process by inserting an incorrect offset in the subsequent fragment. This results in DoS for these flows because the receiver assumes the incorrectly assembled packet as packet error and discards it.

Smurf is a more sophisticated kind of attack, in which an attacker sends a large number of echo request packets to a group of uncompromised nodes in the network as shown in Figure 23.8. The source address of these echo requests is forged to be that of an innocent node in the network. The uncompromised nodes unknowingly flood the innocent node when they generate echo reply packet in response to the spoofed echo request packet.

A DoS attack can be conducted at different levels of the protocol stack as discussed below: at the physical layer, an attacker can jam the wireless channel resulting in denial of service. It can attack the MAC protocol by sending fake de-authentication messages from the compromised MR. This results in service disconnection for the meshclient.

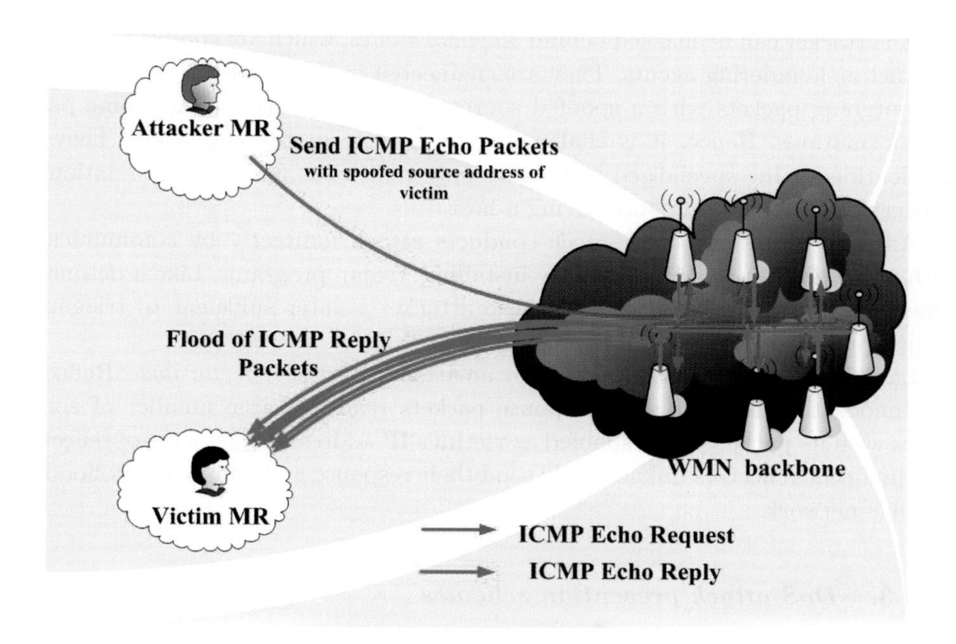

Fig. 23.8. Smurf attack.

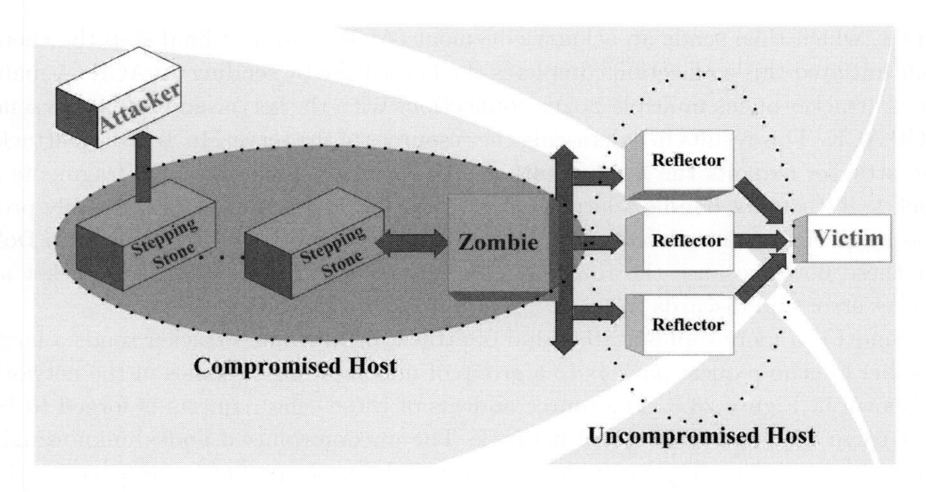

Fig. 23.9. Generalized attack model.

23.4.2. *Generalized DoS attack model*

An attacker perpetrating DoS attack on a distant victim could be buried behind several other entities. Figure 23.9 shows the generalized attack model. An attacker can masquerade as a stepping stone or a zombie or as a reflector to discretely conduct attack and avoid detection.

An attacker can be masked behind *stepping stones,* which are compromised hosts that act as laundering agents. They are engineered in such a way that it overwrites the outgoing packets with a spoofed source address and also applies some packet transformations. Hence, it is challenging to identify a stepping stone. They can be identified using specialized techniques only that looks for causality relationship between packets entering and leaving a host.[28]

An attacker using *zombie node* conducts attack indirectly by communicating with a stepping stones or by directly installing Trojan programs. Like a detonating time bomb, a single command from the attacker is later sufficient to trigger the Trojan program.

Last, in the chain of disguise used by an attacker, is the *reflector node.* Reflectors are innocent nodes that send response packets readily. Large number of zombie nodes with its packet source spoofed as victim's IP address, target a set of reflectors. The innocent reflectors unknowingly send their response and cumulatively flood the victim's network.

23.4.3. *DoS attack prevention schemes*

To combat DoS attacks, there are several schemes in the literature. The baseline of many DoS defense mechanisms includes Traceback schemes, Filter based schemes, and Rate limiting schemes. The defense mechanisms collaboratively identify any deviant behavior in the traffic measurements and take immediate remedial steps to curb the attack. While the first two schemes defend the victim of an attack; a rate based scheme defends the network under DoS attack.

23.4.3.1. *Traceback schemes*

Traceback is the process of reconstructing the path of an attack leading to the true attacker. It works on the theory that the best possible defense against DoS attack lies not only in taking preventive measures *but also in identifying the true origin of the attacker and in blocking further occurrences of such incidents.*[29] It is however, a tough task because of the rampant spoofing of source address. As traceback reveals the exact location of the attacker, an attacker thinks twice before conducting a DoS attack. It also helps in implementing filtering rules closer to the origin of attack.

Figure 23.10 shows the operation of traceback scheme for tracing the source of DoS attack on a MR (MR4) near the IGW. An attacker under MR1 spoofs the source address of an innocent node (MR9) and overwrites this address on attack packets.

There are innumerous ways of conducting traceback. A simple traceback that examines the source IP address of each packet is useless due to the rampant spoofing of source IP address and thus should be designed on different guidelines. In order to facilitate traceback, information on the flow of the packet has to be recorded.

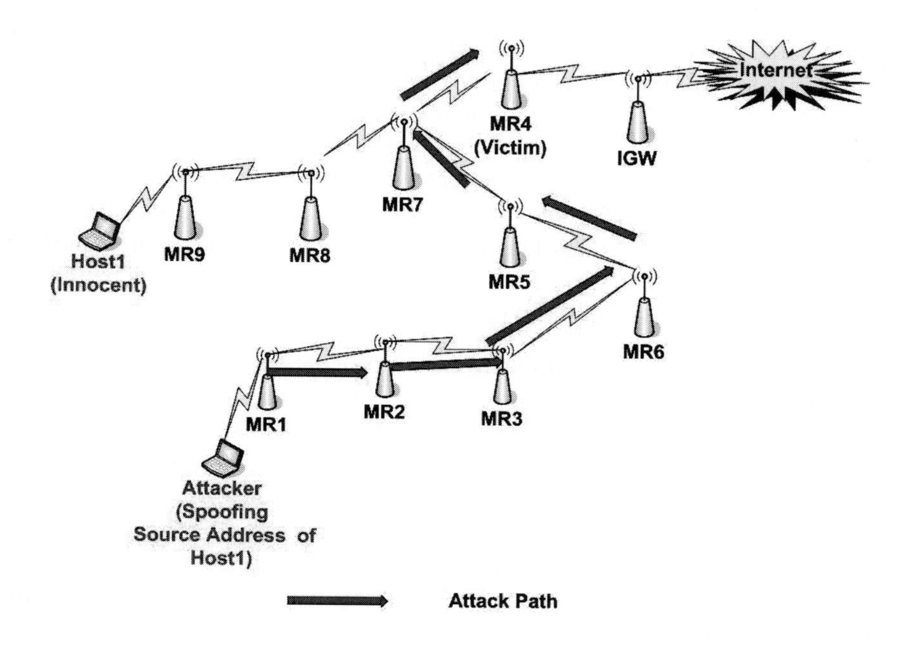

Fig. 23.10. Traceback scheme.

The resources required for storing this trace data can be distributed across various network components which are as listed as follows:

23.4.3.1.1. Inside the packet

Each router stamps its IP address probabilistically or deterministically on some unused fields like ID/Option field of IP header of the packet. As the trace data is carried in-band within the packet, there is limited space available for storing the path of the packet. In the pioneering work of Savage *et al.*,[32] each router probabilistically marks a packet that passes through it with its IP address.

If a router chooses to mark the packet,
It inscribes its IP address in the ***Start Field*** and sets the ***distance field*** (dist) to 0.

Else if a router doesn't chose to mark the packet,
It check if the packet has been already marked,
If Yes, it dumps its IP address in the ***End field*** and increment the value of distance bit.
If No, Just increment the distance field.

On the detection of the DoS attack by an Intrusion Detection System (IDS), the victim node reconstructs an attack graph back using the edge sampled in the packets. It pairs the received packets in the increasing order of the distance element. First, packet with dist = 1 and dist = 0 are paired together. If the value of start filed of dist = 0 matches with the end field of dist = 1 packet, it represents an edge. This is iteratively repeated at each hop by pairing sequential dist packet. However, the complexity of path reconstruction increases exponentially with the increase in the attack path length and accuracy of the scheme depends upon the marking probability. Other schemes mark the packet deterministically.[33] Song *et al.*[34] enhance the basic marking scheme of Savage *et al.*[32] by adding an authentication mechanism to validate the router's marking. This in effect prevents spurious router markings.

23.4.3.1.2. On the forwarding routers

The packet flow is recorded at various routers in the network and the attack path is extrapolated by examining the router logs. But this could consume colossal amount of space on the routers, considering the speed of today's links.

Snoeren *et al.*[35] propose a hash based IP traceback scheme that uses a space efficient data structure called bloom filters to deterministically log critical packet information on the routers. It is called Hash Based scheme as a hash of the invariants fields in the IP header is stored in each router as a 32-bit digest. A trace back manager, on obtaining an alert from IDS, dispatches the signature to all routers which in turn analyze their and report the sub-path of the attack. The scheme is very powerful in tracing even fragmented packets. However, due to the deterministic nature of recording, packet digests are stored only for a limited period of time and hence it can only trace recently delivered packets.

23.4.3.1.3. Emitting a separate trace packet

Bellovin *et al.*[36] propose a scheme in which each intermediate router emits a separate trace packet for every 1 in 20,000 packets that it handles. As most of the DoS attacks are flooding style attacks, this marking probability is sufficient to ensure that victim receives a considerable amount of trace packets. The trace packets like ICMP trace packets collects information of all the routers on its way to its destination. It carries in its payload apart from data, some useful information on the routers such as generating router's id, timestamp of its marking, forward link element along which packet traverses, MAC address pair of the link traversed, etc. However, the traceback incurs additional overhead in terms of the network bandwidth.

A traceback scheme can also be classified as reactive or pro-active depending on when traceback operation was started. In a reactive scheme, traceback is executed in response to an ongoing attack like a stimuli-response mechanism. An IDS raises alert on the occurrence of attack and traceback process is started immediately from the victim, going hop-by-hop backward until the source of the attack. A proactive approach takes a different orientation in pinpointing the source by proactively recording and logging the traffic packets as they flow through the network. These records are useful indicators for the victim in path reconstruction to the actual source and provide timely response on the occurrence of an attack. A complete taxonomy of various traceback approaches in given by Santhanam *et al.*[29]

A traceback scheme for WMN should be designed with following points in mind. The traceback scheme should be scalable and should be feasible to realize in terms of memory and bandwidth requirements. The traceback scheme should be able to handle mobility of meshclient as it moves from one MR to another. Apart from these requirements, in general, any traceback scheme should be able to function even in the presence of source address and MAC address spoofing. It should be remembered that a traceback is only useful after the occurrence of the DoS attack, it is not designed with a goal to limit the infiltration of DoS attack. In the next subsection, we discuss other approaches that limit the perpetuation of a DoS attack.

23.4.3.2. *Packet filtering schemes*

Distributed packet filtering schemes like Ingress filtering[38] can be employed at the edge routers to prevent the generation of spoofed IP address. While the scheme does nothing to prevent the origination of DoS attack from valid IP addresses in the network, it prohibits an attacker from using forged source address that do not conform with the predefined filtering rules. Hence, a traceback scheme can easily trace the attacker to a valid, legitimately reachable source address. However, it is not pragmatic to apply filtering pervasively at all MRs in a large community based network, as it would require considerable degree of cooperation among the independently operated MRs.

23.4.3.3. *Rate limiting schemes*

A rate limiting scheme monitors the network for anomalous high-bandwidth aggressive flows. The main goal of a rate limiting scheme is to penalize aggressive flows and ensure fairness of flows. It penalizes them because, admitting such flows would result in no buffer space for other flows when they eventually arrive. We regulate the traffic rate of high-bandwidth using various heuristics. Rate limiting schemes can be sub-divided as: buffer management schemes and congestion control schemes.

23.4.3.3.1. Buffer management schemes

RED (Random Early Detection[39]) is a buffer management schemes that control bursty flows to ensure fair service for all flows. RED drops packets probabilistically depending on the average queue length at the router. It maintains a moving average of the queue length. The drop probability at a node determines whether a packet should be dropped or admitted in the link layer queue (IFQ).

(1) If the average queue length lies between minimum and maximum threshold, the subsequent packets are dropped with a random probability.
(2) Else, if it exceeds the maximum threshold all subsequent packets are dropped at the IFQ.

However, RED requires extensive tuning of threshold parameters for different traffic profiles. It also suffers from the problem of synchronous backoff in TCP flows when all flows simultaneously reduce their rate. It is quite adaptable to the variable bandwidth of wireless links. Stabilized RED[40] extend the approach of RED for misbehaving sources. Each router maintains a zombie list containing the most recently seen bad flows. It then probabilistically replaces a random packet from the list with a new arriving packet. However, such a naïve control is not sufficient to throttle DoS flows. RED and Stabilized RED require per-flow state information which could be a considerable overhead.

A *most frequently used* cache based mechanism is proposed by Santhanam *et al.*[37] to identify attack flows in mesh networks. Each MR maintains a cache table of fixed size that contains information on all the active flows routed through it. It contains a packet signature (Type of Service — ToS), source input interface identifier, destination address, and IP protocol type) and the frequent count of the packet in the current time window.

When the cache becomes full, the cache entry with the least count is pre-empted to make room for the new flow. In this way, the cache maintains a list of potential high-bandwidth flows at a reduced cost. As attack packets arrive in bulk to flood the victim MR, they are guaranteed to be present in the cache with a large frequency count.

It is possible that the scheme can misclassify bursty flows as an attack flow. In order to avoid such false positives, they compare the frequency count of this flow with the frequency counts of the other flows in the cache. If the frequency count of this flow is greater than half of the maximum frequency count in the cache, the bursty flow is misclassified as an attack flow. But, this labeling is not permanent as in another time frame the bursty flow would behave normally and would no longer be treated as an attack flow.

Once this classification has been done, the throttle controls the attack flows. When a packet belonging to the attack traffic arrives it is dropped even if there is space in IFQ depending on this *drop probability*. Unlike attack flows, an innocent flow has a larger inter-arrival time and hence is not affected greatly by this *drop probability*. As more and more innocent packets are dropped during the attack, the *drop probability* is increased multiplicatively, which in turn acts to control the attack flow. It is however not scalable to handle distributed DoS attack. A Distributed DoS (DDoS) is a virulent attack in which an attack is launched synchronously from several compromised hosts piloted by a remotely hidden attacker. It requires a rigorous control of flows as several small aggregate can converge to flood the victim network.

23.4.3.3.2. Congestion control schemes

The next class of rate limiting scheme is the congestion control mechanism, which are active only when excessive congestion has been detected in a network. Mahajan *et al.*[43] propose an Aggregate Congestion Control (ACC)/pushback mechanism against low bandwidth aggregate attack flows. It is mainly useful for throttling a DDoS attack in which low-bandwidth attack traffic from multiple source clusters together at the victim and overburden the links at the victim. ACC applies a rate limit on a group of similar packets and carries the rate limit backward towards the attacker. The signatures are formed on the basis of source or destination address, an address prefix or a target site or a specific application type like a virulent worm that propagates by emails. But, forming an appropriate congestion signature for the aggregates is very challenging. Each node maintains a drop history over the past K seconds. The ACC agent analyzes these logs and identifies the low bandwidth aggregates responsible for congestion. It then computes a rate limit for each of these aggregate individually such that the drop rate at the output queue is reduced below a certain safe threshold. The limit is adaptively tuned such that the limit for rate-limited aggregates is greater than the arrival rate of the largest non-rate limited aggregate.

Yaar *et al.*[44] propose a scheme in which a sender uses a capability exchange mechanism to acquire a privileged channel from the receiver. The sender initiates a handshake mechanism with the receiver and sends an explorer packet to the server which collects the signature of all routers on its way to the receiver. The aggregate signature of the routers acts as the capability token for the privileged packets. The privileged packets contain capabilities that are verified by each and every intermediate router in the network. The router drops the unprivileged packets in favor of the privileged packets. The privileged channels are time limited and require continuous updates from the server to maintain its privileged state. The use of privileged channel gives the receiver a flexible control to halt the attack flows by not sending updates. The scheme however, assumes the existence of a secure network environment, which is not always readily available.

Stoica *et al.*[45] propose Stateless Fair Queuing (SFQ) at the edge routers, to control high bandwidth attack flows. The edge routers are responsible for estimating per flow arrival rate. It then stamps the packet headers with this arrival rate information and labels them. The core routers probabilistically drop attack packets based on the embedded flow label and using its own estimate of the aggregate. The edge routers are burdened with the task of maintaining per flow estimate. SFQ is more applicable to an enterprise network or an ISP which can manually intervene the traffic entering the network. The applicability of SFQ in WMN is questionable as traffic can originate from any MR.

23.4.4. *Design goals of DoS attack detection scheme*

We can conclude from the above discussions that it is important to mitigate the effect of DoS attacks in WMNs by raising an early alert in the network. The attack traffic must be effectively filtered at every intermediate MR to avoid wastage of valuable network resources. A distributed approach needs to be adopted to progressively control the attack flow within a short time. The scheme should incur minimal overhead at the router and minimal changes in the router hardware. Per-flow state information needs to be avoid as it incur considerable memory overhead which is unsuitable for a large scale network like WMN. The scheme should be robust to multiple sources of attack. The hierarchical nature of WMN makes them vulnerable to DDoS and hence suitable measures should be taken to handle these attacks. The DDoS attack prevention scheme should be capable of handling low-bandwidth aggregates originating from multiple sources.

23.5. Conclusion

In this chapter, we discussed the security threats that are hindering the rapid deployment of WMNs. The inherent characteristics of WMNs such as multi-hop communication, hierarchical architecture, and static nature of WMNs introduce new challenges in securing the network. We analyzed various vulnerabilities that can be exploited by an attacker to conduct attacks such as selfishness, denial of service attack, etc. As MRs form the backbone of the WMN, they need to be protected strongly against various network attacks. Specifically, we discussed the necessity of new authentication mechanism to validate the identity of MRs and meshclients. We also discussed the need for detection mechanism in the event that a malicious intruder compromises an MR. We discussed the potential threat posed by selfish MRs, and discussed the threat of DoS attack on the MRs. We provided an exhaustive analysis of various existing collaboration schemes to stimulate cooperation between MRs. We provided taxonomy of collaboration approaches based on credit based, reputation based, game theory based, and other mechanisms. We discussed the imminent threat imposed by a denial of service attack on the mesh backbone. We highlighted the ease with which such an attack can be conducted in the unlicensed bands used in WMNs. Finally, we provided a detailed analysis of various approaches to combat denial of service attacks in WMNs that are based on Filtering, Traceback, and rate limiting approaches.

Thus, we need to develop a new security framework for WMNs that is scalable, reliable, distributed, autonomous, and fault-tolerant.

Problems

(1) Explain the three categories of WMN, classified based on their operational style.

(2) Differentiate WMNs and ad hoc network and explain how unique characteristics of WMNs lead to new issues in detecting security threats.

(3) An attacker compromises a MR to send fake route request messages to unknown destination in order to disrupt the network activity. Is this:

  (a) Impersonation attack
  (b) Anti-integrity attack
  (c) Denial of Service attack
  (d) None.

(4) Why is it challenging to build a secure authentication protocol for WMN, enumerate the open research challenges.

(5) A selfish MR knowingly drop's other traffic in order to give higher priority to traffic originating from its meshclients. Does this fall under active or passive attack? Justify your answer.

(6) How does a selfish MR exhibit selfishness at MAC layer? How is this different from network layer selfishness?

(7) A reputation based selfish detection scheme is unsuitable for WMNs primarily due to

  (a) Lack of trustable entities in WMNs
  (b) Static nature of meshclients
  (c) Presence of multiple radios at each MR
  (d) Assignment of non-overlapping channels at MRs.

(8) Explain the operation of Smurf attack? Why is it difficult to detect such an attack? Identify strategies that an attack can employ to hide from detection.
(9) What is traceback?
    (a) Identifying the time of attack
    (b) Identifying the compromised host responsible for attack
    (c) Identify the location of the attacker and his/her path
    (d) Identifying the spoofed source address used by the attacker.
(10) Why is Random Early Detection (RED) unsuitable for the detection of DoS attack in WMNs?

Bibliography

1. N. B. Salem and J.-P. Hubaux, Securing wireless mesh networks, *Proceedings of IEEE Wireless Communication Magazine* **13**(2) (2005) 15–55.
2. A. Zimmermann, *Wireless Mesh Networks: An Overview*, Technical Report, Mobile Communication Group, Informatik Research Lab (2006).
3. Förderverein freie Netzwerke e.V. Freifunk.net, available from: http://freifunk.net/.
4. I. F. Akyildiz and X. Wang, A survey on wireless mesh networks, *Proceedings of IEEE Communication Magazine* **43**(9) (2005) S23–30.
5. R. Bruno, M. Conti and E. Gregori, Mesh networks: commodity multi-hop ad hoc networks, *Proceedings of IEEE Communications Magazine* **43**(3) (2005) 123–131.
6. P. Kyasanur and N. H. Vaidya, Capacity of multi-channel wireless networks: impact of number of channels and interfaces, *Proceedings of ACM Mobicom*, Cologne, Germany (2005).
7. B. Xie, A. Kumar, D. P. Agrawal and S. Srinivasan, Secured macro/micro-mobility protocol for multi-hop cellular IP, *Proceedings of Pervasive and Mobile Computing* 111–136 (2006).
8. A. Gupta, A. Mukherjee, B. Xie and D. P. Agrawal, Decentralized key generation scheme for cellular-based heterogeneous wireless ad hoc networks, Under review, *Journal for Parallel and Distributed Computing* (2006).
9. H. Tewari and D. O'. Mahony, Lightweight AAA for cellular IP, *Proceedings of European Wireless* (2002).
10. L. Maccari, R. Fantacci and T. Pecorella, A secure and performant token based authentication for infrastructure and mesh 802.1X networks, *Proceedings of INFOCOMM* (2006).
11. IEEE Std. Local and metropolitan area networks port based network access control (2001).
12. A. Aboba and L. Blunk, Extensible authentication protocol (EAP), *RFC* **3748** (2004).
13. A. Aboba and D. Simon, PPP EAP TLS authentication protocol, *RFC* **2716** (1999).
14. P. Funk and S.-B. Wilson, EAP tunneled TLS authentication protocol version 0, *RFC* **4017** (2005).
15. C. Rigney and A. Rubens, Remote authentications dial in user service (RADIUS), *RFC* **2138** (1997).
16. R. Fantacci, L. Maccari and T. Pecorella, Analysis of secure handover for IEEE 802.1X-based wireless ad hoc networks, *Proceedings of IEEE Wireless Communications Magazine* (2007)

17. E. Perez, 802.11i: how we got here and where are we headed, *Proceedings of SANS Institute* (2004).
18. D. Davis, Compliance defects in public key cryptography, *Proceedings of USENIX Security Symposium* (1996).
19. L. Santhanam, B. Xie and D. P. Agrawal, Selfishness in mesh networks: wired multi-hop MANETS, *Proceedings of IEEE Wireless Communication Magazine* (2007).
20. L. Buttyan and J.-P. Hubaux, Enforcing service availability in mobile ad hoc WANs, *Proceedings of IEEE/ACM MobiHOC Workshop* (2000).
21. S. Zhong, Y. Yang and J. Chen, Sprite: a simple, cheat proof, credit-based system for mobile ad hoc networks, *Proceedings of IEEE INFOCOM* (2003).
22. Y. Yoo, S. Ahn and D. P. Agrawal, A credit-payment scheme for packet forwarding fairness in mobile ad hoc networks, *Proceedings of IEEE ICC* (2005).
23. S. Marti, T. J. Giuli, K. Lai and M. Baker, Mitigating router misbehavior in mobile ad hoc networks, *Proceedings of Mobi-Com* (2000).
24. S. Buchegger and J.-Y. L. Boudec, Performance analysis of the CONFIDANT protocol: cooperation of nodes-fairness in dynamic ad-hoc networks, *Proceedings of MobiHOC* (2002).
25. R. Mahajan, M. Rodrig, D. Wetherall and J. Zahorjan, Sustaining cooperation in multi-hop wireless networks, *Proceedings of NSDI* (2005).
26. V. Srinivasan, P. Nuggehalli, C. Chiasserini and R. R. Rao, Cooperation in wireless ad hoc networks, *Proceedings of IEEE INFOCOM* (2003).
27. H.-Y. Wei and R. D. Gitlin, Incentive mechanism design for selfish hybrid wireless relay networks, *Proceedings of Mobile Networks and Applications* **10** (2005) 929–937.
28. Y. Zhang and V. Paxson, Detecting stepping stones, *Proceedings of the 9th USENIX Security Symposium* (2000) 171–184.
29. L. Santhanam, N. Nandiraju, Y. Yoo and D. P. Agrawal, Distributed self-policing architecture for fostering node cooperation in wireless mesh networks, *Proceedings of PWC* (2006) 147–158.
30. B. Wang, S. Soltani, J. K. Shapiro and P.-N. Tan, Distributed detection of selfish routing in wireless mesh networks, Technical Report MSU-CSE-06-19, University of Michigan (2006).
31. M. Conti, E. Gregori and G. Maselli, Reliable and efficient forwarding in ad hoc networks, *Journal of Ad Hoc Networks* **4** (2006) 398–415.
32. S. Savage, D. Wetherall, A. Karlin and T. Anderson, Practical network support for IP traceback, *Proceedings of ACM SIGCOM* **9**(3) (2001) 226–237.
33. A. Belenky and N. Ansari, IP traceback with deterministic packet marking, *Proceedings of the IEEE Communication Letters* **7**(4) (2003) 162–164.
34. D. X. Song and A. Perrig, Advanced and authenticated marking schemes for IP traceback, *Proceedings of the IEEE INFOCOM* (2001) 878–886.
35. A. C. Snoeren, C. Patridge, L. A. Sanchez, C. E. Jones, F. Tchakountio, S. T. Kent and W. T. Strayer, Hash-based IP traceback, *Journal of IEEE/ACM Transactions on Networking* **10**(6) (2002) 721–734.
36. S. M. Bellovin, ICMP traceback messages, *The Network Working Group Internet Draft* (2000).
37. L. Santhanam, A. Kumar and D. P. Agrawal, Taxonomy of IP traceback, *Journal of Information Assurance and Security* **1** (2006) 79–94.
38. P. Ferguson and D. Senie, Network ingress filtering: defeating denial of service attacks which employ IP source address spoofing, *Internet Engine Task Force RFC 2827*, www.ietf.org/rfc/rfc2827.txt (2000).

39. S. Floyd and V. Jacobson, Random early detection gateways for congestion avoidance, *Proceedings of IEEE/ACM Transactions on Networking* **1**(4) (1993).
40. T. J. Ott, T. V. Lakshman and L. H. Wong, SRED: stabilized RED, *Proceedings of INFOCOM* (1999).
41. B. Suter, T. V. Lakshman, D. Stiliadis and A. K. Choudhary, Design consideration for supporting TCP with per-flow queuing, *Proceedings of INFOCOMM* (1998).
42. L. Santhanam, D. Nandiraju, N. Nandiraju and D. P. Agrawal, Active cache based defense against DoS attack in wireless mesh networks, *Proceedings of ISWPC* (2007).
43. R. Mahajan, S. M. Bellovin, S. Floyd, J. Ioannidis, V. Paxson and S. Shenker, Controlling high bandwidth aggregates in the network, *Proceedings of Computer Communication Review* **32**(3) (2002).
44. A. Yaar, A. Perrig and D. Song, SIFF: a stateless Internet flow filter to mitigate DDoS flooding attacks, *Proceedings of the IEEE Symposium on Security and Privacy* (2004).
45. I. Stoica, S. Shenker and H. Zhang, Core-stateless fair queuing: a scalable architecture to approximate fair bandwidth allocations in high speed networks, *Proceedings of SIFCOMM* (1998).

Solutions

Solutions for Chapter 1

1. In hop-by-hop retransmission, each node is able to retransmit upon packet loss, while only the source can retransmit in end-to-end retransmission. Hop-by-hop retransmission is supported by link layer, while end-to-end retransmission is supported by transportation layer.
2. In end-to-end coding, only the source is able to retransmit, while each node can retransmit in hop-by-hop coding.
3. In multi-path schemes, packets are transmitted through multiple paths simultaneously. Although packets might be transmitted through multiple paths in alternative path schemes, packets are transmitted through a single path at any given time.
4. ETX is the average number of transmissions between two nodes to guarantee transmission reliability. If the cost of a single transmission attempt on each link is 1, ETX will be equal to the cost in the hop-by-hop retransmission model.
5. The ETXs of link (s, x), (x, t), and (t, d) are 12, 6, and 3, respectively.
6. The ETXs of link (s, x), (x, y), and (y, d) are 4, 2, and 4, respectively.
7. In opportunistic routing, candidate receivers should be prioritized so that candidate receivers can sequentially send acknowledgements. The drawback of this scheme is that it introduces additional delay and it cannot guarantee transmission of duplicate packets from multiple candidate receiver.
8. The expected utility is $u = p \cdot v - c$, where p and c denote link reliability and link cost, respectively, and v represents packet benefit.
9. It is a greedy solution that first assigns each link one retry limit and then iteratively selects a link to increase its retry limit. The retry limit of a selected link is always increased one at a time. Each time, the selected link must be the one that can increase its link reliability the most. This process repeats until the reliability constraint is satisfied.
10. In order to increase the reliability of packet delivery, t $(t > k)$ erasure-coded packets will be sent by the source. If any k out of t packets are received by the destination, the original message can be reconstructed. Packet j $(1 \leq j \leq t)$ is

an erasure-coded packet and packet a_i $(1 \leq i \leq t')$ can be any packet j. As long as $t' \geq k$, the message can be decoded. Obviously, the increment of the value of t can increase the reliability of packet delivery, but it is at the expense of the increment of the transmission cost. Hence, a trade-off exists between reliability and cost. The quantity of erasure-coded packets is the key to balancing the trade-off between reliability and cost.

Solutions for Chapter 2

1. Data packets contain sequence numbers that, in conjunction with the address of the source, uniquely identify a given packet. Nodes contain a cache that stores pairs (source address, sequence number) that is used to discard duplicate packets. This is particularly important for protocols where nodes relay packets not regarding the identity of its previous relay.
2. The main problem of having a fixed set of cores is that they become a single point of failure and, therefore, receivers would be unable to join the multicast groups represented by those cores while they are down. The main disadvantages of using a core election protocol are: (a) the overhead induced by the distributed election; and (b) during the time it takes to elect the core, the performance of the routing structure may be degraded.
3. Mesh protocols tend to be more resilient to topology changes induced by node mobility and failures. Multicast trees are very fragile, because a single link breakage in a multicast tree may disconnect a significant part of the routing structure. Mesh protocols also tend to be much simpler, because no extra control packets are needed to force a tree topology. On the other hand, tree-based protocols incur in less data overhead, because packets are forwarded over tree branches.
4. The main reason is that the packets generated by all the sources share a single common structure, which in general is not composed by shortest paths from sources to group members. Moreover, because data packets are targeted to the core, its vicinity is likely to become a contention hot-spot.
5. Source-specific protocols establish a tree per-group per-active source, and hence the order of their control overhead is $O(gs)$, where g is the number of groups and s is the maximum number of concurrent sources per-group. On the other hand, the order of the control overhead of core based-protocols is $O(g)$ because only one mesh is built per-multicast group.
6. In contention-based MAC protocols, the time it takes for a node to access the channel is a function of the current traffic load. Hence, if the channel becomes congested due to control or data packets, the channel access time will also be increased which will have a negative impact on the end-to-end delay. Moreover, this will not only affect the performance of the multicast protocol, but the performance of any other communication that uses the channel.

7. Sequence numbers are used to differentiate stale control packets from the fresh ones. Control packets with a larger sequence number are disseminated along the network to build or refresh the routing structure while packets with smaller sequence numbers are simply discarded.
8. In PUMA, the routing state established by a core that contended and lost an election is discarded, while in Hydra is reutilized.
9. First, control packets for a given source of a multicast group are confined to its region of interest. This way, other parts of the network are not flooded with irrelevant control information. Second, Hydra identifies regions of the network where two or more sources share common sub-graphs and performs routing-state aggregation, so that nodes located inside of those common regions only keep routing state regarding one of the aggregated sources and receive join query and join reply packets only from that source.

Solutions for Chapter 3

1.

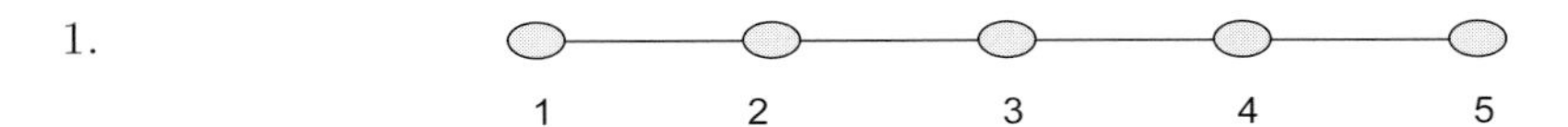

Consider the 4 hop network shown in the figure above. In 802.11 when link 1–2 is active, only link 4–5 may also be active. Link 2–3 cannot be active because node 2 cannot transmit and receive simultaneously. Link 3–4 cannot be active because communication by node 3 will affect the reception at node 2. Thus for the 3 hop case, the capacity $= \frac{C}{i}$ for $1 \leq i \leq 3$ and it is equal to $\frac{C}{3}$ for $i \geq 4$ (approximately). Students should prove the theorem.

2. Its quite possible that at higher speeds, the sender and receiver may come closer in terms of hop-count e.g. may be one hop apart and as a result throughput may increase at higher speeds.
3. Admission control of flows is essential for providing quality of service in multi-hop wireless networks. In order to make an admission decision for a new flow, the expected bandwidth consumption of the flow must be correctly determined. Due to the shared nature of the wireless medium, nodes along a multi-hop path contend among themselves for access to the medium. This leads to intra-flow contention i.e. contention between packets of the same flow being forwarded at different hops along a multi-hop path, causing the actual bandwidth consumption of the flow to become a multiple of its single hop bandwidth requirements. Determining the intra-flow contention therefore is very important in assessing the actual bandwidth requirements of a flow when this flow will be only flow as well as when other flows are also present. Only after determining the intra-flow contention one can assure applications that they will get the guaranteed bandwidth.

4. (A) MAC should aid TCP in differentiating losses due to link failures versus congestion. (B) MAC should also inform TCP to reduce its congestion window when it sees significant link-layer contention. This will reduce the amount of data as well as ACK traffic in a neighborhood. (C) MAC should inform TCP not to increase its back-off timer just freeze it when MAC is dealing with link layer losses using its own back-off timer. (D) MAC should address the unfairness problem that arises due to node captures. Several other MAC related properties could be suggested.
5. Routing protocol should inform TCP if new route discovery is being undertaken so that TCP can adjust its RTO accordingly. Routing protocol should remove stale routes so that invalid routes are not discovered as a result valid routes are found with shorter delays.

Solutions for Chapter 4

1. Transmission power control and the use of Directional Antenna.
2. A node transmits senses if the channel as idle (i.e. the noise power in its surroundings is below its carrier sensing threshold). This termed as physical carrier sensing mechanism. It also checks its Network Allocation vector for any recorded ongoing transmission from its neighbour. This is termed virtual carrier sensing. If both PCS and VCS are clear, a node transmits its packet.
3. The exposed terminal problem and the hidden terminal problem.

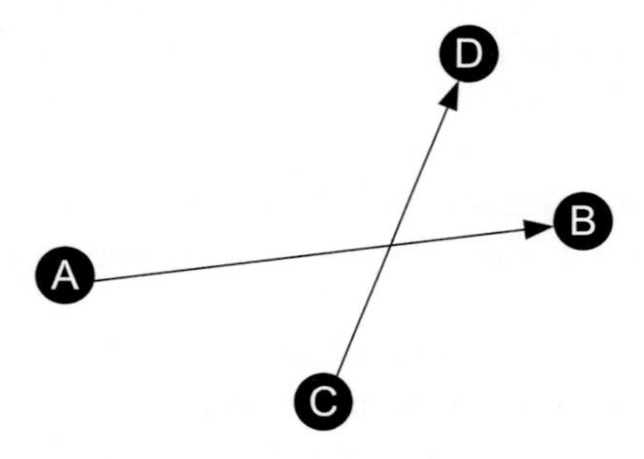

Fig. 4.9. System configuration of Question 4.

4. Consider the network of four Nodes A, B, C, D shown in Figure 4.9. Node A is communicating with Node B and Node C is communicating with Node D. If the distances between terminals are given as d(A,B) = 95 meters, d(A,D) = 98 meters, d(C,B) = 65 meters, and d(C,D) = 57 meters. Given that the target SINR for both connections is 10 dB. Assume that the target SINR is defined as the minimum required SINR when receiving a packet so as the node will be able to correctly receive and decode a packet. The additive average noise power is $-90\,\text{dBw}$ ($10^{-10}\,\text{m}W$). The transmitted power is 24 dB (250 mW) for both

transmitting Nodes A and C. Moreover, each node is equipped with a transceiver of processing gain = 10; the height and the antenna Gains are assumed equal 1. The channel model is a two-ray model.

(a) Assume no power control is applied; check if the transmission from Node A to Node B and from Node C to Node D will be successful.
(b) Repeat Part A but now with applying power control. Assume that the transmitted power of Node A is 20 mW and that of Node C is 100 mW.

Here, the SINR is calculated according to the following equation:

$$SINR = G * \left(\frac{P_r}{P_n + I}\right),$$

where P_r is the good received signal, P_n is the average noise power, I is the interference from other communicating nodes. G is the processing gain and is assumed to be equal to 10. We adopt the two-ray channel model. Thus the received power at a receiver from a transmitter is modeled as:

$$P_r = \frac{P_t^* G_t^* G_r^* h_t^* h_r}{d^4},$$

where P_t is the transmitted power, G_t, G_r, h_t, h_r are the antenna gains and heights of the transmitter and receiver respectively. Now we analyze the case when no power control is implemented. The SINR at Node B when it receives a packet from Node A is calculated as:

$$SINR_B = \frac{10^*(250/(95)^4)}{10^{-10} + (250/(65)^4)} = 2.046$$
$$SINR_B = 3.109\,\text{dB}.$$

Here the interference I results from the ongoing communication between C and D. Since $SINR_B \leq 10\,\text{dB}$, the connection between A and B is not established.
Similarly, the SINR at Node D is:

$$SINR_D = \frac{10^*(250/(57)^4)}{10^{-10} + (250/(98)^4)} = 87.3750$$
$$SINR_D = 19.41\,\text{dB}.$$

Here, the connection is established between C and D since $SINR_D \geq 10\,\text{dB}$. The average transmitting power in this case is 250 mW.

(b) Alternatively, if we apply power control, repeating the same steps as before we obtain:

$$SINR_B = \frac{10^*(100/(95)^4)}{10^{-10} + (20/(65)^4)} = 10.9570$$

$$SINR_B = 24.538\,\text{dB}$$

$$SINR_D = \frac{10^*(20/(57)^4)}{10^{-10} + (100/(98)^4)} = 17.4741$$

$$SINR_D = 10.3969\,\text{dB}.$$

Now, both connections are established with the target SINR. The average power transmitted is 60 mW.

5. Switched Beam and Steerable beam antenna.
6. Four possible scenarios: O-O, D-O, O-D, D-D.
7. Deafness, hidden terminals, exposed terminals, head of line blocking, symmetry gain problems.
8. The example in Section 4.6.1 has presented the benefits of coupling directional antenna with power control. Repeat the gain of using directional antenna for angle of 90 degrees.
   With Directional Antenna (A/2)
   Directional Antenna and Power Control A/16.
9. A generic model of directional antenna for determining the interference is shown in Figure 4.10, where R denotes the maximal permission range of Node A, R” is the maximal range of the side lobe of Node A, R’ is the constraint range of the side lobe, and θ is the beam width of the main lobe. Here, nodes lying in the area of constrained range and in the area formed by the intersection of the node’s main beam (the white region in Figure 4.10) with that of side lobe of radius R”

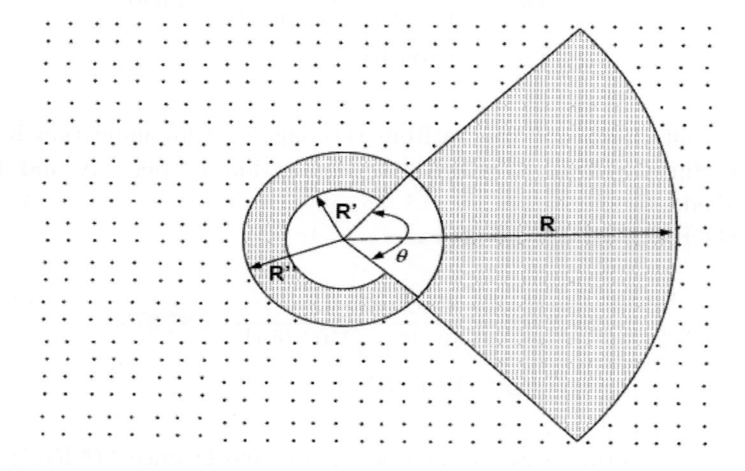

Fig. 4.10. Generic directional antenna interference region.

are refrained from transmission in any direction since their transmission may highly affect the ongoing communication. Two types of interference result from the application of directional antennas, namely the potential interference and the indirect interference. All nodes outside the main lobe and outside the side lobe range (the dotted shaded region in Figure 4.10) are considered potential interferences, and may turn their directional antenna in any direction. All nodes inside the main beam of A within range R and greater than R" or inside the side lobe of A with a range between R' and R" are considered indirect interferences (the shaded region in Figure 4.10) since they will refrain from transmission in the direction of Node A and they will not cause any direct interference to Node A. These nodes are free to be engaged in any communication towards other directions. Let P_t be the transmission power, G_t be the transmitter gain of the main lobe, G_r be the receiving gain. G_s be the gain of the main side lobe represented by R$''$, and h is the antenna height. The receiver is able to receive and decode the packet correctly if the received power P_r of a frame from a transmitter node in its transmission zone is higher than or equal to κ (the reception sensitivity). Furthermore, using the two-ray propagation model and with the exponential attenuation factor equal to 4, determine the relations between R, R' and R".

$$R = \left(\frac{P_t G_t G_r h^2}{\kappa}\right)^{1/4}$$

$$R'' = RG_s^{1/4}$$

$$R' = RG_s^{1/2}.$$

10. DATA rate adaptation, tuning the physical carrier sensing threshold.

Solutions for Chapter 5

1. This is because VANETs have some characteristics that set them apart from MANETs. For example, VANETs consist of a very large number of cars running on very hug size roads while MANETs used to be deployed in a small scale. Also, VANETs have a frequent disconnection between cars specially in sparse traffic while nodes in MANETs usually enjoy some form of connectivity. A third difference is that the deployment cost of MANETs infrastructures is less expensive compared to VANETS, for example, a very few access points are enough to deploy MANET in an airport or a building while we need thousands of these access points to cover streets and highways.
2. Network coding is very efficient when each node, in a group of nodes, has a frame to share in which case a very small number of broadcasting is required to allow all nodes to maintain the set of all frames. On the other hand, consider the

scenario when only one car possesses a file. If this car divided that file into blocks and broadcasted the resulting frames to all requesting cars, the saving is zero as the car has to transmit the entire set of frames which is equivalent to number of transmission needed to send the blocks. Moreover, network coding here is an overhead in coding and decoding file blocks.

3. With infrastructure
   Advantage: (a) guaranteed connectivity for some areas, (b) can perform some secure operations as roadside is more trusted than other cars.
   Disadvantage: (a) cost, (b) compromising a roadside is more dangerous than compromising a car.
4. Relying on the cellular network for file sharing has some the following advantages:
   (a) It involves no routing which is not an easy task in VANETs
   (b) Easy to implement
   (c) It may be a perfect approach for very small files and few cars.

   On the other hand, it has the following disadvantages:

   (a) large files may saturate the networks that has a few kilo bytes of bandwidth
   (b) the network would go down for sufficiently large number of cars
   (c) if the data requires only few hops to be routed, it is much better to rely on car to car communications that has much bandwidth to use.
5. Yes. In highways we can estimate the position of a car after a few seconds from now which is not an easy task in city scenario. Also, in cities, we can install access points at the main intersections and hot spot areas. While in highways, we cannot install access points along the entire highway. A third difference is that dynamics of cars in cities is controlled by traffic lights and stop signs while in highways, we have a different traffic model.
6. GOSSIP would work for VANETs if group of cars stayed connected sufficiently for long time that may happen in congested streets or when most cars maintain the same average speed. While it is not appropriate for a sparse traffic in which cars meet for a very short duration.
7. The first approach is better if many cars are interested in what is being advertised. The second approach is better if the traffic is congested and very few cars are interested in downloading.

Solutions for Chapter 6

1. VANET stands for Vehicular Ad Hoc Network. In VANET large numbers of simple, inexpensive on-board units are embedded in moving vehicles on the road. The vehicles will be capable of exchanging information between vehicles and from road-side units using dedicated short range communications (DSRC) at 5.9 GHz assigned by the Federal Communications Commission (FCC) in the US, or at other frequencies as assigned by regulatory agencies in other countries.

VANET brings significant potential for supporting a wide range of services and applications, including safety, business, and infotainment. Safety applications include collision avoidance and cooperative driving, etc. In addition to safety applications, VANETs will provide convenience and commercial applications to reduce time on the road and to improve the driving experience.

2. Compared to standard cellular scenarios, a V2V system is a mobile-to-mobile system with maximum relative speeds in excess of 90 m/s. The topology is constrained to street maps (cell towers can be off-road), although this varies dynamically with environments. Not only are the transmitter and receiver antenna heights are similar, but the heights of the adjacent vehicles and pedestrians resemble the height of transmit and receive antennas as well. A V2V system exhibits highly dynamic connections as well as a reduced duration of communication links. All of the above make the V2V different from well-understood, conventional cellular scenarios.
3. In V2V scenarios, the vehicles travel through different environments with changing speeds. Consequently, the vehicles pass through many kinds of local scatters, including those represented by other vehicles that are also in movement. Therefore, the mobile V2V channel is time variant because the motion of the transmitter and receiver, as well as the motion of the scatters around the vehicles, results in propagation-path changes.
4. The highway environment showed the largest median RMS delay spread (about 110 ns), the largest median maximum excess delay (about 600 ns) and the smallest median 90% coherence bandwidth (about 900 kHz). In general, the distributions of these quantities for the suburban environment were narrower than those for the rural and highway environments. This is interpreted in terms of the more restricted range of distances to scattering objects in the suburban environment. The average Doppler spread was observed to depend linearly on effective speed, defined as the square root of the sum of the squares of the ground speeds of the two vehicles. The coherence time was observed to vary inversely with effective speed; the highway appears to exhibit the smallest minimum 90% Coherence Time of about 0.3 ms.
5. Yes, the Doppler spread will potentially be impacted by other adjacent vehicles with high speed. The moving vehicle receives signals from multiple paths including reflections from other cars. While the observed frequency of each signal component will be Doppler shifted, the overall effect at the receiver (where all of the received signals are summed together) is a broadened spectrum compared to the transmitted signal.
6. In some environments (e.g. suburban), a correlation is observed between the Doppler spread and vehicle separation. This correlation results from driver behavior: at high speeds, drivers tend to leave greater distances between vehicles. Using the observation that Doppler spread is proportional to the effective speed, an approximation to the Doppler spread for a given distance bin can be constructed using a Speed-Separation diagram.

7. According to Table. 6.1, a particular equalizer setting can be used in a suburban environment for up to 1 ms with the expectation that it would be effective 90% of the time.
8. Using Equation (6.5), the expected Doppler spread is computed to be 385.4 Hz.
9. From Figure 6.7(a), the estimated Doppler spread at 10 m separation is about 8 Hz, at 100 m separation is about 60 Hz.
10. From Figure 6.2, the median values of RMS delay spread ($y = 0.5$) for the rural and suburban environments appear to be very similar (about 70 ns) though the rural distribution is broader. The highway environment also exhibits the largest median value (about 110 ns).

Solutions for Chapter 7

1. *Hint*: See.
2. (a) Solve the problem.

$$(x_1^*, x_2^*) = (1/2, 1/2)$$

(b) Decompose the following problem using Lagrangean relaxation, into two subproblems with x_1 only and x_2 only, respectively.

$$(\text{sub1}) \begin{array}{ll} \min & x_1^2 - \lambda x_1 \\ \text{s.t} & x_1 \geq 0 \end{array} \quad \text{and} \quad (\text{sub2}) \begin{array}{ll} \min & x_2^2 - \lambda x_2 \\ \text{s.t} & x_2 \geq 0 \end{array}.$$

(c) Solve the subproblems.

$$x_1^* = 1/2 \quad \text{and} \quad x_2^* = 1/2.$$

3. *Hint*: See.
4. Proof: The objective function $f_0(\mathbf{x})$ can be rewritten in terms of the rate vector $\mathbf{r}$ as

$$f_1(\mathbf{r}) = \prod_{i=1}^{N-1} (r_i - R_i),$$

where $\mathbf{r} = (r_1, \ldots, r_{N-1})^T$. Given that R_i is zero, the objective function is reduced to

$$f_1(\mathbf{r}) = \prod_{i=1}^{N-1} r_i.$$

Taking advantage of the strictly concave increasing property of a logarithm function, we can transform the problem (P1) into the same problem with the objective function $\log f_1(\mathbf{r})$. Since the optimal solution of (P1), $\mathbf{x}^*$, determines the optimal rate vector $\mathbf{r}^* = (r_1^*, \ldots, r_{N-1}^*)^T$, the following condition holds at $\mathbf{r} = \mathbf{r}^*$:

$$\sum_i \frac{\partial \log f_1(\mathbf{r})}{\partial r_i}\bigg|_{r_i = r_i*} \cdot (r_i - r_i^*) = \sum_i \frac{r_i - r_i^*}{r_i^*} \leq 0.$$

This is because movement along any direction $(r_i - r_i^*)$ at the optimum rate vector $\mathbf{r}^*$ cannot improve the objective function. Thus, the rate allocation $\mathbf{r}^*$ is proportionally fair.

5. Proof: For all x_i's such that $x_i W \log_2(1 + \delta_i/x_i) > R_i$, the denominator of $g_i(x_i)$ above is positive. Thus, it is sufficient for us to show that the numerator is positive for $x_i > x_i^{\min}$, where $x_i^{\min} > 0$ is the unique solution of $x_i W \log_2(1+\delta_i/x_i) = R_i$. Consider a function

$$\rho(z) = \log(1+z) - \frac{z}{1+z}(z > 0).$$

Since the first derivative is positive for $z > 0$, i.e.

$$\frac{d\rho(z)}{dz} = \frac{z}{(1+z)^2} > 0,$$

$\rho(z)$ is strictly increasing $\forall z > 0$, and $\rho(z) > \rho(0) = 0$. Thus, $\forall x_i > 0$, the numerator of $g_i(x_i)$ is reduced to

$$\frac{1}{\log 2}\left\{\log\left(1 + \frac{\delta_i}{x_i}\right) - \frac{\delta_i}{x_i + \delta_i}\right\} = \frac{1}{\log 2} \cdot \rho\left(\frac{\delta_i}{x_i}\right) > 0.$$

This results in the numerator of $g_i(x_i) > 0$ for $x_i > 0$ and $x_i > x_i^{\min}$.

6. Proof: The optimal power allocation vector is obtained by maximizing $\sum_{k\in\mathbf{C}_i} W \log_2(1 + aG_{ij}^k p_i^k)$ subject to $-\overline{p}_i + \sum_{k\in\mathbf{C}_i} p_i^k \leq 0$. Let

$$L(\mathbf{p}_i, \lambda) = \sum_{k\in\mathbf{C}_i} W \log_2(1 + aG_{ij}^k p_i^k) - \lambda \cdot \left(- \overline{p}_i + \sum_{k\in\mathbf{C}_i} p_i^k\right)$$

where $\mathbf{p}_i^* = (p_i^{k*})^T$ and λ is the Lagrange multiplier.
Then

$$\frac{\partial L}{\partial p_i^k} = \frac{W a G_{ij}^k}{\log 2(1 + aG_{ij}^k p_i^k)} - \lambda$$

$$\frac{\partial L}{\partial \lambda} = \left(- \overline{p}_i + \sum_{k\in\mathbf{C}_i} p_i^k\right).$$

Employing KKT conditions yields

$$p_i^{k*} = \left(\frac{1}{|C_i|} \left\{ \overline{p}_i + \sum_{l \in \mathbf{C}_i} \frac{1}{aG_{ij}^l} \right\} - \frac{1}{aG_{ij}^k} \right)^+ . \qquad \text{—}\mathbf{QED}$$

7. Proof: Here, we take logarithm on the objective function of (P3). Let

$$L(\tau, \lambda) = \sum_{j \in \mathbf{A}_i} \log \left(\tau_j \sum_{k \in \mathbf{C}_i} W \log_2(1 + aG_{ij}^k p_i^{k*}) - R_{ij} \right) - \lambda \cdot \left(\sum_{j \in \mathbf{A}_i} \tau_j - 1 \right)$$

where $\tau = (\tau_j)^T$ and λ is the Lagrange multiplier. Then

$$\frac{\partial L}{\partial \tau_j} = \frac{\sum_{k \in \mathbf{C}_i} W \log_2(1 + aG_{ij}^k p_i^{k*})}{\tau_j \sum_{k \in \mathbf{C}_i} W \log_2(1 + aG_{ij}^k p_i^{k*}) - R_{ij}} - \lambda$$

$$\frac{\partial L}{\partial \lambda} = \left(\sum_{j \in \mathbf{A}_i} \tau_j - 1 \right) .$$

Employing KKT conditions [5] yields

$$\tau_j^* = \frac{1}{\lambda^*} + \frac{R_{ij}}{\sum_{k \in \mathbf{C}_i} W \log_2(1 + aG_{ij}^k p_i^{k*})}$$

where

$$\frac{1}{\lambda^*} = \frac{1}{|\mathbf{A}_i|} \left\{ 1 - \sum_{l \in \mathbf{A}_i} \frac{R_{il}}{\sum_{k \in \mathbf{C}_i} W \log_2(1 + aG_{ij}^k p_i^{k*})} \right\} .$$

8. Proof: We have

$$\begin{aligned}
\overline{I} - I_{no} &= I \cdot \left(1 - \frac{R_0^2}{R^2} \right) + I_0 \cdot \frac{R_0^2}{R^2} - \int_0^R I_0(a) \cdot \frac{2a}{R^2} da \\
&= I \cdot \left(1 - \frac{R_0^2}{R^2} \right) - \int_{R_0}^R I_0(a) \cdot \frac{2a}{R^2} da \\
&= \int_{R_0}^R \{ \overline{I}(a) - I_0(a) \} \cdot \frac{2a}{R^2} da \\
&= \int_{R_0}^R \{ (\phi_0 I_0(a) + \phi_1 I_1(a) + (1 - \phi_1 - \phi_2) I_2(a)) - I_0(a) \} \cdot \frac{2a}{R^2} da.
\end{aligned}$$

Note that the second relaying agent is chosen so that the SNR may increase, i.e. $I_1(a) \leq I_2(a)$. From this, we have

$$\begin{aligned}\overline{I} - I &\geq \int_{R_0}^{R} \{(\phi_0 I_0(a) + \phi_1 I_1(a) + (1 - \phi_1 - \phi_2) I_1(a)) - I_0(a)\} \cdot \frac{2a}{R^2} da \\ &= \int_{R_0}^{R} \{(I_1(a) - I_0(a))(1 - \phi_0)\} \cdot \frac{2a}{R^2} da \\ &\geq 0 \; (\because I_1(a) \geq I_0(a)).\end{aligned}$$

It is obvious that the equality holds if $\phi_0 = 1$ or if $I_0 = I_1 = I_2$.

Solutions for Chapter 8

1. **Pure Nash Equilibrium and Dominant Strategy Calculation:** For column player, Left is the dominated by Middle. There is no dominated strategy for the row player. The only pure Nash equilibrium in the game is the pair (N, Middle).
2. **Pure Strategy Equilibrium and Dominant Strategies:** See the paper.
3. **Mixed Strategy Equilibrium Calculation:** See the paper.
4. **Nash Equilibrium Calculation:** No pure strategy exists in this game. There exists one mixed Nash equilibrium where row player uses (1/3, 2/3) and column player uses (1/2, 1/2, 0) .
5. **PoA and PoS calculation:** There are 3 Nash equilibria. Hence, PoA = 2, PoS =1.
6. **Routing and Selfishness (Braess's paradox):** Nash: $S \to a \to b \to T$. Delay is 2. It is not the optimal solution. After deleting one link, the expected delay decreases to 1.5.
7. **Supermodularity:** See the paper.
8. **Supermodularity:** This is not a supermodular game. By checking increasing difference property (to establish supermodularity), we need $\alpha\lambda_i + \lambda_{-i} \geq \mu$ for $i = 1, 2$. Since $\alpha < 1$ and we know $\lambda_1 + \lambda_2 < \mu$, the condition will never hold.
9. **Truthful Auction:** See the paper.
10. **Truthful Auction (general):** kth price sealed auction only ensures $k - 1$ items/winners truthfulness.

Solutions for Chapter 9

1. Wireless sensor networks applications include those that require monitoring some particular parameters in environment etc. where the connection among sensors are built easier or more inexpensive compared to wired sensor networks. Some potential applications are: wildlife monitoring, health and human related

monitoring, agriculture monitoring such as soil moisture, consumption monitoring and retail management, for example, in supermarkets, and industry and home building applications.

2. Four main functions of the nodes in a wireless sensor network are:

   (a) Event detection: Source detects the events that match the query and reports back to the sink with the sensed information.
   (b) Function approximation: To map the area approximately, sensor nodes are used to approximate a function of location that estimates the change of the physical value from one point to another.
   (c) Periodic measurement: Sources report the sensed value to the destination sink periodically according to the set interval or as per-application requirements.
   (d) Tracking: As an object of interest, information is obtained regarding the mobility of a sensor node. In fact, sensor nodes generate meaningful information by interacting with each other.

3. It is likely that the provider/operator of the WSN intends to operate the network at least for the time necessary to fulfill the given task. Particularly the lifetime will be application specific and of course will depend on the energy efficiency of various mechanisms employed (such as routing protocol, sleep strategy, aggregation mechanism etc.). A topology change due to the mobility or due to the death of a sensor node (resulting in network partitioning) must not act as a deterrent for the regular service in the network. A WSN must be robust and adaptive enough to tackle such a situation and therefore be sustainable.
4. Flat routing architecture refers to the situation where all sensor nodes have the same level of functionality in the network. In these networks usually each sensor node has a direct communication to the sink. An example of flat routing is the minimum transmit efficiency (MTE) scheme.

   Hierarchical routing refers to the networks where some sensor nodes (usually with higher capability and functionality) act as intermediate point of connection for a group of normal sensor nodes. Normal nodes need to transmit their information to those intermediate nodes and then they will direct the information to the sink. An example of this network is the low energy adaptive clustering hierarchy (LEACH) scheme.
5. Direct diffusion (DD) is a hybrid data centric routing approach for WSN. Here, sinks generate interests that propagate through the network looking for nodes with matching event records. The events are forwarded to the originators of the interests through multiple paths. The interests contain the interval attribute field that indicates the frequency with which data related to these interests should be sent. The longevity of the communication allows the protocol to discover the good paths. The network reinforces one or a small number of these paths during the communications.

6. Random walk is a routing techniques used in wireless sensor networks where load balancing is considered. This can be used in a sensor network with very limited mobility while randomly turning the sensor nodes on and off. The protocol uses multi-path routing and achieves load balancing in a statistical sense. Although the topology can be irregular, the nodes are so arranged that each node falls in the intersection of a grid on a plane. This intersection i.e. the lattice coordinate is obtained by computing distances between the nodes using distributed asynchronous Bellman-Ford algorithm. Nodes in the neighborhood are selected as next hop according to the probability in relation to the closeness of the node to the destination.
7. The difference is in the selection of cluster heads (CHs), that is the cluster formation. To address the irregular formation of clusters in LEACH (a decentralized approach), LEACH-C (a centralized cluster formation scheme) was proposed, where the CH selection is carried out by the sink. Here, the information about the location and energy of the sensor nodes is delivered to the sink prior to the CH selection. Using this information the sink then calculates the average energy of the nodes. Nodes having energy level below the average energy are not considered as a prospective CH. Accordingly the cluster is formed in such a way that the total amount of energy needed to transmit data from the non-CHs to the CHs is minimized. Apart from this cluster formation scheme, the aggregation and forwarding in LEACH-C is same as LEACH.
8. LEACH and TEEN are different types of protocols where the former collects the periodic data proactively while the latter reports time critical data reactively. LEACH sends data to the sink even if there is no significant importance of the sensed values thereby resulting in wastage of resources. Conversely, if the thresholds are not reached, TEEN may not deliver data to the sink for a long period of time. To resolve these issues, APTEEN, a hybrid protocol, has been protocol that can respond to the queries on the historic data, forward the periodic data and also respond to any sudden change on the sensed value.
9. Power efficient gathering in sensor information systems (PEGASIS) addresses the energy saving problem by constructing a near optimal chain. The chain is constructed so that the nodes can only communicate with their nearest neighbors. PEGASIS reduces the communication cost efficiently; as a result it increases the life of the network. But this encouraging result comes with the cost of increasing delay in data delivery (approximately 4.3 times more than LEACH). Instead of this very high delay sometimes a moderate delay is required for the applications.
10. Geographic routing approaches are based on the location of the nodes in the networks. Greedy forwarding is one of such routing approaches that provide a simple and efficient routing mechanism. Geographic adaptive fidelity (GAF) and geographic and energy aware routing (GEAR) along with a group of protocols

that handle the routing hole in the geographical forwarding arena are included in this category of routing protocols.

11. In order to extend the network lifetime through conservation of battery energy and distribution of data traffic across the sensor field in an equitable manner, new energy-efficient routing protocol designs are warranted. This will be even more important as the network size grows beyond the capabilities of existing proposals. Add to this the node mobility, and the problem with offering energy-efficient routing becomes manifold. Although traditional proposals can address some of the issues to some extent, a lot more is desired. Also, single network architecture may not suffice and protocol may need to support both flat and hierarchical networks, unlike the traditional support of a specific architecture. Similarly well referenced routing protocols such as LEACH family of protocols suffer from scalability problem where they limit their corresponding network size due to the adaptation of single long hops. On the other hand, deploying a high cost CH is also undesirable for the high monitory cost involved.
12. In principle, energy and power limitation, processing power, short range transmission range, un-rechargeable or un-replaceable nodes, and large number of nodes are the main features of a wireless sensor network which are not the case for ad hoc networks. Therefore completely different or well-modified routing protocols are needed in the case of wireless sensor networks compared to those of ad hoc networks.

Solutions for Chapter 10

1. Real-time target tracking in battlefield environments, event triggering in monitoring applications.
2. Mobility of the target or the sink node places more demand on the network for supporting QoS traffic.
3. CEDAR depends on using core nodes (dominating sets). If a node in the core breaks down, which is highly probable in the constrained sensor networks, it will cost too much in terms of resources to reconstruct the core. In addition, CEDAR is not energy-aware, which is a key concern in sensor networks.
4. SAR, Energy-aware QoS Routing Protocol, SPEED.
5. By dividing the network into cells and requiring that all nodes within a cell hear each other.
6. It depends on using velocity-monotonic scheduling, where packets are put in different queues based on their requested velocity, i.e. the deadline and closeness to the sink.
7. Because it assumes an ideal flow model, where infinitely small amount of data can be served from each queue in a round robin fashion. This is not the case with the discrete size of packets, where packets should be sent as a whole.

8. Bandwidth limitation, removal of redundancy, energy and delay trade-off, buffer size limitation, support of multiple traffic types.
9. Since the typical applications of QoS in sensor networks involve image and video streams, which require significant amount of processing power.
10. Buffer size will increase the delay variation, and hence reduce the QoS. In addition, the large packet size of typical image and video traffic will require large buffers, which are costly for the resource constrained sensor nodes.
11. Mapping between the different QoS classes in the two heterogeneous networks to support end-to-end quality of service.
12. By using hybrid metrics where the energy cost and QoS metric costs are included.

Solutions for Chapter 11

1. The common types of mobility based data collection approaches proposed for WSNs are:

   (a) Mobile sink based approach.
   (b) Mobile data collector based approach.

2. Mobile base station can increase the network lifetime by balancing the energy consumption of nodes, thereby reducing the occurrence of hot spots that affects the connectivity of the network.
3. The disadvantage of using a mobile base station is that the sink's position needs to be informed regularly to all sensor nodes, with modifications of routing tables at the sensor nodes changed accordingly. This can become a serious overhead undermining the advantages of using a mobile node.
4. The following are the main challenges to be dealt with for designing a system with a mobile base station:

   - Selecting the GW node's path and the stopping points on the path.
   - Time spend at each stopping point.
   - Updating routing tables.
   - Should the path be changed dynamically, such as the possible creation of a hotspot or detection of an important event in one network region.
   - Use of multiple mobile base stations.

5. For a circular deployment area, the optimum mobility strategy would be along the periphery of the network. For routing, the circle is divided into an inner circle and the annulus around this circle. For nodes within the inner circle, shortest path routing is used; for nodes in the annulus, the concept of "round routing", where a path in parallel with the circle's periphery until it is in line with the sink location (at the periphery), followed by shortest-path routing to the sink.

6. The dual-sink based approach reduces the overhead for updating the routing table as the mobile sink's location changes. In this architecture, the static sink broadcasts its location information at the beginning, where as the mobile sink broadcasts its location information only to a subset of nodes around it. To handle multiple mobile and static nodes, either the mobile nodes can move around randomly or co-operatively such that the same area in the network is not served by more than one mobile sink.
7. Advantages: Prevention of hotspots and hence increased network life time, it can efficiently deal with the limited buffer size of sensor nodes.
   Disadvantages: Increased overhead of updating sink position to the entire network and communicating network events to the sink.
   Challenges in implementing this approach are: updating the sink position to the entire network, informing network events to the sink in regular intervals, dynamically changing the mobile sink path.
8. The main theoretical result discussed in this paper is: the optimal network lifetime for an un-constrained mobile base station movement is $(1- \in)$ times that of a constrained mobile base station movement.
9. This is a hybrid approach where both the residual energy of the nodes and their respective distance from the sink is jointly considered for deciding the mobile sink path. The objective of increasing network lifetime is achieved by reducing the battery usage at each node. The data disseminators are selected not just by considering their estimated residual energy, but also their distance from the mobile sink. The selection of data disseminator is formulated as an optimization problem where the search space of coordinates is pruned and the mobile sink is assumed to follow the Random Way Point (RWP) model. In RWP model, the mobile sink randomly selects a point in the sensor field as its destination, known as waypoint, and moves towards it with a random constant speed. After reaching the destination the sink stays there for a particular amount of time and randomly selects the next waypoint. A weighted entropy value is calculated based on Shannon's entropy to guide the selection of data disseminators. Entropy, as defined in the laws of thermodynamics, increases as the system spends more energy; a system with high entropy enters a disordered state. Hence, we see a direct correlation between the concept of entropy and energy efficient data dissemination being discussed. The residual energy and the position of the sink are calculated from the previously known accurate values. After calculating these, the values are used to find a weighted entropy value that decides the forwarding node selected for data dissemination to the mobile sink.
10. Simulations show that any of the mobility models can increase packet success rate by around 140% as compared to Directed Diffusion and a 40% reduction in energy consumption. For networks with higher delay tolerance, the random walk and biased random walk approaches are more suitable; with the second

approach, the delay can be reduced at the cost of higher energy multi-hop paths; the deterministic walk is suitable for networks with limited mobility capabilities and higher loss tolerance. One-step and multi-step moving scheme both prolong the network lifetime and are adaptive to different network topologies.

11. Design challenges:
    (a) Number of mobile nodes to get maximum coverage.
    (b) Determining the path taken by the mobile node.
    (c) Buffer size of the mobile node.
    (d) Mobile element's speed will affect the performance.
    (e) Possibility of rendezvous points; helps data aggregation.
12. A three-tier approach is used: (i) the top tier that consists of access points or gateway nodes; (ii) the middle tier that consists of the mobile data nodes, called data MULEs, that execute a random walk, and (iii) the lower tier that consists of the sensor nodes. The data MULE node traverses the network, collects the data from sensor nodes using single-hop communication and reports the data to the GW node. This method is preferred over mobile base station based approach, when data delivery delay (latency) is the major constraint for the system.
13. The main advantages of mobile data collector based approach are: energy efficiency, no routing overhead, robustness, scalability and simplicity in design.
14. If the mobile data collector does not stop at any point, then the latency will be reduced. However, packet loss will be high since the communication time between the data collector and the sensor node is less and also the sensor node has limited buffer size. If it stops at regular intervals to collect data, the latency will be high, but the packet delivery ratio will be high.
15. If the buffer size at the sensor node is less, then the MDCs have to wait at each stopping point for more time so as to collect as much data as possible from the nodes and there by preventing the buffer overflow. If the buffer size of the MDC is less then it has to reach the base station fast, so as to transfer its buffer contents and start the next round of data collection; this essentially implies less stopping time at each node.
16. Multiple mobile node based data collector approach works the same way as a single mobile node based approach, with the difference that each mobile data collector will cater to the data collection of a particular subset of nodes in the network. This increases the network lifetime, while making sure that latency is not high.
17. Advantages of this approach:
    (a) Decreased latency.
    (b) Data aggregation is possible.
    (c) Can effectively deal with the limited buffer size of sensor nodes.

    This approach is useful in systems, which are data intensive and where latency should be preferably less.

Solutions for Chapter 12

1. There are five major characteristics of DTMSN: nodal mobility, sparse connectivity, delay tolerability, fault tolerability, and limited message buffer at sensor node.
2. Because most of the mainstream approaches of data delivery in traditional sensor networks assume that end-to-end connectivity exists, which is not true in DTMSN. Meanwhile, they do not consider some other unique characteristics of DTMSN, such as frequent topology change, delay/fault tolerability, and limited buffer size at sensor node.
3. The direct transmission has minimal transmission cost (each message will be only transmitted once). But its delivery ratio may be low and delivery delay is high. On the contrary, while flooding may present a better delivery probability and lower delay, its overhead is high.
4. The major consideration is the tradeoff between delivery delay/ratio and transmission overhead/energy.
5. In ZebraNet approach, each node maintains a hierarchy level, which is the likelihood of this node being in the transmission range of the base station. When a node meets other sensor nodes with higher hierarchy level, it sends its collected data to the neighboring node with the highest hierarchy level, until the message reaches the base station.

   There are two major issues of this approach. First, it assumes each node can reach the sink node, which may not be true. Second, the energy consumption can be unnecessarily high due to the lack of proper overhead control.
6. SWIM assumes each node has the same nodal delivery probability. When a message is generated, it's associated with a TTL. Each time the message is replicated to other nodes, the TTL decreases by one. The initial TTL value is calculated so that the message can be delivered successfully to the sink node with a predetermined high probability according to the current nodal delivery probability.
7. The nodal delivery probability in DFT-MSN is calculated in a cascaded way. If a node i transmits a message to node j, node i's delivery probability is updated according to node j's probability.
8. In DFT-MSN, only the minimal number of message copies are generated in order to achieve a desired message delivery probability.
9. In FAD data delivery scheme, both the source node and the intermediate nodes need to calculate and generate the necessary number of message copies locally and dynamically. While in RED scheme, only the source node generates block messages according to its local information. All the intermediate nodes only do simple forwarding.
10. In the erasure coding approach, a message is first split into b blocks with equal size. Erasure coding is then applied to these b blocks, producing $S \times b$ small

messages (which are referred as block messages), where S is the replication overhead. The gain of erasure coding stems from its ability of recovering the original message based on any b block messages.

11. They differ in many aspects, such as how to calculate nodal delivery probability, where to replicate message, how to identify receivers, and how to control overhead. Please refer to Table 1.1 for details.

Solutions for Chapter 13

1. Radio Frequency Identification is a means of storing and retrieving data through electromagnetic transmission to an RF compatible integrated circuit. It usually consists of a batch of tags and readers which are used to read data from tags.
2. There are three types of tags: passive tags, semi-passive tags and active tags. Passive tags operate without any battery. Semi-passive tags are equipped batteries as power supply for additional functions. However, like passive tags, they use the radio waves of senders as an energy source for their transmissions. Active tags use batteries to power their transceiver and all other components.
3. Active tags are more powerful than passive tags and semi-passive tags. Therefore they have a larger range and memory and more functions. Active tags are more expensive than passive tags and semi-passive tags.
4. There are three frequency ranges which are used in RFID: low frequency (LF, 30–500 kHz), high frequency (HF, 10–15 MHz), and ultra high frequency (UHF, 850–950 MHz, 2.4–2.5 GHz, 5.8 GHz).
5. Tags with higher frequencies have higher transmission rates and longer transmission ranges. However they are also more expensive and severely affected by fluids and metals when compared with tags with lower frequencies.
6. A wireless sensor network is composed of a large number of sensor nodes that can be deployed on the ground, in the air, in vehicles, inside buildings or even on bodies. WSNs are widely employed in environmental monitoring, biomedical observation, surveillance, security, etc.
7. The difference between RFID and WSN is illustrated in Table 1.
8. Based on manners of integration, there are four types of integration: integrating tags with sensors, integrating tags with wireless sensor nodes and wireless devices, integrating readers with wireless sensor nodes and wireless devices, and mix of RFID and WSNs. Based on the roles that RFID and WSN play in the integration, current applications of integration of RFID and WSN can be classified as follows: RFID provides identification, RFID provides localization, WSN provides sensing abilities, WSN provides advanced communications and multi-hop communications and finally WSN provides localization.
9. RFID is usually used to detect presence and location of interested objects. RFID tags are required to be attached to the objects.

10. Generally speaking, active tags have larger power supply and memory size, use higher frequency range, have more flexible programmability than that of passive tags and semi-passive tags. The active tags could be integrated with variant sensors or be attached with external sensors.

Solutions for Chapter 14

1. Data sources on WWW include, for example, webpages, MP3, image files, video stream files, database services, blogs, forums, and etc.
2. Sensor networks will enhance the WWW with new types of real-time data that were not available before, and expand WWW services to such new applications like real-time health monitoring, security surveillance, and so on. On the other hand, WWW provides the means to promote access to and sharing of sensor network resources similar to any other kinds of information resources, resulting in higher information availability and more efficient sensor resource utilization.
3. Similar to other web services, it is expected that new sensors and sensor network services will be developed and attached to the web at any time. In order to support the progressive nature of the sensor augmentation to the web, common standards must be defined which will serve as a spanning layer to handle the heterogeneity of existing and emerging sensor systems in order to provide a seamless integration to the worldwide web services. The idea of designing such a spanning layer is similar to the Internet design strategy which uses IP (Internet Protocol) to operate over various physical communication networks, e.g. Ethernet, WiFi, optical, satellite and etc. For sensor system designers, the standards serve as guidelines for integrating their products into the web. For users and application designers, the use of such a layered design is also particularly important because it provides common interfaces for accessing various sensor systems and eases the technical demand for domain specific knowledge and also makes the high-level constructs reusable.
4. For concurrent executions — more users and more a higher throughput can be supported by the same sensor server, because the server has to access sensors to collect data, i.e. I/O intensive, and data processing can run concurrently.

   Against concurrent executions — problems of interference and competition between concurrent operations. For example, one application may require a video sensor to collect data at one view angle while another application may require a different one.

   To overcome interference use some form of concurrency control. For example, for a Java server use synchronized operations such as credit and debit.
5. The OCG SWE initiative has adopted a layered and service-oriented architecture. As an enhancement to the existing WWW, sensor web adopts a similar layered architecture as the WWW. Compared to various physical networks that Internet is built upon, sensor networks may display even higher level of heterogeneity

and complexity due to its highly specific needs and properties. For example, besides many benefit of the layered architecture in general, one of its particular advantages in sensor web design is the capability to support the deployment of various types of sensor networks that are designed for different special tasks, while at the same time enable inter-operation between them through spanning layers. A layered design also has the benefit of hiding the complexity of the physical sensor networks from the end users as well as application designer, and making sensor web application design similar to other web applications.

6. The OCG SWE makes use of SOA for discovering a sensor service attached to the web. The Service Oriented Architecture (SOA) provides an approach to describe, discover, and invoke services from heterogeneous platforms using XML and SOAP standards. Sensor services described in SensorML can be discovered and executed. Each service defines its input, output, parameters, and method, as well as provides relevant metadata for discovery.
7. In sensor networks, such metadata may represent the properties of and relationships between sensors, useful derived properties of sensor data. By using formal semantic descriptions of sensors, sensor data and processing agents, automatic planning techniques can be developed to compose them in legitimate and efficient ways so that they collectively produce meaningful end results.

   Secondly, formal semantic description makes it possible to share sensor sources and processing agents among different applications. By tapping into the growing reservoir of data sources and processing agents, increasingly powerful applications can be built progressively. More specifically, metadata will specify the characteristics of service providers, e.g. sensor, and processing agents. A service registry uses these characteristics to categorize service providers, and a service planner uses them to discover matched sensing and processing resources to its requirements. Metadata will also specify non-functional characteristics which may be used to help a service planner find an appropriate service provider, e.g. availability or reliability measures (e.g. mean time to fail, mean time to repair); description of needed resources (CPU, network bandwidth, memory) for the component to work properly; or trust-based information on other components. Finally, metadata describes the interfaces used to access its service. The interface description includes its signature, allowed operations, data typing, and access protocols. A service planner can then use this information to bind to the service provider and invoke its service using the published interfaces.
8. Building a system capable of semantic brokering and dynamic task composition requires the use of a universal language to represent the metadata. This is achieved by composing metadata using a standard ontology to describe the sensing and processing resources in the network.
9. An upper ontology is limited to concepts that are meta, generic, abstract and philosophical, and therefore are general enough to address a broad range of

domain areas. Concepts specific to given domains will not be included; however, this standard will provide a structure and a set of general concepts upon which domain ontologies (e.g. medical, financial, engineering, etc.) could be constructed.

For example, the NASA SWEET ontologies (Semantic Web for Earth and Environmental Terminology) provide an upper-level ontology for Earth system science. As an upper ontology it also defines ontologies in support of the sensor systems, which can be used as an alternative to the OntoSensor ontology. SWEET ontologies include several thousands of terms, spanning a broad extent of Earth system science and related concepts (such as data characteristics) using the OWL language. Other than sensors, it includes ontologies, e.g. earth, space, time, data, units, biosphere, human activity etc.

Domain and domain task ontologies specify terms, relationships between terms and generic tasks and activities that are relevant in a particular domain. Domain concepts provide further specialization of concepts from the upper ontology. Moreover, sensor data must be described in terms of the phenomenon being measured, the time and space over which it is measured and in terms of the larger data set to which this data belongs. For example, the SWEET ontologies provide support for representing time and space.

10. When the user accesses a sensor web, he/she makes an invocation of services in a sensor web server. The following can affect responsiveness:

    (i) sensor server overload;
    (ii) latency in exchanging request and reply messages (due to layers of OS and middleware software in user and server);
    (iii) load on network.

    Replication of the processing service helps with (i). The use of lightweight communication protocols helps with (ii). The use of multicast helps with (iii).

Solutions for Chapter 15

1. Wireless Sensor Networks (WSNs) have enabled data gathering from a vast geographical region and present unprecedented opportunities for a wide range of tracking and monitoring applications from both civilian and military domains. In these applications, WSNs are expected to process, store, and provide the sensed data to the network users upon their demands.
2. A digital signature algorithm is a cryptographic tool for generating non-repudiation evidence, authenticating the integrity as well as the origin of a signed message. In a digital signature algorithm, a signer keeps a private key secret and publishes the corresponding public key. The private key is used by the signer to generate digital signatures on messages and the public key is used

by anyone to verify signatures on messages. The digital signature algorithms mostly used are RSA and DSA.

3. A Bloom filter is a simple space-efficient randomized data structure for representing a set in order to support membership queries. A Bloom filter for representing a set $S = S_1, S_2, \ldots, S_n$ of n elements is described by a vector v of m bits, initially all set to 0. A Bloom filter uses k independent hash functions $h_1, \ldots, h_k$ with range $0, \ldots, m-1$ which map each item in the universe to a random number uniform over $[0, \ldots, m-1]$. For each element $s \in S$, the bits $h_i(s)$ are set to 1 for $1 \leq i \leq k$. Note that a bit of v can be set to 1 multiple times. To check if an item x is in S, we check whether all bits $h_i(x)$ are set to 1. If not, x is not a member of S for certain, that is, no false negative error. If yes, x is assumed to be in S.
4. A Merkle Tree is a construction introduced by Merkle in 1979 to build secure authentication schemes from hash functions. It is a tree of hashes where the leaves in the tree are hashes of the authentic data values $n_1, n_2, \ldots, n_w$. Nodes further up in the tree are the hashes of their respective children.
5. A Bloom filter may yield a false positive. It may suggest that an element x is in S even though it is not. The probability of a false positive for an element not in the set can be calculated as follows. After all the elements of S are hashed into the Bloom filter, the probability that a specific bit is still 0 is $\left(1 - \frac{1}{m}\right)^{kn} \approx e^{-kn/m}$. The probability of a false positive f is then $\left(1 - \left(1 - \frac{1}{m}\right)^{kn}\right)^k \approx (1 - e^{-kn/m})^k$.
6. Simply speaking, counting bloom filter is employed to mitigate the storage space for the user ID and public key. A novel signature scheme is applied for shortening message length. In such signature scheme, the signature is appended to a truncated message but the signature authenticates the whole message. Merkle hash tree is used for supporting more users.
7. We point out a much more serious vulnerability of μTESLA-like schemes when they are applied in multi-hop WSNs. Since sensor nodes buffer all the messages received within one time interval, an adversary can hence food the whole network arbitrarily. All he has to do is to claim that the flooding messages belong to the current time interval which should be buffered for authentication until the next time interval. Since wireless transmission is very expensive in WSNs, and WSNs are extremely energy constrained, the ability to food the network arbitrarily could cause devastating Denial of Service (DoS) attacks. Moreover, this type of energy-depletion DoS attacks become more devastating in multi-user scenario as the adversary now can have more targets and hence more chances to generate bogus messages without being detected. Obviously, all these attacks are due to delayed authentication of the broadcast messages.
8. $f = (0.6185)^{\frac{m}{N}}$ (see Theorem 1). f decreases sharply as $\frac{m}{N}$ increases. When $\frac{m}{N}$ increases from 8 to 96 bits, f decreases from $2.1{*}10^{-2}$ to $9.3{*}10^{-21}$. f determines the security strength. For example, when $\frac{m}{N} = 92$ bits, the adversary has to generate around $2^{63.8}$ public/private key pairs on average before finding a valid

one to pass the Bloom filter. This is almost computationally infeasible, at least within the lifetime of the WSN (usually at most several years).

9. Simply speaking, in partial message recovery digital signature scheme, the signature is appended to a truncated message and the discarded bytes are recovered by the verification algorithm. Also, the signature authenticates the whole message.
10. For BAS, the signature size is still the same as that of ECDSA, but only part of the message now has to be transmitted, with the saving of up to 10 bytes. Therefore, the per-message overhead of BAS is 54 bytes, which is 10 bytes less than that of DAS. The message size affects the energy consumption in communication in a WSN, and the energy consumption in communication is the critical cost for WSN. Figure 15.5 shows that BAS consumes a much lower energy as compared to others. For example, when $W = 15,000$, CAS always costs 2.20 KJ, while BAS costs only 1.18 KJ. The energy saving for a single broadcast can be more than 1000 J between BAS and CAS.
11. To revoke a user, say U_{ID_j}, the sink follows the steps below: (1) First, it hashes $h_l(U_{IDj})\|PK_{U_{ID_j}}) = i$ and decreases $\bar{v}_i$ by 1. It repeats this operation for all $h_l, l \in [1, k]$. (2) From the updated counting Bloom filter $\bar{v}$, the sink obtains the corresponding updated Bloom filter v' with $v' = v'_0 v'_1 \ldots v'_{m-1}$. Here, $v'_i = 1$ only when $\bar{v}_i \geq 1$, and $vi' = 0$ otherwise. (3) The sink further calculates $v_\Delta = v' \oplus v$ and deletes v afterwards. Here $\oplus$ denotes bitwise exclusive OR operation. Obviously, v_Δ is an m-bit vector with at most k bits set to 1. Hence v_Δ can be simply represented by enumerating its 1-valued bits, requiring $\overline{k}\lceil \log 2m \rceil$ bits for indexing ($\overline{k} \leq k$). This representation is efficient for a small $\overline{k}$ as will be analyzed in Section VI.B. (4) The sink finally broadcasts v_Δ after signing it. The message format follows (III) but with the sink's public key omitted, as every sensor already has it. (5) Upon receiving and successfully authenticating the broadcast message, every sensor node updates its own Bloom filter accordingly, that is, if $v_{\Delta,i} = 1$, then $v_i = 0$, $i \in [0, m-1]$.

Solutions for Chapter 16

1. Six principles that are considered for security of any system are collectively known as the "*Philosophy of Mistrust*". They are:

    - *Don't talk to any one you don't know.*
    - *Accept nothing without a guarantee.*
    - *Take everyone as an enemy until proved otherwise.*
    - *Don't trust your friend for long.*
    - *Use well-tried solutions.*
    - *Watch the ground you are standing on for cracks.*

2. Mainly three aspects:

   - Constrained Resources of Sensors.
   - Nature of Work of Wireless Sensor Networks.
   - Wireless Communications.

3. Most of the encryption-decryption techniques devised for traditional wired networks are not fit for direct use in wireless networks because WSNs consist of tiny low-cost devices which possess very scarce processing, memory, and battery power. Applying any kind of encryption scheme requires transmission of extra bits, and thus it needs extra processing, memory, and battery power which are very important resources for the sensors' longevity. Applying the encryption and decryption operations can also increase delay, jitter, and packet loss in wireless sensor networks. Hence, the lightweight versions of the schemes could be made suitable for WSNs.
4. Steganography is the art of covert communication by embedding a message into the multi-media data (image, sound, video, etc.). The main objective of steganography is to modify the carrier in a way so that it is not perceptible and hence, looks just like ordinary. It hides the existence of the covert channel, and furthermore, if we want to send a secret data without sender information or want to distribute secret data publicly, it is very useful.

   Securing wireless sensor networks is not directly related to steganography and processing multi-media data (like audio, video) with the inadequate resources of the sensors is difficult. Applying steganography in wireless sensor networks still remains as an open research issue.
5. **Type I**

   Attacks on the Basic Mechanism (e.g. attacks against routing in the network).
   Attacks on the Security Mechanisms (e.g. against cryptographic scheme or against key management scheme).

   Type II

   Passive Attack — It typically means eavesdropping of data. In this case, the attacker passively listens to the transmitted data in the network and can use the collected information later for launching other types of attacks.
   Active Attack — It means any type of direct attack caused by an adversary. The attacker actively participates in the collection, modification, and fabrication of data. Sometimes, the information collected by passive attacks can be used for active attacks.

   Type III

   External Attack — In an external attack, an outsider is involved. These attacks can cause denial of service situation, congestion, propagation of wrong routing information, etc. Typically external attacks can be resisted using firewalls, encryption mechanisms, good security management policy, and other available techniques.

Internal Attack — An Internal attack sometimes could be very harmful for the network as any node within the network works as an attacker in this case. Often it is difficult to detect an internal attacker within the network which shows a legitimate identity. Various kinds of authentication schemes, intrusion detection schemes, membership verification schemes, etc. can be used for preventing internal attacks.

6. DoS (Denial of Service) is basically a given formal name of a particular condition or state of the network but when it occurs as a result of an intentional attempt of an adversary, it is called DoS attack. In general, "*Denial of Service* (*DoS*)" is an umbrella term that can indicate many kinds of events in the network in which legitimate nodes are deprived of getting of expected services for some reasons (intentional attempts or unintentional incidents).

   When DoS situation is created because of intentional attempts by the attackers, it is called a DoS attack.

7. **Jamming** — Jamming means the deliberate interference with radio reception to deny a target's use of a communication channel. For single-frequency networks, it is simple and effective, causing the jammed node unable to communicate or coordinate with others in the network. Due to their very nature, wireless sensor networks are probably the category of wireless networks most vulnerable to "radio channel jamming"-based Denial of Service (DoS) attacks. Mainly two types of jamming could be possible; constant and sporadic. In case of constant jamming, attacker interferes with the signals of a legitimate node continuously for a certain period of time while in case of sporadic jamming, the attacker intermittently causes jamming. Sporadic jamming in the network is often more difficult to detect than detecting constant jamming.

   Tampering — Due to the unattended feature of wireless sensor networks, an attacker can physically damage/replace sensors, parts of computational and sensitive hardware, even can extract cryptographic keys to gain unrestricted access to higher communication layers. Tampering is actually any type of physical attack on sensors in the network. Success in tampering depends on:

   - how accurately and efficiently the designer considered the potential threats at design time,
   - resources available for design, construction, and test
   - attacker's cleverness and determination.

8. In case of Sybil attack, when a malicious device takes several identities, the additional identities of that malicious device are called the *Sybil nodes*. Each of the extra identities is called a Sybil node.

9. There is no difference between blackhole and sinkhole attack, rather these are the two names of the same type of attack. In this attack, a malicious node acts as a blackhole (or sinkhole) to attract all the traffic in the network. Especially in a flooding based protocol, the attacker listens to the route request and then replies to the target node saying that it has a high quality or shortest path

to the base station. A victim node is thus lured to select it as a forwarder of its packets. Once the malicious device is able to insert itself between the communicating entities (between the base station and sensor node), it is able to do whatever it wishes with the packets that pass through it. The blackhole (i.e. malicious node or the attacker) can drop the packets, selectively forward those to the base station or to the next node, or even can change the content of the packets. This type of attack could be very harmful for those nodes that are considerably far from the base station.

10. Many protocols require broadcasting of HELLO packets for neighbor discovery in the network. All of those protocols are vulnerable to HELLO Flood attack where the attacker convinces the victim that it is the immediate neighbor of the victim.
11. Wormhole attack is a significant threat to wireless sensor networks because this is possible even if the attacker has not compromised any node, and even if all communications provide authenticity and confidentiality. It could be performed even at the initial phase when the sensors start discovering the neighborhood information.
12. (i) Unknown scalability of the network.
    (ii) Unknown topological distribution of sensors in the network.
    (iii) Limited available resources of the sensors.
    (iv) Node compromise.
    (v) Re-keying.
13. Compromising a node means, convincing a legitimate node to help the attacker or persuading a node in the network to work on behalf of the attacking entity.
14. (i) The most suitable way to tackle physical attack is to use "self-destruction" mechanism. In this case, a sensor detects a physical attack and quickly deletes all of its hidden information to become non-functional. For a large-scale sensor network, this could be a feasible solution as their might be several backups of the sensors' data, cryptographic keys, codes, and other secret information. Also if a part of the network is attacked, the sensors in other parts can be ready to destroy themselves before getting captured. Though this sort of self-destruction mechanism is expensive to incorporate with the sensor's physical package, it is not impossible.
    (ii) An alternate solution could be using a mechanism where each sensor monitors the status of its neighboring sensors. Any suspicious behavior or lack of response of a neighbor for a certain period of time might trigger a warning. Consequently, the other neighbors can get ready for hiding all of their secret information.
    (iii) Analyzing the deployment policy and detailed mapping of the network could also be effective for reducing the probability of physical attacks. However, in many applications, such thorough study of the deployment area might not be possible.

(iv) Camouflaging of sensors could be efficient in some deployment scenarios. Say, for example, a wireless sensor network is to be deployed over a rocky hilly area. In that case, the sensors could be colored like rocks or could be given the shapes of rocks (with some outer coverings!), which can make the task of physically locating them more difficult.

(v) Sensors might have some sort of protective shields that can save the internal hardware from external pressure or other environmental conditions.

15. Ensuring a good level of QoS and a good level of security at the same time is always very difficult and often contradictory! Not only for sensor networks but also for other types of networks this statement is true. This is because, any sort of security operation requires some processing time. If the level of security is increased, the processing delay also increases causing degradation of quality of service. For real time multi-media applications, it poses a great research challenge.

Solutions for Chapter 17

1. Asymmetric cryptography can hardly be used in WSN because of the induced storage and computation overheads. Indeed signature footprints require relatively high storage capacities, and the calculation and verification of digital signatures induce important calculations and hence high energy consumption. Nevertheless, some experiments show that Elliptic Curve Cryptography is a promising asymmetric crypto-system which provides the same security level of classical asymmetric systems while inducing relatively acceptable overheads.
2. This is due to the fact that compromising a sensor leads to compromising its key ring (m keys) and hence compromising all the secure links based on those compromised keys. Thus, the greater is the number (m) of the preloaded keys in a sensor, the worse is the resiliency.
3. SEDAN and SDAP scale better to large networks because the data aggregation verification and revocation mechanisms are completely distributed; each node contributes to the verification process using local information. However, the verification process in SAWN and SecureDAV rely on keys shared between each node and the sink node or the cluster head node respectively. This may induce bottlenecks at these specific nodes. Moreover, the revocation process in these two solutions relies on the assumption that all nodes are in the radio range of the sink or a cluster head respectively. This means that the scalability of these solutions is limited by the radio range of these specific nodes again.
4. In SEDAN and SecureDAV, the use of A pair-wise key between a node and its upstream node allows data origin authentication, and rejects any message coming from unauthenticated nodes. In SAWN, when a node detects an invalid MAC, it must exclude the two downward nodes (child and grandchild) from the sensor network. However, there is no mechanism in SAWN that enables to verify the

origin of a packet. This enables an intruder to launch an impersonation attack to remove legitimate nodes from the network.

5. In SDAP, a malicious node can send a faulty data to one selected parent, using the identity of one chosen member node belonging to the same clique of the selected parent. To lunch such an attack, the malicious node must be positioned in the neighbourhood of the parent. In this case, the parent calculates a false aggregation value and hence will be considered as a malicious node by its child members.
6. No. It depends on the cryptographic mechanism used to verify data aggregation integrity. Indeed, protocols based on end-to-end encrypted data suffer from the lack of localization of the intruder, since the sink receives and verifies only the final result. However, the localization of malicious nodes in the protocols based on hop-by-hop encryption is possible because intermediate nodes have access to payload data and thus can detect the malicious nodes that falsify the aggregation.
7. During the initial phase of main parent selection, SeRINS simply use the hop count value as the solely metric of selection, without any security check. Therefore, an intruder can easily inject a bogus RREQ message with a very low hop count, which will increase his chance to be chosen as a main parent by a set of neighbors. Each one of these neighbor will use this falsified hop count as a referential to verify the correctness of the received sub-sequent hop counts. Consequently, a valid hop count from a legitimate node i will be detected as erroneous information. In addition, when an alarm is sent to the base station and the latter collects the received hop counts from the neighbors of node i, the base station will conclude that i is a legitimate node and that the sender of the alert, which is also legitimate, will be detected as a comprised node advertising false alerts.
8. RREQ messages are relayed using successive broadcasts, i.e. one-to-many communications. A broadcast key is a symmetric key shared between a node and all its neighbors. Therefore, if a node will use a broadcast key to authenticate itself to its neighbors, every neighbor will be able to use the same key to generate the same authentication credentials. Hence, any neighbor can launch a spoofing attack by endorsing the identity of one neighbor and using the broadcast key of this target to compute the required one hop authentication certificates.
9. Energy exhaustion attacks aim at breaking down nodes by draining all their energy resources uselessly. As radio reception is a major energy consuming operation, the fact of receiving a message even without processing it drains significant amounts of energy. Therefore, an efficient solution to energy exhaustion attacks should be implemented at lower layers: channel access and/or physical layers. Coping with such attacks is hard because there is no efficient way to avoid receiving unsolicited messages. Even filtering out messages from blacklisted nodes at the access layer according to their addresses is inefficient because an intelligent attacker may spoof valid addresses.

Solutions for Chapter 18

1. WLANs are broadly categorized into infrastructure WLANs and Mobile Ad Hoc Networks (MANETs). Infrastructure WLANs are structured networks consisting of Access Points (APs) and the client-stations, or the subscriber units. The client stations communicate via the AP which is connected to the wired network. MANETs are characterized by absence of any infrastructure such as an AP and is formed by a set of nodes which communicate over wireless medium.

   WMNs are a combination of infrastructure WLANs and MANETs. They comprise of a set of Mesh Routers (MRs) and Mesh Clients (MCs). The majority of MRs form the wireless backbone and deliver relaying services to the MCs while a subset of them are connected directly to the wired network. The MRs that are connected to the wired network providing Internet are called Internet Gateways (IGWs). WMNs leverage the benefits of both the infrastructure connectivity and the ad hoc nature of the MRs so that a large number of user community can be supported with minimal planning and infrastructure support required to deploy them.

   Refer Section 18.2 for more details.
2. The main functional components in a typical WMN architecture are
   (a) Mesh Topology Learning, Routing and Forwarding — Responsible for route discovery, routing packets to their respective destinations.
   (b) Channel State Monitor — Keeps track of the underlying channel conditions.
   (c) Mesh Medium Access Coordination — Responsible for efficient medium access by the nodes in a neighborhood, congestion control, priority control.
   (d) Mesh Security — Targets secure data transmission over the network.
   (e) Interworking — Responsible for interconnection of networks.
   (f) Mesh Configuration and Management — Performs the automatic setting of parameters like frequency channel, transmit power etc.
3. The IEEE 802.11 family of protocols standardizes WLAN technology and includes the three well known standards: 802.11a, 802.11b, and 802.11g. IEEE has setup a task group 802.11s for specifying the PHY and MAC standards for WMNs.
4. Challenges at Physical Layer
   (a) Frequently changing link quality in the network.
   (b) Omni-directional antennas exposed to interference resulting in reduced network capacity.

   Challenges at MAC Layer

   (a) Multi-hop flow unfairness and performance degradation.
   (b) Multi-radio, multi-channel assignment with minimal or no interference, efficient use of the multiple channels.

Challenges at Routing Layer

(a) Design of a routing protocol which takes care of congested paths, frequent route oscillations.
(b) Efficient traffic distribution policy over multiple paths in the network.

5. In multi-hop wireless networks, flows spanning multiple hops experience dismal performance due to the problems such as spatial bias and the underlying link layer buffer management policy at a node. Achieving fairness in wireless networks could be obtained by provisioning fairness at the node level or at the flow level. In other words, regardless of the number of hops, a flow travels in reaching its destination; it should receive throughput proportional to any other flow in the network.
6. Queue management approaches for wired networks randomly drop packets at the intermediate nodes in the network. In a multi-hop WMN, such a policy would penalize any flow, irrespective of the fact that it has traversed many hops en route towards its destination before being dropped.
7. QMMN, CBTR and DQSD. Refer to Section 18.6 for a detailed description.
8. Achieving fairness in wireless networks could be obtained by provisioning fairness at the node level or at the flow level. Per-node or per-user fairness involves guaranteeing a fair access of the communication medium and/or ensuring a fair degree of performance for every node in the network. In per-flow fairness, individual flows are monitored and their traffic rates are regulated so that every flow in the network obtains a fair access to the network resources.
9. Refer to Figure 18.3.
10. Equipping a node with multiple radios leads to capacity improvement with reduced interference and simultaneous communication with peer nodes. Also, the concurrent communication paradigm can aid in QoS provisioning.

Solutions for Chapter 19

1. A wireless local area network supports one-hop wireless communication directly between stations or between stations and an access point (AP). It needs a distribution system in order to allow communication between stations connected to different APs. A wireless mesh network (WMN) supports multi-hop forwarding among network nodes. A WMN may be implemented using traditional layer-three routing protocols or using IEEE 802.11s layer-two forwarding mechanisms, which provide multi-hop communication making a WMN appear as a WLAN.
2. Two examples of quality-aware metrics are ETX and ETT. ETX (Expected Transmission Count) represents the expected number of transmissions a node needs to successfully transmit a packet to a neighbor. ETT (Expected Transmission Time) represents the time a data packet needs to be successfully transmitted to a neighbor. Basically, ETT adjusts ETX to different PHY rates and

data-packet sizes. Other quality-aware routing metrics include ML, WCETT, NBLC, MIC, mETX, ENT, DBETX, iAWARE and Airtime Link metric.

3. WMN ad hoc-based protocols adapt ad hoc routing protocols to deal with link-quality variations. Therefore, routers continually update their metrics and disseminate them to other routers. Controlled-flooding protocols implement algorithms to reduce control traffic overhead, avoiding the flooding of the network with routing updates. Traffic-aware, or tree-based, protocols explore the traffic matrix typical of WMNs. They assume that backhaul access is the common-case application and, therefore, consider a network topology similar to a tree, rooted at the gateway. Opportunistic protocols improve classical routing by exploring cooperative diversity schemes. Instead of computing a sequence of hops to the destination before sending a packet, opportunistic protocols exploit the broadcast nature of radio-frequency transmissions to use multiple paths towards a destination.
4. Mesh Access Points or MAPs are mesh nodes extended with access point capabilities. Acting like a regular AP, a MAP is able to provide connectivity to non-mesh devices, allowing them to communicate with other nodes in a mesh cloud. A mesh portal point or MPP is a multi-interface device that connects a mesh cloud to another network, as an infrastructured network or even another mesh network (for instance, operating in a different channel). Hence, both MAPs and MPPs are specialized gateways proposed in IEEE 802.11s.
5. The mandatory protocol is the Hybrid Wireless Mesh Protocol (HWMP), which was inspired in the Ad Hoc On Demand Distance Vector (AODV) protocol. It is mostly a reactive (on-demand) protocol with proactive extensions for a tree-based (traffic-aware) approach. It is a truly hybrid protocol since both modes (proactive and reactive) can be used concurrently.
6. The mandatory metric is the Airtime Link metric, which is a quality-aware metric that tries to grasp the expected time needed to transmit a MAC frame. It takes into account not only the error probability for transmitting a frame but also the data rate used.
7. Yes. The standard is extensible and different protocols and metrics may be implemented without losing compatibility. Since the design of protocols and metrics for MANETs is an active research field, new approaches will probably appear in a near future. For instance, the draft had proposed an optional proactive (controlled-flooding) protocol (RA-OLSR), which was removed in recent versions of the proposal.
8. Although routing is typically associated to the network layer (due mainly to the ISO/OSI model), there are some benefits of implementing path discovery on layer two. The most obvious is the fact that it is easier to gather inherently layer-two data (radio-specific parameters like noise and signal levels, transmitting power and data rates) and to calculate quality-aware link metrics. Other advantage is allowing non-IP devices to implement multi-hop forwarding. On the

other hand, one disadvantage is that layer-two deals with MAC addresses and forwarding tables must have one entry for each destination, so hierarchical addresses cannot be used, what can limit the size of a layer-two mesh network. Another fact is that link-layer functions are usually implemented on hardware while installing a layer-three routing protocol is generally a user task, so users will probably depend on what is provided by their device vendors.

9. No. Quality-aware link metrics may have different values for each link direction, thus path discovery mechanisms may result in different reverse and forward paths (with different intermediate mesh points) between two nodes in a mesh network. In IEEE 802.11s, whenever a mesh point receives a Path Request (PREQ) from a source mesh point, it learns a path back to it. This path is the reverse path and it may be used later (in case this mesh point is in the selected path) to forward Path Reply (PREP) frames in the opposite direction. In order to deal with asymmetric links, the destination mesh point can issue a new PREQ to the source. Therefore, source and destination nodes run path discovery separately and may create asymmetrical paths.
10. Data frames in IEEE 802.11s are still defined as type-2 frames, exactly like the original IEEE 802.11 but they have both Distribution System flags, FromDS and ToDS, activated. They also include a Mesh Header in the Body field of the IEEE 802.11 data frame. The mesh header defines, among other fields, additional addresses that complete the set of up to six necessary MAC addresses, which may include: source and destination STAs, source and destination MAPs and intermediate transmitter and receiver MPs.

Solutions for Chapter 20

1. The capacity per-MR in a WMN degrades with an increase in the number of MRs as $\Theta(\frac{1}{\sqrt{n}})$ (Equation (20.1)). By implementing multi-channel multi-radio in a WMN, MR can receive and transmit packets simultaneously and more simultaneous transmission pairs can exist within an interference region. Hence, the capacity per-node can be improved. Channel assignment can assign the channels to multiple radios so that the network is able to explore such benefits. However, multiple channels have to be appropriately assigned to the radios of MRs, otherwise the network may have serious interference problem which may lead to a worst case performance than in a single channel system.

2. The number is 20 (without counting the MR at the center).

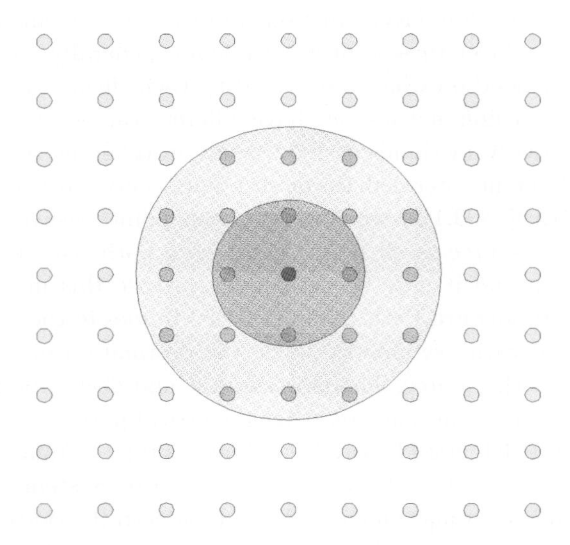

3. The number is 24.

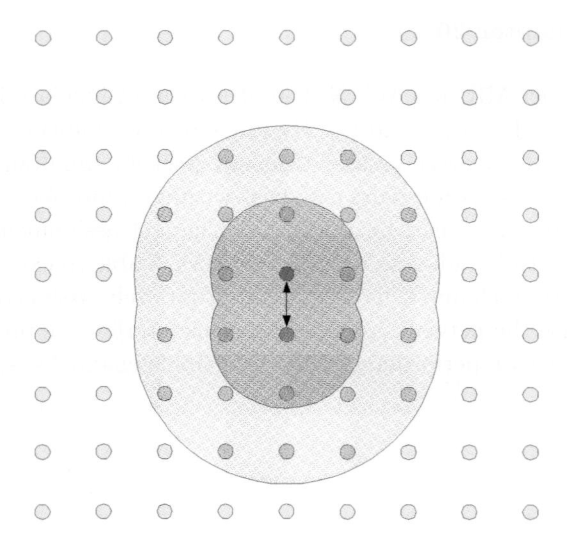

4. The number is 9. An example is shown below.

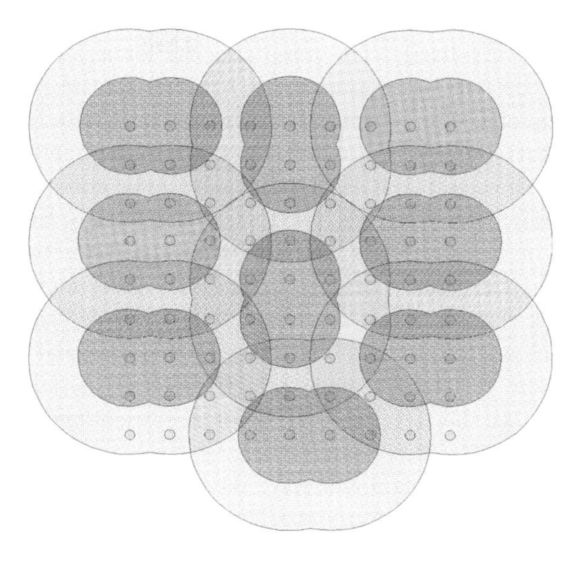

5. There are many possible answers. One of the solutions is shown as follows (suppose nodes A, B, C, and D use Ch_1 and Ch_2, and nodes E, F, G, H, and I use Ch_2 and Ch_3 in ICA).

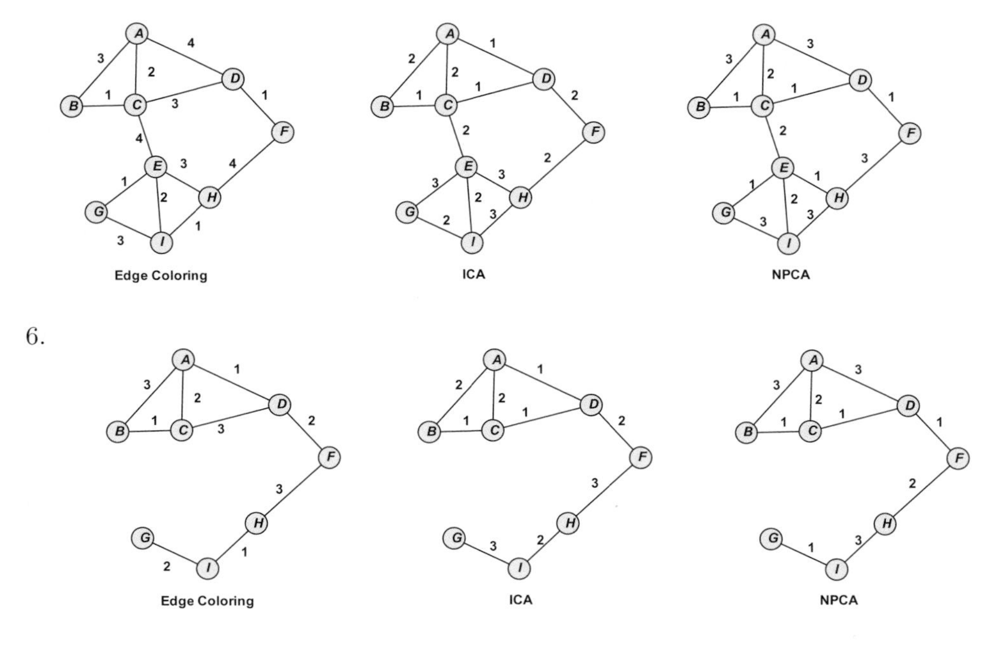

7. Four hopping schedules are needed so that there is no interference with each other.

8. Channel assignment in a WMN is mainly to assign channels among MRs. The assignment is managed by negotiation among MRs (distributed) or the IGW in the network (centralized). The main concern for assignment includes many factors such as the network connectivity, channel spatial reuse, time schedule, and so on. Channel assignment in a cellular system is mainly to assign channels by the Based Station (BS) to mobile clients. The assignment is managed by the BSs. The main concern in the assignment is to have adequate spatial reuse.
9. Centralized channel assignment is able to achieve global optimization. The channel assignment involves all MRs in the network, so the network size affects the performance and the time required for channel assignment. It is suitable for a stable network that does not need to change channel assignment often. Distributed channel assignment involves a set of MRs in a local area, which achieve local optimization. It achieves scalability. It has a potential to track the traffic or other changes in the network and sequentially change channel assignment quickly to accommodate any change.

   Centralized channel assignment may have better performance for a WMN with IGWs and the Internet dominated traffic. In contrast, in distributed channel assignment, IGW is not necessary to be involved in the process and it is suitable for the network with peer-to-peer dominated traffic.
10. Depends on the simulation.

Solutions for Chapter 21

1. Besides having the gateway and repeater functionalities of conventional routers, mesh routers are usually equipped with multiple wireless interfaces built on same or different wireless access technologies.
2. Route Stability, good performance for minimum weight paths, efficient algorithms to calculate minimum weight paths and loop free routing.
3. ETX is defined as the expected number of packet transmissions (including retransmissions) required to successfully deliver a packet along a wireless link. The total weight of a path is calculated by summing up the ETXs for all the links that form the path. Paths with the minimum ETX are expected to have the highest throughput.
4. WCETT extends the ETT concept for the inclusion of intra-flow interference in the metric determination process. With WCETT, the paths that have more diversified channel assignments on their links are given a lower weight; hence when it comes to the optimal path selection, these paths are chosen over other paths that have less diversified channels.
5. Scalability, Robustness against link failures, Load Balancing and Support for heterogeneous nodes.
6. HWMP and RA-OLSR.

7. The HWMP protocol uses on-demand (reactive) routing for dynamic topologies (with mobile nodes) and a tree-based proactive routing for the predominantly fixed backbone networks. If the root portal does not exist, the routing protocol that is used is an adaptation of AODV and is called Radio Metric AODV (RM-AODV). In the case of RA-OLSR, there is a provision for using OLSR instead of AODV. It is a proactive link-state routing protocol based on the specifications of OLSR and optional features from Fisheye State Routing (FSR). Instead of IP addresses, MAC addresses are used and it can work with any link metric, including the airtime metric.
8. Load balancing, increase in fault tolerance in case of link failures, support of applications with high QoS and high bandwidth requirements.
9. Robustness, easy maintenance, low overheads and processing delays, link sensitive routing and forwarding.
10. Since the protocol assumes that all participating nodes are stationary, it is more suitable for the backbone mesh routers.

Solutions for Chapter 22

1. Wireless Mesh Network (WMN) and Mobile and Ad Hoc Network (MANET) are very different in many respects. The primary difference between these two types of networks lies in node mobility. MANET focuses on ad hoc capability for devices with high mobility while WMN mainly focuses on reliable communication and practical deployment for static nodes. Compared to the dynamic topology of MANET which is changing over time, the topology of WMN is relatively static. Moreover, MANETs are infrastructure-less with no planning required at deployment, while WMNs usually have fixed infrastructure and require some planning at deployment. Finally, most nodes in WMNs are connected to a power source, thus removing the energy constraint of the mostly battery powered MANETs.
2. The architecture of WMNs can be classified into three main groups: Infrastructure wireless mesh networks, Client mesh networks and Hybrid mesh networks. *Infrastructure wireless mesh networks* consist of mesh routers and access points forming an infrastructure for mesh clients to connect to. The mesh client does not participate in the operation of the WMN. *Client mesh networks* constitute peer-to-peer networks among client devices such as PDAs, smart phones and laptops. The client nodes themselves constitute the actual network which routes data and carries out configuration functions *Hybrid mesh networks* is a combination of infrastructure mesh networks and client mesh networks.
3. Mobility management can be divided into two main components: location management and handoff management. Location management is a two-stage process that enables the network to discover the current point of attachment of the mobile client for packet and call delivery. The two stages are location registration stage and packet/call delivery stage. Handoff management enables the network

to maintain a client's connections as it continues to move and change its point of access to the network. It is a three-stage process which includes initiation stage, new connection generation stage and data flow control stage.

4. Vertical handoff refers to a mobile client changing the type of radio access technology used to access the Internet. Horizontal handoff refers to handoff within networks that uses the same radio access technology.
5. In 802.11, as a mobile client reaches the coverage limit of its current access point, it must temporarily abandon its connection and actively probe the network to discover alternative access points and to find the one with the best signal quality to reconnect to. This approach minimizes the management overhead, but has a high handoff latency and even worse, can produce "gaps" in connectivity for a duration up to 1 second. This process can be improved by removing the "gaps" and reducing the scanning overhead. By using the current access point as a conduit to other APs, 802.11r allows a mobile client to establish a QoS and security state at a new AP before making the transition, thus removing the "gap" in connectivity. Synscan synchronizes the announcements of beaconing packets on each channel so a mobile client only has to switch to a particular channel at a predetermined synchronized time for all the beacon packets on that channel and there is therefore no waiting required. Moreover, instead of scanning only at handoff, the mobile client regularly scans the existence of other APs while still being associated with the current AP thus reduces the handoff latency by one order of magnitude.
6. Any protocol is fine. "Group Mobility Management in MANETs" proposes a scheme to predict network partitioning and topology changes, and for applying the prediction to routing protocols to minimize disruptions, hence achieving seamless mobility. More specifically, it creates a group mobility model; a group-id based clustering algorithm and a partition prediction scheme. The mobility model is based on parameters such as distance, velocity and acceleration rate and is comparable to real user mobility patterns. The clustering algorithm forms mobility groups where the mobile nodes in each group have the same characteristics. With the grouping and information from the mobility model, the network changes and partitioning can be predicted. Thus, if a destination is unreachable, packets in the network can be rerouted to predicted locations, thereby minimizing the connection disruption and achieving seamless mobility management.
7. Mobile IP (MIP) describes a global mobility solution that provides host mobility for a diverse array of applications and devices on the Internet. With MIP, each mobile client is identified by a static network address from its home network, regardless of its current point of attachment. When a mobile client moves into a foreign network, it obtains a temporary network care-of-address from its current point of attachment and updates its home agent with this temporary address. Thus, the home agent can intercept any packets destined for the mobile client and forward them to the mobile client's current location (most commonly using IP in IP tunneling). The schemes for improving the handoff latency of MIP can

be classified into two groups. The first group aims to reduce the address resolution time through address pre-configuration (FMIPv6) while the second group aims to reduce the network registration time by using a hierarchical network management structure (HMIPv6). There are also schemes combining the two approaches such as (S-MIP).

8. Both IEEE 802.11r and IEEE 802.21 are developing mobility management standards. IEEE 802.11r will facilitate *fast handoff* from one access point to another managed in a seamless manner by refining the transition process as a mobile client moves between access points. IEEE 802.21 will enable seamless handoff between networks of the same type as well as handover between different network types. More specifically, IEEE 802.21 enables *vertical handoff* between different types of networks which includes but is not limited to GSM, GPRS, 802.11, 802.15 and 802.16 networks.

Solutions for Chapter 23

1. The WMNs can be classified into 3 categories based on the management style as: fully managed, semi-managed, and unmanaged.

   - Fully managed WMN: They exist in enterprises such as university, office where the entire network is administered by one single authority.
   - Semi-managed WMN: The ISP has a partial control over few critical MRs that are further connected to other independent MRs installed at public locations such as rooftops, street lamps, public parks, etc.
   - Unmanaged WMN: They are temporary WMN formed in public locations such as conferences, airports, etc. wherein some MRs provide external connectivity to WLAN, GSM.

2. Though WMNs follow the paradigm of ad hoc networking by connecting together group of MRs, certain characteristics of WMN complicate the design of security solutions such as:

   Uni-directional traffic: The traffic pattern in WMN is either towards or away from IGW, and hence MRs near IGW are vulnerable to DoS attack.

   Static nature: As mesh backbone is static, a selfish node attack leads to persistent disruption of the network activity. The static nature introduces route dependencies leading to severity of selfish behavior.

   Multi-radio design: As adjacent radios operate on non-overlapping channels, it is not possible to apply reputation based techniques for detection of selfish nodes.

3. b.

4. In the absence of authentication mechanism, an attacker can conduct wide variety of bogus attacks such as false registration attack, route poisoning attack. Thus, an authentication protocol is very much essential.
   - The hierarchical nature of WMN necessitates a chain of authentication to be performed from the meshclients to MRs to IGW.
   - The multi-hop nature of meshclients requires network layer authentication mechanism instead of a port based authentication like 802.1X framework.
   - A newly joining meshclient needs to validate the authenticity of MR and it is equally important for the MR to validate the identity of the client by suitable authentication protocols. A light weight authentication mechanism should be used here to reduce communication overhead in distributing and validating the keys between different entities.
   - As WMNs are large scale network, scalability of the scheme is very much important.
   - As a meshclients moves from one MR to another, re-authentication overhead has to be reduced.
5. Selfishness falls under passive attack as the primary motivation of a selfish MR is to favor its own client's traffic. A selfish MR drops relay traffic only when the residual bandwidth available for its own clients falls below a threshold. Thus, in order to increase its own client's Internet throughput, it partially or fully drops others traffic. Also, some MRs drops relay traffic in order to avoid congestion in the paths leading to the IGW. An active attack like DoS, on the other hand, intentionally overwhelms a victim router and cripples the network activity.
6. A selfish MR can exhibit MAC layer selfishness in several ways such as:

   Radio Selfishness: A multi-radio capable MR predominantly utilizes its radio for sending its own clients traffic and declares a busy radio when requested to send others traffic.

   Channel Riding Selfishness: A selfish MR fails to abide by the dynamic channel negotiation protocol and refuses to agree to a common channel for communication with its adjacent neighbor.

   Exponential Backoff: A selfish MR tries to grab the channel for itself by choosing a smaller backoff interval.

   Network layer selfishness, on the other hand, fails to abide by the routing protocol and gives preferential treatment to its own client's traffic.
7. d.
8. An attacker conducts Smurf attack by first spoofing the address of an innocent node. Then it broadcasts ICMP echo request packets to a bunch of innocent reflector nodes. The reflector nodes, inturn, compose an ICMP echo reply packet back to the sender, thereby unknowingly flooding the victim's network.

   The attacker employs innocent reflector nodes to flood the victim network and uses a spoofed source address thereby making it difficult to identify the exact source of attack. This requires special traceback techniques that analyze the traffic pattern to determine the inflow of attack packets in the network.

To hide its location, the attacker could be further buried behind stepping stones that are laundering agents overwriting a packet's source address repeatedly with a series of spoofed source addresses. The attacker can also use zombie node that trigger the inflow of ICMP echo packets at timed interval using a remote command from the attacker.

9. c.
10. RED is a rate limiting scheme that ensures fairness among flows in a wired network. Its applicability to wireless network is questionable as it applies rate limit based on the average queue length at the MR. A DoS attack detection scheme requires more stringent measures of control to throttle the infiltration of attack packets in the network. Use of application layer information, such as SYN packet would be useful in applying a heavy filtration on the attack flows. Also, an attack traffic can be composed using multiple low bandwidth attack flows that converge at the victim MR.

Biography

Dharma P. Agrawal is the Ohio Board of Regents Distinguished Professor of Computer Science and the founding director for the Center for Distributed and Mobile Computing in the Department of ECECS, University of Cincinnati, OH. He was a Visiting Professor of ECE at the Carnegie Mellon University, on sabbatical leave during the Autumn 2006 and Winter 2007 Quarters. He has been a faculty member at the N.C. State University, Raleigh, NC (1982–1998) and the Wayne State University, Detroit (1977–1982). He has been a consultant to the General Dynamics Land Systems Division, Battelle, Inc., and the US Army. He has held visiting appointments at AIRMICS, Atlanta, GA, and the AT&T Advanced Communications Laboratory, Whippany, NJ. Dr. Agrawal received a B.E. in Electrical Engineering from Ravishanker University, Raipur, India, in 1966, M.E. (Honors) in Electronics and Communication Engineering from the University of Roorkee, Roorkee (now IIT Roorkee), India, in 1968, and the D.Sc. Technology degree in Electrical Engineering from the Swiss Federal Institute of Technology, Lausanne, Switzerland, in 1975.

His recent research interests include resource allocation and security in mesh networks, efficient query processing and security in sensor networks, and heterogeneous wireless networks. He has published over 500 papers in Journals and international meetings. His co-authored introductory text book on *Wireless and Mobile Computing* has been widely accepted throughout the world and a second edition was published in 2006. The book has been reprinted both in China and India and translated in to Korean and Chinese languages. His second co-authored book on *Ad hoc and Sensor Networks* published in spring of 2006 has been named as the best seller by the publisher. He has given tutorials and extensive training courses in various conferences in USA, and numerous institutions in Taiwan, Korea, Jordan, Malaysia, and India in the areas of ad hoc and Sensor Networks and Mesh Networks. He has been a Tutorial presenter entitled, "Security Vulnerabilities and Preventive Measures in Wireless Mesh Networks," at the MILCOM 2007, October 29–31, 2007, Orlando, FL.

He is an editor for the *Journal of Parallel and Distributed Systems, International Journal on Distributed Sensor Networks, International Journal of Ad Hoc and Ubiquitous Computing (IJAHUC)*, and *International Journal of Ad Hoc & Sensor Wireless Networks*. He has served as an editor of the IEEE *Computer magazine*, the *IEEE Transactions on Computers,* and the *International Journal of High Speed Computing.* He has been the Program Chair and General Chair for many international conferences and meetings. He has received numerous certificates and awards from the IEEE Computer Society and been elected as a core member. He was awarded a "*Third Millennium Medal*," by the IEEE for his outstanding contributions. He has also delivered keynote speech for ten international meetings. He also has five patents in wireless networking area. He has also been named as an **ISI Highly Cited Researcher** in Computer Science. He is a Fellow of the IEEE, the ACM, the AAAS, and the World Innovation Foundation.

Bin Xie received his BSc degree from Central South University, Changsha, China, MSc and PhD degrees (with honors) in Computer Science and Computer Engineering from the University of Louisville, Kentucky, USA. As a research associate, he is currently with Department of Computer Science, University of Cincinnati. He is the author of the book titled *Heterogeneous Wireless Networks — Networking Protocol to Security*, and published 40+ papers in the international conferences and journals.

His research interests are focused on ad hoc networks, sensor networks, wireless mesh networks, integrated WLAN/MANET/cellular with Internet, in particular, on the fundamental aspects of mobility management, performance evaluation, Internet/ wireless infrastructure security, and wireless network capacity. In addition to his academic experience, he has six years of industry experience including ISDN, 3G, and Lucent Excel programmable switching systems.

He is an IEEE senior member and the vice chair of the IEEE TCSIM (Computer Society Technical Committee on Simulation).